Student's Solutions Manual

to accompany

Basic Technical Mathematics

Ninth Edition

AND

Basic Technical Mathematics with Calculus

Ninth Edition

by

Allyn J. Washington

Dutchess Community College

Prepared by

John R. Martin

Tarrant County College, Northeast Campus

PEARSON
Prentice Hall

Upper Saddle River, New Jersey
Columbus, Ohio

Editor in Chief: Vernon Anthony
Senior Acquisitions Editor: Gary Bauer
Development Editor: Linda Cupp
Editorial Assistant: Megan Heintz
Project Manager: Rex Davidson
Operations Specialist: Laura Weaver
Senior Operations Supervisor: Pat Tonneman
Art Director: Diane Ernsberger
Director of Marketing: David Gesell
Marketing Manager: Leigh Ann Sims
Senior Marketing Coordinator: Alicia Wozniak

This book was set in Times Roman by John R. Martin and John Garlow, and was printed and bound by Bradford and Bigelow. The cover was printed by Phoenix Color Corp.

Pearson Education Ltd., London
Pearson Education Singapore Pte. Ltd.
Pearson Education Canada, Inc.
Pearson Education—Japan
Pearson Education Australia Pty. Limited
Pearson Education North Asia Ltd. , Hong Kong
Pearson Educación de Mexico, S.A. de C.V.
Pearson Education Malaysia Pte. Ltd.

10 9 8 7 6 5 4 3 2 1

PEARSON
Prentice Hall

ISBN-13: 978-0-13-814563-7
ISBN-10: 0-13-814563-6

CONTENTS

CHAPTER 1 BASIC ALGEBRAIC OPERATIONS

CHAPTER 2 GEOMETRY

CHAPTER 3 FUNCTIONS AND GRAPHS

CHAPTER 4 THE TRIGONOMETRIC FUNCTIONS

CHAPTER 5 SYSTEMS OF LINEAR EQUATIONS; DETERMINANTS

CHAPTER 9 VECTORS AND OBLIQUE TRIANGLES

CHAPTER 10 GRAPHS OF THE TRIGONOMETRIC FUNCTIONS

CHAPTER 11 EXPONENTS AND RADICALS

CHAPTER 12 COMPLEX NUMBERS

CHAPTER 13 EXPONENTIAL AND LOGARITHMIC FUNCTIONS

CHAPTER 20 ADDITIONAL TOPICS IN TRIGONOMETRY

CHAPTER 21 PLANE ANALYTIC GEOMETRY

CHAPTER 22 INTRODUCTION TO STATISTICS

CHAPTER 23 THE DERIVATIVE

CHAPTER 24 APPLICATIONS OF THE DERIVATIVE

CHAPTER 25 INTEGRATION

CHAPTER 26 APPLICATIONS OF INTEGRATION

CHAPTER 27 DIFFERENTIATION OF TRANSCENDENTAL FUNCTIONS

CHAPTER 28 METHODS OF INTEGRATION

CHAPTER 31 DIFFERENTIAL EQUATIONS

Chapter 1

BASIC ALGEBRAIC OPERATIONS

1.1 Numbers

1. The numbers -3 and 14 are integers. They are also rational numbes since they can be written as $\frac{-3}{1}$ and $\frac{14}{1}$.

5. 3: integer, rational $\left(\frac{3}{1}\right)$, real

$\sqrt{-4}$: imaginary

$-\frac{\pi}{6}$: irrational, real

9. $6 < 8$

13. $-|-3| = -3 \Rightarrow -4 < -3 = -|-3|$

17. The reciprocal of $3 = \frac{1}{3}$. The reciprocal of $-\frac{4}{\sqrt{3}}$ is $\frac{1}{-\frac{4}{\sqrt{3}}} = -\frac{\sqrt{3}}{4}$.

The reciprocal of $\frac{y}{b}$ is $\frac{1}{\frac{y}{b}} = \frac{b}{y}$.

21. An absolute value is not always positive, $|0| = 0$ which is not positive.

25. Since $0.13 = \frac{13}{100} \cdot \frac{\frac{3}{13}}{\frac{3}{13}} = \frac{3}{\frac{300}{13}} = \frac{3}{23.0769} \cdots <$

$\frac{3}{23} = \frac{3}{23} \cdot \frac{\frac{14}{3}}{\frac{14}{3}} = \frac{14}{107.3} < \frac{14}{100} = 0.14$

$\frac{3}{23}$ is a rational number between 0.13 and 0.14 with numerator 3 and an integer, 23, in the denominator.

29. (a) $b - a$; $b > a$, positive integer

(b) $a - b; b > a$, negative integer

(c) $\frac{b-a}{b+a}$, positive rational number less than 1

33. (a) x is a positive number located to the right of 0.

(b) x is a negative number located to the left of -4.

37. $a + bj = a + b\sqrt{-1}$ which for $b = 0$ is $a + bj = a$, a real number. The complex number $a + bj$ is a real number for all values of a and $b = 0$.

41. $\text{N} = \frac{a \text{ bits}}{\text{byte}} \cdot \frac{1000 \text{ bytes}}{\text{kilobytes}} \cdot n \text{ kilobytes}$

$= 1000\, an \text{ bits}$

1.2 Fundamental Operations of Algebra

1. $16 - 2 \times (-3) = 16 - (-6) = 22$

5. $8 + (-4) = 8 - 4 = 4$

9. $-19 - (-16) = -19 + 16 = -3$

13. $-7(-5) = 35$

17. $-2(4)(-5) = (-8)(-5) = 40$

21. $-9 - |2 - 10| = -9 - 1|-8| = -9 - 8 = -17$

25. $8 - 3(-4) = 8 - (-12) = 8 + 12 = 20$

29. $30(-6)(-2) \div (0 - 40) = 360 \div (-40) = -9$

33. $-7 - \frac{|-14|}{2(2-3)} - 3|6 - 8| = -7 - \frac{14}{2(-1)} - 3|-2|$

$= -7 - \frac{14}{-2} - 3(2)$

$= -7 - (-7) - 6$

$= 0 - 6$

$= -6$

37. $6(7)=7(6)$ demonstrates the commutative law of multiplication.

41. $3+(5+9)=(3+5)+9$ demonstrates the associative law of addition.

45. $-a+(-b)=-a-b$ which is expression (d).

49. (a) The product of an even number of negative numbers is positive.
(b) The product of an odd number of negative numbers is negative.

53. (a) $-xy=1 \Rightarrow y=-\frac{1}{x}$ which shows the numbers are negative reciprocals.
(b) $\frac{x-y}{x-y}=1$ provided $x \neq y$ (to prevent division by zero). The numbers cannot be equal.

57. $\frac{7+3+(-2)+(-3)+(-1)+4+6}{7}=2^{\circ}\text{C}$

61. 100 m + 200 m = 200 m + 100 m illustrates the commutative law of addition.

1.3 Calculators and Approximate Numbers

1. Yes, 0.390 has three significant digits since the 0 after the 9 is not needed to locate the decimal.

5. 8 cylinders is exact because they can be counted. 55 mi/h is approximate since it is measured.

9. 107 has 3 significant digits. 3004 has 4 significant digits.

13. 3000 has 1 significant digit. 3000.1 has 5 significant digits.

17. (a) Both numbers have the same precision with digits in the tenths place.
(b) 78.0 with 3 significant digits is more accuate than 0.1 with 1 significant digit.

21. (a) 4.936 = 4.94 rounded to 3 significant digits.
(b) 4.936 = 4.9 rounded to 2 significant digits.

25. (a) 9549 = 9550 rounded to 3 significant digits.
(b) 9549 = 9500 rounded to 2 significant digits.

29. (a) Calculator: $12.78+1.0495-1.633=12.1965$
$=12.20$
(b) Estimate: $13+1-2=12$

33. (a) Calculator: $\frac{23.962\times 0.01537}{10.965-8.249}=0.1356$
(b) Estimate: $\frac{20\times 0.02}{10-8}=0.2$

37. 0.9788 + 14.9 = 15.8788 since 4 is the number of decimal places in the least precise number.

41. 2.745 MHz and 2.755 MHz are the least possible and greatest possible frequencies respectively.

45. (a) $2.2+3.8\times 4.5=19.3$
(b) $(2.2+3.8)\times 4.5=27$

49.

```
61812311→X
            61812311
11381216→Y
            11381216
(X-Y)/9
             5603455
```

53. (a) $\frac{1}{3}=0.333\cdots$
(b) $\frac{5}{11}=0.454545\cdots$
(c) $\frac{2}{5}=0.4000\cdots$
0 is the repeating part

57. $1\text{ K}=1024$ bytes
$256\text{ K}=256\times 1024=262{,}144$ bytes

60. $\frac{50.45(9.80)}{1+\frac{100.9}{23}}=92\text{ N}$

1.4 Exponents

1. $(-x^3)^2 = [(-1)x^3]^2 = (-1)^2(x^3)^2 = x^{3(2)} = x^6$

5. $x^3 \cdot x^4 = x^{3+4} = x^7$

9. $\frac{m^5}{m^3} = m^{5-3} = m^2$

13. $(P^2)^4 = P^{2\cdot4} = P^8$

17. $(aT^2)^{30} = a^{30}T^{2(30)} = a^{30}T^{60}$

21. $\left(\frac{x^2}{2}\right)^4 = \frac{x^{2\cdot4}}{2^4} = \frac{x^8}{16}$

25. $-3x^0 = (-3)(1) = -3$

29. $\frac{1}{R^{-2}} = R^2$

33. $(2v^2)^6 = (2^6)(v^{2\cdot6}) = 64v^{12}$

37. $\frac{2v^4}{(2v)^4} = \frac{2v^4}{2^4v^4} = \frac{1}{2^3} = \frac{1}{8}$

41. $(\pi^0x^2a^{-1})^{-1} = (1\cdot x^2\cdot a^{-1})^{-1} = x^{-2(1)}(a^{-1(-1)})$

$= \frac{a}{x^2}$

45. $\left(\frac{4x^{-1}}{a^{-1}}\right)^{-3} = \frac{4^{-3}x^{-1(-3)}}{a^{-1(-3)}} = \frac{x^3}{4^3a^3} = \frac{x^3}{64a^3}$

49. $7(-4)-(-5)^2 = -28-25 = -53$

53. $\frac{3.07(-1.86)}{(-1.86)^4+1.596} = \frac{-5.71}{11.97+1.596} = \frac{-5.71}{13.57} = -0.421$

57. $\left(\frac{1}{x^{-1}}\right)^{-1} = \frac{1^{-1}}{x^{-1(-1)}} = \frac{1}{x} =$ reciprocal of x, yes.

61. $(x^a\cdot x^{-a})^5 = (x^{a+(-a)})^5 = (x^0)^5$

$= (1)^5,\ x \neq 0$

$= 1$

65. $\pi\left(\frac{r}{2}\right)^3\left(\frac{4}{3\pi r^2}\right) = \pi\cdot\frac{r^3}{8}\cdot\frac{4}{3\pi r^2} = \frac{r}{6}$

1.5 Scientific Notation

1. $8.06\times10^3 = 8060$

5. $2.01\times10^{-3} = 0.00201$; move decimal point 3 places to the left by adding 2 zeros.

9. $1.86\times10 = 18.6$; move decimal point 1 place to the right.

13. $0.0087 = 8.7\times10^{-3}$; move decimal point to the right 3 places.

17. $0.063 = 6.3\times10^{-2}$; move decimal point 2 places to the right.

21. $28{,}000(2{,}000{,}000{,}000) = 2.8\times10^4(2\times10^9)$

$= 5.6\times10^{13}$

25. $2\times10^{-35}+3\times10^{-34} = 0.2\times10^{-34}+3\times10^{-34}$

$= 3.2\times10^{-34}$

29. $1280(865{,}000)(43.8) = 4.85\times10^{10}$

33. $(3.642\times10^{-8})(2.736\times10^5) = 9.965\times10^{-3}$

37. $6{,}500{,}000\text{ kW} = 6.5\times10^6\text{ kW}$

41. $2{,}000{,}000{,}000\text{ Hz} = 2\times10^9\text{ Hz}$

45. (a) $2300 = 2.3\times10^3$

(b) $0.23 = 230\times10^{-3}$

(c) $23 = 23\times10^0$

49. $\frac{7.5\times10^{-15}s}{\text{addition}}\cdot5.6\times10^6\text{ additions} = 4.2\times10^{-8}s$

53. mass

$= 125{,}000{,}000 \text{ atoms}\left(\dfrac{16 \text{ amu}}{\text{atom}}\right)\left(\dfrac{1.66\times 10^{-27}}{\text{amu}}\right)$

$= 3.32\times 10^{-18}$ kg

1.6 Roots and Radicals

1. $-\sqrt[3]{64} = -4$ since $(-4)^3 = -64$

5. $\sqrt{81} = 9$

9. $-\sqrt{49} = -7$

13. $\sqrt[3]{125} = 5$

17. $\left(\sqrt{5}\right)^2 = 5$

21. $\left(\sqrt[4]{53}\right)^4 = 53$

25. $2\sqrt{84} = 2\sqrt{4\cdot 21} = 2\cdot 2\sqrt{21} = 4\sqrt{21}$

29. $\sqrt[3]{8^2} = \sqrt[3]{64} = 4$

33. $\sqrt{36+64} = \sqrt{100} = 10$

37. $\sqrt{85.4} = 9.24$

41. (a) $\sqrt{1296+2304} = \sqrt{3600} = 60.00$

(b) $\sqrt{1296} = \sqrt{2304} = 36.00 + 48.00 = 84.00$

45. $\sqrt{24s} = \sqrt{(24)(150)} = \sqrt{3600} = 60$ mi/h

49. $d = \sqrt{w^2+h^2} = \sqrt{3.6+20.5} = 42.0$ in.

53.

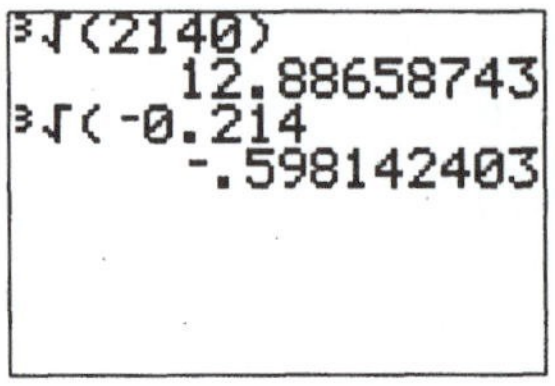

1.7 Addition and Subtraction of Algebraic Expressions

1. $3x+2y-5y = 3x-3y$

5. $5x+7x-4x = 12x-4x = 8x$

9. $2F-2T-2+3F-T = 5F-3T-2$

13. $s+3a-4-2 = s+(2a-4) = 3s-4$

17. $2-3-(4-5a) = -1-4+5a = -5+5a = 5a-5$

21. $-(t-2u)+(3u-t) = -t+2u+3u-t = 5u-2t$

25. $-7(6-3j)-2(j+4) = -42+21j-2j-8$

$= 19j-50$

29. $2\left[4-\left(t^2-5\right)\right] = 2\left[4-t^2+5\right] = 2\left[9-t^2\right] = 18-2t^2$

33. $aZ-\left[3-(aZ+4)\right] = aZ-\left[3-aZ-4\right]$

$aZ-\left[-1-aZ\right] = aZ+1+aZ$

$= 2aZ+1$

37. $5p-(q-2p)-\left[3q-(p-q)\right]$

$= 5p-q+2p-\left[3q-p+q\right]$

$= 7p-q-\left[4q-p\right]$

$= 7p-q-4q+p$

$= 8p-5q$

41. $5V^2-\left(6-\left(2V^2+3\right)\right) = 5V^2-\left(6-2V^3-3\right)$

$= 5V^2-\left(3-2V^2\right)$

$= 5V^2-3+2V^2$

$= 7V^2-3$

45. $-4\left[4R-2.5(Z-2R)-1.5(2R-Z)\right]$

$= -4\left[4R-2.5Z+5R-3R+1.5Z\right]$

$= -4\left[6R-Z\right]$

$= -24R+4Z$

$= 4Z-24R$

49. $\left[\left(B+\frac{4}{3}\alpha\right)+2\left(B-\frac{2}{3}\alpha\right)\right]$

$-\left[\left(B+\frac{4}{3}\alpha\right)-\left(B-\frac{2}{3}\alpha\right)\right]$

$=\left[B+\frac{4}{3}\alpha+2B-\frac{4}{3}\alpha\right]-\left[B+\frac{4}{3}\alpha-B+\frac{2}{3}\alpha\right]$

$=[3B]-\left[\frac{6}{3}\alpha\right]=3B-2\alpha$

53. (a) $2x^2-y+2a+3y-x^2-b=x^2+2y+2a-b$

(b) $2x^2-y+2a-\left(3y-x^2-b\right)$

$=2x^2-y+2a-3y+x^2+b$

$=3x^2-4y+2a+b$

1.8 Multiplication of Algebraic Expressions

1. $2s^3\left(-st^4\right)^3\left(4s^2t\right)=2s^3\left(-s\right)^3\left(t^4\right)^3\left(4s^2t\right)$

$=-2s^6t^{12}\left(4s^2t\right)$

$=-8s^8t^{13}$

5. $\left(a^2\right)\left(ax\right)=a^{2+1}x=a^3x$

9. $\left(2ax^2\right)^2\left(-2ax\right)=4a^2x^4\left(-2ax\right)=-8a^3x^5$

13. $-3s\left(s^2-5t\right)=-3s\left(s^2\right)+3s\left(5t\right)=-3s^2+15st$

17. $3M\left(-M-N+2\right)=3M\left(-M\right)+3M\left(-N\right)+3M\left(2\right)$

$=-3M^2-3MN+6M$

21. $(x-3)(x+5)=x^2+5x-3x-15=x^2+2x-15$

25. $(2a-b)(3a-2b)=6a^2-4ab-3ab+2b^2$

$=6a^2-7ab+2b^2$

29. $\left(x^2-1\right)(2x+5)=2x^3+5x^2-2x-5$

33. $2(a+1)(a-9)=2\left(a^2-8a-9\right)=2a^2-16a-18$

37. $2L(L+1)(4-L)=2L\left(4L-L^2+4-L\right)$

$=2L\left(-L^2+3L+4\right)$

$=-2L^3+6L^2+8L$

41. $\left(x_1+3x_2\right)^2=\left(x_1+3x_2\right)\left(x_1+3x_2\right)$

$=x_1^2+3x_1x_2+3x_1x_2+9x_2^2$

$=x_1^2+6x_1x_2+9x_2^2$

45. $\left(b-2x^2\right)^2=b^2-4bx^2+4x^4$

49. $3T(T+2)(2T-1)=\left(3T^2+6T\right)(2T-1)$

$=6T^3-3T^2+12T^2-6T$

$=6T^3+9T^2-6T$

(b) For $x=3,\ y=4$,

$(x-y)^2=(3-4)^2=(-1)^2=1$

$x^2-y^2=3^2-4^2=9-16=-7$

which shows $(x-y)^2\neq x^2-y^2$.

52. $(98)(102)=(100-2)(100+2)=100^2-2^2$

$=10,000-4=9996$

53. Let $1<n<9$, $n^2-1=(n-1)(n+1)$ which shows the square of the integer minus $1=$ product of the integer before n and the integer after n.

57. $P(1+0.01r)^2=P\left(1+0.02r+0.0001r^2\right)$

$=P+0.02\,\mathrm{Pr}+0.0001\,\mathrm{Pr}^2$

61. $(n+100)^2=n^2+200n+100^2$

$=n^2+200n+10,000$

1.9 Division of Algebraic Expressions

1. $\dfrac{-6a^2xy^2}{-2a^2xy^5}=3y^{2-5}=3y^{-3}=\dfrac{3}{y^3}$

5. $\dfrac{8x^3y^2}{-2xy}=-4x^2y$

9. $\dfrac{(15x^2)(4bx)(2y)}{30bxy} = 4x^2$

13. $\dfrac{3a^2x+6xy}{3x} = \dfrac{3a^2x}{3x}+\dfrac{6xy}{3x} = a^2+2y$

17. $\dfrac{4pq^3+8p^2q^2-16pq^5}{4pq^2} = \dfrac{4pq^3}{4pq^2}+\dfrac{8p^2q^2}{4pq^2}-\dfrac{16pq^5}{4pq^2}$
$= q+2p-4q^3$

21. $\dfrac{3ab^2-6ab^3+9a^2b^2}{9a^2b^2} = \dfrac{3ab^2}{9a^2b^2}-\dfrac{6ab^3}{9a^2b^2}+\dfrac{9a^2b^2}{9a^2b^2}$
$= \dfrac{1}{3a}-\dfrac{2b}{3a}+1$

25.
$$\begin{array}{r} 2x+1 \\ x+3\overline{)2x^2+7x+3} \\ \underline{2x^2+6x} \\ x+3 \\ \underline{x+3} \\ 0 \end{array}$$

29.
$$\begin{array}{r} 4x^2-x-1 \\ 2x-3\overline{)8x^3-14x^2+x+0} \\ \underline{8x^3-12x^2} \\ -2x^2+x \\ \underline{-2x^2+3x} \\ -2x+0 \\ \underline{-2x+3} \\ -3 \end{array}$$

33.
$$\begin{array}{r} x^2+x-6 \\ x+2\overline{)x^3+3x^2-4x-12} \\ \underline{x^3+2x^2} \\ x^2-4x \\ \underline{x^2+2x} \\ -6x-12 \\ \underline{-6x-12} \\ 0 \end{array}$$

37.
$$\begin{array}{r} x^2-2x+4 \\ x+2\overline{)x^3+0x^2+0x+8} \\ \underline{x^3+2x^2} \\ -2x^2+0x \\ \underline{-2x^2-4x} \\ 4x+8 \\ \underline{4x+8} \\ 0 \end{array}$$

41.
$$\begin{array}{r} x-y+z \\ x+y-z\overline{)x^2+0xy+0xz-y^2+2yz-z^2} \\ \underline{x^2+xy-xz} \\ -xy+xz-y^2+2yz-z^2 \\ \underline{-xy \quad -y^2+yz} \\ +xz \quad +yz-z^2 \\ \underline{+xz \quad +yz-z^2} \\ 0 \end{array}$$

$\dfrac{x^2-y^2+2yz-z^2}{x+y-z} = x-y+z+\dfrac{0}{x+y-z}$

45.
$$\begin{array}{r} x^3-x^2+x-1 \\ x+1\overline{)x^4 \qquad\qquad +1} \\ \underline{x^4+x^3} \\ -x^3 \\ \underline{-x^3-x^2} \\ x^2 \\ \underline{x^2+x} \\ -x+1 \\ \underline{-x-1} \\ 2 \end{array}$$

$\dfrac{x^4+1}{x+1} = x^3-x^2+x-1+\dfrac{2}{x+1} \neq x^3$

49. $\dfrac{GMm[(R+r)-(R-r)]}{2rR} = \dfrac{GMm[R+r-R+r]}{2rR}$
$= \dfrac{GMm[2r]}{2rR} = \dfrac{GMm}{R}$

1.10 Solving Equations

1. (a)
$$x-3=-12$$
$$x-3+3=-12+3$$
$$x=-9$$

(b)
$$x+3=-12$$
$$x+3-3=-12-3$$
$$x=-15$$

(c)
$$\frac{x}{3}=-12$$
$$3\left(\frac{x}{3}\right)=3(-12)$$
$$x=-36$$

(d)
$$3x=-12$$
$$\frac{3x}{3}=\frac{-12}{3}$$
$$x=-4$$

5.
$$x-2=7$$
$$x=7+2$$
$$x=9$$

9.
$$\frac{t}{2}=-5$$
$$t=2(-5)$$
$$t=-10$$

13.
$$3t+5=-4$$
$$3t=-4-5$$
$$3t=-9$$
$$t=-3$$

17.
$$3x+7=x$$
$$3x-x=-7$$
$$2x=-7$$
$$x=\frac{-7}{2}$$

21.
$$6-(r-4)=2r$$
$$6-r+4=2r$$
$$-3r=-10$$
$$r=\frac{10}{3}$$

25.
$$0.1x-0.5(x-2)=2$$
$$x-5(x-2)=20$$
$$x-5x+10=20$$
$$-4x=10$$
$$x=-2.5$$

29.
$$\frac{4x-2(x-4)}{3}=8$$
$$4x-2(x-4)=24$$
$$4x-2x+8=24$$
$$2x=16$$
$$x=8$$

33.
$$5.8-0.3(x-6.0)=0.5x$$
$$5.8-0.3x+1.8=0.5x$$
$$7.6=0.8x$$
$$x=9.5$$

37.
$$\frac{x}{2.0}=\frac{17}{6.0}$$
$$x=\frac{34}{6.0}$$
$$x=5.7$$

41. (a)
$$2x+3=3+2x$$
$$2x+3=2x+3, \text{ identity}$$

(b)
$$2x-3=3-2x$$
$$2x-3+3=3+3-2x$$
$$2x=6-2x$$
$$4x=6$$
$$x=\frac{3}{2}, \text{ conditional}$$

45.
$$2.0v+40=2.5(v+5.0)$$
$$2.0v+40=2.5v+12.5$$
$$-0.5v=-27.5$$
$$v=55 \text{ mi/h}$$

49.
$$0.14n+0.06(2000-n)=0.09(2000)$$
$$0.14n+120-0.06n=180$$
$$0.08n=180-120$$
$$n=750 \text{ gal}$$

1.11 Formulas and Literal Equations

1.
$$v=v_0+at$$
$$v-v_0=at$$
$$a=\frac{v-v_0}{t}$$

5.
$$E=IR$$
$$\frac{E}{I}=\frac{IR}{I}$$
$$R=\frac{E}{I}$$

9.
$$Q=SLd^2$$
$$L=\frac{Q}{Sd^2}$$

13.
$$A=\frac{Rt}{PV}$$
$$Rt=APV$$
$$t=\frac{APV}{R}$$

17. $T = \dfrac{c+d}{v}$

$Tv = c + d$

$d = Tv - c$

21. $a = \dfrac{2mg}{M+2m}$

$aM + 2ma = 2mg$

$aM = 2mg - 2ma$

$M = \dfrac{2mg - 2ma}{a}$

25. $N = r(A - s)$

$N = rA - rs$

$rs = rA - N$

$s = \dfrac{rA - N}{r}$

29. $Q_1 = P(Q_2 - Q_1)$

$Q_1 = PQ_2 - PQ_1$

$Q_1(1+P) = PQ_2$

$Q_2 = \dfrac{Q_1 + PQ_1}{P}$

33. $L = \pi(r_1 + r_2) + 2x_1 + x_2$

$L = \pi r_1 + \pi r_2 + 2x_1 + x_2$

$\pi r_1 = L - \pi r_2 - 2x_1 - x_2$

$r_1 = \dfrac{L - \pi r_2 - 2x_1 - x_2}{\pi}$

37. $C = \dfrac{2eAk_1k_2}{d(k_1 + k_2)}$

$Cd(k_1 + k_2) = 2eAk_1k_2$

$e = \dfrac{Cd(k_1 + k_2)}{2Ak_1k_2}$

41. $e = \dfrac{T_1}{T_1 + T_2}$

$T_1 + T_2 = \dfrac{T_2}{e}$

$T_1 = \dfrac{T_2}{e} - T_2 = \dfrac{850}{0.45} - 850$

$T_1 = 1040 \text{ K}$

45. $V_1 = \dfrac{VR_1}{R_1 + R_2}$

$R_1 + R_2 = \dfrac{VR_1}{V_1}$

$R_2 = \dfrac{VR_1}{V_1} - R_1 = \dfrac{12.0(3.56)}{6.30} - 3.56$

$R_2 = 3.22\Omega$

1.12 Applied Word Problems

1. x = number of 1.5Ω resistors

$34 - x$ = number of 2.5Ω resistors

$1.5x + 2.5(34 - x) = 56$

$1.5x + 85 - 2.5x = 56$

$-1.0x = -29$

$x = 29$

$34 - x = 5$

29 1.5Ω resistors 5 2.5Ω resistors

5. x = cost 6 years ago

$x + 5000$ = cost today

$x + (x + 5000) = 49{,}000$

$2x + 5000 = 49{,}000$

$2x = 44{,}000$

$x = 22{,}000$

$x + 5000 = 27{,}000$

$22,000 six years ago $27,000 is the cost today

$\dfrac{x}{4} = 2000$, 3rd day

9. Let x = number of acres @ $200

$200 \cdot x + 300 \cdot (140 - x) = 37{,}000$

$2x + 420 - 3x = 370$

$x = 50$ acres @ $200

$140 - x = 90$ acres @ $300

13. Let x = 15 m girders

$x - 4$ = 18 m girders

$15x = 18(x - 4)$

$3x = 72$

$x = 24$,

so there are twenty 18 m girders needed.

17. Let x = the main pipeline

$x+2.6$ = the smaller pipeline

$$3(x+2.6)+x=35.4$$
$$3x+7.8+x=35.4$$
$$4x=27.6$$
$x=6.9$ km for the main pipeline;

9.5 km for the smaller pipeline

21. Let d = slope length

$$d=vt,\ t=\frac{d}{v},\ \frac{d}{v_1}=24=\frac{d}{v_2}$$
$$\frac{d}{50}=24-\frac{d}{140};\ \frac{d}{50}+\frac{d}{150}=24$$
$0.0267d=24;\ d=900$ m

25. Let $x-30$ = time first car started race;

x = time second car started race

$$260(x-30)=240(x)$$
$$260x-7800=240x$$
$$20x=7800$$
$x=390$ s

8 laps $=250\times8=20$ mi, 105 600 ft, total distance

$d=vt=260\times390=101{,}400$ ft $<105{,}600$ ft

So, the first car is ahead after 8 laps.

29. Let x = number of liters of pure antifreeze added

x = number of liters of 50% antifreeze drained

$$0.25(12-x)+x=0.50(12)$$
$$3.0-0.25x+x=6.0$$
$$0.75x=3.0$$
$x=4.0$ L

Chapter 1 Review Exercises

1. $(-2)+(-5)-3=-7-3=-10$

5. $-5-|2(-6)|+\frac{-15}{3}=-5-|-12|+(-5)$

$$=-5-12-5$$
$$=-17-5$$
$$=-22$$

9. $\sqrt{16}-\sqrt{64}=4-8=-4$

13. $\left(-2rt^2\right)^2=4r^2\left(t^2\right)^2=4r^2t^4$

17. $\dfrac{-16N^{-2}\left(NT^2\right)}{-2N^0T^{-1}}=8N^{-2+1}T^{2-(-1)}$

$$=8N^{-1}T^3$$
$$=\frac{8T^3}{N}$$

21. (a) 8840 has 3 significant digits

(b) 8840 rounded to 2 significant digits is 8800

25. $37.3-16.92(1.067)^2=18.03676612$

on a calculator; 18.0

29. $a-3ab-2a+ab=a-2a-3ab+ab=-a-2ab$

33. $(2x-1)(x+5)=2x^2+10x-x-5$

$$=2x^2+9x-5$$

37. $\dfrac{2h^3k^2-6h^4k^5}{2h^2k}=\dfrac{2h^3k^2}{2h^2k}-\dfrac{6h^4k^5}{2h^2k}$

$$=hk-3h^2k^4$$

41. $2xy-\{3z-[5xy-(7z-6xy)]\}$

$$=2xy-\{3z-[5xy-7z+6xy]\}$$
$$=2xy-\{3z-[11xy-7z]\}$$
$$=2xy-\{3z-11xy+7z\}$$
$$=2xy-\{10-11xy\}$$
$$=2xy-10z+11xy$$
$$=13xy-10z$$

45. $-3y(x-4y)^2=-3y\left(x^2-8xy+16y^2\right)$

$$=-3x^2y+24xy^2-48y^3$$

49. $\dfrac{12p^3q^2-4p^4q+6pq^5}{2p^4q}=\dfrac{12p^3q^2}{2p^4q}-\dfrac{4p^4q}{2p^4q}+\dfrac{6pq^5}{2p^4q}$

$$=\frac{6q}{p}-2+\frac{3q^4}{p^3}$$

53.
$$\begin{array}{r} x^2-2x+3 \\ 3x-1\overline{)3x^3-7x^2+11x-3} \\ \underline{3x^3-\ x^2} \\ -6x^2+11x \\ \underline{-6x^2+\ 2x} \\ 9x-3 \\ \underline{9x-3} \end{array}$$

57.
$$\begin{aligned} &-3\{(r+s-t)-2[(3r-2s)-(t-2s)]\} \\ &=-3\{(r+s-t)-2[3r-2s-t+2s]\} \\ &=-3\{r+s-t-2[3r-t]\} \\ &=-3\{r+s-t-6r+2t\} \\ &=-3\{-5r+s+t\} \\ &=15r-3s-3t \end{aligned}$$

61.
$$\begin{aligned} 3x+1&=x-8 \\ 3x-x&=-8-1 \\ 2x&=-9 \\ x&=\frac{-9}{2} \end{aligned}$$

65.
$$\begin{aligned} 6x-5&=3(x-4) \\ 6x-5&=3x-12 \\ 3x&=-7 \\ x&=\frac{-7}{3} \end{aligned}$$

69.
$$\begin{aligned} 3t-2(7-t)&=5(2t+1) \\ 3t-14+2t&=10t+5 \\ 5t-14&=10t+5 \\ -5t&=19 \\ t&=\frac{-19}{5} \end{aligned}$$

73. $60,000,000,000=6\times10^{10}$

77. 2.53×10^{13} mi $=25,300,000,000,000$ mi

81. 1.5×10^{-1} Bq/L $=0.15$ Bq/L

85.
$$\begin{aligned} P&=\frac{\pi^2EI}{L^2} \\ E&=\frac{PL^2}{\pi^2I} \end{aligned}$$

89.
$$\begin{aligned} d&=(n-1)A \\ d&=nA-A \\ na&=d+A \\ n&=\frac{d+A}{A} \end{aligned}$$

93.
$$\begin{aligned} R&=\frac{A(T_2-T_1)}{H} \\ RH&=AT_2-AT_1 \\ AT_2&=RH+AT_1 \\ T_2&=\frac{RH+AT_1}{A} \end{aligned}$$

97. $\dfrac{5.25\times10^{10}}{6.4\times10^4}=8.2\times10^5$

101. $\dfrac{R_1R_2}{R_1+R_2}=\dfrac{0.0275(0.0590)}{0.0275+0.0590}=0.0188\Omega$

105.
$$\begin{aligned} 4(t+h)-2(t+h)^2&=4t+4h-2(t^2+2th+h^2) \\ &=4t+4h-2t^2-4th-2h^2 \end{aligned}$$

109.
$$\begin{aligned} x-(3-x)&=2x-3 \\ x-3+x&=2x-3 \\ 2x-3&=2x-3, \text{the equation is an identity} \end{aligned}$$

113. $\dfrac{8\times10^{-3}}{2\times10^4}=4\times10^{-7}$

117. Let $x=$ the # of liters of 0.50% mixture

$100-x=$ the # of liters of 0.75% mixture

$$\begin{aligned} 0.50\%\cdot x+0.75\%(1000-x)&=0.65\%\cdot1000 \\ 50x+75(1000-x)&=65,000 \\ 50x+75,000-75x&=65,000 \\ 25x&=10,000 \\ x&=400 \text{ L - 0.50\% mixture} \\ 1000-x&=600 \text{ L - 0.75\% mixture} \end{aligned}$$

121. $$\frac{12h}{450m} = \frac{x}{250m}$$
$$x = 6.67h$$

125. $$x + y = 1000$$
$$0.0050x + 0.0075y = 0.0065(1000)$$
from which $x = 400L$

$y = 600L$

129. $$P = P_0 + P_0rt$$
$$P_0rt = P - P_0$$
$$r = \frac{P - P_0}{P_0t}$$
$$r = \frac{7625 - 6250}{6250(4.000)} = 0.055$$
$$r = 5.5\%$$
On a calculator $r = (7625 - 6250)/(6250(4.000))$.

Chapter 2

GEOMETRY

2.1 Lines and Angles

1. $\angle ABE = 90^\circ$

5. $\angle EBD$ and $\angle DBC$ are acute angles.

9. The complement of $\angle CBD = 65^\circ$ is 25°.

13. $\angle AOB = 90^\circ + 50^\circ = 140^\circ$

17. $\angle 1 = 180^\circ - 145^\circ = 35^\circ = \angle 2 = \angle 4$

21. $\angle 3 = 90^\circ - 62^\circ = 28^\circ$

25. $\angle DEB = 44^\circ$

29. $\dfrac{a}{4.75} = \dfrac{3.05}{3.20} \Rightarrow a = 4.75 \cdot \dfrac{3.05}{3.20} = 4.53 \text{ m}$

33. $\angle BCD = 180^\circ - 47^\circ$
$= 133^\circ$

37. $\measuredangle 1 + \measuredangle 2 + \measuredangle 3 = 180^\circ$,
($\measuredangle 1$, $\measuredangle 2$, and $\measuredangle 3$ form a straight line)

2.2 Triangles

1. $\angle 5 = 45^\circ \Rightarrow \angle 3 = 45^\circ$
$\angle 2 = 180^\circ - 70^\circ - 45^\circ = 65^\circ$

5. $\angle A = 180^\circ - 84^\circ - 40^\circ = 56^\circ$

9. $A = \dfrac{1}{2}bh = \dfrac{1}{2}(7.6)(2.2) = 8.4 \text{ ft}^2$

13. $A = \dfrac{1}{2}bh = \dfrac{1}{2}(3.46)(2.55) = 4.41 \text{ ft}^2$

17. $p = 205 + 322 + 415$
$p = 942 \text{ cm}$

21. $c = \sqrt{13.8^2 + 22.7^2} = 26.6 \text{ ft}$

25. $\angle B = 90^\circ - 23^\circ = 67^\circ$

29.

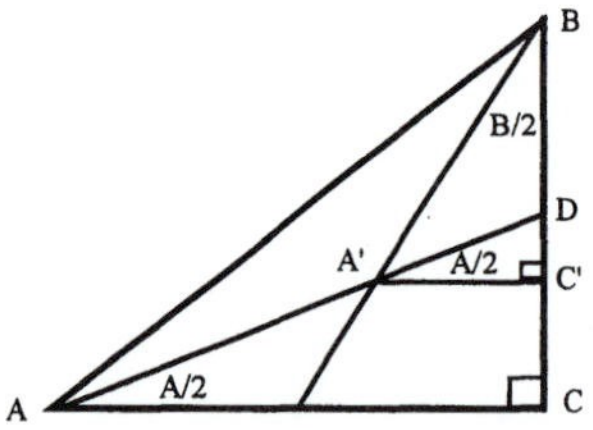

$\Delta ADC \sim \Delta A'DC' \Rightarrow \measuredangle DA'C' = A/2$

$\measuredangle$ between bisectors $= \measuredangle BA'D$

$\Delta BA'C'$, $\dfrac{B}{2} + (\measuredangle BA'D + A/2) = 90^\circ$

from which $\measuredangle BA'D = 90^\circ - \left(\dfrac{A}{2} + \dfrac{B}{2}\right)$

or $\measuredangle BA'D = 90^\circ - \left(\dfrac{A+B}{2}\right)$

33.

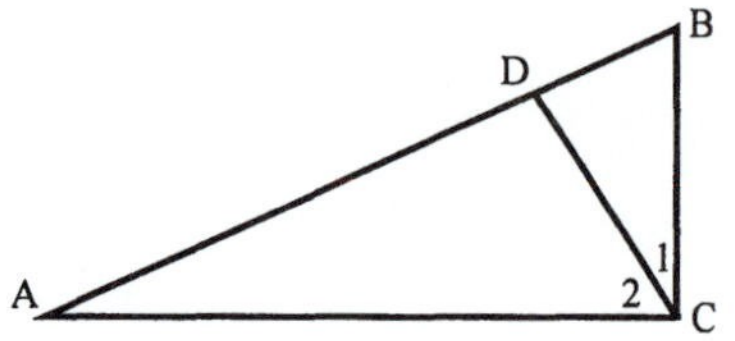

$\measuredangle A + \measuredangle B = 90^\circ$

$\measuredangle 1 + \measuredangle B = 90^\circ$

$\Rightarrow \quad \measuredangle A = \measuredangle 1$

redraw ΔBDC as

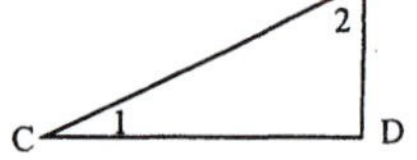

$\measuredangle 1 + \measuredangle 2 = 90^\circ$

$\measuredangle 1 + \measuredangle B = 90^\circ$

$\Rightarrow \quad \measuredangle 2 = \measuredangle B$

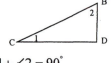

and ΔADC as

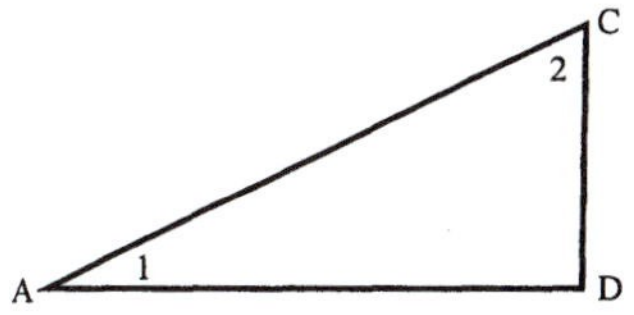

ΔBDC and ΔADC are similar.

37. Since $\triangle MKL \sim \triangle MNO$; $KN = KM - MN$; $15 - 9 = 6$

$= KM$; $\frac{KM}{MN} = \frac{LM}{MO}$; $\frac{6}{9} = \frac{LM}{12}$; $9LM = 72$; $LM = 8$

41. $s = \frac{2(76.6) + 30.6}{2} = 91.9$

$A = \sqrt{91.9(91.9 - 76.6)^2 (91.6 - 30.6)}$

$A = 1150 \text{ cm}^2$

45.

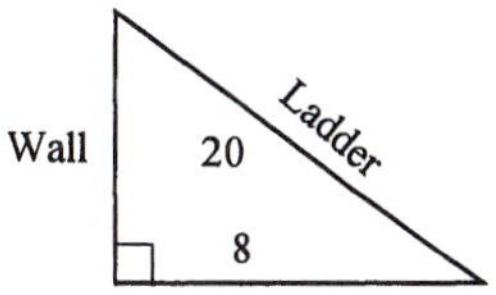

distance up wall $= \sqrt{20^2 - 8^2} = \sqrt{336} = 18.3$ ft

49.

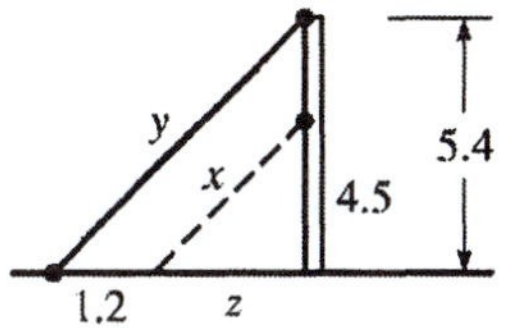

$\frac{4.5}{z} = \frac{5.4}{1.2 + z}$

$z = 6.0$ m

$x^2 = z^2 + 4.5^2$

$x = 7.5$ m

$y^2 = (1.2 + 6)^2 + 5.4^2$

$y = 9.0$ m

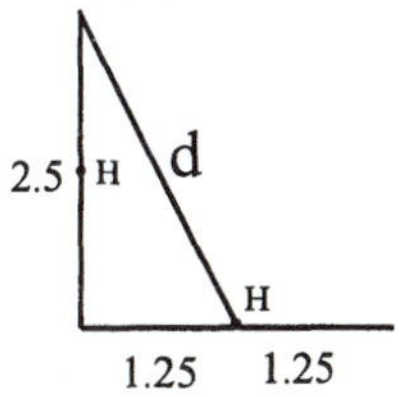

53. Redraw ΔBCP as

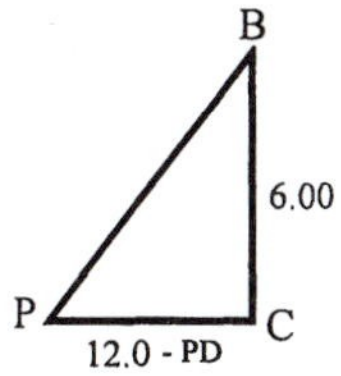

ΔAPD is

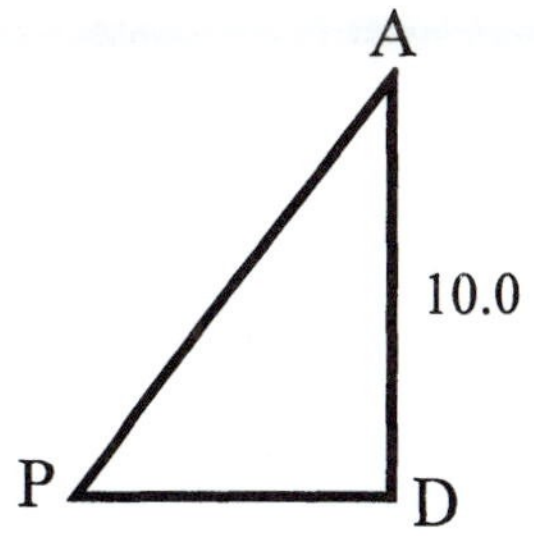

from which $\Delta BCP \sim \Delta ADP$, so $\frac{6.00}{12.0 - PD} = \frac{10.0}{PD}$

$\Rightarrow PD = 7.50$ and $PC = 12.0 - PD = 4.50$

$l = PB + PA = \sqrt{4.50^2 + 6.00^2} + \sqrt{7.50^2 + 10.0^2}$

$l = 20.0$ ft

2.3 Quadrilaterals

1.

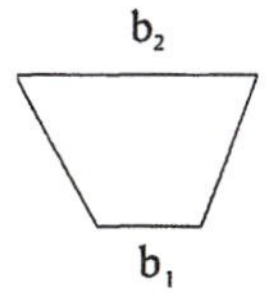

trapezoid

5. $p = 4s = 4(65) = 260$ m

9. $p = 2l + 2w = 2(3.7) + 2(2.7) = 12.8$ m

13. $A = s^2 = 2.7^2 = 7.3 \text{ mm}^2$

17. $A = bh = 3.7(2.5) = 9.3 \text{ m}^2$

21. $p = 2b + 4a$

25. The parallelogram is a rectangle.

29. The diagonal always divides the rhombus into two congruent triangles. All outer sides are always equal.

33.

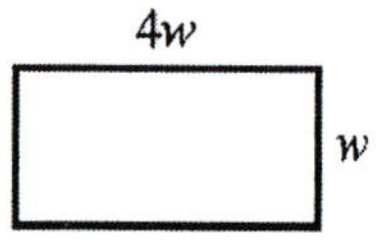

$$w + 2.5 = 4w - 4.7$$
$$w = 2.4 \text{ ft}$$
$$4w = 9.6 \text{ ft}$$

37.

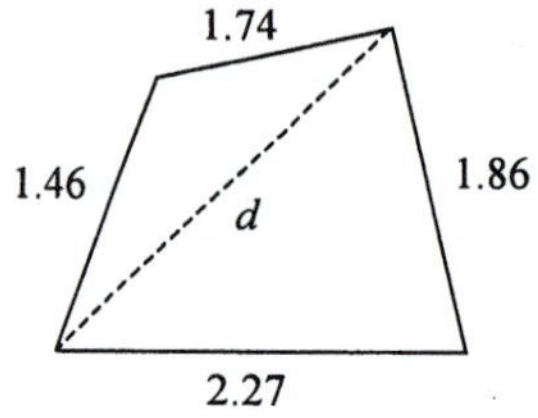

$$d = \sqrt{2.27^2 + 1.86^2}$$

For right triangle, $A = \frac{1}{2}(2.27)(1.86)$

For obtuse triangle, $s = \dfrac{1.46 + 1.74 + d}{2}$

and $A = \sqrt{s(s-1.46)(s-d)(s-1.74)}$

A of quadrilateral = Sum of areas of two triangles,

$$A = \frac{1}{2}(2.27)(1.86) + \sqrt{s(s-1.46)(s-d)(s-1.74)}$$

$$A = 3.04 \text{ km}^2$$

2.4 Circles

1. $\angle OAB + OBA + \angle AOB = 180^\circ$

$$\angle OAB + 90^\circ + 72^\circ = 180^\circ$$
$$\angle OAB = 18^\circ$$

5. (a) AD is a secant line.
(b) AF is a tangent line.

9. $c = 2\pi r = 2\pi(275) = 1730 \text{ ft}$

13. $A = \pi r^2 = \pi(0.0952^2) = 0.0285 \text{ yd}^2$

17. $\angle CBT = 90^\circ - \angle ABC = 90^\circ - 65^\circ = 25^\circ$

21. ARC $\text{BC} = 2(60^\circ) = 120^\circ$

25. $022.5^\circ\left(\dfrac{\pi}{180^\circ}\right) = 0.393 \text{ rad}$

29. $P = \dfrac{1}{4}(2\pi r) + 2r = \dfrac{\pi r}{2} + 2r$

33. All are on the same diameter.

37. $C = 2\pi r = 2\pi(3960) = 24{,}900 \text{ mi}$

41. $c = 112;\ c = \pi d;\ d = c/\pi = 112/\pi = 35.7 \text{ in.}$

45. A of room $= A$ of rectangle $+ \dfrac{3}{4} A$ of circle

$$A = 24(35) + \frac{3}{4}\pi(9.0)^2$$
$$A = 1000 \text{ ft}^2$$

2.5 Measurement of Irregular Areas

1. The use of smaller intervals improves the approximation since the total omitted area or the total extra area is smaller.

5. $A_{\text{trap}} = \dfrac{2.0}{2}\left[0.0 + 2(6.4) + 2(7.4) + 2(7.0) + 2(6.1)\right]$

$$\left[+2(5.2) + 2(5.0) + 2(5.1) + 0.0\right]$$

$A_{\text{trap}} = 84.4 = 84 \text{ m}^2$ to two significant digits

9. $A_{\text{trap}} = \dfrac{0.5}{2}\left[0.6 + 2(2.2) + 2(4.7) + 2(3.1) + 2(3.6)\right]$

$$\left[+2(1.6) + 2(2.2) + 2(1.5) + 0.8\right]$$

$A_{\text{trap}} = 9.8 \text{ m}^2$

13. $A_{\text{trap}} = \frac{45}{2}[170 + 2(360) + 2(420) + 2(410) + 2(390)$
$+ 2(350) + 2(330) + 2(290) + 230]$

$A_{\text{trap}} = 120,000 \text{ ft}^2$

17. A_{trap}

$= \frac{0.500}{2}[0.0 + 2(1.732) + 2(2.000) + 2(1.732)$
$+ 0.0] = 2.73 \text{ in.}^2$

This value is less than 3.14 in.^{2} because all of the trapezoids are inscribed.

2.6 Solid Geometric Figures

1. $V_1 = lwh_1,\ V_2 = (2l)(w)(2h) = 4lwh = 4V_1$

The volume is four times as much.

5. $V = e^3 = 7.15^3 = 366 \text{ ft}^3$

9. $V = \frac{4}{3}\pi r^3 = \frac{4}{3}\pi(0.877^3) = 2.83 \text{ yd}^3$

13. $V = \frac{1}{3}Bh = \frac{1}{3}(76^2)(130) = 250,000 \text{ in.}^3$

17. $V = \frac{1}{2}\left(\frac{4}{3}\pi r^3\right) = \frac{2}{3}\pi\left(\frac{0.83}{2}\right)^3 = 0.15 \text{ yd}^3$

21. $V = \frac{4}{3}\pi r^3 = \frac{4}{3}\pi\left(\frac{d}{2}\right)^3 = \frac{4}{3}\pi\frac{d^3}{8}$

$V = \frac{1}{6}\pi d^3$

25. $\frac{\text{final surface area}}{\text{original surface area}} = \frac{4\pi(2r)^2}{4\pi r^2} = \frac{4}{1}$

29. $V = \pi r^2 h = \pi(d/2)^2 h = \pi(4.0/2)^2(3,960,000)$
$= 5.0 \times 10^7 \text{ ft}^3 \text{ or } 0.00034 \text{ mi}^3$

33. $V = \frac{4}{3}\pi r^3 = \frac{4}{3}\pi(d/2)^3$
$= \frac{4}{3}\pi(165/2)^3$
$= 2.35 \times 10^6 \text{ ft}^3$

37. $c = 2\pi r = 29.8 \Rightarrow r = \frac{29.8}{2\pi}$

$V = \frac{4}{3}\pi r^3 = \frac{4}{3}\pi\left(\frac{29.8}{2\pi}\right)^3$

$V = 447 \text{ in.}^3$

Chapter 2 Review Exercises

1. $\angle CGE = 180^\circ - 148^\circ = 32^\circ$

5. $c = \sqrt{9^2 + 40^2} = 41$

9. $c = \sqrt{6.30^2 + 3.80^2} = 7.36$

13. $P = 3s = 3(8.5) = 25.5 \text{ mm}$

17. $C = \pi d = \pi(98.4) = 309 \text{ mm}$

21. $V = Bh = \frac{1}{2}(26.0)(34.0)(14.0) = 6190 \text{ cm}^3$

25. $A = 6e^2 = 6(0.520) = 1.62 \text{ m}^2$

29. $\angle BTA = \frac{50^\circ}{2} = 25^\circ$

33. $\angle ABE = 90^\circ - 37^\circ = 53^\circ$

37. $P = b + \sqrt{b^2 + (2a^2)^2} + \frac{1}{2}\pi(2a) = b + \sqrt{b^2 + 4a^2} + \pi a$

41. A square is a rectangle with four equal sides and a rectangle is a parallelogram with perpendicular intersecting sides so a square is a parallelogram. A rhombus is a parallelogram with four equal sides and since a square is a parallelogram, a square is a rhombus.

45.

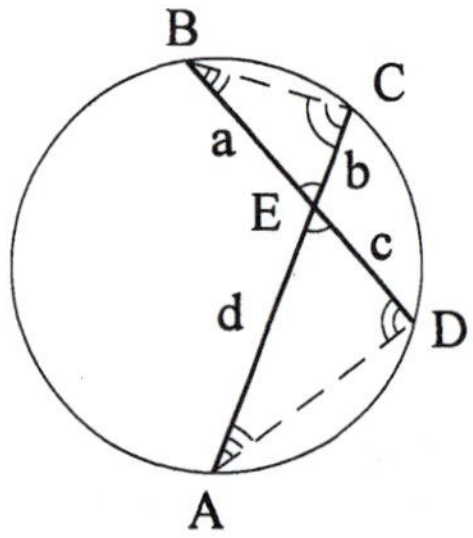

$\measuredangle BEC = \measuredangle AED$, vertical $\measuredangle$'s.

$\measuredangle BCA = \measuredangle ADB$, both are inscribed in $\overset{\frown}{AB}$

$\measuredangle CBE = \measuredangle CAD$, both are inscribed in $\overset{\frown}{CD}$

which shows $\Delta AED \sim \Delta BEC \Rightarrow \dfrac{a}{d} = \dfrac{b}{c}$

49. $L = \sqrt{1.2^2 + 7.8^2} = 7.9 \text{ m}$

53. $\dfrac{AB}{38} = \dfrac{42}{54}$

$AB = \dfrac{38(42)}{54} = 30 \text{ m}$

57. The longest distance in inches between points on the photograph is,

$\sqrt{8.00^2 + 10.0^2} = 12.8$ in. from which

$\dfrac{x}{12.8} = \dfrac{18{,}450}{1}$

$x = (12.8)(18{,}450) \text{ in.} \left(\dfrac{1 \text{ ft}}{12 \text{ in.}}\right)\left(\dfrac{\text{mi}}{5280 \text{ ft}}\right)$

$x = 3.73 \text{ mi}$

61. $A = (4.0)(8.0) - 2\pi \cdot \dfrac{1.0^2}{4} = 30 \text{ ft}^2$

65. $V = \pi r^2 h = \pi \left(\dfrac{4.3}{2}\right)^2 (13) = 190 \text{ m}^3$

69. $V = \pi r^2 h + \dfrac{1}{2} \cdot \dfrac{4}{3} \pi r^3$

$= \left[\pi \left(\dfrac{2.50}{2}\right)^2 \left(4.75 - \dfrac{2.50}{2}\right) + \dfrac{1}{2} \cdot \dfrac{4}{3} \cdot \pi \left(\dfrac{2.50}{2}\right)^3\right]$

$\left(\dfrac{7.48 \text{ gal}}{\text{ft}^3}\right)$

$= 159 \text{ gal}$

73. Label the vertices of the pentagon ABCDE. The area is the sum of the areas of three triangles, one with sides 921, 1490, and 1490 and two with sides 921, 921, and 1490. The semi-perimeters are given by;

$s_1 = \dfrac{921 + 921 + 1490}{2} = 1666$ and

$s_2 = \dfrac{921 + 1490 + 1490}{2} = 1950.5.$

$A = 2\sqrt{1666(1666 - 921)(1666 - 921)(1666 - 1490)}$

$+ \sqrt{1950.5(1950.5 - 1490)(1950.5 - 1490)}$

$+ \sqrt{(1950.5 - 921)}$

$= 1{,}460{,}000 \text{ ft}^2$

Chapter 3

FUNCTIONS AND GRAPHS

3.1 Introduction to Functions

1. $f(x)=3x-7$
$f(-2)=3(-2)-7=-13$

5. (a) $A(r)=\pi r^2$

(b) $A(d)=\pi\left(\frac{d}{2}\right)^2=\frac{1}{4}\pi d^2$

9. From geometry, $A=s^2$

$$\sqrt{A}=\sqrt{s^2}$$
$$s=\sqrt{A}$$

13. $f(x)=2x+1;\ f(1)=2(1)+1=3;$
$f(-1)=2(-1)+1=-1$

17. $\phi(x)=\frac{6-x^2}{2x}$

$$\phi(2\pi)=\frac{6-(2\pi)^2}{2(2\pi)}=\frac{6-4\pi^2}{4\pi}$$
$$\phi(2\pi)=\frac{3-2\pi^2}{2\pi}$$
$$\phi(-2)=\frac{6-(-2)^2}{2\cdot(-2)}=\frac{2}{-4}=-\frac{1}{2}$$

21. $K(s)=3s^2-s+1;$
$K(-s)=3(-s)^2-(-s)+6=3s^2+s+6$
$K(2s)=3(2s)^2-2s+6=12s^2-2s+6$

25. $f(x)=5x^2-3x;\ f(3.86)=5\cdot 3.86^2-3\cdot 3.86=62.9$
$f(-6.92)=5\cdot(-6.92)^2-3\cdot(-6.92)=260$

29. $f(x)=x^2+2;$ square x and add 2 to the result.

33. Take 3 times the sum of twice the independent variable and 5, then subtract 1.

37. $A=5e^2$
$f(e)=5e^2$

41. $s=f(t)=17.5-4.9t^2;\ f(12)=17.5-4.9\cdot(1.2)^2$
$s=10.4$ m

45. $d=f(t)=55t$

3.2 More About Functions

1. $f(x)=-x^2+2$ is defined for all real values of x; the domain is all real numbers. Since $-x^2+2\le 2$ the range is all real numbers $f(x)\le 2$.

5. The domain and range of $f(x)=x+5$ is all real numbers.

9. The domain of $f(s)=\frac{2}{s^2}$ is all real numbers except zero since it gives a division by zero. The range is all positive real numbers because $\frac{2}{s^2}$ is always positive.

13. The domain of $y=|x-3|$ is all real numbers and the range is all nonnegative numbers, $y\ge 0$.

17. $f(D)=\frac{D}{D-2}+\frac{4}{D+4}$; since division by zero is undefined, the domain must be restricted to exclude any value for which $D-2,\ D+4$ are equal to zero. In this case, $D\ne 2,\ -4$. So the domain is the set of all real numbers except $2,\ -4$.

21. $f(1)=\sqrt{1+3}=\sqrt{4}=2$ (since $1\ge 1$)
$f(-0.25)=-0.25+1=0.75$ (since $-0.25<1$)

25. $w=f(t)=5500-2t$

29. $C=f(l)=500+5(l-50)=5l+250$

33. $2750 = 15x + 25y \Rightarrow y = f(x) = \dfrac{2750 - 15x}{25}$

37. $d = f(h)$

$120^2 + h^2 = d^2$

$d = \sqrt{14,400 + h^2}$

Range: $d \geq 120$ m since $d = 120$ when $h = 0$.

Domain: $h \geq 0$ since distance above ground is nonnegative.

41. The domain of $f = \dfrac{1}{2\pi\sqrt{C}}$ is $C > 0$ because C must be ≥ 0 to avoid taking the square root of a negative and > 0 to prevent division by zero.

45. (a) $V = lwh$

$V = (2w - 4)(w - 4)(2)$

$V = (2w^2 - 12w + 16)(2)$

$V = 4w^2 - 24w + 32$

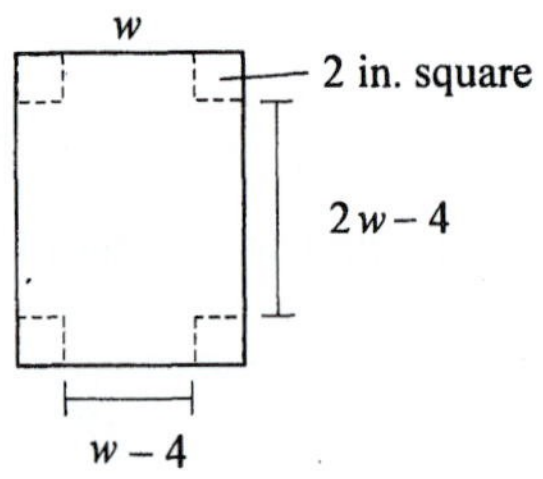

(b) Domain: $w > 4$ in.

49. $f(x) = |x| + |x - 2|$, Since $|x| \geq 0$ and $|x - 2| \geq 2$, $f(x) \geq 0 + 2 = 2$ the range of $f(x)$ is all real numbers $y \geq 2$.

3.3 Rectangular Coordinates

1. $A(-1, -2), (4, -2), C(4, 1)$

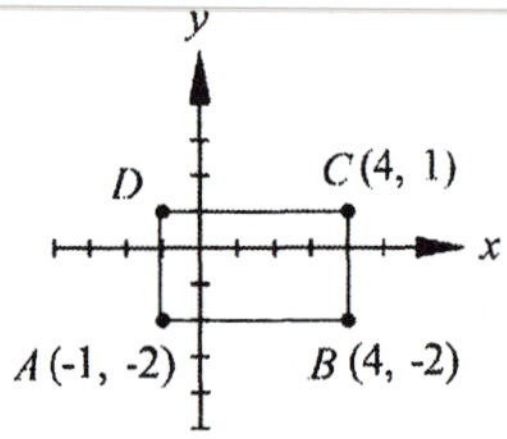

5.

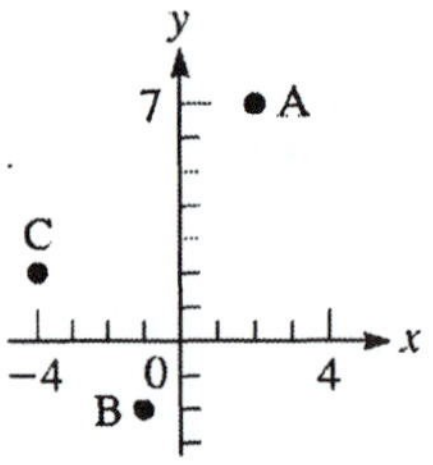

9. Rectangle

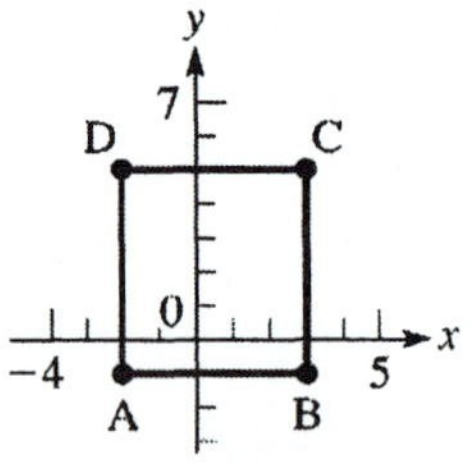

13. In order for the x-axis to be the perpendicular disector of the line segment join P and Q, Q must be $(3, -2)$.

17. All points $(x, 3)$, where x is any real number, are points on a line parallel to the x-axis, 3 units above it.

21. Abscissas are x-coordinates; thus the abscissa of all points on the y-axis is zero.

25. All points which lie to the left of a line that is parallel to the y-axis, one unit to the left have $x < -1$.

29. $xy = 0$ for $x = 0$ (y-axis) or $y = 0$ (x-axis)
$xy = 0$ on either axis.

33. (a) $d = 3-(-5) = 8$
(b) $d = 4-(-2) = 6$

3.4 The Graph of a Function

1. $f(x) = 3x+5$

x	y
−3	−4
−2	−1
−1	2
0	5
1	8

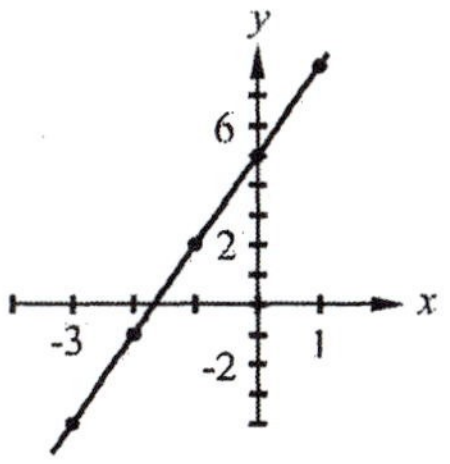

5. $y = 3x$

x	y
−1	−3
0	0
1	3

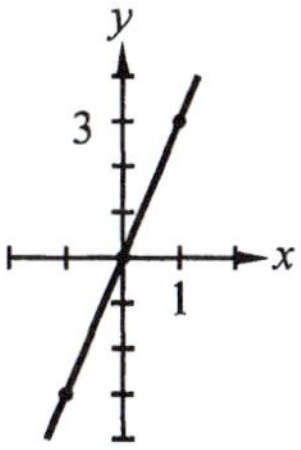

9. $s = 7-2t$

s	t
−1	9
0	7
1	5

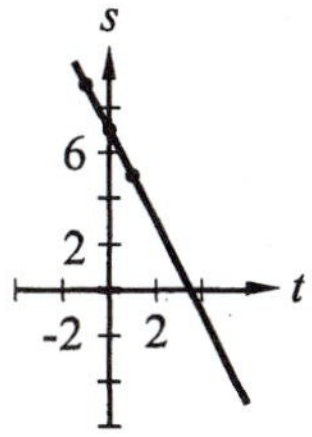

13. $y = x^2$

x	y
−2	4
−1	1
0	0
1	1
2	4

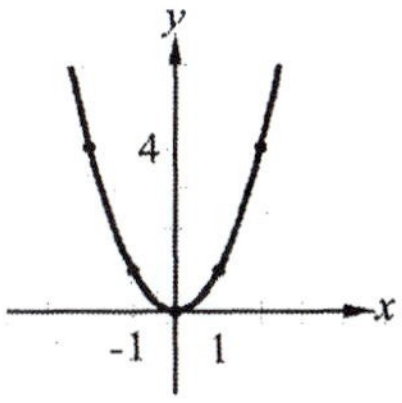

17. $y = \frac{1}{2}x^2 + 2$

x	y
−4	10
−2	4
0	2
2	4
4	10

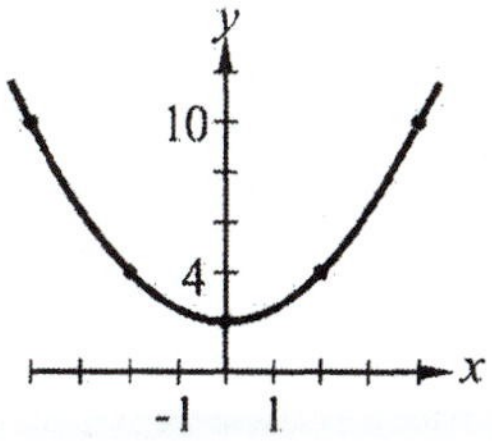

21. $y = x^2 - 3x + 1$

x	y
3	1
2	−1
1.5	−1.25
1	−1
0	1

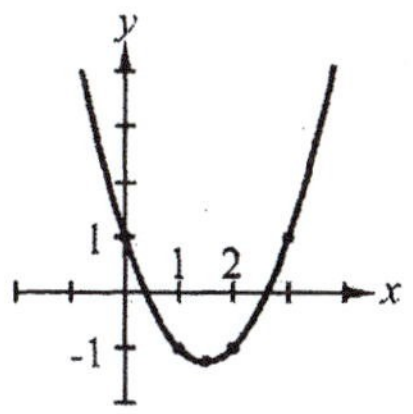

25. $y = x^3 - x^2$

x	y
−2	−12
−1	−2
0	0
2/3	−4/27
1	0
2	4

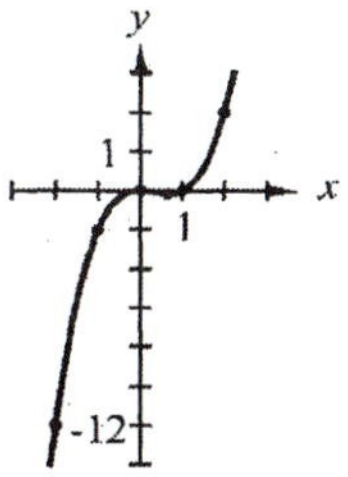

29. $P = \frac{1}{V} + 1$

V	P
−3	2/3
−2	1/2
−1	0
1	2
2	3/2
3	4/3

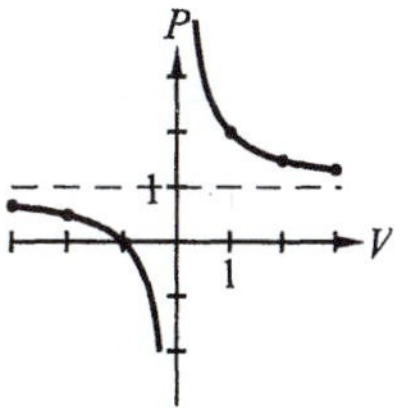

33. $y=\sqrt{x}$

x	y
0	0
1	1
4	2
9	3
16	4

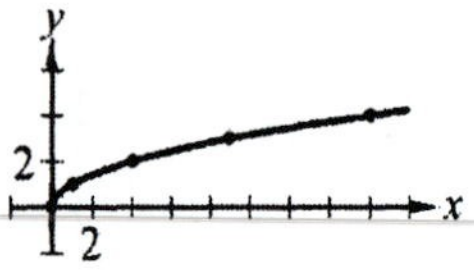

37. $n=0.40m$

m	n
10	4
50	20
80	32

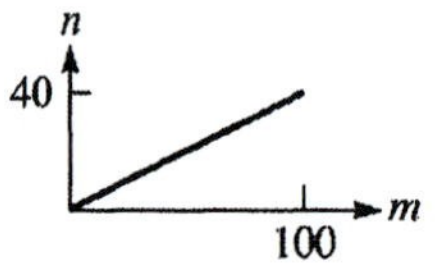

41. $H=240I^2$

I	H
0	0
0.2	9.6
0.4	38.4
0.6	86.4
0.8	153.6

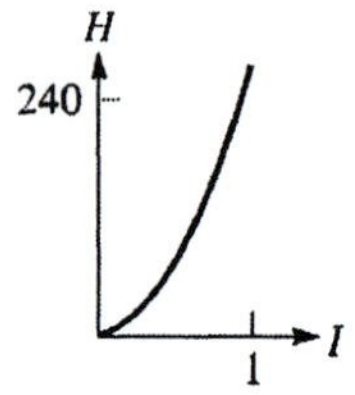

45. $P=0.004v^3$

v	P
0	0
5	0.5
10	4.0
15	13.5
20	32.0

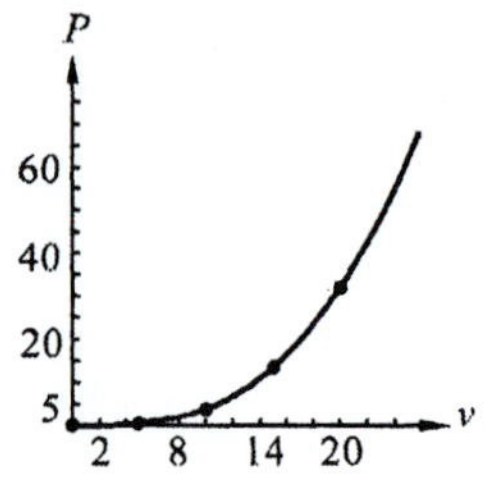

49. $P=2l+2w=200 \Rightarrow l=100-w$

$A=lw=(100-w)w=100w-w^2$ for $30\le w\le 70$

w	30	40	50	60	70
A	2100	2400	2500	2400	2100

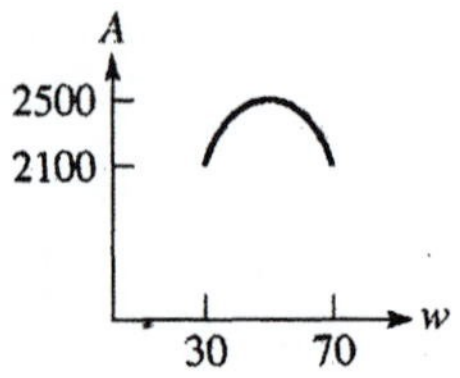

53. $S=\dfrac{5n}{4+n}$

n	S
0	0
2	5/3
4	5/2
6	3
8	10/3

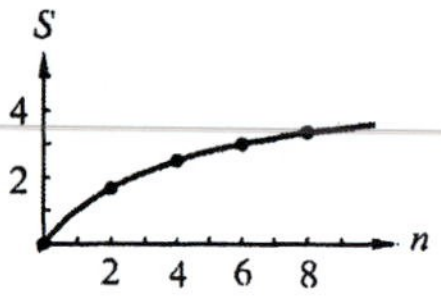

57.

x	y
−2	2
−1	1
0	0
1	1
2	2

$y=x$ is the same as $y=|x|$ for $x\ge 0$.
$y=|x|$ is the same as $y=-x$ for $x<0$.

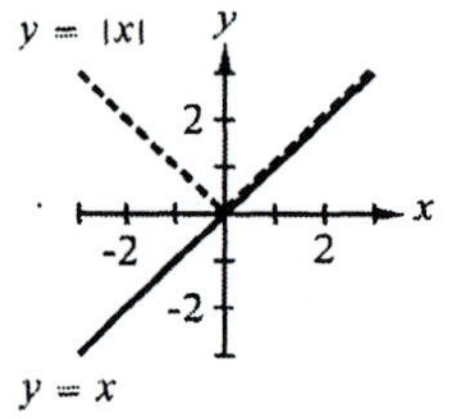

For negative values of x, $y=|x|$ becomes $y=-x$.

61. (a) $y=x+2$

x	y
−3	−1
−2	0
−1	1
0	2
1	3
2	4

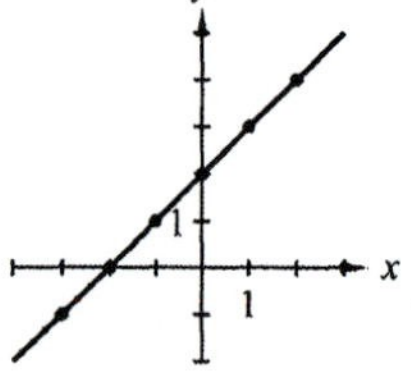

(b) $y = \dfrac{x^2 - 4}{x - 2}$

x	y
−3	−1
−2	0
−1	1
0	2
1	3
2	undefined

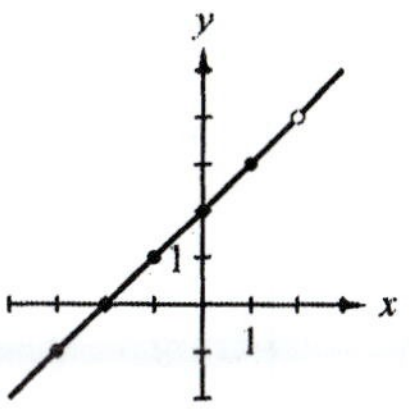

65. No. Some vertical lines will intercept the graph at multiple points.

3.5 Graphs on the Graphing Calculator

1. $x^2 + 2x = 1 \Rightarrow x^2 + 2x - 1 = 0$. Graph $y = x^2 + 2x - 1$.

x	y
−4	7
−3	2
−2	−1
−1	−2
0	−1
1	2
2	7

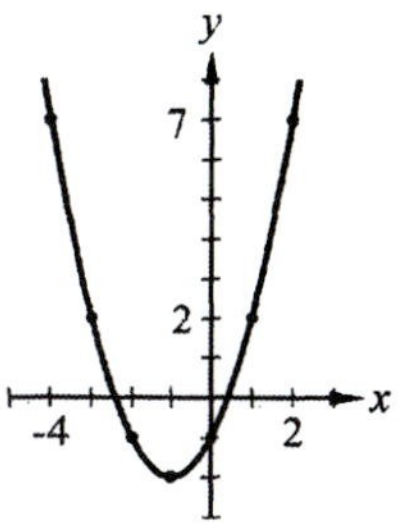

$x = -2.4$, $x = 0.4$

5. $y = 3x - 1$

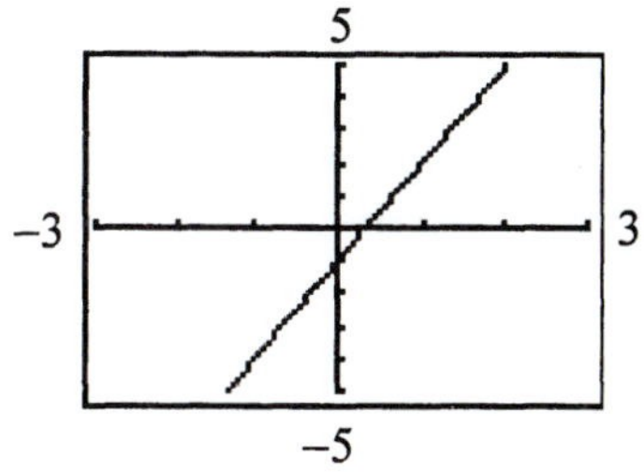

9. $y = 6 - x^3$

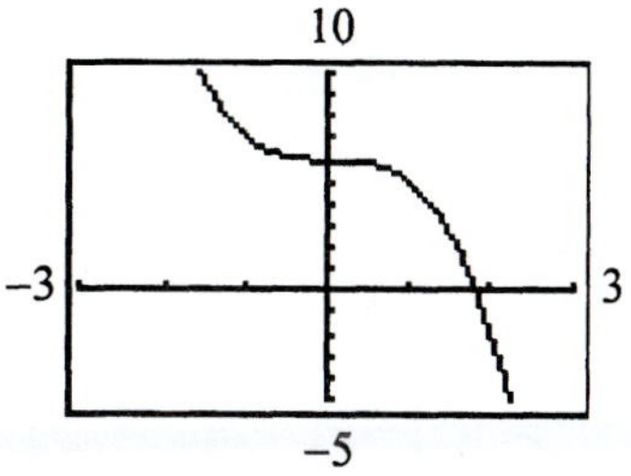

13. $y = \dfrac{2x}{x - 2}$

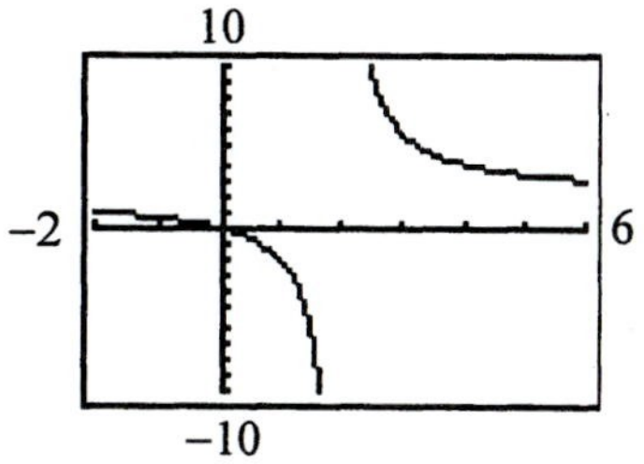

17. $y = 3 + \dfrac{2}{x}$

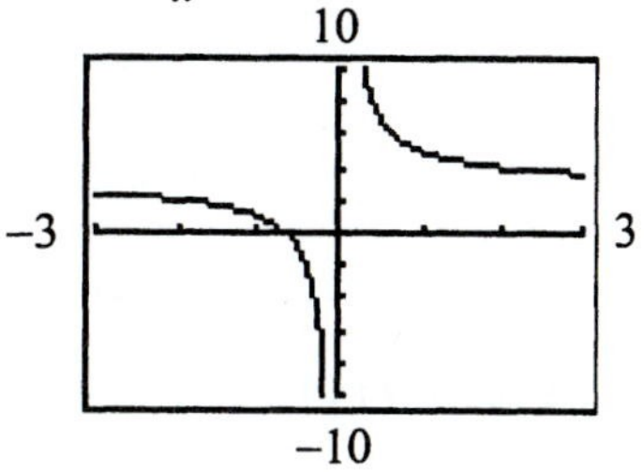

21. $x^3 - 3 = 3x \Rightarrow x^3 - 3x - 3 = 0$. Graph $y = x^3 - 3x - 3$ and use the zero feature to solve.

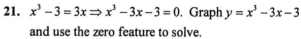

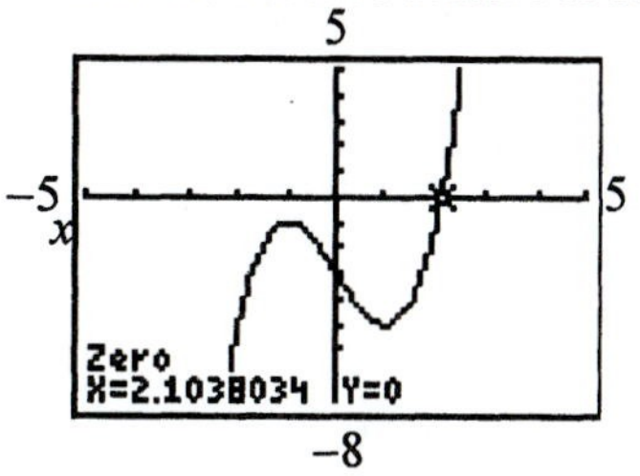

25. $\sqrt{5R+2}=3 \Rightarrow \sqrt{5R+2}-3=0$

Graph $y=\sqrt{5x+2}-3$ and use the zero feature to solve.

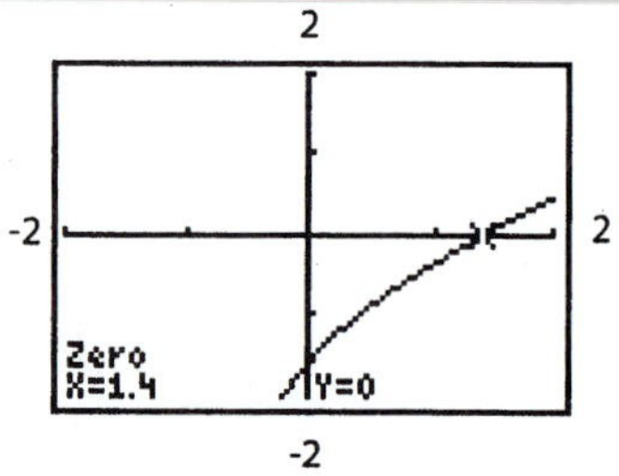

$x=1.400$

29. From the graph, $y=\dfrac{4}{x^2-4}$ has a range $y \le -1$ or $y > 0$.

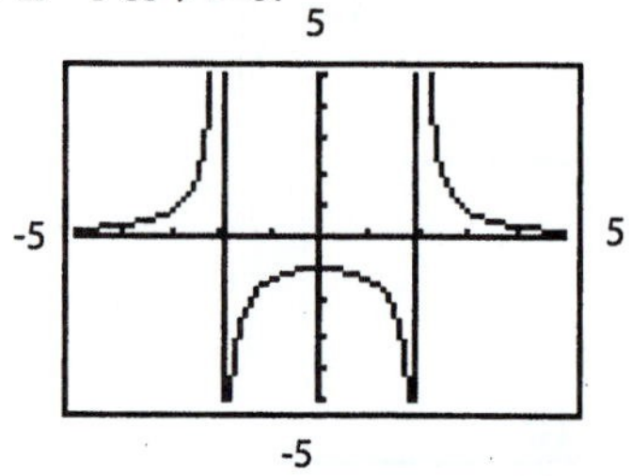

33. Graph $y=\dfrac{x+1}{\sqrt{x-2}}$ on the graphing calculator and use the minimum feature, then from the graph, $Y(y)=\dfrac{y+1}{\sqrt{y-2}}$ has range $Y(y) \ge 3.464$.

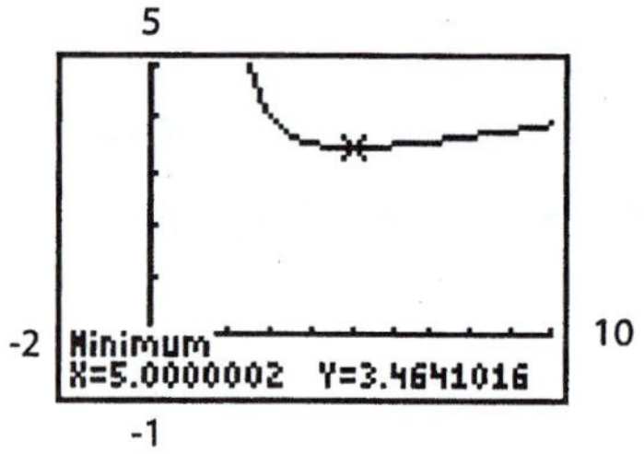

37. function: $y=3x$

function shifted up 1: $y=3x+1$

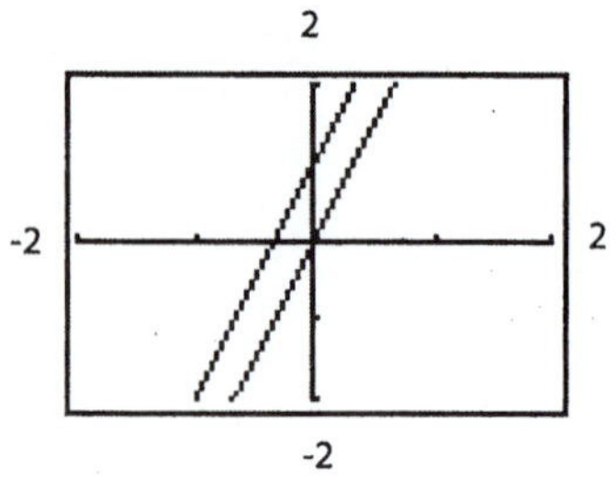

41. function: $y=-2x^2$

function shifted down 3, left 2: $y=-2(x+2)^2-3$

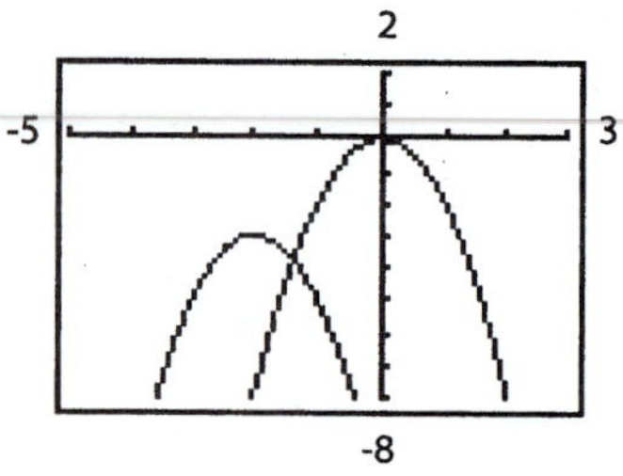

45. $i=0.1v-0.06$, graph $y=0.01x-0.06$

From the graph, $v=6.0\ V$ for $i=0$.

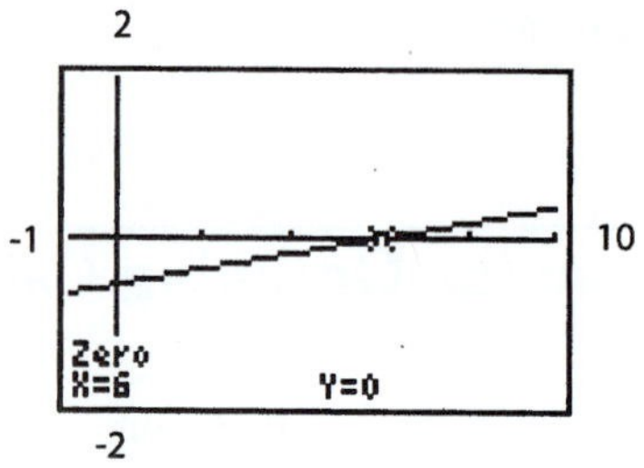

49. $A=520=lw=(x+12)w=w^2+12w \Rightarrow w^2+12w-520=0$

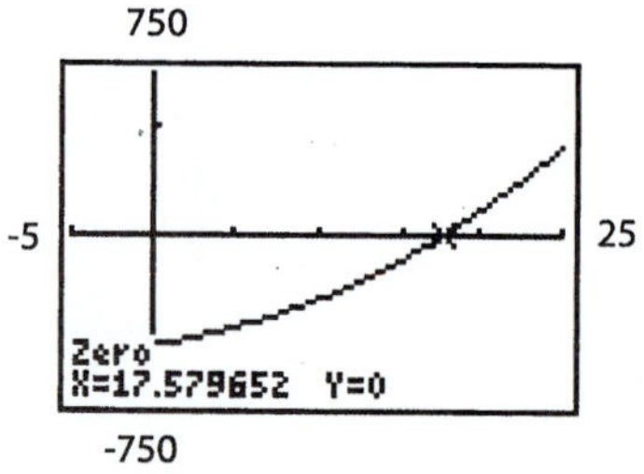

The approximate dimensions, in. cm, to 2 significant digits are $w \approx 18$ cm, $l \approx 30$ cm.

53. Graph $y=x^3-2.000x^2-30.00$ and use the zero feature to solve.

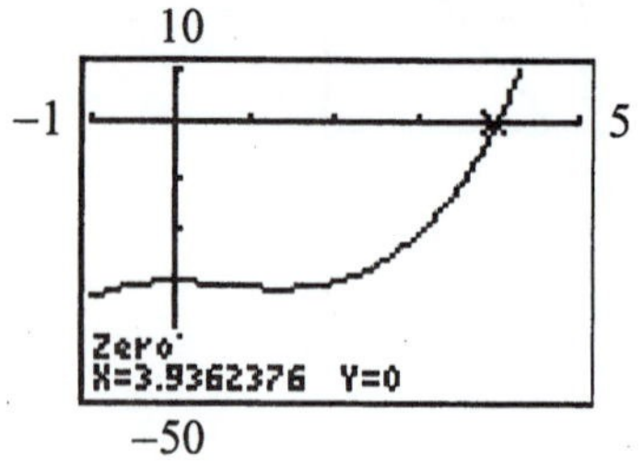

From the graph $x = 3.936$. The dimensions are 3.936 cm, 2.936 cm, and 1.936 cm.

57. (a) Since V is increasing at a constant rate and $V = \frac{4}{3}\pi r^3$ it seems reasonable that r^3 should be increasing at a constant rate. Thus, r should be proportional to the cube root of time, $r = k\sqrt[3]{t}$.

(b) $r = \sqrt[3]{3t}$ would be a typical situation.

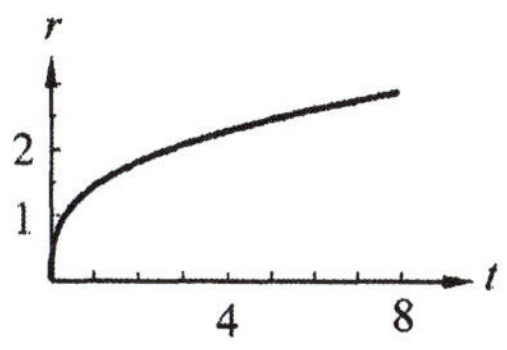

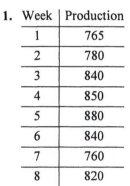

3.6 Graphs of Functions Defined by Tables of Data

1.

Week	Production
1	765
2	780
3	840
4	850
5	880
6	840
7	760
8	820

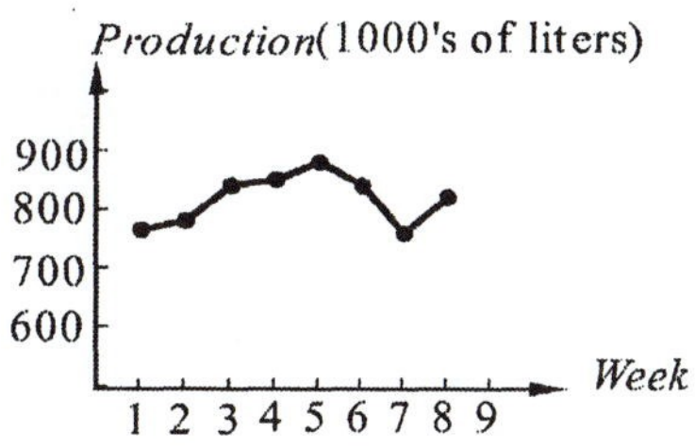

5.

Dist. (cm)	M. ind. (H)
0.0	0.77
2.0	0.75
4.0	0.61
6.0	0.49
8.0	0.38
10.0	0.25
12.0	0.17

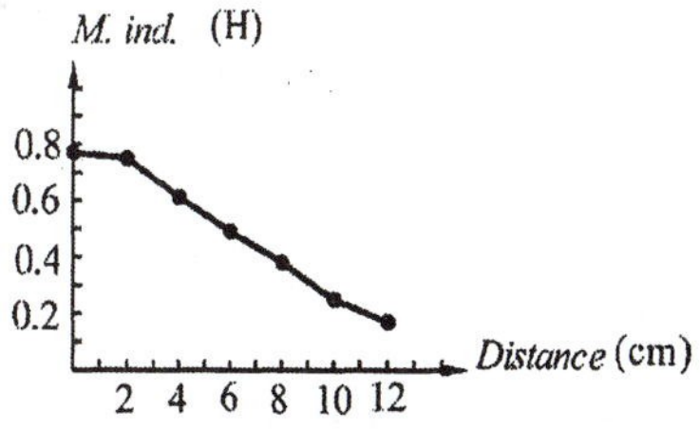

9. (a) Reading from the graph, $T = 132^\circ$C for $t = 4.3$ min.

(b) Reading from the graph, $t = 0.7$ min for $T = 145.0^\circ$C.

13. $$2\left[\begin{array}{c} 1.2\begin{bmatrix} 8.0 & 0.38 \\ 9.2 & ? \end{bmatrix} x \\ 10.0 \quad 0.25 \end{array}\right] -0.13$$

$\frac{1.2}{2} = \frac{x}{-0.13}$, $x = -0.78$

Therefore, $M = 0.38 - 0.078 = 0.30$ H

17.

Height (ft)	Rate (ft^3 / s)
0	0
1.0	10
2.0	15
4.0	22
6.0	27
8.0	31
12	35

(a) For $R = 20\ \text{ft}^3/\text{s}$, $H = 3.4$ ft

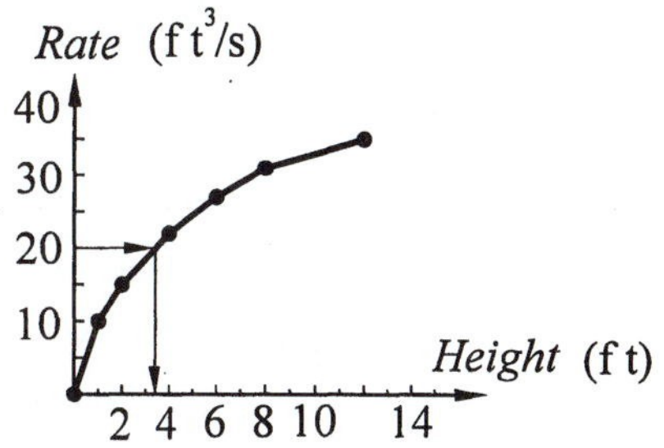

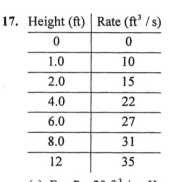

(b) For $H = 2.5$ ft, $R = 17\ \text{ft}^3/\text{s}$

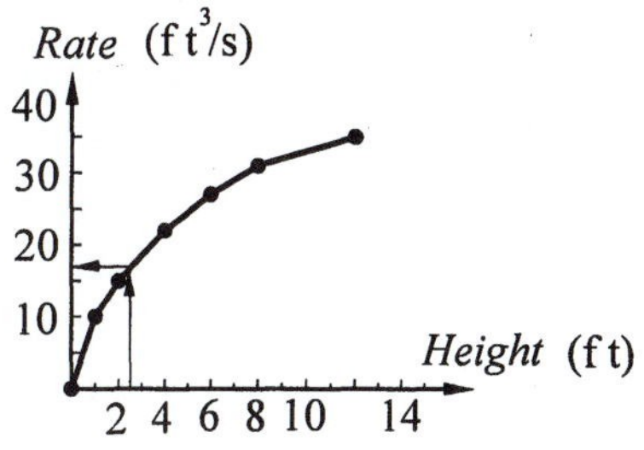

21. $10\left[6\begin{bmatrix}30 & 0.30\\ 46 & ?\end{bmatrix}x \atop \begin{matrix}40 & 0.37\end{matrix}\right]0.07$

$\frac{6}{10}=\frac{x}{0.07}$, $x=0.042$

Therefore, $f=0.30+0.042=0.34$

25. The graph is extended using a straight line segment.

$T\approx 130.3^{\circ}\text{C}$ for $t=5.3$ min

Chapter 3 Review Exercises

1. $A=\pi r^2=\pi(2t)^2$

$A=4\pi t^2$

5. $f(x)=7x-5$

$f(3)=7(3)-5=21-5=16$

$f(-6)=7(-6)-5=-42-5=-47$

9. $F(x)=x^3+2x^2-3x$

$$\begin{aligned}F(3+h)-F(3)&=(3+h)^3+2(3+h)^2-3(3+h)\\&\quad-\left(3^3+2\left(3^2\right)-3(3)\right)\\&=27+27h+9h^2+h^3+18+12h\\&\quad+2h^2-9-3h-27-18+9\\&=h^3+11h^2+36h\end{aligned}$$

13. $f(x)=8.07-2x$

$f(5.87)=8.07-2\cdot 5.87=-3.67$

$f(-4.29)=8.07-2(-4.29)=16.65\approx 16.7$

17. The domain of $f(x)=x^4+1$ is $-\infty<x<\infty$. The range is $y\geq 1$.

21. $f(n)=1+\frac{2}{(n-5)^2}$ has domain $n\neq 5$ and range $f(n)>1$.

25. $s=4t-t^2$

t	s
-1	-5
0	0
1	3
2	4
3	3
4	0
5	-5

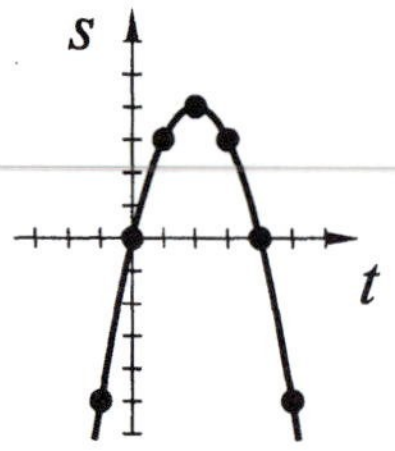

29. The graph of $A=2-s^4$ is

A	s
-2	-14
-1	1
0	2
1	1
0	-14

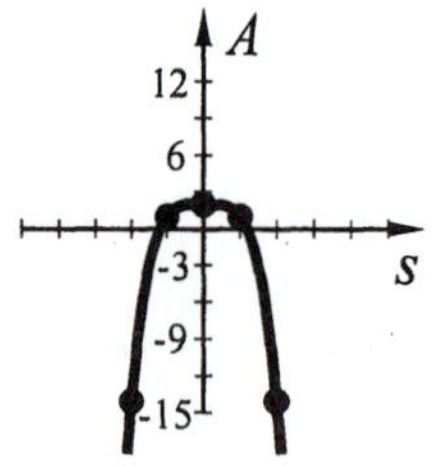

33. Graph $y_1=7x-3$ and use the zero feature to solve.

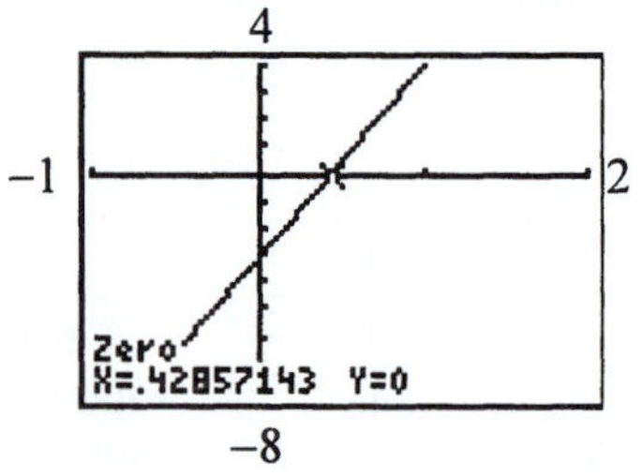

37. $x^3-x^2=2-x\Rightarrow x^3-x^2+x-2=0$

Graph $y_1=x^3-x^2+x-2$ and use the zero feature to solve.

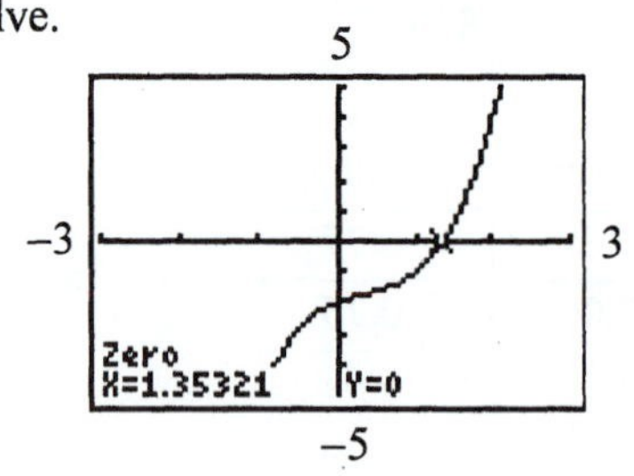

41. Graph $y_1 = x^4 - 5x^2$ and use the minimum feature, then from the graph the range is $y \geq -6.25$.

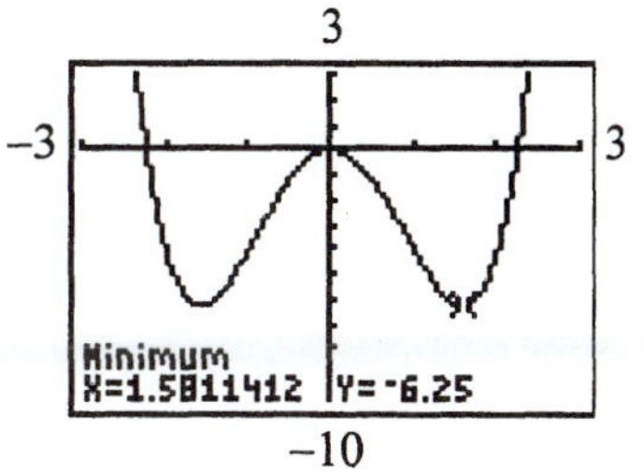

45. $A(a,\ b) = A(2,\ -3)$ is in QIV which $B(b,\ a)$ $= B(-3,\ 2)$ is in QII.

49. $\left|\frac{y}{x}\right| > 0$ for $\frac{y}{x} \neq 0 \Rightarrow$ all $(x,\ y)$ not on x-axis or y-axis.

53. $y = \sqrt{x-1}$ shifted left 2 and up 1 is

$y = \sqrt{(x+2)-1}+1$

$y = \sqrt{x+1}+1$

57. $(2,\ 3),\ (5,\ -1),\ (-1,\ 3)$ do not lie on a straight line. However, they do all lie on a parabola of the form.

$f(x) = ax^2 + bx + c$

$f(2) = 4a + 2b + c = 3$

$f(5) = 25a + 5b + c = -1$

$f(-1) = a - b + c = 3$

The last three equations have $a = -\frac{2}{9},\ b = \frac{2}{9}$

$c = \frac{31}{9}$ as the solution. All three points lie on the graph of $f(x) = \frac{-2x^2}{9} + \frac{2x}{9} + \frac{31}{9}$.

61. Graph $y_1 = 8.0 + 12x^2 - 2x^3$ for $0 \leq x \leq 6$ and use the maximum feature.

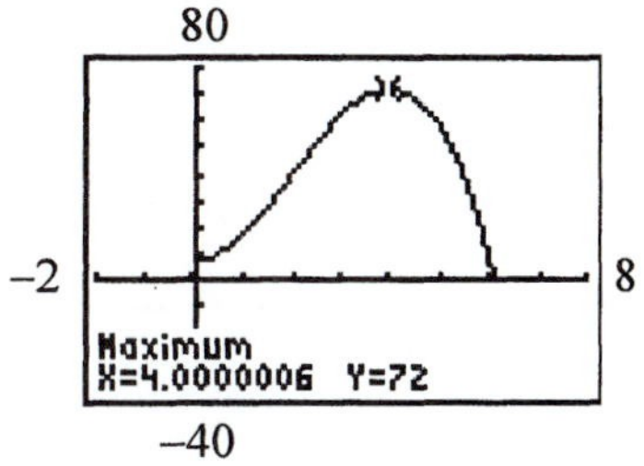

65. $P = 50(p-50) \qquad 30 \leq p \leq 150$

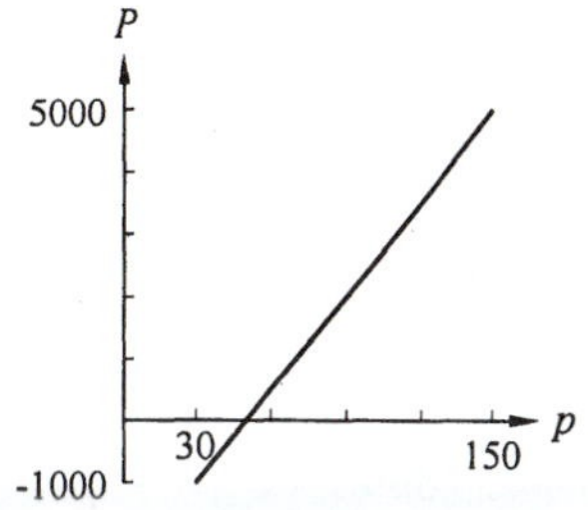

69. $e = f(T) = \dfrac{100\left(T^4 - 307^4\right)}{307^4}$

$e = f(309) = \dfrac{100\left(309^4 - 307^4\right)}{307^4} = 2.63$

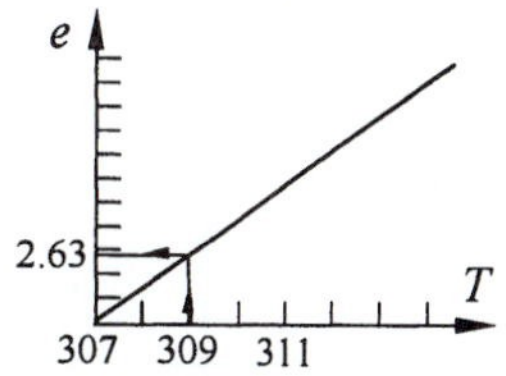

73.

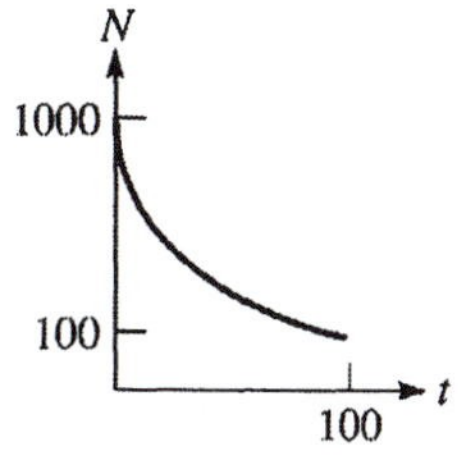

77. $20\left[\ 7\begin{bmatrix} 40 & 11.1 \\ 47 & ? \end{bmatrix} x \ \right] 0.6$
$\qquad\quad 60 \quad 11.7$

$\frac{x}{0.6} = \frac{7}{20} \Rightarrow x = \frac{0.6(7)}{20}$

$f(47) = 11.1 + \frac{0.6(7)}{20} = 11.3$ ft

81. $$s = 135 + 4.9T + 0.19T^2$$

$$500 = 135 + 4.9T + 0.19T^2$$

$$0.19T^2 + 4.9T - 365 = 0$$

Graph $y_1 = 0.19x^2 + 4.9x - 365$ and use the zero feature to solve.

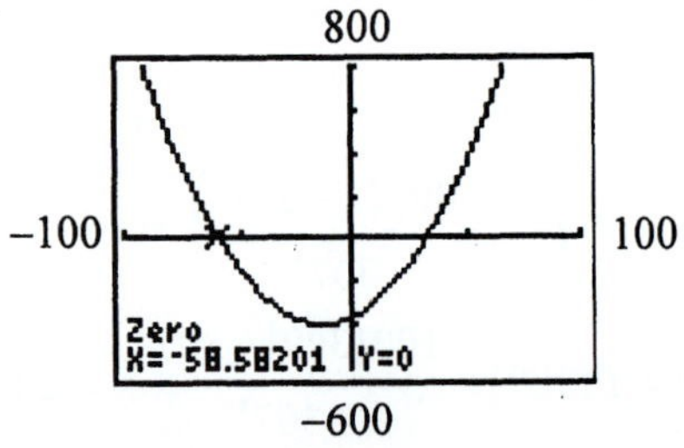

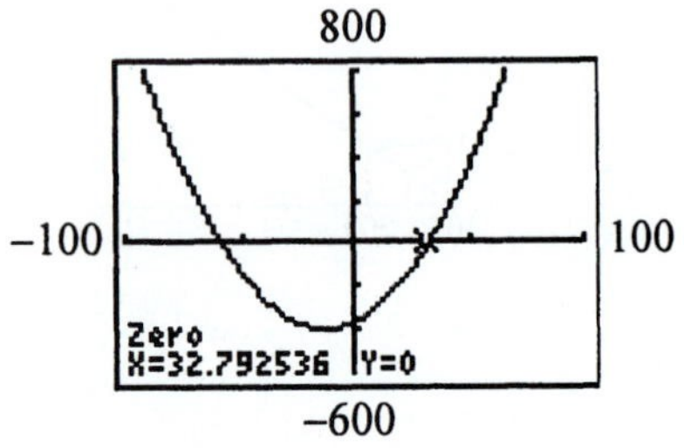

85. $T = \dfrac{4t^2}{t+2} - 20,\ t \geq 0$

$$0 = \frac{4t^2}{t+2} - 20 \Rightarrow 4t^2 - 20(t+2) = 0$$

$$4t^2 - 20t - 40 = 0$$

$$t^2 - 5t - 10 = 0$$

Graph $y_1 = x^2 - 5x - 10$ for $x \geq 0$ and use the zero feature to solve.

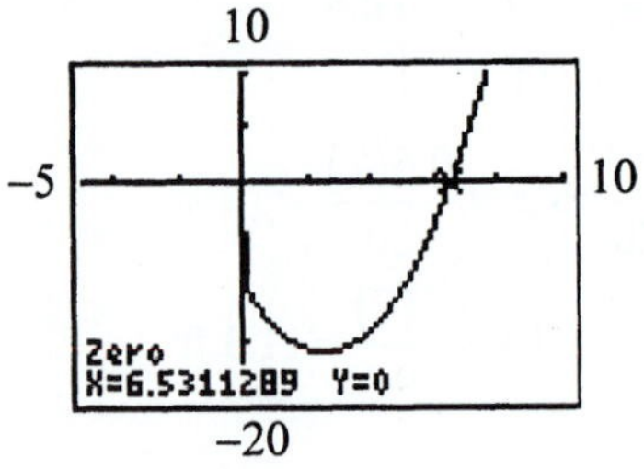

Chapter 4

THE TRIGONOMETRIC FUNCTIONS

4.1 Angles

1. $145.6^\circ + 2(360^\circ) = 865.6^\circ$

5.

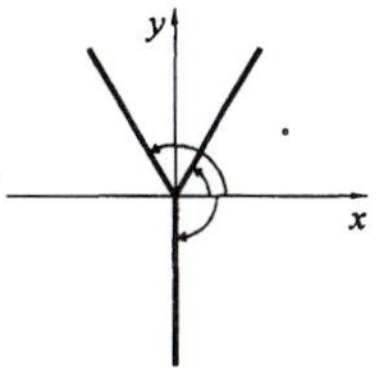

9. positive: $45^\circ + 360^\circ = 405^\circ$

negative: $45^\circ - 360^\circ = -315^\circ$

13. positive: $70^\circ 30' + 360 = 430^\circ 30'$

negative: $70^\circ 30' - 360^\circ = -289^\circ 30'$

17. To change 0.265 rad to degrees multiply by $\frac{180}{\pi}$,

$0.265 \text{ rad}\left(\frac{180^\circ}{\pi \text{ rad}}\right) \approx 15.18^\circ$

21. $0.329 \text{ rad} \approx 18.85^\circ$

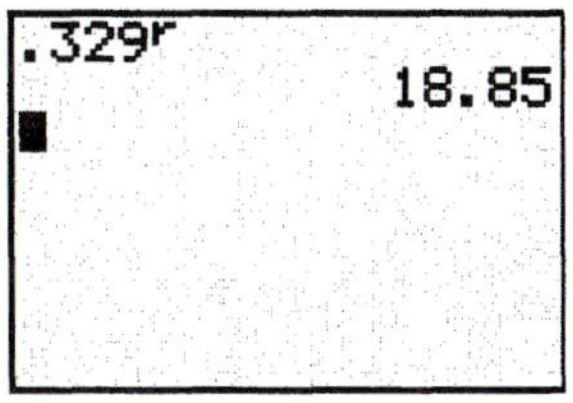

25. $56.0^\circ = 0.977$ rad to three significant digits

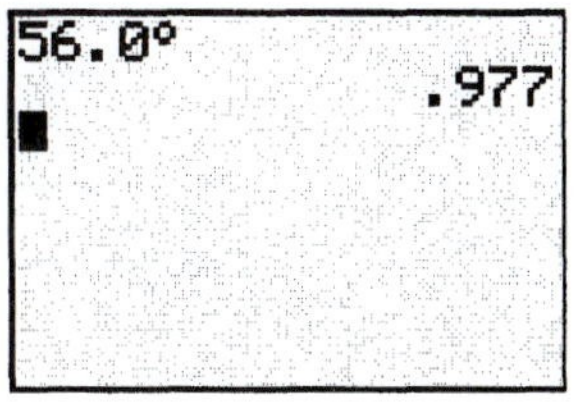

29. $47^\circ + 0.5^\circ \cdot \frac{60'}{1^\circ} = 47^\circ + 30' = 47^\circ 30'$

33. $15^\circ 12' = 15^\circ + 12' \cdot \frac{1^\circ}{60'} = 15.2^\circ$

37. Angle in standard position terminal side passing through $(4, 2)$.

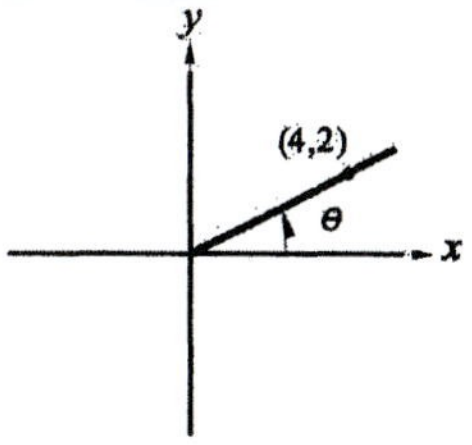

41. Angle in standard position terminal side passing through $(-7, 5)$.

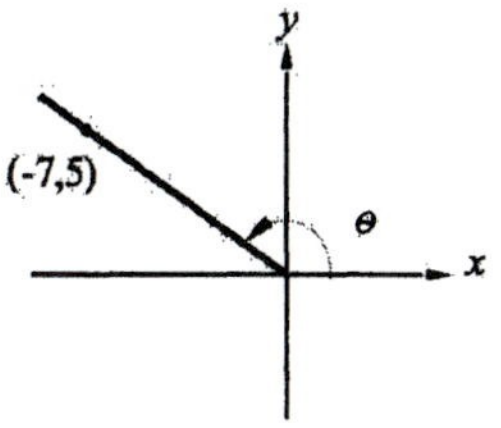

45. 31°, *QI*; 310°, *QIV*

49. 1 rad, QI; 2 rad, QII

53. $21^\circ 42' 36''$

$= 21^\circ + 42' \cdot \frac{1^\circ}{60'} + 36'' \cdot \frac{1'}{60''} \cdot \frac{1^\circ}{60'} = 21.710^\circ$

4.2 Defining the Trigonometric Functions

1. $r = \sqrt{4^2 + 3^2} = \sqrt{25} = 5$

$\sin\theta = \frac{3}{5}$

$\cos\theta = \frac{4}{5}$

$\tan\theta = \frac{3}{4}$

$\csc\theta = \frac{5}{3}$

$\sec\theta = \frac{5}{4}$

$\cot\theta = \frac{4}{3}$

5. $r = \sqrt{15^2 + 8^2} = \sqrt{289} = 17$

$\sin\theta = \frac{y}{r} = \frac{8}{17}$

$\cos\theta = \frac{x}{r} = \frac{15}{17}$

$\tan\theta = \frac{y}{x} = \frac{8}{15}$

$\csc\theta = \frac{r}{y} = \frac{17}{8}$

$\sec\theta = \frac{r}{x} = \frac{17}{15}$

$\cot\theta = \frac{x}{y} = \frac{15}{8}$

9. $r = \sqrt{1+15} = \sqrt{16} = 4$

$\sin\theta = \frac{y}{r} = \frac{\sqrt{15}}{4}$

$\cos\theta = \frac{x}{r} = \frac{1}{4}$

$\tan\theta = \frac{y}{x} = \sqrt{15}$

$\csc\theta = \frac{r}{y} = \frac{4}{\sqrt{15}}$

$\sec\theta = \frac{r}{x} = 4$

$\cot\theta = \frac{x}{y} = \frac{1}{\sqrt{15}}$

13. $r = \sqrt{50^2 + 20^2} = \sqrt{2900} = 10\sqrt{29}$

$\sin\theta = \frac{y}{r} = \frac{20}{10\sqrt{29}} = \frac{2}{\sqrt{29}}$

$\cos\theta = \frac{x}{r} = \frac{50}{10\sqrt{29}} = \frac{5}{\sqrt{29}}$

$\tan\theta = \frac{y}{x} = \frac{20}{50} = \frac{2}{5}$

$\csc\theta = \frac{r}{y} = \frac{10\sqrt{29}}{20} = \frac{\sqrt{29}}{2}$

$\sec\theta = \frac{r}{x} = \frac{10\sqrt{29}}{50} = \frac{\sqrt{29}}{5}$

$\cot\theta = \frac{x}{y} = \frac{50}{20} = \frac{5}{2}$

17. $\cos\theta = \frac{12}{13} \Rightarrow x = 12$ and $r = 13$ with θ in QL

$r^2 = x^2 + y^2 \Rightarrow 169 = 144 + y^2 \Rightarrow y^2 = 25$

$y = 5$

$\sin\theta = \frac{y}{r} = \frac{5}{13}$, $\cot\theta = \frac{x}{y} = \frac{12}{5}$.

21. $\sin\theta = 0.750 \Rightarrow y = 0.750$ and $r = 1$ with θ in QI.

$r^2 = x^2 + y^2 \Rightarrow 1^2$

$= x^2 + 0.750^2 \Rightarrow x^2$

$= 0.4375 \Rightarrow x = 0.661$

$\cot\theta = \frac{x}{y} = \frac{0.661}{0.750} = 0.881.$

$\csc\theta = \frac{r}{y} = \frac{1}{0.750} = 1.333.$

25. For $(3, 4)$, $r = 5$,

$\sin\theta = \frac{y}{r} = \frac{4}{5}$ and $\tan\theta = \frac{y}{x} = \frac{4}{3}$.

For $(6, 8)$, $r = 10$,

$\sin\theta = \frac{y}{r} = \frac{8}{10} = \frac{4}{5}$ and $\tan\theta = \frac{y}{x} = \frac{8}{6} = \frac{4}{3}$.

For $(4.5, 6)$, $r = 7.5$,

$\sin\theta = \frac{y}{r} = \frac{6}{7.5} = \frac{4}{5}$ and $\tan\theta = \frac{y}{x} = \frac{6}{4.5} = \frac{4}{3}$.

29. $\sin^2\theta + \cos^2\theta = \left(\frac{3}{5}\right)^2 + \left(\frac{4}{5}\right)^2$

$= \frac{9}{25} = \frac{16}{25}$

$= \frac{25}{25} = 1$

33. $\tan\theta = \frac{y}{x} = \frac{6}{-2} = \frac{4}{x+1}$

$6x + 6 = -8$

$6x = -14$

$x = -\frac{7}{3}$

4.3 Values of the Trigonometric Functions

1. $\sin\theta = 0.3527$

$\theta = 20.65^\circ$

```
sin⁻¹(0.3527)
         20.65254838
■
```

5. Answers may vary in exercises 5, 6, 7, 8. One set of measurements gives $x = 7.6$ and $y = 6.5$.

$\sin 40^\circ = \frac{6.5}{10} = 0.65$

$\cos 40^\circ = \frac{7.6}{10} = 0.76$

$\tan 40^\circ = \frac{6.5}{7.6} = 0.86$

$\csc 40^\circ = \frac{10}{6.5} = 1.54$

$\sec 40^\circ = \frac{10}{7.6} = 1.32$

$\cot 40^\circ = \frac{7.6}{6.5} = 1.17$

9. $\sin 22.4^\circ = 0.381$

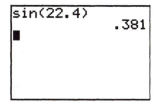

```
sin(22.4)
                .381
■
```

13. $\cos 15.71^\circ = 0.9626$

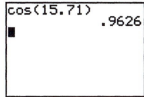

```
cos(15.71)
               .9626
■
```

17. $\cot 67.78^\circ = 0.4085$

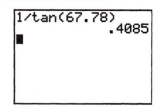

```
1/tan(67.78)
               .4085
■
```

21. $\csc 0.49^\circ = 116.9$

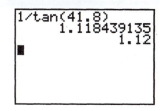

```
1/tan(41.8)
         1.118439135
                1.12
■
```

25. $\cos\theta = 0.3261$

$\theta = 70.97^\circ$

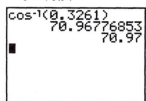

```
cos⁻¹(0.3261)
         70.96776853
               70.97
■
```

29. $\tan\theta = 0.207$

$\theta = 11.7^\circ$

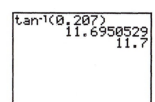

```
tan⁻¹(0.207)
          11.6950529
                11.7
■
```

33. $\csc\theta = 1.245$

$\theta = 53.44^\circ$

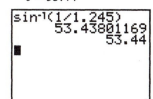

```
sin⁻¹(1/1.245)
         53.43801169
               53.44
■
```

37. $\sec\theta = 3.65$

$\theta = 74.1^\circ$

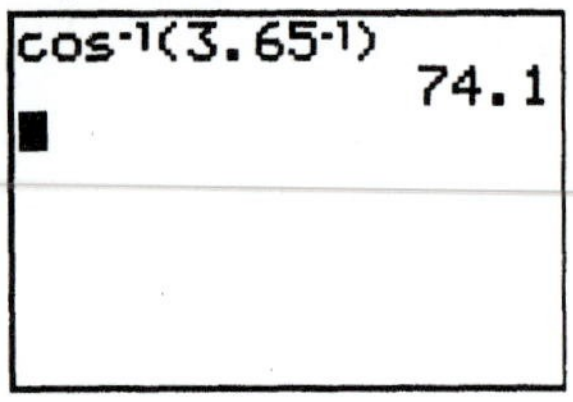

41. $\dfrac{\sin 43.7^\circ}{\cos 43.7^\circ} = \tan 43.7^\circ$

$= 0.96$

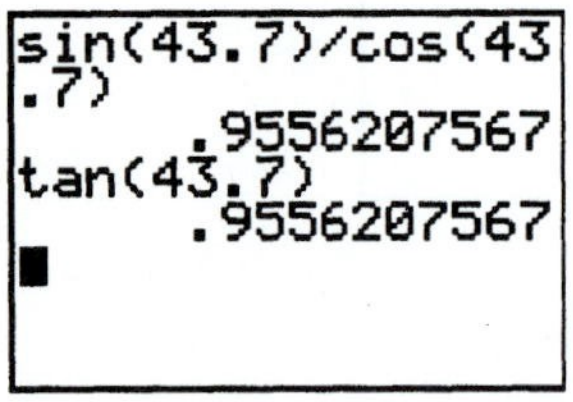

45. y is always less than r.

49. Given $\tan\theta = 1.936$

$\sin\theta = \sin\left(\tan^{-1} 1.936\right)$

$\sin\theta = 0.8885$

53.

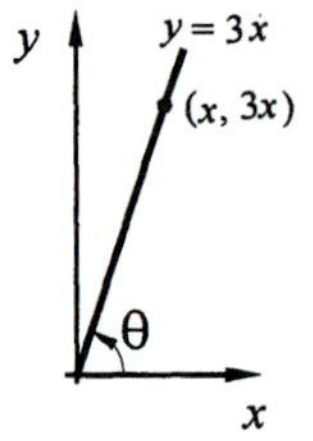

$\theta = \tan^{-1}\dfrac{3x}{x}$

$\theta = \tan^{-1} 3$

$\theta = 71.7^\circ$

57. $\sin\theta = \dfrac{1.5\lambda}{d} = \dfrac{1.5(200)}{400}$

$\theta = \sin^{-1}\dfrac{1.5(200)}{400}$

$\theta = 48.6^\circ$

4.4 The Right Triangle

1.

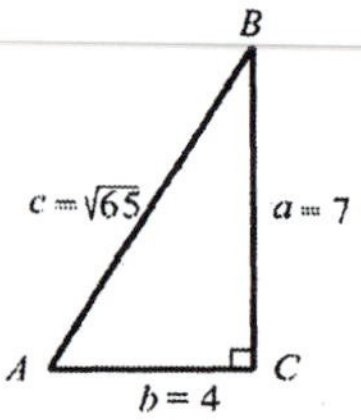

$\sin A = \dfrac{7}{\sqrt{65}} = 0.868$

$\cos A = \dfrac{4}{\sqrt{65}} = 0.496$

$\tan A = \dfrac{7}{4} = 1.75$

$\sin B = \dfrac{4}{\sqrt{65}} = 0.496$

$\cos B = \dfrac{4}{\sqrt{65}} = 0.868$

$\tan B = \dfrac{4}{7} = 0.571$

5. A 60° angle between sides of 3 in. and 6 in. determines the unique triangle shown in the figure below.

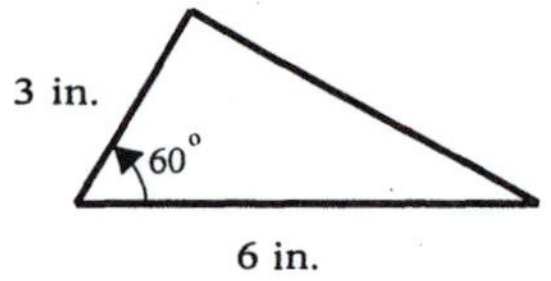

9. $\sin 77.8^\circ = \dfrac{6700}{c} \Rightarrow c = 6850$

$\angle B = 90^\circ - 77.8^\circ = 12.2^\circ$

$\tan 77.8^\circ = \dfrac{6700}{b} \Rightarrow b = 1450$

13. $\angle A = 90^\circ - 32.1^\circ = 57.9^\circ$,

$\sin 32.1^\circ = \dfrac{b}{23.8} \Rightarrow b = 12.6$

$\cos 32.1^\circ = \dfrac{a}{23.8} \Rightarrow a = 20.2$

17. $\angle B = 90^\circ - 32.1^\circ = 57.9^\circ$,

$$\sin 32.1^\circ = \frac{a}{56.85} \Rightarrow a = 30.21$$

$$\cos 32.1^\circ = \frac{b}{56.85} \Rightarrow b = 48.16$$

21. $\angle A = 90^\circ - 37.5^\circ = 52.5^\circ$,

$$\tan 37.5^\circ = \frac{b}{0.862} \Rightarrow b = 0.661$$

$$\cos 37.5^\circ = \frac{0.862}{c} \Rightarrow c = 1.09$$

25. $\tan A = \frac{591.87}{264.93} \Rightarrow A = 65.89^\circ$;

$$\tan B = \frac{264.93}{591.87} \Rightarrow B = 24.11^\circ$$

$$c = \sqrt{264.93^2 + 591.87^2} = 648.46$$

29. $A = 90.0^\circ - 9.56^\circ = 70.44^\circ$

$$\sin B = \frac{b}{c}$$

$$c = \frac{b}{\sin B} = \frac{1.849}{\sin 74.18^\circ} = 1.922$$

$$\tan B = \frac{b}{a}$$

$$\sin 9.56^\circ = \frac{b}{0.00973}$$

$$b = 0.00162$$

$$\cos B = \frac{a}{c}$$

$$\cos 9.56^\circ = \frac{a}{0.00973}$$

$$a = 0.00959$$

33. $\sin 61.7^\circ = \frac{3.92}{x} \Rightarrow x = \frac{3.92}{\sin 61.7^\circ} = 4.45$

37.

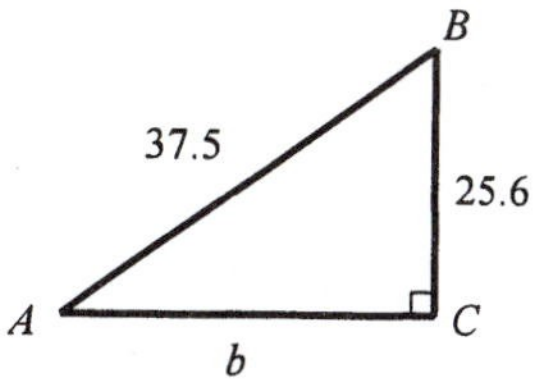

$$\sin A = \frac{25.6}{37.5}$$

$$A = 43.1^\circ < B = 90^\circ - 43.1^\circ = 47.9^\circ$$

41. $\angle B = 90^\circ - \angle A$,

$$A = \frac{a}{c} \Rightarrow a = c \sin A,$$

$$\cos A = \frac{b}{c} \Rightarrow b = c \cos A$$

4.5 Applications of Right Triangles

1. $90^\circ - 62.1^\circ = 27.9^\circ$

$$d = 2090 \text{ ft}$$

5. $\tan 62.6^\circ = \frac{h}{22.8}$

$$h = 22.8 \tan 62.6^\circ = 44.0 \text{ ft}$$

9. $\cos 76.67^\circ = \frac{196.0}{h}$

$$h = \frac{196.0}{\cos 76.67^\circ} = 850.1 \text{ cm}$$

13.

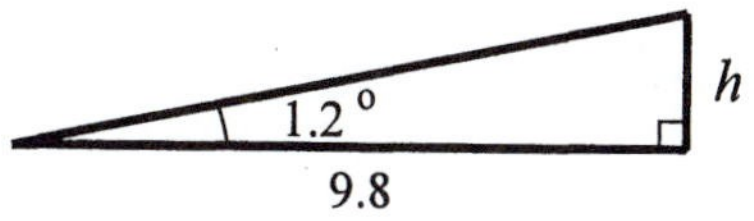

$$\tan 1.2^\circ = \frac{h}{9.8}$$

$$h = 9.8 \tan 1.2^\circ$$

$$h = 0.21 \text{ mi}$$

17. $\theta = \tan^{-1} \frac{6.0}{100} = 3.4^\circ$

21. $\theta = \sin^{-1}\dfrac{12.0}{85.0} = 8.1^\circ$

25. Each angle of the pentagon is $\dfrac{360^\circ}{5} = 72^\circ$. Radii drawn from the center of the pentagon (which is also the center of the circle) through adjacent vertices of the pentagon outward to the fence form an isosceles triangle with base 92.5 and equal sides x. A ⊥ bisector from the center of the pentagon to the base this isosceles triangle forms a right triangle with hypotenuse x and base 46.25. The base angle of this right triangle is $\dfrac{180^\circ - 72^\circ}{2} = 54^\circ$.

Thus,

$$\cos(54^\circ) = \frac{46.25}{x} \Rightarrow x = \frac{46.25}{\cos 54^\circ}$$

and $C = 2\pi(x+25)$

$$C = 2\pi\left(\frac{46.25}{\cos 54^\circ} + 25\right) = 651 \text{ ft}$$

29.

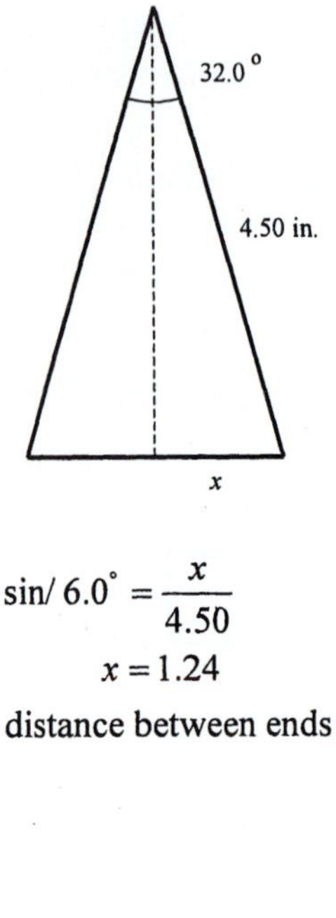

$$\sin/\ 6.0^\circ = \frac{x}{4.50}$$

$$x = 1.24$$

distance between ends $= 2x = 2.48$ in.

33.

$$\sin D = \frac{5380}{6680}$$

$$D = \sin^{-1}\left(\frac{5380}{6680}\right) = 53.7^\circ$$

$$\tan 53.7^\circ = \frac{2860}{h}$$

$$h = \frac{2860}{\tan 53.7^\circ} = 2100 \text{ ft}$$

37.

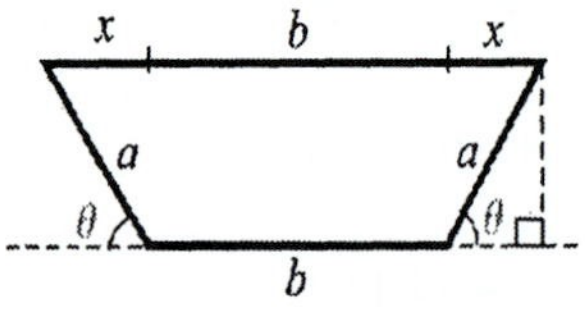

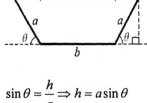

$$\sin\theta = \frac{h}{a} \Rightarrow h = a\sin\theta$$

$$\cos\theta = \frac{x}{a} \Rightarrow x = a\cos\theta$$

$$A = \frac{b+(b+2x)}{2}\cdot h$$

$$= \frac{2b+2a\cos\theta}{2}\cdot a\sin\theta$$

$$A = (b+a\cos\theta)\cdot a\sin\theta$$

Chapter 4 Review Exercises

1. $17.0^\circ + 360.0^\circ = 377.0^\circ$, $17.0^\circ - 360.0^\circ = -343.0^\circ$

5. $31^\circ + 54'\left(\dfrac{1^\circ}{60'}\right) = 31.9^\circ$

9. $17.5^\circ = 17^\circ + 0.5^\circ\left(\dfrac{60'}{1^\circ}\right) = 17^\circ 30'$

13. $r = \sqrt{x^2+y^2} = \sqrt{24^2+7^2} = \sqrt{625} = 25$

$\sin\theta = \dfrac{y}{r} = \dfrac{7}{25}$, $\csc\theta = \dfrac{r}{y} = \dfrac{25}{7}$

$\cos\theta = \dfrac{x}{r} = \dfrac{24}{25}$, $\sec\theta = \dfrac{r}{x} = \dfrac{25}{24}$

$\tan\theta = \dfrac{y}{x} = \dfrac{7}{24}$, $\cot\theta = \dfrac{x}{y} = \dfrac{24}{7}$

17. $r^2 = x^2 + y^2 \Rightarrow 13^2 = x^2 + 5^2 \Rightarrow x = 12$

$$\cos\theta = \frac{x}{r} = \frac{12}{13} = 0.923$$

$$\cot\theta = \frac{x}{y} = \frac{12}{5} = 2.40$$

21. $\sin 72.1^\circ = 0.952$ from the calculator

25. $\sec 18.4^\circ = 1.05$ from the calculator

29. $\cos\theta = 0.950 \Rightarrow \theta = \cos^{-1}(0.950)$

$\theta = 18.2^\circ$ from the calculator

33. $\csc\theta = 4.713 \Rightarrow \frac{1}{\sin\theta} = 4.13 \Rightarrow \sin\theta = \frac{1}{4.713}$

$\theta = \sin^{-1}\frac{1}{4.713} = 12.25^\circ$ from the calculator

37. $\cot\theta = \frac{1}{\tan\theta} = 7.117$

$$\tan\theta = \frac{1}{7.117}$$

$$\theta = \tan^{-1}\frac{1}{7.117}$$

$\theta = 8.00^\circ$ from the calculator

41. $\angle B = 90.0^\circ - 17.0^\circ = 73.0^\circ$

$$\tan 17.0^\circ = \frac{a}{6.00} \Rightarrow a = 1.83$$

$$\cos 17.0^\circ = \frac{6.00}{c} \Rightarrow c = \frac{6.00}{\cos 17.0^\circ} = 6.27$$

45. $\angle B = 90.0^\circ - 37.5^\circ = 52.5^\circ$

$$\tan 37.5^\circ = \frac{12.0}{b} \Rightarrow b = 15.6$$

$$\sin 37.5^\circ = \frac{12.0}{c} \Rightarrow c = \frac{12.0}{\sin 37.5^\circ} = 19.7$$

49. $\angle B = 90.0^\circ - 49.67^\circ = 40.33^\circ$

$$\sin 49.67^\circ = \frac{a}{0.8253} \Rightarrow a = 0.6292$$

$$\cos 49.67^\circ = \frac{b}{0.8253} \Rightarrow b = 0.5341$$

53.

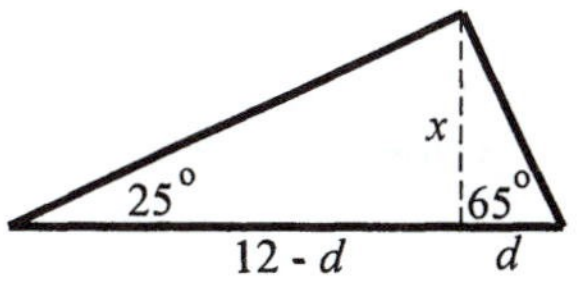

$$90^\circ - 25^\circ = 65^\circ$$

$$\tan 65^\circ = \frac{x}{d} \Rightarrow d = \frac{x}{\tan 65^\circ}$$

$$\tan 25^\circ = \frac{x}{12-d} = \frac{x}{12 - \frac{x}{\tan 65^\circ}}$$

$$x = \left(12 - \frac{x}{12-d}\right)\tan 25^\circ$$

$$x = 12\tan 25^\circ - x\cdot\frac{\tan 25^\circ}{\tan 65^\circ}$$

$$x\left(1 + \frac{\tan 25^\circ}{\tan 65^\circ}\right) = 12\tan 25^\circ$$

$$x = \frac{12\tan 25^\circ}{1 + \frac{\tan 25^\circ}{\tan 65^\circ}}$$

$$x = 4.6$$

57.

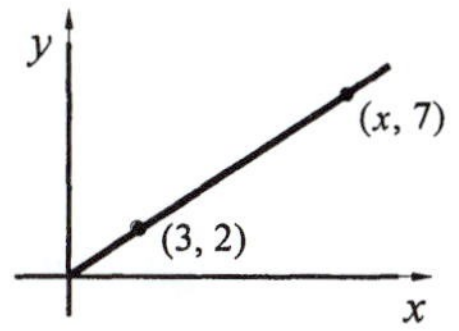

$$\tan\theta = \frac{2}{3} = \frac{7}{x}$$

$$2x = 21$$

$$x = 10.5$$

61.

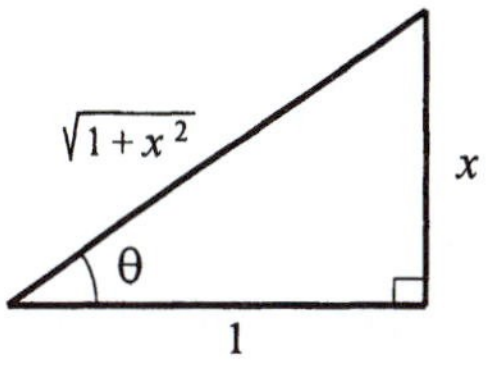

$$\tan\theta = x = \frac{x}{1}$$

$$\csc\theta = \frac{\sqrt{x^2+1}}{x}$$

65. $e = E\cos\alpha \Rightarrow 56.9 = 339\cos\alpha \Rightarrow \alpha$

$$= \cos^{-1}\frac{56.9}{339} = 80.3^\circ$$

69. (a) A triangle with angle θ included between sides a and b, the base, has an altitude of $a\sin\theta$. The area, A, is $A = \frac{1}{2}\cdot b\cdot a\sin\theta$.

(b) The area of the tract is $A = \frac{1}{2}\cdot 31.96\cdot 47.25$ $\sin 64.09^\circ = 679.2\ \text{m}^2$.

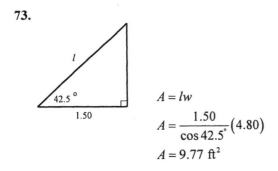

73.

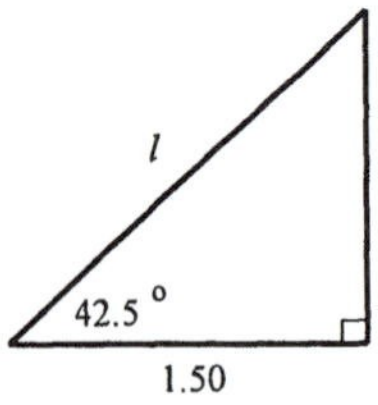

$$A = lw$$

$$A = \frac{1.50}{\cos 42.5^\circ}(4.80)$$

$$A = 9.77\ \text{ft}^2$$

77. $d = a + b = \dfrac{1.85}{\tan 28.3^\circ} + \dfrac{1.85}{\tan(90.0^\circ - 28.3^\circ)}$

$= 4.43$ m

81. Let x = length of window through which sun does not shine.

$\tan 65^\circ = \dfrac{x + 2.5}{2.0} \Rightarrow x = 2.0\tan 65^\circ - 2.5$

percent window shaded $= \dfrac{2.0\tan 65^\circ - 2.5}{3.2} = 56\%$

85. I. Line of sight perpendicular to end of span

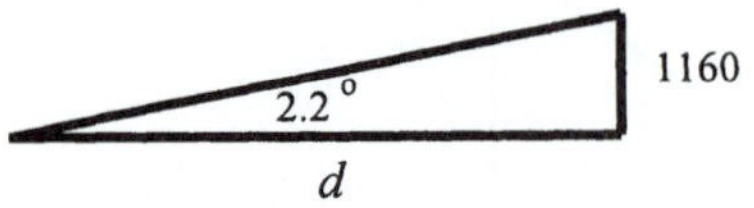

d = distance from helicopter to end of span

$\tan 2.2^\circ = \dfrac{1160}{d} \Rightarrow d = 30{,}200$ m

II. Line of sight perpendicular to middle of span

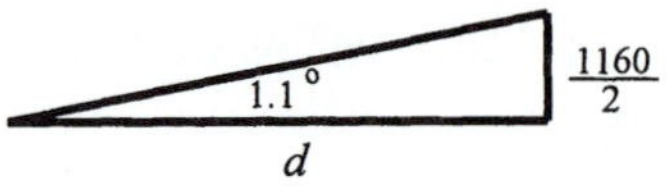

$\tan 1.1^\circ = \dfrac{\frac{1160}{2}}{d} \Rightarrow d = 30{,}200$ m

89. Each angle of a regular pentagon is $\dfrac{(5-2)\cdot 180}{5} = 108^\circ$. A regular pentagon iwth a side of 45.0 mm consists of 5 triangles of base 45.0 and altitude of 22.5 tan 54°. The area of 12 such pentagons is $12\cdot 5\cdot\frac{1}{2}\cdot 45.0\cdot 22.5\tan 54^\circ$ or $41{,}807.60083\ \text{mm}^2$.

Each angle of a regular hexagon is $\dfrac{(6-2)\cdot 180}{6} = 120^\circ$. A regular hexagon with a side of 45.0 mm consists of 6 triangles of base 45.0 and an altitude of 22.5 tan 60°. The area of 20 such hexagons is $20\cdot 6\cdot\frac{1}{2}\cdot 45\cdot 22.5\tan 60^\circ = 105{,}222.0866\ \text{mm}^2$.

Thus, the area of 12 regular pentagons of side 45.0 mm and 20 regular hexagons of side 45.0 mm is $147{,}029.6874\ \text{mm}^2$ $(147{,}000\ \text{mm}^2$ rounded off$)$. Since this is the area of a flat surface it approximates the area of the spherical soccer ball which is given by

$$4\pi r^2 = 4\pi\cdot\left(\frac{222}{2}\right)^2 = 154{,}830.2523\ \text{mm}^2$$

$(155{,}000\ \text{mm}^2$ rounded off$)$.

93.

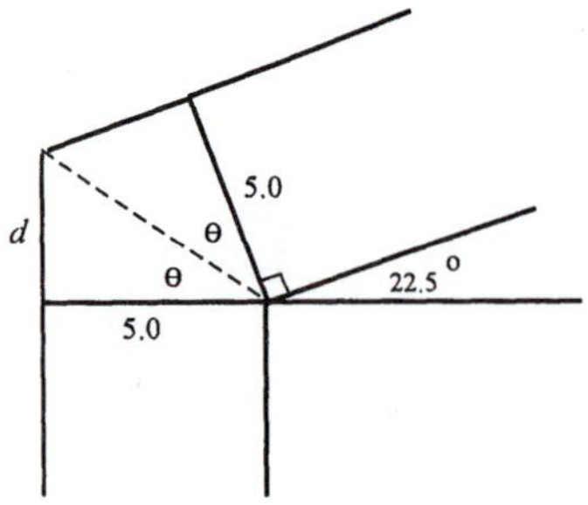

$$2\theta + 90^\circ + 22.5^\circ = 180^\circ$$

$$\theta = \frac{67.5^\circ}{2}$$

$$\tan\frac{67.5^\circ}{2} = \frac{d}{5.0}$$

$$d = 5.0\tan\frac{67.5^\circ}{2}$$

$$l = 5.0 + 65.0 + 5.0\tan\frac{67.5^\circ}{2} = 73.3 \text{ cm}$$

Chapter 5

Systems of Linear Equations; Determinants

5.1 Linear Equations

1. $x-\dfrac{y}{6}+z-4w=7 \Rightarrow x-\dfrac{1}{6}y+z-4w=7$ is linear.

5. The coordinates of the point $(3, 1)$ do not satisfy the equation since $2(3)+3(1)=6+3=9$.
The coordinates of the point $(5, 1/3)$ do not satisfy the equation since $2(5)+3\left(\dfrac{1}{3}\right)=10+1=11\neq 9$.

9. $3(2)-2y=12;\ 2y=6-12=-6;\ y=-3$

$3(-3)-2y=12;\ 2y=-9-12=-21;\ y=-\dfrac{21}{2}$

13. $24(2/3)-9y=16;\ 9y=16-16=0;\ y=0$

$24(-1/2)-9y=16;\ 9y=-12-16=-28;$

$y=(-28/9)$

17. If the values $A=-2$ and $B=1$ satisfy both equations, they are a solution.

$-2+5(1)=3\neq 7;\ 3(-2)-4(1)=-10\neq 4$

Since the given values do not satisfy both equations they are not a solution.

21. If the values $x=0.6$ and $y=-0.2$ satisfy both equations, they are a solution.

$3(0.6)-2(-0.2)=2.2;\ 5(0.6)-0.2=2.8$

Therefore the given values are a solution.

25. For $x=-2$, $3x+b=0$ becomes $3(-2)+b=0$

$b=6$

29. If $F_1=45$ N and $F_2=28$ N, then

$0.80(45)+0.50(28)=50;$

$0.60(45)-0.87(28)=2.64\neq 12$

Since both values do not satisfy both equations, they are not a solution.

5.2 Graphs of Linear Functions

1. $(-1,-2),(3,-1)$

$m=\dfrac{-1-(-2)}{3-(-1)}=\dfrac{1}{4}$. The line rises 1 unit for each 4 units in going from left to right.

5. By taking $(3, 8)$ as (x_2, y_2) and $(1, 0)$ as (x_1, y_1)

$m=\dfrac{8-0}{3-1}=\dfrac{8}{2}=4$

9. By taking $(-2,-5)$ as (x_2, y_2) and $(5,-3)$ as (x_1, y_1)

$m=\dfrac{-5-(-3)}{-2-5}=\dfrac{-5+3}{-7}=\dfrac{2}{7}$

13. $m=2,\ (0,-1)$

Plot the y-intercept point $(0,-1)$. Since the slope is 2/1, from this point, go over 1 unit and up 2 units, and plot a second point. Sketch the line between the 2 points.

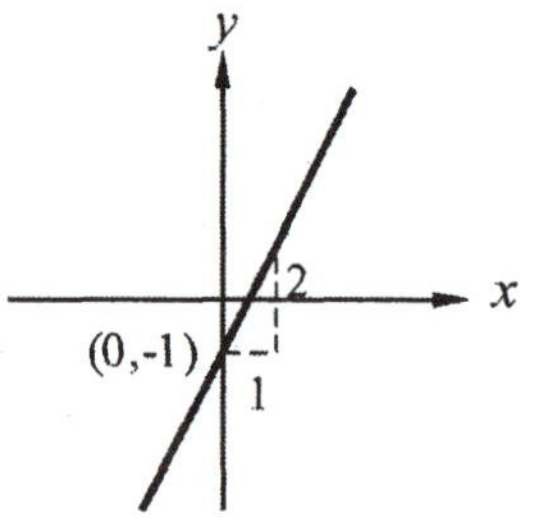

17. $m = \dfrac{1}{2}, (0, 0)$

Plot the y-intercept point $(0, 0)$. Since the slope is 1/2, from this point, go over 2 units and up 1 unit, and plot a second point. Sketch the line between the 2 points.

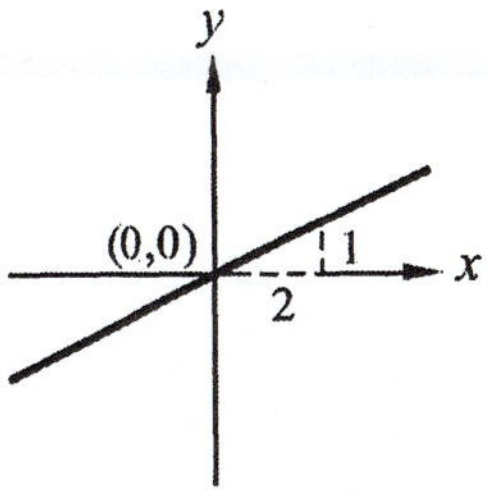

21. $m = -2x + 1,\ m = -2,\ b = 1$

Plot the y-intercept point $(0, 1)$. Since the slope is $-2/1$, from this point, go over 1 unit and down 2 units, and plot a second point. Sketch the line between the 2 points.

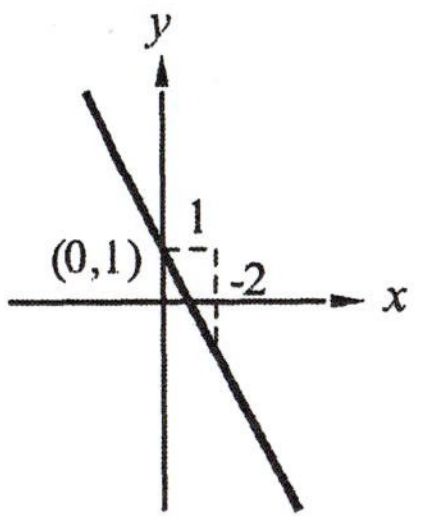

25. $5x - 2y = 40 \Rightarrow y = \dfrac{5}{2}x - 20,\ m = \dfrac{5}{2},\ b = -20$

Plot the y-intercept point $(0, 20)$. Since the slope is 5/2, from this point, go over 2 units and up 5 units, and plot a second point. Sketch the line between these 2 point.

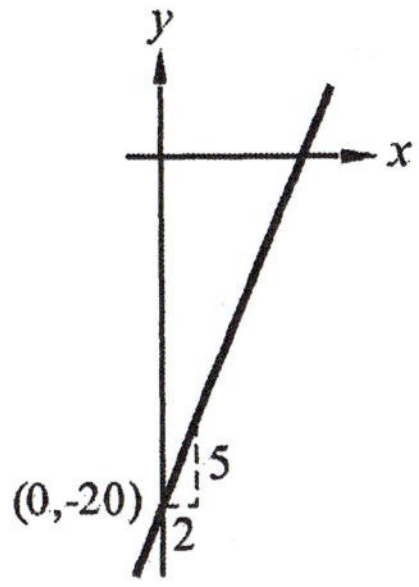

29. $x + 2y = 4\big|_{x=0} \Rightarrow 0 + 2y = 4 \Rightarrow y\text{-int} = 2$

$x + 2y = 4\big|_{y=0} \Rightarrow x + 2 \cdot 0 = 4 \Rightarrow x\text{-int} = 4$

Plot the y-intercept point $(4, 0)$ and the y-intercept point $(0, 2)$. Sketch the line between these 2 pts. A third point is found as a check. Let $x = -2$, $-2 + 2y = 4,\ 2y = 6,\ y = 3$. Therefore the point $(-2, 3)$ should lie on the line.

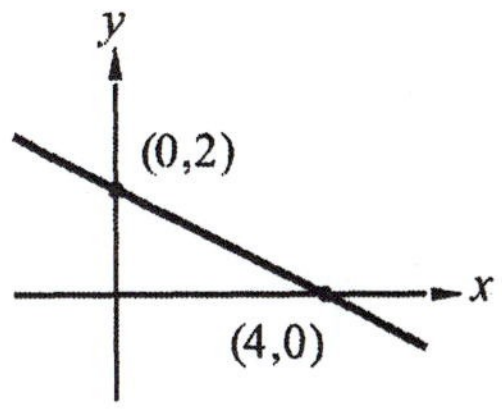

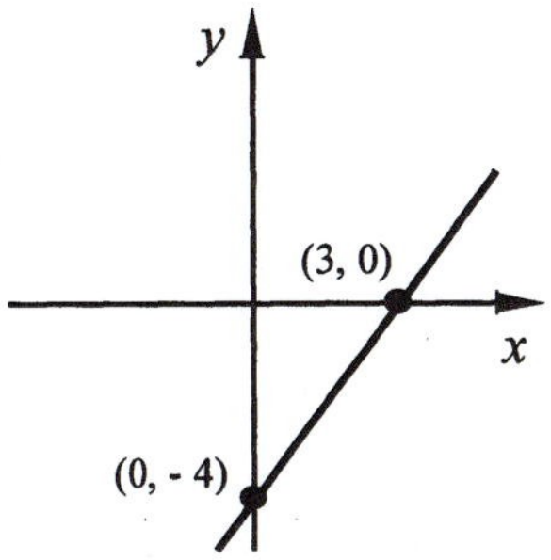

33. $y = 3x + 6\big|_{x=0} \Rightarrow y = 3 \cdot 0 + 6 \Rightarrow y\text{-int} = 6$

$y = 3x + 6\big|_{y=0} \Rightarrow 0 = 3x + 6 \Rightarrow x\text{-int} = -2$

Plot the x-intercept point $(-2, 0)$ and the y-intercept point $(0, 6)$. Sketch the line between these 2 pts. A third point is found as a check. Let $x = 1$, $y = 3(1) + 6 = 9$. Therefore the point $(1, 9)$ should lie on the line.

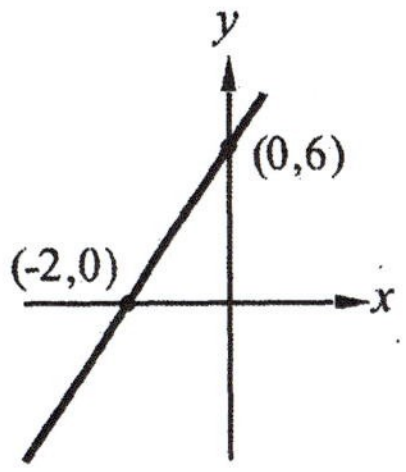

37. $kx - 2y = 9$

$$2y = kx - 9$$

$$y = \frac{k}{2}x - \frac{9}{2}$$

$$m = \frac{k}{2} = 3 \Rightarrow k = 6$$

41. $4I_1 - 5I_2 = 2;\ I_2 = -\frac{2}{5} + \frac{4}{5}I_1$

The y-intercept is $(0, -2/5)$; $m = 4/5$. Plot $(0, -2/5)$. From this point go up 4 units and over 5 units to the right.

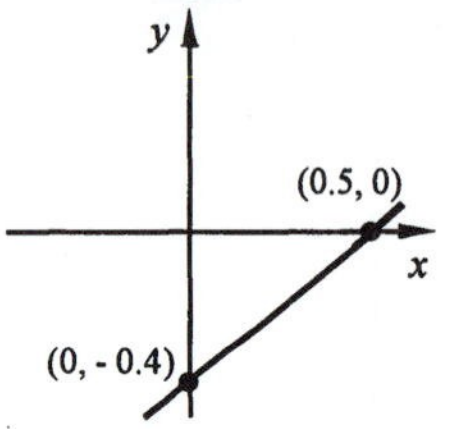

5.3 Solving Systems of Two Linear Equations in Two Unknowns Graphically

1. $2x + 5y = 10$. Let $x = 0,\ y = 0$ to find intercepts of $(0, 2), (5, 0)$. A third point is $\left(1, \frac{8}{5}\right)$.

$3x + y = 6$. Let $x = 0,\ y = 0$ to find intercepts of $(0, 6), (2, 0)$. A third point is $(1, 3)$. From the graph the solution is approximately $x = 1.5,\ y = 1.4$.

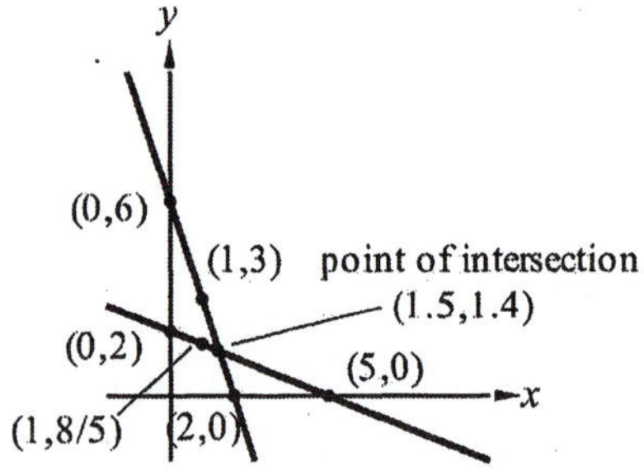

5. $y = 2x - 6;\ y = -\left(\frac{1}{3}\right)x + 1$

The slope of the first line is 2, and the y-intercept is -6. The slope of the second line is $-1/3$ and the y-intercept is 1. From the graph, the point of intersection is $(3.0, 0.0)$. Therefore, the solution of the system of equations is $x = 3.0,\ y = 0.0$.

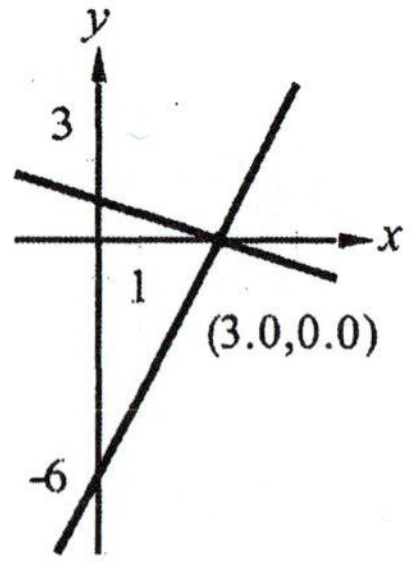

9. $2x - 5y = 10;\ 3x + 4y = -12$

The intercepts of the first line are $(5, 0), (0, -2)$. A third point is $\left(1, -\frac{8}{5}\right)$. The intercepts of the second line are $(-4, 0), (0, -3)$. A third point is $\left(\frac{4}{3}, 4\right)$. From the graph, the point of intersection is $x = -0.9,\ y = -2.3$.

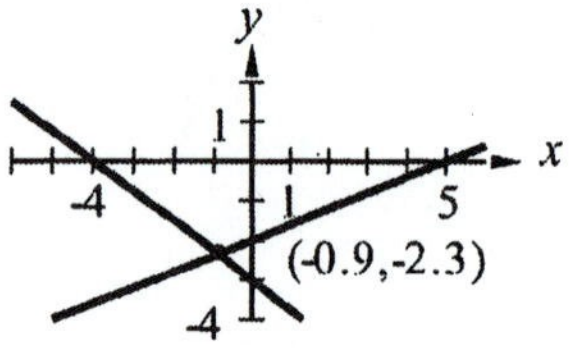

13. $y = -x + 3;\ y = -2x + 3$

The intercepts of the first line are $(3, 0), (0, 3)$. A third point is $(2, 1)$. The intercepts of the second line are $\left(\frac{3}{2}, 0\right), (0, 3)$. A third point is $(1, 1)$. From the graph, the point of intersection is $(0.0, 3.0)$. The solution of the system of equations is $x = 0.0,\ y = 3.0$.

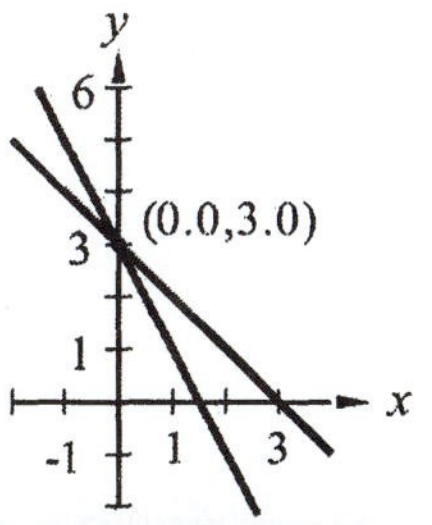

17. $-2r_1 + 2r_2 = 7;\ 4r_1 - 2r_2 = 1$

The intercepts of the first line are $\left(-\frac{7}{2}, 0\right), \left(0, \frac{7}{2}\right)$. A third point is $\left(-2, \frac{3}{2}\right)$. The intercepts of the second line are $\left(\frac{1}{4}, 0\right), \left(0, -\frac{1}{2}\right)$. A third point is $\left(1, \frac{3}{2}\right)$. From the graph, the point of intersection is $(4.0, 7.5)$, and the solution of the system of equations is $r_1 = 4.0,\ r_2 = 7.5$.

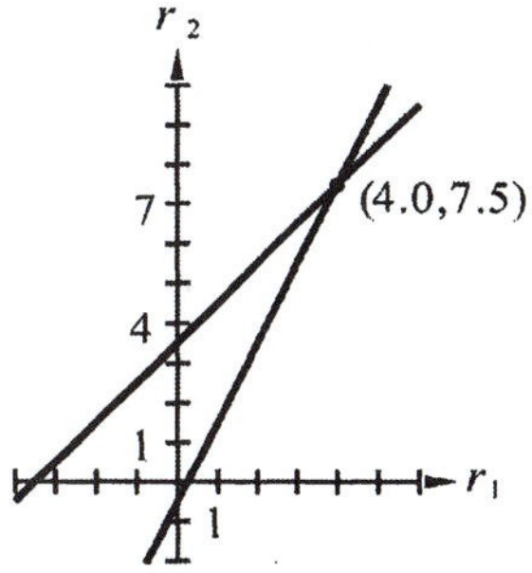

21. $x = 4y + 2 \Leftrightarrow y = \dfrac{x-2}{4}$ and

$3y = 2x + 3 \Leftrightarrow y = \dfrac{2x+3}{3}$

On a graphing calculator let $y_1 = \dfrac{x-2}{4}$ and $y_2 = \dfrac{2x+3}{3}$. Using the intersect feature, the point of intersection is $(-3.6, -1.4)$, and the solution of the system of equations is $x = -3.600$, $y = -1.400$.

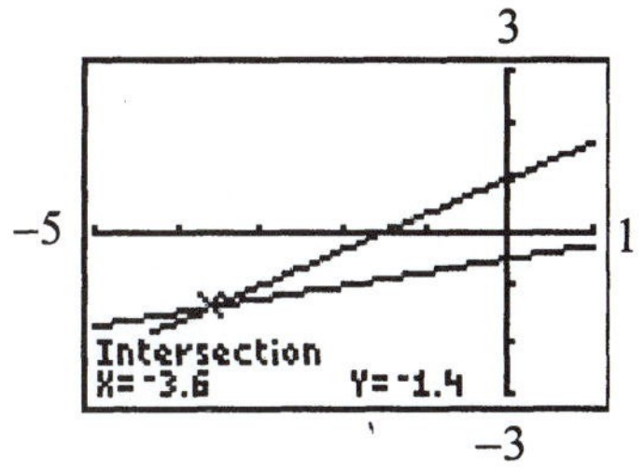

25. $x - 5y = 10 \Leftrightarrow y = \dfrac{x-10}{5}$ and

$2x - 10y = 20 \Leftrightarrow y = \dfrac{x-10}{5}$

On the graphing calculator let $y_1 = \frac{x-10}{5}$ and $y_2 = \frac{x-10}{5}$. From the graph the lines are the same. The system is dependent.

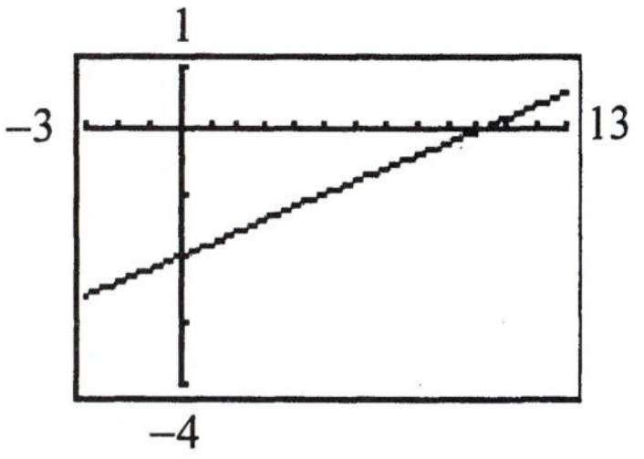

29. $5x - y = 3 \Leftrightarrow y = 5x - 3$ and

$4x = 2y - 3 \Leftrightarrow y = \dfrac{4x+3}{2}$

On a graphing calculator let $y_1 = 5x - 3$ and $y_2 = \frac{4x+3}{2}$. Using the intersect feature, the point of intersection is $(1.5, 4.5)$, and the solution to the system of equations is $x = 1.500,\ y = 4.500$.

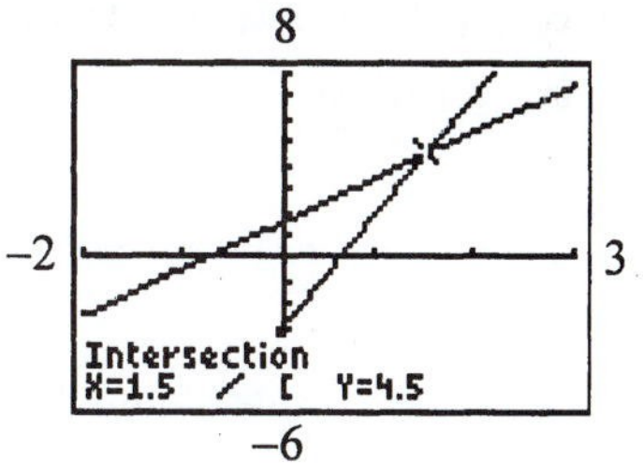

33. $0.8T_1 - 0.6T_2 = 12 \Leftrightarrow$

$$T_2 = \frac{0.8T_1 - 12}{0.6} \Rightarrow y = \frac{0.8x - 12}{0.6}$$

$0.6T_1 + 0.8T_2 = 68 \Leftrightarrow$

$$T_2 = \frac{68 - 0.6T_1}{0.8} \Rightarrow y = \frac{68 - 0.6x}{0.8}$$

On the graphing calculator let $y_1 = \frac{0.8x-12}{6}$ and $y_2 = \frac{68-0.6x}{0.8}$. Using the intersect feature, the point of intersection is $(50, 47)$. The tensions are $T_1 = 50\text{ N},\ T_2 = 47\text{ N}$.

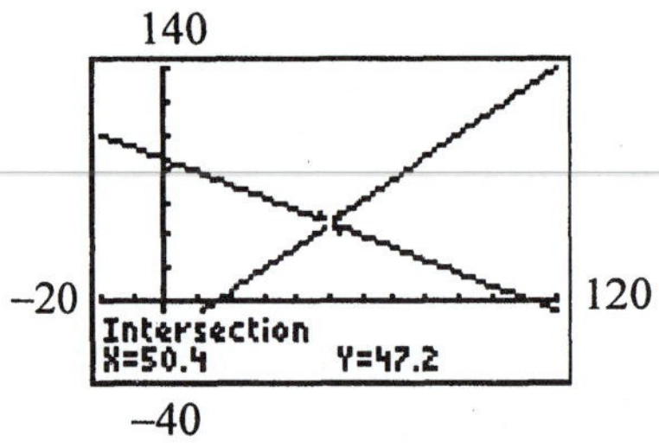

37. $x+y=120 \Rightarrow y=120-x$

$0.7x+0.4y=60 \Rightarrow 0.7x+0.4(20-x)=60$

$0.7x+48-0.4x=60 \Rightarrow 0.3x=12$

$x=40$

$x=40$ kg of 70% lead alloy

$y=120-x=120-40$

$=80$ kg of 40% lead alloy

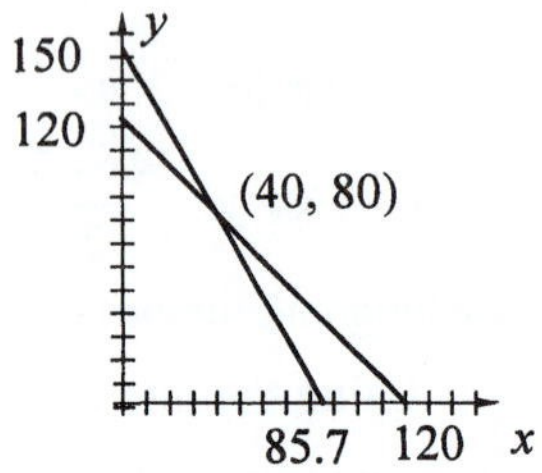

5.4 Solving Systems of Two Linear Equations in Two Unknowns Algebraically

1. (1) $x-3y=6 \Rightarrow x=3y+6$

(2) $2x-3y=3$

$2(3y+6)-3y=3$ substitute x from (1) into (2)

$6y+12-3y=3$

$3y=-9$

$y=-3$

$x-3(3)=6$ substitute -3 for y in (1)

$x+9=6$ $x=-3$

5. (1) $x=y+3$

(2) $x-2y=5$

$(y+3)-2y=5$ substitute x from (1) into (2)

$-y=2$

$y=-2$ substitute -2 for y in (1)

$x=-2+3=1$

9. (1) $x+y=-5,\ y=-x-5$

(2) $2x-y=2$

$2x-(-x-5)=2$ substitute y from (1) into (2)

$3x=-3$

$x=-1$

$-1+y=-5$ substitute -1 for x in (1)

$y=-4$

13. (1) $33x+2y=34 \Rightarrow y=-\frac{33}{2}x+17$

(2) $40y=9x+11$

$40\left(-\frac{33}{2}x+17\right)=9x+11$ substitute y from (1) in (2)

$-660x+680=9x+11$

$-669x=-669$

$x=1$

$33(1)+2y=34$ substitute 1 for x in (1)

$2y=1$

$y=\frac{1}{2}$

17. (1) $2x-3y=4$

(2) $\underline{2x+\ y=-4}$

$-4y=8$ subtract (1) and (2)

$y=-2$

$2x-3(-2)=4$ substitute for x in (1)

$2x+6=4$

$2x=-2$

$x=-1$

21. (1) $v+2t=7$

(2) $2v+4t=9$

(3) $2v+4t=14$ multiply (1) by 2

(4) $\underline{2v+4t=9}$ recopy (2)

$0=5$ inconsistent

25. (1) $2x - y = 5, y = 2x - 5$
(2) $6x + 2y = -5$
$6x + 2(2x - 5) = -5$ substitute y from (1) into (2)
$10x = 5$
$x = \frac{1}{2}$
$y = 2\left(\frac{1}{2}\right) - 5 = -4$ substitute $\frac{1}{2}$ for x in (1)

$x = \frac{1}{2}$, from (1) $y = 2x - 5 = 2 \cdot \frac{1}{2} - 5 = 1 - 5 = -4$

$\left(\frac{1}{2}, -4\right)$

29. (1) $15x + 10y = 11 \Rightarrow 75x + 50y = 55$ m
(2) $20x - 25y = 7 \Rightarrow 40x - 50y = 14$ add m
$115x = 69$
$x = \frac{3}{5}$
$15\left(\frac{3}{5}\right) + 10y = 11$ substitute $\frac{3}{5}$ for x in (1)
$10y = 2$
$y = \frac{1}{5}$

33. (1) $44A = 1 - 15B \Rightarrow 44A + 15B = 1$
(2) $5B = 22 + 7A \Rightarrow 7A - 5B = -22$

(1) $44A + 15B = 1$
(2) $21A - 15B = -66$ multiply (2) by 3
$65A = -65$
$A = -1$

$5B = 22 + 7(-1)$ substitute (-1) for A in (2)
$5B = 15$
$B = 3$

37. $0.3x - 0.7y = 0.4$ $\quad 3x - 7y = 4$ $\quad 15x - 35y = 30$
$0.2x + 0.5y = 0.7$ $\quad 2x + 5y = 7$ $\quad 14x + 35y = 49$
$29x = 69$
$x = \frac{69}{29}$

$2\left(\frac{69}{29}\right) + 5y = 7$
$5y = \frac{65}{29}$
$y = \frac{13}{29}$

41. $V_1 + V_2 = 15$
$V_1 - V_2 = 3$
$2V_1 = 18$
$V_1 = 9\ V$

$9 + V_2 = 15 \Rightarrow V_2 = 6$, the solution is $V_1 = 9$ and $V_2 = 6\ V$

45. x = number of regular email messages
y = number of spam messages
$x + y = 79$ $\quad x + 2x + 4 = 79$
$y = 2x + 4$ $\quad 3x = 75$
$x = 25$ reg. messages
$y = 2(25) + 4 = 54$ spam messages

49. (1) $2000t_1 = 3200t_2$

(2) $t_1 - 12 = t_2$

$2000t_1 = 3200(t_1 - 12) \Rightarrow 2000t_1 = 3200t_1 - 38,400 \Rightarrow -1200t_1 = -38,400$

$t_1 = 32\ s$ and $t_2 = 32 - 12 = 20\ s$

53. w = capacity of windmill in kW,
g = capacity of generator in kW

(1) $0.45w(10 \cdot 24) + g(10 \cdot 24) = 3010$ for first 10 day period
(2) $108w + 240g = 3010$
(3) $54w + 120g = 1505$
(4) $0.72w(10 \cdot 24) + g(10 \cdot 24 - 60) = 2900$ for second 10 day period
(5) $172.8w + 180g = 2900$

Solving (3) and (5) gives $w = 7.00$ kW, $g = 9.39$ kW

57. $ax + y = c \Rightarrow y = -ax + c$, slope $= -a$
$bx + y = d \Rightarrow y = -bx + d$, slope $= -b$

System will have unique solution if slopes of lines are different which requires $a \neq b$.

5.5 Solving Systems of Two Linear Equations in Two Unknowns by Determinants

1. $\begin{vmatrix} 4 & -6 \\ 3 & 17 \end{vmatrix} = 4(17) - 3(-6) = 68 + 18 = 86$

5. $\begin{vmatrix} 2 & 4 \\ 3 & 1 \end{vmatrix} = (2)(1) - (3)(4) = 2 - 12 = -10$

9. $\begin{vmatrix} 8 & -10 \\ 0 & 4 \end{vmatrix} = (8)(4) - (0)(-10) = 32 - 0 = 32$

13. $\begin{vmatrix} 0.75 & -1.32 \\ 0.15 & 1.18 \end{vmatrix} = 0.75(1.18) - (0.15)(-1.32)$

$= 0.885 + 0.198 = 1.083$

17. $x + 2y = 5$
$x - 2y = 1$

$$x = \frac{\begin{vmatrix} 5 & 2 \\ 1 & -2 \end{vmatrix}}{\begin{vmatrix} 1 & 2 \\ 1 & -2 \end{vmatrix}} = \frac{(5)(-2) - (1)(2)}{(1)(-2) - (1)(2)} = \frac{-10 - 2}{-2 - 2} = 3$$

$$y = \frac{\begin{vmatrix} 1 & 5 \\ 1 & 1 \end{vmatrix}}{-4} = \frac{(1)(1) - (1)(5)}{-4} = \frac{1 - 5}{-4} = 1$$

21. Rewrite the system with both equations in standard form.

$12t + 9y = 14$
$6t - 7y = -16$

$$t = \frac{\begin{vmatrix} 14 & 9 \\ -16 & -7 \end{vmatrix}}{\begin{vmatrix} 12 & 9 \\ 6 & -7 \end{vmatrix}} = \frac{46}{-138} = -\frac{1}{3},$$

$$y = \frac{\begin{vmatrix} 12 & 14 \\ 6 & -16 \end{vmatrix}}{-138} = \frac{-276}{-138} = 2$$

25. -3 Rewrite the system with both equations in standard form.

$2x - 3y = 4$
$3x - 2y = -2$

$$x = \frac{\begin{vmatrix} 4 & -3 \\ -2 & -2 \end{vmatrix}}{\begin{vmatrix} 2 & -3 \\ 3 & -2 \end{vmatrix}} = -2.8 = -\frac{14}{5}$$

$$y = \frac{\begin{vmatrix} 2 & 4 \\ 3 & -2 \end{vmatrix}}{\begin{vmatrix} 2 & -3 \\ 3 & -2 \end{vmatrix}} = -3.2 = \frac{-16}{5}$$

29. $$s = \frac{\begin{vmatrix} 60 & 30 \\ -50 & 40 \end{vmatrix}}{\begin{vmatrix} 40 & 30 \\ 20 & 40 \end{vmatrix}} = \frac{3900}{1000} = 3.9,$$

two significant digits

$$t = \frac{\begin{vmatrix} 40 & 60 \\ 20 & -50 \end{vmatrix}}{1000} = \frac{-3200}{1000} = 3.2$$

two significant digts

33. $1.2y + 8.4x = -10.8$
$3.5x + 4.8y = -12.9$

$$x = \frac{\begin{vmatrix} -10.8 & 1.2 \\ -12.9 & 4.8 \end{vmatrix}}{\begin{vmatrix} 8.4 & 1.2 \\ 3.5 & 4.8 \end{vmatrix}} = \frac{-51.84 + 15.48}{40.32 - 4.2} = -1.0$$

$$y = \frac{\begin{vmatrix} 8.4 & -10.8 \\ 3.5 & -12.9 \end{vmatrix}}{\begin{vmatrix} 8.4 & 1.2 \\ 3.5 & 4.8 \end{vmatrix}} = \frac{-108.36 + 37.8}{40.32 - 4.2} = -2.0$$

37. If $a = kb$, $c = kd$, $\begin{vmatrix} a & b \\ c & d \end{vmatrix} = \begin{vmatrix} kb & b \\ kd & d \end{vmatrix} = kbd - kbd = 0$

41. $x + y = 36.0$
$0.250x + 0.375y = 11.2$

$$x = \frac{\begin{vmatrix} 36.0 & 1 \\ 11.2 & 0.375 \end{vmatrix}}{\begin{vmatrix} 1 & 1 \\ 0.250 & 0.375 \end{vmatrix}} = \frac{13.5 - 11.2}{0.375 - 0.250}$$
$$= 18.4 \text{ gal}$$

$$y = \frac{\begin{vmatrix} 1 & 36.0 \\ 0.250 & 11.2 \end{vmatrix}}{\begin{vmatrix} 1 & 1 \\ 0.250 & 0.375 \end{vmatrix}} = \frac{11.2 - 9}{0.375 - 0.250}$$
$$= 17.6 \text{ gal}$$

45. x = number of phones, y = number of detectors

(1) $x + y = 320$
(2) $110x + 160y = 40,700$

Solving (1) and (2) gives $x = 210$ phones, $y = 110$ detectors

49. Convert 24 minutes to hours.
24 min = 0.4 h

$t_2 = t_1 - 0.4 \Leftrightarrow t_1 - t_2 = 0.4$

$42t_1 = 50t_2 \Leftrightarrow 42t_1 - 50t_2 = 0$

$$t_1 = \frac{\begin{vmatrix} 0.4 & -1 \\ 0 & -50 \end{vmatrix}}{\begin{vmatrix} 1 & -1 \\ 42 & -50 \end{vmatrix}} = 2.5 \text{ h}$$

$$t_2 = \frac{\begin{vmatrix} 1 & 0.4 \\ 42 & 0 \end{vmatrix}}{\begin{vmatrix} 1 & -1 \\ 42 & -50 \end{vmatrix}} = 2.1 \text{ h}$$

5.6 Solving Systems of Three Linear Equations in Three Unknowns Algebraically

1.

(1)	$4x + y + 3z = 1$	
(2)	$2x - 2y + 6z = 12$	
(3)	$-6x + 3y + 12z = -14$	
(4)	$8x + 2y + 6z = 2$	(1) multiplied by 2
(2)	$2x - 2y + 6z = 12$	add
(5)	$10x + 12z = 14$	
(6)	$12x + 3y + 9z = 3$	(1) multiplied by 3
(3)	$-6x + 3y + 12z = -14$	subtract
(7)	$18x - 3z = 17$	
(8)	$72x - 12z = 68$	
(5)	$10x + 12z = 14$	add
(9)	$82x = 82$	
(10)	$x = 1$	
(11)	$18(1) - 3z = 17$	substituting $x = 1$ into (7)
(12)	$-3z = -1$	
(13)	$z = \frac{1}{3}$	
(14)	$4(1) + y + 3\left(\frac{1}{3}\right) = 1$	substitute $x = 1$ and $z = \frac{1}{3}$ into (1)
(15)	$4 + y + 1 = 1$	
(16)	$y = -4$	

Thus, the solution is $x = 1$, $y = -4$, $z = \frac{1}{3}$

5.

(1) $2x + 3y + z = 2$	
(2) $-x + 2y + 3z = -1$	
(3) $-3x - 3y + z = 0$	
(4) $-2x + 4y + 6z = -2$	multiply (2) by 2
(5) $7y + 7z = 0$	add (1) and (4)
(6) $y = -z$	
(7) $-x + 2(-z) + 3z = -1$	substitute (5) in (2)
(8) $-x - z = -1$	
(9) $-3x - 3(-z) + z = 0$	substitute (5) in (3)
(10) $-3x + 4z = 0$	
$3x = 4z$	
$x = \frac{4}{3}z$	
(11) $2\left(\frac{4}{3}z\right) + 3(-z) + z = 2$	substitute (10), (6) in (1)
$\frac{8}{3}z - 3z + z = 2$	
$\left(\frac{8}{3} - \frac{9}{3} + \frac{3}{3}\right) z = 2$	
$\frac{2}{3}z = 2$	
$z = 2$	

The solution is $x = 4, y = -3, z = 3$.

9.

(1) $2x - 2y + 3z = 5$	
(2) $2x + y - 2z = -1$	
(3) $4x - y - 3z = 0$	
(4) $-3y + 5z = 6$	subtract (1) and (2)
(5) $4x + 2y - 4z = -2$	multiply (2) by 2
(6) $-3y + z = 2$	subtract (5) and (3)
$4z = 4$	
$z = 1$	
(7) $2x - 2y + 3(1) = 5$	substitute (6) in (1)
$2x - 2y = 2$	
(8) $2x + y - 2(1) = -1$	substitute (6) in (2)
$2x + y = 1$	
(9) $-3y = 1$	subtract (7) and (8)
$y = -\frac{1}{3}$	
(10) $2x - 2\left(-\frac{1}{3}\right) + 3(1) = 5$	substitute (9), (6) into (1)
$2x + \frac{2}{3} + 3 = 5$	
$2x = \frac{15}{3} - \frac{11}{3}$	
$2x = \frac{4}{3}$	
$x = \frac{2}{3}$	

The solution is $x = 2/3, y = -1/3, z = 1$.

13. (1) $10x+15y-25z=35$

(2) $40x-30y-20z=10$

(3) $16x-2y+8z=6$

(4) $20x+30y-50z=70$ multiply (1) by 2

(2) $40x-30y-20z=10$ add

(5) $60x-70z=80$

(2) $40x-30y-20z=10$

(6) $-240x+30y-120z=-90$ multiply (3) by -15, add

(7) $-200x-140z=-80$

(8) $-120x+140z=-160$ multiply (5) by -2, add

$-320x=-240$

$x=\frac{3}{4}$

(5) $60\left(\frac{3}{4}\right)-70z=80 \Rightarrow z=-\frac{1}{2}$

(1) $10\left(\frac{3}{4}\right)+15y-25\left(-\frac{1}{2}\right)=35 \Rightarrow y=1$

The solution is $x=\frac{3}{4}$, $y=1$, $z=-\frac{1}{2}$

17. $Ax+By+Cz=D$

(1) $2A+4B+4C=12$

(2) $3A-2B+8C=12$

(3) $-A+8B+6C=12$

(4) $-2A+16B+12C=24$ multiply (3) by 2

(1) $2A+4B+4C=12$ add

(5) $20B+16C=36 \Rightarrow 5B+4C=9$

(6) $-3A+24B+18C=36$

(2) $3A-2B+8C=12$ add

(7) $22B+26C=48 \Rightarrow 11B+13C=24$

(8) $55B+44C=99$ multiply (5) by 11

(9) $-55B-65C=-21$ multiply (7) by -5, add

$-21C=-21$

$C=1$

(5) $5B+4(1)=9 \Rightarrow B=1$

(1) $2A+4(1)+4(1)=12 \Rightarrow A=2$

The constants are $A=2$, $B=1, C=1$ and the equation is $2x+y+z=12$.

21. (1) $0.707F_1-0.800F_2=0$

(2) $0.707F_1+0.600F_2-F_3=10.0$

(3) $3.00F_2-3.00F_3=20.0$

(4) $-1.4F_2+F_3=-10.0$ subtract (1) and (2)

(5) $-4.2F_2+3.00F_3=-30.0$ multiply (4) by 3

(6) $-1.2F_2=-10.0$ add (3) and (5)

(7) $F_2=8.33$

(8) $-1.4(8.33)+F_3=-10.0$ substitute (7) into (4)

$F_3=-10.0+11.662$

(9) $F_3=1.67$

(10) $0.707F_1-0.800(8.33)=0$ substitute (7) into (1)

(11) $F_1=\frac{6.66}{0.707}=9.43$

The solution is $F_1=9.43$ N, $F_2=8.33$ N, $F_3=1.67$ N.

25. Letting $t = 1, 3, 5$ and $\theta = 19, 30.9, 19.8$ in $at^3 + bt^2 + ct = \theta$ gives

(1) $a + b + c = 19$
(2) $27a + 9b + 3c = 30.9$
(3) $125a + 25b + 5c = 19.8$. Solving (1) for $c = 19 - a - b$ and substituting into (2) and (3) gives
(4) $24a + 6b = -26.1$ and (5) $120a + 20b = -75.2$. From (4), (6) $b = \dfrac{-26.1 - 24a}{6}$.

Substituting in (5) gives

(7) $120a + 20\left(\dfrac{-26.1 - 24a}{6}\right) = -75.2 \Rightarrow a = 0.295$

(6) $b = \dfrac{-26.1 - 24(0.295)}{6} = -5.53$

From (1) $c = 19 - 0.295 - (-5.53) = 24.235$.
The solution is $a = 0.295$, $b = -5.53$, and $c = 24.235$.
$\theta = 0.295t^3 - 5.53t^2 + 24.2t$

29. (1) $x - 2y - 3z = 2 \Rightarrow x = 2y + 3z + 2$ substituted in (2) and (3) gives
(2) $x - 4y - 13z = 14 \Rightarrow 2y + 3z + 2 - 4y - 13z = 14 \Rightarrow -2y - 10z = 12 \Rightarrow$
(4) $y + 5z = -6$
(3) $-3x + 5y + 4z = 0 \Rightarrow -3(2y + 3z + 2) + 5y + 4z = 0 \Rightarrow y + 5z = -6$ which is (4).
Thus, $y = -6 - 5z$ and from (1) $x - 2(-6 - 5z) - 3z = 2 \Rightarrow x = -7z - 10$. The solution is $x = -7z - 10, y = -5z - 6, z = z$. Letting $z = 0, x = -10, y = -6, z = 0$.

5.7 Solving Systems of Three Linear Equations in Three Unknowns by Determinants

1.

$$\begin{array}{|rrr|rr} -2 & 3 & -1 & -2 & 3 \\ 1 & 5 & 4 & 1 & 5 \\ 2 & -1 & 5 & 2 & -1 \end{array}$$

$$= -2(5)(5) + 3(4)(2) + (-1)(1)(-1) - 2(5)(-1) - (-1)(4)(-2) - 5(1)(3)| = -38$$

5.

$$\begin{array}{|rrr|rr} 8 & 9 & -6 & 8 & 9 \\ -3 & 7 & 2 & -3 & 7 \\ 4 & -2 & 5 & 4 & -2 \end{array}$$

$$= 280 - (-32) + 72 - (-135) + (-36) - (-168)$$
$$= 651$$

9.

$$\begin{array}{|rrr|rr} 4 & -3 & -11 & 4 & -3 \\ -9 & 2 & -2 & -9 & 2 \\ 0 & 1 & -5 & 0 & 1 \end{array}$$

$$= -40 - (-8) + 0 - (-135) + 99 - 0 = 202$$

13.

$$\begin{array}{|rrr|rr} 0.1 & -0.2 & 0 & 0.1 & -0.2 \\ -0.5 & 1 & 0.4 & -0.5 & 1 \\ -2 & 0.8 & 2 & -2 & 0.8 \end{array}$$

$$= 0.2 - 0.032 + 0.16 - 0.2 + 0 - 0 = 0.128$$

17.

$$x = \frac{\begin{array}{|rrr|rr} 2 & 1 & 1 & 2 & 1 \\ 1 & 0 & -1 & 1 & 0 \\ 1 & 1 & 0 & 1 & 1 \end{array}}{\begin{array}{|rrr|rr} 1 & 1 & 1 & 1 & 1 \\ 1 & 0 & -1 & 1 & 0 \\ 1 & 1 & 0 & 1 & 1 \end{array}}$$

$$= \frac{0 - (-2) + (-1) - 0 + 1 - 0}{0 - (-1) + (-1) - 0 + 1 - 0} = \frac{2}{1} = 2$$

$$y = \frac{\begin{array}{|rrr|rr} 1 & 2 & 1 & 1 & 2 \\ 1 & 1 & -1 & 1 & 1 \\ 1 & 1 & 0 & 1 & 1 \end{array}}{1}$$

$$= \frac{0 - (-1) + (-2) - 0 + 1 - 1}{1} = \frac{-1}{1} = -1$$

$$z = \frac{\begin{array}{|rrr|rr} 1 & 1 & 2 & 1 & 1 \\ 1 & 0 & 1 & 1 & 0 \\ 1 & 1 & 1 & 1 & 1 \end{array}}{1}$$

$$= \frac{0 - 1 + 1 - 1 + 2 - 0}{1} = \frac{1}{1} = 1$$

21.

$$l = \frac{\begin{array}{|rrr|rr} 6 & 6 & -3 & 6 & 6 \\ -3 & -7 & -2 & -3 & -7 \\ 1 & 1 & -7 & 1 & 1 \end{array}}{\begin{array}{|rrr|rr} 5 & 6 & -3 & 5 & 6 \\ 4 & -7 & -2 & 4 & -7 \\ 3 & 1 & -7 & 3 & 1 \end{array}}$$

$$= \frac{294 + 12 - 12 - 126 + 9 - 21}{245 + 10 - 36 + 168 - 12 - 63}$$

$$= \frac{156}{312} = \frac{1}{2}$$

$$w = \frac{\begin{array}{|rrr|rr} 5 & 6 & -3 & 5 & 6 \\ 4 & -3 & -2 & 4 & -3 \\ 3 & 1 & -7 & 3 & 1 \end{array}}{312}$$

$$= \frac{105 + 10 - 36 + 168 - 12 - 27}{312}$$

$$= \frac{208}{312} = \frac{2}{3}$$

$$h = \frac{\begin{array}{|rrr|rr} 5 & 6 & -3 & 5 & 6 \\ 4 & -3 & -2 & 4 & -3 \\ 3 & 1 & -7 & 3 & 1 \end{array}}{312}$$

$$= \frac{-35 + 15 - 54 - 224 + 24 + 126}{312}$$

$$= \frac{52}{312} = \frac{1}{6}$$

25.

$$x=\frac{\begin{vmatrix}6&-7&3\\1&3&6\\5&-5&2\end{vmatrix}\begin{matrix}6&-7\\1&3\\5&-5\end{matrix}}{\begin{vmatrix}3&-7&3\\3&3&6\\5&-5&2\end{vmatrix}\begin{matrix}3&-7\\3&3\\5&-5\end{matrix}}$$

$$=\frac{36+180-210+14-15-45}{18+90-210+42-45-45}$$

$$=\frac{-40}{-150}=\frac{4}{15}$$

$$y=\frac{\begin{vmatrix}3&1&6\\5&5&2\end{vmatrix}\begin{matrix}3&1\\5&5\end{matrix}}{-150}$$

$$=\frac{6-90+180-36+45-15}{-150}$$

$$=\frac{90}{-150}=-\frac{3}{5}$$

$$z=\frac{\begin{vmatrix}3&-7&6\\3&3&1\\5&-5&5\end{vmatrix}\begin{matrix}3&-7\\3&3\\5&-5\end{matrix}}{-150}$$

$$=\frac{45+15-35+105-90-90}{-150}$$

$$=\frac{-50}{-150}=\frac{1}{3}$$

29.

$$x=\frac{\begin{vmatrix}10.5&4.5&-7.5\\1.2&-3.6&-2.4\\1.5&-0.5&2.0\end{vmatrix}\begin{matrix}10.5&4.5\\1.2&-3.6\\1.5&-0.5\end{matrix}}{\begin{vmatrix}3.0&4.5&-7.5\\4.8&-3.6&-2.4\\4.0&-0.5&2.0\end{vmatrix}\begin{matrix}3.0&4.5\\4.8&-3.6\\4.0&-0.5\end{matrix}}$$

$$=\frac{-75.6-12.6-16.2-10.8+4.5-40.5}{-21.6-3.6-43.2-43.2+18-108}$$

$$=\frac{-151.2}{-201.6}$$

$$=\frac{3}{4}$$

$$y=\frac{\begin{vmatrix}3.0&10.5&-7.5\\4.8&1.2&-2.4\\4.0&1.5&2.0\end{vmatrix}\begin{matrix}3.0&10.5\\4.8&1.2\\4.0&1.5\end{matrix}}{-201.6}$$

$$=\frac{7.2+10.8-100.8-100.8-54+36-201.6}{-201.6}$$

$$=1$$

$$z=\frac{\begin{vmatrix}3.0&4.5&10.5\\4.8&-3.6&1.2\\4.0&-0.5&1.5\end{vmatrix}\begin{matrix}3.0&4.5\\4.8&-3.6\\4.0&-0.5\end{matrix}}{-201.6}$$

$$=\frac{-16.2+1.8+21.6-32.4-25.2+151.2}{-201.6}$$

$$=\frac{100.8}{-201.6}=-\frac{1}{2}$$

33. $\begin{vmatrix}4&2&1\\7&8&6\\7&9&8\end{vmatrix}=19$, using calculator

the value does not change.

37.

$$s_o+2v_o+2a=20$$
$$s_o+4v_o+8a=54$$
$$s_o+6v_o+18a=104$$

$$s_o=\frac{\begin{vmatrix}20&2&2\\54&4&8\\104&6&18\end{vmatrix}}{\begin{vmatrix}1&2&2\\1&4&8\\1&6&18\end{vmatrix}}=\frac{16}{8}=2\text{ ft}$$

$$v_o=\frac{\begin{vmatrix}1&20&2\\1&54&8\\1&104&18\end{vmatrix}}{8}=\frac{40}{8}=5\text{ ft/s}$$

$$a=\frac{\begin{vmatrix}1&2&20\\1&4&54\\1&6&104\end{vmatrix}}{8}=\frac{32}{8}=4\text{ ft/s}^2$$

41.

x = percent of nickel
y = percent of iron
z = percent of molybdenum

$$x + y + z = 100$$
$$x - 5y = -1$$
$$y - 3z = 1$$

$$x = \frac{\begin{vmatrix} 100 & 1 & 1 \\ -1 & -5 & 0 \\ 1 & 1 & -3 \end{vmatrix}}{\begin{vmatrix} 1 & 1 & 1 \\ 1 & -5 & 0 \\ 0 & 1 & -3 \end{vmatrix}} = \frac{1501}{19} = 79\% \text{ nickel}$$

$$y = \frac{\begin{vmatrix} 1 & 100 & 1 \\ 1 & -1 & 0 \\ 0 & 1 & -3 \end{vmatrix}}{19} = \frac{304}{19} = 16\% \text{ iron}$$

$$z = \frac{\begin{vmatrix} 1 & 1 & 100 \\ 1 & -5 & -1 \\ 0 & 1 & 1 \end{vmatrix}}{19} = \frac{95}{19} = 5\% \text{ molybdenum}$$

Chapter 5 Review Exercises

1. $\begin{vmatrix} -2 & 5 \\ 3 & 1 \end{vmatrix} = (-2)(1) - (3)(5) = -2 - 15 = -17$

5. $m = \dfrac{y_2 - y_1}{x_2 - x_1} = \dfrac{-8-0}{4-2} = \dfrac{-8}{2} = -4$

9. Comparing $y = -2x + 4$ to $y = mx + b$ gives a slope of -2 and y-intercept of 4.

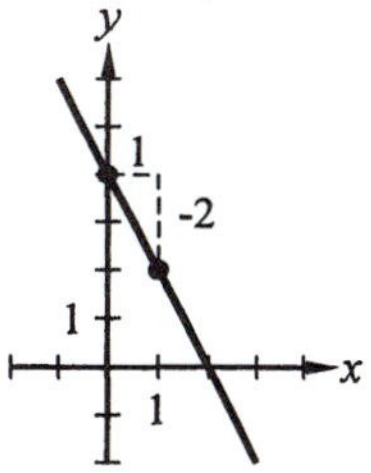

13.

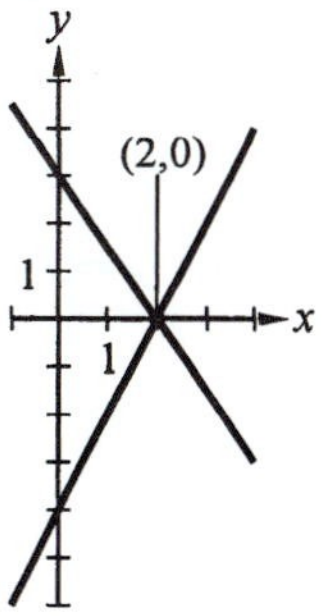

17. $7x = 2y + 14 \Rightarrow y = \dfrac{7x - 14}{2}$

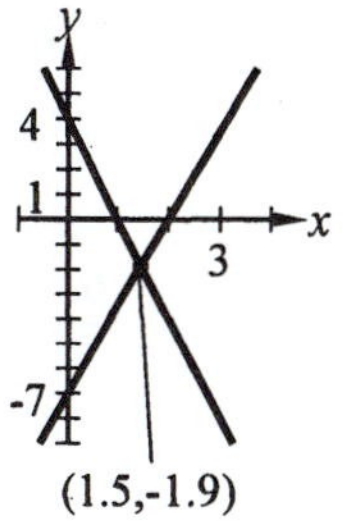

21. (1) $x + 2y = 5 \Rightarrow x = 5 - 2y$ which substitutes into (2) $x + 3y = 7$ to give $5 - 2y + 3y = 7 \Rightarrow y = 2$. From (1) $x = 5 - 2 \cdot 2 = 1$. The solution is (1, 2).

25. $i = \dfrac{\begin{vmatrix} 29 & -27 \\ 69 & 33 \end{vmatrix}}{\begin{vmatrix} 10 & -27 \\ 40 & 33 \end{vmatrix}} = \dfrac{2820}{1410} = 2$

$$v = \frac{\begin{vmatrix} 10 & 29 \\ 40 & 69 \end{vmatrix}}{1410} = \frac{-470}{1410} = -\frac{1}{3}$$

The solution is $i = 2$, $v = -\dfrac{1}{3}$.

29. (1) $90x-110y=40 \Rightarrow x=\dfrac{11y+4}{9}$ which substitutes into (2) $30x-15y=25$ to give $30\dfrac{11y+4}{9}-15y=25 \Rightarrow y=\dfrac{7}{13}$. From (1) $x=\dfrac{11\cdot\frac{7}{13}+4}{9}=\dfrac{43}{39}$. The solution is $\left(\dfrac{43}{39},\dfrac{7}{13}\right)$.

33. $4x+3y=-4 \Rightarrow 4x+3y=-4$

$y=2x-3 \qquad 2x-y=3$

$$x=\frac{\begin{vmatrix}-4 & 3\\ 3 & -1\end{vmatrix}}{\begin{vmatrix}4 & 3\\ 2 & -1\end{vmatrix}}=\frac{-5}{-10}=\frac{1}{2}$$

$$y=\frac{\begin{vmatrix}4 & -4\\ 2 & 3\end{vmatrix}}{-10}=\frac{20}{-10}=-2$$

37. $7x=2y-6 \Rightarrow 7x-2y=-6$

$7y=12-4x \Rightarrow 4x+7y=12$

$$x=\frac{\begin{vmatrix}-6 & -2\\ 12 & 7\end{vmatrix}}{\begin{vmatrix}7 & -2\\ 4 & 7\end{vmatrix}}=\frac{-18}{57}=-\frac{6}{19}$$

$$y=\frac{\begin{vmatrix}7 & -6\\ 4 & 12\end{vmatrix}}{57}=\frac{108}{57}=\frac{36}{19}$$

41. Exercise 33 is most easily solved by substitution because the second equation is already solved for y.

45.
$$\left|\begin{matrix}4 & -1 & 8\\ -1 & 6 & -2\\ 2 & 1 & -1\end{matrix}\right|\begin{matrix}4 & -1\\ -1 & 6\\ 2 & 1\end{matrix}$$

$=4(6)(-1)+(-1)(-2)(2)+8(-1)(1)-2(6)(8)$
$-1(-2)(4)-(-1)(-1)(-1)=-115$

49. (1) $2x+\ y+z=4$

(2) $\underline{x-2y-z=3}$ add

(4) $3x-\ y \quad =7$

(1) $4x+2y+2z=8$

(3) $\underline{3x+3y-2z=1}$ add

(5) $7x+5y \quad =9$

(4) $\underline{15x-5y=35}$ add

$22x \quad =44$

$x=2$

(4) $3x-y=7 \Rightarrow 3(2)-y=7 \Rightarrow y=-1$

(1) $2x+y+z=4 \Rightarrow 2(2)-1+z=4 \Rightarrow z=1$

53. Multiply both sides of all three equations by 10 to clear decimals.

(1) $36x+52y-10z=-22$ solve for z:

(4) $z=\dfrac{36x+52y+22}{10}$

(2) $32x-48y+39z=81$

(3) $64x+41y+23z=51$

(5) $32x-48y+39\cdot\dfrac{36x+52y+22}{10}=81$

(2) with z from (4) which simplifies to

(5) $1724x+1548y=-48 \Rightarrow y=\dfrac{-48-1724x}{1548}$

(6) $64x+41y+23\cdot\dfrac{36x+52y+22}{10}=51$

(3) with z from (4) which simplifies to

(6) $1468x+1606\cdot\dfrac{-48-1724x}{1548}=4 \Rightarrow$

$x=-0.1678084952$. From (5) $y=\dfrac{-48-1724x}{1548}$ $=0.1558797453$. From (4) $z=\dfrac{36x+52y+22}{10}=2.406464093$. The solution is $(-0.17, 0.16, 2.4)$.

57.
$$r=\frac{\begin{vmatrix}8 & 1 & 2\\ 5 & -2 & -4\\ -3 & 3 & 4\end{vmatrix}}{\begin{vmatrix}2 & 1 & 2\\ 3 & -2 & -4\\ -2 & 3 & 4\end{vmatrix}}=\frac{42}{14}=3$$

$$s = \frac{\begin{vmatrix} 2 & 8 & 2 \\ 3 & 5 & -4 \\ -2 & -3 & 4 \end{vmatrix}}{14} = \frac{-14}{14} = -1$$

$$t = \frac{\begin{vmatrix} 2 & 1 & 8 \\ 3 & -2 & 5 \\ -2 & 3 & -3 \end{vmatrix}}{14} = \frac{21}{14} = \frac{3}{2}$$

61. $\begin{vmatrix} 2 & 5 \\ 1 & x \end{vmatrix} = 3 \Rightarrow 2x - 5 = 3 \Rightarrow 2x = 8 \Rightarrow x = 4$

65. (1) $\frac{1}{x} - \frac{1}{y} = \frac{1}{2} \Rightarrow u - v = \frac{1}{2}$

(2) $\frac{1}{x} + \frac{1}{y} = \frac{1}{4} \Rightarrow u + v = \frac{1}{4}$

Adding (1), (2) $2u = \frac{3}{4} \Rightarrow u = \frac{3}{8} \Rightarrow x = \frac{8}{3}$, from (2)

$v = \frac{1}{4} - \frac{3}{8} = -\frac{1}{8}$, $y = -8$. $y = -8$. The solution is $\left(\frac{8}{3}, -8\right)$.

69. (1) $3x - ky = 6$

(2) $x + 2y = 2$. Multiplying (2) by 3 gives $3x + 6y = 6$ which is (1) with $k = -6$. A k-value of -6 makes the system dependent.

73. $$F_1 = \frac{\begin{vmatrix} 280 & 2.0 & 0 \\ 0 & 0 & -1 \\ 600 & -4.0 & 0 \end{vmatrix}}{\begin{vmatrix} 1 & 2.0 & 0 \\ 0.87 & 0 & -1 \\ 3.0 & -4.0 & 0 \end{vmatrix}} = \frac{-2320}{-10} = 232 = 230 \text{ lb}$$

$$F_2 = \frac{\begin{vmatrix} 1 & 280 & 0 \\ 0.87 & 0 & -1 \\ 3.0 & 600 & 0 \end{vmatrix}}{-10} = \frac{-240}{-10} = 24 \text{ lb}$$

$$F_3 = \frac{\begin{vmatrix} 1 & 2.0 & 280 \\ 0.87 & 0 & 0 \\ 3.0 & -4 & 600 \end{vmatrix}}{-10} = \frac{-2018.4}{-10} = 201.84 = 200 \text{ lb}$$

77. $I = 0.105$

$I = 2400 + 0.045$

$0.105 = 2400 + 0.045 \Rightarrow S = 40,000$

$I = 0.01(40,000) = 4000$

Both plans produce an income of \$4000 for sales of \$40,000.

81. $T = \frac{a}{x+100} + b$

$\frac{a}{0+100} + b = 14$

$\frac{a}{900+100} + b = 10$

$$a = \frac{\begin{vmatrix} 14 & 1 \\ 10 & 1 \end{vmatrix}}{\begin{vmatrix} \frac{1}{100} & 1 \\ \frac{1}{1000} & 1 \end{vmatrix}} = 440 \text{ m} \cdot {}^\circ\text{C} \text{ using calculator}$$

$$b = \frac{\begin{vmatrix} \frac{1}{100} & 14 \\ \frac{1}{1000} & 10 \end{vmatrix}}{\begin{vmatrix} \frac{1}{100} & 1 \\ \frac{1}{1000} & 1 \end{vmatrix}} = 9.6^\circ\text{C} \text{ using calculator}$$

85. $I^2 R = P$

$1.0^2 R_1 + 3.0^2 R_2 = 14.0$

$3.0^2 R_1 + 1.0^2 R_2 = 6.0$

$$R_1 = \frac{\begin{vmatrix} 14.0 & 3.0^2 \\ 6.0 & 1.0^2 \end{vmatrix}}{\begin{vmatrix} 1.0^2 & 3.0^2 \\ 3.0^2 & 1.0^2 \end{vmatrix}} = 0.50\ \Omega \text{ using calculator}$$

$$R_2 = \frac{\begin{vmatrix} 1.0^2 & 14.0 \\ 3.0^2 & 6.0 \end{vmatrix}}{\begin{vmatrix} 1.0^2 & 3.0^2 \\ 3.0^2 & 1.0^2 \end{vmatrix}} = 1.5\ \Omega \text{ using calculator}$$

89. (1) $A+B+C=180$

(2) $A \quad = 2B-55 \Rightarrow 2B = A+55$

substitute into (1)

(3) $\quad C = B-25$

(4) $A+B+B-25=180 \Rightarrow$

$\underline{2B=-A+205}$ add

$4B=260 \Rightarrow B=65^\circ$

(3) $C=65-25 \Rightarrow C=40^\circ$

(1) $A+65^\circ+40^\circ=180 \Rightarrow A=75^\circ$

93. x = weight of gold in air

y = weight of silver in air

$x+y=6.0 \Rightarrow y=6.0-x$

$0.947x+0.9y=5.6 \Rightarrow 0.947x+0.9(6.0-x)=5.6$

$x=4.3\text{ N}$

$y=6.0-4.3=1.7\text{ N}$

Chapter 6

FACTORING AND FRACTIONS

6.1 Special Products

1. $(3r-2s)(3r+2s)=(3r)^2-(2s)^2$
$=9r^2-4s^2$

5. $40(x-y)=40x-40y$

9. $(T+6)(T-6)=T^2-6^2=T^2-36$

13. $(4x-5y)(4x+5y)=(4x)^2-(5y)^2$
$=16x^2-25y^2$

17. $(5f+4)^2=(5f)^2+2(5f)(4)+4^2$
$=25f^2+40f+16$

21. $(L^2-1)^2=(L^2)^2-2\cdot L^2\cdot 1+1^2$
$=L^4-2L^2+1$

25. $(0.6s-t)^2=(0.6s)^2-2(0.6s)(t)+t^2$
$=0.36s^2-1.2st+t^2$

29. $(3+C^2)(6+C^2)=18+(3C^2+6C^2)+(C^2)^2$
$=18+9C^2+C^4$

33. $(10v-3)(4v+15)=40v^2+138v-45$

37. $2(x-2)(x+2)=2(x^2-4)=2x^2-8$

41. $6a(x+2b)^2=6a(x^2+4bx+4b^2)$
$=6ax^2+24abx+24ab^2$

45. $[(2R+3r)(2R-3r)]^2=[4R^2-9r^2]^2$
$=16R^4-72R^2r^2+81r^4$

49. $\left[3-(x+y)^2\right]=9-6(x+y)+(x+y)^2$
$=9-6x-6y+x^2+2xy+y^2$

53. $(3L+7R)^3$
$=(3L)^3+3(3L)^2(7R)+3(3L)(7R)^2+(7R)^3$
$=27L^3+189L^2R+441LR^2+343R^3$

57. $(x+2)(x^2-2x+4)=x^3-2x^2+4x+2x^2-4x+8$
$=x^3+8$

61. $(x+y)^2(x-y)^2=(x^2+2x+y^2)(x^2-2xy+y^2)$
$=x^4-2x^3y+x^2y^2+2x^3y-4x^2y^2$
$+2xy^3+x^2y^2-2xy^3+y^4$
$=x^4-2x^2y^2+y^4$

65. $4(p+DA)^2=4(p^2+2pDA+D^2A^2)$
$=4p^2+8pDA+4D^2A^2$

69. $\frac{L}{6}(x-a)^3=\frac{L}{6}\left[x^3-3x^2a+3xa-a^3\right]$
$=\frac{L}{6}x^3-\frac{L}{2}ax^2+\frac{L}{2}a^2x-\frac{L}{6}a^3$

73. $(49)(51)=(50-1)(50+1)=50^2-1^2$
$=2500-1$
$=2499$

77. (a) $A=(x+y)^2=x^2+2xy+y^2$
(b) $A=x^2+xy+xy+y^2=x^2+2xy+y^2$

6.2 Factoring: Common Factor and Difference of Squares

1. $4ax^2-2ax=2ax(2x-1)$

5. $6x+6y=6(+y)$
(6 is a common monomial factor, c.m.f.)

9. $3x^2-9x=3x(x-3)$ (3x is a c.m.f.)

13. $288n^2 + 24n = 24n(12n+1)$ ($24n$ is a c.m.f.)

17. $3ab^2 - 6ab + 12ab^3 = 3ab(b - 2 + 4b^2)$
($3ab$ is a c.m.f.)

21. $2a^2 - 2b^2 + 4c^2 - 6d^2 = 2(a^2 - b^2 + 2c^2 - 3d^2)$
(2 is a c.m.f.)

25. $100 - 9A^2 = (10 - 3A)(10 + 3A)$
(because $-30A + 30A = 0A = 0$)

29. $162s^2 - 50t^2 = 2(81s^2 - 25t^2) = 2(9s - 5t)(9s + 5t)$
(because $-45st + 45st = 0st = 0$)

33. $(x+y)^2 - 9 = (x + y - 3)(x + y + 3)$

37.
$$\begin{aligned} 300x^2 - 2700z^2 &= 300(x^2 - 9z^2) \\ &= 300(x - 3z)(x + 3z) \end{aligned}$$

41.
$$\begin{aligned} x^4 - 16 &= (x^2 - 4)(x^2 + 4) \\ &= (x-2)(x+2)(x^2+4) \end{aligned}$$

45. Solve $2a - b = ab + 3$ for a.
$$\begin{aligned} 2a - ab &= b + 3 \\ a(-b) &= b + 3 \\ a &= \frac{b+3}{2-b} \end{aligned}$$

49.
$$\begin{aligned} (x+2k)(x-2) &= x^2 3x - 4k \\ x^2 - 2x + 2kx - 4k &= x^2 + 3x - 4k \\ 2kx &= 5x \\ k &= \frac{5}{2} \end{aligned}$$

53.
$$\begin{aligned} a^2 + ax - ab - bx &= (a^2 + ax) - (ab + bx) \\ &= a(a+x) - b(a+x) \\ &= (a+x)(a-b) \\ &= (a-b)(a+x) \end{aligned}$$

57.
$$\begin{aligned} x^2 - y^2 + x - y &= (x^2 - y^2) + (x - y) \\ &= (x+y)(x-y) + (x-y) \\ &= (x+y+1)(x-y) \\ &= (x-y)(x+y+1) \end{aligned}$$

61. $u^2 + n = n(n+1)$, the product of two consecutive integers of which one must be even. Therefore, the product is even.

65. $Rv + Rv^2 + Rv^3 = Rv(1 + v + v^2)$

69.

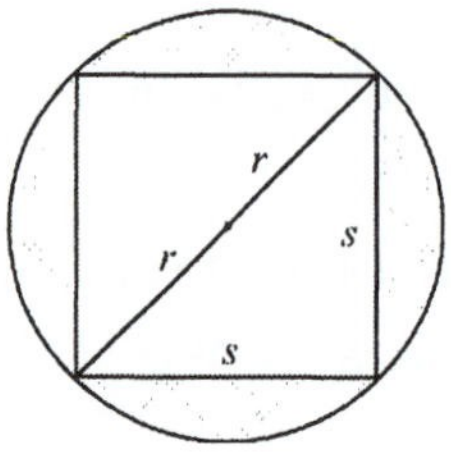

$$\begin{aligned} s^2 + s^2 &= (2r)^2 \\ 2s^2 &= 4r^2 \\ s^2 &= 2r^2 = \text{area of square} \end{aligned}$$
$$\begin{aligned} \text{Area left} &= \text{Area of circle} - \text{Area of square} \\ &= \pi r^2 - 2r^2 \\ &= r^2(\pi - 2) \end{aligned}$$

73.
$$\begin{aligned} 3BY + 5Y &= 9BS \\ 3BY - 9BS &= -5Y \\ B(3Y - 9S) &= -5Y \\ B &= \frac{-5Y}{3Y - 9S} = \frac{5Y}{3(3S - Y)} \end{aligned}$$

6.3 Factoring Trinomials

1. $x^2 + 4x + 3 = (x+3)(x+1)$

5. $2x^2 + 6x - 36 = 2(x^2 + 3x - 18) = 2(x+6)(x-3)$

9. $s^2 - s - 42 = (s-7)(s+6)$
(because $-s = -7s + 6s$)

13. $x^2 + 2x + 1 = (x+1)(x+1) = (x+1)^2$
(because $2x = x + x$)

17. $3x^2 - 5x - 2 = (3x+1)(x-2)$
(because $-5x = -6x + x$)

21. $2s^2 + 13s + 11 = (2s+11)(s+1)$
(because $13s = 2s + 11s$)

25. $2t^2 + 7t - 15 = (2t-3)(t+5)$
(because $7t = 10t - 3t$)

29. $4x^2 - 3x - 7 = (4x-7)(x+1)$
(because $-3x = 4x - 7x$)

33. $4m^2 + 20m + 25 = (2m+5)(2m+5) = (2m+5)^2$
(because $20m = 10m + 10m$)

37. $9t^2 - 15t + 4 = (3t-4)(3t-1)$
(because $-15t = -3t - 12t$)

41. $4p^2 - 25pq + 6q^2 = (4p-q)(p-6q)$
(because $-25pq = -24pq - pq$)

45. $2x^2 - 14x + 12 = 2(x^2 - 7x + 6) = 2(x-1)(x-6)$
(because $-7x = -6x - x$)

49. $ax^3 + 4a^2x^2 - 12a^3x = ax(x^2 + 4ax - 12a^2)$
$= ax(x+6a)(x-2a)$

53. $25a^2 - 25x^2 - 10xy - y^2$
$= 25a^2 - (25x^2 + 10xy + y^2)$
$= 25a^2 - (5x+y)^2$
$= (5a+5x+y)(5a-5x-y)$

57. $3p^2 + 9p - 54 = 3(p^2 + 3p - 18)$
$= 3(p+6)(p-3)$

61. $200n^2 - 2100n - 3600 = 100(2n^2 - 21 - 36)$
$= 100(2x+3)(n-12)$

65. $wx^4 - 5wLx^3 + 6wL^2x^2 = wx^2(x^2 - 5Lx + 6L^2)$
$= wx^2(x-3L)(x-2L)$

69. Write $4x^2 + 4x - k$ as
$(2x)^2 + 2(2x)(1) + (-k)$ and comparing to
$(2x)^2 + 2(2x)(1) + 1^2$, which is a perfect square,
gives $-k = 1^2 \Rightarrow k = -1$ from which
$4x^2 + 4x - k$ becomes $4x^2 + 4x + 1$ which factors
as $4x^2 + 4x + 1 = (2x+1)^2$

73. $36x^2 + 9 = 9(4x^2 + 1)$

6.4 The Sum and Difference of Cubes

1. $x^3 - 8 = x^3 - 2^3 = (x-2)(x^2 + 2x + 2^2)$
$= (x-2)(x^2 + 2x + 4)$

5. $8 - t^3 = 2^3 - t^3 = (2-t)(4 + 2t + t^2)$

9. $4x^3 + 32 = 4(x^3 + 8) = 4(x+2)(x^2 - 2x + 4)$

13. $54x^3y - 6x^3y^4 = 6x^3y(9 - y^3)$

17. $8a^6 - 8a^2 = 8a^2(a^4 - 1)$
$= 8a^2(a^2 - 1)(a^2 + 1)$
$= 8a^2(a+1)(a-1)(a^2 + 1)$

21. $27L^6 + 216L^3 = 27L^3(L^3 + 8)$
$= 27L^3(L+2)(L^2 - 2L + 4)$

25. $64 - x^6 = 4^3 - (x^2)^3 = (4 - x^2)(16 + 4x^2 + x^4)$
$= (2+x)(2-x)(16 + 4x^2 + x^4)$

29. $D^4 - d^3D = D(D^3 - d^3)$
$= D(D-d)(D^2 + Dd + d^2)$

33.
$$
\begin{array}{r|l}
 & x^4+x^3y+x^2y^2+xy^3+y^4 \\
x-y & x^5 \qquad\qquad\qquad -y^5 \\
 & x^5-x^4y \\
 & x^4y \\
 & x^4y-x^3y^2 \\
 & x^3y^2 \\
 & x^3y^2-x^2y^3 \\
 & x^2y^3 \\
 & x^2y^3-xy^4 \\
 & xy^4-y^5 \\
 & xy^4-y^5
\end{array}
$$

$$(x^5-y^5)\div(x-y)=x^4+x^3y+x^2y^2+xy^3+y^4$$
$$x^5-y^5=(x-y)(x^4+x^3y+x^2y^2+xy^3+y^4)$$
$$(x^7-y^7)\div(x-y)=x^6+x^5y+x^4y^2+x^3y^3+x^2y^4+xy^5+y^6$$
$$x^7-y^7=(x-y)(x^6+x^5y+x^4y^2+x^3y^3+x^2y^4+xy^5+y^6)$$

37. For $n=1$, $n^3+1=(n+1)(n^2-n+1)$ becomes

$$1^3+1=(1+1)(1^2-1+1)$$
$$2=2\cdot 1 \text{ which is prime}$$

For $n=2,3,4,\ldots$, $n^3+1=(n+1)(n^2-n+1)$

$\Rightarrow (n+1)$ is a factor of $n^3+1 \Rightarrow n^3+1$ is not prime.

6.5 Equivalent Fractions

1. $\dfrac{18abc^6}{24ab^2c^5}=\dfrac{6abc^5(3c)}{6abc^5(4b)}=\dfrac{3c}{4b}$

5. $\dfrac{2}{3}\cdot\dfrac{7}{7}=\dfrac{14}{21}$

9. $\dfrac{2}{(x+3)}\cdot\dfrac{(x-2)}{(x-2)}=\dfrac{2(x-2)}{(x+3)(x-2)}=\dfrac{2x-4}{x^2+x-6}$

13. $\dfrac{28}{44}=\dfrac{\frac{28}{4}}{\frac{44}{4}}=\dfrac{7}{11}$

17. $\dfrac{2(R-1)}{(R-1)(R+1)}=\dfrac{\frac{2(R-1)}{(R-1)}}{\frac{(R-1)(R+1)}{(R-1)}}=\dfrac{2}{R+1}$

21. $\dfrac{A}{6y^2}=\dfrac{3x}{2y}\cdot\dfrac{3y}{3y}=\dfrac{9xy}{6y^2}\Rightarrow A=9xy$

25. $\dfrac{A}{x^2-1}=\dfrac{2x^3+2x}{x^4-1}=\dfrac{2x(x^2+1)}{(x^2+1)(x^2-1)}$

$$A=\frac{2x}{x^2-1}=2x$$

29. $\dfrac{2a}{8a}=\dfrac{2a}{2a\cdot 4}=\dfrac{1}{4}$

33. $\dfrac{a+b}{5a^2+5ab}=\dfrac{(a+b)}{5a(a+b)}=\dfrac{1}{5a}$

37. $\dfrac{4x^2+1}{4x^2-1}=\dfrac{4x^2+1}{(2x-1)(2x+1)}$

Since no cancellations can be made the fraction cannot be reduced.

41. $\dfrac{3+2y}{4y^3+6y^2}=\dfrac{(2y+3)}{2y^2(2y+3)}=\dfrac{1}{2y^2}$

45. $\dfrac{2w^45w^2-3}{w^4+11w^2+24}=\dfrac{(2w^2-1)(w^2+3)}{(w^2+8)(w^2+3)}=\dfrac{2w^2-1}{w^2+8}$

49. $\dfrac{N^4-16}{8N-16}=\dfrac{(N^2+4)(N^2-4)}{8(N-2)}$

$$=\frac{(N^2+4)(N+2)(N-2)}{8(N-2)}$$
$$=(N^2+4)(N+2)$$

53. $\dfrac{(x-1)(3+x)}{(3-x)(1-x)}=\dfrac{(x-1)(3+x)}{-(3-x)(x-1)}=\dfrac{3+x}{-(3-x)}=\dfrac{x+3}{x-3}$

57. $$\frac{x^3+x^2-x-1}{x^3-x^2-x+1}=\frac{x^3-1+x^2-x}{x^2+1-x^2-x}$$
$$=\frac{(x-1)(x^2+x+1)+x(x-1)}{(x+1)(x^2-x+1)-x(x+1)}$$
$$=\frac{(x-1)(x^2+x+1+x)}{(x+1)(x^2-x+1-x)}$$
$$=\frac{(x-1)(x^2+2x+1)}{(x+1)(x^2-2x+1)}$$
$$=\frac{(x-1)(x+1)(x+1)}{(x+1)(x-1)(x-1)}$$
$$=\frac{x+1}{x-1}$$

61. $$\frac{x^3+y^3}{2x+2y}=\frac{(x+y)(x^2-xy+y^2)}{2(x+y)}=\frac{x^2-xy+y^2}{2}$$

65. (a) $\frac{x^2(x+2)}{x^2+4}$ will not reduce further since x^2+4 does not factor.

(b) $$\frac{x^4+4x^2}{x^4-16}=\frac{x^2(x^2+4)}{(x^2+4)(x^2-4)}=\frac{x^2}{x^2-4}$$
$$=\frac{x^2}{(x+2)(x-2)}$$

69. $$\frac{mu^2-mv^2}{mu-mv}=\frac{m(u^2-v^2)}{m(u-v)}$$
$$=\frac{(u-v)(u+v)}{(u-v)}=u+v$$

6.6 Multiplication and Division of Fractions

1. $$\frac{4x+6y}{(x-y)^2}\times\frac{(x^2-y^2)}{6x+9y}=\frac{2(2x+3y)(x+y)(x-y)}{(x-y)(x-y)\cdot 3(2x+3y)}$$
$$=\frac{2(x+y)}{3(x-y)}$$

5. $$\frac{3}{8}\times\frac{2}{7}=\frac{3}{4}\times\frac{1}{7}=\frac{3}{28}$$
(divide out a common factor of 2)

9. $$\frac{2}{9}\div\frac{4}{7}=\frac{2}{9}\times\frac{7}{4}=\frac{1}{9}\times\frac{7}{2}=\frac{7}{18}$$
(divide out a common factor of 2)

13. $$\frac{4x+12}{5}\times\frac{15t}{3x+9}=\frac{4(x+3)}{5}\times\frac{5(3t)}{3(x+3)}=4t$$
(divide out common factors of $15(x+3)$)

17. $$\frac{2a+8}{15}\div\frac{a^2+8a+16}{125}=\frac{2(a+4)}{3\times 5}\times\frac{5\times 5\times 5}{(a+4)(a+4)}$$
$$=\frac{50}{3(a+4)}$$
(divide out a common factor of $5(a+4)$)

21. $$\frac{3ax^2-9ax}{10x^2+5x}\times\frac{2x^2+x}{a^2x-3a^2}=\frac{3ax(x-3)}{5x(2x+1)}\times\frac{x(2x+1)}{a^2(x-3)}$$
$$=\frac{3x}{5a}$$
(divide out a common factor of $ax(2x+1)(x-3)$)

25. $$\frac{\frac{x^2+ax}{2b-cx}}{\frac{a^2+2ax+x^2}{2bx-cx^2}}=\frac{x(x+a)}{(2b-cx)}\times\frac{x(2b-cx)}{(a+x)(a+x)}=\frac{x^2}{a+x}$$
(divide out a common factor of $(a+x)(2b-cx)$)

29. $$\frac{x^2-6x+5}{4x^2-17x-15}\times\frac{6x+21}{2x^2+5x-7}$$
$$=\frac{(x-5)(x-1)}{(4x+3)(x-5)}\times\frac{3(2x+7)}{(2x+7)(x-1)}$$
$$=\frac{3}{4x+3}$$
(divide out common factor $(x-5)(x-1)(2x+7)$)

33. $$\frac{7x^2}{3a}\div\left(\frac{a}{x}\times\frac{a^2x}{x^2}\right)=\frac{7x^2}{3a}\div\frac{a^3}{x^2}=\frac{7x^2}{3a}\times\frac{x^2}{a^3}=\frac{7x^4}{3a^4}$$
(divide out a common factor of x)

37. $\frac{x^3-y^3}{2x^2-2y^2}\times\frac{y^2+2xy+x^2}{x^2+xy+y^2}$

$=\frac{(x-y)(x^2+xy+y^2)}{2(x-y)(x+y)}\times\frac{(x+y)(x+y)}{(x^2+xy+y^2)}=\frac{x+y}{2}$

(divide out common factors of

$(x-y),(x+y),(x^2+xy+y^2)$)

41. $\frac{x}{2x+4}\times\frac{x^2-4}{3x^2}=\frac{x(x+2)(x-2)}{2(x+2)(3x^2)}=\frac{x-2}{6x}$

45. $\frac{d}{2}\div\frac{v_1d+v_2d}{4v_1v_2}=\frac{d}{2}\times\frac{4v_1v_2}{d(v_1+v_2)}=\frac{2v_1v_2}{v_1+v_2}$

(divide out a common factor $2d$)

6.7 Addition and Subtraction of Fractions

1. $4a^2b=2\cdot2\cdot a\cdot a\cdot b$

$6ab^3=2\cdot3\cdot a\cdot b\cdot b\cdot b$

$4a^2b^2=2\cdot2\cdot a\cdot a\cdot b\cdot b$

$L.C.D.=2^2\cdot3\cdot a^2\cdot b^3=12a^2b^3$

5. $\frac{3}{5}+\frac{6}{5}=\frac{3+6}{5}=\frac{9}{5}$

9. $\frac{1}{2}+\frac{3}{4}=\frac{2}{4}+\frac{3}{4}=\frac{2+3}{4}=\frac{5}{4}$

13. $\frac{a}{x}-\frac{b}{x^2}=\frac{ax}{x^2}-\frac{b}{x^2}=\frac{ax-b}{x^2}$

17. $\frac{2}{5a}+\frac{1}{a}-\frac{a}{10}=\frac{4}{10a}+\frac{10}{10a}-\frac{a^2}{10a}=\frac{4+10-a^2}{10a}$

$=\frac{14-a^2}{10a}$

21. $L.C.D.=2(2x-1)$

$\frac{3}{2x-1}+\frac{1}{4x-2}=\frac{3}{(2x-1)}\times\frac{2}{2}+\frac{1}{2(2x-1)}$

$=\frac{6+1}{2(2x-1)}=\frac{7}{2(2x-1)}$

25. $L.C.D.=4(s-3)$

$\frac{s}{2s-6}+\frac{1}{4}-\frac{3s}{4s-12}=\frac{s}{2(s-3)}\times\frac{2}{2}+\frac{1}{4}\times\frac{(s-3)}{(s-3)}$

$-\frac{3s}{4(s-3)}=\frac{2s+(s-3)-3s}{4(s-3)}=\frac{-3}{4(s-3)}$

29. $L.C.D.=(x-4)(x-4)=(x-4)^2$

$\frac{3}{x^2-8x+16}-\frac{2}{4-x}$

$=\frac{3}{(x-4)(x-4)}+\frac{2}{(x-4)}\times\frac{(x-4)}{(x-4)}$

$=\frac{3+2(x-4)}{(x-4)(x-4)}=\frac{3+2x-8}{(x-4)(x-4)}$

$=\frac{2x-5}{(x-4)(x-4)}=\frac{2x-5}{(x-4)^2}$

$=\frac{10N^2+N-1}{2N^2(N-2)}$

33. $L.C.D.=(3x-1)(x-4)$

$\frac{x-1}{3x^2-13x+4}-\frac{3x+1}{4-x}$

$=\frac{(x-1)}{(3x-1)(x-4)}+\frac{(3x+1)}{(x-4)}\times\frac{(3x-1)}{(3x-1)}$

$=\frac{x-1+9x^2-1}{(3x-1)(x-4)}=\frac{9x^2+x-2}{(3x-1)(x-4)}$

37. $L.C.D.=(w+1)(w^2-w+1)$

$\frac{1}{w^3+1}+\frac{1}{w+1}-2$

$=\frac{1}{(w+1)(w^2-x+1)}+\frac{(w^2-w+1)}{(w+1)(w^2-w+1)}$

$-\frac{2(w+1)(w^2-w+1)}{(w+1)(w^2-w+1)}$

$=\frac{1+w^2-w+1-2(w+1)(w^2-w+1)}{(w+1)(w^2-w+1)}$

$=\frac{w^2-w+2-2w^2-2}{(w+1)(w^2-w+1)}$

$=\frac{-2w^3+w^2-w}{(w+1)(w^2-w+1)}$

41. $\dfrac{\frac{x}{y}-\frac{y}{x}}{1+\frac{y}{x}}\times\dfrac{xy}{xy}=\dfrac{x^2-y^2}{xy+y^2}=\dfrac{(x+y)(x-y)}{y(x+y)}=\dfrac{x-y}{y}$

45.
$$f(x)=\frac{x}{x+1},\ f(x+h)-f(x)$$
$$=\frac{x+h}{x+h+1}-\frac{x}{x+1}$$
$$f(x+h)-f(x)=\frac{(x+h)}{(x+h+1)}\times\frac{(x+1)}{(x+1)}\times\frac{(x+h+1)}{(x+h+1)}$$
$$=\frac{x^2+x+hx+h-x^2-xh-x}{(x+1)(x+h+1)}$$
$$=\frac{h}{(x+1)(x+h+1)}$$

49.
$$\tan\theta\times\cot\theta(\sin\theta)^2=\frac{y}{x}\times\frac{x}{y}+\left(\frac{y}{r}\right)^2-\frac{x}{r}$$
$$=1+\frac{y^2}{r^2}-\frac{x}{r}=\frac{r^2+y^2-rs}{r^2}$$

53.
$$f(x)=x-\frac{2}{x}$$
$$f(a+1)=a+1-\frac{2}{a+1}=\frac{(a+1)^2-2}{a+1}$$
$$=\frac{a^2+2a+1-2}{a+1}$$
$$=\frac{a^2+2a-1}{a+1}$$

57.
$$\frac{y^2-x^2}{y^2+x^2}=\frac{\left(\frac{mn}{m-n}\right)^2-\left(\frac{mn}{m+n}\right)^2}{\left(\frac{mn}{m-n}\right)^2+\left(\frac{mn}{m+n}\right)^2}\times\frac{(m-n)^2(m+n)^2}{(m-n)^2(m+n)^2}$$
$$=\frac{m^2n^2(m+n)^2-m^2n^2(m-n)^2}{m^2n^2(m+n)^2+m^2n^2(m-n)^2}$$
$$=\frac{m^2n^2\left(m^2+2mn+n^2-m^2+2mn-n^2\right)}{m^2n^2\left(m^2+2mn+n^2+m^2-2mn+n^2\right)}$$
$$=\frac{4mn}{2m^2+2n^2}$$
$$=\frac{2mn}{m^2n^2}$$

61.
$$\frac{2n^2-n-4}{2n^2+2n-4}+\frac{1}{n-1}=\frac{2n^2-n-4+2(n+2)}{2(n-1)(n+2)}$$
$$=\frac{2n^2-n-4+2n+4}{2(n-1)(n+2)}$$
$$=\frac{2n^2+n}{2(n-1)(n+2)}$$
$$=\frac{n(2n+1)}{2(n-1)(n+2)}$$

65.
$$\frac{\frac{L}{C}+\frac{R}{sC}}{sL+R+\frac{1}{sC}}=\frac{\frac{Ls+R}{sC}}{\frac{(sL+R)sC+1}{sC}}=\frac{\frac{IS+R}{sC}}{\frac{s^2CL+CRs+1}{sC}}$$
$$=\frac{LsR}{Cs}\times\frac{Cs}{CLs^2+CRs+1}$$
$$=\frac{Ls+R}{CLs^2+CRs+1}$$

6.8 Equations Involving Fractions

1.
$$\frac{x}{2}-\frac{1}{b}=\frac{x}{2b}$$
$$xb-2=x$$
$$x(b-1)=2$$
$$x=\frac{2}{b-1}$$

5.
$$\frac{x}{2}+6=2x$$
$$x+12=4x$$
$$3x=12$$
$$x=4$$

9.
$$1-\frac{t-5}{6}=\frac{3}{4}$$
Multiply both sides by the L.C.D. = 12
$$12-2(t-5)=9$$
$$12-2t+10=9$$
$$2t=13$$
$$t=\frac{13}{2}$$

13. $\frac{3}{T}+2=\frac{5}{3}$

Multiply both sides by the L.C.D. $= 3T$

$9+6T=5T$

$T=-9$

17. $\frac{2y}{y-1}=5$

Multiply both sides by the L.C.D. $= y-1$

$2y=5y-5$

$3y=5$

$y=\frac{5}{3}$

21. $\frac{5}{2x+4}+\frac{3}{6x+12}=2$

Multiply both sides by the L.C.D. $= 6(x+2)$

$\frac{5}{2(x+2)}+\frac{3}{6(x+2)}=2$

$15+3=2\times 6(x+2)$

$18=12x+24$

$12x=-6$

$x=-\frac{1}{2}$

25. $\frac{1}{4x}+\frac{3}{2x}=\frac{2}{x+1}$

Multiply both sides by L.C.D. $= 4x(x+1)$

$(x+1)+6(x+1)=2\times 4x$

$x+1+6x+6=8x$

$7x+7=8x$

$x=7$

29. $\frac{1}{x^2-x}-\frac{1}{x}=\frac{1}{x-1}$

Multiply both sides by L.C.D. $= x(x-1)$

$\frac{1}{x(x-1)}-\frac{1}{x}=\frac{1}{(x-1)}$

$1-(x-1)=x$

$1-x+1=x$

$2x=2$

$x=1$, no solution

33. $2-\frac{1}{b}+\frac{3}{c}=0$, for c

Multiply both sides by L.C.D. $= bc$

$2bc-c+3b=0$

$c(2b-1)=-3b$

$c=\frac{3b}{1-2b}$

37. $\frac{s-s_\circ}{t}=\frac{v+v_\circ}{2}$ for v

Multiply both sides by L.C.D. $= 2t$

$2(s-s_\circ)=t(v+v_\circ)$

$2(s-s_\circ)=tv+tv_\circ$

$2(s-s_\circ)-tv_\circ=tv$

$v=\frac{2(s-s_\circ)-t_\circ v_\circ}{t}$

41. $z=\frac{1}{g_m}-\frac{jX}{g_m R}$ for R

$g_m Rz=R-jX$

$g_m Rz-R=-jX$

$R(g_m z-1)=-jX$

$R=\frac{jX}{1-g_m z}$

45. $\frac{1}{C}=\frac{1}{C_2}+\frac{1}{C_1+C_3}$

$C_2(C_1+C_3)=C(C_1+C_3)+CC_2$

$C_1C_2+C_2C_3=CC_1+CC_3+CC_2$

$C_1(C_2-C)=CC_3+CC_2-C_2C_3$

$C_1=\frac{CC_3+CC_2-C_2C_3}{C_2-C}$

49. $\frac{1}{5.0}\times t+\frac{1}{8.0}\times t=1$

$8.0t+5.0t=40$

$13t=40$

$t=\frac{40}{13}$

$t=0.33$ h

53. $d = 2.0t_1$ for trip up

$d = 2.2t_2$ for trip down

$$t_1 + t_2 + 90 = 5.0(60)$$

$$\frac{d}{2.0} + \frac{d}{2.2} + 90 = 5(60)$$

$$d = 220 \text{ m}$$

57. $$\frac{V}{R_1} + \frac{V}{R_2} = i$$

$$\frac{V}{2.7} + \frac{V}{6.0} = 1.2$$

$$V = 2.2V$$

Chapter 6 Review Exercises

1. $3a(4x + 5a) = 12ax + 15a^2$

5. $(2a+1)^2 = 4a^2 + 4a + 1$

9. $(2x+5)(x-9) = 2x^2 - 13x - 45$

13. $3s + 9t = 3(s + 3t)$

17. $$W^2b^{x+2} - 144b^x = b^x(W^2b^2 - 144)$$
$$= b^x(Wb+12)(Wb-12)$$

21. $4(36t^2 - 24t + 4) = 4(3t-1)(3t-1) = 4(3t-1)^2$

25. $x^2 + x - 56 = (x+8)(x-7)$

29. $2k^2 - k - 36 = (2k-9)(k+4)$

33. $10b^2 + 23b - 5 = (5b-1)(2b+5)$

37. $$250 - 16y^6 = 2(125 - 8y^6)$$
$$= 2\left(5^3 - (2y^2)^3\right)$$
$$= 2(5 - 2y^2)(25 + 10y^2 + 4y^4)$$

41. $$ab^2 - 3b^2 + a - 3 = b^2(a-3) + (a-3)$$
$$= (a-3)(b^2+1)$$

45. $$\frac{48ax^3y^6}{9a^3xy^6} = \frac{16x^2}{3a^2}$$

49. $$\frac{4x+4y}{35x^2} \times \frac{28x}{x^2-y^2} = \frac{4(x+y)}{35x^2} \times \frac{28x}{(x+y)(x-y)}$$

53. $$\frac{\frac{3x}{7x^2+13x-3}}{\frac{6x^2}{x^2+4x+4}} = \frac{3x}{(7x-1)(x+2)} \times \frac{(x+2)(x+2)}{3 \times 2x^2}$$
$$= \frac{x+2}{2x(7x-1)}$$

57. $$\frac{4}{9x} - \frac{5}{12x^2} = \frac{4}{9x} \times \frac{4x}{4x} - \frac{5}{12x^2} \times \frac{3}{3}$$
$$= \frac{16x-15}{36x^2}$$

61. $$\frac{a+1}{a+2} - \frac{a+3}{a} = \frac{(a+1)}{(a+2)} \times \frac{a}{a} - \frac{(a+3)}{a} \times \frac{(a+2)}{(a+2)}$$
$$= \frac{a(a+1) - (a+3)(a+2)}{a(a+2)}$$
$$= \frac{a^2 + a - a^2 - 5a - 6}{a(a+2)}$$
$$= \frac{-4a-6}{a(a+2)} = \frac{-2(a+3)}{a(a+2)}$$

65. $$\frac{3x}{2x^2-2} - \frac{2}{4x^2-5x+1} = \frac{3x}{2(x+1)(x-1)} \times \frac{(4x-1)}{(4x-1)}$$
$$- \frac{2}{(4x-1)(x-1)} \times \frac{2(x+1)}{2(x+1)}$$
$$= \frac{3x(4x-1) - 4(x+1)}{2(4x-1)(x+1)(x-1)} = \frac{12x^2 - 3x - 4x - 4}{2(4x-1)(x+1)(x-1)}$$
$$= \frac{12x^2 - 7x - 4}{2(4x-1)(x+1)(x-1)}$$

69. $$\frac{6x^2-7x-3}{4x^2-8x+3} = \frac{(2x-3)(3x+1)}{(2x-1)(2x-3)} = \frac{3x+1}{2x-1}$$

Graph $y_1 = \dfrac{6x^2-7x-3}{4x^2-8x+3}$

and $y_2 = \dfrac{3x+1}{2x-1}$. The graphs are the same.

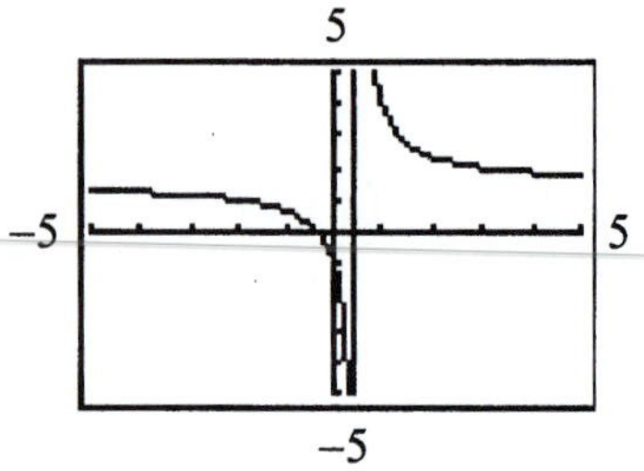

73. $x^2 - 5 = \left(x + \sqrt{5}\right)\left(x - \sqrt{5}\right)$

77. $\dfrac{x}{2} - 3 = \dfrac{x - 10}{4}$

81.
$$\frac{2x}{2x^2 - 5x} - \frac{3}{x} = \frac{1}{4x - 10}$$
$$\frac{2x}{x(2x-5)} - \frac{3}{x} = \frac{1}{2(2x-5)}$$
$$4x - 6(2x-5) = x$$
$$4x - 12x + 30 = x$$
$$9x = 30$$
$$x = \frac{10}{3}$$

85. (a) Changing an odd number of signs changes the sign of the fraction.
(b) Changing an even number of signs leaves the sign of the fraction unchanged.

89. $2zS(S+1) = 2zS^2 + 2zS$

93.
$$cT_2 - cT_1 + RT_2 - RT_1 = c(T_2 - T_1) + R(T_2 - T_1)$$
$$= (T_2 - T_1)(c + R)$$

97. $(n+1)^3(2n+1)^3$
$$= \left(n^3 + 3n^2 + 3n + 1\right)\left(8n^3 + 12n^2 + 6n + 1\right)$$
$$= 8n^6 + 12n^5 + 6n^4 + n^3 + 24n^5 + 36n^4 + 18n^3 + 3n^2$$
$$+ 24n^4 + 36n^3 + 18n^2 + 3n + 8n^3 + 12n^2 + 6n + 1$$
$$= 8n^6 + 36n^5 + 66n^4 + 63n^3 + 33n^2 + 9n + 1$$

101. Increase in volume $= (x+4)^3 - x^3$
$$= x^3 + 12x^2 + 48x + 64 - x^2$$
$$= 12x^2 + 48x + 64$$
$$= 4\left(3x^2 + 12x + 16\right)$$

105.
$$\frac{\frac{\pi ka}{2}\left(R^4 - r^4\right)}{\pi ka\left(R^2 - r^2\right)} = \frac{1\left(R^2 + r^2\right)\left(R^2 - r^2\right)}{2\left(R^2 - r^2\right)} = \frac{R^2 + r^2}{2}$$

109.
$$\frac{4k-1}{4k-4} + \frac{1}{2k} = \frac{(4k-1)}{4(k-1)} \times \frac{k}{k} + \frac{1}{2k} \times \frac{2(k-1)}{2(k-1)}$$
$$= \frac{4k^2 - k + 2k - 2}{4k(k-1)}$$
$$= \frac{4k^2 + k - 2}{4k(k-1)}$$

113.
$$\frac{\frac{u^2 2g}{} - x}{\frac{1}{2gc^2} - \frac{u^2}{2g} + x} \times \frac{2gc^2}{2gc^2} = \frac{u^2c^2 - 2gc^2x}{1 - u^2c^2 + 2gc^2x}$$

117.
$$R = \frac{wL}{H(w+L)}$$
$$RHw + RHL = wL$$
$$wL - RHL = RHw$$
$$L(w - RH) = RHw$$
$$L = \frac{RHw}{w - RH}$$

121.
$$s^2 + \frac{cs}{m} + \frac{kL^2}{mb^2} = 0$$
$$s^2mb^2 + csb^2 + kL^2 = 0$$
$$csb^2 = -s^2mb^2 - kL^2$$
$$c = \frac{-s^2mb^2 - kL^2}{sb^2}$$

125.
$$\frac{1}{4} \times t + \frac{1}{24} \times t = 1$$
$$\frac{7}{24} \times t = 1$$
$$t = 3.4 \text{ lb}$$

129. $d = \dfrac{w_a}{w_a - w_w} = \dfrac{1.097w_w}{1.097w_w - w_w} = \dfrac{1.097}{1.097 - 1} = 11.3$

133. $\dfrac{\left(1+\frac{1}{s}\right)\left(1+\frac{1}{s/2}\right)}{3+\frac{1}{s}+\frac{1}{s/2}}$

When you "cancel" the basic operation being performed is division.

$$=\frac{\left(\frac{s+1}{s}\right)\left(\frac{s/2+1}{s/2}\right)}{\left(3+\frac{1}{s}+\frac{1}{s/2}\right)}\times\frac{s(s/2)}{s(s/2)}$$

$$=\frac{(s+1)(s/2+1)}{3s(s/2)+\frac{s}{2}+s}\times\frac{2}{2}$$

$$=\frac{(s+1)(s+2)}{3s^2+s+2s}=\frac{(s+1)(s+2)}{3s(s+1)}$$

$$=\frac{s+2}{3s}$$

Chapter 7

Quadratic Equations

7.1 Quadratic Equations; Solution By Factoring

1. $2N^2-7N-4=0$

$(2N+1)(N-4)=0$ factor

$2N+1=0$ or $N-4=0$

$2N=-1$ $\quad N=4$

$N=-\frac{1}{2}$ $\quad N=4$

The roots are $N=-\frac{1}{2}$ and $N=4$.

5. $x^2=(x+2)^2$

$x^2=x^2+4x+4$

$4x+4=0$, no x^2 term, not quadratic

9. $x^2-4=0$

$(x+2)(x-2)=0$

$x+2=0$ or $x-2=0$

$x=-2$ $\quad x=2$

13. $x^2-8x-9=0$

$(x-9)(x+1)=0$

$x-9=0$ or $x+1=0$

$x=9$ $\quad x=-1$

17. $40x-16x^2=0$

$2x^2-5x=0$

$x(2x-5)=0$

$x=0$ or $2x-5=0$

$2x=5$

$x=\frac{5}{2}$

21. $3x^2-13x+4=0$

$(3x-1)(x-4)=0$

$3x-1=0$ or $x-4=0$

$3x=1$ $\quad x=4$

$x=\frac{1}{3}$

25. $6x^2=13x-6$

$6x^2-13x+6=0$

$(3x-2)(2x-3)=0$

$3x-2=0$ or $2x-3=0$

$3x=2$ $\quad 2x=3$

$x=\frac{2}{3}$ $\quad x=\frac{3}{2}$

29. $6y^2+by=2b^2$

$6y^2+by-2b^2=0$

$(2y-b)(3y+2b)=0$

$2y-b=0$ or $3y-2b=0$

$y=\frac{b}{2}$ $\quad y=\frac{2b}{3}$

33. $(x+2)^3=x^3+8$

$x^3+6x^2+12x+8=x^3+8$

$6x^2+12x=0$

$6x(x+2)=0$

$6x=0$ or $x+2=0$

$x=0$ $\quad x=-2$

37. $x^2+2ax=b^2-a^2$

$x^2+2ax+(a^2-b^2)=0$

$(x+(a+b))(x+(a-b))=0$

$x+a+b=0$ or $x+a-b=0$

$x=-a-b$ $\quad x=b-a$

41. $V = \alpha I + \beta I^2$

$$2I + 0.5I^2 = 6$$
$$I^2 + 4I - 12 = 0$$
$$(I+6)(I-2) = 0$$
$$I + 6 = 0 \quad \text{or} \quad I = 0$$
$$I = -6 \qquad I = 2$$

The current is -6 A or 2 A.

45.

$$x^3 - x = 0$$
$$x(x^2 - 1) = 0$$
$$x(x+1)(x-1) = 0$$
$$x = 0 \quad \text{or} \quad x + 1 = 0 \quad \text{or} \quad x - 1 = 0$$
$$x = -1 \qquad x = 1$$

The three roots are -1, 0, 1.

49.

$$\frac{1}{2x} - \frac{3}{4} = \frac{1}{2x+3}$$
$$4(2x+3) - 3(2x)(2x+3) = 2x(4)$$
$$8x + 12 - 12x^2 - 18x = 8x$$
$$-12x^2 - 18x + 12 = 0$$
$$-6(2x-1)(x+2) = 0$$
$$2x - 1 = 0 \quad \text{or} \quad x + 2 = 0$$
$$2x = 1 \qquad x = -2$$
$$x = \frac{1}{2}$$

(2) $R_{\text{series}} = 16 = R_1 + R_2$; $R_1 = 16 - R_2$

$$3R_2 + 3(16 - R_2) = (16 - R_2)R_2 \text{ substitute (2)}$$
$$3R_2 + 48 - 3R_2 = 16R_2 - R_2^2$$
$$R_2^2 - 16R_2 + 48 = 0$$
$$(R_2 - 12)(R_2 - 4) = 0$$
$$R_2 - 12 = 0 \quad \text{or} \quad R_2 - 4 = 0$$
$$R_2 = 12 \qquad R_2 = 4$$
$$R_1 = 16 - 12 = 4\Omega \qquad R_1 = 16 - 4 = 12\Omega$$

53. (1) $d = v_1 t_1$, going

(2) $d = v_2 t_2$, returning

(3) $2d = 120 \Rightarrow d = 60$

(4) $t_1 + t_2 = 3.5 \Rightarrow t_1 = 3.5 - t_2$

(5) $v_1 + 10 = v_2$

add (1) and (2) $2d = 120 = v_1 t_1 + v_2 t_2$

$$120 = v_1(3.5 - t_2) + (v_1 + 10)t_2$$
$$120 = v_1\left(3.5 - \frac{d}{v_2}\right) + (v_1 + 10) \times \frac{d}{v_2}$$
$$120 = v_1\left(3.5 - \frac{60}{v_1 + 10}\right) + (v_1 + 10) \times \frac{60}{(v_1 + 10)}$$
$$120 = 3.5v_1 - \frac{60v_1}{v_1 + 10} + 60$$
$$120(v_1 + 10) = 3.5v_1(v_1 + 10) - 60v_1 + 60(v_1 + 10)$$
$$120v_1 + 1200 = 3.5v_1^2 + 35v_1 - 60v_1 + 60v_1 + 600$$
$$3.5v_1^2 - 85v_1 - 600 = 0 \text{ from which}$$
$$v_1 = 30 \text{ km/h}$$
$$v_2 = v_1 + 10 = 40 \text{ km/h}$$

7.2 Completing the Square

1. $x^2 + 6x - 8 = 0$

$$x^2 + 6x = 8$$
$$x^2 + 6x + 9 = 8 + 9$$
$$(x+3)^2 = 17$$
$$x + 3 = \pm\sqrt{17}$$
$$x = -3 \pm \sqrt{17}$$

5. $x^2 = 7$

$$\sqrt{x^2} = \pm\sqrt{7}$$
$$x = \sqrt{7} \text{ or } x = -\sqrt{7}$$

9. $(x+3)^2 = 7$

$$\sqrt{(x+3)^2} = \pm\sqrt{7}$$
$$x + 3 = \pm\sqrt{7} \text{ or } x = -3 \pm \sqrt{7}$$

13. $D^2 + 3D + 2 = 0$

$$D^2 + 3D = -2$$
$$D^2 + 3D + \frac{9}{4} = -2 + \frac{9}{4}$$
$$\left(D + \frac{3}{2}\right)^2 = \frac{1}{4}$$

$$D+\frac{3}{2}=-\frac{1}{2} \quad \text{or} \quad D+\frac{3}{2}=\frac{1}{2}$$
$$D=-\frac{4}{2} \qquad D=-\frac{2}{2}$$
$$=-2 \qquad =-1$$

17.
$$v(v+2)=15$$
$$v^2+2v=15$$
$$v^2+2v-15=0$$
$$v^2+2v=15$$
$$v^2+2v+1=15+1$$
$$(v+1)^2=16$$
$$v+1=\pm 4$$
$$v=-4-1=-5$$
$$v=4-1=3$$

21.
$$3y^2=3y+2$$
$$3y^2-3y-2=0$$
$$y^2-y=\frac{2}{3}$$
$$y^2-y+\frac{1}{4}=\frac{2}{3}+\frac{1}{4}$$
$$\left(y-\frac{1}{2}\right)^2=\frac{11}{12}$$
$$y-\frac{1}{2}=\pm\sqrt{\frac{11}{12}}$$
$$y=\frac{1}{2}\pm\frac{1}{2}\sqrt{\frac{11}{3}}=\frac{1}{2}\pm\frac{1}{2}\frac{\sqrt{33}}{3}$$
$$=\frac{1}{2}\pm\frac{1}{6}\sqrt{33}=\frac{1}{6}\left(3\pm\sqrt{33}\right)$$

25. $5T^2-10T+4=0$
$$5\left(T^2-2T+1\right)=-4+5$$
$$5(T-1)^2=1$$
$$(T-1)^2=\frac{1}{5}$$
$$T=1\pm\frac{\sqrt{5}}{5}$$

29. $x^2+2bx+c=0$
$$x^2+2bx=-c$$
$$x^2+2bx+b^2=-c+b^2$$
$$(x+b)^2=-c+b$$
$$x+b=\pm\sqrt{b^2-c}$$
$$x=-b\pm\sqrt{b^2-c}$$

33.

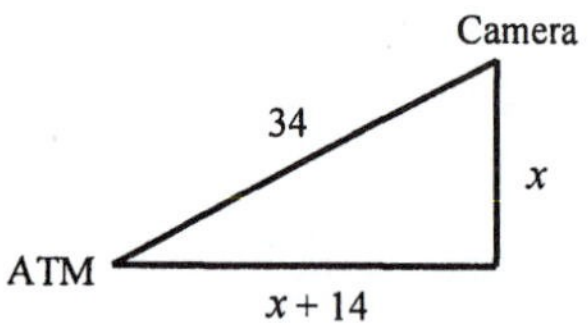

$$34^2=(x+14)^2+x^2$$
$$1156=x^2+28x+196+x^2$$
$$2x^2+28x-960=0$$
$$x^2+14x-480=0$$
$$(x-16)(x+30)=0$$
$$x-16=0 \quad \text{or} \quad x+30=0$$
$$x=16 \qquad x=-30, \text{ reject}$$

The camerais 16 ft above the ATM.

7.3 The Quadratic Formula

1. $x^2+5x+6=0;\ a=1,\ b=5,\ c=6$
$$x=\frac{-5\pm\sqrt{5^2-4(1)(6)}}{2(1)}=\frac{-5\pm\sqrt{1}}{2}=\frac{-5\pm 1}{2}$$
$$x=\frac{-5+1}{2}=-2 \quad \text{or} \quad x=\frac{-5-1}{2}=-3$$

5. $x^2+2x-8=0;\ a=1,\ b=2,\ c=-8$
$$x=\frac{-2\pm\sqrt{2^2-4(1)(-8)}}{2(1)}=\frac{-2\pm\sqrt{36}}{2}$$
$$=\frac{-2\pm 6}{2}$$
$$x=2 \quad \text{or} \quad x=-4$$

9. $x^2-4x+2=0;\ a=1,\ b=-4,\ c=2$

$$x=\frac{-(-4)\pm\sqrt{(-4)^2-4(1)(2)}}{(2)}$$
$$=\frac{4\pm\sqrt{8}}{2}$$
$$=\frac{4\pm2\sqrt{2}}{2}$$
$$=2\pm\sqrt{2}$$

13. $2s^2+5s=3$

$2s^2+5s-3=0;\ a=2,\ b=5,\ c=-3$

$$s=\frac{-5\pm\sqrt{5^2-4(2)(-3)}}{2(2)}$$
$$=\frac{-5\pm\sqrt{49}}{4}$$
$$=\frac{-5\pm7}{4}$$
$$s=\frac{1}{2}\text{ or }s=-3$$

17. $y+2=2y^2$

$2y^2-y-2=0;\ a=2,\ b=-1,\ c=-2$

$$y=\frac{-(-1)\pm\sqrt{(-1)^2-4(2)(-2)}}{2(2)}$$
$$=\frac{1\pm\sqrt{17}}{4}$$

21. $8t^2+61t=-120$

$8t^2+61t+120=0;\ a=8,\ b=61,\ c=120$

$$t=\frac{-61\pm\sqrt{61^2-4(8)(120)}}{2(8)}$$
$$=\frac{-61\pm\sqrt{-119}}{16}$$

25. $25y^2-121=0;\ a=25,\ b=0,\ c=-121$

$$y=\frac{-0\pm\sqrt{0^2-4(25)(-121)}}{2(25)}$$
$$=\frac{\pm\sqrt{12{,}100}}{50}$$
$$=\frac{\pm110}{50}$$
$$=\frac{\pm100}{50}=\pm\frac{11}{5}$$

29. $x^2-0.20x-0.40=0;\ a=1,\ b=-0.20,\ c=-0.40$

$$x=\frac{-(-0.20)\pm\sqrt{(-0.20)^2-4(1)(-0.40)}}{2(1)}$$
$$=\frac{0.2\pm\sqrt{1.64}}{2}$$
$$=-0.54\quad\text{or}\quad x=0.74$$

33. $x^2+2cx-1=0$

$$x=\frac{-2c\pm\sqrt{(2c)^2-4(1)(-1)}}{2(1)}$$
$$=\frac{-2c\pm\sqrt{4c^2+4}}{2}$$
$$=\frac{-2c\pm2\sqrt{c^2+1}}{2}$$
$$=-c\pm\sqrt{c^2+1}$$

37. $2x^2-7x=-8$

$2x^2-7x+8=0;\ a=2;\ b=-7;\ c=8$

$D=\sqrt{(-7)^2-4(2)(8)}=\sqrt{-15},$

unequal imaginary roots

41. $x^2+4x+k=0$ will have a double root if

$$b^2-4ac=0\Rightarrow4^2-4(1)(k)=0$$
$$k=0$$

45. For $D=3.625,\ D_0^2-DD_0-0.25D^2=0$ is

$D_0^2-3.625D_0-0.25(3.625)^2=0$

$D_0^2-3.625D_0-3.28515625=0;$

$a=1,\ b=-3.625,\ c=-3.28515625$

$$D_0=\frac{-(-3.625)\pm\sqrt{(-3.625)^2-4(1)(-3.28515625)}}{2}$$

$D_0 = 4.376$ cm or $D_0 = -0.75$, reject since $D_0 > 0$.

49. $Lm^2+Rm+\frac{1}{C}=0;\ a=L,\ b=R,\ c=\frac{1}{C}$

$$m=\frac{-R\pm\sqrt{R^2-4(L)\left(\frac{1}{C}\right)}}{2(L)}$$

$$=\frac{-R\pm\sqrt{R^2-\frac{4L}{C}}}{2L}$$

53. $A=l\times w=262$

$(w+12.8)\times w=262$

$w^2+12.8w-262=0$,

using the quadratic formula $w=11.0$ m or

$w=-24$, reject since $w>0$.

$l=w+12.8=23.8$ m. The dimension of the rectangle are $l=23.8$ m and $w=11.0$ m

57. v = truck speed

$v+15$ = car speed

From $d=vt$

$77=vt$, car

$77=v\left(t+\frac{18}{60}\right)$, truck

$77=v\left(\frac{77}{v+15}+\frac{18}{60}\right)$

$77(v+15)=77v+\frac{18}{60}v(v+15)$

$\frac{18}{60}v^2+\frac{18}{4}v-1155=0;\ a=\frac{18}{60},\ b=\frac{18}{4},\ c=-1155$

$$v=\frac{-\frac{18}{4}\pm\sqrt{\left(\frac{18}{4}\right)^2-4\left(\frac{18}{60}\right)(-1155)}}{2\left(\frac{18}{60}\right)}$$

$v=-70,\ 55$

The truck speed is 55 mi/h and the car speed is 70 mi/h.

7.4 The Graph of the Quadratic Function

1. $y=2x^2+8x+6;\ a=2,\ b=8,\ c=6$

x-coordinate of vertex $=\frac{-b}{2a}$

$=\frac{-8}{2(2)}=-2$

y-coordinate of vertex $=2(-2)^2+8(-2)+6$

$=-2$

The vertex is $(-2,\ -2)$ and since $a>0$, it is a minimum. Since $c=6$, the y-intercept is (0, 6) and the check is:

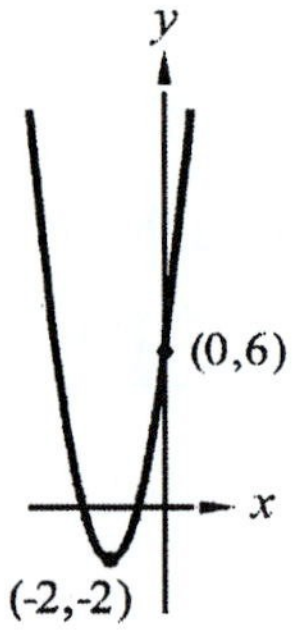

5. $y=-3x^2+10x-4;\ a=-3,\ b=10.$

This means that the x-coordinate of the extreme is

$$\frac{-b}{2a}=\frac{-10}{2(-3)}=\frac{10}{6}=\frac{5}{3}$$

and the y-coordinate is

$$y=-3\left(\frac{5}{3}\right)^2+10\left(\frac{5}{3}\right)-4=\frac{13}{3}.$$

Thus the extreme point is $\left(\frac{5}{3},\frac{13}{3}\right)$.

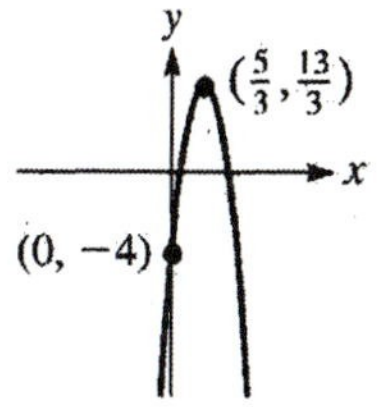

Since $a < 0$, it is a maximum point.
Since $c = -4$, the y-intercept is $(0, -4)$. Use the maximum point $\left(\frac{5}{3}, \frac{13}{3}\right)$, and the y-intercept $(0, -4)$, and the fact that the graph is a parabola, to sketch the graph.

9. $y = x^2 - 4 = x^2 + 0x - 4$; $a = 1$, $b = 0$, $c = -4$
The x-coordinate of the extreme point is
$\frac{-b}{2a} = \frac{-0}{2(1)} = 0$, and the y-coordinate is
$y = 0^2 - 4 = -4$.
The extreme point is $(0, -4)$.

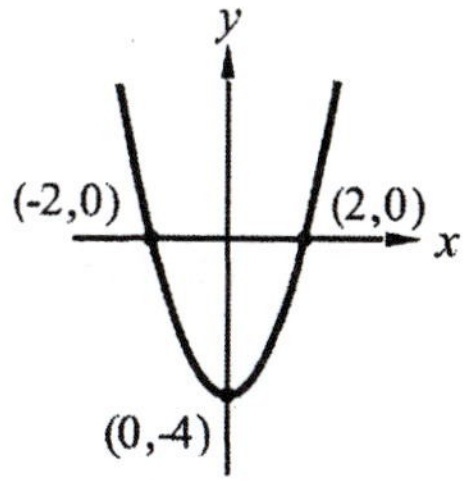

Since $a > 0$, it is a minimum point.
Since $c = -4$, the y-intercept is $(0, -4)$.
$x^2 - 4 = 0$, $x^2 = 4$, $x = \pm 2$ are the x-intercepts.
Use the minimum points and intercepts to sketch the graph.

13. $y = 2x^2 + 3 = 2x^2 + 0x + 3$; $a = 2$, $b = 0$, $c = 3$
The x-coordinate of the extreme point is
$\frac{-b}{2a} = \frac{-0}{2(2)} = 0$, and the y-coordinate is
$y = 2(0)^2 + 3 = 3$.
The extreme point is $(0, 3)$. Since $a > 0$ it is a minimum point.

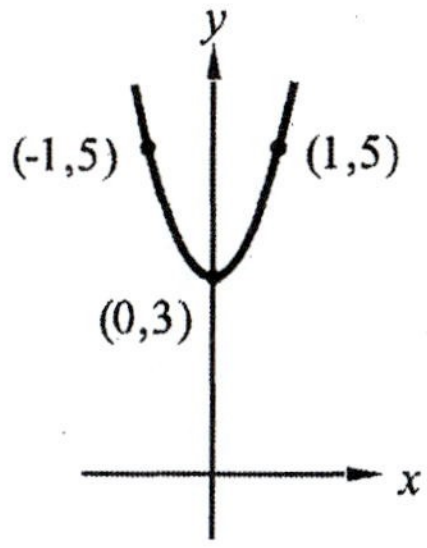

Since $c = 3$, the y-intercept is $(0, 3)$ there are no x-intercepts, $b^2 - 4ac = -24$. $(-1, 5)$ and $(1, 5)$ are on the graph. Use the three points to sketch the graph.

17.
$2x^2 - 3 = 0$. Graph $y = 2x^2 - 3$ on a graphing calculator and use the zero feature to find the roots. $x = -1.2$ and $x = 1.2$.

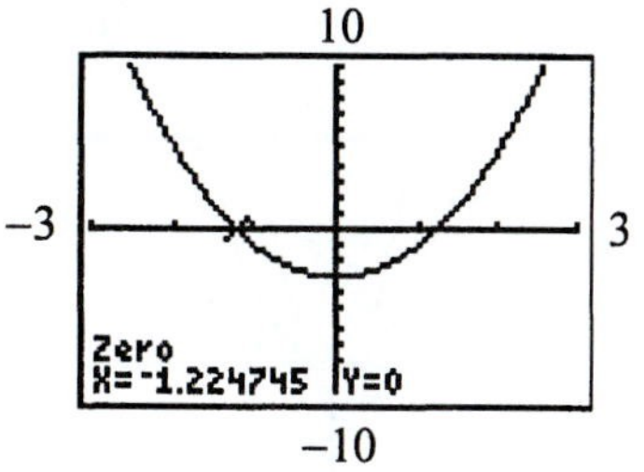

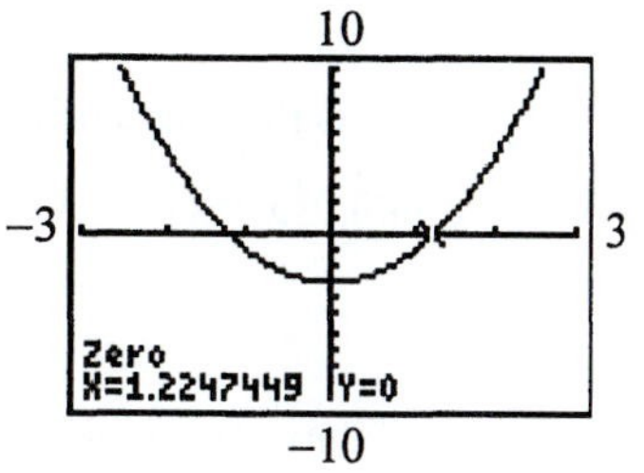

21. $x(2x-1) = -3$.
Graph $y_1 = x(2x-1) + 3$ and use the zero feature.

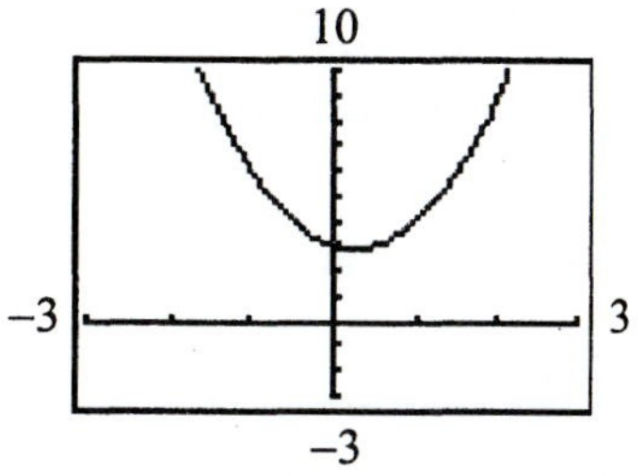

As the graph shows there are no real solutions.

25. (a) $y = x^2$ (b) $y = x^2 + 3$ (c) $y = x^2 - 3$

The parabola $y = x^2 + 3$ is shifted up +3 units (minimum point $(0, 3)$).

The parabola $y = x^2 - 3$ is shifted down -3 units (minimum point $(0, -3)$).

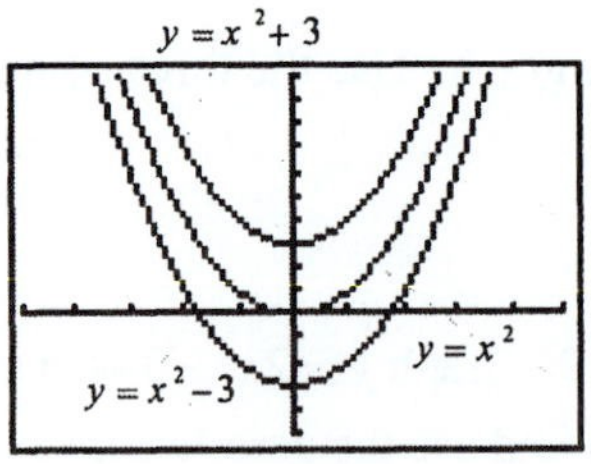

29. (a) $y = x^2$ (b) $y = 3x^2$ (c) $y = \frac{1}{3}x^2$

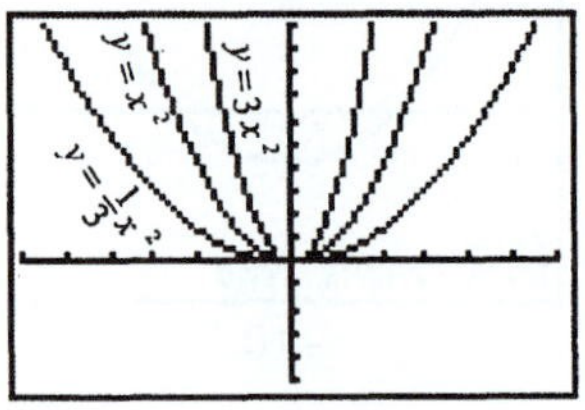

The graph of $y = 3x^2$ is the graph of $y = x^2$ narrowed. The graph of $y = \frac{1}{3}x^2$ is the graph of $y = x^2$ broadened.

33. $y = 2x^2 - 4x - c$ will have two real roots if

$$(-4)^2 - 4(2)(-c) > 0$$
$$16 + 8c > 0$$
$$c > -2$$

-1 is the smallest integral value of c such that $y = 2x^2 - 4x - c$ has two real roots.

37. Graph $y_1 = x(8 - x)$ for $0 > x > 8$.

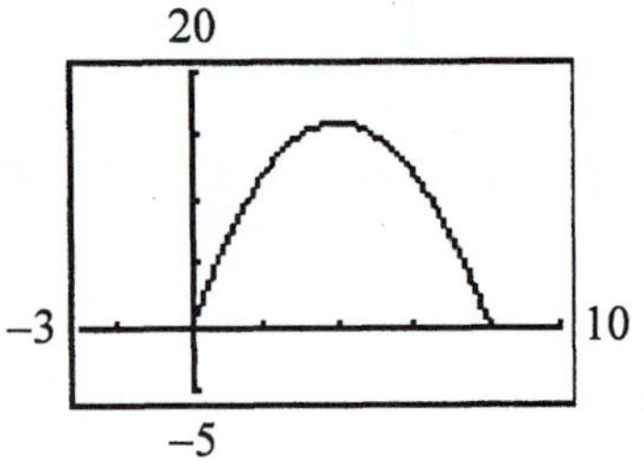

41. Graph $y_1 = 150 + 250x - 16x^2$ and use the zero and maximum feature.

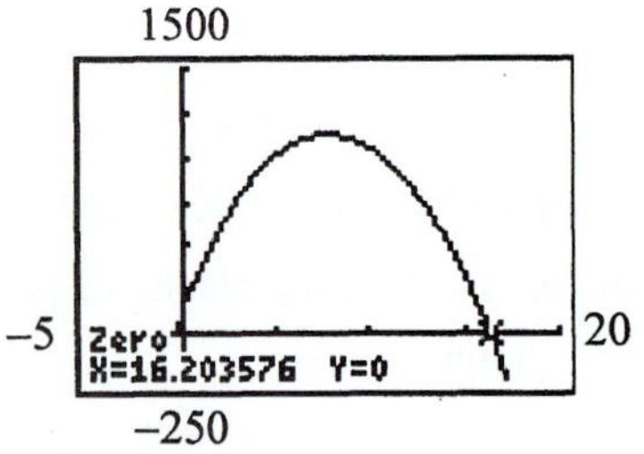

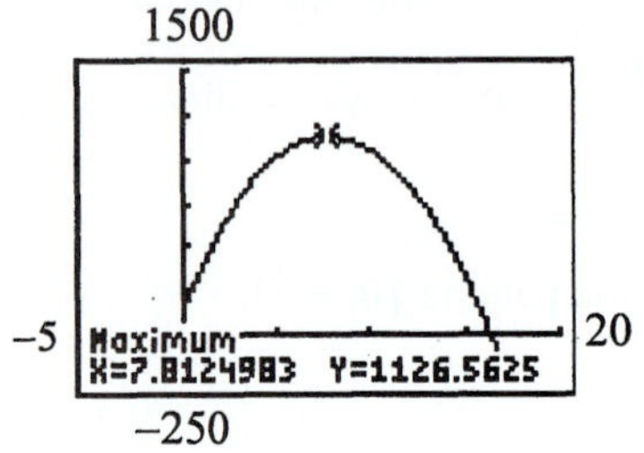

Graph $y_1 = 800$ and use the intersect feature.

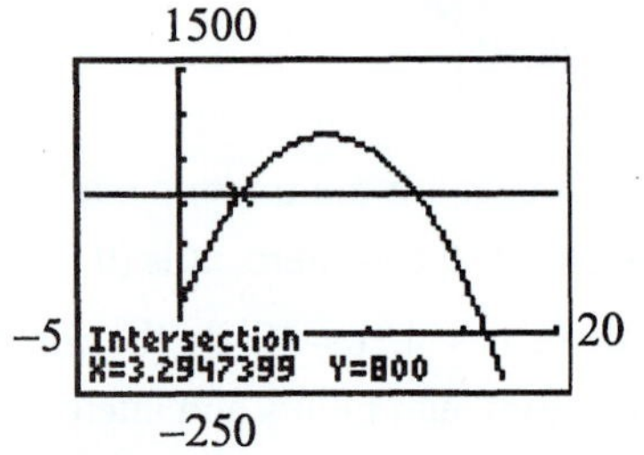

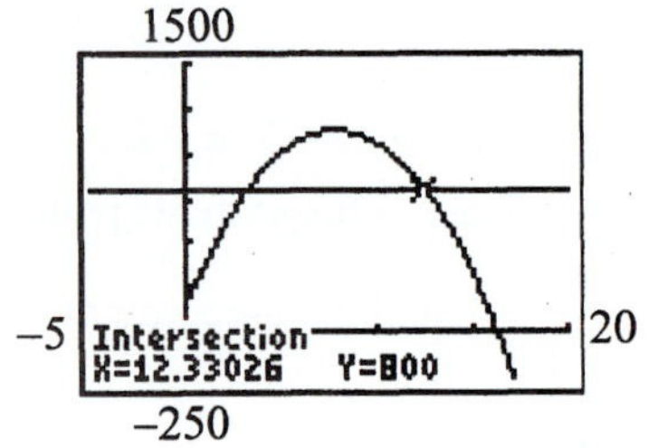

45. $A = xy = 20{,}000,\ y = \dfrac{20{,}000}{x}$

cost $= 30x + 20y = 7500$

$$30x + 20\left(\frac{20{,}000}{x}\right) = 7500$$

$$30x^2 + 400{,}000 = 7500x$$

$$3x^2 - 750x + 40{,}000 = 0$$

Graph $y_1 = 3x^2 - 750x + 40{,}000$ and use the zero feature to solve.

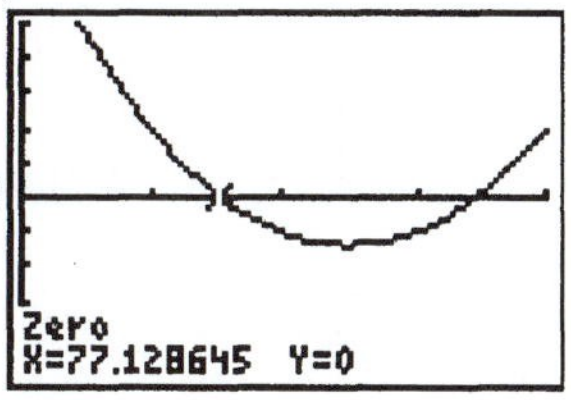

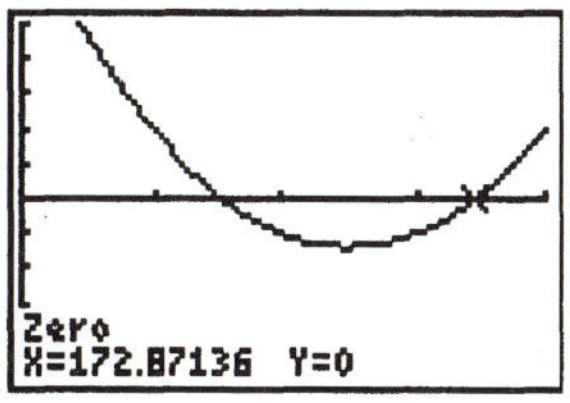

$x_1 = 77.128465,\ x_2 = 172.87136$

$$y_1 = \frac{20{,}000}{x_1} = 259.3076369$$

$$y_2 = \frac{20{,}000}{x_2} = 115.692938$$

The dimensions are 77 ft by 259 ft, or 173 ft by 116 ft.

Chapter 7 Review Exercises

1. $x^2 + 3x - 4 = 0$

$(x+4)(x-1) = 0$

$x + 4 = 0$ or $x - 1 = 0$

$x = -4 \qquad x = 1$

5. $3x^2 + 11x = 4$

$3x^2 + 11x - 4 = 0$

$(3x-1)(x+4) = 0$

$3x - 1 = 0$ or $x + 4 = 0$

$3x = 1 \qquad x = -4$

$x = \dfrac{1}{3}$

9. $6s^2 = 25s$

$6s^2 - 25s = 0$

$s(6s - 25) = 0$

$s = 0$ or $6s - 25 = 0$

$6s = 25$

$s = \dfrac{25}{6}$

13. $x^2 - x - 110 = 0$

$$x = \frac{-(-1) \pm \sqrt{(-1)^2 - 4(1)(-110)}}{2(1)}$$

$$= \frac{\pm 21}{2}$$

$x = -10$ or $x = 11$

17. $2x^2 - x = 36$

$2x^2 - x - 36 = 0$

$$x = \frac{-(-1) \pm \sqrt{(-1)^2 - 4(2)(-36)}}{2(2)}$$

$$= \frac{1 \pm \sqrt{289}}{4}$$

$$= \frac{1 \pm 17}{4}$$

$x = \dfrac{9}{2}$ or $x = -4$

21. $2.1x^2 + 2.3x + 5.5 = 0$

$$x = \frac{-2.3 \pm \sqrt{2.3^2 - 4(2.1)(5.5)}}{2(2.1)}$$

$$= \frac{\pm\sqrt{\ }}{42} = \frac{\pm\sqrt{\ }}{42}$$

there are two unequal imaginary roots.

25. $x^2+4x-4=0$

$$x=\frac{-4\pm\sqrt{4^2-4(1)(-4)}}{2(1)}$$

$$=\frac{-4\pm\sqrt{32}}{2}$$

$$=\frac{-4\pm4\sqrt{2}}{2}$$

$$x=-2\pm2\sqrt{2}$$

29. $4v^2=v+5$

$$4v^2-v-5=0$$

$$(v+1)(4v-5)=0$$

$$v+1=0 \quad \text{or} \quad 4v-5=0$$

$$v=-1 \qquad 4v=5$$

$$v=\frac{5}{4}$$

33. $a^2x^2+2ax+2=0$

$$x=\frac{-2a\pm\sqrt{(2a)^2-4(a^2)(2)}}{2(a^2)}$$

$$x=\frac{-2a\pm\sqrt{-4a^2}}{2a^2} \text{ and for } a>0,$$

$$x=\frac{-2a\pm2a\sqrt{-1}}{2a^2} \qquad x=\frac{-1\pm\sqrt{-1}}{a}$$

37. $x^2-x-30=0$

$$x^2-x+\frac{1}{4}=30+\frac{1}{4}=\frac{121}{4}$$

$$\left(x-\frac{1}{2}\right)^2=\frac{121}{4}$$

$$x-\frac{1}{2}=\frac{\pm11}{2}$$

$$x=\frac{1}{2}\pm\frac{\pm11}{2}$$

41. $\dfrac{x-4}{x-1}=\dfrac{2}{x}$

$$x(x-4)=2(x-1)$$

$$x^2-4x=2x-2$$

$$x^2-6x+2=0$$

$$x=\frac{-(-6)\pm\sqrt{(6)^2-4(1)(2)}}{2(1)}$$

$$=\frac{6\pm\sqrt{28}}{2}$$

$$=\frac{6\pm\sqrt{4\times7}}{2}$$

$$=\frac{6\pm\sqrt{7}}{2}$$

$$x=3\pm\sqrt{7}$$

45. $y=2x^2-x-1;\ a=2,\ b=-1,\ c=-1$

$c=-1\Rightarrow y\text{-intercept}=-1$

$2x^2-x-1=0\Rightarrow x=-\frac{1}{2},\ x=1$, the x-intercepts

$$x \text{ vertex}=\frac{-b}{2a}=\frac{-(-1)}{2(2)}=\frac{1}{4}$$

$$y \text{ vertex}=2\left(\frac{1}{4}\right)^2-\left(\frac{1}{4}\right)-1=-\frac{9}{8}$$

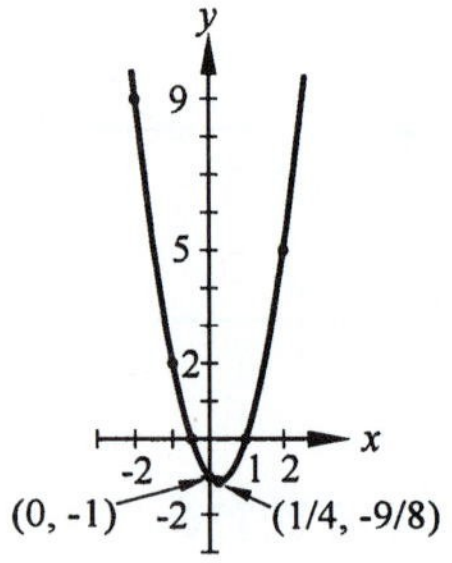

49. Graph $y_1=2x^2+x-4$ and use the zero feature.

$(x=-1.7,\ 1.2)$

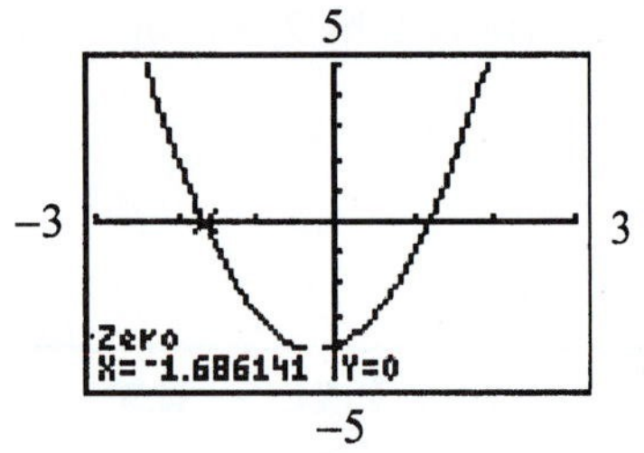

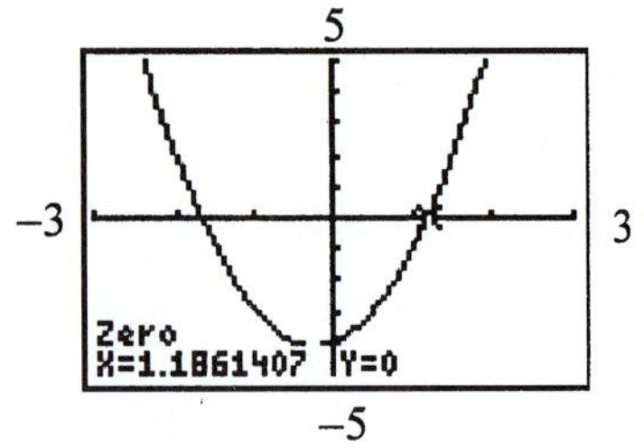

53. The roots are equally spaced on either side of $x = 2$.

$$\frac{x_1 + x_2}{2} = 2$$

$$x_1 + x_2 = 4$$

$$2 + x^2 = 4$$

$x^2 = 2$ is the other solution.

57. $0.01x^2 + 0.8x + 7 = 50$

$0.01x^2 + 0.8x - 43 = 0$ from which

$x = 17$ units

61. $h = vt\,\sin\theta - 16t^2$

$18 = 44t\,\sin\theta\text{-}16t^2$

$16t^2 - 44\left(\sin 65^\circ\right)t + 18 = 0$ from which

$t = 0.6,\ 1.9$

$h = 18$ ft when $t = 0.6$ s and 1.9 s

65. $A = 2\pi r^2 + 2\pi rh$

$2\pi r^2 + 2\pi hr - A = 0$

$$r = \frac{-2\pi h \pm \sqrt{(2\pi)^2 - 4ac(2\pi)(-A)}}{2(2\pi)}$$

$$r = \frac{-2\pi h + \sqrt{4\pi^2 h^2 + 8\pi A}}{4\pi}$$

$$r = \frac{-2\pi + \sqrt{4\left(\pi^2 h^2 + 2\pi A\right)}}{4\pi}$$

$$r = \frac{-\pi h + \sqrt{\pi^2 h^2 + 2\pi A}}{2\pi}$$

69.

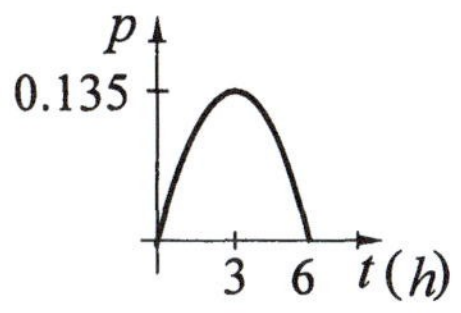

73. $V = e^3$

$V - 29 = (e - 0.10)^3$

$= e^3 - 3e^2(0.10) + 3e(0.10)^2 - (0.10)^3$

$e^3 - 29 = e^3 - 3(0.10)e^2 + 3(0.10)^2 e - (0.10)^3$

$3(0.10)e^2 - 3(0.10)^2 e - 29 + (0.10)^3 = 0$

from which $e = 9.9$ cm

77. $(h + 5.7)^2 + h^2 = 27.0^2$

$h^2 + 2(5.7)h + 5.7^2 + h^2 = 27.0^2$

$2h^2 + 2(.7)h + 5.7^2 - 27.0^2 = 0$ from which

$h = 16.0$

$h + 5.7 = 21.7$

The dimensions of the screen are 16.0 cm, 21.7 cm

81. From the graph $p = 0.205$ at 6 h and 18 h.

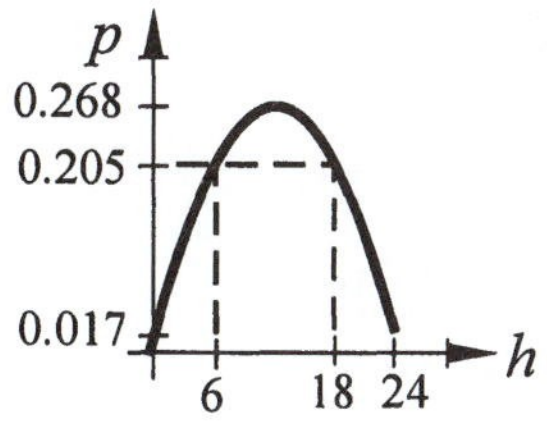

Chapter 8

TRIGONOMETRIC FUNCTIONS OF ANY ANGLE

8.1 Signs of the Trigonometric Functions

1. (a) $\sin(150^\circ + 90^\circ)$ is $-$

$\cos(290^\circ + 90^\circ)$ is $+$

$\tan(190^\circ + 90^\circ)$ is $-$

$\cot(260^\circ + 90^\circ)$ is $-$

$\sec(350^\circ + 90^\circ)$ is $+$

$\csc(100^\circ + 90^\circ)$ is $-$

(b) $\sin(300^\circ + 90^\circ)$ is $+$

$\cos(150^\circ + 90^\circ)$ is $-$

$\tan(100^\circ + 90^\circ)$ is $+$

$\cot(300^\circ + 90^\circ)$ is $+$

$\sec(200^\circ + 90^\circ)$ is $+$

$\csc(250^\circ + 90^\circ)$ is $-$

5. csc 98° is positive since 98° is in QII, where csc θ is positive.

cot 82° is positive since 82° is in QI, where cot θ is positive.

9. cos 348° is positive since 348° is in QIII, where csc θ is negative.

csc 238° is negative since 238° is in QIII, where csc θ is negative.

13. cot -2° is negative since -2° is in QIV, where cot θ is negative.

$\cos 710^\circ = \cos(710^\circ - 360^\circ) = \cos 350^\circ$ which is positive since 350° is in QIV, where cos θ is positive.

$$\sec\theta = \frac{r}{x} = \frac{\sqrt{2}}{-1} = -\sqrt{2}$$

$$\cot\theta = \frac{x}{y} = \frac{-1}{1} = -1$$

17. $(-2, -3)$, $x = -2$, $y = -3$, $r = \sqrt{x^2 + y^2} = \sqrt{13}$

$$\sin\theta = \frac{y}{r} = \frac{-3}{\sqrt{13}}$$

$$\cos\theta = \frac{x}{r} = \frac{-2}{\sqrt{13}}$$

$$\tan\theta = \frac{y}{x} = \frac{-3}{-2} = \frac{3}{2}$$

$$\csc\theta = \frac{r}{y} = \frac{\sqrt{13}}{3}$$

$$\sec\theta = \frac{r}{x} = \frac{\sqrt{13}}{2}$$

$$\cot\theta = \frac{x}{y} = \frac{-2}{-3} = \frac{2}{3}$$

21. $(20, -8)$; $x = 20$, $y = -8$, $r = \sqrt{x^2 + y^2} = 4\sqrt{29}$

$$\sin\theta = \frac{y}{r} = \frac{-8}{4\sqrt{29}} = \frac{-2}{\sqrt{29}}$$

$$\cos\theta = \frac{x}{r} = \frac{20}{4\sqrt{29}} = \frac{5}{\sqrt{29}}$$

$$\tan\theta = \frac{y}{x} = \frac{-8}{20} = -\frac{2}{5}$$

$$\csc\theta = \frac{r}{y} = \frac{4\sqrt{29}}{-8} = -\frac{\sqrt{29}}{2}$$

$$\csc\theta = \frac{r}{y} = \frac{4\sqrt{29}}{-8} = -\frac{\sqrt{29}}{2}$$

$$\sec\theta = \frac{r}{x} = \frac{4\sqrt{29}}{20} = \frac{\sqrt{29}}{5}$$

$$\cot\theta = \frac{x}{y} = \frac{20}{-8} = -\frac{5}{2}$$

25. $\tan\theta = 1.500 \Rightarrow$ QI, QIII

29. $\sin\theta$ is positive and $\cos\theta$ is negative
$\sin\theta$ is positive in QI and QII
$\cos\theta$ is negative in QII and QIII. The terminal side of θ is in QII.

33. $\csc\theta$ is negative and $\tan\theta$ is negative
$\csc\theta$ is negative in QIII and QIV
$\tan\theta$ is negative in QII and QIV. The terminal side of θ is in QIV.

37. $\sin\theta$ is positive and $\cot\theta$ is negative
$\sin\theta$ is positive in QI and QII
$\cot\theta$ is negative in QII and QIV. The terminal side of θ is in QII.

41. For (x, y) in QIV, $x(+)$ and $y(-) \Rightarrow \frac{y}{x} = \frac{(-)}{(+)} = (-)$

8.2 Trigonometric Functions of Any Angle

1. $\sin 200^\circ = -\sin 20^\circ = -0.3420$
$\tan 150^\circ = -\tan 30^\circ = -0.5774$
$\cos 265^\circ = -\cos 85^\circ = -0.0872$
$\cot 300^\circ = -\cot 60^\circ = -0.5774$
$\sec 344^\circ = \sec 16^\circ = 1.040$
$\sin 397^\circ = \sin 37^\circ = 0.6018$

5. $\sin 160^\circ = \sin\left(180^\circ - 160^\circ\right) = \sin 20^\circ$
$\cos 220^\circ = \cos\left(180^\circ + 40^\circ\right) = -\cos 40^\circ$

9.

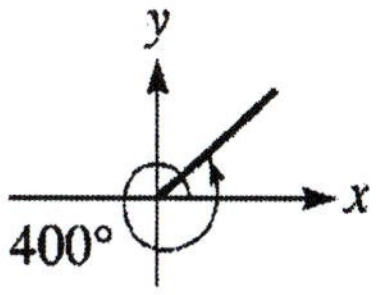

$$\cos 400^\circ = \cos\left(360^\circ + 40^\circ\right) = \cos 40^\circ$$

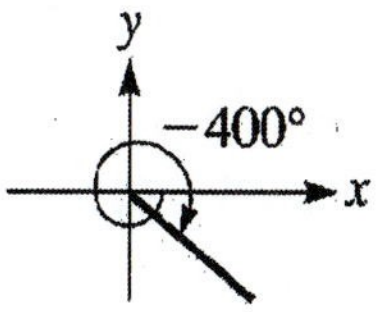

13. $\cos 106.3^\circ = -\cos 73.7^\circ = -0.281$

17. $\tan\left(-31.5^\circ\right) = -\tan 31.5^\circ = -0.613$

21. $\sin 310.36^\circ = -0.7620$

25. $\cos\left(-72.61^\circ\right) = 0.2989$

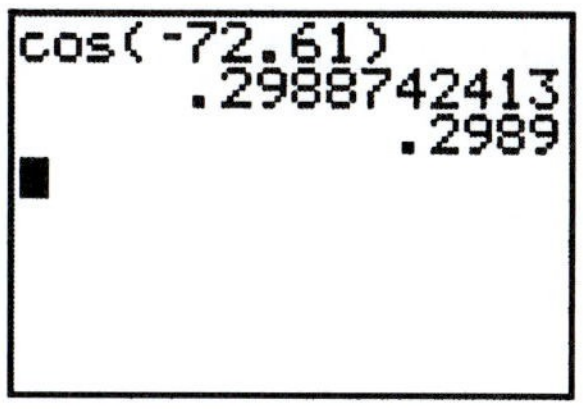

29. $\cos\theta = 0.4003$; $\theta_{ref} = \cos^{-1} 0.4003 = 66.4^\circ$
Since $\cos\theta$ is positive, θ is in QI or QIV.
Therefore, $\theta = 66.4^\circ$
or $\theta = 293.6^\circ$
$0^\circ \le \theta < 360^\circ$

33. $\sin\theta = 0.870$; $\theta_{ref} = \sin^{-1} 0.870 = 60.4^\circ$
Since $0^\circ \le \theta < 360^\circ$, $\cos\theta < 0$, θ is in QII and $\theta = 119.5^\circ$.

37. $\tan\theta = -1.366$; $\theta_{ref} = \tan^{-1}\left(-1.366\right) = -53.8^\circ$
Since $0^\circ \le \theta < 360^\circ$, $\cos\theta > 0$, θ is in QIV and $\theta = 306.2^\circ$.

41. $\sin\theta = -0.5736;\ \theta_{\text{ref}} = \sin^{-1}(-0.5736) = -35^\circ$
Since $\cos\theta > 0$, θ is in QIV.
$\theta = 325^\circ + k\times 360^\circ$ where $k = 0, \pm 1, \pm 2, \cdots$
$\tan\theta = -0.7003$

45. $\sin 90^\circ = 1,\ 2\sin 45^\circ = 2\times\frac{\sqrt{2}}{2} = \sqrt{2}$
$1 < \sqrt{2}$
$\sin 90^\circ < 2\sin 45^\circ$

49. $\theta = 195^\circ$ has $\theta_{\text{ref}} = 15^\circ \Rightarrow \cos 195^\circ = -\cos 15^\circ$
$= -\sin 75^\circ$
$= -0.9659^\circ$

53. $i = i_m \sin\theta$
$i = 0.0259\times\sin 495.2^\circ$
$i = 0.0183$ A
$y = \frac{x\sin\beta}{\sin\alpha}$
$= \frac{(6.78)\sin 104.7^\circ}{\sin 31.3^\circ}$
$= \frac{(6.78)(0.9673)}{(0.5195)}$
$= 12.6$ in.

8.3 Radians

1. $2.80 = \left(\frac{180^\circ}{\pi}\right)(2.80) = 160^\circ$

5. $15^\circ = \frac{\pi}{180}(15) = \frac{\pi}{12}$
$150^\circ = \frac{\pi}{180}(150) = \frac{5\pi}{6}$

9. $210^\circ = \frac{\pi}{180}(210) = \frac{7\pi}{6}$
$27^\circ = \frac{\pi}{180}(27) = \frac{3\pi}{20}$

13. $\frac{2\pi}{5} = \frac{180^\circ}{\pi} = \left(\frac{2\pi}{5}\right) = 72^\circ$
$\frac{3\pi}{2} = \frac{180^\circ}{\pi} = \left(\frac{3\pi}{2}\right) = 270^\circ$

17. $\frac{7\pi}{18} = \frac{180^\circ}{\pi} = \left(\frac{7\pi}{18}\right) = 70^\circ$
$\frac{5\pi}{3} = \frac{180^\circ}{\pi} = \left(\frac{5\pi}{3}\right) = 300^\circ$

21. $23^\circ = \frac{\pi}{180}(23) = 0.401$

25. $333.5^\circ = \frac{\pi}{180}(333.5) = 5.821$

29. $0.750 = \frac{180^\circ}{\pi} = (0.750) = 43.0^\circ$

33. $12.4 = \frac{180^\circ}{\pi} = (12.4) = 710^\circ$

37. $\sin\frac{\pi}{4} = \sin\left[\left(\frac{\pi}{4}\right)\left(\frac{180}{\pi}\right)\right] = \sin 45^\circ = 0.7071$

41. $\cos\frac{5\pi}{6} = \cos\left[\left(\frac{5\pi}{6}\right)\left(\frac{180}{\pi}\right)\right] = \cos 150^\circ = -0.8660$

45. $\tan 0.7359 = 0.9056$

49. $\sec 2.07 = \frac{1}{\cos 2.07} = -2.1$

53. $\sin\theta = 0.3090,\ \theta = 0.3141$
$\sin\theta$ is positive, θ is in QI and QII.
QI, $\theta = 0.3141$
$QII,\ \theta = \pi - 0.3141 = 2.827$

57. $\cos\theta = 0.6742,\ \theta = 0.8309$
$\cos\theta$ is positive, θ is in QI and QIV.
QI, $\theta = 0.8309$
$QIV,\ \theta = 2\pi - 0.8309 = 5.452$

61. $\theta = \frac{s}{r} = \frac{15 \text{ cm}}{12 \text{ cm}} = \frac{5}{4}$

65. For $0 < \theta < \frac{\pi}{2}, \frac{\pi}{2} < \frac{\pi}{2} + \theta < \pi \Rightarrow \frac{\pi}{2} + \theta$ is in QII.

$\frac{\pi}{2} + \theta$ has $\theta_{\text{ref}} = \frac{\pi}{2} - \theta \Rightarrow \tan\left(\frac{\pi}{2} + \theta\right)$

$= -\tan\left(\theta - \frac{\pi}{2}\right)$

69. $1.75(2\pi) = \frac{7\pi}{2}$ rad $= 11.0$ rad

73. $h = 1200 \tan \frac{5t}{3t + 10}$ $t < 10$ s. Let $t = 8.0$ s

$h = 1200 \tan \frac{5(8.0)}{3(8.0) + 10} = 1200 \tan \frac{40.0}{34.0}$

$= 1200 \tan 2.403 = 2900$ m

8.4 Applications of Radian Measure

1. $s = \left(\frac{\pi}{4}\right)(3.00) = 2.36$ in.

5. $s = r\theta = (3.30)\left(\frac{\pi}{3}\right) = 3.46$ in.

9. $\theta = \frac{s}{r} = \frac{0.3913}{0.9449} = 0.4141 = 23.73^\circ$

$A = \frac{1}{2}\theta r^2 = \frac{1}{2}(0.4141)(0.9449)^2 = 0.1849 \text{ mi}^2$

13. $A = \frac{1}{2}r^2\theta \Rightarrow r = \sqrt{\frac{2A}{\theta}} = \sqrt{\frac{2(0.0119)}{326^\circ \frac{\pi}{180}}}$

$r = 0.0647$ ft

17. $r = \frac{s}{\theta} = \frac{0.203}{\frac{3}{4}(2\pi)}$

$r = 0.0431$ mi

21. From $\theta = wt$,

hour hand: $\theta = \frac{2\pi}{12}t$

minute hand: $\theta + \pi = \frac{2\pi}{1}t$, t in hours

$\frac{\pi}{6}t + \pi = 2\pi t \Rightarrow t = \frac{6}{11}$ hour $= 32.73$ minutes

$t = 32$ minutes 44 seconds

at 32 minutes and 44 seconds after noon the hour and minute hands will be at 180°.

25. $w = \frac{\theta}{t} = \frac{\pi}{6.0}$ rad/s $= 0.52$ rad/s

29.

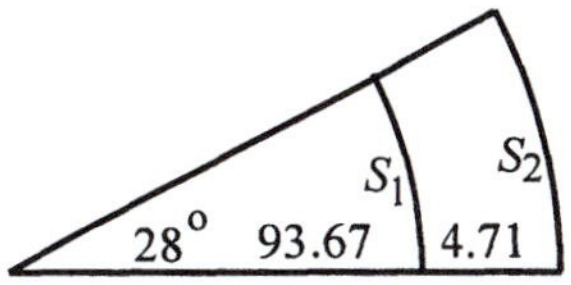

From $S = r\theta$

$S_1 = 93.67(28.0^\circ)\frac{\pi}{180^\circ}$

$S_1 = 45.78$

$S_2 = (93.67 + 4.71)(28.0^\circ)\frac{\pi}{180^\circ}$

$S_2 = 48.08$

$S_2 - S_1 = 2.30$ ft. Outer rail is 2.30 ft longer.

33. $V = At = \left[\frac{1}{2}r_1^2\theta - \frac{1}{2}r_2^2\theta\right]t = \frac{1}{2}\theta\left(r_1^2 - r_2^2\right)t$

$V = \frac{1}{2}(15.6^\circ)\frac{\pi}{180^\circ}\left((285 + 15.2)^2 - 285^2\right)(0.305)$

$V = 369 \text{ m}^3$

37. $v = rw = 8.5(20)(2\pi) = 1070$ ft/min

41. $v = wr = \frac{1 \text{ rev}}{2.88 \text{ day}} \cdot \frac{2\pi}{\text{rev}} \cdot 3{,}500{,}000 \text{ mi} \cdot \frac{1 \text{ day}}{24 \text{ h}}$

$v = 318{,}000$ mi/h

45.

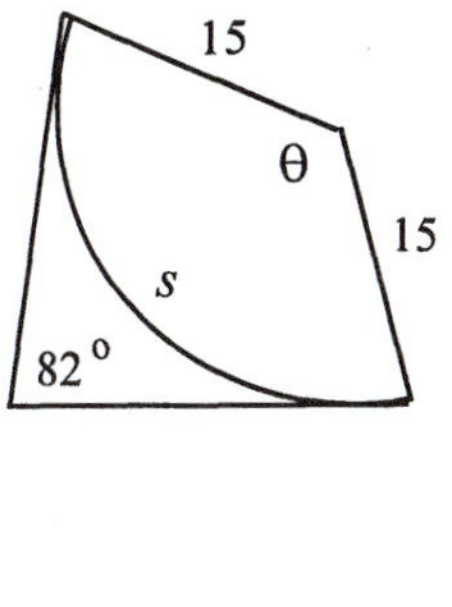

$$82.0^\circ + 2(90^\circ) + \theta = 360^\circ$$
$$\theta = 98.0^\circ$$
$$s = r\theta$$
$$s = 15.0(98.0^\circ)\frac{\pi}{180^\circ}$$
$$s = 25.7 \text{ ft}$$

49. $w = \frac{v}{r} = \frac{\frac{1}{4}(6.5)}{3.75} = 0.433 \text{ rad/s}$

53. $\theta = wt = 2400 \cdot \frac{\text{r}}{\text{min}} \cdot \frac{2\pi \text{ rad}}{\text{r}} \cdot \frac{\text{min}}{60 \text{ s}}(1 \text{ s})$

$\theta = 80\pi \text{ rad} = 250 \text{ rad}$

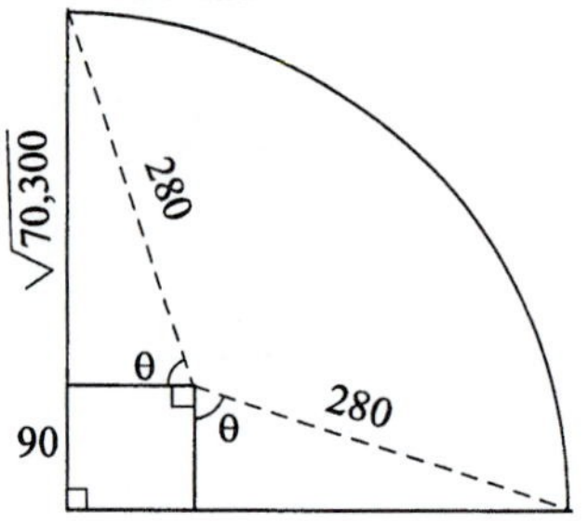

$$A = 90^2 + 2\left(\frac{1}{2}\right)(90)\sqrt{70{,}300} + \frac{1}{2}(280)^2$$
$$\left(2\pi - \frac{\pi}{2} - 2\tan^{-1}\frac{\sqrt{70{,}300}}{90}\right)$$
$$A = 119{,}000 \text{ ft}^2$$

57. $x = \sqrt{1.10^2 - 0.74^2} = 0.81;$

$$\theta = \sin^{-1}\left(\frac{0.74}{1.10}\right) = 42.3^\circ;$$
$$2\theta = 84.6^\circ \frac{\pi}{180^\circ} = 1.477$$
$$A_{\text{sector}} = \frac{1}{2}(1.10)^2(1.477) = 0.89$$
$$A_{\text{triangle}} = \frac{1}{2}(1.48)^2(0.81) = 0.60$$
$$A_{\text{segment}} = 0.89 - 0.60 = 0.29$$
$$A_{\text{circle}} = \pi(1.10)^2 = 3.80$$
$$A_{\text{tank}} = 3.80 - 0.29 = 3.51$$
$$A = 3.51 \times 4.25 = 14.9 \text{ m}^3$$

61. $0.2'' = \frac{0.2^\circ}{3600}$

$$= (5.556 \times 10)^\circ \frac{\pi}{180^\circ}$$
$$= 9.696 \times 10^{-7} \text{ rad}$$

12.5 light years $= (12.5)(5.88 \times 10^{12})\text{min}$

$$\tan 0.2'' = \frac{x}{12.5 \times 5.88 \times 10^{12}}$$
$$x = (12.5)(5.88 \times 10^{12})(9.696 \times 10^{-5})$$
$$= 7.13 \times 10^7 \text{ mi}$$

Chapter 8 Review Exercises

1. $r = \sqrt{6^2 + 8^2} = 10$ for $(6, 8)$

$$\sin\theta = \frac{y}{r} = \frac{8}{10} = \frac{4}{5}$$
$$\cos\theta = \frac{x}{r} = \frac{6}{10} = \frac{3}{5}$$
$$\tan\theta = \frac{y}{x} = \frac{8}{6} = \frac{4}{3}$$
$$\csc\theta = \frac{r}{y} = \frac{5}{4}$$
$$\sec\theta = \frac{r}{x} = \frac{5}{3}$$
$$\cot\theta = \frac{x}{y} = \frac{3}{4}$$

5. $\cos 132^\circ = -\cos(180^\circ - 132^\circ) = -\cos 48^\circ$

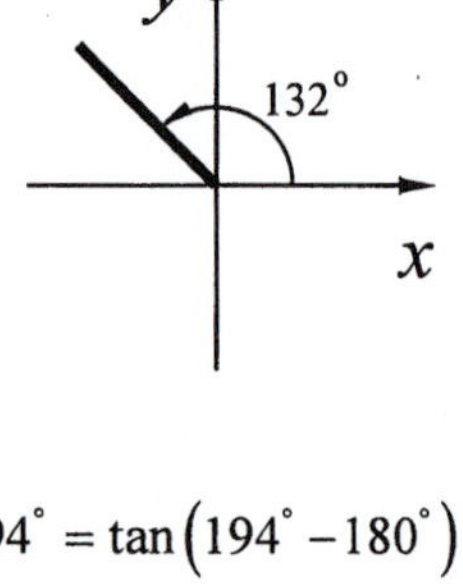

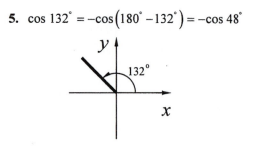

$\tan 194^\circ = \tan(194^\circ - 180^\circ) = \tan 14^\circ$

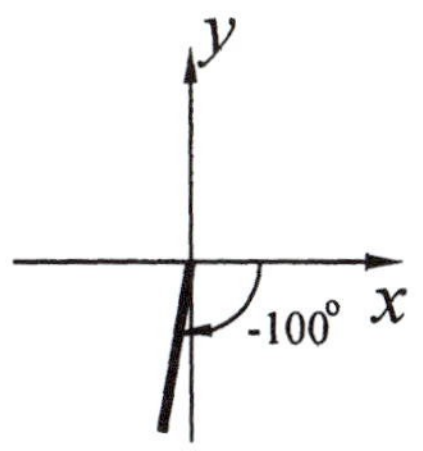

9. $40^\circ \cdot \frac{\pi}{180^\circ} = \frac{2\pi}{9}$

$153^\circ \cdot \frac{\pi}{180^\circ} = \frac{7\pi}{20}$

13. $\frac{7\pi}{5} \cdot \frac{180^\circ}{\pi} = 252^\circ$; $\frac{13\pi}{18} \cdot \frac{180^\circ}{\pi} = 130^\circ$

17. $0.560 \cdot \frac{180^\circ}{\pi} = 32.1^\circ$

21. $102^\circ \cdot \frac{\pi}{180^\circ} = 1.78$

25. $262.05^\circ \cdot \frac{\pi}{180^\circ} = 4.5736$

29. $\cos 245.5^\circ = -0.415$

33. $\csc 247.82^\circ = -1.080$

37. $\tan 301.4^\circ = -1.64$

41. $\sin \frac{9\pi}{4} = -0.5878$

45. $\sin 0.5906 = 0.5569$

49. $\tan \theta = 0.1817, 0 \le \theta < 360^\circ$

$\theta = \tan^{-1}(0.1817) = 10.3^\circ$ in QI

$\theta = 180^\circ + 10.3^\circ = 190.3^\circ$ in QIII

53. $\cos \theta = 0.8387, 0 \le \theta < 2\pi$

$\theta = \cos^{-1}(0.8387) = 0.5759$ in QI

$\theta = 2\pi - 0.5759 = 5.707$ in QIV

57. $\cos \theta = -0.7222$, $\sin \theta < 0$ for $0^\circ \le \theta < 360^\circ \Rightarrow \theta$ in QIII

$\theta = \cos^{-1}(-0.7222) = 136.2364165$ from calculator reference angle $= 180^\circ - \theta = 43.76^\circ$

QIII angle $= 180^\circ +$ reference angle $= 223.76^\circ$.

61. $r = \frac{s}{\theta} = \frac{20.3 \text{ in.}}{107.5^\circ \cdot \frac{\pi}{180^\circ}} = 10.8 \text{ in.}$

65. $s = r\theta = r\frac{2A}{r^2} = \frac{2A}{r}$

$s = \frac{2(32.8)}{4.62} = 14.2 \text{ m}$

69.

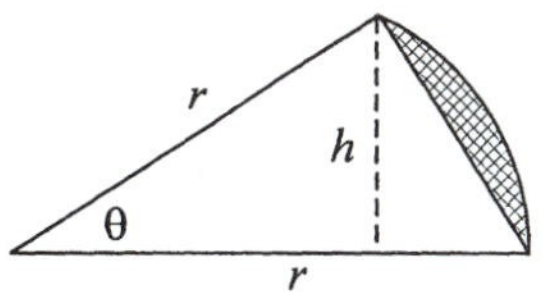

$\sin \theta = \frac{h}{r} \Rightarrow h = r \sin \theta$

$$A_{\text{segment}} = A_{\text{sector}} - A_{\text{triangle}}$$
$$= \frac{1}{2}r^2\theta - \frac{1}{2}rh$$
$$= \frac{1}{2}r^2\theta - \frac{1}{2}r(r\sin\theta)$$
$$= \frac{1}{2}r^2(\theta - \sin\theta)$$

73. $p = P_m \sin^2 377 \cdot t = 0.120 \sin^2(377 \cdot 2 \cdot 10^{-3})$

$p = 0.0562W$

77. $w = \frac{v}{r} = \frac{3.5 \text{ mi/h}}{2.4 \text{ ft}} \cdot \frac{5280 \text{ ft}}{\text{mi}} \cdot \frac{\text{r}}{2\pi} \cdot \frac{\text{h}}{60 \text{ min}}$

$w = 20.4$ r/min

81. (a) $3960 \cdot 60^\circ \cdot \frac{\pi}{180^\circ} = 3960 \cdot \frac{\pi}{3}$ over with pole, 4150 mi

(b) $3960 \cdot \sin 30^\circ \cdot \pi = 3960 \cdot \frac{\pi}{2}$ along 60°N latitude arc, 6220 mi

The distance over the north pole is shorter.

85.

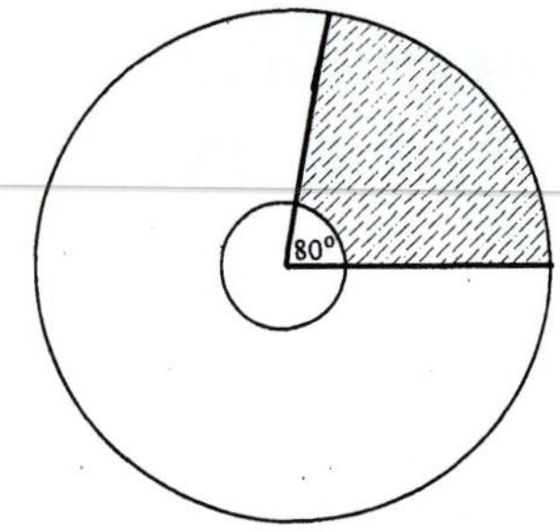

$$A_{\text{circle}} = \pi \cdot 3.25^2 = A_1$$
$$A_{\text{hole}} = \pi \cdot 0.75^2 = A_2$$
$$A_{\text{hatched}} = \frac{1}{2} \cdot 80^\circ \cdot \frac{\pi}{180^\circ}\left(3.25^2 - 0.75^2\right) = A_3$$
$$A_{\text{hood}} = A - A_2 - A_3 = 24.4 \text{ ft}^2$$

89. $A = \underbrace{\frac{1}{2} \cdot 15^2 \cdot 60^\circ \cdot \frac{\pi}{180^\circ}}_{\text{area of sector formed by one arc}} +$

$$\left(\frac{1}{2} \cdot 15^2 \cdot 60^\circ \cdot \frac{\pi}{180^\circ} - \underbrace{\frac{1}{2} \cdot 15 \cdot 15 \sin 60^\circ}_{\text{area of equilateral triangle inside one sector}}\right)$$

$A = 138 \text{ m}^2$

93. $v = r \cdot w = \left(1080 \text{ mi} + 70.0 \text{ mi}\right) \cdot \frac{1 \text{ r}}{1.95 \text{ h}} \cdot \frac{2\pi \text{ rad}}{\text{r}}$

$v = 3705.5 \text{ mi/h}$

Chapter 9

VECTORS AND OBLIQUE TRIANGLES

9.1 Introduction to Vectors

1.

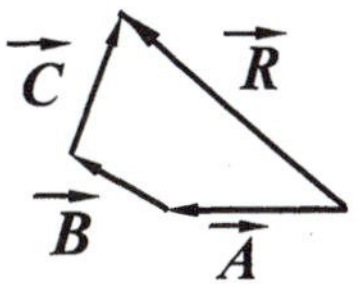

5. (a) scalar, no direction given
   (b) vector, magnitude and direction both given

9.

13.

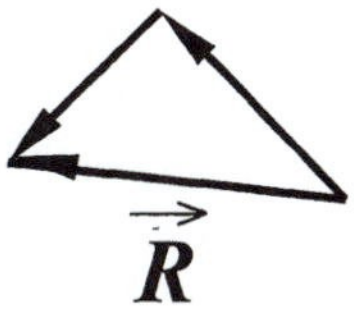

17. 4.3 cm, 156°

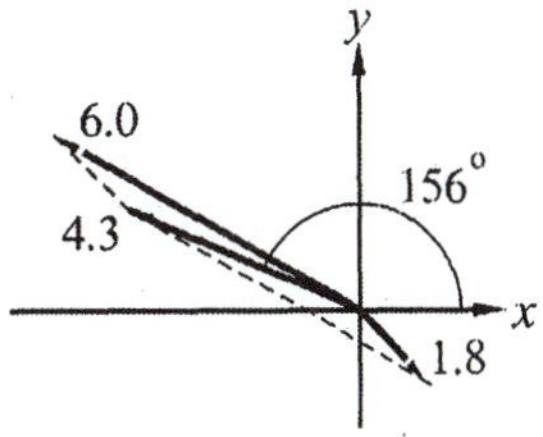

21.

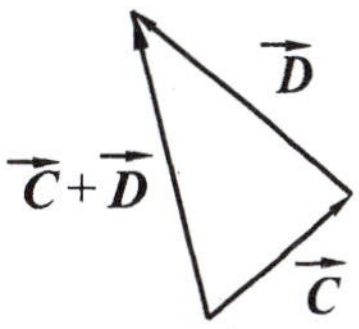

25.

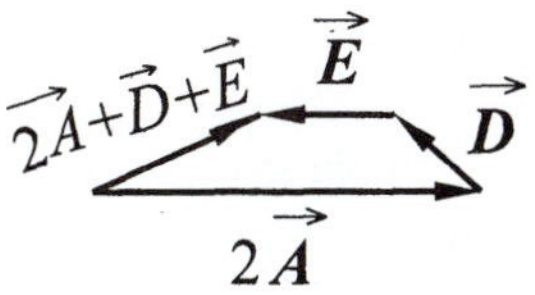

29.

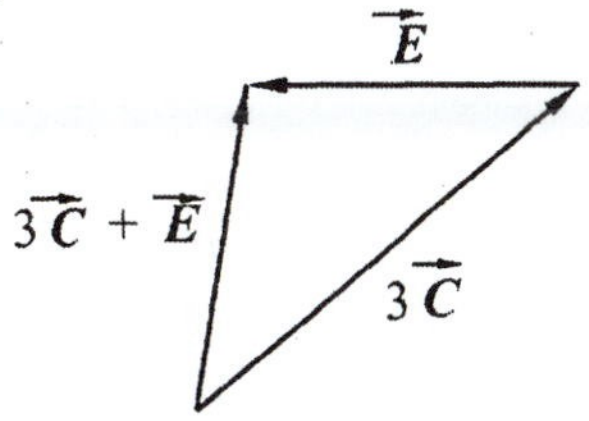

33.

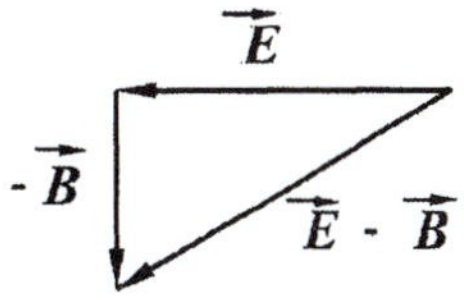

37.

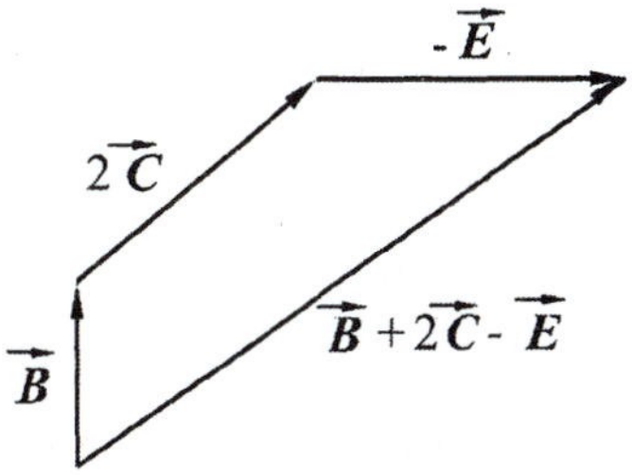

41.

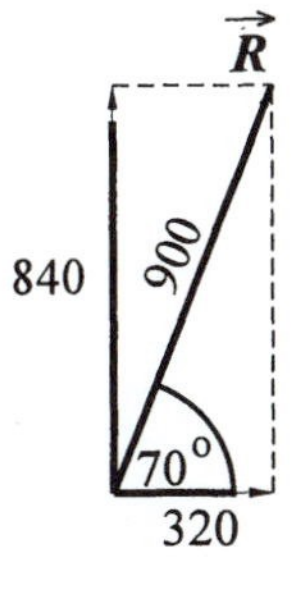

From drawing $\vec{R}$ is approximately 900 lb at 70°

45.

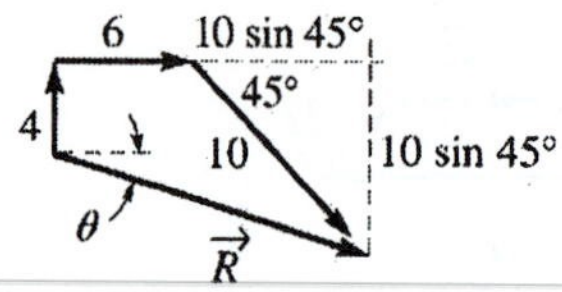

From the drawing,

$R = 13$ mi

$\theta = 13^\circ$

9.2 Components of Vectors

1.

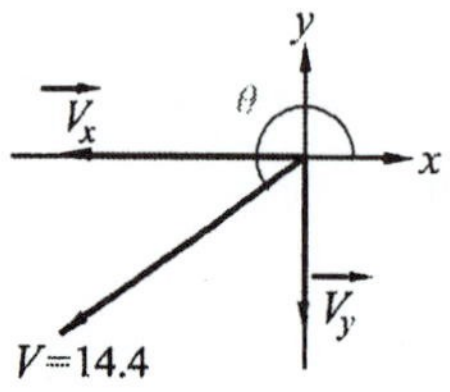

$V_x = V\cos\theta$

$V_x = 14.4\cos 216^\circ = -11.6$

$V_y = V\sin\theta$

$V_y = 14.4\sin 216^\circ = -8.46$

5. $V_x = 750\cos 28^\circ = 662$

$V_y = 750\sin 28^\circ = 352$

9. $V_x = -750$

$V_y = 0$

13. Let $V = 76.8$

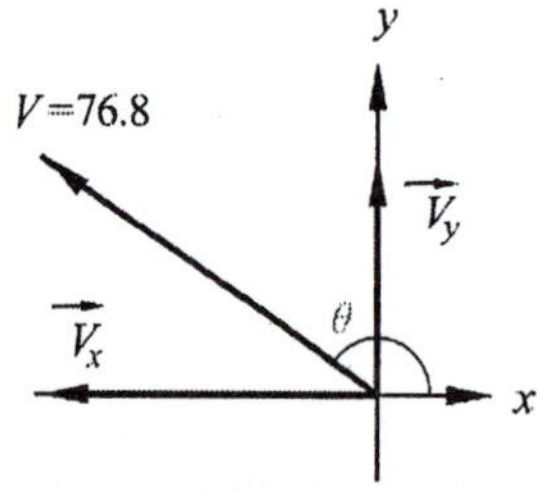

$V_x = V\cos 145.0^\circ = 76.8(-0.819) = -62.9$ m/s

$V_y = V\sin 145.0^\circ = 76.8(0.574) = 44.1$ m/s

17. Let $V = 2.65$

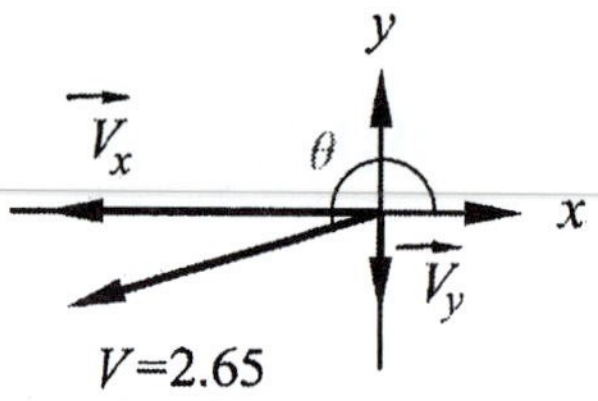

$V_x = V\cos 197.3^\circ = 2.65(-0.955) = -2.53$ mN

$V_y = V\sin 197.3^\circ = 2.65(-0.297) = -0.788$ m/N

21.

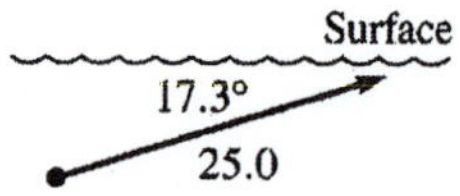

$V_x = 25.0\cos 17.3^\circ = 23.9$ km/h

$V_y = 25.0\sin 17.3^\circ = 7.43$ km/h

25.

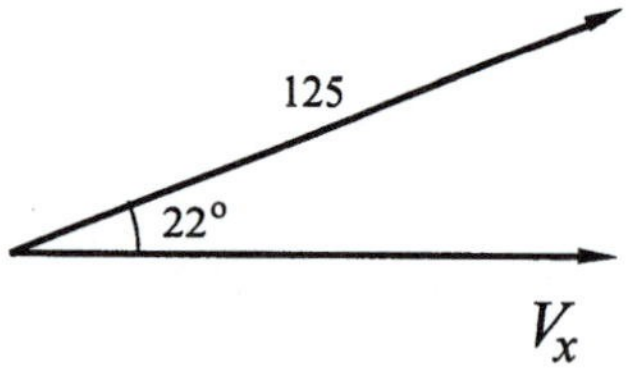

$V_x = 125\cos 22^\circ = 116$ km/h

29.

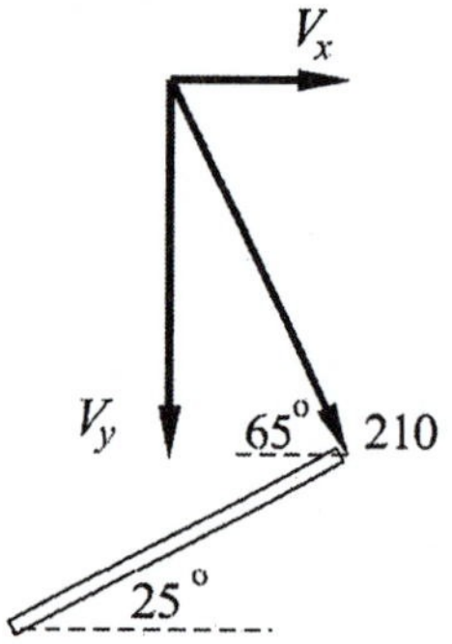

$V_x = 210\cos 65^\circ = 89$ N

$V_y = -210\sin 65^\circ = -190$ N

33.

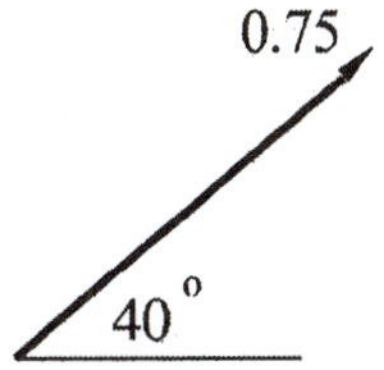

horizontal component $= 0.75\cos 40^\circ$

$= 0.57$ (km/h)/m

vertical component $= 0.75\sin 40^\circ$

$= 0.48$ (km/h)/m

9.3 Vector Addition by Components

1. $A = 1200 = A_x$, $B = 1750$

$Ay = 0$

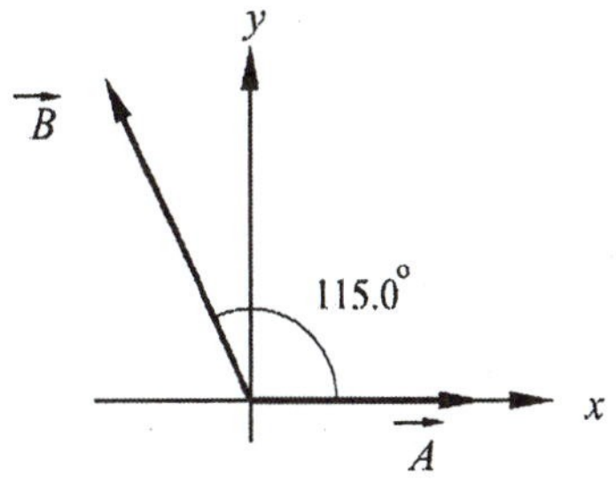

$$R_x = A_x + B_x = 1200 + 1750\cos 115^\circ = 460.4$$
$$R_y = A_y + B_y = 0 + 1750\sin 115^\circ = 1586$$
$$R = \sqrt{R_x^2 + R_y^2} = \sqrt{460.4^2 + 1586^2} = 1650$$
$$\theta = \tan^{-1}\frac{R_y}{R_x} = \tan^{-1}\frac{1586}{460.4} = 73.8^\circ$$

5.

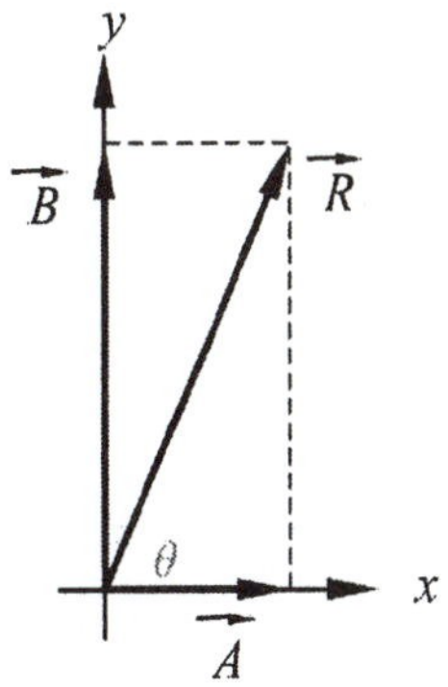

$$R = \sqrt{3.086^2 + 7.143^2} = \sqrt{60.54} = 7.781$$
$$\tan\theta = \frac{7.143}{3.086} = 2.315$$
$$\theta = 66.63^\circ \text{ (with } \vec{A}\text{)}$$

9.

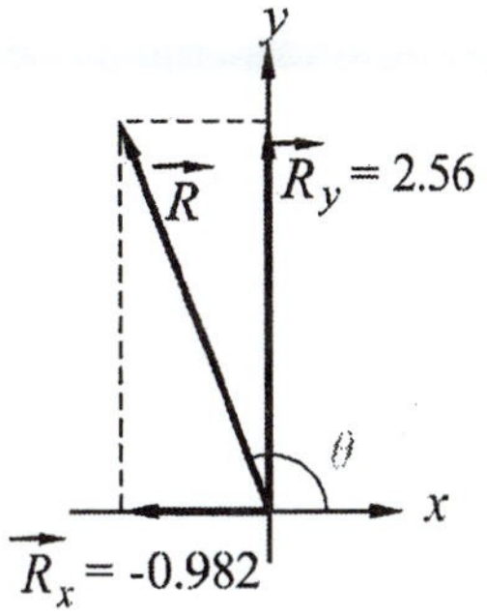

$$R_x = -0.982, R_y = 2.56$$
$$R = \sqrt{R_x^2 + R_y^2} = \sqrt{(-0.982)^2 + 2.56^2} = 2.74$$
$$\tan\theta_{\text{ref}} = \left|\frac{2.56}{-0.982}\right| = 2.61$$
$$\theta_{\text{ref}} = 69.0^\circ$$
$$\theta = 180^\circ - 69.0^\circ = 111.0^\circ$$

(θ is in Quad II since R_x is negative and R_y is positive)

13.

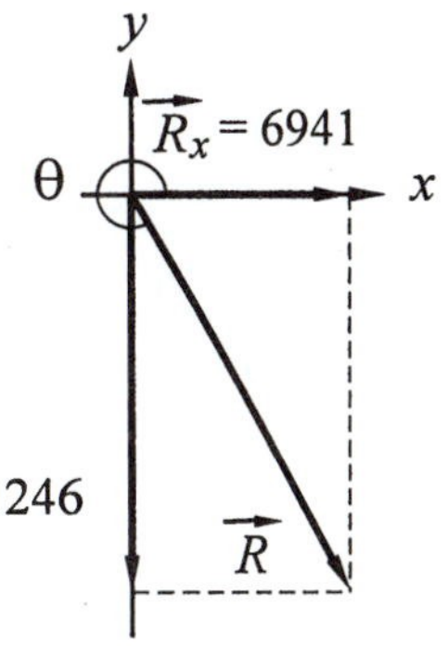

$R_x = 6941,\ R_y = -1246$

$R = \sqrt{6941^2 + (-1246)^2} = 7052$

$\tan\theta_{ref} = \left|\frac{-1246}{6941}\right| = 1.795$

$\theta_{ref} = 60.88^\circ$

$\theta = 360^\circ - 60.88^\circ = 299.12^\circ$

(θ is in QIV since R_x is positive and R_y is negative)

17.

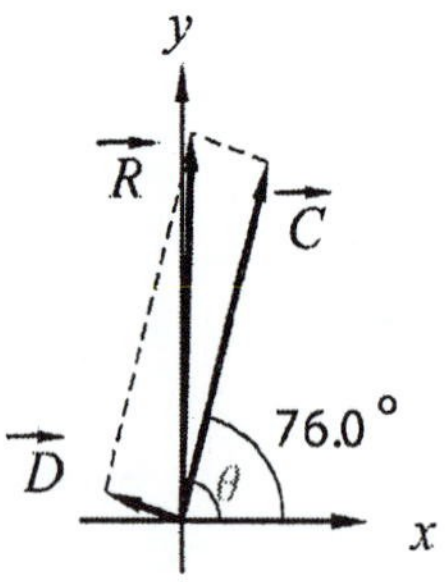

$C = 5650, \theta_C = 76.0°$
$C_x = 5650 \cos 76.0° = 1370$
$C_y = 5650 \sin 76.0° = 5480$
$D = 1280, \theta_D = 160.0°$
$D_x = 1280 \cos 160.0° = -1200$
$D_y = 1280 \sin 160.0° = 438$
$R_x = 1370 - 1200 = 170$
$R_y = 5480 + 438 = 5920$
$R = \sqrt{170^2 + 5920^2} = 5920$

$\tan\theta = \frac{R_y}{R_x} = \frac{5920}{170}, \theta = 88.4°$

21.

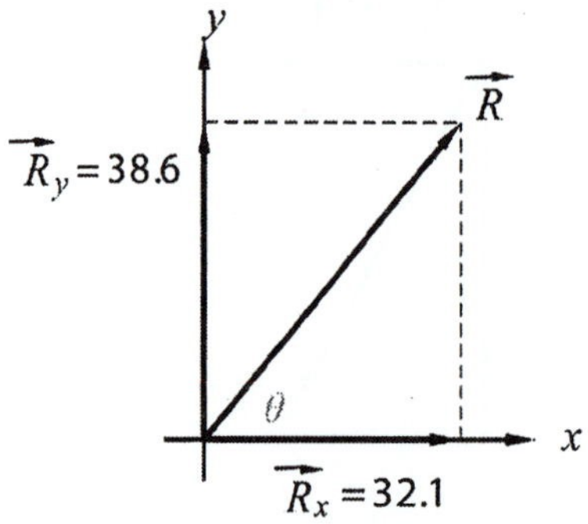

$R_x = A_x + B_x + C_x$
$R_x = 21.9 \cos 236.2° + 96.7 \cos 11.5° + 62.9 \cos 143.4°$
$R_y = A_y + B_y + C_y$
$R_y = 21.9 \sin 236.2° + 96.7 \sin 11.5° + 62.9 \sin 143.4°$
$R = \sqrt{R_x^2 + R_y^2} = 50.2$

$\theta = \tan^{-1}\frac{R_y}{R_x} = 50.3°$

25.

Vector	Magnitude	Ref. Angle
A	318	67.5°
B	245	16.3°

	x-component	y-component
	$-318\cos 67.5^\circ = -122$	$318\sin 67.5^\circ = 294$
	$245\sin 16.3^\circ = 68.8$	$245\cos 16.3^\circ = 235$
R	-53.2	529

$\theta_{ref} = \tan^{-1}\left|\frac{529}{-53.2}\right| = 84.3^\circ$

$\theta = 180^\circ - 84.3^\circ = 95.7^\circ$

$R = \sqrt{(-53.2)^2 + 529} = 531$

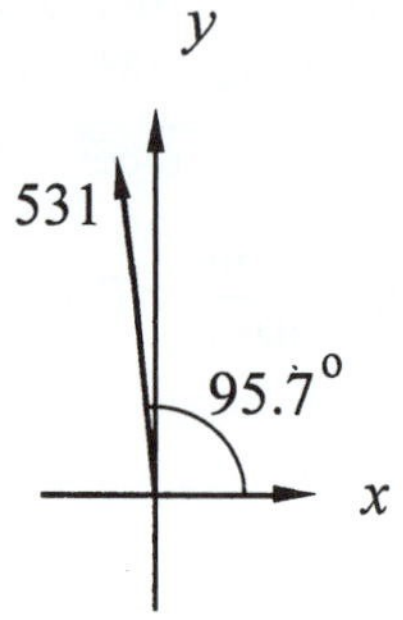

29.

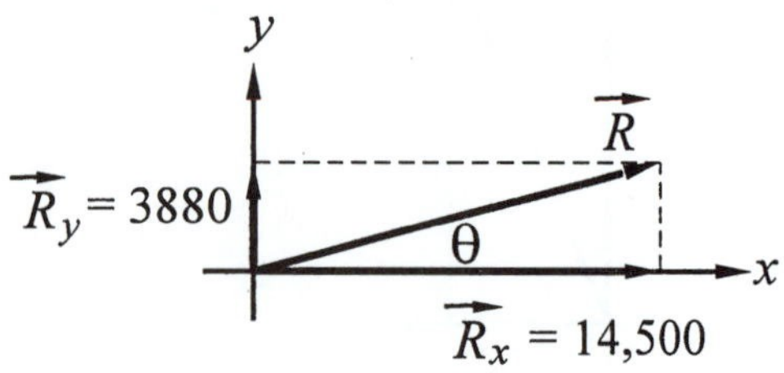

$$R_x = 5500 + 6500\cos 15.5° + 3500\cos 37.7°$$
$$= 14{,}500$$
$$R_y = 6500\sin 15.5° + 3500\sin 37.7°$$
$$= 3880$$
$$R = \sqrt{14{,}500^2 + 3880^2} = 15{,}000 \text{ lb}$$
$$\theta = \tan^{-1}\frac{3880}{14{,}500} = 15.0° \text{ down from 5500 lb force}$$

33.

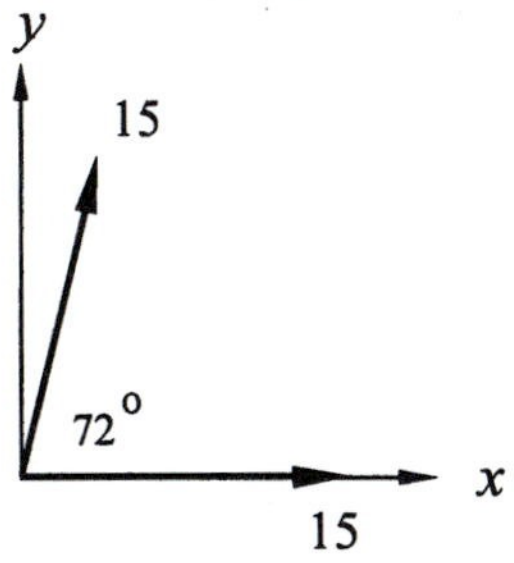

$$R = \sqrt{R_x^2 + R_y^2}$$
$$= \sqrt{(15\cos 72° + 15)^2 + (15\sin 72°)^2}$$
$$= 20 \text{ kg}\cdot\text{m/s}$$

9.4 Applications of Vectors

1.

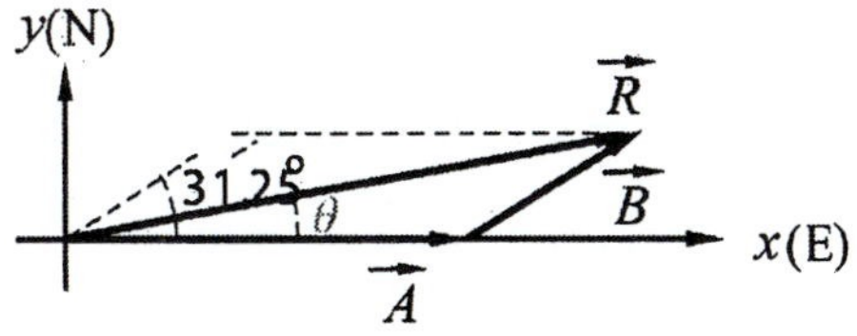

$$R_x = A + B_x = 32.50 + 16.18\cos 31.25° = 46.33$$
$$R_y = 16.18\sin 31.25° = 8.394$$
$$R = \sqrt{46.33^2 + 8.394^2} = 47.08 \text{ mi}$$
$$\theta = \tan^{-1}\frac{8.394}{46.33} = 10.27°$$

The ship is 47.08 mi from start in direction 10.27° N of E.

5.

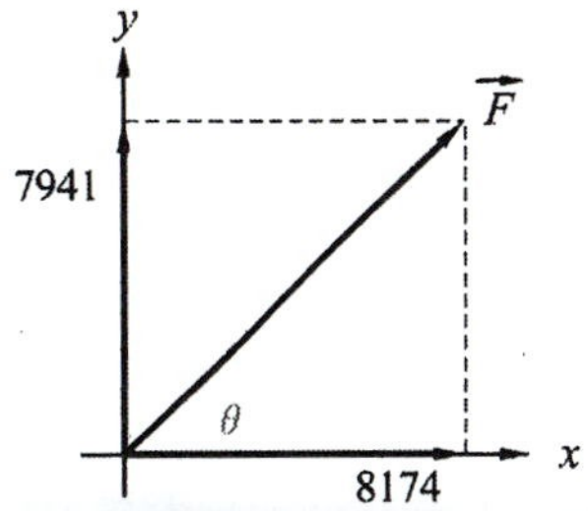

$$F_x = 8300\cos 10.0°$$
$$= 8174 \text{ N}$$
$$F_y = 8300\sin 10.0° + 6500$$
$$= 7941 \text{ N}$$
$$F = \sqrt{8174^2 + 7941^2}$$
$$= 11{,}000 \text{ N}$$

$$\tan\theta = \frac{7941}{8174} = 0.97, \theta = 44° \text{ above horizontal}$$

9.

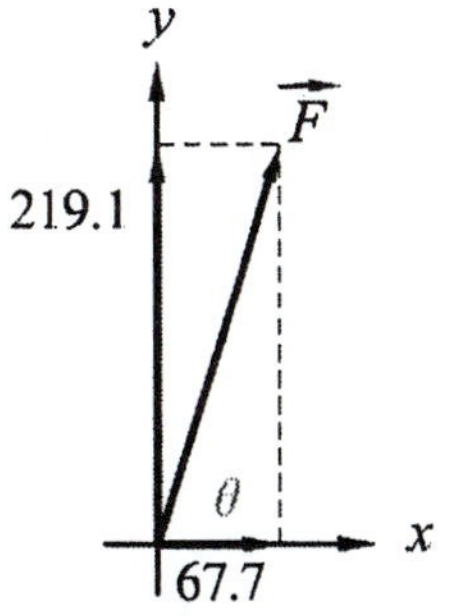

$$F_x = 358.2\cos 37.72° - 215.6 = 67.7$$
$$F_y = 358.2\sin 37.72° = 219.1$$
$$F = \sqrt{67.7^2 + 219.1^2}$$
$$= 229.4 \text{ ft}$$

$$\tan\theta = \frac{219.1}{67.74} = 3.234,\ \theta = 72.82°, \text{ N of E}$$

13.

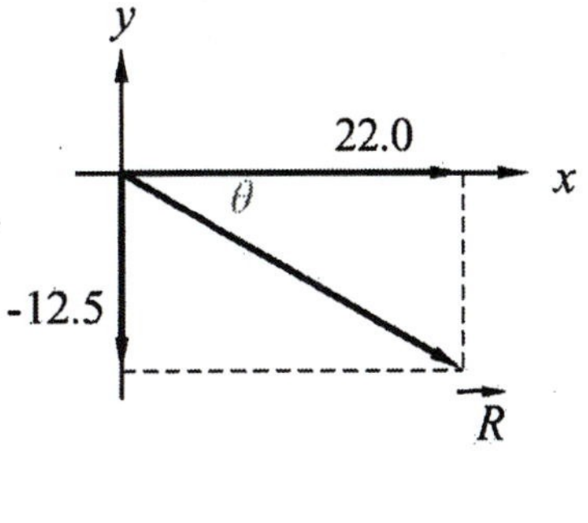

$$R = \sqrt{22.0^2 + 12.5^2} = 25.3 \text{ km/h}$$

$$\theta = \tan^{-1}\frac{12.5}{22.0} = 29.6°$$

17.

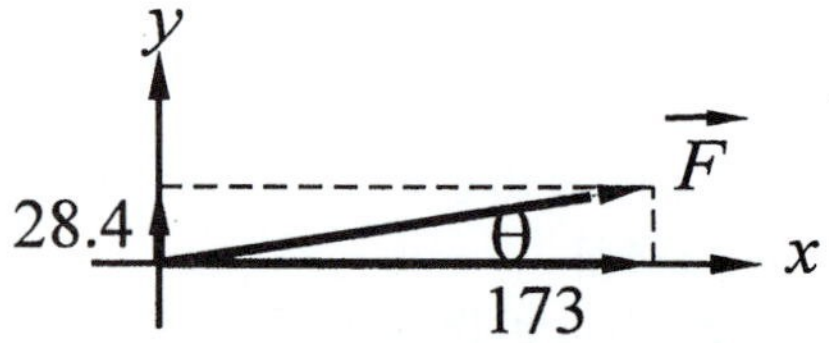

$$F_x = 95.0 + 83.0\cos 20.0° = 173 \text{ lb}$$
$$F_y = 83.0\sin 20.0° = 28.4 \text{ lb}$$
$$F = \sqrt{173^2 + 28.4^2} = 175 \text{ lb}$$

$$\tan\theta = \frac{28.4}{173} = 0.163$$

$\theta = 9.3°$, above horizontal and to right.

21. $v_{sx} = 18,250 - 120\cos 5.20° = 18,130$ mi/h
$v_{sy} = 120\sin 5.20° = 10$ mi/h
$v = \sqrt{18,130^2 + 10^2} = 18,130$ mi/h

$$\tan\theta = \frac{10}{18,310} = 0.000055$$

$\theta = 0.03°$ from direction of shuttle

25. Assume that as the smoke pours our of the funnel it immediately takes up the velocity of the wind. Let $\vec{w}$ = velocity of wind, $\vec{u}$ = velocity of boat, $\vec{v}$ = velocity of smoke as seen by passenger.

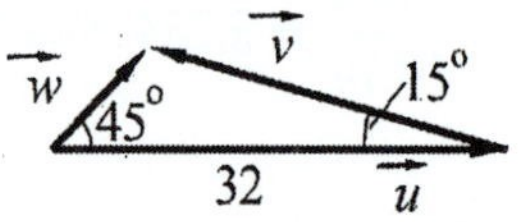

$$w\cos 45° + v\cos 15° = 32$$
$$w\sin 45° = v\sin 15°$$
$$v = \frac{w\sin 45°}{\sin 15°}$$
$$w\cos 45° + \frac{w\sin 45°}{\sin 15°}\cos 15° = 32$$
$$w = 9.6 \text{ km/h}$$

29. $r = \frac{d}{2} = \frac{8.20}{2} = 4.10$

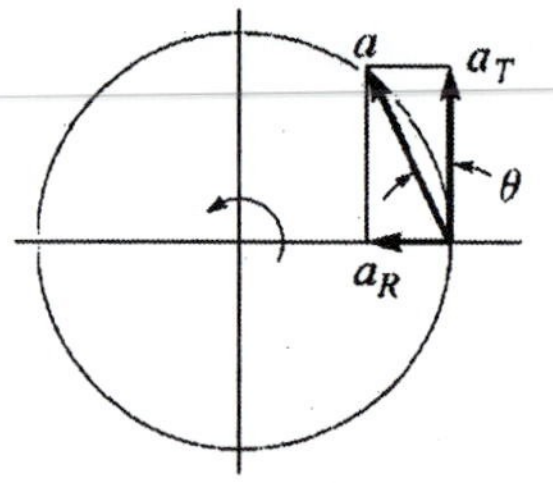

$$a = \sqrt{a_T^2 + a_R^2} = \sqrt{(\alpha r)^2 + (w^2 r)^2}$$
$$a = \sqrt{(318(4.10))^2 + (212^2 \cdot 4.10)^2}$$
$$a = 184,000 \text{ in/min}^2$$

$$\theta = \tan^{-1}\frac{a_R}{a_T} = \tan^{-1}\frac{w_r^2 r}{\alpha r}$$
$$= \tan^{-1}\frac{212^2}{318}$$
$$\theta = 89.6°$$

33. top view of plane

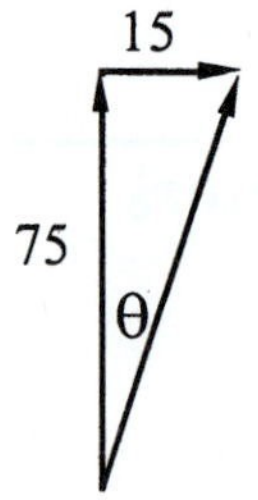

$$V_H = \sqrt{75.0^2 + 15.0^2} = 76.5$$
$$\theta = \tan^{-1}\frac{15.0}{75.0} = 11.3°$$
$$V_v = 9.80(2.00) = 19.6$$

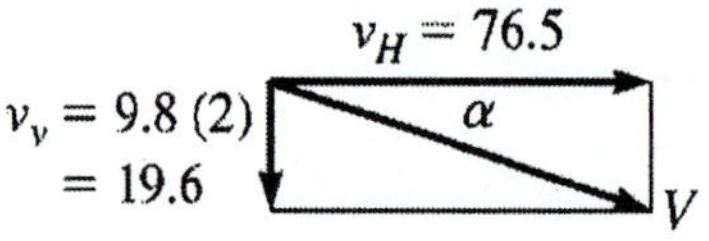

$$V = \sqrt{76.5^2 + 19.6^2} = 79.0 \text{ m/s}$$

$$\alpha = \tan^{-1}\frac{19.6}{76.5} = 14.4°, \ 75.6° \text{ from vertical}$$

9.5 Oblique Triangles, the Law of Sines

1.

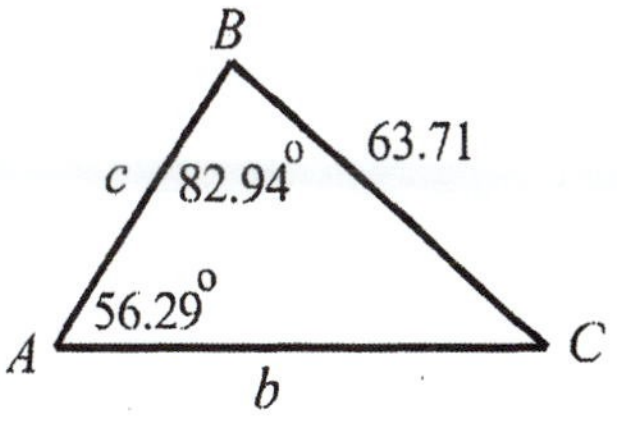

$C = 180° - (56.29° - 82.94°) = 40.77°$

$$\frac{b}{\sin 82.94°} = \frac{63.71}{\sin 56.29°} = \frac{c}{\sin 40.77°}$$

$$b = \frac{63.71 \sin 82.94°}{\sin 56.29°} = 76.01$$

$$c = \frac{63.71 \sin 40.77°}{\sin 56.29°} = 50.01$$

5.

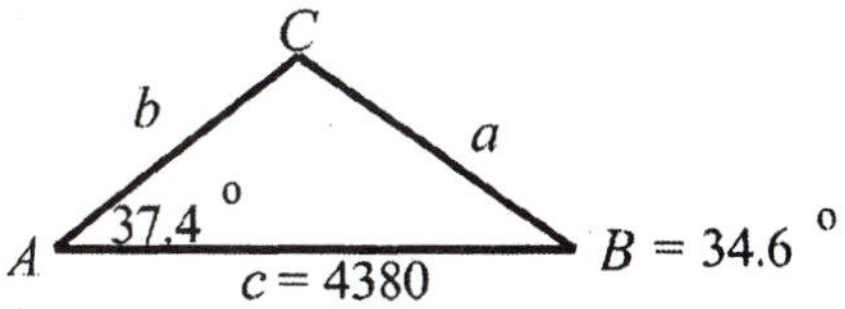

$c = 4380,\ A = 37.4°,\ B = 34.6°$

$C = 180.0° - (37.4° + 34.6°) = 108.0°$

$$\frac{b}{\sin B} = \frac{c}{\sin C};\ \frac{b}{\sin 34.6°} = \frac{4380}{\sin 108.0°}$$

$$b = \frac{4380 \sin 34.6°}{\sin 108.0°} = 2620$$

$$\frac{a}{\sin A} = \frac{c}{\sin C};\ \frac{a}{\sin 37.4°} = \frac{4380}{\sin 108.0°}$$

$$a = \frac{4380 \sin 37.4°}{\sin 108.0°} = 2800$$

9.

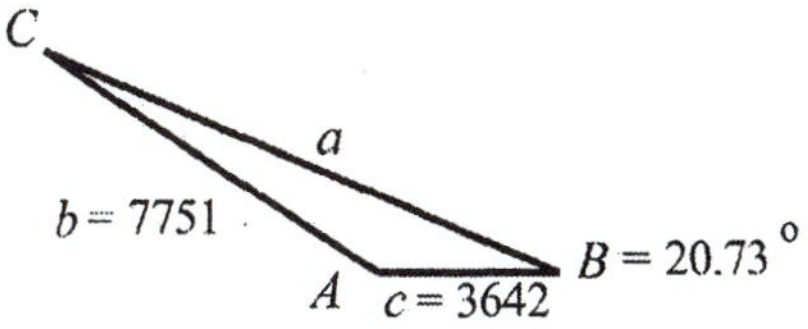

$b = 7751,\ c = 3642,\ B = 20.73°$

$$\frac{b}{\sin B} = \frac{c}{\sin C};\ \frac{7751}{\sin 20.73°} = \frac{3642}{\sin C}$$

$$\sin C = \frac{3642 \sin 20.73°}{7751} = 0.1663$$

$C = 9.57°$

$A = 180.0° - (20.73° - 9.57°) = 149.70°$

$$\frac{a}{\sin A} = \frac{b}{\sin B};\ \frac{a}{\sin 149.70°} = \frac{7751}{\sin 20.73°}$$

$$a = \frac{7751 \sin 149.70°}{\sin 20.73°} = 11{,}050$$

13.

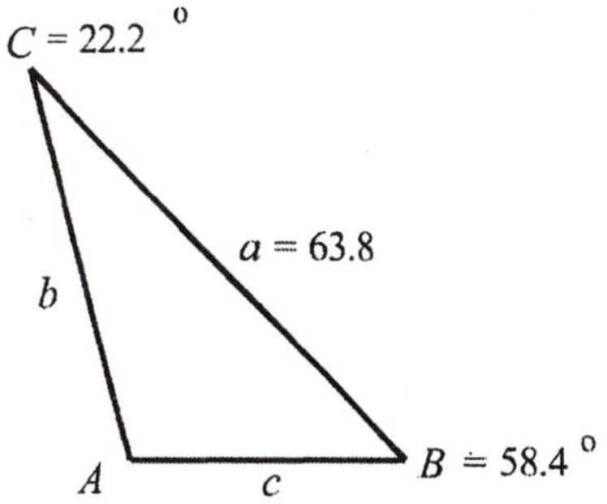

$a = 63.8, B = 58.4°, C = 22.2°$

$A = 180.0° - 58.4° - 22.2° = 99.4°$

$$\frac{a}{\sin A} = \frac{b}{\sin B};\ \frac{63.8}{\sin 99.4°} = \frac{b}{\sin 58.4°}$$

$$b = \frac{63.8 \sin 58.4°}{\sin 99.4°} = 55.1$$

$$\frac{a}{\sin A} = \frac{c}{\sin C};\ \frac{63.8}{\sin 99.4°} = \frac{c}{\sin 22.2°}$$

$$c = \frac{63.8 \sin 22.2°}{\sin 99.4°} = 24.4$$

17.

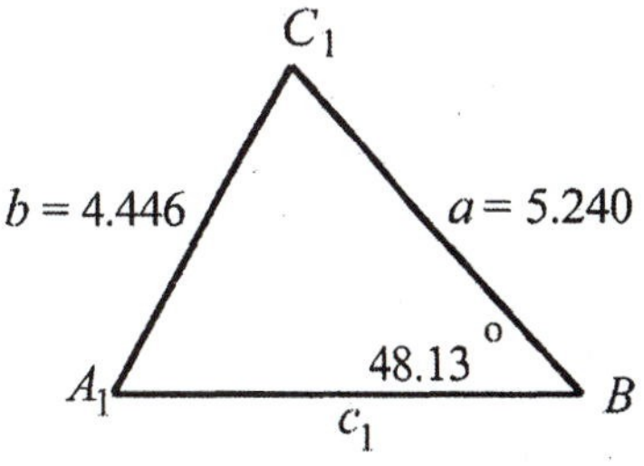

21.

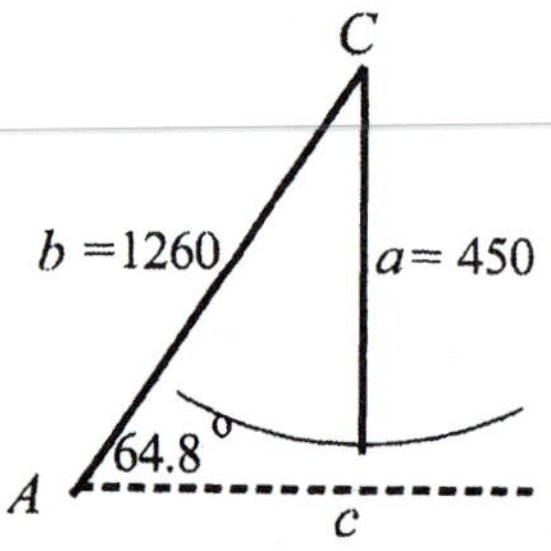

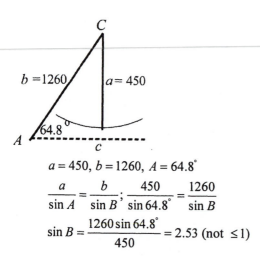

$a = 450,\ b = 1260,\ A = 64.8^\circ$

$$\frac{a}{\sin A} = \frac{b}{\sin B};\ \frac{450}{\sin 64.8^\circ} = \frac{1260}{\sin B}$$

$$\sin B = \frac{1260 \sin 64.8^\circ}{450} = 2.53\ (\text{not } \le 1)$$

Therefore, no solution.

25.

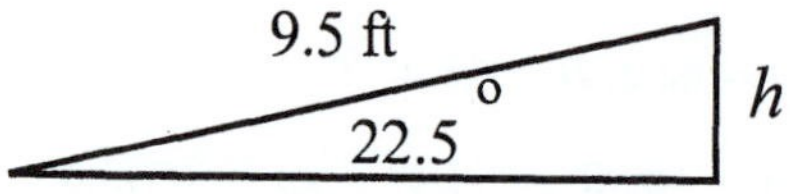

$h = 9.5 \sin 22.5^\circ$

$$\theta = \sin^{-1} \frac{9.5 \sin 22.5^\circ}{22.5}$$

$\theta = 9.3^\circ$

29.

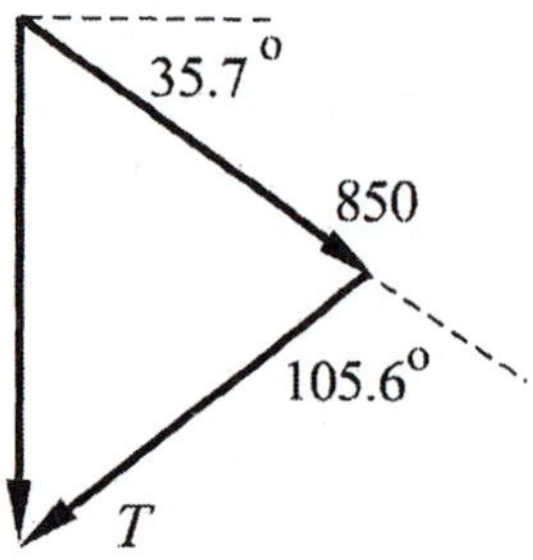

$$\frac{T}{\sin 54.3^\circ} = \frac{850}{\sin 51.3^\circ};\ T = \frac{850 \sin 54.3^\circ}{\sin 51.3^\circ} = 880\ \text{N}$$

33.

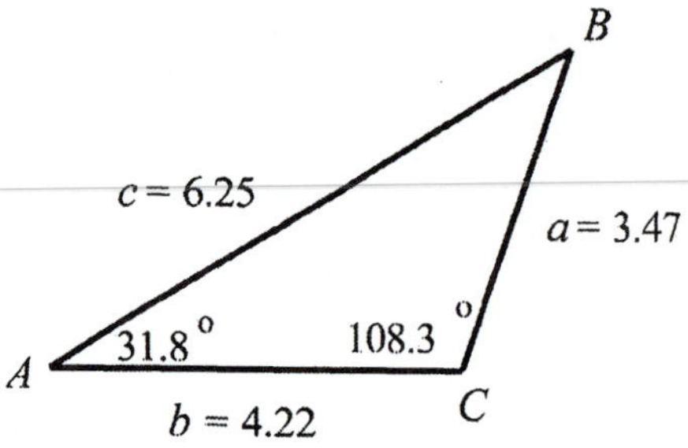

$$\frac{6.25}{\sin 108.3^\circ} = \frac{a}{\sin 31.8^\circ}$$

$$a = \frac{6.25 \sin 31.8^\circ}{\sin 108.3^\circ} = 3.47\ \text{cm}$$

$$B = 180^\circ - 108.3^\circ - 31.8^\circ = 39.9^\circ$$

$$\frac{6.25}{\sin 108.3^\circ} = \frac{b}{\sin 39.9^\circ}$$

$$b = \frac{6.25 \sin 39.9^\circ}{\sin 108.3^\circ} = 4.22\ \text{cm}$$

$$\text{Perimeter} = 6.25 + 3.47 + 4.22 = 13.94\ \text{cm}$$

37.

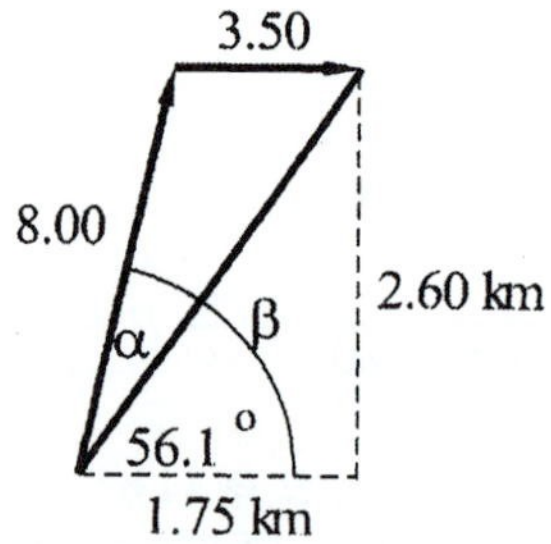

$$\theta = \tan^{-1} \frac{2.60}{1.75} = 56.1^\circ$$

$$\frac{8.00}{\sin 56.1^\circ} = \frac{3.50}{\sin \alpha}$$

$\alpha = 21.3^\circ,$

$\beta = 56.1^\circ + 21.3^\circ = 77.4^\circ$

with bank downstream

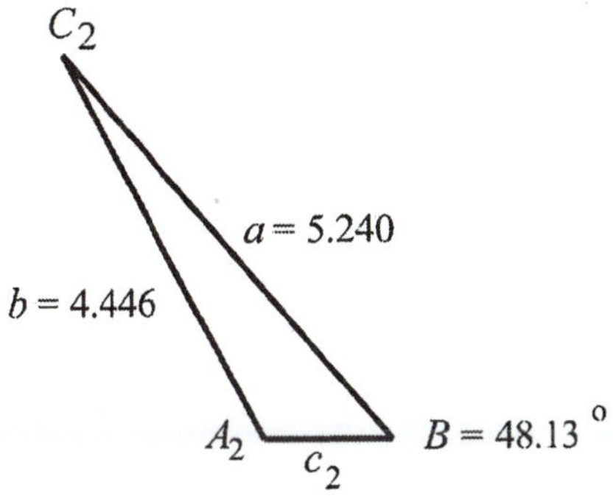

$a = 5.240,\ b = 4.446,\ B = 48.13^\circ$

$$\frac{a}{\sin A} = \frac{b}{\sin B};\ \frac{5.240}{\sin A} = \frac{4.446}{\sin 48.13^\circ}$$

$$\sin A = \frac{5.240 \sin 48.13^\circ}{4.446} = 0.8776$$

$A_1 = 61.36^\circ$

$C_1 = 180.0^\circ - 48.13^\circ - 61.36^\circ = 70.51^\circ$

or

$A_2 = 180.0^\circ - 61.36^\circ = 118.64^\circ$

and

$C_2 = 180.0^\circ - 48.13^\circ - 118.64^\circ = 13.23^\circ$

$$\frac{b}{\sin B} = \frac{c_1}{\sin C_1};\ \frac{4.446}{\sin 48.13^\circ} = \frac{c}{\sin 70.51^\circ}$$

$$c_1 = \frac{4.446 \sin 70.51^\circ}{\sin 48.13^\circ} = 5.628$$

or

$$\frac{b}{\sin B} = \frac{c_2}{\sin C_2};\ \frac{4.446}{\sin 48.13^\circ} = \frac{c_2}{\sin 13.23^\circ}$$

$$c_2 = \frac{4.446 \sin 13.23^\circ}{\sin 48.13^\circ} = 1.366$$

9.6 The Law of Cosines

1.

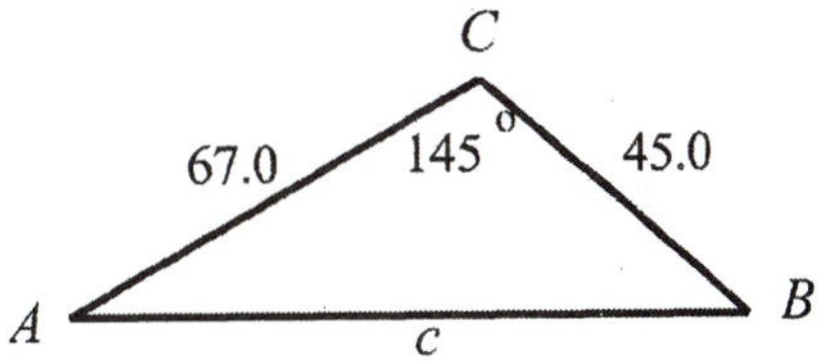

$$c = \sqrt{45.0^2 + 67.0^2 - 2(45.0)(67.0)\cos 145^\circ}$$
$$= 107$$

$$\frac{45.0}{\sin A} = \frac{c}{\sin 145^\circ} = \frac{67.0}{\sin B}$$

$$A = \sin^{-1}\frac{45.0 \sin 145^\circ}{c} = 14.0^\circ$$

$$B = \sin^{-1}\frac{67.0 \sin 145^\circ}{c} = 21.0^\circ$$

5.

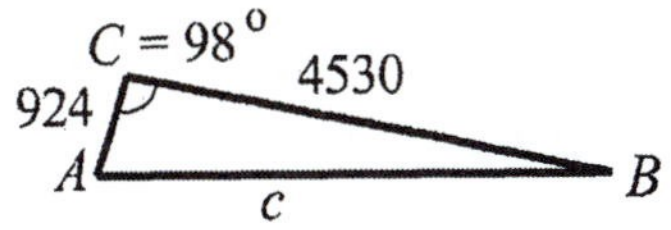

$a = 4530,\ b = 924,\ C = 98.0^\circ$

$$c = \sqrt{4530^2 + 924^2 + 2(4530)(924)(\cos 98.0^\circ)}$$
$$= 4750$$

$$\frac{c}{\sin C} = \frac{b}{\sin B};\ \frac{4750}{\sin 98.0^\circ} = \frac{924}{\sin B}$$

$$\sin B = \frac{924 \sin 98.0^\circ}{4750} = -0.193$$

$B = 11.1^\circ$

$A = 180^\circ - 98.0^\circ - 11.1^\circ = 70.9^\circ$

9.

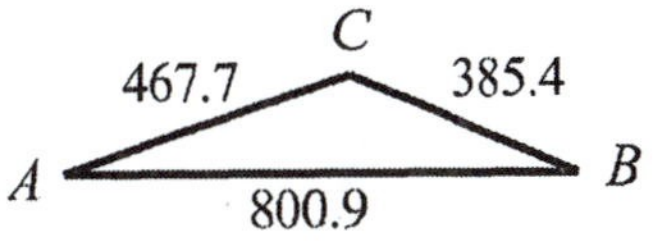

$a = 385.4,\ b = 467.7,\ c = 800.9$

$$\cos A = \frac{467.7^2 + 800.9^2 - 385.4^2}{2(467.7)(800.9)} = 0.9499$$

$A = 18.21^\circ$

$$\cos B = \frac{385.4^2 + 800.9^2 - 467.7^2}{2(385.4)(800.9)} = 0.9253$$

$B = 22.28^\circ$

$C = 180^\circ - 18.21^\circ - 22.28^\circ = 139.51^\circ$

13.

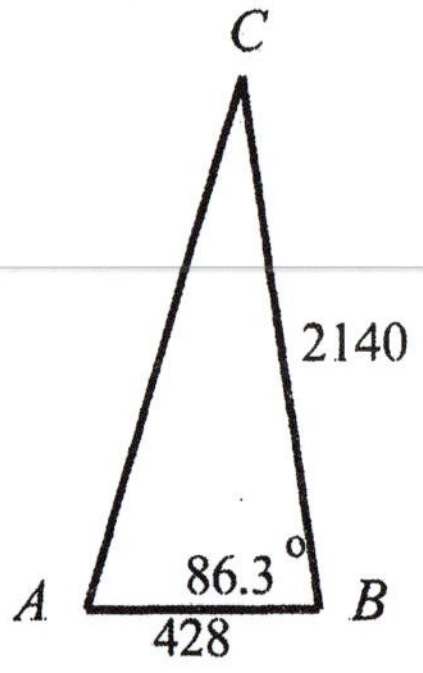

$a = 2140, c = 428, B = 86.3°$

$$b = \sqrt{2140^2 + 428^2 - 2(2140)(428)(\cos 86.3°)}$$
$$= 2160$$
$$\frac{b}{\sin B} = \frac{c}{\sin C}; \frac{2160}{\sin 86.3°} = \frac{428}{\sin C}$$
$$\sin C = \frac{428 \sin 86.3°}{2160} = 0.198$$
$$C = 11.4°$$
$$A = 180° - 86.3° - 11.4° = 82.3°$$

17.

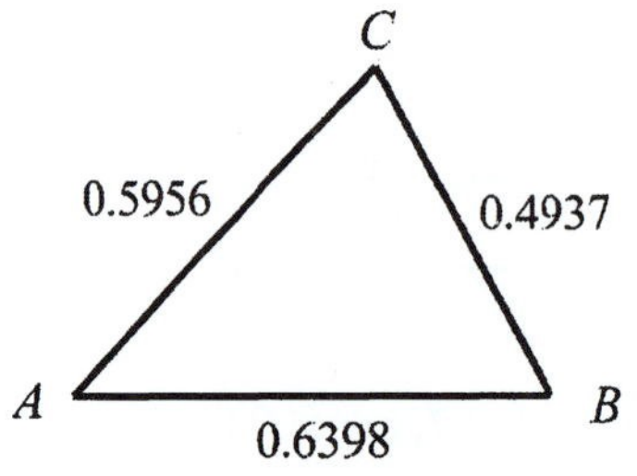

$a = 0.4937, b = 0.5956, c = 0.6398$

$$\cos A = \frac{0.5956^2 + 0.6398^2 - 0.4937^2}{2(0.5956)(0.6398)} = 0.6827$$
$$A = 46.94°$$
$$\cos B = \frac{0.4937^2 + 0.6398^2 - 0.5956^2}{2(0.4937)(0.6398)} = 0.4723$$
$$B = 61.82°$$
$$C = 180° - 46.94° - 61.82° = 71.24°$$

21.

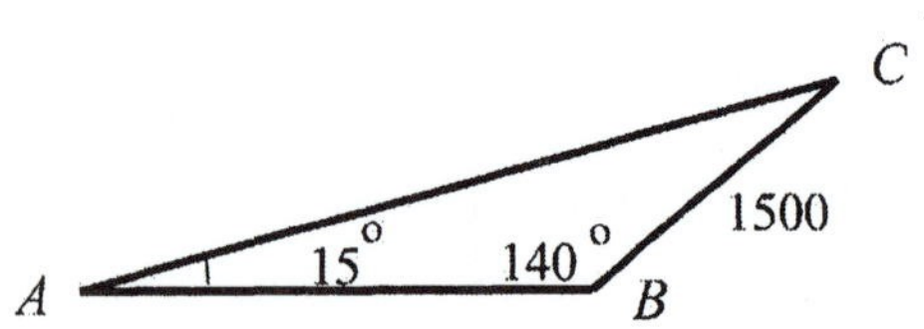

$a = 1500, A = 15°, B = 140°$

$$\frac{a}{\sin A} = \frac{b}{\sin B}; b = \frac{1500 \sin 140°}{\sin 15°} = 3700$$
$$C = 180° - 15° - 140° = 25°$$
$$c = \sqrt{1500^2 + 3700^2 - 2(1500)(3700)(\cos 25°)}$$
$$= 2400$$

25. Case 3: Two sides and the included angle always determine a unique triangle which will have a solution. The triangle may be constructed by drawing the angle and then measuring the two sides along the sides of the angle. The third side is then uniquely determined and the triangle may be solved. Case 4: Three sides determines a unique triangle provided the sum of the lengths of any two sides is greater than the length of the third side. The triangle may be constructed by drawing one side as the horizontal base and the swinging circular arcs from each end. The intersection of these arcs determines the other two sides of the triangle.

29. $19^2 = 21^2 + 25^2 - 2(21)(25)\cos\theta$

$\theta = 47.7°$

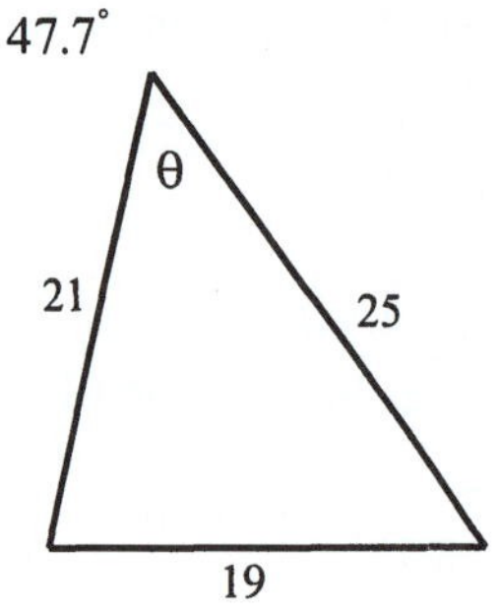

33. $x = \sqrt{9.53^2 + 9.53^2 - 2(9.53)(9.53)(\cos 120°)}$

$= 16.5$ mm

37. $A = 26.4° - 12.4° = 14.0°$

$a = \sqrt{15.8^2 + 32.7^2 - 2(15.8)(32.7)\cos 14.0°}$

$= 17.8$ mi

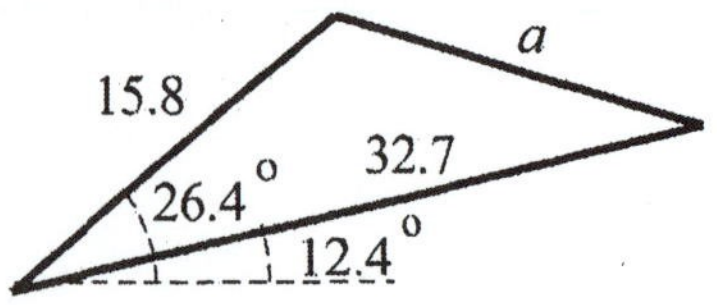

The distance between the planes is
$31.94 - 14.15 = 17.8$ mi

Chapter 9 Review Exercises

1.

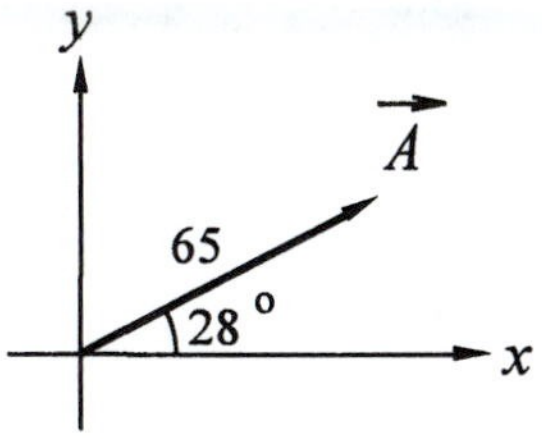

y-component $= 65.0\cos 28.0^\circ = 57.4$

x-component $= 65.0\sin 28.0^\circ = 30.5$

5.

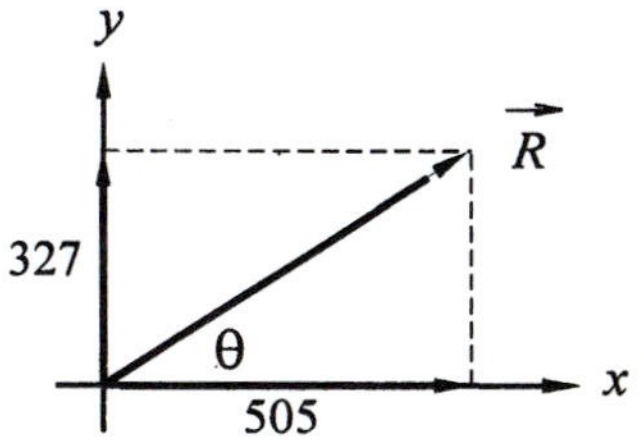

$R = \sqrt{327^2 + 505^2} = 602$

$\theta = \tan^{-1}\dfrac{327}{505} = 32.9^\circ$

9.

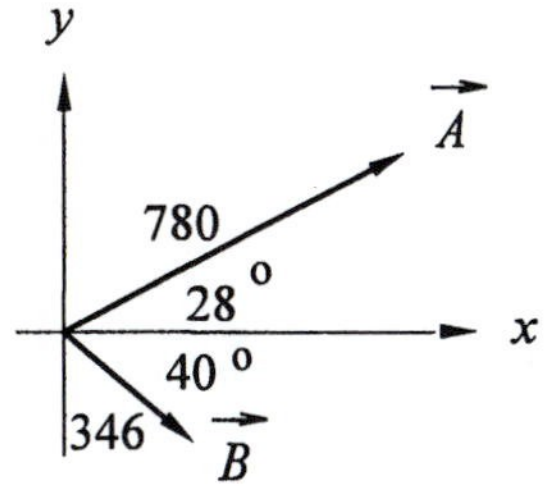

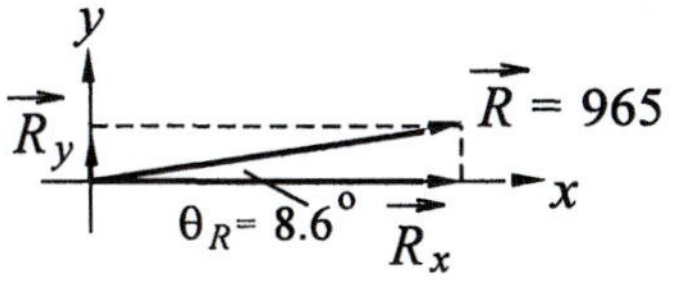

$R_x = 780\cos 28.0^\circ + 346\cos 40.0^\circ$

$R_y = 780\sin 28.0^\circ - 346\sin 40.0^\circ$

$R = \sqrt{R_x^2 + R_y^2} = \sqrt{954^2 + 1442} = 965$

$\theta_R = \tan^{-1}\dfrac{144}{954} = 8.6^\circ$

13.

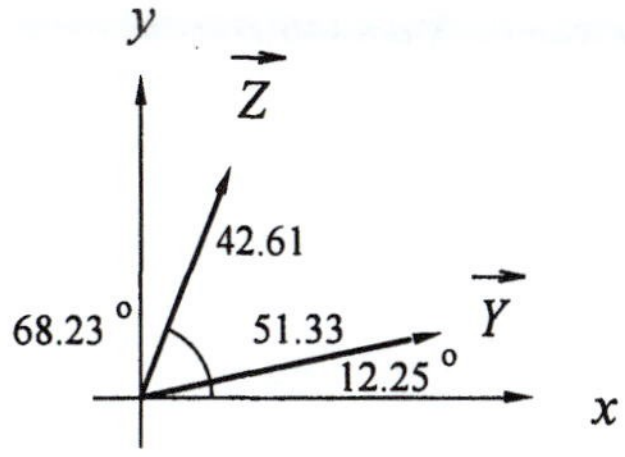

$Y_x = 51.33\cos 12.25^\circ = 5016$

$Y_y = 51.33\sin 12.25^\circ = 10.89$

$Z_x = 42.61\cos 68.23^\circ = 15.80$

$Z_y = -42.61\sin 68.23^\circ = -39.57$

$R_x = 50.16 + 15.80 = 65.98$

$R_y = 10.89 - 39.57 = -28.68$

$R = \sqrt{R_x^2 + R_y^2} = \sqrt{65.98^2 + (-28.68)^2} = 71.94$

$\theta_R = \tan^{-1}\left(\dfrac{-28.68}{65.98}\right) = 336.5^\circ$

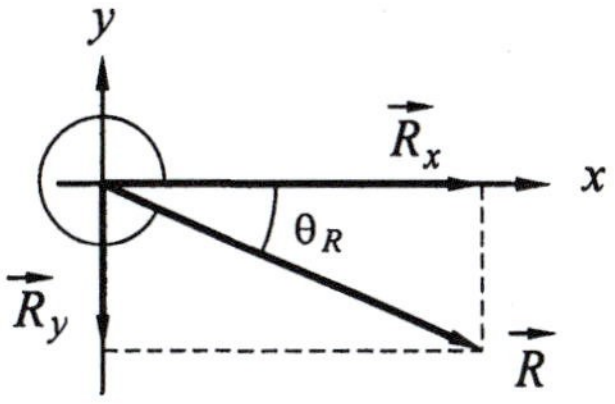

$A_x = 0.750\cos 15.0^\circ = 0.724$

$B_x = 0.265\cos 192.4^\circ = -0.259$

$C_x = 0.548\cos 344.7^\circ = 0.529$

$R_x = 0.994$

$A_y = 0.750\sin 15.0^\circ = 0.194$

$B_y = 0.265\sin 192.4^\circ = -0.0569$

$C_y = 0.548\sin 344.7^\circ = -0.145$

$R_y = -0.00790$

$R = \sqrt{R_x^2 + R_y^2} = 0.994$

$\theta_R = \tan^{-1}\dfrac{R_y}{R_x} = 359.5^\circ$

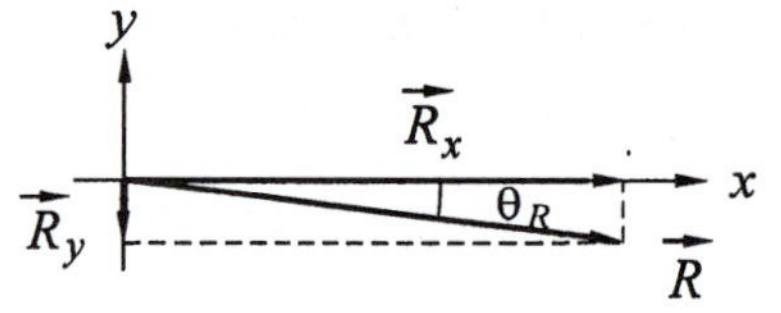

17.

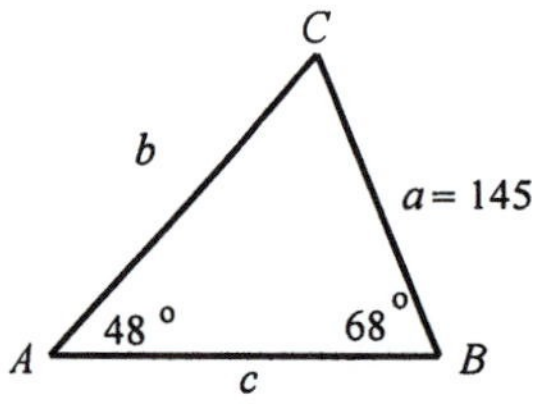

$$C = 180^\circ - 48.0^\circ - 68.0^\circ = 64.0^\circ$$

$$\frac{145}{\sin 48.0^\circ} = \frac{b}{\sin 68.0^\circ} = \frac{c}{\sin 64.0^\circ}$$

$$b = \frac{145\sin 68.0^\circ}{\sin 48.0^\circ} = 181,$$

$$c = \frac{145\sin 64.0^\circ}{\sin 48.0^\circ} = 175$$

21.

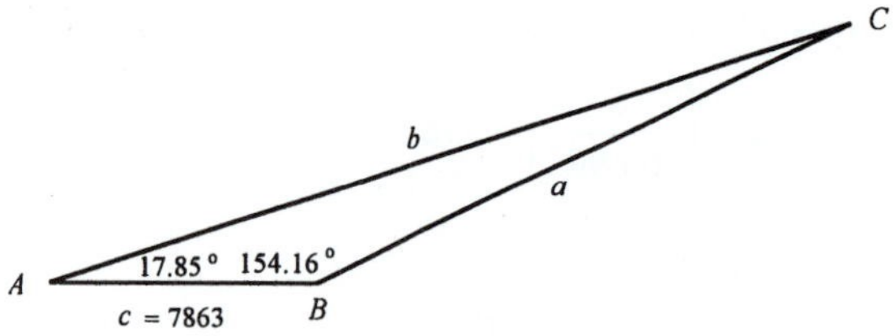

$$C = 180.0^\circ - 17.85^\circ - 154.16^\circ = 7.99^\circ$$

$$\frac{a}{\sin 17.85^\circ} = \frac{b}{\sin 154.16^\circ} = \frac{7863}{\sin 7.99^\circ}$$

$$b = \frac{7863\sin 154.16^\circ}{\sin 7.99^\circ} = 24,660$$

$$a = \frac{7863\sin 17.85^\circ}{\sin 7.99^\circ} - 17,340$$

25.

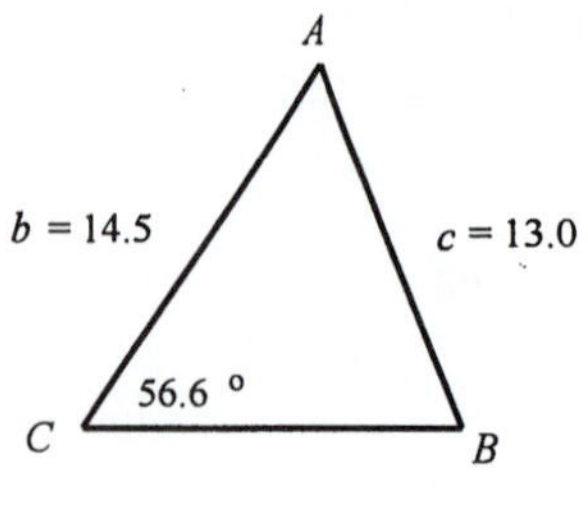

$$\frac{a}{\sin A} = \frac{14.5}{\sin B} = \frac{13.0}{\sin 56.6^\circ}$$

$$\sin B = \frac{14.5\sin 56.6^\circ}{13.0}$$

$$B = 68.6^\circ \text{ or } 111.4^\circ$$

Case I:

$$B = 68.6^\circ,\ A = 180^\circ - 68.6^\circ - 56.6^\circ = 54.8^\circ$$

$$\frac{a}{\sin 54.8^\circ} = \frac{13.0}{\sin 56.6^\circ} \Rightarrow a = 12.7$$

Case II:

$$B = 111.4^\circ,\ A = 180^\circ - 111.4^\circ - 56.6^\circ = 12.0^\circ$$

$$\frac{a}{\sin 12.0^\circ} = \frac{13.0}{\sin 56.6^\circ} \Rightarrow a = 3.24$$

29.

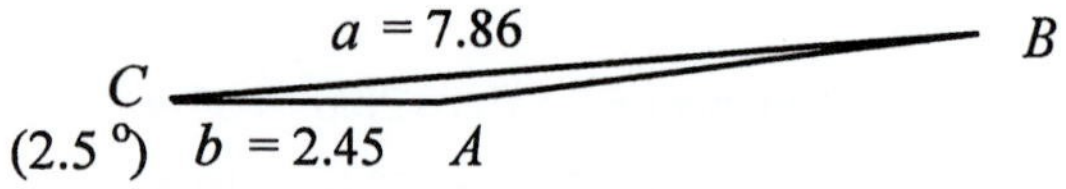

$$c^2 = a^2 + b^2 - 2ab\cos C$$

$$c^2 = 7.86^2 + 2.45^2 - 2(7.86)(2.45)\cos 2.5^\circ$$

$$c = 5.41$$

$$a^2 = b^2 + c^2 - 2bc\cos A$$

$$7.86^2 = 2.45^2 + 5.41^2 - 2(2.45)(5.41)\cos A$$

$$\cos A = 0.9986$$

$$A = 177.0^\circ$$

$$B = 180^\circ = C - A = 180^\circ = 2.5^\circ - 177.0^\circ$$

$$B = 0.5^\circ$$

33.

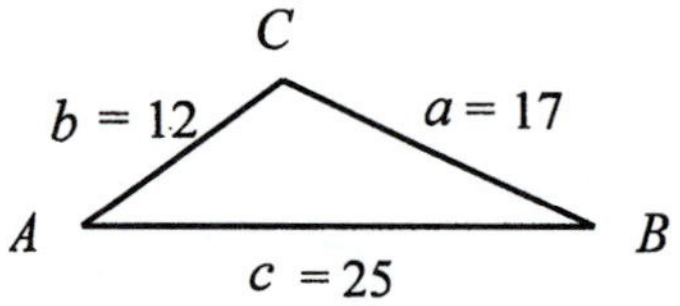

$$a^2 = b^2 + c^2 - 2bc\cos A$$

$$17^2 = 12^2 + 25^2 - 2(12)(25)\cos A$$

$$\cos A = 0.8$$

$$A = 37^\circ$$

$$b^2 = a^2 + c^2 - 2ac\cos B$$

$$12^2 = 17^2 + 25^2 - 2(17)(25)\cos B$$

$$\cos B = 0.9059$$

$$B = 25^\circ$$

$$C = 180^\circ - 37^\circ - 25^\circ = 118^\circ$$

37.

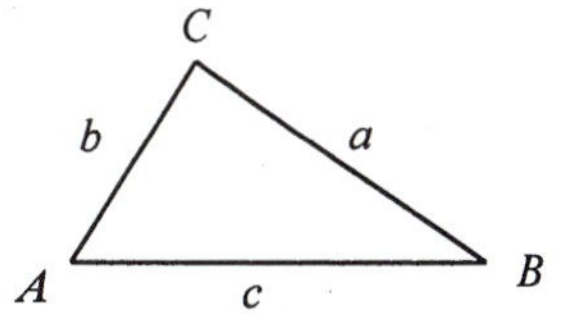

$$a^2 = b^2 + c^2 - 2bc\cos A$$
$$b^2 = a^2 + c^2 - 2ac\cos B$$
$$\underline{c^2 = a^2 + b^2 - 2ab\cos C} \quad \text{add}$$

$$a^2 + b^2 + c^2 = 2a^2 + 2b^2 + 2c^2 - 2bc\cos A - 2ac\cos B - 2ab\cos C$$
$$a^2 + b^2 + c^2 = 2bc\cos A + 2ac\cos B + 2ab\cos C$$
$$\frac{a^2 + b^2 + c^2}{2abc} = \frac{\cos A}{a} + \frac{\cos}{b} + \frac{\cos C}{c}$$

41.

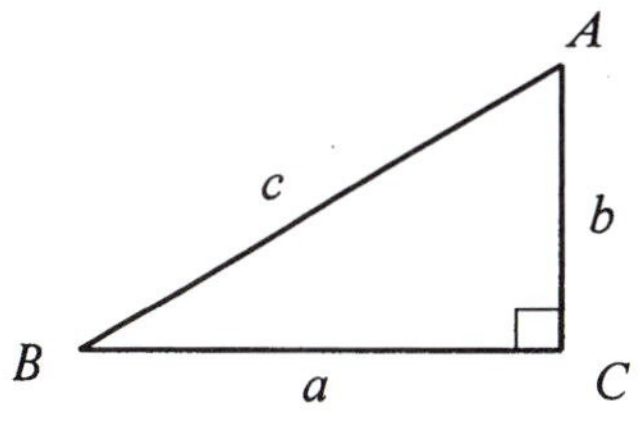

$$\frac{a}{\sin A} = \frac{b}{\sin B} \Rightarrow b = \frac{a\sin B}{\sin A}$$
$$A_t = \frac{1}{2}ab = \frac{1}{2}\cdot a \cdot \frac{a\sin B}{\sin A}$$
$$A_t = \frac{a^2\sin B}{2\sin A}$$

45.

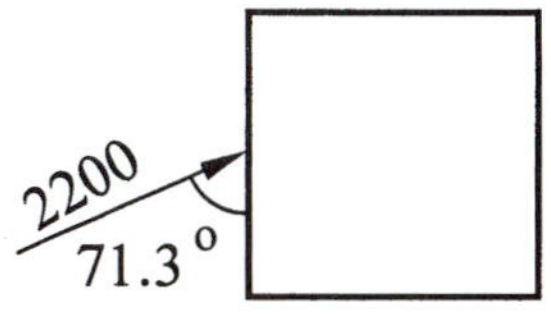

$$v_\perp = 2200\sin 71.3^\circ$$
$$= 2100 \text{ ft/s}$$

component toward tower $= 15\cos 15^\circ$
$= 14$ mi/h

49.

vector	x-component	y-component
1300	$-1300\cos 54^\circ$	$1300\sin 54^\circ$
3200	$3200\sin 32^\circ$	$3200\cos 32^\circ$
2100	$-2100\cos 35^\circ$	$-2100\sin 35^\circ$
	-788.6	2561

$$R = \sqrt{(-788.6)^2 + (2561)^2} = 2700 \text{ lb}$$
$$\theta_{\text{ref}} = \tan^{-1}\left|\frac{2561}{-788.6}\right|$$
$$\theta = 180^\circ - \theta_{\text{ref}} = 107^\circ$$

53. upward force $= 0.15\sin 22.5^\circ + 0.20\sin 15.0^\circ$
$= 0.11$ N

57.

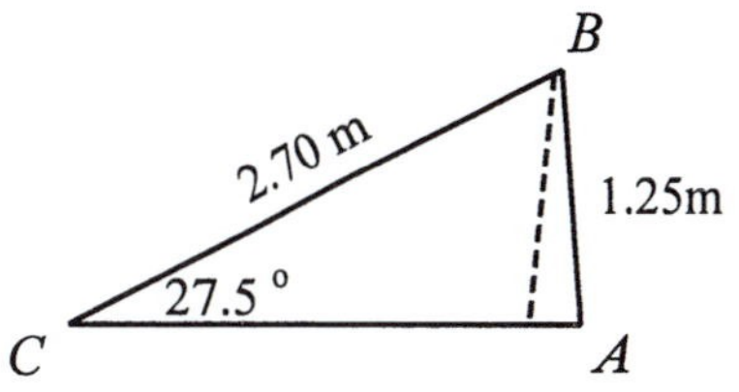

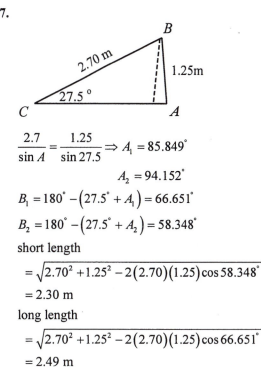

$$\frac{2.7}{\sin A} = \frac{1.25}{\sin 27.5} \Rightarrow A_1 = 85.849^\circ$$
$$A_2 = 94.152^\circ$$
$$B_1 = 180^\circ - (27.5^\circ + A_1) = 66.651^\circ$$
$$B_2 = 180^\circ - (27.5^\circ + A_2) = 58.348^\circ$$

short length
$$= \sqrt{2.70^2 + 1.25^2 - 2(2.70)(1.25)\cos 58.348^\circ}$$
$$= 2.30 \text{ m}$$

long length
$$= \sqrt{2.70^2 + 1.25^2 - 2(2.70)(1.25)\cos 66.651^\circ}$$
$$= 2.49 \text{ m}$$

61.

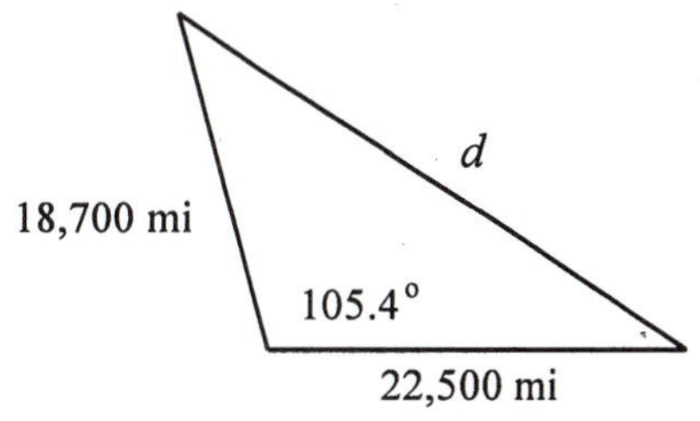

$$d = \sqrt{18,700^2 + 22,500^2 - 2(18,700)(22,500)\cos 105.4^\circ}$$
$$= 32,900 \text{ mi}$$

65.

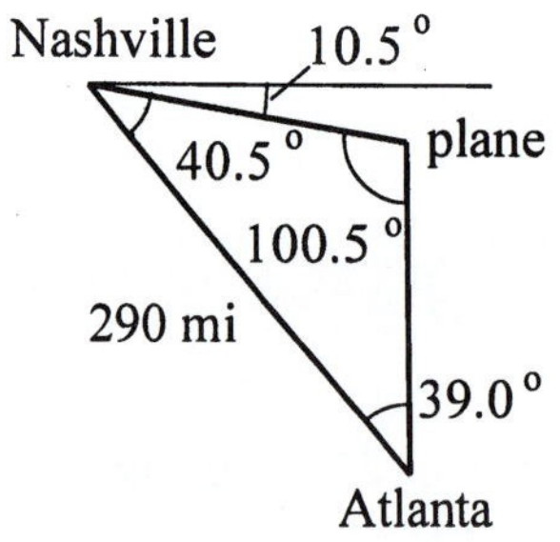

distance from plane to Nashville = d

$$\frac{d}{\sin 39.0^\circ} = \frac{290 \text{ mi}}{\sin 100.5^\circ}$$
$$d = 190 \text{ mi}$$

69.

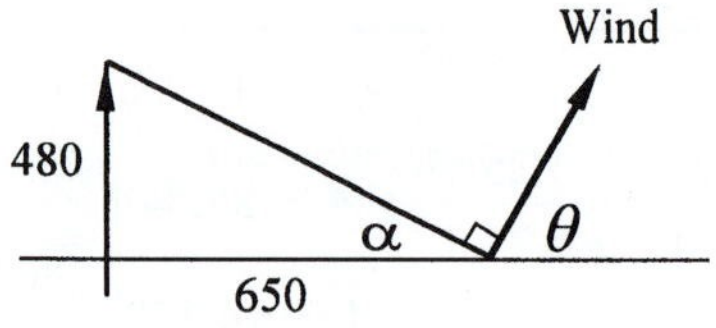

$$\tan\alpha = \frac{480}{650}$$
$$\alpha = 36.4^\circ \text{ N of E}$$
$$F = \sqrt{F_x^2 + F_y^2} = \sqrt{650^2 + 480^2}$$
$$= 810 \text{ N}$$

Chapter 10

GRAPHS OF THE TRIGONOMETRIC FUNCTIONS

10.1 Graphs of $y = a \sin x$ and $y = a \cos x$

1. $y = 3 \cos x$

x	0	$\frac{\pi}{6}$	$\frac{\pi}{3}$	$\frac{\pi}{2}$	$\frac{2\pi}{3}$	$\frac{5\pi}{6}$
y	0	2.6	1.5	0	−1.5	−2.6

x	π	$\frac{7\pi}{6}$	$\frac{4\pi}{3}$	$\frac{3\pi}{2}$	$\frac{5\pi}{3}$	$\frac{11\pi}{6}$	2π
y	−3	−2.6	−1.5	0	1.5	2.6	3

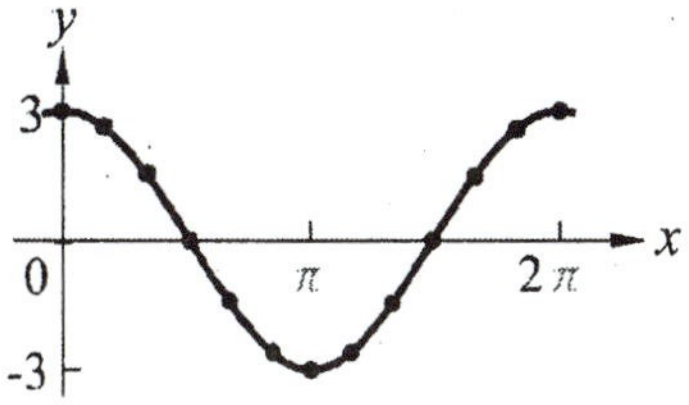

5. $y = 3 \cos x$

x	$-\pi$	$-\frac{3\pi}{4}$	$-\frac{\pi}{2}$	$-\frac{\pi}{4}$	0	$\frac{\pi}{4}$	$\frac{\pi}{2}$	$\frac{3\pi}{4}$	π
y	−3	−2.1	0	2.1	3	2.1	0	−2.1	3

x	$\frac{5\pi}{4}$	$\frac{3\pi}{4}$	$\frac{7\pi}{4}$	2π	$\frac{9\pi}{4}$	$\frac{5\pi}{2}$	$\frac{11\pi}{4}$	3π
y	−2.1	0	2.1	3	2.1	0	−2.1	−3

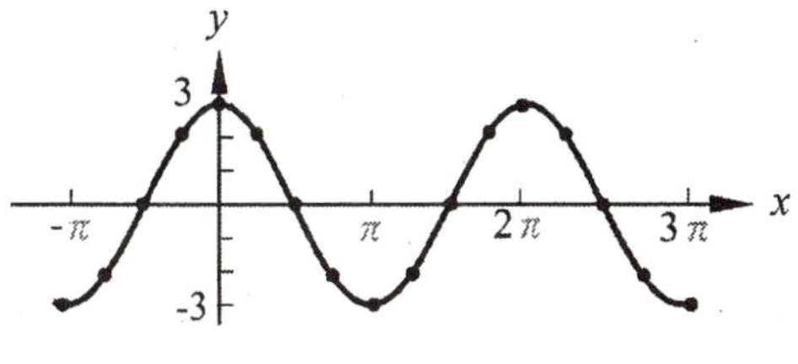

9. $y = \dfrac{5}{2} \sin x$; $\sin x$ has its amplitude value at $x = \dfrac{\pi}{2}$ and $x = \dfrac{3\pi}{2}$ and has intercepts at $x = 0$, $x = \pi$, and $x = 2\pi$. The graph can be sketched with these values.

x	0	$\frac{\pi}{2}$	π	$\frac{3\pi}{2}$	2π
y	0	$\frac{5}{2}$	0	$-\frac{5}{2}$	0

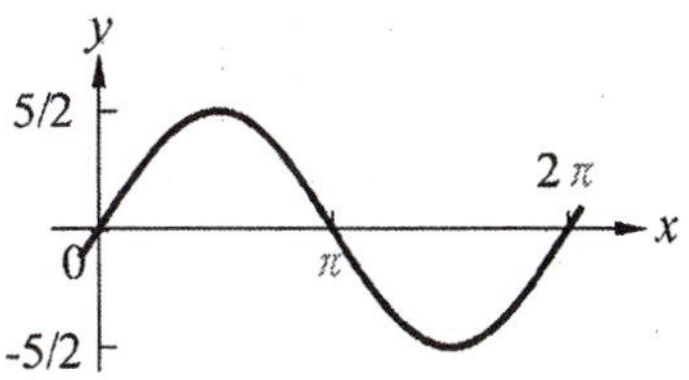

13. $y = 0.8 \cos x$; $\cos x$ has its amplitude value at $x = 0$ and $x = \pi$, and $x = 2\pi$, and has intercepts at $x = \dfrac{\pi}{2}$, and $x = \dfrac{3\pi}{2}$. The graph can be sketched with these values.

x	0	$\frac{\pi}{2}$	π	$\frac{3\pi}{2}$	2π
y	0.8	0	−0.8	0	0.8

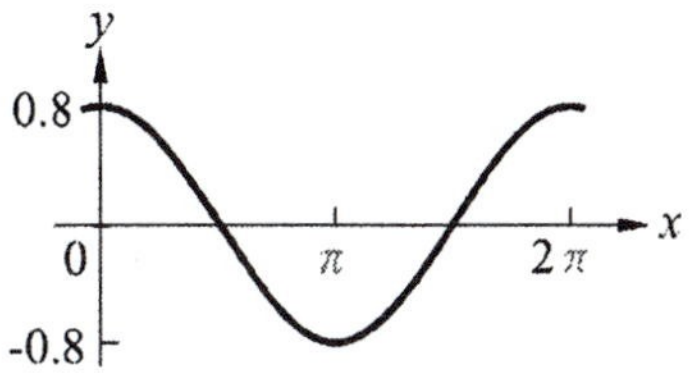

17. $y = -1500 \sin x = -1500(\sin x)$; $\sin x$ has its amplitude value at $x = \frac{\pi}{2}$ and $x = \frac{3\pi}{2}$ and has intercepts at $x = 0$, $x = \pi$, and $x = 2\pi$.
(The negative sign will invert the graph values.)

x	0	$\frac{\pi}{2}$	π	$\frac{3\pi}{2}$	2π
y	0	−1500	0	1500	0

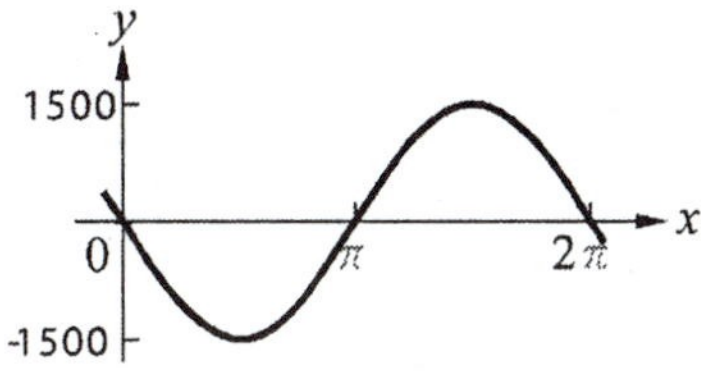

21. $y = -50 \cos x = -50(\cos x)$; $\cos x$ has its amplitude value at $x = 0$, $x = \pi$, and $x = 2\pi$. and has intercepts at $x = \frac{\pi}{2}$, $x = \frac{3\pi}{2}$.
(The negative sign will invert the graph values.)

x	0	$\frac{\pi}{2}$	π	$\frac{3\pi}{2}$	2π
y	−50	0	50	0	−50

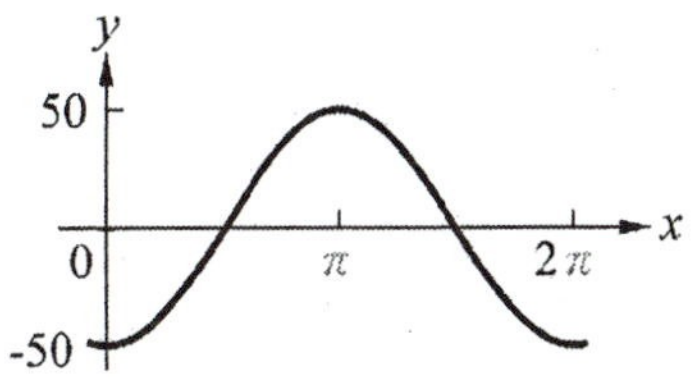

25. Sketch $y = 12\cos x$ for $x = 0, 1, 2, 3, 4, 5, 6, 7$

x	0	1	2	3	4
$12\cos x$	12	6.5	5.0	11.9	7.8

x	5	6	7
$12\cos x$	3.4	11.5	9.0

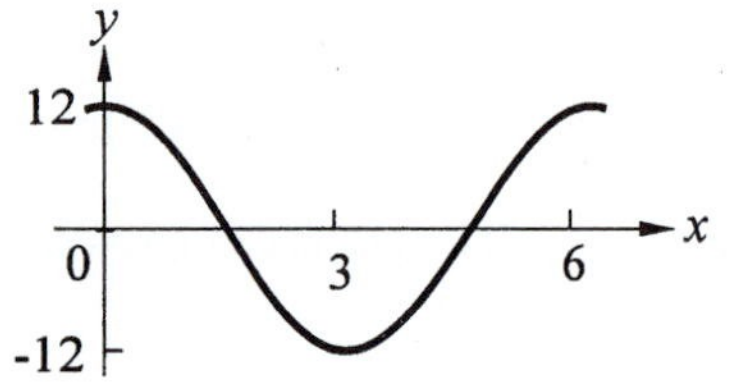

29. $y = a\cos x, (\pi, 2)$

$2 = a\cos x \Rightarrow a = -2$

$y = -2\cos x$

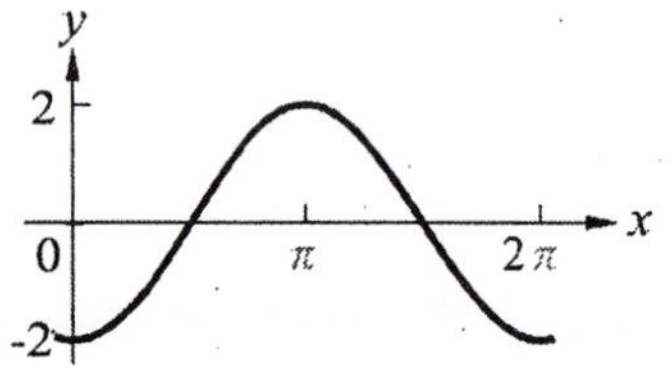

33. The graph passes through $(0,0), (\pi,0)$, and $(2\pi,0)$ with amplitude 4. The graph is $y = 4\sin x$.

37.

x	$\pm 2.5\sin x$	$\pm 2.5\cos x$
0.67	± 1.55	± 1.96

The function is $y = -2.5\sin x$

10.2 Graphs of $y = a\sin bx$ and $y = a\cos bx$

1. $y = 3\sin 6x$, amplitude $= 3$, period $= \frac{2\pi}{6} = \frac{\pi}{3}$

x	0	$\frac{\pi}{12}$	$\frac{\pi}{6}$	$\frac{\pi}{4}$	$\frac{\pi}{3}$
y	0	3	0	-3	0

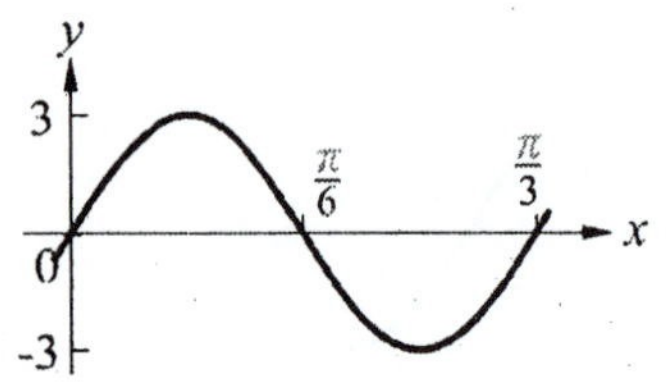

5. Since $\cos bx$ has period $\frac{2\pi}{b}$, $y = 3\cos 8x$ has a period of $\frac{2\pi}{8}$, or $\frac{\pi}{4}$.

9. $y = -\cos 16x$ has period of $\frac{2\pi}{16}$, or $\frac{\pi}{8}$.

13. $y = 3\cos 4\pi x$ has period of $\frac{2\pi}{4\pi}$, or $\frac{2}{4} = \frac{1}{2}$.

17. $y = -\frac{1}{2}\cos\frac{2}{3}x$ has period of $\frac{2\pi}{\frac{2}{3}} = \frac{2\pi}{1} \times \frac{3}{2} = 3\pi$.

21. $y = 3.3\cos\pi^2 x$ has period of $\frac{2\pi}{\pi^2} = \frac{2}{\pi}$.

25. $y = 3\cos 8x$ has amplitude of 3 and period $\frac{\pi}{4}$.

x	0	$\frac{\pi}{16}$	$\frac{\pi}{8}$	$\frac{3\pi}{16}$	$\frac{\pi}{4}$
y	3	0	-3	0	3

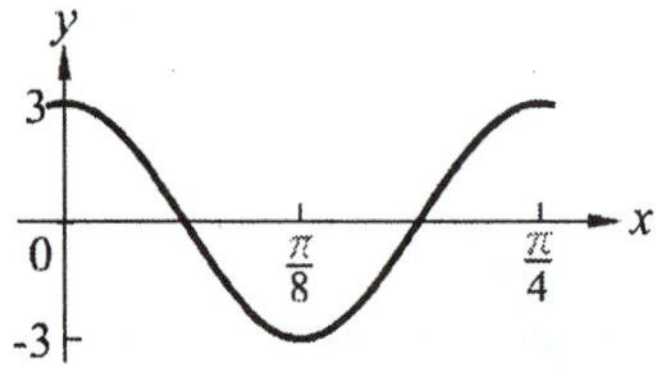

29. $y = -\cos 16x$ has amplitude of $|-1| = 1$, and period of $\frac{\pi}{8}$.

x	0	$\frac{\pi}{32}$	$\frac{\pi}{16}$	$\frac{3\pi}{32}$	$\frac{\pi}{8}$
y	-1	0	1	0	-1

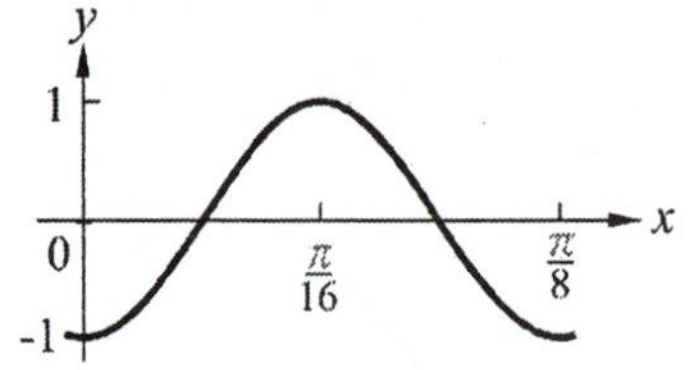

33. $y = 3\cos 4\pi x$ has amplitude of 3 and period of $\frac{1}{2}$.

x	0	$\frac{1}{8}$	$\frac{1}{4}$	$\frac{3}{8}$	$\frac{1}{2}$
y	3	0	-3	0	3

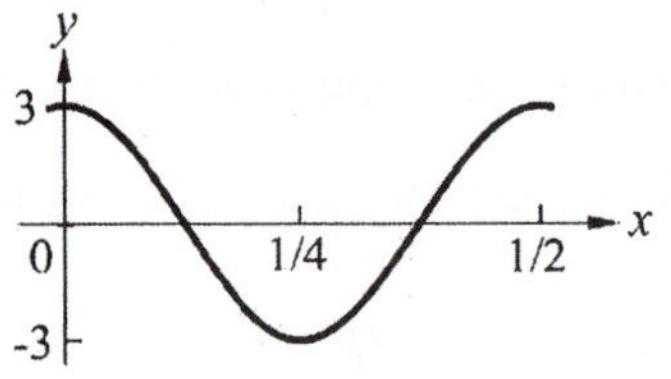

37. $y=-\frac{1}{2}\cos\frac{2}{3}x$ has amplitude of $\left|-\frac{1}{2}\right|=\frac{1}{2}$, and period of 3π.

x	0	$\frac{3\pi}{4}$	$\frac{3\pi}{2}$	$\frac{9\pi}{4}$	3π
y	$-\frac{1}{2}$	0	$\frac{1}{2}$	0	$-\frac{1}{2}$

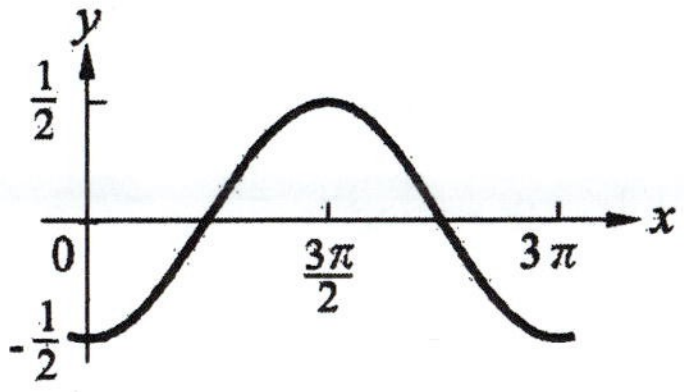

41. $y=3.3\cos\pi^2x$ has amplitude of 3.3 and period of $\frac{2}{\pi}$.

x	0	$\frac{1}{2\pi}$	$\frac{1}{\pi}$	$\frac{3}{2\pi}$	$\frac{2}{\pi}$
π^2x	0	$\frac{\pi}{2}$	π	$\frac{3\pi}{2}$	2π
$\cos\pi^2x$	1	0	-1	0	1
$3.3\cos\pi^2x$	3.3	0	-3.3	0	3.3

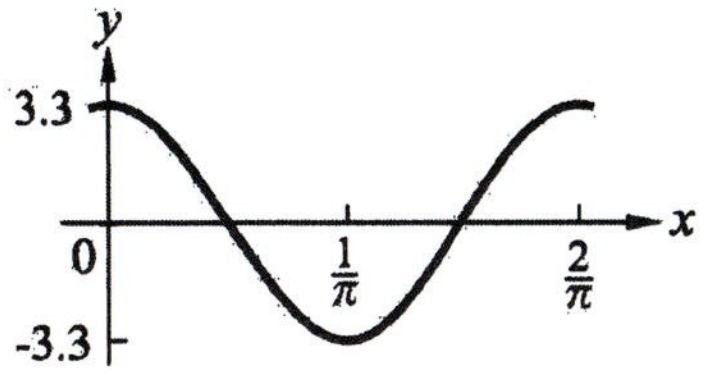

45. $b=\frac{2\pi}{1/3}=6$; $y=\sin 6\pi x$

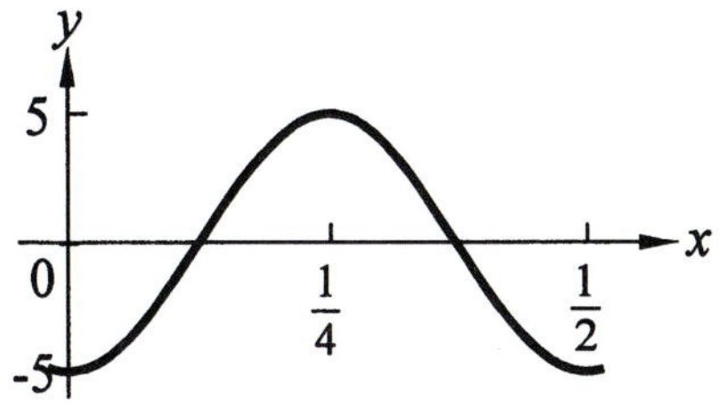

49. $y=8\left|\cos\frac{\pi}{2}x\right|$

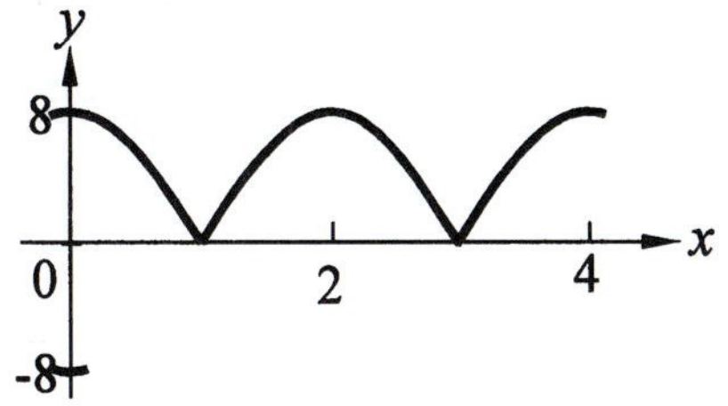

53. $y=-2\sin bx,\left(\frac{\pi}{4},-2\right), b>0$

$-2=-2\sin b\cdot\frac{\pi}{4}$

$\sin\frac{b\pi}{4}=1\Rightarrow\frac{b\pi}{4}=\frac{\pi}{2}+2\pi n$

$b=2+8n$ of which

the smallest is $b=2$

$y=-2\sin 2x$ is the function.

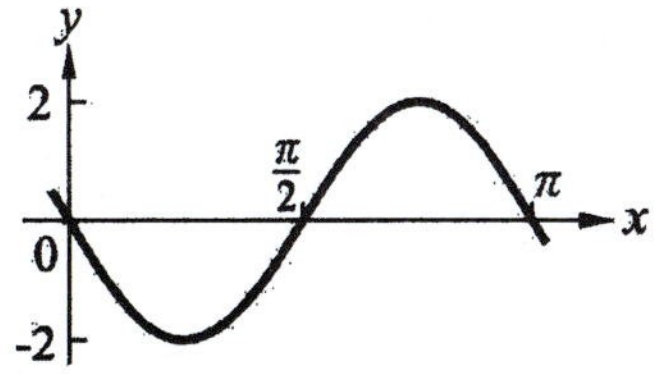

57. $V=170\sin 120\pi t$

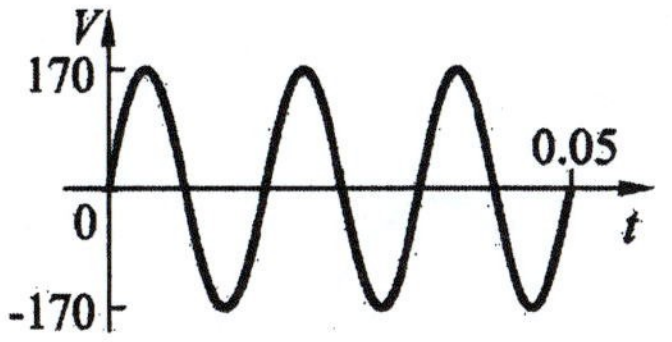

61. $y=\frac{1}{2}\cos 2x$ period of π, amplitude 0.5

10.3 Graphs of $y=a\sin(bx+c)$ and $y=a\cos(bx+c)$

1. $y=-\cos\left(2x-\frac{\pi}{6}\right)$

(1) the amplitude is 1

(2) the period is $\frac{2\pi}{2}=\pi$

(3) the displacement is $-\frac{\frac{-\pi}{6}}{2}=\frac{\pi}{12}$

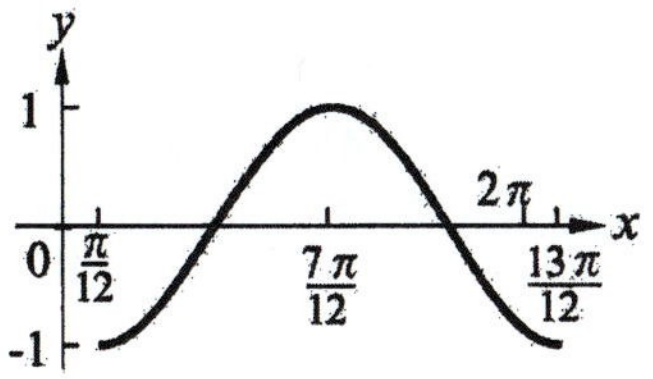

5. $y=\cos\left(x+\dfrac{\pi}{6}\right)$; $a=1$, $b=1$, $c=-\dfrac{\pi}{6}$

Amplitude is $|a|=1$; period is $\dfrac{2\pi}{b}=2\pi$;

displacement is $-\dfrac{c}{b}=\dfrac{\pi}{6}$

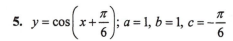

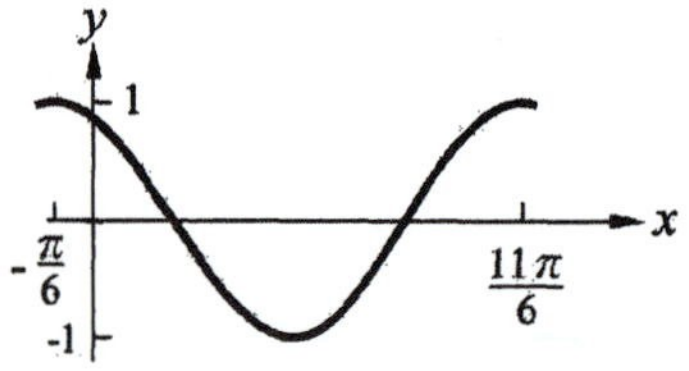

9. $y=-\cos(2x-\pi)$; $a=-1$, $b=2$, $c=-\pi$

Amplitude is $|a|=1$; period is $\dfrac{2\pi}{b}=\dfrac{2\pi}{1}=\pi$;

displacement is $-\dfrac{c}{b}=-\left(\dfrac{-\pi}{2}\right)=\dfrac{\pi}{2}$

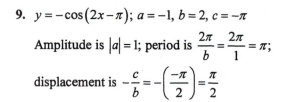

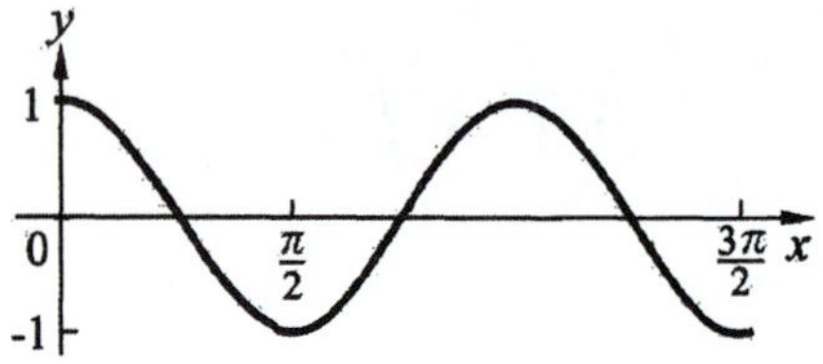

13. $y=30\cos\left(\dfrac{1}{3}x+\dfrac{\pi}{3}\right)$; $a=30$, $b=\dfrac{1}{3}$, $c=\dfrac{\pi}{3}$

Amplitude is $|a|=30$; period is $\dfrac{2\pi}{b}=\dfrac{2\pi}{1/3}=6\pi$;

displacement is $-\dfrac{c}{b}=\dfrac{-\pi/3}{1/3}=-\pi$

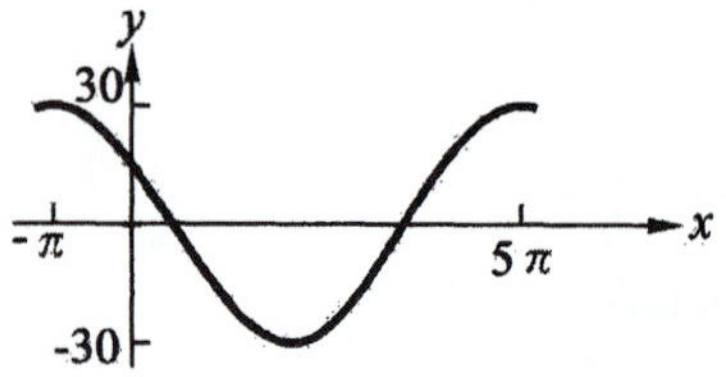

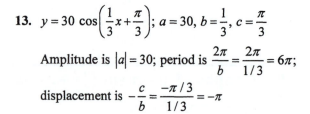

17. $y=0.08\cos\left(4\pi x-\dfrac{\pi}{5}\right)$; $a=0.08$, $b=4\pi$, $c=-\dfrac{\pi}{5}$

Amplitude is $|a|=0.08$; period is $\dfrac{2\pi}{b}=\dfrac{2\pi}{4\pi}=\dfrac{1}{2}$;

displacement is $-\dfrac{c}{b}=-\left(-\dfrac{\pi/5}{4\pi}\right)=\dfrac{1}{20}$

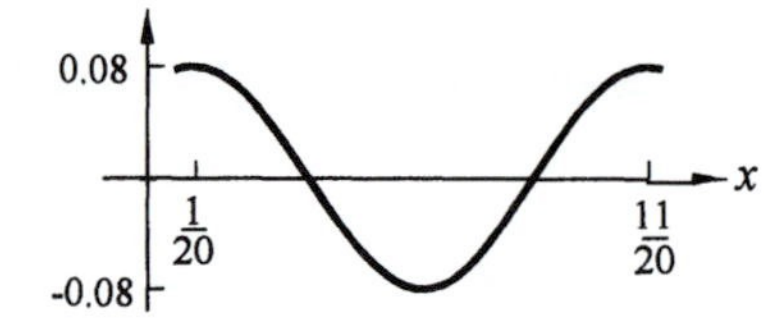

21. $y=40\cos(3\pi x+2)$; $a=40$, $b=3\pi$, $c=2$

Amplitude is $|a|=40$; period is $\dfrac{2\pi}{b}=\dfrac{2\pi}{3\pi}=\dfrac{2}{3}$;

displacement is $-\dfrac{c}{b}=-\dfrac{2}{3\pi}$

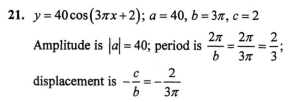

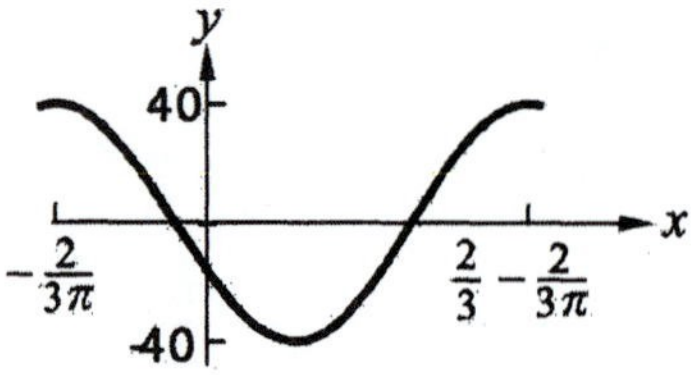

25. $y=-\dfrac{3}{2}\cos\left(\pi x-\dfrac{\pi^2}{6}\right)$; $a=-\dfrac{3}{2}$, $b=\pi$, $c=-\dfrac{\pi^2}{6}$

Amplitude is $|a|=\dfrac{3}{2}$; period is $\dfrac{2\pi}{b}=\dfrac{2\pi}{\pi}=2$;

displacement is $-\dfrac{c}{b}=-\dfrac{\pi^2/6}{\pi}=-\dfrac{\pi}{6}$

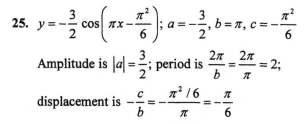

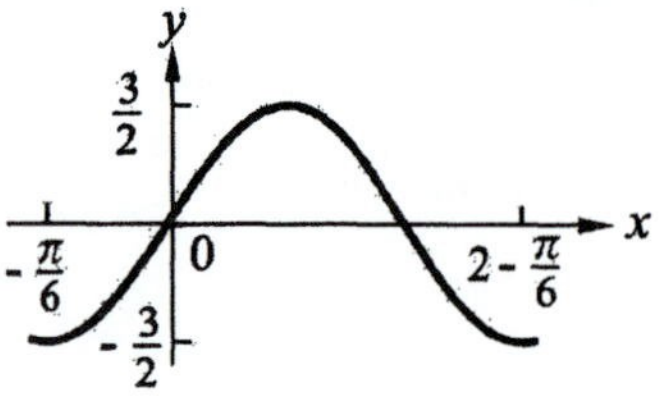

29. cosine, 12, $\dfrac{1}{2}$, $\dfrac{1}{8}$

$a=12$; period $=\dfrac{1}{2}=\dfrac{2\pi}{b}\Rightarrow b=4\pi$

displacement $=\dfrac{1}{8}=-\dfrac{c}{4\pi}\Rightarrow c=-\dfrac{\pi}{2}$

$y=12\cos\left(4\pi x-\dfrac{\pi}{2}\right)$

33. Graph $y_1=\sin\left(\dfrac{x}{2}-\dfrac{3\pi}{4}\right)$ and $y_2=-\sin\left(\dfrac{3\pi}{4}-\dfrac{x}{2}\right)$.

Graphs are the same.

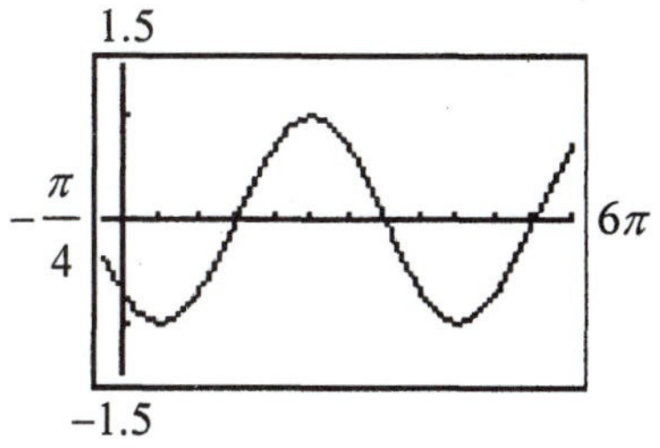

37. $y = 2.00 \sin 2\pi\left(\dfrac{t}{0.100} - \dfrac{5.00}{20.0}\right)$; $a = 2.00$,

$b = \dfrac{2\pi}{0.100}$, $c = \dfrac{-500(2\pi)}{20.0}$

Amplitude $= |a| = 2.00$,

period $= \dfrac{2\pi}{b} = 0.100$,

displacement $= -\dfrac{c}{b} = 0.025$

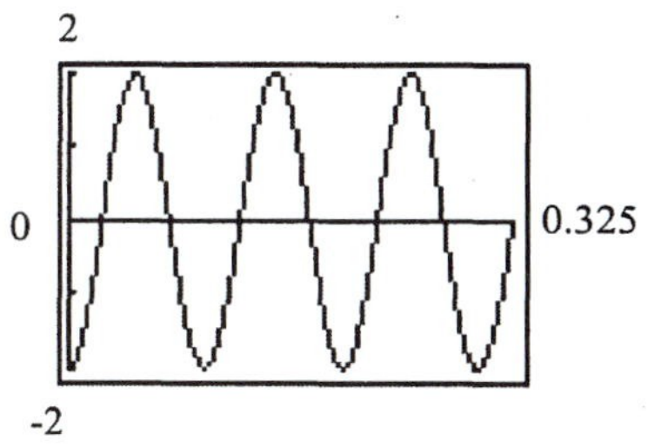

41. $y = a \sin(bx + c)$. Amplitude $= 5$,

period $= \dfrac{2\pi}{b} = 16$, $b = \dfrac{\pi}{8}$

displacement $= -\dfrac{c}{b} = -1$, $c = \dfrac{\pi}{8}$

$y = 5 \sin\left(\dfrac{\pi}{8}x + \dfrac{\pi}{8}\right)$

10.4 Graphs of $y = \tan x$, $y = \cot x$, $y = \sec x$, $y = \csc x$

1. $y = 5.0 \cot 2x$. Graph $y_1 = \dfrac{5.0}{\tan 2x}$.

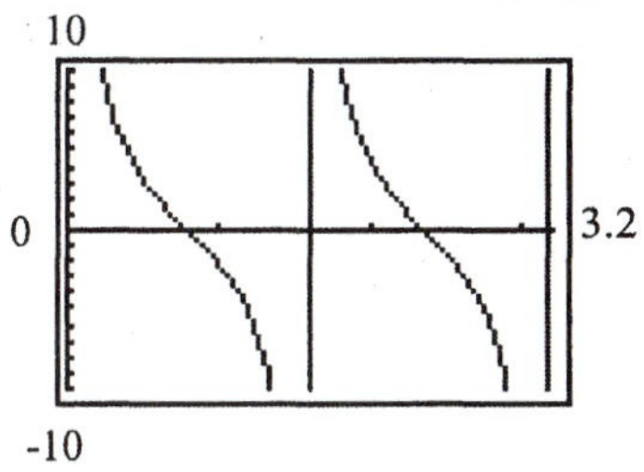

5.

x	$-\frac{\pi}{2}$	$-\frac{\pi}{3}$	$-\frac{\pi}{4}$	$-\frac{\pi}{6}$	0	$\frac{\pi}{6}$	$\frac{\pi}{4}$
sec x	*	2	1.4	1.2	1	1.2	1.4

x	$\frac{\pi}{3}$	$\frac{\pi}{2}$	$\frac{2\pi}{3}$	$\frac{3\pi}{4}$	$\frac{5\pi}{6}$	π
sec x	2	*	−2	−1.4	−1.2	−1

(* = undefined)

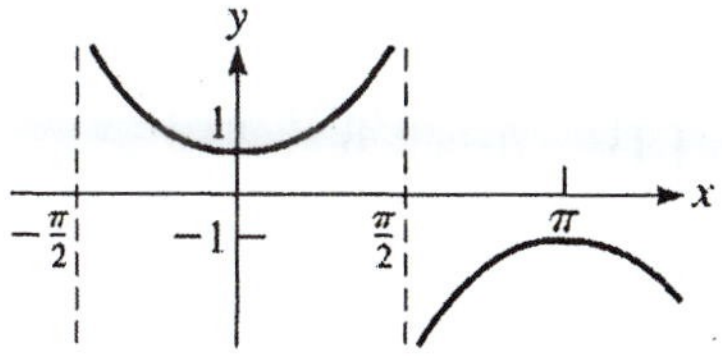

9. For $y = \dfrac{1}{2} \sec x$, first sketch the graph of $y = \sec x$, then multiply the y – values of the secant function by $\dfrac{1}{2}$ and graph.

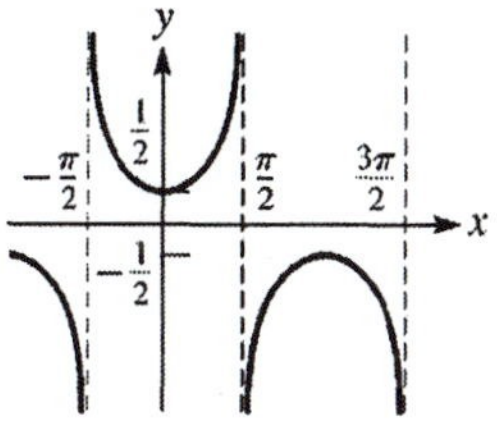

13. For $y = -3 \csc x$, sketch the graph of $y = \csc x$, then multiply the y – values by -3, and resketch the graph. It will be inverted.

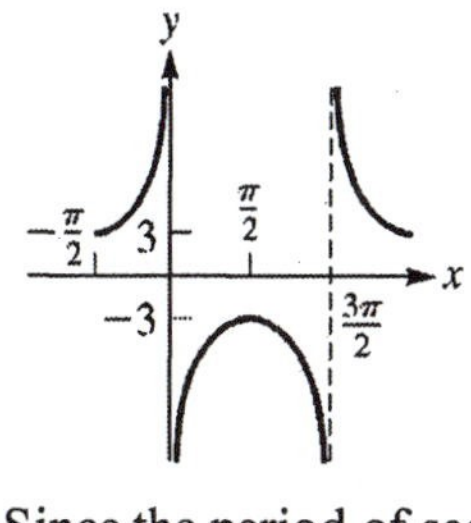

17. Since the period of sec x is 2π, the period of $y = \dfrac{1}{2}\sec 3x$ is $\dfrac{2\pi}{3}$. Graph $y_1 = 0.5(\cos 3x)^{-1}$.

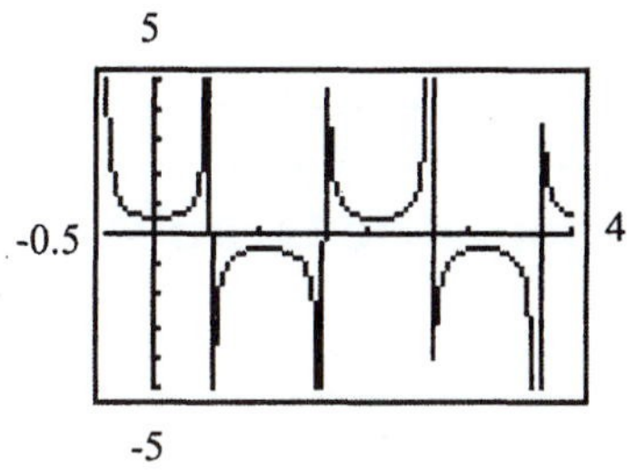

21. Since the period of $\csc x$ is 2π, the period of $y = 18\csc\left(3x - \frac{\pi}{3}\right)$ is $\frac{2\pi}{3}$. The displacement is $-\left(-\frac{\pi/3}{3}\right) = \frac{\pi}{9}$. Graph $y_1 = 18\left(\sin\left(3x - \frac{\pi}{3}\right)\right)^{-1}$.

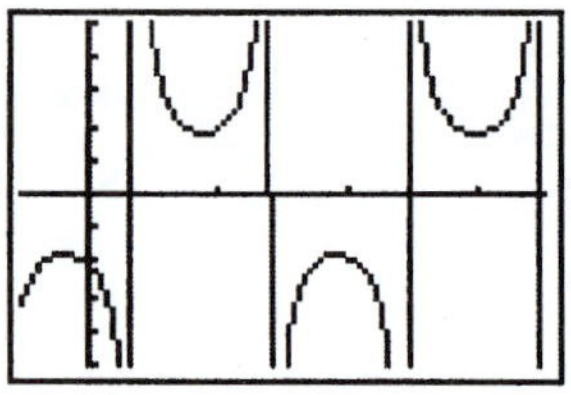

25. $y = \sec(bx) + k$ is a secant function with zero displacement, period 4π and passing through $(0, -3)$ if $\frac{2\pi}{b} = 4\pi \Rightarrow b = \frac{1}{2}$ and

$1 + k = -3 \Rightarrow k = -4.$

$y = \sec\left(\frac{1}{2}x\right) - 4$ is the required function

29. $b = (a\sin B)\csc A$

$= \left(4.00\sin\frac{\pi}{4}\right)\csc A$

$= 2.83\csc A$

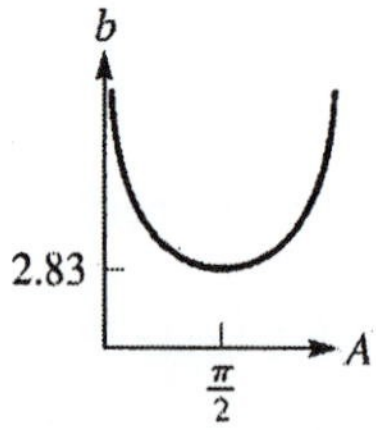

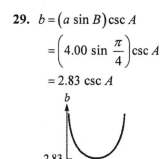

10.5 Applications of the Trigonometric Graphs

1. The displacement of the projection on the y-axis is d and is given by $d = R\cos wt$.

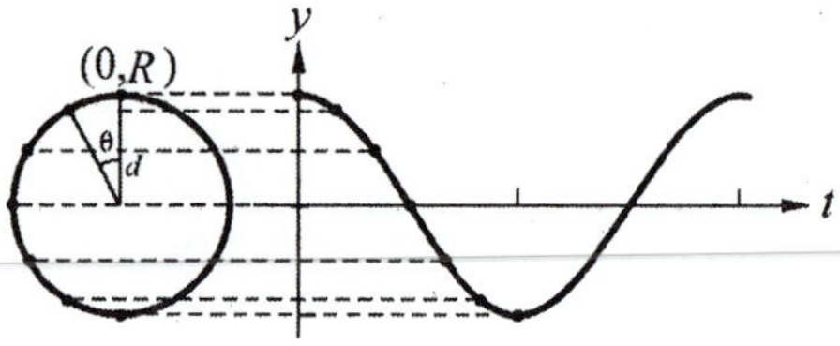

5. $y = R\cos\omega t$

$= 8.30\cos[(3.20)(2\pi)]t$

Amplitude is 8.30 cm;

period is $\frac{1}{3.20} = 0.3125$ s,

0.625 s for 2 cycles;

displacement is 0 s.

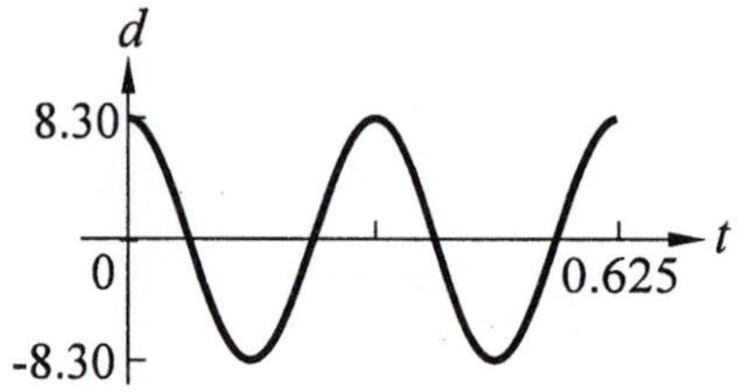

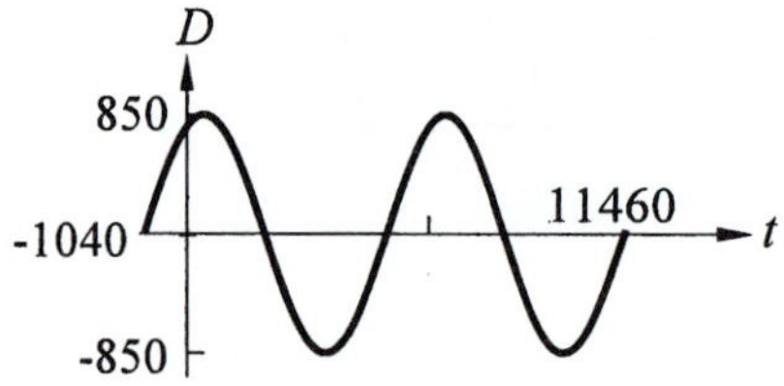

9. $e = E\cos(\ )$

$= 170\cos\left[2\pi(60.0)t - \frac{\pi}{3}\right]$

Amplitude is 170 V

period is $\frac{2\pi}{2\pi(60.0)} = 0.016$ s, 0.033 s

for 2 cycles; displacement is $\frac{\pi/3}{2\pi(60.0)} = \frac{1}{360}$ s

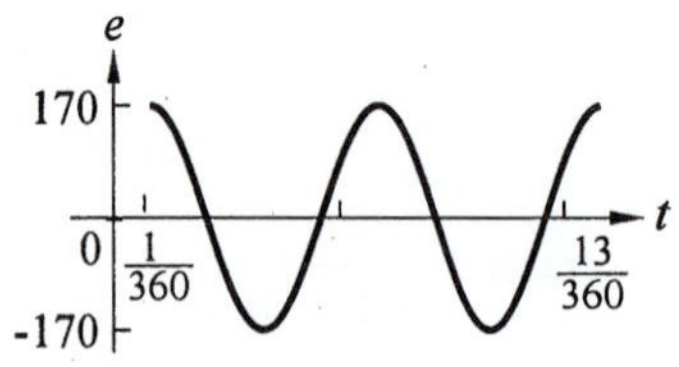

13. $p = p_0 \sin 2\pi ft$

$= 2.80 \sin\left[2\pi(2.30)\right]t$

$= 2.80 \sin 14.45t$

Amplitude is 2.80 lb/in.^{2}, period is $\dfrac{2\pi}{14.45} = 0.435$ s for 1 cycle, 0.87 s for 2 cycles; displacement is 0 s

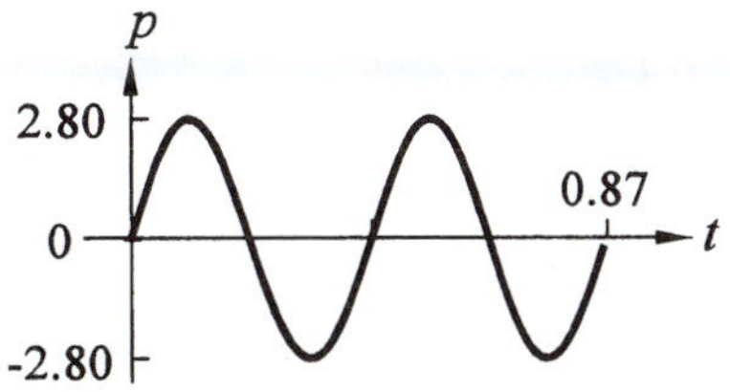

17. $e = 0.014 \cos\left(2\pi ft + \dfrac{\pi}{4}\right)$

$= 0.014 \cos\left[2\pi(0.950)t + \dfrac{\pi}{4}\right]$

Amplitude is 0.014 V

period is $\dfrac{2\pi}{2\pi(0.950)} = 1.05$ s, 2.10 s

for 2 cycles; displacement is $\dfrac{-\pi/4}{2\pi(0.950)} = -0.13$ s

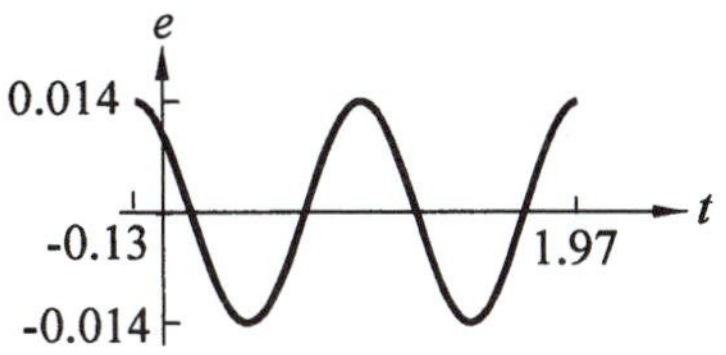

21. $a = 12$

$\omega = 18 \text{ r/min} \times \dfrac{2\pi}{r} = 36\pi/\text{min}$

$y = 12 \sin(36\pi t)$

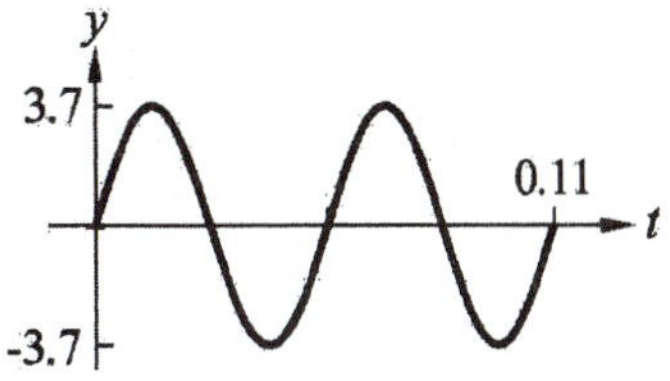

10.6 Composite Trigonometric Curves

1. $y = 1 + \sin x$

x	-2π	$-\frac{3\pi}{2}$	$-\pi$	$-\frac{\pi}{2}$	0	$\frac{\pi}{2}$	π	$\frac{3\pi}{2}$	2π
y	1	2	1	0	1	2	1	0	1

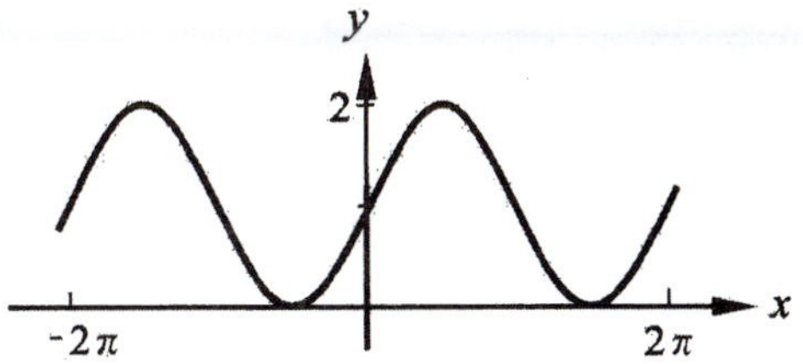

5. $y = \dfrac{1}{10}x^2 - \sin \pi x$

x	−4	−3.43	−2.55	−1.88	−1.47
y	1.60	0.20	1.64	0	−0.78

x	−1.03	−0.51	0	0.49	0.97	1.53
y	0	1.03	0	−0.98	0	1.23

x	2.15	2.45	2.73	0
y	0	−0.39	4	1.6

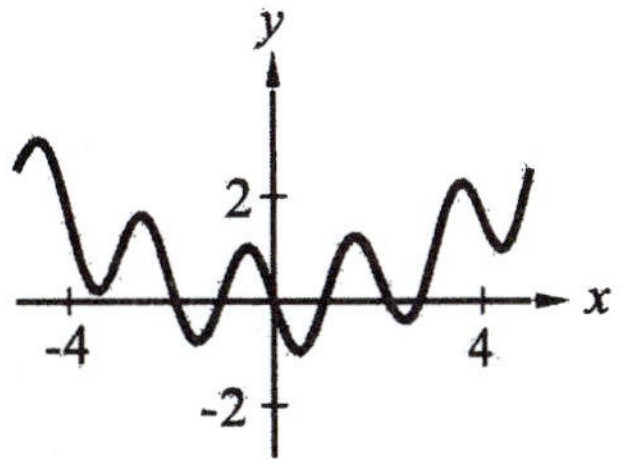

7. $y = \sin x + \cos x$

x	−5.5	−3.9	−2.4	−0.8
y	1.41	0	−1.41	0

x	0	0.8	2.4	3.9	5.5
y	1	1.41	0	−1.41	0

9. Graph $y_1 = x^3 + 10 \sin 2x$.

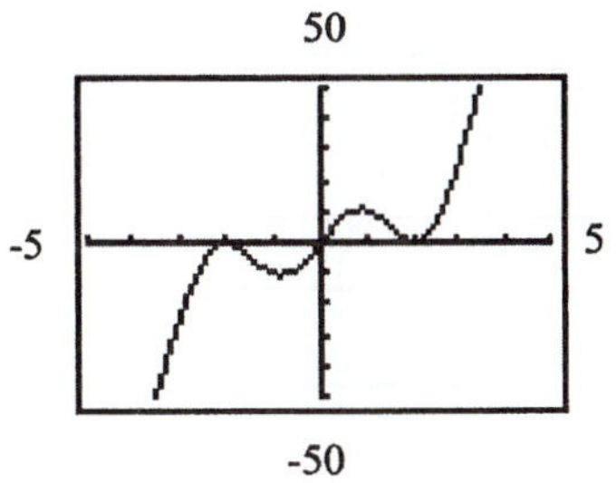

13. Graph $y_1 = 20 \cos 2x + 30 \sin x$.

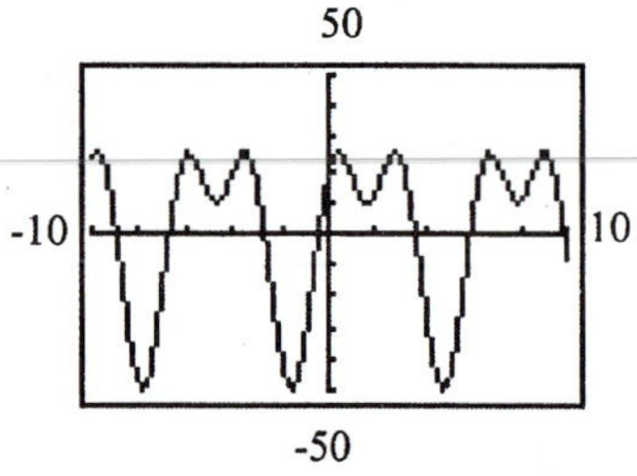

17. Graph $y_1 = \sin \pi x - \cos 2x$.

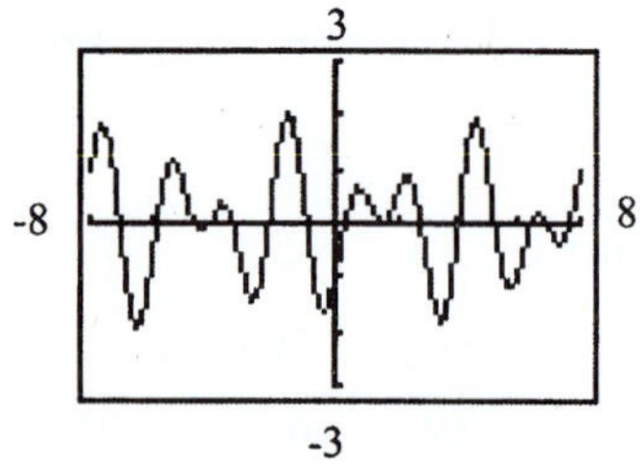

21. In parametric mode graph

$x_{IT} = 3 \sin t$, $y_{IT} = 2 \sin t$

t	x	y
$-\frac{\pi}{2}$	-3	-2
$-\frac{\pi}{4}$	-2.12	-1.41
0	0	0
$\frac{\pi}{4}$	2.12	1.41
$\frac{\pi}{2}$	3	2

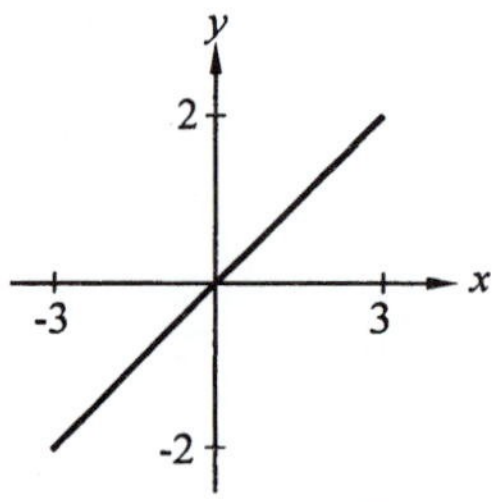

25. In parametric mode graph

$x_{IT} = \cos \pi\left(t + \frac{1}{6}\right)$, $y_{IT} = 2 \sin \pi t$

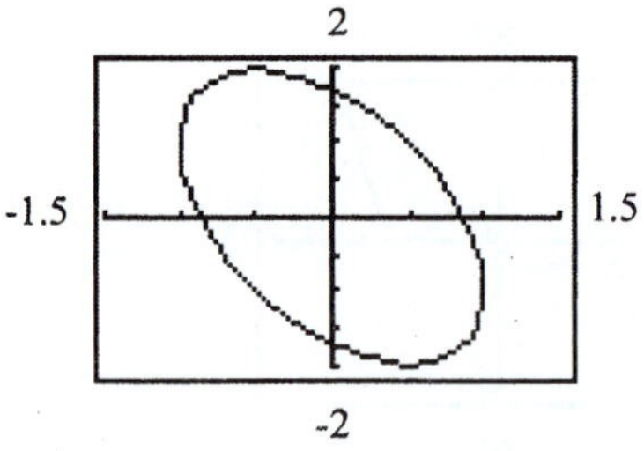

29. In parametric mode graph

$x_{IT} = \sin(t+1)$, $y_{IT} = \sin 5t$.

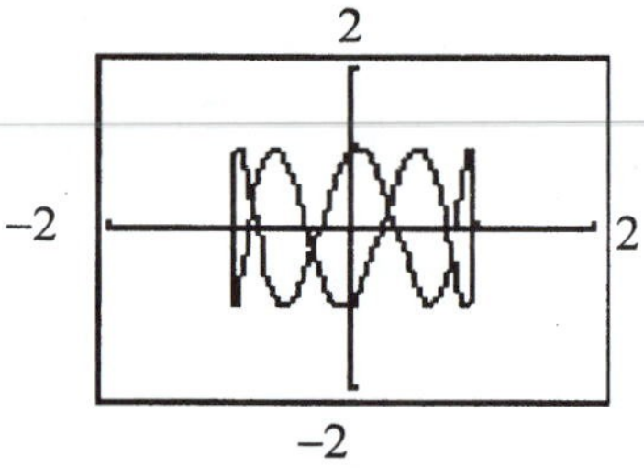

33. $y = 0.4 \sin 4t + 0.3 \cos 4t$

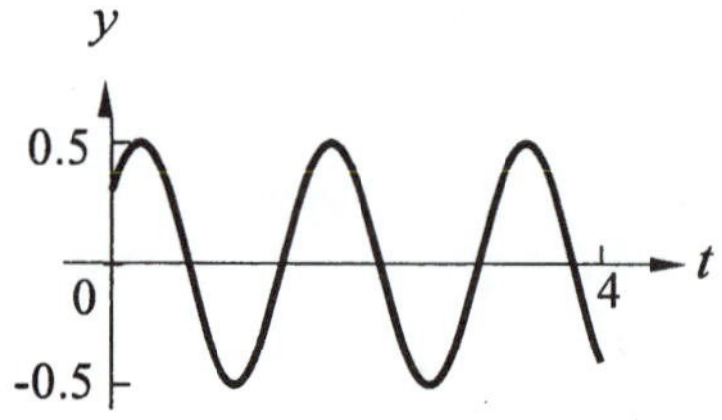

37. $60 \text{ beats/min} \times \frac{\text{min}}{60 \text{ s}} = 1 \text{ beat/s}$ from which $\omega = 2\pi$

$p = 100 + 20 \cos(2\pi t)$

Graph $y_1 = 100 + 20 \cos(2\pi x)$.

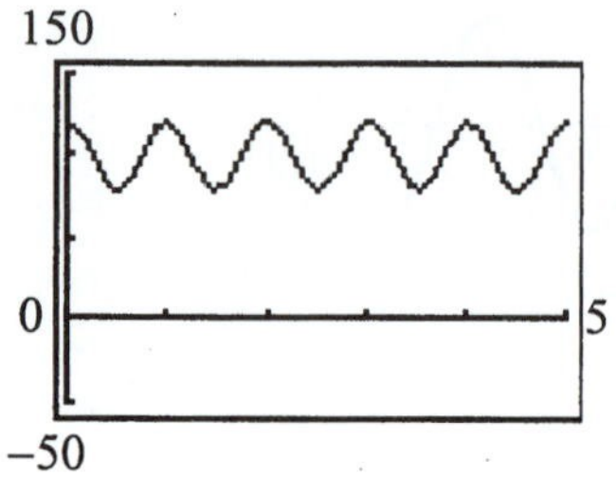

41. $x = 4 \cos \pi t$, $y = 2 \sin 3\pi t$

t	0	$\frac{\pi}{4}$	$\frac{\pi}{2}$	$\frac{3\pi}{4}$	π
x	4	-3.12	0.88	1.75	-3.61
y	0	1.80	1.57	-0.43	-1.94

t	$\frac{5\pi}{4}$	$\frac{3\pi}{2}$	$\frac{7\pi}{4}$	2π
x	3.90	-2.48	-0.028	2.52
y	-1.27	0.84	2.0	0.91

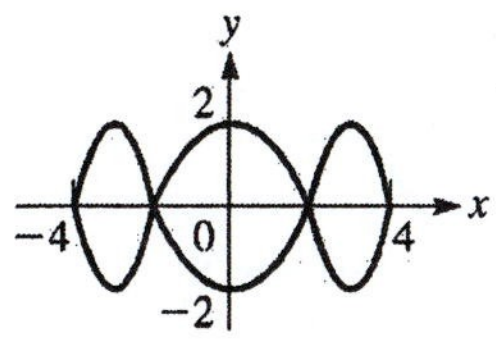

Chapter 10 Review Exercises

1. $y = \frac{2}{3}\sin x$

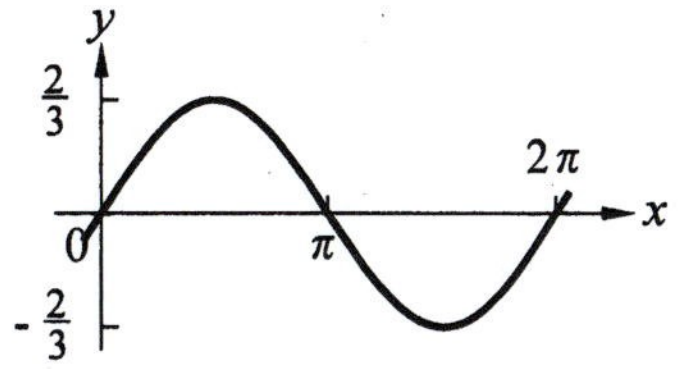

9. $y = 3\cos\frac{1}{3}x$

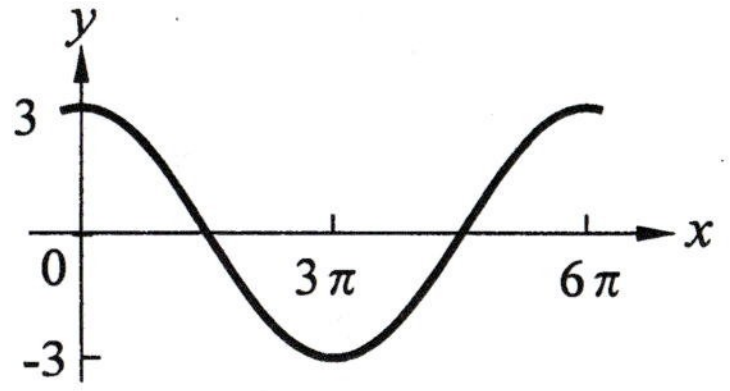

13. $y = 5\cos\left(\frac{\pi x}{2}\right)$

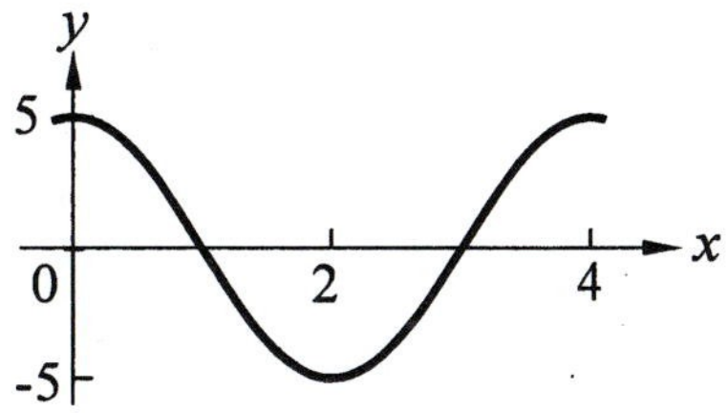

17. $y = 2\sin\left(3x - \frac{\pi}{2}\right)$

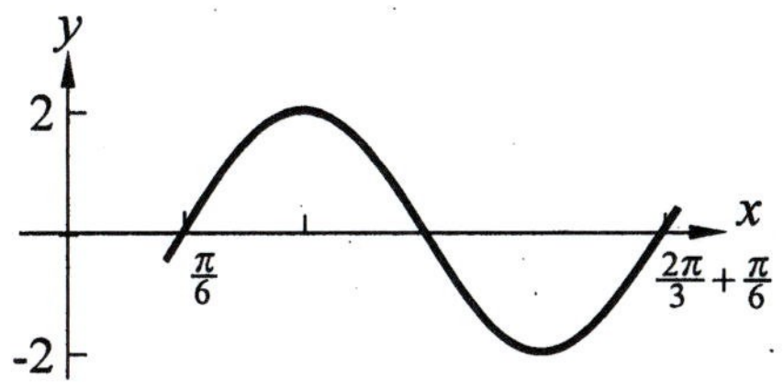

21. $y = -\sin\left(\pi x + \frac{\pi}{6}\right)$

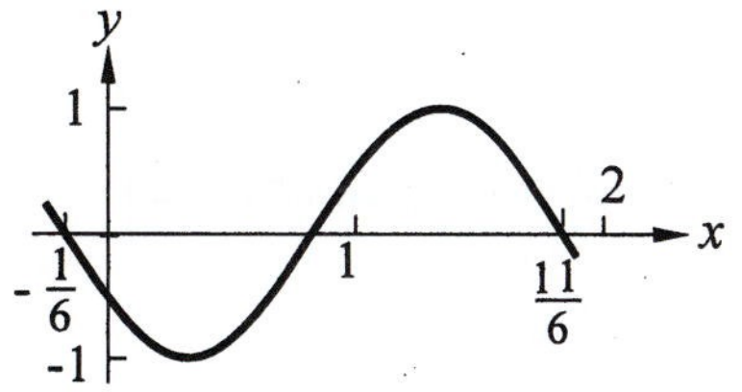

25. $y = 0.3\tan 0.5x$

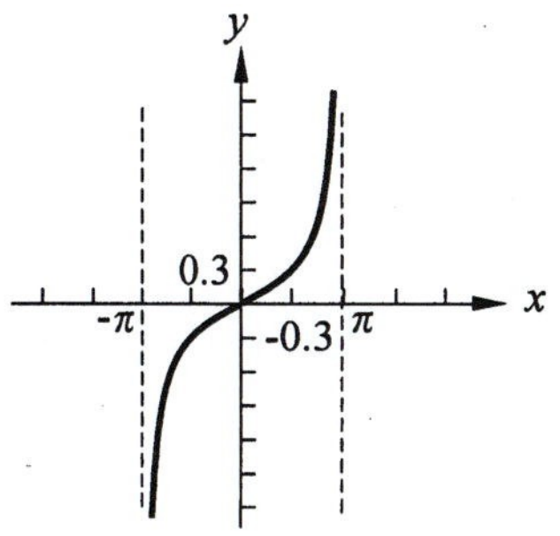

29. $y = 2 + \frac{1}{2}\sin 2x$

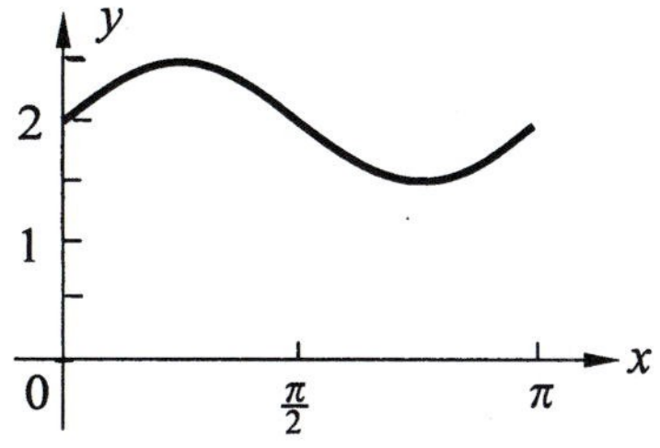

33. Graph $y_1 = 2\sin x - \cos 2x$.

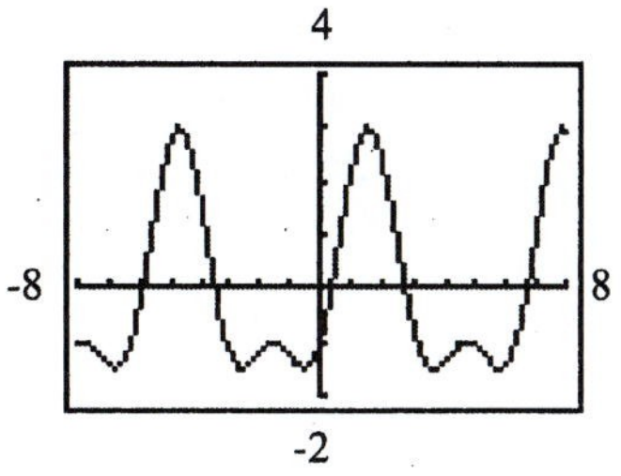

37. Graph $y_1 = \frac{\sin x}{x}$.

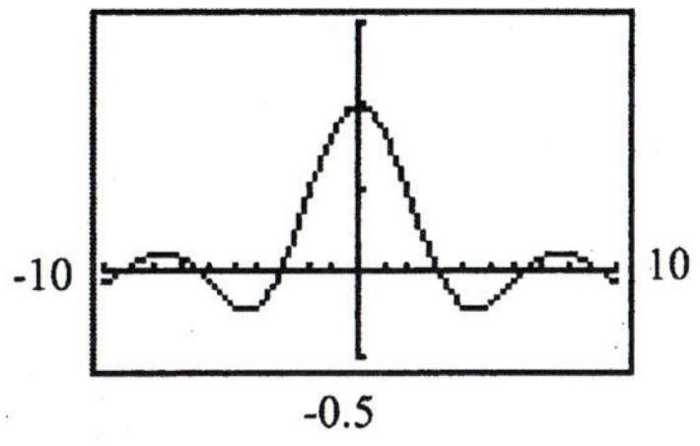

41. From the graph, $a = 2$, period $= \pi = \dfrac{2\pi}{b} \Rightarrow b = 2$,

and displacement $= 0 = -\dfrac{c}{b} = -\dfrac{\pi}{4} \Rightarrow c = \dfrac{\pi}{2}$.

$y = a\sin(bx + c)$ is $y = 2\sin\left(2x + \dfrac{\pi}{2}\right)$

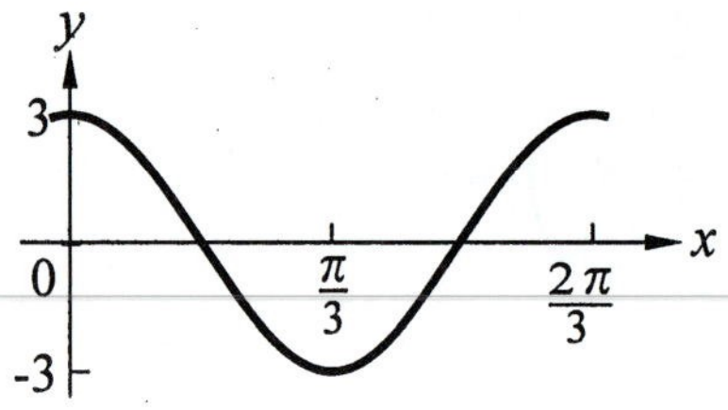

45. In parametric mode graph

$x_1 = -\cos 2\pi t,\ y_1 = 2\sin \pi t$

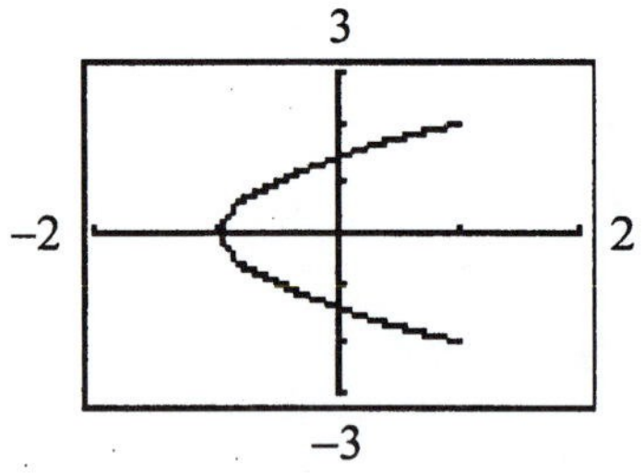

49. Graph $y_1 = 2|2\sin 0.2\pi x| - |\cos 0.4\pi x|$

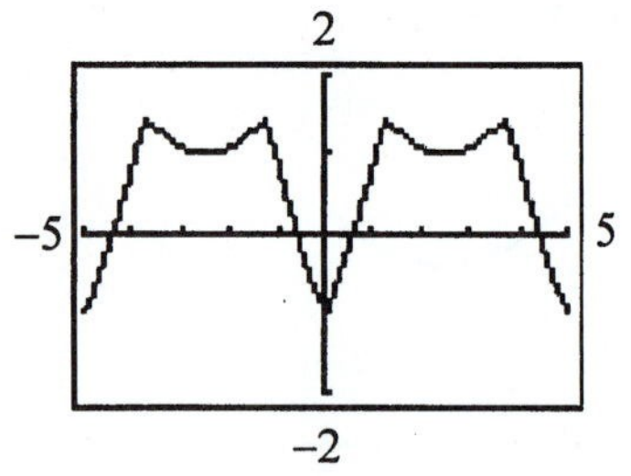

53. The period of $\cos 0.5x$ is $\dfrac{2\pi}{0.5} = 4\pi$.

The period of $\sin 3x$ is $\dfrac{2\pi}{3}$.

The period of $y = 2\cos 0.5x + \sin 3x$ is the least common multiple of 4π and $\dfrac{2\pi}{3}$; 4π.

57. $y = 3\cos bx, \left(\dfrac{\pi}{3}, -3\right), b > 0$

$$-3 = 3\cos\left(b\cdot\frac{\pi}{3}\right) \Rightarrow b\cdot\frac{\pi}{3} = \pi + 2\pi\cdot n$$

$$b = 3 + 6\cdot n$$

of which the smallest is $b = 3$. Graph $y = 3\cos 3x$.

61.

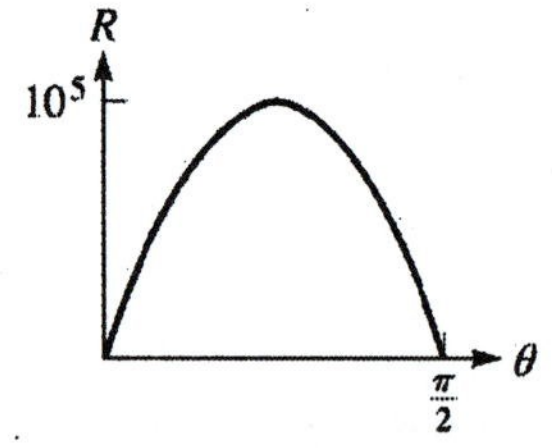

65.

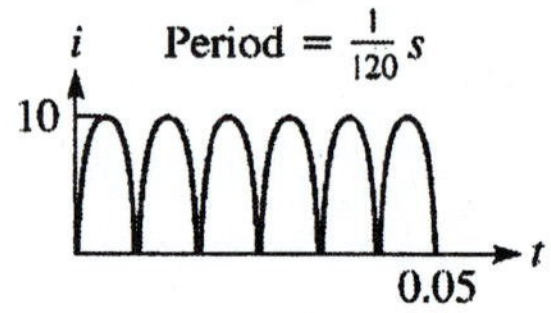

69.

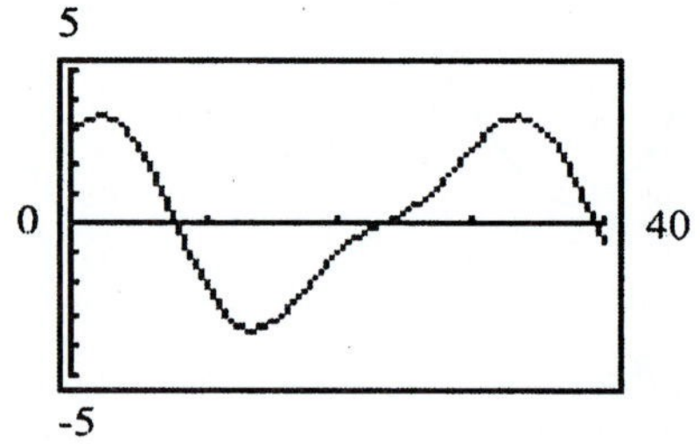

73.

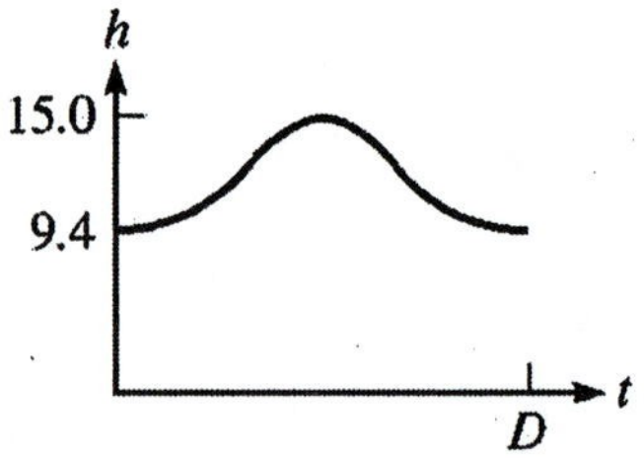

77.

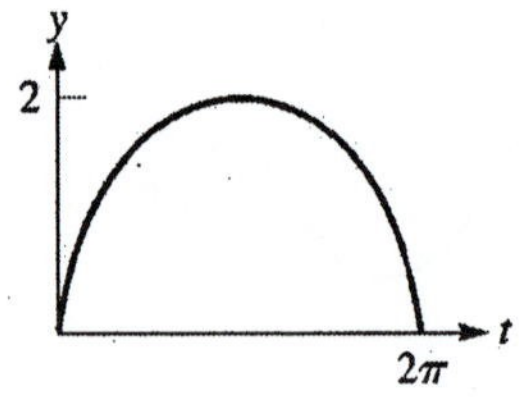

81.

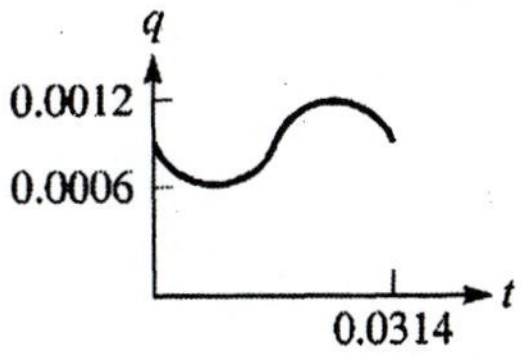

Chapter 11

EXPONENTS AND RADICALS

11.1 Simplifying Expressions with Integral Exponents

1. $(x^{-2}y)^2\left(\frac{2}{x}\right)^2 = \frac{x^{-4}y^2}{\left(\frac{2}{x}\right)^2} = \frac{x^{-4}y^2}{\frac{4}{x^2}}$

$= \frac{x^{-4}y^2}{1} \times \frac{x^2}{4}$

$= \frac{x^{-2}y^2}{4}$

$= \frac{y^2}{4x^2}$

5. $x^7 \cdot x^{-4} = x^{7+(-4)} = x^3$

9. $5^0 \times 5^{-3} = 5^{0+(-3)} = 5^{-3} = \frac{1}{5^3} = \frac{1}{125}$

13. $2\left(5an^{-2}\right)^{-1} = 2\times 5^{-1}a^{-1}n^{(-2)(-1)} = \frac{2n^2}{5a}$

17. $-7x^0 = -7\times 1 = -7$

21. $\left(7a^{-1}x\right)^3 = 7^{-3}a^3x^{-3} = \frac{a^3}{7^3x^3} = \frac{a^3}{343x^3}$

25. $\left(\frac{a}{b^{-2}}\right)^{-3} = \frac{3a^{-3}}{\left(b^{-2}\right)^{-3}} = \frac{\frac{3}{a^3}}{b^{(-2)(-3)}} = \frac{\frac{3}{a^3}}{b^6} = \frac{3}{a^3b^6}$

29. $3x^{-2} + 2y^{-2} = \frac{3}{x^2} + \frac{2}{y^2} = \frac{2x^2 + 3y^2}{x^2y^2}$

33. $\left(\frac{3a^2}{4b}\right)^{-3}\left(\frac{4}{a}\right)^{-5} = \frac{3^{-3}a^{-6}}{4^{-3}b^{-3}} \times \frac{4^{-5}}{a^{-5}}$

$= \frac{4^3b^3}{3^3a^6} \times \frac{a^5}{4^5} = \frac{b}{432a}$

37. $2a^{-2} + \left(2a^{-2}\right)^4 = \frac{2}{a^2} + 2^4a^{-8} = \frac{2}{a^2} + \frac{16}{a^8}$

$= \frac{2a^6 + 16}{a^8}$

41. $\left(R_1^{-1} + R_2^{-1}\right)^{-1} = \frac{1}{\frac{1}{R_1} + \frac{1}{R_2}} = \frac{R_1R_2}{R_1 + R_2}$

45. $\frac{6^{-1}}{4^{-2} + 2} = \frac{\frac{1}{6}}{\frac{1}{4^2} + 2} = \frac{\frac{1}{6}}{\frac{1}{16} + 2} \times \frac{48}{48}$

$= \frac{8}{3 + 96} = \frac{8}{99}$

49. $2t^{-2} + t^{-1}(t+1) = \frac{2}{t^2} + t^0 + t^{-1}$

$= \frac{2}{t^2} + 1 + \frac{1}{t}$

$= \frac{2 + t^2 + t}{t^2}$

53. If $x < 0$, then $x^2 > 0$ and $x^{-2} > 0$

If $x < 0$, then $\frac{1}{x} < 0$ and $x^{-1} < 0$.

$x^{-2} > 0 > x^{-1}$ or $x^{-2} > x^{-1}$.

Is it ever true that $x^{-2} < x^{-1}$? No.

57. (a) $\left(\frac{a}{b}\right)^{-n} = \frac{1}{\left(\frac{a}{b}\right)^n} = \frac{1}{\frac{a^n}{b^n}} = 1\times\frac{b^n}{a^n} = \left(\frac{b}{a}\right)^n$

(b) $\left(\frac{3.576}{8.091}\right) = (0.4419725)^{-7} = 303.55182$

$\left(\frac{8.091}{3.576}\right) = (2.26258)^7 = 303.55182$

61. $2^{5x} = 2^7\left(2^{2x}\right)^2 = 2^7\left(2^{4x}\right)$

$= 2^{7+4x} \Rightarrow 5x = 7 + 4x \Rightarrow x = 7$

65.

$$\begin{aligned}1\,\text{J} &= 1\ \text{kg}\times\left(\text{m}\times\text{s}^{-1}\right)^2\\ &= 1\ \text{kg}\times\text{m}^2\times\text{s}^{-2}\end{aligned}$$

$$\begin{aligned}\text{kg}\times\text{s}^{-1}\left(\text{m}\times\text{s}^{-2}\right)^2 &= \text{kg}\times\text{s}^{-1}\times\text{m}^2\times\text{m}^2\times\text{s}^{-4}\\ &= \text{kg}\times\text{m}^2\times\text{s}^{-2}\left(\text{s}^{-3}\right)=\text{J/s}^3\end{aligned}$$

69.

$$\begin{aligned}\frac{p(1+i)^{-1}\left[(1+i)^{-n}-1\right]}{(1+i)^{-1}-1} &= \frac{p(1+i)^{-n-1}-p(1+i)^{-1}}{(1+i)^{-1}-1}\\
&= \frac{p\left[(1+i)^{-n-1}-(1+i)^{-1}\right]}{\left(\frac{1}{1+i}-1\right)}\\
&= \frac{p\left[(1+i)^{-n-1}-(1+i)^{-1}\right]}{\frac{1-1-i}{1+i}}\\
&= \frac{p\left(\frac{1}{(1+i)^{n+1}}-\frac{1}{(1+i)^{1}}\right)}{\frac{-i}{1+i}}\\
&= \frac{p\left[(1+i)-(1+i)^{n+1}\right]}{(1+i)^{n+1}(1+i)}\times\frac{i+1}{-i}\\
&= \frac{p\left[(1+i)-(1+i)^{n+1}\right]}{-i(1+i)^{n+1}}\\
&= \frac{p\left[(1+i)-(1+i)^{n}(1+i)\right]}{-i(1+i)^{n+1}}\\
&= \frac{p(1+i)\left[1-(1+i)^{n}\right]}{-i(1+i)^{n}(1+i)}\\
&= \frac{p\left[1-(1+i)^{n}\right]}{-i(1+i)^{n}}\\
&= \frac{p\left[(1+i)^{n}-1\right]}{i(1+i)^{n}}\end{aligned}$$

11.2 Fractional Exponents

1. $8^{4/3}=\left(8^{1/3}\right)^4=\left(\sqrt[3]{8}\right)^4=2^4=16$

5. $25^{1/2}=\sqrt{25}=5$

9. $100^{25/2}=\left(100^{1/2}\right)^{25}=\left(\sqrt{100}\right)^{25}=10^{25}$

13. $64^{-2/3}=\dfrac{1}{\left(64^{1/3}\right)^2}=\dfrac{1}{\left(\sqrt[3]{64}\right)^2}=\dfrac{1}{4^2}=\dfrac{1}{16}$

17. $\left(3^6\right)^{2/3}=(3)^{12/3}=3^4=81$

21. $\dfrac{15^{2/3}}{5^2\times 15^{-1/3}}=\dfrac{15^{2/3+1/3}}{5^2}=\dfrac{15^1}{25}=\dfrac{3}{5}$

25.

$$\begin{aligned}125^{-2/3}-100^{-3/2} &= \frac{1}{\left(125^{1/3}\right)^2}-\frac{1}{\left(100^{1/2}\right)^3}\\
&= \frac{1}{\left(\sqrt[3]{125}\right)^2}-\frac{1}{\left(\sqrt{100}\right)^3}\\
&= \frac{1}{5^2}-\frac{1}{10^3}=\frac{1}{25}-\frac{1}{1000}\\
&= \frac{39}{1000}\end{aligned}$$

29. $17.98^{1/4}=2.059$

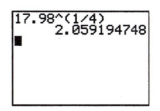

```
17.98^(1/4)
          2.059194748
■
```

33. $B^{2/3}\times B^{1/2}=B^{2/3+1/2}=B^{7/6}$

37. $\dfrac{x^{3/10}}{x^{-1/5}x^2}=x^{3/10+1/5-2}=x^{-3/2}=\dfrac{1}{x^{3/2}}$

41.

$$\begin{aligned}\left(16a^4b^3\right)^{-3/4} &= 16^{-3/4}a^{4(-3/4)}b^{3(-3/4)}\\
&= \frac{1}{16^{3/4}a^3b^{9/4}}=\frac{1}{8a^3b^{9/4}}\end{aligned}$$

45. $\dfrac{1}{2}\left(4x^2+1\right)^{-1/2}(8x)=\dfrac{4x}{\left(4x^2+1\right)^{1/2}}$

49. $$\left(T^{-1}+2T^{-2}\right)^{-1/2} = \frac{1}{\left(\frac{1}{T}+\frac{2}{T^2}\right)^{1/2}}$$
$$= \frac{1}{\left(\frac{T+2}{T^2}\right)} = \frac{1}{\frac{(T+2)^{1/2}}{\left(T^2\right)^{1/2}}}$$
$$= \frac{T}{(T+2)^{1/2}}$$

53. $$\left[\left(a^{1/2}-a^{-1/2}\right)^2+4\right]^{1/2} = \left[\left(a^{1/2}-\frac{1}{a^{1/2}}\right)^2+4\right]^{1/2}$$
$$= \left[\left(\frac{a-1}{a^{1/2}}\right)^2+4\right]^{1/2}$$
$$= \left[\frac{(a-1)^2}{a}+4\right]^{1/2}$$
$$= \left[\frac{a^2-2a+1+4a}{a}\right]^{1/2}$$
$$= \left[\frac{a^2+2a+1}{a}\right]^{1/2}$$
$$= \left[\frac{(a+1)^2}{a}\right]^{1/2} = \frac{a+1}{a^{1/2}}$$

57. $f(x) = 3x^{1/2}$

x	0	1	2	4	9
y	0	3	4.24	6	9

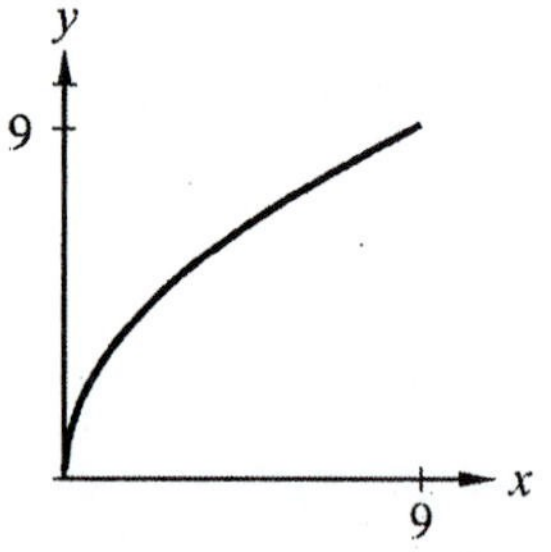

61. $$\left(x^{n-1} \div x^{n-3}\right)^{1/3} = \left(x^{n-1-(n-3)}\right)^{1/3}$$
$$= \left(x^2\right)^{1/3}$$
$$= \sqrt[3]{x^2}$$

65. $$T^2 = kR^3\left(1+\frac{d}{R}\right)^3$$
$$T^2 = kR^3\left(\frac{R+d}{R}\right)^3$$
$$= kR^3\frac{(R+d)^3}{R^3}$$
$$(R+d)^3 = \frac{T^2}{k} \Rightarrow R = -d\left(\frac{T^2}{k}\right)^{1/3}$$
$$= \frac{T^{2/3}}{k^{1/3}} - d$$

11.3 Simplest Radical Form

1. $\sqrt{a^3b^4} = \sqrt{a^2\left(b^2\right)^2 \cdot a} = ab^2\sqrt{a}$

5. $\sqrt{24} = \sqrt{4 \cdot 6} = \sqrt{4} \cdot \sqrt{6} = 2\sqrt{6}$

9. $\sqrt{x^2y^5} = \sqrt{x^2y^4y} = \sqrt{x^2}\sqrt{y^4}\sqrt{y} = xy^2\sqrt{y}$

13. $$\sqrt{18R^5TV^4} = \sqrt{9R^4V^4 \cdot 2RT}$$
$$= 3R^2V^2\sqrt{2RT}$$

17. $\sqrt[5]{96} = \sqrt[5]{32 \cdot 3} = \sqrt[5]{32}\sqrt[5]{3} = 2\sqrt[5]{3}$

21. $$\sqrt[4]{64r^3s^4t^5} = \sqrt[4]{16 \cdot 4s^4t^4r^3t}$$
$$= \sqrt[4]{16}\sqrt[4]{s^4}\sqrt[4]{t^4}\sqrt[4]{4r^3t}$$
$$= 2st\sqrt[4]{4r^3t}$$

25. $\sqrt[3]{P^3}\sqrt[3]{P^2V} = P\sqrt[3]{P^2V}$

29. $\sqrt[3]{\frac{3}{4}} = \sqrt[3]{\frac{48}{64}} = \frac{\sqrt[3]{48}}{\sqrt[3]{64}} = \frac{\sqrt[3]{8 \cdot 6}}{4} = \frac{2\sqrt[3]{6}}{4} = \frac{\sqrt[3]{6}}{2}$

33. $\sqrt[4]{400} = \sqrt[4]{2^4 \cdot 5^2} = 2 \cdot \sqrt[4]{25} = 2\sqrt{5}$

37. $\sqrt{4\times10^4} = \sqrt{4}\times = \sqrt{10^4} = 2\times10^2 = 200$

41. $\sqrt[4]{4a^2} = \left(4a^2\right)^{1/4} = \left(2^2\right)^{1/4}\left(a^2\right)^{1/4} = 2^{1/2}a^{1/2} = \sqrt{2a}$

45. $$\sqrt[4]{\sqrt[3]{16}} = \sqrt[4]{16^{1/3}} = \left(16^{1/3}\right)^{1/4}$$
$$= \left(16^{1/4}\right)^{1/3} = 2^{1/3} = \sqrt[3]{2}$$

49. $\sqrt{28u^3v^{-5}} = \sqrt{\frac{4u^2 \cdot 7u \cdot v}{v^6}} = \frac{2u\sqrt{7uv}}{v^3}$

53. $\sqrt{\frac{2x}{3c^4}} = \sqrt{\frac{2x}{3c^4} \cdot \frac{3}{3}} = \frac{\sqrt{6x}}{3c^2}$

57. $\sqrt{xy^{-1} + x^{-1}y} = \sqrt{\frac{x}{y} + \frac{y}{x}} = \sqrt{\frac{x^2 + y^2}{xy}}$

$= \sqrt{\frac{x^2 + y^2}{xy} \cdot \frac{xy}{xy}} = \sqrt{\frac{xy\left(x^2 + y^2\right)}{x^2y^2}}$

$= \frac{\sqrt{xy\left(x^2 + y^2\right)}}{xy}$

61. $\sqrt{a^2 + b^2}$ cannot be simplified any further.

65. $\sqrt{a} = a^{1/2} = a^{3/6}$

$\sqrt[3]{b} = b^{1/3} = b^{2/6}$

$\sqrt[6]{c} = c^{1/6}$

69. $a\sqrt{\frac{2g}{a}} = a\sqrt{\frac{2g}{a} \cdot \frac{a}{a}} = a\sqrt{\frac{2ag}{a^2}}$

$= \frac{a}{a}\sqrt{2ag} = \sqrt{2ag}$

11.4 Addition and Subtraction of Radicals

1. $3\sqrt{125} - \sqrt{20} + \sqrt{45} = 3\sqrt{25(5)} - \sqrt{4(5)} + \sqrt{9(5)}$

$= 15\sqrt{5} - 2\sqrt{5} + 3\sqrt{5}$

$= 15\sqrt{5} + \sqrt{5}$

$= 16\sqrt{5}$

5. $\sqrt{28} + \sqrt{5} - 3\sqrt{7} = 2\sqrt{7} + \sqrt{5} - 3\sqrt{7}$

$= \sqrt{5} + (2-3)\sqrt{7}$

$= \sqrt{5} - \sqrt{7}$

9. $2\sqrt{3t^2} - 3\sqrt{12t^2} = 2t\sqrt{3} - 3t\sqrt{3(4)}$

$= 2t\sqrt{3} - 3(2)t\sqrt{3} = (2-6)t\sqrt{3}$

$= -4t\sqrt{3}$

13. $2\sqrt{28} + 3\sqrt{175}$

$= 2\sqrt{4(7)} + 3\sqrt{25(7)} + 2(2)\sqrt{7} + 3(5)\sqrt{7}$

$(4+15)\sqrt{7} = 19\sqrt{7}$

17. $3\sqrt{75R} + 2\sqrt{48R} - 2\sqrt{18R}$

$= 3\sqrt{25(3)R} + 2\sqrt{16(3)R} - 2\sqrt{9(2)R}$

$= 3(5)\sqrt{3R} + 2(4)\sqrt{3R} - 2(3)\sqrt{2R}$

$= (15+8)\sqrt{3R} - 6\sqrt{2R}$

$= 23\sqrt{3R} - 6\sqrt{2R}$

21. $\sqrt{\frac{1}{2}} + \sqrt{\frac{25}{2}} - 4\sqrt{18} = \sqrt{\frac{1(2)}{2(2)}} + \sqrt{\frac{25(2)}{4}} - 4\sqrt{(9)2}$

$= \sqrt{\frac{2}{4}} - \sqrt{\frac{25(2)}{4}} - 12\sqrt{2}$

$= \left(\frac{1}{2} + \frac{5}{2} - 12\right)\sqrt{2} = -9\sqrt{2}$

25. $\sqrt[4]{32} - \sqrt[8]{4} = \sqrt[4]{16(2)} - \sqrt[8]{2(2)}$

$= 2\sqrt[4]{2} - 2^{2/8} = 2\sqrt[4]{2} - 2^{1/4}$

$= 2\sqrt[4]{2} - \sqrt[4]{2} = (2-1)\sqrt[4]{2}$

$= \sqrt[4]{2}$

29. $\sqrt{6}\sqrt{5}\sqrt{3} - \sqrt{40a^2} = \sqrt{90} - \sqrt{4a^2(10)}$

$= \sqrt{9(10)} - 2a\sqrt{10}$

$= (3-2a)\sqrt{10}$

33. $\sqrt{\frac{a}{c^5}} - \sqrt{\frac{c}{a^3}} = \sqrt{\frac{a(c)}{c^5(c)}} - \sqrt{\frac{ca}{a^3(a)}}$

$= \sqrt{\frac{ac}{c^6}} - \sqrt{\frac{ac}{a^4}}$

$= \left(\frac{1}{c^3} - \frac{1}{a^2}\right)\sqrt{ac}$

$= \frac{\left(a^2 - c^3\right)\sqrt{ac}}{a^2c^3}$

37. $\sqrt{\frac{T-V}{T+V}}-\sqrt{\frac{T+V}{T-V}}$

$=\sqrt{\frac{(T-V)(T+V)}{(T+V)^2}}-\sqrt{\frac{(T+V)(T-V)}{(T-V)^2}}$

$=\frac{\sqrt{T^2-V^2}}{T+V}-\frac{\sqrt{T^2-V^2}}{T-V}$

$=\frac{(T-V)\sqrt{T^2-V^2}-(T+V)\sqrt{T^2-V^2}}{(T+V)(T-V)}$

$=\frac{(T-V-T-V)\sqrt{T^2-V^2}}{T^2-V^2}$

$=\frac{-2V\sqrt{T^2-V^2}}{T^2-V^2}$

$=\frac{-2V\sqrt{T^2-V^2}}{T^2-V^2}$

$=\frac{2V\sqrt{T^2-V^2}}{V^2-T^2}$

41. $2\sqrt{\frac{2}{3}}+\sqrt{24}-5\sqrt{\frac{3}{2}}=2\sqrt{\frac{2(3)}{3(3)}}+\sqrt{4(6)}-5\sqrt{\frac{3(2)}{2(2)}}$

$=2\sqrt{\frac{6}{9}}+2\sqrt{6}-5\sqrt{\frac{6}{4}}$

$=\frac{2}{3}\sqrt{6}+2\sqrt{6}-\frac{5}{2}\sqrt{6}$

$=\frac{1}{6}\sqrt{6}=0.40824829$

45. $x^2-2x-2=0$ has roots

$x=\frac{2\pm\sqrt{(-2)^2-4(1)(-2)}}{2}=\frac{2\pm\sqrt{12}}{2}$

$x=\frac{2\pm2\sqrt{3}}{2}=1\pm\sqrt{3}$

$x=1+\sqrt{3}$ is the positive root

$x^2+2x-11=0$ has roots

$x=\frac{-2\pm\sqrt{2^2-4(1)(-11)}}{2(1)}=\frac{-2\pm\sqrt{48}}{2}$

$x=\frac{-2\pm4\sqrt{3}}{2}$

$x=-1+2\sqrt{3}$ is the positive root

sum of positive roots $=1+\sqrt{3}-1+2\sqrt{3}=3\sqrt{3}$

49.

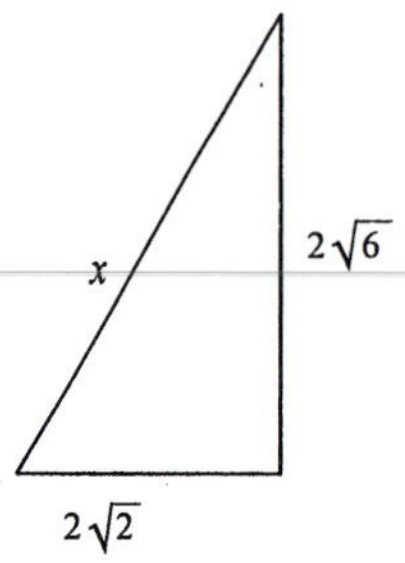

$x^2=\left(2\sqrt{2}\right)^2+\left(2\sqrt{6}\right)^2$

$x^2=8+24$

$x^2=32$

$x=\sqrt{32}$

Perimeter $=\sqrt{32}+2\sqrt{2}+2\sqrt{6}$

$=\sqrt{16(2)}+2\sqrt{2}+2\sqrt{6}$

$=4\sqrt{2}+2\sqrt{2}+2\sqrt{6}$

$=6\sqrt{2}+2\sqrt{6}$ units

11.5 Multiplication and Division of Radicals

1. $\sqrt{2}\left(3\sqrt{5}-4\sqrt{8}\right)=3\sqrt{10}-4\sqrt{16}$

$=3\sqrt{10}-4(4)$

$=3\sqrt{10}-16$

5. $\sqrt{3}\sqrt{10}=\sqrt{3(10)}=\sqrt{30}$

9. $\sqrt[3]{4}\cdot\sqrt[3]{2}=\sqrt[3]{4(2)}=\sqrt[3]{8}=2$

13. $\sqrt{8}\cdot\sqrt{\frac{5}{2}}=\sqrt{\frac{40}{2}}=\sqrt{20}=\sqrt{4\cdot5}=2\sqrt{5}$

17. $\left(2-\sqrt{5}\right)\left(2+\sqrt{5}\right)=2^2-\sqrt{5}^2=4-5=-1$

21. $\left(3\sqrt{11}-\sqrt{x}\right)\left(2\sqrt{11}+5\sqrt{x}\right)$

$=6\sqrt{11}^2+15\sqrt{11x}-2\sqrt{11x}-5\sqrt{x}^2$

$=6\cdot11+13\sqrt{11x}-5x$

$=66+13\sqrt{11x}-5x$

25. $\dfrac{\sqrt{6}-3}{\sqrt{6}}=\dfrac{\sqrt{6}-3}{\sqrt{6}}\cdot\dfrac{\sqrt{6}}{\sqrt{6}}\dfrac{\sqrt{6}^2-3\sqrt{6}}{\sqrt{6}^2}$

$=\dfrac{6-3\sqrt{6}}{6}=\dfrac{2-\sqrt{6}}{2}$

29. $\sqrt{2}\sqrt[3]{3}=2^{1/2}3^{1/3}=2^{3/6}3^{2/6}=\left(2^3 3^2\right)^{1/6}$

$=\sqrt[6]{2^3 3^2}=\sqrt[6]{72}$

33. $\dfrac{\sqrt{2}-1}{\sqrt{7}-3\sqrt{2}}=\dfrac{\sqrt{2}-1}{\sqrt{7}-3\sqrt{2}}\cdot\dfrac{\sqrt{7}+3\sqrt{2}}{\sqrt{7}+3\sqrt{2}}$

$=\dfrac{\sqrt{14}+3\sqrt{2}^2-\sqrt{7}-3\sqrt{2}}{\sqrt{7}^2-3^2\sqrt{2}^2}$

$=\dfrac{\sqrt{14}+3\cdot 2-\sqrt{7}-3\sqrt{2}}{7-9\cdot 2}$

$=\dfrac{\sqrt{14}+6-\sqrt{7}-3\sqrt{2}}{7-18}$

$=-\dfrac{\sqrt{14}+6-\sqrt{7}-3\sqrt{2}}{11}$

37. $\dfrac{2\sqrt{x}}{\sqrt{x}-\sqrt{5}}=\dfrac{2\sqrt{x}}{\sqrt{x}-\sqrt{5}}\cdot\dfrac{\sqrt{x}+\sqrt{5}}{\sqrt{x}+\sqrt{5}}$

$=\dfrac{2\sqrt{x}^2+2\sqrt{x}\sqrt{5}}{\sqrt{x}^2-\sqrt{5}^2}=\dfrac{2x+2\sqrt{5x}}{x-5}$

41. $\left(\sqrt{\dfrac{2}{R}}+\sqrt{\dfrac{R}{2}}\right)\left(\sqrt{\dfrac{2}{R}}-2\sqrt{\dfrac{R}{2}}\right)$

$=\dfrac{2}{R}-2\sqrt{\dfrac{2R}{2R}}+\sqrt{\dfrac{2R}{2R}}-2\left(\dfrac{R}{2}\right)$

$=\dfrac{2}{R}-2+1-R=\dfrac{2-2R+R-R^2}{R}$

$=\dfrac{2-R-R^2}{R}$

45. $\dfrac{\sqrt{a}+\sqrt{a-2}}{\sqrt{a}-\sqrt{a-2}}\cdot\dfrac{\sqrt{a}+\sqrt{a-2}}{\sqrt{a}+\sqrt{a-2}}$

$=\dfrac{a+2\sqrt{a}\sqrt{a-2}+a-2}{a-(a-2)}$

$=\dfrac{2a+2\sqrt{a}\sqrt{a-2}-2}{2}$

$=a+\sqrt{a}\sqrt{a-2}-1$

$=a-1+\sqrt{a(a-2)}$

49. $\dfrac{2\sqrt{6}-\sqrt{5}}{3\sqrt{6}-4\sqrt{5}}=\dfrac{2\sqrt{6}-\sqrt{5}}{3\sqrt{6}-4\sqrt{5}}\cdot\dfrac{3\sqrt{6}+4\sqrt{5}}{3\sqrt{6}+4\sqrt{5}}$

$=\dfrac{6(6)+8\sqrt{30}-3\sqrt{30}-4(5)}{9(6)-16(5)}$

$=-\dfrac{16+5\sqrt{30}}{26}=-1.6686962$

53. $\dfrac{x^2}{\sqrt{2x+1}}+2x\sqrt{2x+1}$

$=\dfrac{x^2}{\sqrt{2x+1}}+\dfrac{2x\sqrt{2x+1}\sqrt{2x+1}}{\sqrt{2x+1}}$

$=\dfrac{x^2+2x\sqrt{2x+1}}{\sqrt{2x+1}}=\dfrac{x^2+4x^2+2x}{\sqrt{2x+1}}$

$=\dfrac{5x^2+2x}{\sqrt{2x+1}}$

57. $\dfrac{\sqrt{x+h}-\sqrt{x}}{h}=\dfrac{\sqrt{x+h}-\sqrt{x}}{h}\cdot\dfrac{\sqrt{x+h}+\sqrt{x}}{\sqrt{x+h}+\sqrt{x}}$

$=\dfrac{x+h-x}{h\sqrt{x+h}+h\sqrt{x}}$

$=\dfrac{h}{h\left(\sqrt{x+h}+\sqrt{x}\right)}$

$=\dfrac{1}{\sqrt{x+h}+\sqrt{x}}$

61. $\dfrac{-b+\sqrt{b^2-4ac}}{2a}=\dfrac{1}{\dfrac{-b-\sqrt{b^2-4ac}}{2a}}$

$\dfrac{-b+\sqrt{b^2-4ac}}{2a}=\dfrac{2a}{-b-\sqrt{b^2-4ac}}$

$\left(-b+\sqrt{b^2-4ac}\right)\left(-b-\sqrt{b^2-4ac}\right)=4a^2$

$b^2b\sqrt{b^2-4ac}-b\sqrt{b^2-4ac}-\left(b^2-4ac\right)=4a^2$

$b^2-b^2+4ac=4a^2$

$c=a$

65. $m^2+bm+k^2=0;\ m=\frac{1}{2}\left(\sqrt{b^2-4k^2}-b\right)$

$$\left(\frac{1}{2}\sqrt{b^2-4k^2}-b\right)^2+b\left(\frac{1}{2}\sqrt{b^2-4k^2}-b\right)+k^2$$
$$=\frac{1}{4}\left(b^2-4k^2\right)-\frac{b}{2}\sqrt{b^2-4k^2}$$
$$+\frac{1}{4}b^2+\frac{b}{2}\sqrt{b^2-4k^2}-\frac{1}{2}b^2+k^2=0$$

69. $$\frac{2Q}{\sqrt{\sqrt{2}-1}}=\frac{2Q}{\sqrt{\sqrt{2}-1}}\cdot\frac{\sqrt{\sqrt{2}+1}}{\sqrt{\sqrt{2}+1}}$$
$$=\frac{2Q\sqrt{\sqrt{2}+1}}{\sqrt{2}-1}$$
$$=2Q\sqrt{\sqrt{2}+1}$$

Chapter 11 Review Exercises

1. $2a^{-2}b^0=2a^{-2}\cdot 1=2\cdot\frac{1}{a^2}=\frac{2}{a^2}$

5. $3(25)^{3/2}=3\left[(25)^{1/2}\right]^3=3[25]^3$
$$=3[5]^3=3\cdot 125=375$$

9. $$\left(\frac{3}{t^2}\right)^{-2}=\frac{1}{\left(\frac{3}{t^2}\right)^2}=\frac{1}{\frac{3^2}{\left(t^2\right)^2}}=\frac{1}{\frac{9}{t^4}}=\frac{t^4}{9}$$

13. $\left(2a^{1/3}b^{5/6}\right)^6=2^6\cdot\left(a^{1/3}\right)^6\cdot\left(b^{5/6}\right)^6$
$$=64\cdot a^{6/3}\cdot b^{5/6\cdot 6}$$
$$=64a^2b^5$$

17. $2L^{-2}-4C^{-1}=\frac{2}{L^2}-\frac{4}{C}=\frac{2C-4L^2}{L^2C}$

21. $\left(a-3b^{-1}\right)^{-1}=\frac{1}{\left(a-3b^{-1}\right)}=\frac{1}{a-\frac{3}{b}}\cdot\frac{b}{b}=\frac{b}{ab-3}$

25. $\left(W^2+2WH+H^2\right)^{-1/2}=\frac{1}{\left((W+H)^2\right)^{1/2}}=\frac{1}{W+H}$

29. $\sqrt{68}=\sqrt{4\cdot 17}=\sqrt{4}\cdot\sqrt{17}=2\sqrt{17}$

33. $\sqrt{9a^3b^4}=\sqrt{9a^2b^4\cdot a}=3ab^2\sqrt{a}$

37. $\frac{5}{\sqrt{2s}}=\frac{5}{\sqrt{2s}}\cdot\frac{\sqrt{2s}}{\sqrt{2s}}=\frac{5\sqrt{2s}}{2s}$

41. $\sqrt[4]{8m^6n^9}=\sqrt[4]{8m^4\cdot m^2\cdot n^8\cdot n}=mn^2\sqrt[4]{8m^2n}$

45. $\sqrt{36+4}=-2\sqrt{10}=\sqrt{4(10)}-2\sqrt{10}$
$$=2\sqrt{10}-2\sqrt{10}$$
$$=0$$

49. $a\sqrt{2x^3}+\sqrt{8a^2x^3}=a\sqrt{2x^2\cdot x}+\sqrt{4\cdot 2\cdot a^2x^2\cdot x}$
$$=ax\sqrt{2x}+2ax\sqrt{2x}=3ax\sqrt{2x}$$

53. $5\sqrt{5}\left(6\sqrt{5}-35\right)=30\sqrt{5}^2-5\sqrt{5}\sqrt{35}$
$$=30\cdot 5-5\sqrt{175}=150-5\sqrt{25(7)}$$
$$=150-25\sqrt{7}$$

57. $\left(2-3\sqrt{17B}\right)\left(3+\sqrt{17B}\right)$
$$=6+2\sqrt{17B}-9\sqrt{17B}-3\sqrt{17B}$$
$$=6-51B-7\sqrt{17B}$$

61. $$\frac{\sqrt{3x}}{2\sqrt{3x}-\sqrt{y}}=\frac{\sqrt{3x}}{\left(2\sqrt{3x}-\sqrt{y}\right)}\cdot\frac{\left(2\sqrt{3x}+\sqrt{y}\right)}{\left(2\sqrt{3x}+\sqrt{y}\right)}$$
$$=\frac{2\sqrt{3x}^2+\sqrt{3x}\cdot\sqrt{y}}{4\sqrt{3x}^2-\sqrt{y}^2}=\frac{2\cdot 3x+\sqrt{3xy}}{4\cdot 3x-y}$$
$$=\frac{6x+\sqrt{3xy}}{12x-y}$$

65. $$\frac{\sqrt{7}-\sqrt{5}}{\sqrt{5}+3\sqrt{7}}=\frac{\left(\sqrt{7}-\sqrt{5}\right)}{\left(\sqrt{5}+3\sqrt{7}\right)}\cdot\frac{\left(\sqrt{5}-3\sqrt{7}\right)}{\left(\sqrt{5}-3\sqrt{7}\right)}$$
$$=\frac{\sqrt{35}-3\cdot 7-5+3\sqrt{35}}{5-9\cdot 7}$$
$$=\frac{4\sqrt{35}-26}{5-63}=\frac{4\sqrt{35}-26}{-58}$$

69. $\sqrt{4b^2+1}$ is in simplest form

73. $\left(1+6^{1/2}\right)\left(3^{1/2}+2^{1/2}\right)\left(3^{1/2}-2^{1/2}\right)$

$=\left(1+6^{1/2}\right)\left(\left(3^{1/2}\right)^2-\left(2^{1/2}\right)^2\right)$

$=\left(1+6^{1/2}\right)(3-2)$

$=1+6^{1/2}$

77. $\sqrt{3+n}\left(\sqrt{3+n}-\sqrt{n}\right)^{-1}=\dfrac{\sqrt{3+n}}{\sqrt{3+n}-\sqrt{n}}\cdot\dfrac{\sqrt{3+n}+\sqrt{n}}{\sqrt{3+n}+\sqrt{n}}$

81. $\sqrt{\sqrt{2}-1}\left(\sqrt{2}+1\right)=\sqrt{\sqrt{2}-1}\left(\sqrt{2}+1\right)\cdot\dfrac{\sqrt{\sqrt{2}+1}}{\sqrt{\sqrt{2}+1}}$

$=\dfrac{\sqrt{2}+1}{\sqrt{\sqrt{2}+1}}\cdot\dfrac{\sqrt{\sqrt{2}+1}}{\sqrt{\sqrt{2}+1}}$

$=\dfrac{\left(\sqrt{2}+1\right)\sqrt{\sqrt{2}+1}}{\left(\sqrt{2}+1\right)}$

$=\sqrt{\sqrt{2}+1}$

85. $3x^2-2x+5=3\left(\frac{1}{2}\left(2-\sqrt{3}\right)\right)^2-2\left(\frac{1}{2}\left(2-\sqrt{3}\right)\right)+5$

$=\frac{3}{4}\left(4-4\sqrt{3}+3\right)-2+\sqrt{3}+5$

$=3-3\sqrt{3}+\frac{9}{4}+3+\sqrt{3}$

$=\frac{33}{4}-2\sqrt{3}$

89. (a) $v=k\sqrt[3]{\dfrac{P}{W}}=k\left(\dfrac{P}{W}\right)^{1/3}$

(b) $v=k\sqrt[3]{\dfrac{P}{W}\cdot\dfrac{W^2}{W^2}}=\dfrac{k}{W}\sqrt[3]{PW^2}$

93. $\left(1-\dfrac{v^2}{c^2}\right)^{-1/2}=\dfrac{1}{\left(1-\dfrac{v^2}{c^2}\right)^{1/2}}\cdot\dfrac{\left(1-\dfrac{v^2}{c^2}\right)^{1/2}}{\left(1-\dfrac{v^2}{c^2}\right)^{1/2}}$

$=\dfrac{\left(1-\dfrac{v^2}{c^2}\right)^{1/2}}{1-\dfrac{v^2}{c^2}}\cdot\dfrac{c^2}{c^2}$

$=\dfrac{c^2\left(1-\dfrac{v^2}{c^2}\right)^{1/2}}{c^2-v^2}$

97. $\sqrt{3^2+3^2}+\sqrt{2^2+2^2}+\sqrt{1^2+1^2}=\sqrt{18}+\sqrt{8}+\sqrt{2}$

$=\sqrt{9\cdot2}+\sqrt{4\cdot2}+\sqrt{2}$

$=3\sqrt{2}+2\sqrt{2}+\sqrt{2}$

$=6\sqrt{2}$ cm

Chapter 12

COMPLEX NUMBERS

12.1 Basic Definitions

1. $-\sqrt{-3}\sqrt{-12} = -j\sqrt{3}\left(j\sqrt{12}\right) = -j^2\sqrt{36}$
$= -(-1)6$
$= 6$

5. $\sqrt{-81} = \sqrt{81(-1)} = \sqrt{81}\sqrt{-1} = 9j$

9. $\sqrt{-0.36} = \sqrt{0.36(-1)} = \sqrt{0.36}\sqrt{-1} = 0.6j$

13. $\sqrt{-\frac{7}{4}} = \frac{\sqrt{-7}}{\sqrt{4}} = \frac{\sqrt{7(-1)}}{2} = \frac{\sqrt{7}\sqrt{-1}}{2} = j\frac{\sqrt{7}}{2}$

17. (a) $\left(\sqrt{-7}\right)^2 = \left(\sqrt{7(-1)}\right)^2 = \left(\sqrt{7}\cdot\sqrt{-1}\right) = \left(\sqrt{7}\cdot j\right)^2$
$= \sqrt{7}^2\cdot j^2 = 7(-1) = -7$

(b) $\left(\sqrt{-7}\right)^2 = \sqrt{49} = 7$

21. $\sqrt{-\frac{1}{15}}\sqrt{-\frac{27}{5}} = j\sqrt{\frac{1}{15}}j\sqrt{\frac{27}{5}} = j^2\sqrt{\frac{27}{75}}$
$= -\sqrt{\frac{9(3)}{25(3)}} = -\frac{3}{5}$

25. (a) $-j^6 = -\left(j^2\right)^3 = -(-1)^3 = -(-1) = 1$

(b) $(-j)^6 = \left((-j)^2\right)^3 = (-1)^3 = -1$

29. $j^{15} - j^{13} = j^{12}\cdot j^3 - j^{12}\cdot j = (1)(-j) - (1)(j)$
$= -j - j = -2j$

33. $2 + \sqrt{-9} = 2 + \sqrt{9(-1)} = 2 + 3j$

37. $\sqrt{-4j^2} + \sqrt{-4} = \sqrt{(-4)(-1)} + 2j = 2 + 2j$

41. $\sqrt{18} - \sqrt{-8} = \sqrt{9\cdot 2} - \sqrt{4\cdot 2}j$
$= 3\sqrt{2} - 2j\sqrt{2}$

45. (a) the conjugate of $6 - 7j$ is $6 + 7j$
(b) the conjugate of $8 + j$ is $8 - j$

49. $7x - 2yj = 14 + 4j \Rightarrow 7x = 14$
$x = 2$
and $-2y = 4$
$y = -2$

53. $x - 2j^2 + 7y = yj + 2xj^3$
$x - 2(-1) + 7j = yj + 2x(-j)$
$x + 2 + 7j = 0 + (y - 2x)j$
$x + 2 = 0$ and $7 = y - 2x = y - 2(-2)$
$x = -2$ $\quad 7 = y + 4$
$y = 3$

57. No. $x^4 + 16 = 0$
$2j$ is not a solution since
$(2j)^4 + 16 = 0$
$16j^4 + 16 = 0$
$16(1) + 16 = 0$
$32 \neq 0.$
$-2j$ is not a solution since
$(-2j)^4 + 16 = 0$
$16j^4 + 16 = 0$
$16(1) + 16 = 0$
$32 \neq 0.$

61. For a complex number and its conjugate to be equal, it must be a real number. $a + 0j = a - 0j.$

12.2 Basic Operations with Complex Numbers

1. $(7 - 9j) - (6 - 4j) = 7 - 9j - 6 + 4j = 1 - 5j$

5. $(3 - 7j) + (2 - j) = (3 + 2) + (-7 - 1)j = 5 - 8j$

9. $0.23-(0.46-0.9j)+0.67j$
$=0.23-0.46+0.19j+0.67j$
$=-0.23+0.86j$

13. $(7-j)(j)=49j-7j^2=49j-7(-1)=7+49j$

17. $\sqrt{-18}\sqrt{-4}(3j)=(3\sqrt{2}j)(2j)(3j)$
$=18\sqrt{2}j^3=18\sqrt{2}j^2j$
$=18\sqrt{2}(-1)j=-18j\sqrt{2}$

21. $j\sqrt{-7}-j^6\sqrt{112}+3j=j\sqrt{7(-1)}-j^6\sqrt{16(7)}+3j$
$=j^2\sqrt{7}-4j^6\sqrt{7}+3j$
$=(-1)\sqrt{7}-4j^4(j^2)\sqrt{7}+3j$
$=-\sqrt{7}-4(1)(-1)\sqrt{7}+3j$
$=-\sqrt{7}+4\sqrt{7}+3j$
$=3\sqrt{7}+3j$

25. $(1-j)^3=(1-j)(1-j)^2=(1-j)(1-2j+j^2)$
$=1-2j+j^2-j+2j^2-j^3$
$=1-3j+3j^2-j^3$
$=1-3j+3(-1)-(-1)j$
$=-2-2j$

29. $\frac{1-j}{3j}\cdot\frac{3j}{3j}=\frac{3j-3j^2}{9(j^2)}=\frac{3j-3(-1)}{9(-1)}=\frac{3j+3}{-9}$
$=-\frac{1}{3}(1+j)$

33. $\frac{j^2-j}{2j-j^8}=\frac{-1-j}{2j-1}\cdot\frac{-2j-1}{-2j-1}=\frac{1+3j+2j^2}{1^2+2^2}$
$=\frac{1+3j-2}{5}=\frac{-1+3j}{5}$

37. $(4j^5-5j^4+2j^3-3j^2)^2=(4j-5-2j+3)^2$
$=(-2+2j)^2=4-8j-4$
$=-8j$

41. $\frac{(2-j^3)^4}{(j^8-j^6)}+j=\frac{(2+j)^4}{(1+1)^3}+j=\frac{(2+j)^4}{8}+j$
$=\frac{(2+j)(2+j)^2}{8}+j$
$=\frac{(3+4j)(3+4j)}{8}+j=\frac{-7+24i}{8}+j$
$=-\frac{7}{8}+4j$

45. Multiply $-3+j$ by its conjugate.
$(-3+j)(-3-j)=9-j^2$
$=9-(-1)=10$

49. $j^{-2}+j^{-3}=\frac{1}{j^2}+\frac{1}{j^3}=\frac{1}{-1}+\frac{1}{-j}\cdot\frac{j}{j}$
$=-1+\frac{j}{-j^2}$
$=-1+j$

53. $f(x)=x+\frac{1}{x}$
$f(1+3j)=1+3j+\frac{1}{1+3j}\cdot\frac{1-3j}{1-3j}$
$=1+3j+\frac{1}{10}-\frac{3}{10}j$
$=\frac{11}{10}+\frac{27}{10}j$

57. $E=85+74j;\ Z=2500-1200j$
$I=\frac{85+74j}{2500-1200j}\cdot\frac{2500+1200j}{2500+1200j}$
$=\frac{212500+88800j^2+(102000+185000)j}{6250000-1440000j^2}$
$=\frac{212500+88800(-1)+287000j}{6250000-1440000(-1)}$
$=\frac{123700+287000j}{7690000}$

61. $(a+bj)(a-bj)=a^2-b^2j^2=a^2+b^2$, a positive real number.

12.3 Graphical Representation of Complex Numbers

1. Add $5-2j$ and $-2+j$ graphically.

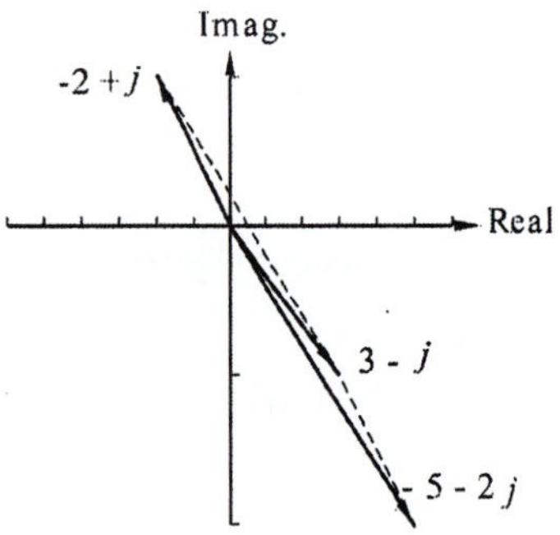

$5-2j+(-2+j)=3-j$

$2+3j-(4-2j)=-2+5j$

5. $-4-3j$

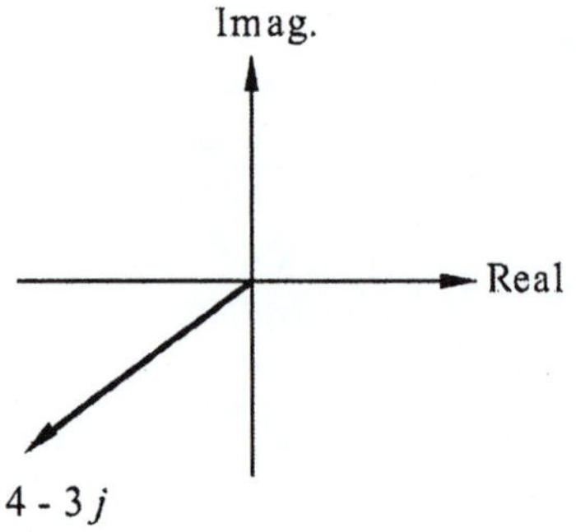

9. $2+3+4j=5+4j$

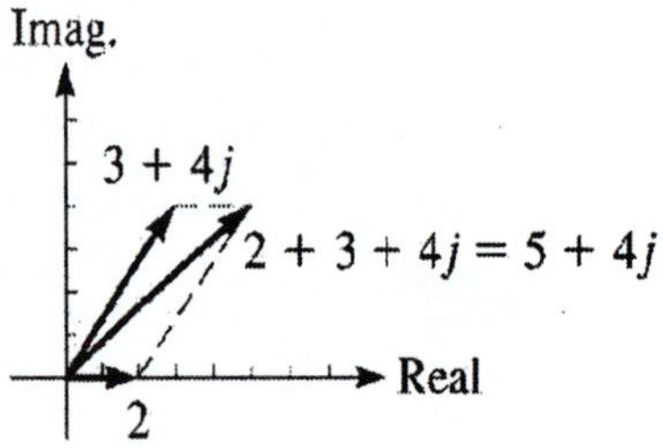

13. $5j-1(1-4j)=5j-1+4j=-1+9j$

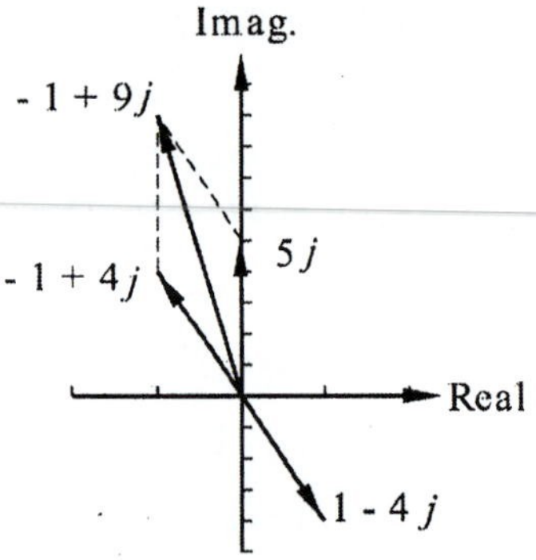

17. $(3-2j)-(4-6j)=3-2j-4+6j=-1+4j$

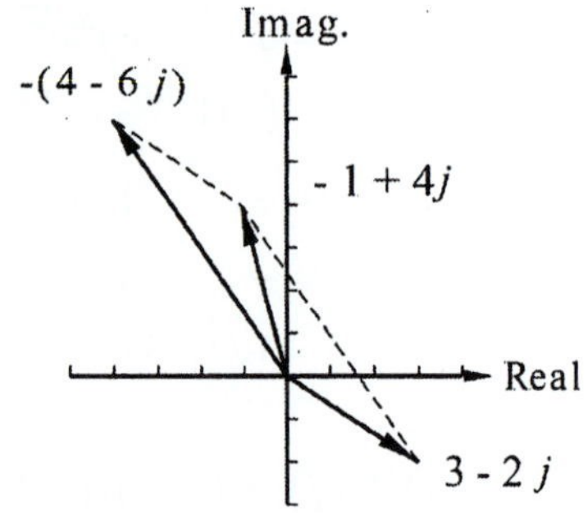

21. $(1.5-0.5j)+(3.0+2.5j)=1.5-0.5j+3.0+2.5j$

$=4.5+2.0j$

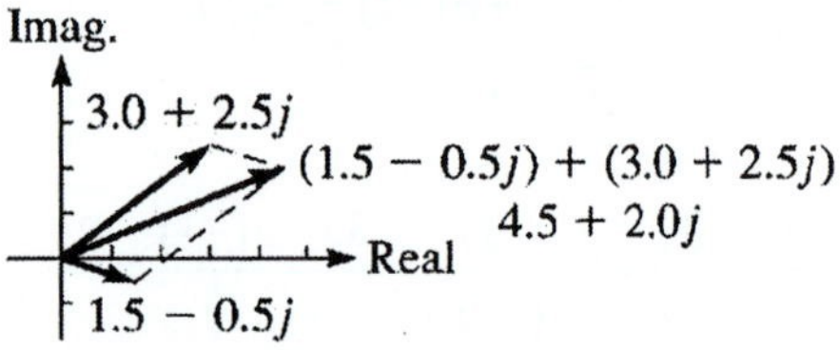

25. $(2j+1)-3j-(j+1)=(2j+1)+(-3j)+(-(j+1))$

$=2j+1-3j-j-1$

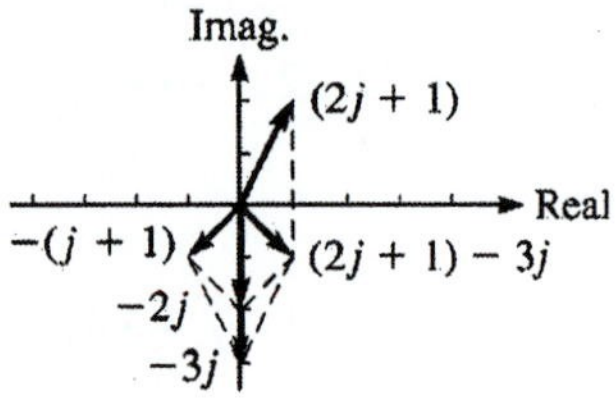

29. $3+2j$

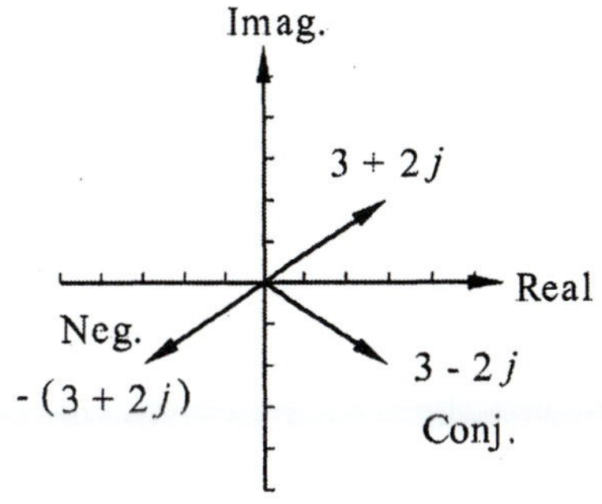

33.
$$a+bj=3-j$$
$$3(a+bj)=9-3j$$
$$-3(a+bj)=-9+3j$$

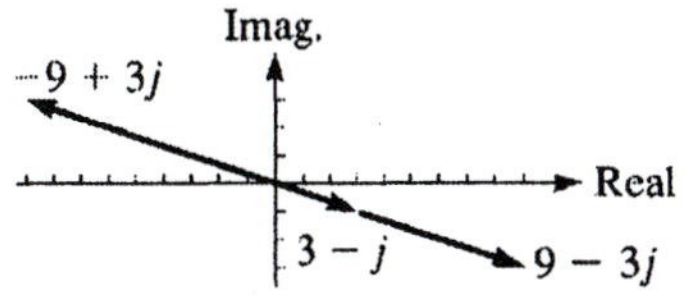

37. $2+4j-(2-4j)=2+4j-2+4j=2(4j)=8j$
The difference of a complex number and its conjugate is twice the imaginary part times the imaginary unit.

12.4 Polar Form of a Complex Number

1. $-3+4j \Rightarrow x=-3,\ y=4 \Rightarrow r=\sqrt{(-3)^2+4^2}=5$

$\tan\theta_{ref}=\frac{4}{3},\ \theta_{ref}=53.1^\circ,$

$\theta=180^\circ-53.1^\circ=126.9^\circ$

$5(\cos 126.9^\circ+j\sin 126.9^\circ)$

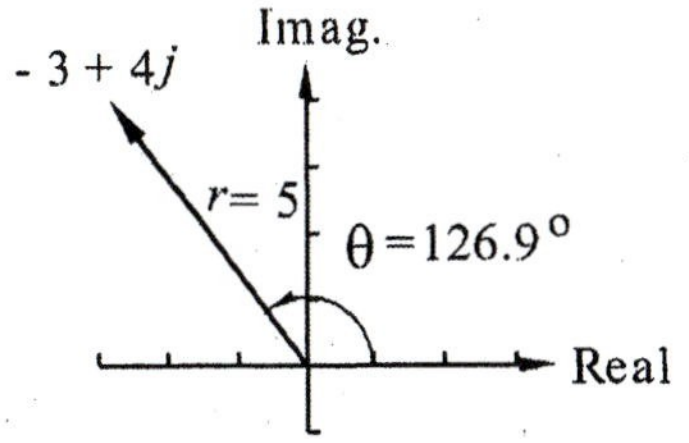

5. $30-40j$

$$r=\sqrt{30^2+(-40)^2}$$
$$=50$$
$$\tan\theta_{ref}=\left|\frac{-40}{30}\right|$$
$$=1.33$$
$$\theta_{ref}=53.1^\circ$$
$$\theta=306.9^\circ$$
$$50(\cos 306.9^\circ+j\sin 306.9^\circ)$$

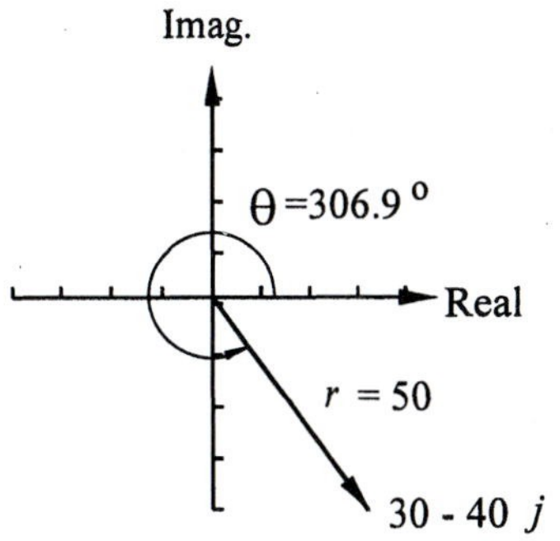

9. $-0.55-0.24j \Rightarrow x=-0.55,\ y=-0.24$

$r=\sqrt{(-0.55)^2+(-0.24)^2}=0.60$

$\tan\theta_{ref}=\frac{0.24}{0.55},\ \theta_{ref}=24^\circ, \theta=180^\circ+24^\circ=204^\circ$

$0.60(\cos 204^\circ+j\sin 204^\circ)$

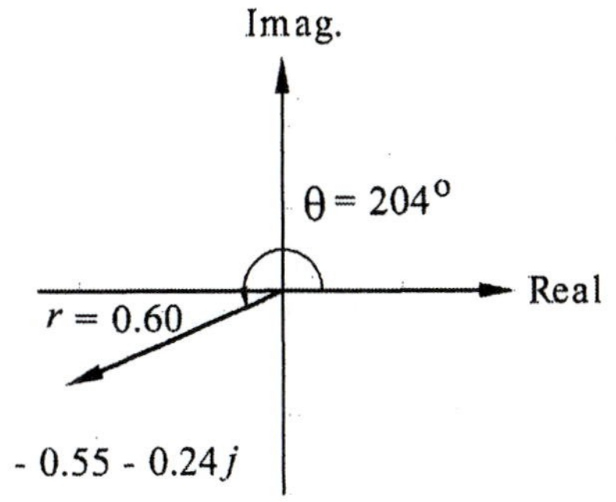

13. $3.514 - 7.256j$

$r = \sqrt{(3.514)^2 + (-7.256)^2} = 8.062$

$\tan\theta_{\text{ref}} = \left|\frac{-7.256}{3.514}\right| = 2.065$

$\theta_{\text{ref}} = 64.16^\circ$

$\theta = 295.84^\circ$

$8.062(\cos 295.84^\circ + j \sin 295.84^\circ)$

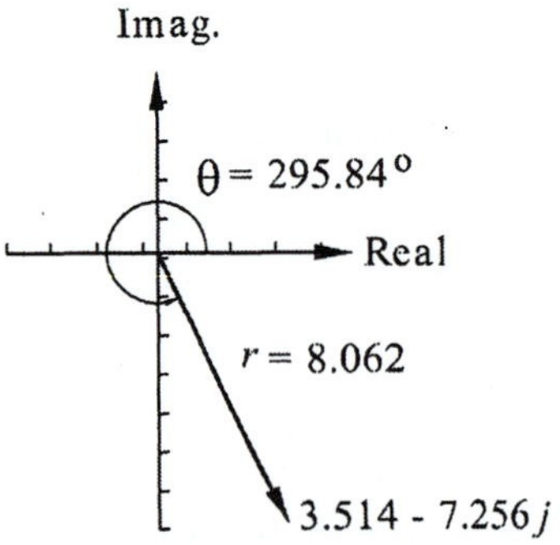

17. $9j = 0 + 9j$

$r = \sqrt{0^2 + 9^2} = 9$

$\theta = 90^\circ$ since y is negative

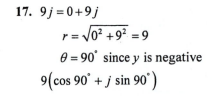

$9(\cos 90^\circ + j \sin 90^\circ)$

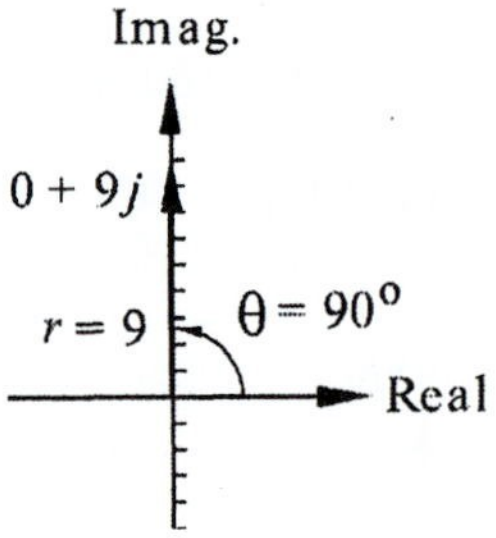

21. $160(\cos 150.0^\circ + j \sin 150.0^\circ)$

$x = 160 \cos 150.0^\circ = 160(0.866) = -139$

$y = 160 \sin 150.0^\circ = 160(0.5) = 80$

$-139 + 80j$

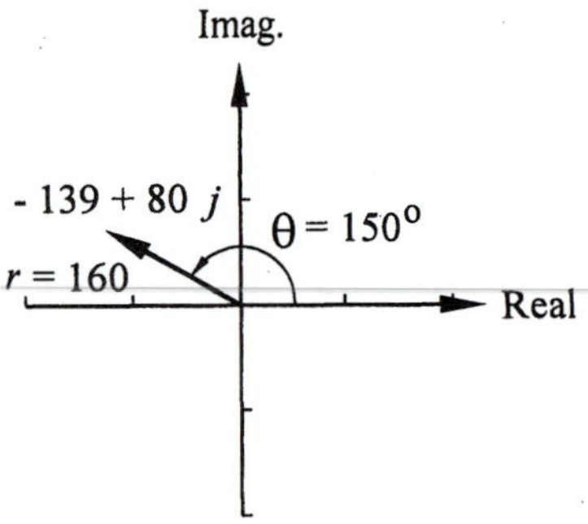

25. $0.08(\cos 360^\circ + j \sin 360^\circ)$

$x = 0.08 \cos 360^\circ = 0.088(1) = 0.08$

$y = 0.088 \sin 360^\circ = 0.088(0) = 0$

0.08

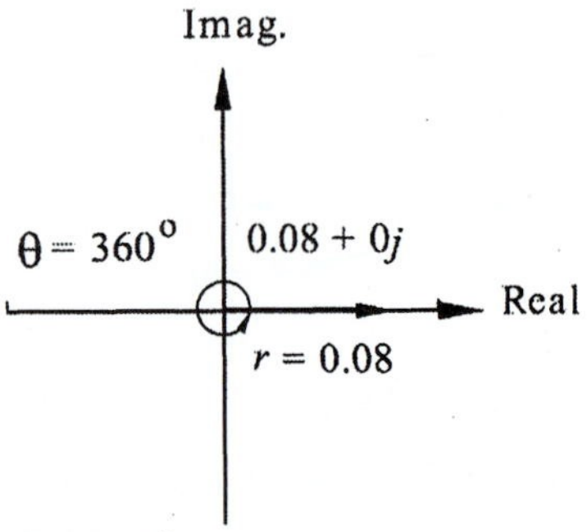

29. $4.75\angle 172.8^\circ$

$x = 4.75(\cos 172.8^\circ + j \sin 172.8^\circ)$

$y = -4.71 + 0.595j$

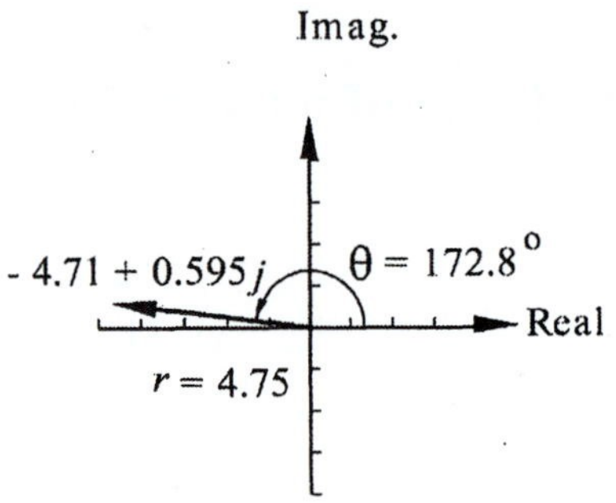

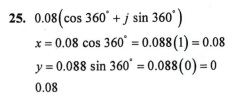

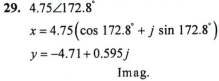

33. $7.32\angle -270^\circ = 7.32\left(\cos\left(-270^\circ\right) + j\,\sin\left(-270^\circ\right)\right)$

$= 0 + 7.32j$

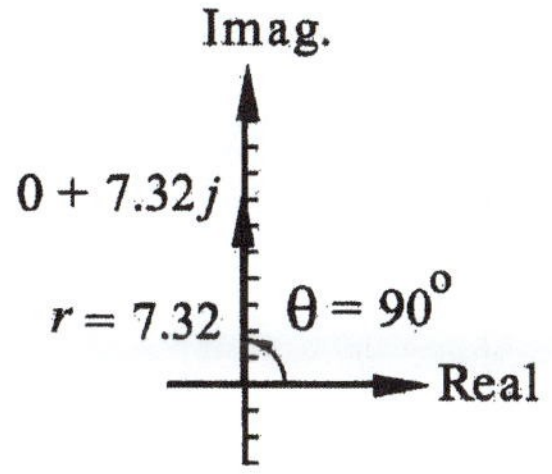

37. The argument for any negative real number is 180°.

41. $r = \sqrt{2.84^2 + (-1.06)^2} = 3.03$

$\tan\theta_{\text{ref}} = \dfrac{-1.06}{2.84},\ \theta_{\text{ref}} = -20.5^\circ,\ \theta = 339.5^\circ$

$2.84 - 1.06j = 3.03\left(\cos 339.5^\circ + j\,\sin 339.5^\circ\right)$

$3.03\angle 339.5^\circ$ kV

12.5 Exponential Form of a Complex Number

1. $8.50\angle 226.3^\circ,\ r = 8.50,$

$\theta = 226.3^\circ\left(\dfrac{\pi}{180^\circ}\right)$

$\theta = 3.95$

$8.50\angle 226.3^\circ = 8.50e^{3.97j}$

5. $3.00\left(\cos 60.0^\circ + j\,\sin 60.0^\circ\right);$

$r = 3.00,\ \theta = 60.0^\circ \cdot \left(\dfrac{\pi}{180^\circ}\right) = 1.05 \text{ rad}$

$3.00e^{1.05j}$

9. $375.5\left(\cos\left(-95.46^\circ\right) + j\,\sin\left(-95.46^\circ\right)\right);\ r = 375.5$

$\theta = -95.46^\circ \cdot \dfrac{\pi}{180^\circ} = -1.666 \text{ rad}$

$375.5e^{-1.666j} = 375.5e^{4.617j}$

13. $4.06\angle -61.4^\circ;\ r = 4.06;$

$\theta = -61.4^\circ = -1.07 \text{ rad}$

$4.06e^{-1.07j} = 4.06e^{5.21j}$

17. $3 - 4j;\ r = \sqrt{3^2 + (-4)^2} = 5,$

$\theta_{\text{ref}} = \tan^{-1}\dfrac{-4}{3} = -0.9273$

$\theta = \theta_{\text{ref}} + 2\pi = 5.36 \text{ rad}$

$3 - 4j = 5e^{5.36j}$

21. $5.90 + 2.40j;\ r = \sqrt{5.90^2 + 2.40^2} = 6.37$

$\theta = \tan^{-1}\dfrac{2.40}{5.90} = 0.386$

$5.90 + 2.40j = 6.37e^{0.386j}$

25. $3.00e^{0.500j};\ r = 3,\ \theta = 0.500 \text{ rad} = 28.6^\circ$

$3.00e^{0.500j} = 3\left(\cos 28.6^\circ + j\,\sin 28.6^\circ\right)$

$= 2.63 + 1.44j$

29. $3.20e^{5.41j};\ r = 3.20,\ \theta = 5.41 \text{ rad} = 310.0^\circ$

$3.20e^{5.41j} = 3.20\left(\cos 310.0^\circ + j\,\sin 310.0^\circ\right)$

$= 2.06 - 2.45j$

33. $\left(4.55e^{1.32j}\right)^2 = 20.7e^{2.64j};$

$r = 20.7,\ \theta = 2.64 \text{ rad} = 151.3^\circ$

$\left(4.55e^{1.32j}\right)^2 = 20.7\left(\cos 151.3^\circ + j\,\sin 151.3^\circ\right)$

$= -18.2 + 9.94j$

37.

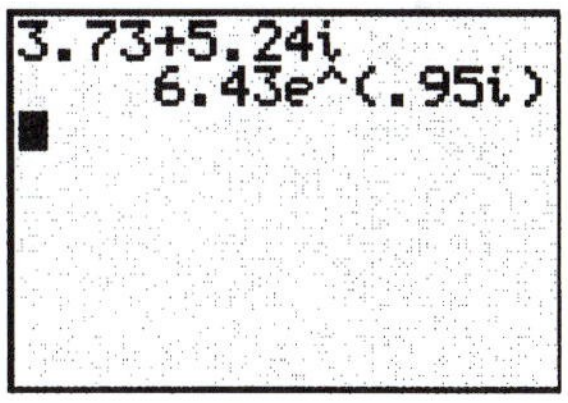

41. $2800-1450j;\ r=\sqrt{2800^2+(-1450)^2}=3153$

$\tan\theta=\dfrac{-1450}{2800}=-0.5179$

$\theta=-27.38^\circ=\dfrac{-27.38\pi}{180^\circ}=-0.478$

$\left(3153e^{-0.478j}\right)^{-1}=3.17\times10^{-4}e^{0.478j}\,1/\Omega$

12.6 Products, Quotients, Powers, and Roots of Complex Numbers

1. $2+3j$: $r_1=\sqrt{2^2+3^2}=3.61$, $\tan\theta_1=\dfrac{3}{2}\Rightarrow\theta_1=56.3^\circ$

$1+j$: $r_2=\sqrt{1^1+1^2}=1.41$, $\tan\theta_2=\dfrac{1}{1}\Rightarrow\theta_2=45.0^\circ$

$$\begin{aligned}(2+3j)(1+j)&=3.61(\cos 56.3^\circ+j\sin 56.3^\circ)(1.41)(\cos 45.5^\circ+j\sin 45.0^\circ)\\&=5.09(\cos 101.3^\circ+j\sin 101.3^\circ)\end{aligned}$$

5. $$\begin{aligned}[4(\cos 60^\circ+j\sin 60^\circ)][2(\cos 20^\circ+j\sin 20^\circ]&=4\cdot 2[(\cos(60^\circ+20^\circ)+j\sin(60^\circ+20^\circ)]\\&=8(\cos 80^\circ+j\sin 80^\circ)\end{aligned}$$

9. $\dfrac{8(\cos 100^\circ+j\sin 100^\circ)}{4(\cos 65^\circ+j\sin 65^\circ)}=\dfrac{8}{4}(\cos(100^\circ-65^\circ)+j\sin(100^\circ-65^\circ))=2(\cos 35^\circ+j\sin 35^\circ)$

13. $[2(\cos 35^\circ+j\sin 35^\circ)]^3=2^3(\cos(3\cdot 35^\circ)+j\sin(3\cdot 35^\circ))=8(\cos 105^\circ+j\sin 105^\circ)$

17. $\dfrac{(50\angle 236^\circ)(2\angle 84^\circ)}{125\angle 47^\circ}=\dfrac{100\angle 320^\circ}{125\angle 47^\circ}=\dfrac{4}{5}\angle 273^\circ$

21. $$\begin{aligned}2.78\angle 56.8^\circ+1.37\angle 207.3^\circ&=2.78(\cos 56.8^\circ+j\sin 56.8^\circ)+1.37(\cos 207.3^\circ+j\sin 207.3^\circ)\\&=1.5222+2.3262j-1.2174-0.6283j=0.3048+1.6979j.\end{aligned}$$

$r=\sqrt{0.3048^2+1.6979^2}=1.73$, $\theta=\tan^{-1}\dfrac{1.6979}{0.3048}=79.8^\circ$.

25. $3+4j=\sqrt{3^2+4^2}=\left(\cos\left(\tan^{-1}\frac{4}{3}\right)+j\sin\left(\tan^{-1}\frac{4}{3}\right)\right)=5(\cos 53.1^\circ+j\sin 53.1^\circ)$

$5-12j=\sqrt{5^2+(-12)^2}\left(\cos\left(\tan^{-1}\frac{-12}{5}+360^\circ\right)+j\sin\left(\tan^{-1}\frac{-12}{5}+360^\circ\right)\right)$

$=13(\cos 292.6^\circ-j\sin 292.6^\circ)$

polar form:

$$\begin{aligned}(3+4j)(5-12j)&=[5(\cos 53.1^\circ+j\sin 53.1^\circ)][13(\cos 292.6^\circ+j\sin 292.6^\circ)]\\&=65(\cos(53.1^\circ+292.6^\circ)+j\sin(53.1^\circ+292.6^\circ))\\&=65(\cos(345.7^\circ)+j\sin(345.7^\circ))\\&=63.0-16.1j\end{aligned}$$

rectangular form:

$(3+4j)(5-12j)=15-36j+20j-48j^2=15-16j+48=63-16j$

29. $\frac{7}{1-3j}=\frac{7(\cos 0^\circ+j\sin 0^\circ)}{\sqrt{1^2+(-3)^2}(\cos\left(\tan^{-1}\frac{-3}{1}+360^\circ\right)+j\sin\left(\tan^{-1}\frac{-3}{1}+360^\circ\right))}=\frac{7(\cos 0^\circ+j\sin 0^\circ)}{3.16(\cos 288.4^\circ+j\sin 288.4^\circ)}$

$=\frac{7}{3.16}(\cos(0^\circ-288.4^\circ)+j\sin(0-288.4^\circ))=2.22(\cos(-288.4^\circ)+j\sin(-288.4^\circ))$

$=2.22(\cos 71.6^\circ+j\sin 71.6^\circ)=0.7+2.1j$

Rectangular form:

$$\frac{7}{1-3j}=\frac{7}{(1-3j)}\cdot\frac{(1+3j)}{(1+3j)}=\frac{7+21j}{1^2+3^2}=\frac{7}{10}+\frac{21}{10}j$$

33. $(3+4j)=\sqrt{3^2+4^2}\left(\cos\left(\tan^{-1}\frac{4}{3}\right)+j\sin\left(\tan^{-1}\frac{4}{3}\right)\right)=5(\cos 53.1^\circ+j\sin 53.1^\circ)$

$$\begin{aligned}(3+4j)^4&=[5(\cos 53.1^\circ+j\sin 53.1^\circ)]^4 5^4(\cos(4\cdot 53.1^\circ)+j\sin(4\cdot 53.1^\circ))\\&=625(\cos 212.5^\circ+j\sin 212.5^\circ)=-527-336j\end{aligned}$$

rectangular form:

$$\begin{aligned}(3+4j)^4&=[(3+4j)^2]^2=(9+24j+(16j^2))^2=(9+24j-16)^2=(-7+24j)^2\\&=49-336j+576j^2 49-336j-576=-527-336j\end{aligned}$$

37. The two square roots of $4(\cos 60^\circ+j\sin 60^\circ)$ are

$r_1=\sqrt{4}\left(\cos\frac{60^\circ+0\cdot 360^\circ}{2}+j\sin\frac{60^\circ+0\cdot 360^\circ}{2}\right)$

$r_1=2(\cos 30^\circ+j\sin 30^\circ)=\sqrt{3}+j$

and

$r_2=\sqrt{4}\left(\cos\frac{60^\circ+1\cdot 360^\circ}{2}+j\sin\frac{60^\circ+1\cdot 360^\circ}{2}\right)$

$r_2=2(\cos 210^\circ+j\sin 210^\circ)=-\sqrt{3}-j$

41. The two square roots of $1+j=\sqrt{2}(\cos 45° + j\sin 45°)$ are

$$r_1 = \sqrt{\sqrt{2}}\left(\cos\frac{45° + 0\cdot 360°}{2} + j\sin\frac{4.5° + 0\cdot 360°}{2}\right)$$

$$r_1 = 2^{1/4}(\cos 22.5° + j\sin 22.5°) = 1.0987 + 0.4551j$$

and

$$r_2 = 2^{1/4}\left(\cos\frac{45° + 1\cdot 360°}{2} + j\sin\frac{4.5° + 1\cdot 360°}{2}\right)$$

$$r_2 = 2^{1/4}(\cos 202.5° + j\sin 202.5°) = -1.0987 - 0.4551j$$

45. $-27j$; $x=0$, $y=-27$; $r=\sqrt{0^2+27^2}=27$; $\theta=270°$

First root: $(0-27j)^{1/3} = [27(\cos 270° + j\sin 270°)]^{1/3} = 3(\cos 90° + j\sin 90°) = 3j$

Second root: $\theta = 360° + 270° = 630°$

$(0+27j)^{1/3} = [27(\cos 630° + j\sin 630°)]^{1/3} = 3(\cos 210° + j\sin 210°) = \dfrac{-3\sqrt{3}}{2} - \dfrac{3}{2}j$

Third root: $\theta = 720° + 270° = 990°$

$(0+27j)^{1/3} = [27(\cos 990° + j\sin 990°)]^{1/3} = 3(\cos 330° + j\sin 330°) = \dfrac{3\sqrt{3}}{2} - \dfrac{3}{2}j$

49. $-125 = 125(\cos 180° + j\sin 180°)$

$$(-125)^{1/3} = 125^{1/3}\left[\cos\left(\frac{1}{3}\cdot 180°\right) + j\sin\left(\frac{1}{3}\cdot 180°\right)\right] = \frac{5}{2} + \frac{5\sqrt{3}}{2}j = 2.500 + 4.330j$$

$$(-125)^{1/3} = 125^{1/3}\left[\cos\left(\frac{1}{3}\cdot 540°\right) + j\sin\left(\frac{1}{3}\cdot 540°\right)\right] = -5$$

$$(-125)^{1/3} = 125^{1/3}\left[\cos\left(\frac{1}{3}\cdot 900°\right) + j\sin\left(\frac{1}{3}\cdot 900°\right)\right] = \frac{5}{2} - \frac{5\sqrt{3}}{2}j = 2.500 - 4.330j$$

53.

$$x^3 + 1 = 0$$

$$(x+1)(x^2 - x + 1) = 0$$

$$x + 1 = 0 \quad \text{or} \quad x^2 - x + 1 = 0$$

$$x = -1 \qquad x = \frac{-(-1) \pm \sqrt{(-1)^2 - 4(1)(1)}}{2(1)} = \frac{1 \pm j\sqrt{3}}{2}$$

The roots are the same as in Example 7.

57. $\dfrac{(8.66\angle 90.0°)(50.0\angle 135.0°)}{10.0\angle 60.0°} = \dfrac{(8.66)(50.0)}{10.0}\angle 90.0° + 135.0° - 60.0°$

$= 43.3\angle 165.0°, V = 43.3\text{ V}$

12.7 An Application to Alternating-Current (ac) Circuits

1. $V_R = IR = 2.00(12.0) = 24.0\text{ V}$

$Z = R + j(X_L - X_C) = 12.0 + j(16.0)$

$|Z| = \sqrt{12.0^2 + 16.0^2} = 20.0\,\Omega$

$V_L = IX_L = 2.00(16.0) = 32.0\text{ V}$

$V_{RL} = IZ = 2.00(20.0) = 40.0\text{ V}$

$\theta = \tan^{-1}\dfrac{X_L}{R} = \tan^{-1}\dfrac{16.0}{12.0} = 53.1^\circ$, voltage leads current.

5. (a) $|Z| = \sqrt{R^2 + X_L^2} = \sqrt{2250^2 + 1750^2} = 2850\,\Omega$

(b) $\tan\theta = \dfrac{1750}{2250};\ \theta = 37.9^\circ$

(c) $V_{RLC} = IZ = (0.005\ 75)(2850) = 16.4\text{ V}$

9. (a) $X_R = 45.0\,\Omega$

$X_L = 2\pi fL = 2\pi(60)(0.0429) = 16.2\,\Omega$

$Z = 45.0 + 16.2j$

$|Z| = \sqrt{45.0^2 + 16.2^2} = 47.8\,\Omega$

(b) $\tan\theta = \dfrac{16.2}{45.0};\ \theta = 19.8^\circ$

13. $R = 25.3\,\Omega$

$X_C = 1/(2\pi fC)$

$= 1/\left(2\pi\left(1.2\times10^6\right)\left(2.75\times10^{-9}\right)\right) = 48.2\,\Omega$

$= f = 1200\text{ kHz} = 1.2\times10^6\text{ Hz}$

$Z = R - X_{Cj} = 25.3 - 48.2j$

$|Z| = \sqrt{25.3^2 + (-48.2)^2} = 54.4\,\Omega$

$\tan\theta = \dfrac{-48.2}{25.3};\ \theta = -62.3^\circ$

17. $L = 12.5\times10^{-6}\text{ H}$

$C = 47.0\times10^{-9}\text{ F}$

$X_L = X_C$

$2\pi fL = \dfrac{1}{2\pi fC}$

$$f = \sqrt{\frac{1}{4\pi^2 LC}}$$

$$= \sqrt{\frac{1}{4\pi^2\left(12.5\times10^{-6}\right)\left(4.70\times10^{-9}\right)}}$$

$= 208\text{ kHz}$

21. $P = \text{VI}\cos\theta$

$V = 225\text{ mV}$

$\theta = -18.0^\circ = 342^\circ$

$Z = 47.3\,\Omega$

$V = IZ$

$I = \dfrac{225\times10^{-3}}{47.3} = 0.00476\text{ A}$

$P = \left(225\times10^{-3}\right)(0.00476)\cos 342^\circ$

$= 0.00102\text{ W} = 1.02\text{ mW}$

Chapter 12 Review Exercises

1. $(6-2j)+(4+j) = 6-2j+4+j = 10-j$

5. $(2+j)(4-j) = 8-2j+4j-j^2 = 8+2j+1 = 9+2j$

9. $\dfrac{3}{7-6j} = \dfrac{3}{(7-6j)}\cdot\dfrac{(7+6j)}{(7+6j)} = \dfrac{21+18j}{7^2+6^2} = \dfrac{21}{85}+\dfrac{18}{85}j$

13. $\dfrac{5j-(3-j)}{4-2j} = \dfrac{(-3+6j)}{(4-2j)}\cdot\dfrac{(4+2j)}{(4+2j)}$

$= \dfrac{-12-6j+24j+12j^2}{4^2+2^2}$

$= \dfrac{-12+18j-12}{16+4}$

$= \dfrac{-24+18j}{20}$

$= -\dfrac{6}{5}+\dfrac{9}{10}j$

17. $3x-2j = yj-9 \Rightarrow 3x = -9,\ -2 = y$

$x = -\dfrac{9}{3}\quad y = -2$

$x = -3$

$x = 3,\ y = -2$

21.

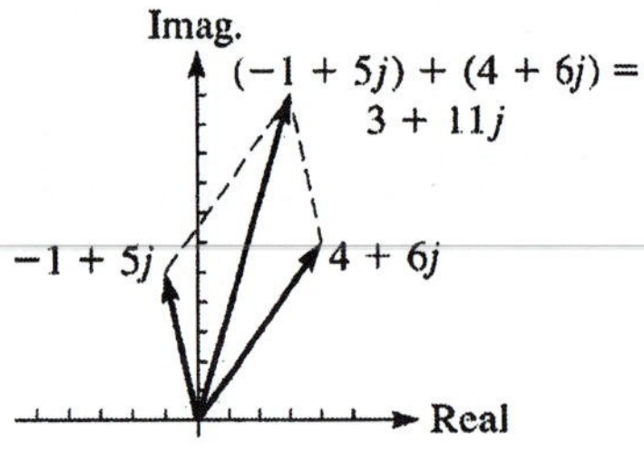

algebraically:

$$(-1+5j)+(4+6j)=-1+5j+4+6j$$
$$=-1+4+5j+6j$$
$$=3+11j$$

25. $1-j \Rightarrow r=\sqrt{1^2+1^2}=\sqrt{2},\ \theta=\tan^{-1}\frac{-1}{1}+360^\circ$

$$=315^\circ=\frac{7\pi}{4}\text{ rad}$$

polar: $1-j=\sqrt{2}\left(\cos 315^\circ + j\sin 315^\circ\right)$

exponential: $1-j=\sqrt{2}e^{7\pi/4j}$

29. $1.07+4.55j \Rightarrow r=\sqrt{1.07^2+4.55^2}=4.67,$

$\theta=\tan^{-1}\frac{4.55}{1.07}=76.8^\circ=1.34\text{ rad}$

polar: $1.07+4.55j=4.67\left(\cos 76.8^\circ + j\sin 76.8^\circ\right)$

exponential: $1.07+4.55j=4.67e^{1.34j}$

33. $2\left(\cos 225^\circ + j\sin 225^\circ\right)=-\sqrt{2}-\sqrt{2}j$

37. $0.62\angle-72^\circ=0.62\left(\cos(-72^\circ)+j\sin(-72^\circ)\right)$

$$=0.19-0.59j$$

41. $200e^{0.25j}=2.00(\cos 0.25+j\sin 0.25)$

$$=1.94+0.495j$$

45. $\left[3\left(\cos 32^\circ + j\sin 32^\circ\right)\right]\cdot\left[\left(\cos 52^\circ + j\sin 52^\circ\right)\right]$

$$=3\cdot 5\left(\cos\left(32^\circ+52^\circ\right)+j\sin\left(32^\circ+52^\circ\right)\right)$$
$$=15\left(\cos 84^\circ + j\sin 84^\circ\right)$$

49. $\dfrac{24\left(\cos 165^\circ + j\sin 165^\circ\right)}{3\left(\cos 106^\circ + j\sin 106^\circ\right)}$

$$=\frac{24}{3}\cos\left(165^\circ-106^\circ\right)+j\sin\left(165^\circ-106^\circ\right)$$
$$=8\left(\cos 59^\circ + j\sin 59^\circ\right)$$

53. $0.983\angle 47.2^\circ + 0.366\angle 95.1^\circ$

$$=0.983\left(\cos 47.2^\circ + j\sin 47.2^\circ\right)$$
$$+0.366\left(\cos 95.1^\circ + j\sin 95.1^\circ\right)$$
$$=0.6679+0.7213j-0.03254+0.3646j$$
$$=0.6354+1.0859j$$

in polar form $r=\sqrt{0.6354^2+1.0859^2}=1.26$

$$\theta=\tan^{-1}\frac{1.0859}{0.6354}=59.7^\circ,\ 1.26\angle 59.7^\circ$$

57. $\left[2\left(\cos 16^\circ + j\sin 16^\circ\right)\right]^{10}$

$$=2^{10}\left[\cos\left(10\cdot 16^\circ\right)+j\sin\left(10\cdot 16^\circ\right)\right]$$
$$=1024\left(\cos 160^\circ + j\sin 160^\circ\right)$$

61. $1-j=\sqrt{2}\left(\cos 315^\circ + j\sin 315^\circ\right)$ from prob. 25.

$$(1-j)^{10}=\left[\sqrt{2}\left(\cos 315^\circ + j\sin 315^\circ\right)\right]^{10}$$
$$=\sqrt{2}^{10}\left(\cos\left(10\cdot 315^\circ\right)+j\sin\left(10\cdot 315^\circ\right)\right)$$
$$=32\left(\cos 3150^\circ + j\sin 3150^\circ\right),\text{ polar form}$$
$$=0-32j,\text{ rectangular form}$$

$$(1-j)^{10}=\left((1-j)^2\right)^5=\left(1-2j+j^2\right)^5=\left(1-2j-1\right)^5$$
$$=(-2j)^5=(-2)^5\cdot j^5$$
$$=-32\cdot j^4\cdot j=-32j$$

65. $-8=-8+0j=8\left(\cos 180^\circ + j\sin 180^\circ\right)$

$$r_1=\sqrt[3]{8}\left(\cos\frac{180^\circ+360^\circ}{3}+j\sin\frac{180^\circ+0\cdot 360^\circ}{3}\right)$$
$$=2\left(\cos 60^\circ + j\sin 60^\circ\right)$$
$$=2\left(\frac{1}{2}+j\cdot\frac{\sqrt{3}}{2}\right)=1+j\sqrt{3}$$

$$r_2 = \sqrt[3]{8}\left(\cos\frac{180^\circ + 360^\circ}{3} + j\sin\frac{180^\circ + 1\cdot 360^\circ}{3}\right)$$
$$= 2\left(\cos 180^\circ + j\sin 180^\circ\right) = 2\left(-1 + j(0)\right) = -2$$
$$r_3 = \sqrt[3]{8}\left(\cos\frac{180^\circ + 2\cdot 360^\circ}{3} + j\sin\frac{180^\circ + 2\cdot 360^\circ}{3}\right)$$
$$= 2\left(\cos 300^\circ + j\sin 300^\circ\right) = 2\left(\frac{1}{2} - \frac{\sqrt{3}}{2}j\right)$$
$$= 1 - j\sqrt{3}$$

69. Rectangular: $40 + 9j$ from the graph

polar: $40 + 9j \Rightarrow r = \sqrt{40^2 + 9^2} = 41,\ \theta = \tan^{-1}\frac{9}{40}$
$= 12.7^\circ$

$40 + 9j = 41\left(\cos 12.7^\circ + j\sin 12.7^\circ\right)$

73. $x^2 - 2x + 4\Big|_{x=5-2j} = (5-2j)^2 - 2(5-2j) + 4$
$= 25 - 20j + 4j^2 - 10 + 4j + 4$
$= 19 - 16j - 4$
$= 15 - 16j$

77. $x = 2 + j,\ x = 2 - j$
$x - (2 + j) = 0,\ x - (2 - j) = 0$
$(x - (2 + j))(x - (2 - j)) = 0$
$x^2 - (2 + j)x - (2 - j)x + (+j)(2 - j) = 0$
$x^2 - 2x - jx - 2x + jx + 4 - j^2 = 0$
$x^2 - 4x + 4 + 1 = 0$
$x^2 - 4x + 5 = 0$

81. $(1 + jx)^2 = 1 + j - x^2$
$1 + 2jx - x^2 = 1 + j - x^2$
$2jx = j$
$x = \frac{1}{2}$

85.

```
abs(5-3i)
          5.830951895
angle(5-3i)
         -30.96375653
```

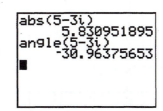

$5 - 3j = 5.83\left(\cos 329 + j\sin 329^\circ\right)$

89. Since $\sqrt{R^2 + X_L^{\ 2}} = \sqrt{6250^2 + 1320^2}$
$= 6390 < 6720 = Z$, Z must be in QIV.

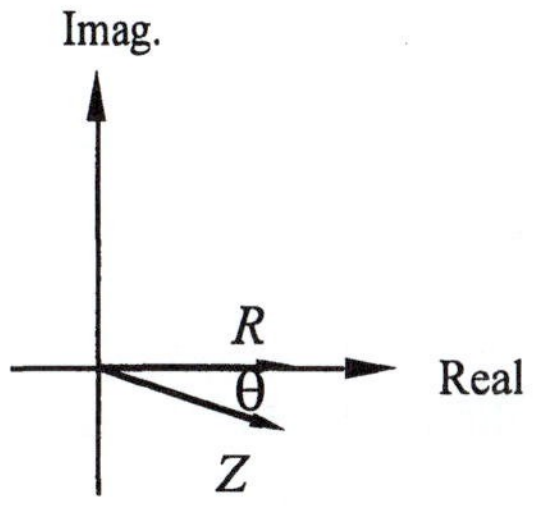

93. resultant $= 2100 - 1200j + 1200 + 5600j$
$= 3300 + 4400j$

$r = \sqrt{3300^2 + 4400^2} = 5500;\ \theta = \tan^{-1}\frac{4400}{3300} = 53^\circ$

resultant $= 5500\angle 53^\circ$ N

97. $re^{j\theta} = r(\cos\theta + j\sin\theta)$
$e^{j\pi} = \cos\pi + j\sin\pi = -1$

Chapter 13

EXPONENTIAL AND LOGARITHMIC FUNCTIONS

13.1 Exponential Functions

1. For $x=-\frac{3}{2}$, $y=-2\left(4^x\right)=-2\left(4^{-3/2}\right)=-\frac{1}{4}$

5. (a) $y=-7(-5)^{-x}$, $-5<0$, not an exponential function.

(b) $y=-7\left(5^{-x}\right)$ is a real number multiple of an exponential function and therefore an exponential function.

9. $y=9^x$; $x=-2$, $=9^{-2}=\frac{1}{9^2}=\frac{1}{81}$

13. $y=4^x$

x	-3	-2	-1	0	1	2	3
y	$\frac{1}{64}$	$\frac{1}{16}$	$\frac{1}{4}$	1	4	16	64

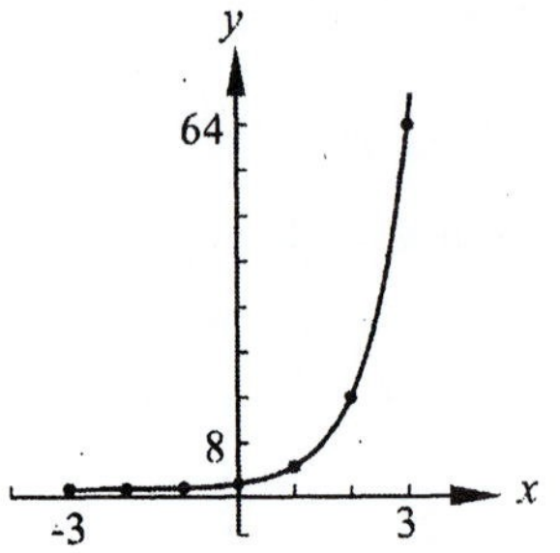

17. $y=-0.5\pi^x$

x	y
-3	27.000
-2	9.000
-1	3.000
0	1.000
1	0.333
2	0.111
3	0.037

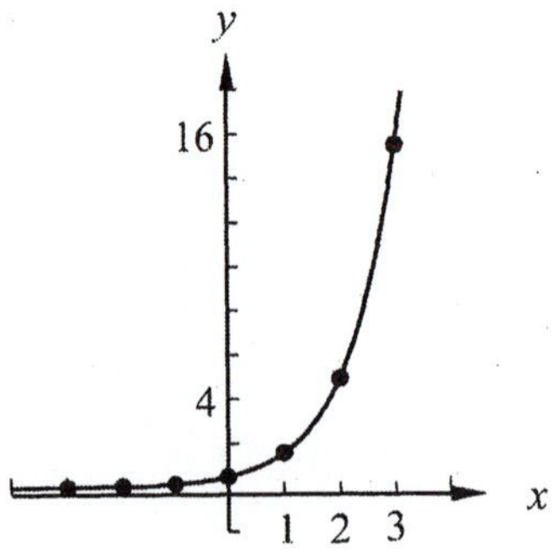

21. $y_1=0.1(0.25)^{2x}$

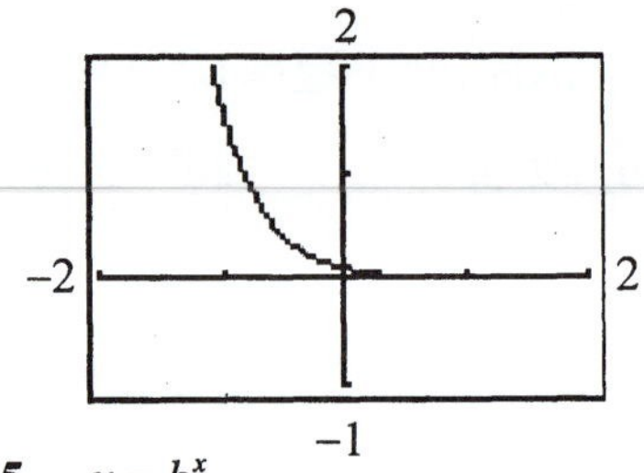

25.
$$y=b^x$$
$$64=b^3$$
$$4^3=b^3$$
$$b=4$$

29. $y=2^{|x|}$

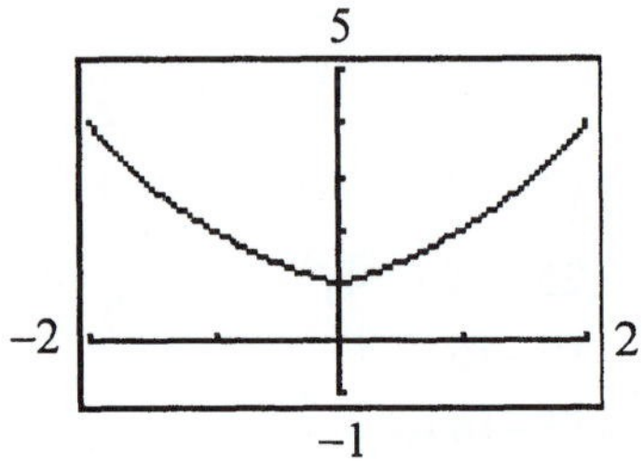

33. $V=250(1.0500)^t=250(1.0500)^4=\$303.88$

37. $q=100e^{-10t}$, graph $y_1=100e^{-10x}$.

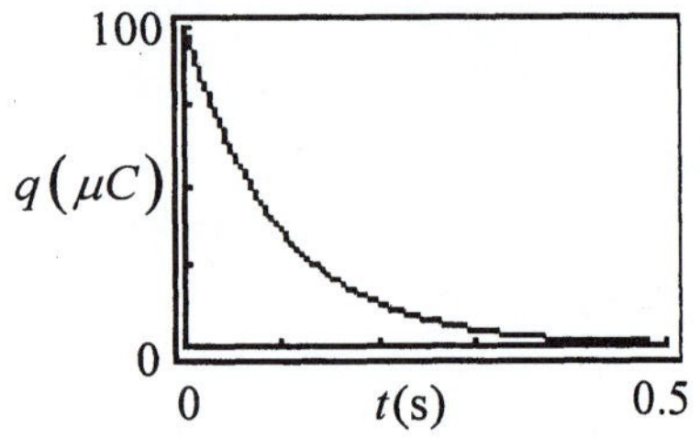

13.2 Logarithmic Functions

1. $32^{4/5}=16$ in logarithmic form is $\frac{4}{5}=\log_{32}16$.

5. $3^3 = 27$ has base 3, exponent 3, and number 27.

$\log_3 27 = 3$

9. $7^{-2} = \frac{1}{49}$ has base 7, exponent -2, and number $\frac{1}{49}$.

$\log_7 \frac{1}{49} = -2$

13. $8^{1/3} = 2$ has base 8, exponent $\frac{1}{3}$, and number 2.

$\log_8 2 = \frac{1}{3}$

17. $\log_3 81 = 4$ has base 3, exponent 2, and number 81.

$3^4 = 81$

21. $\log_{25} 5 = \frac{1}{2}$ has base 25, exponent $\frac{1}{2}$, and number 5.

$25^{1/2} = 5$

25. $\log_{10} 0.1 = -1$ has base 10, exponent -1, and number 0.1.

$0.1 = 10^{-1}$

29. $\log_4 16 = x$ has base 4, exponent x, and number 16.

$4^x = 16$

$4^x = 4^2$

$x = 2$

33. $\log_7 y = 3$ has base 7, exponent 3, and number 3.

$7^3 = y,\ y = 343$

37. $\log_b 5 = 2$ has base b, exponent 2, and number 5.

$b^2 = 5,\ b = \sqrt{5}$

41. $\log_{10} 10^{0.2} = x$ has base 10, exponent x, and number $10^{0.2}$.

$10^x = 10^{0.2},\ x = 0.2$

45. Write $y = \log_3 x$ as $3^y = x$ to find values in table.

x	y
0.19	−1.5
0.58	−0.5
1.00	0.0
1.93	0.6
3.00	1.0
5.20	1.5
9.00	2.0

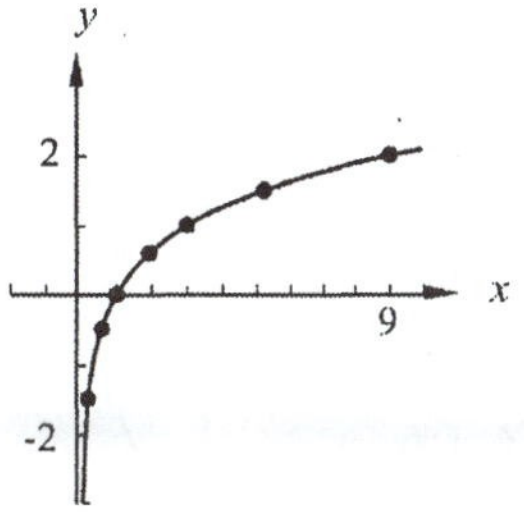

49. $N = 0.2\log_4 v;\ \frac{N}{0.2} = \log_4 v;\ 4^{N/0.2} = v$

v	N
0.2	−0.232
0.6	−0.0737
0.8	−0.0322
1	0
2	0.1
3	0.159
4	0.2

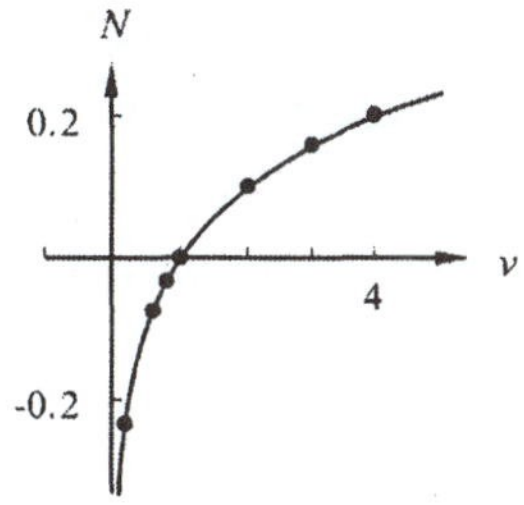

53. Graph $y_1 = -\log_{10}(-x)$.

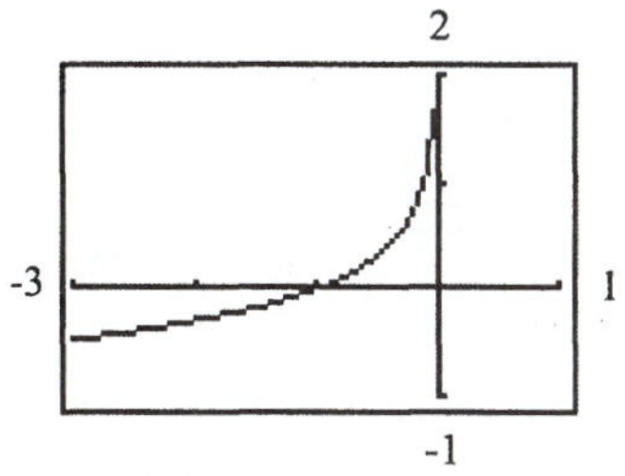

57.

$f(\sqrt{5}) = \log_5 \sqrt{5} = \log_5 5^{1/2}$

$= \frac{1}{2}\log_5 5 = \frac{1}{2}$

(b) $f(0)$ does not exist.

61.

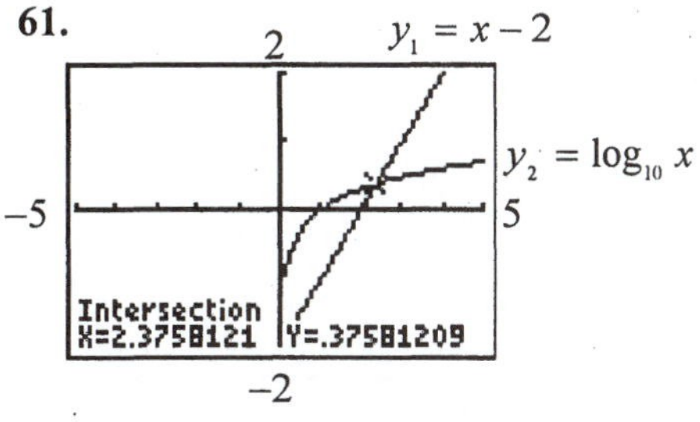

$x = 2.38$ is the solution of $\log_{10} x = x - 2$

65. $\log_e\left(\frac{N}{N_0}\right) = -kt \Rightarrow e^{-kt} = \frac{N}{N_0}$, $N = N_0 e^{-kt}$

69. $t = N + \log_2 N$ where $n > 0$ and $t > 0$.

N	t
1	1
2	3
4	6
8	11

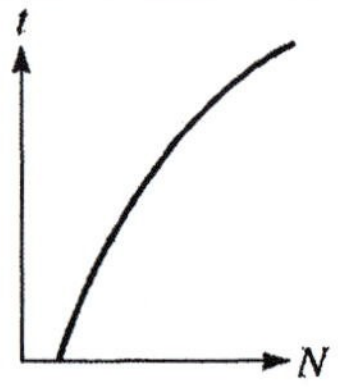

73. Solve $y = 10^{x/2}$ for $x = 2\log_{10} y$. Interchange x and y; $y = 2\log_{10} x$, which is the inverse function. Graph $y_1 = 10^{x/2}$, $y = 2\log_{10} x$.
For each graph to be the mirror image of the other across $y = x$ the calculator window must be "square." One way to do this is ZOOM 5: Square

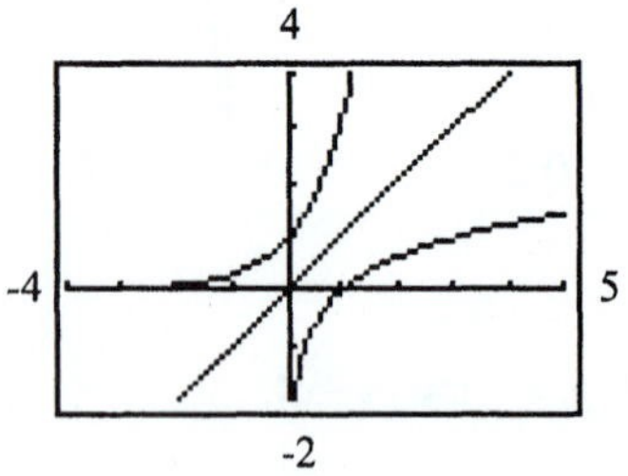

13.3 Properties of Logarithms

1. $\log_4 21 = \log_4(3(7)) = \log_4 3 + \log_4 7$

5. $\log_3 27 = \log_3 3^3 = 3\log_3 3 = 3(1) = 3$

9. $\log_5 33 = \log_5(3\cdot 11) = \log_5 3 + \log_5 11$

13. $\log_2(a^3) = 3\log_2 a$

17. $8\log_5 \sqrt[4]{y} = 8\log_5 y^{1/4} = 2\log_5 y$

21. $\log_b a + \log_b c = \log_b(ac)$

23. $\log_5 9 - \log_5 3 = \log_5\left(\frac{9}{3}\right) = \log_5 3$

25. $-\log_b \sqrt{x} + \log_b x^2 = \log_b \frac{x^2}{x^{1/2}} = \log_b x^{3/2}$

29. $\log_2\left(\frac{1}{32}\right) = \log_2\left(\frac{1}{2^5}\right) = \log_2 2^{-5} = -5\log_2 2 = -5$

33. $6\log_7\sqrt{7} = 6\log_7 7^{1/2} = 3\log_7 7 = 3$

37. $\log_3 18 = \log_3(9\cdot 2) = \log_3 9 + \log_3 2$
$= \log_3 3^2 + \log_3 2 = 2\log_3 3 + \log_3 2$
$= 2 + \log_3 2$

41. $\log_3\sqrt{6} = \log_3(3\cdot 2)^{1/2} = \frac{1}{2}\log_3(3\cdot 2)$
$= \frac{1}{2}\cdot[\log_3 3 + \log_3 2]$
$= \frac{1}{2}\cdot[1 + \log_3 2]$

45. $\log_b y = \log_b 2 + \log_b x$
$\log_b y = \log_b(2x)$
$y = 2x$

49. $\log_{10} y = 2\log_{10} 7 - 3\log_{10} x$
$= \log_{10} 7^2 - \log_{10} x^3$
$= \log_{10} 49 - \log_{10} x^3$
$\log_{10} y = \log_{10}\frac{49}{x^3}$
$y = \frac{49}{x^3}$

53. $\log_2 x + \log_2 y = 1$
$\log_2(xy) = 1 \Leftrightarrow 2^1 = xy$
$y = \frac{2}{x}$

57. $\log_{10}(x+3) = \log_{10} x + \log_{10} 3 = \log_{10}(3x)$
$x + 3 = 3x$
$2x = 3$
$x = \frac{3}{2}$ is the only value for which
$\log_{10}(x+3) = \log_{10} x + \log_{10} 3$ is true

For any other x-value $\log_{10}(x+3) = \log_{10}x + \log_{10}3$ is false and thus not true in general. This can be also be seen from the following graph.

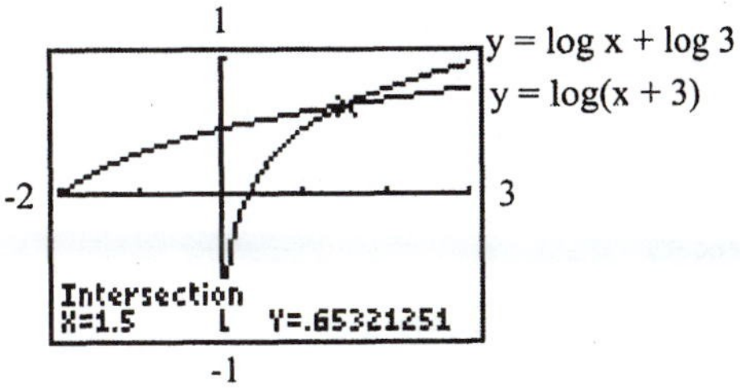

61. $\log_b \sqrt{x^2y^4} = \log_b\left(x^2y^4\right)^{1/2}$

$= \log_b\left(xy^2\right)$

$= \log_b x + 2\log_b y$

$= 2 + 2(3) = 8$

65.

$$\log_e D = \log_e a - br + cr^2$$

$$\log_e D - \log_e a = cr^2 - br$$

$$\log_e \frac{D}{a} = cr^2 - br$$

$$\frac{D}{a} = e^{cr^2 - br}$$

$$D = ae^{cr^2 - br}$$

13.4 Logarithms to the Base 10

1. $\log 0.3654 = -0.4372$

5. $\log 9.24 \times 10^6 = 6.966$

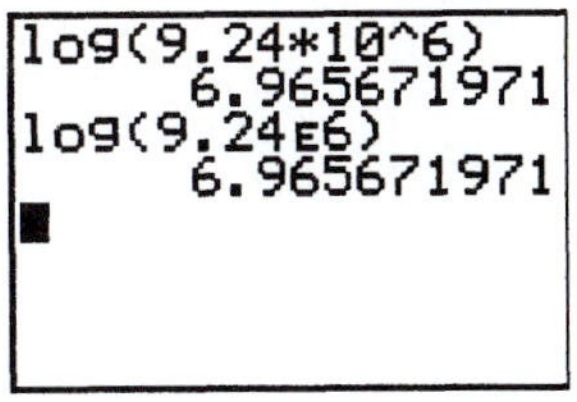

9. $\log\left(\cos 12.5^\circ\right) = -0.0104$

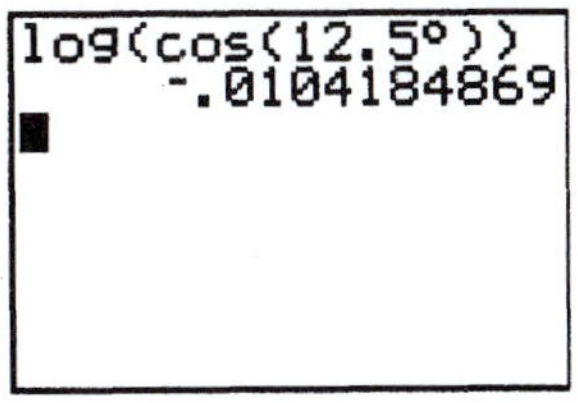

13. $10^{4.437} = 27{,}400$

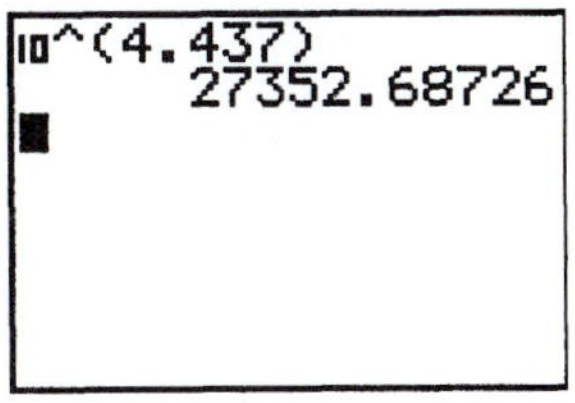

17. $10^{3.30112} = 2000.4$

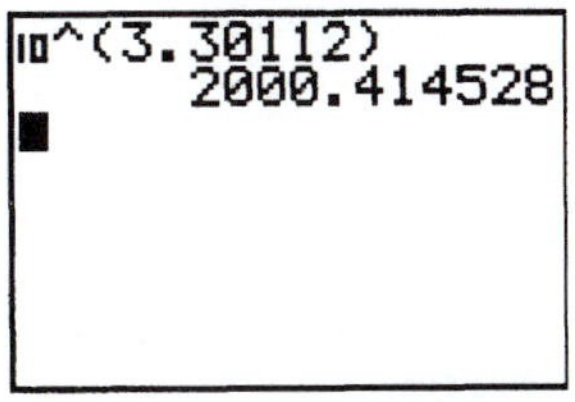

21. $(5.98)(14.3) = 85.5$

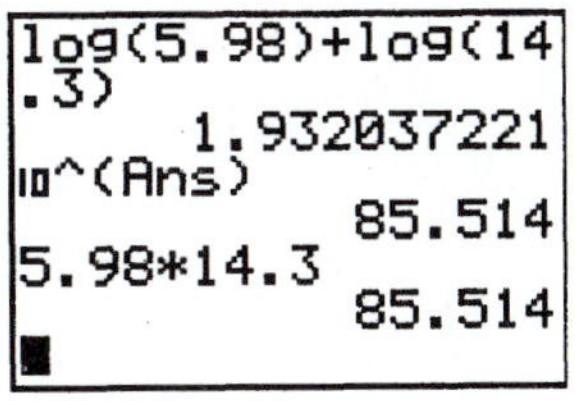

25. $\log 14 + \log 0.5 = \log 7$

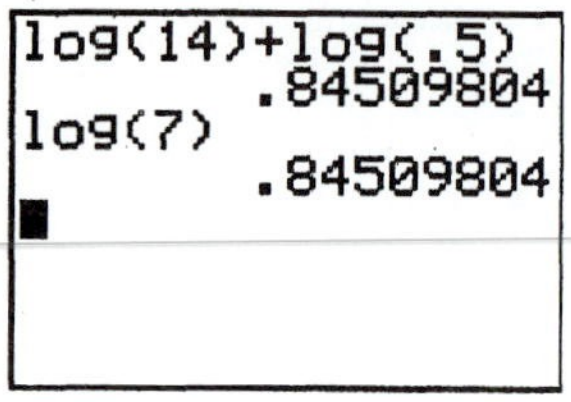

29. $\log 9 \times 10^8 = 8.9542$

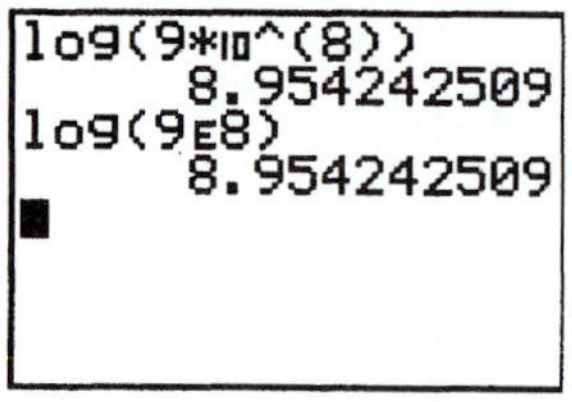

33. $\log T = 8$

$T = 10^8$ K

37. $\frac{\log_b x^2}{\log 100} = \frac{2\log_b x}{\log 10^2} = \frac{2\log_b x}{2\log 10} = \log_b x$

41. $R = \log\left(\frac{I}{I_0}\right);\ I = 20{,}000{,}000 I_0$

$R = \log \frac{20{,}000{,}000 I_0}{I_0}$

$= \log 20{,}000{,}000 = 7.3$

13.5 Natural Logarithms

1. $\ln 200 = \frac{\log 200}{\log e} = 5.298$

5. $\ln 1.562 = \frac{\log 1.562}{\log e} = \frac{0.1937}{0.4343} = 0.4460$

9. $\log_7 42 = \frac{\log 42}{\log 7} = \frac{1.6232}{0.8451} = 1.92$

13. $\log_{40} 750 = \frac{\log 750}{\log 40} = \frac{2.875}{1.6021} = 1.795$

17. $\ln 1.394 = 0.3322$

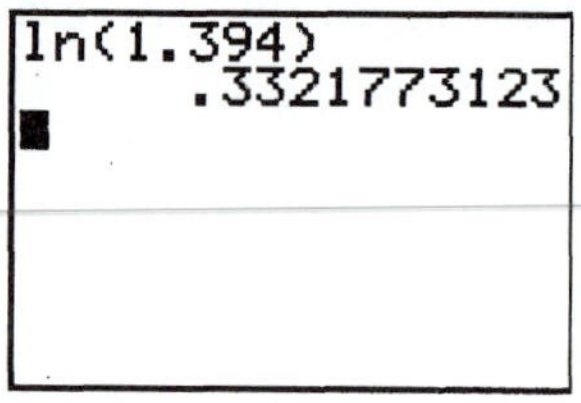

21. $\ln 0.012937^4 = -17.39066$

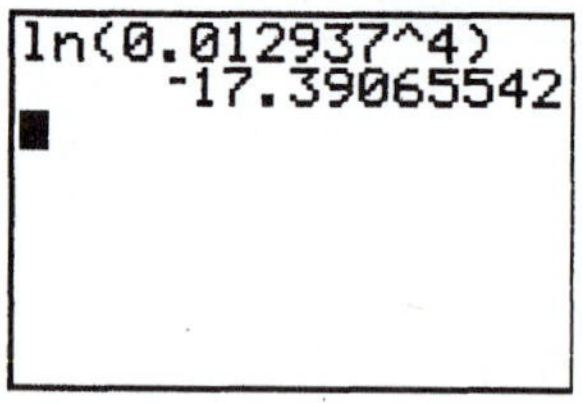

25. $\log 0.68528 = \frac{\ln 0.68528}{\ln 10} = -0.16413$

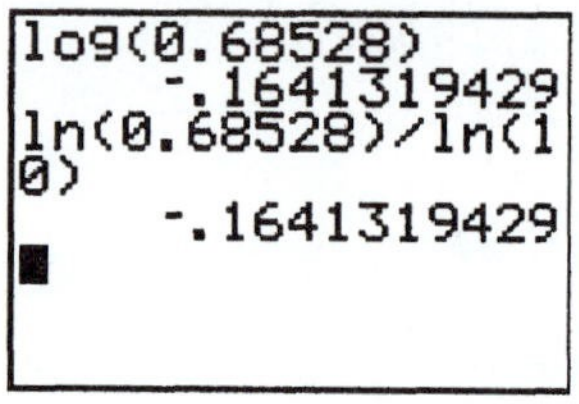

29. $e^{0.0084210} = 1.0085$

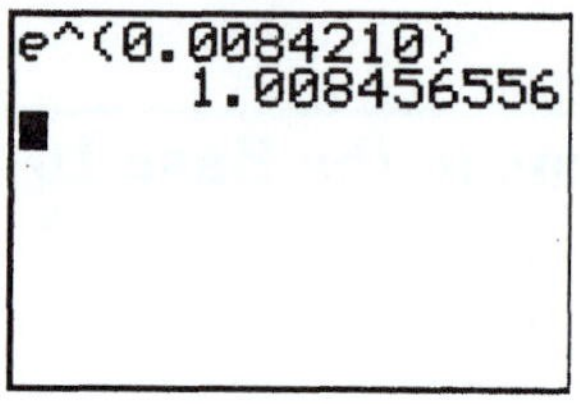

33. $e^{-23.504} = 6.20 \times 10^{-11}$

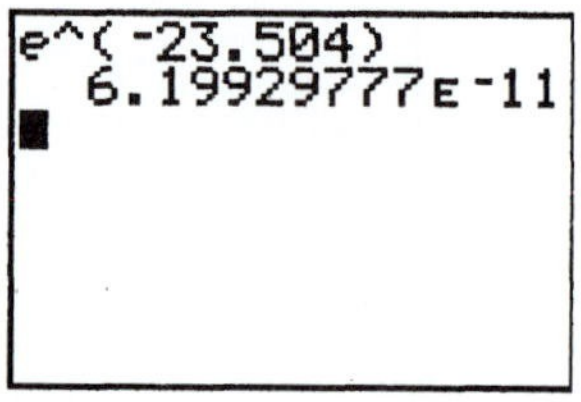

37. $y = 2^x \Leftrightarrow x = \log_2 y \Rightarrow y = \log_2 x$ is the inverse function.

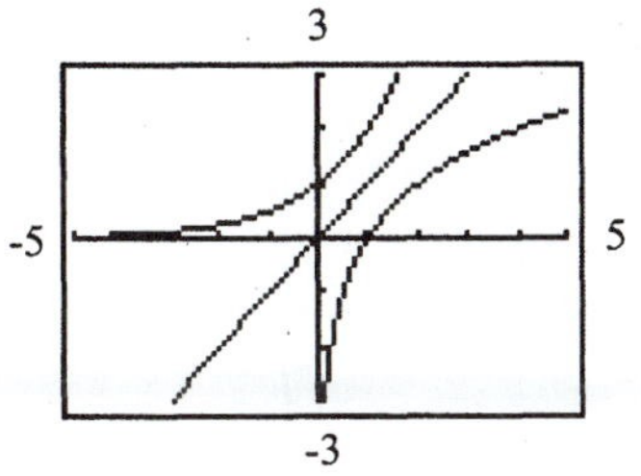

41. $4\ln 3 = \ln 81$

```
4ln(3)
        4.394449155
ln(81)
        4.394449155
```

45. $\ln x(\log x) = 0$

$$e^{\ln(\log x)}e^{0} = 1$$
$$\log x = 1$$
$$10^{\log x} = 10^{1}$$
$$x = 10$$

49. $\ln f = 21.619$

$$f = e^{21.619} = 2.45 \times 10^{9} \text{ Hz}$$

53. $t = \dfrac{L \cdot \ln\left(\dfrac{i}{I}\right)}{R} = -\dfrac{1.25\ln\left(\dfrac{0.1I}{I}\right)}{7.5}$

$t = 0.384$ s

$$t = \frac{e^{kx} - 1}{kv_0} = \frac{e^{(6.80\times10^{-3})(150)} - 1}{(6.80\times10^{-3})(12.0)}$$

$$t = \frac{e^{1.02} - 1}{0.0816} = \frac{1.773}{0.0816} = 21.7 \text{ s}$$

13.6 Exponential and Logarithmic Equations

1.

$$3^{x+2} = 5$$
$$\log 3^{x+2} = \log 5$$
$$(x+2)\log 3 = \log 5$$
$$x + 2 = \frac{\log 5}{\log 3}$$
$$x = \frac{\log 5}{\log 3} - 2$$
$$x = -0.535$$

5. $5^x = 0.3$; $\log 5^x = \log 0.3$; $x\log 5 = \log 0.3$

9.

$$6^{x+1} = 78$$
$$\ln 6^{x+1} = \ln 78$$
$$(x+1)\cdot \ln 6 = \ln 78$$
$$x + 1 = \frac{\ln 78}{\ln 6}$$
$$x = \frac{\ln 78}{\ln 6} - 1$$
$$x = 1.432$$

13.

$$0.6^x = 2^{x^2}$$
$$\ln\left(0.6^x\right) = \ln 2^{x^2}$$
$$x \cdot \ln 0.6 = x^2 \cdot \ln 2$$
$$x^2 \cdot \ln 2 - x\ln 0.6 = 0$$
$$x(x \cdot \ln 2 - \ln 0.6) = 0$$
$$x = 0 \quad \text{or} \quad x \cdot \ln 2 - \ln 0.6 = 0$$
$$x = \frac{\ln 0.6}{\ln 2} = -0.737$$

17. $\log x^2 = (\log x)^2$

$$2\log x = (\log x)^2$$
$$(\log x)^2 - 2\log x = 0$$
$$\log x(\log x - 2) = 0$$
$$\log x = 0 \quad \text{or} \quad \log x = 2$$
$$x = 1 \qquad x = 10^2 = 100$$

21. $2\log(3-x)=1$

$$\log(3-x)=\frac{1}{2}$$
$$3-x=10^{1/2}=3.162$$
$$-x=3.162-3$$
$$=0.162$$
$$x=-0.162$$

25. $3\ln 2+\ln(x-1)=\ln 24$

$$\ln 2^3+\ln(x-1)=\ln 24$$
$$\ln 8+\ln(x-1)=\ln 24$$
$$\ln[8(x-1)]=\ln 24$$
$$8(x-1)=24$$
$$x-1=3$$
$$x=4$$

29. $\log(2x-1)+\log(x+4)=1$

$$\log[(2x-1)(x+4)]=1$$
$$(2x-1)(x+4)=10$$
$$2x^2+7x-4=10$$
$$2x^2x^2 7x-14=0$$

Use the quadratic formula to solve for x:

$$x=\frac{-7\pm\sqrt{49-4(2)(-14)}}{2(2)}=\frac{-7\pm\sqrt{161}}{4}$$
$$=\frac{-7\pm 12.689}{4}=-4.92,\ 142$$

$x=1.42$ (Since logs are not defined on negatives.)

33. $4(3^x)=5$. Graph $y_1=4(3^x)-5$ and use the zero feature to solve.

$x=0.2031$

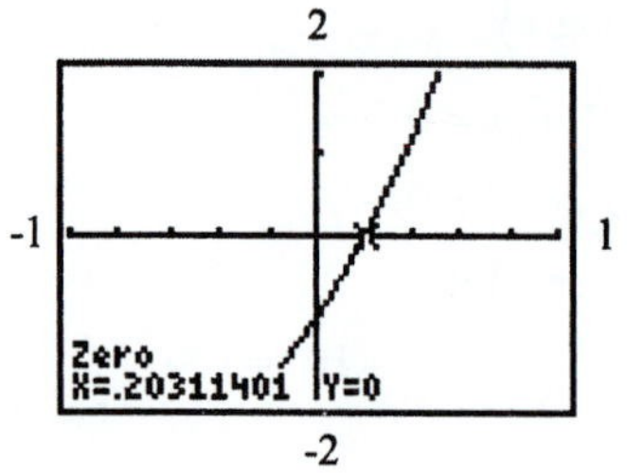

37. $2\ln 2-\ln x=-1$. Graph $y_1=2\ln 2-\ln x+1$ and use the zero feature to solve.

$x=10.87$

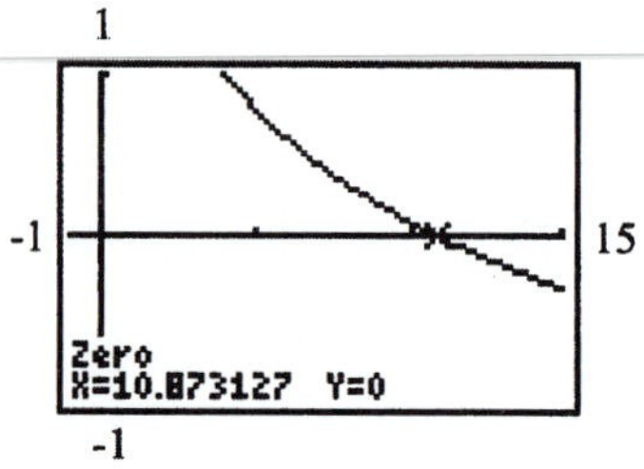

41. $y=1.5e^{-0.90x}=1.5e^{-0.90(7.1)}$

$y=0.0025$

45. $N=2^x\Rightarrow 2.6\times10^8=2^x$

$$\log 2^x=\log 2.6\times10^8$$
$$x\cdot\log 2=\log 2.6\times10^8$$
$$x=\frac{\log 2.6\times10^8}{\log 2}$$
$$=27.95393638\cdots$$
$$x=28$$

49. $pH=-\log(H^+)$

$$4.764=-\log(H^+)$$
$$-4.764=\log(H^+)$$
$$H^+=10^{-4.764}$$
$$H^+=1.772\times10^{-5}$$

53. $\ln c=\ln 15-0.20t$

$$\ln c-\ln 15=-0.20t$$
$$\ln\frac{c}{15}=-0.20t$$
$$\frac{c}{15}=e^{-0.20t}$$
$$c=15e^{-0.20t}$$

57. $2^x + 3^x = 50$. Graph $y_1 = 2^x + 3^x - 50$ and use the zero feature to solve.

$x = 3.353$

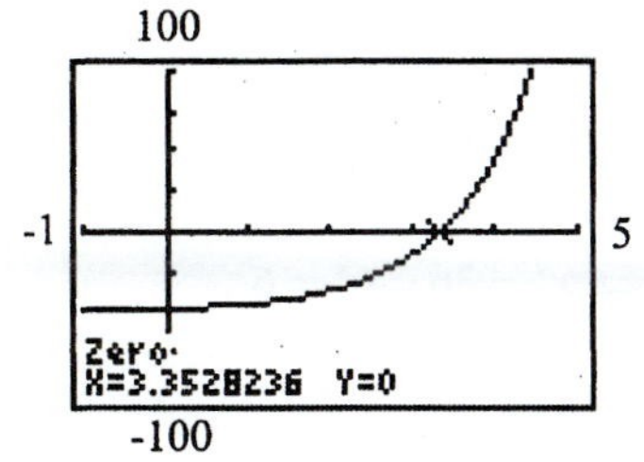

13.7 Graphs on Logarithmic and Semilogarithmic Paper

1. $y = 2(3^x)$

x	−1	0	2	3	4	5
y	0.67	2	18	54	162	486

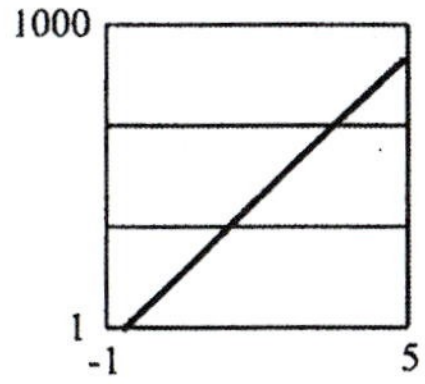

5. $y = 5(4^x)$

x	0	1	2	3	4	5
y	5	20	80	320	1280	5120

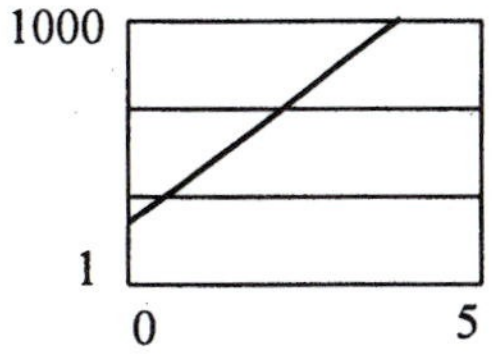

9. $y = 2x^3 + 6x$

x	0	1	2	4	6	8
y	0	8	28	152	468	1072

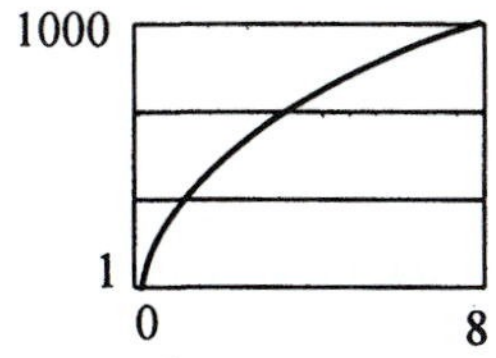

13. $y = x^{2/3}$

x	1	5	10	50	100	500	1000
y	1	2.9	4.6	13.6	21.5	63.0	100

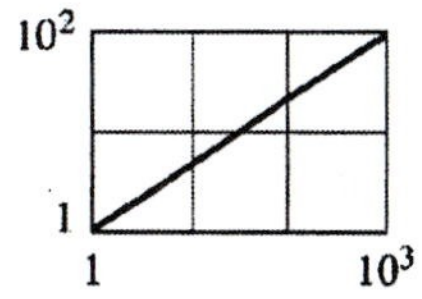

17. $x^2y^3 = 25$

x	0.1	0.5	1	10	50
y	50	10	5	0.5	0.1

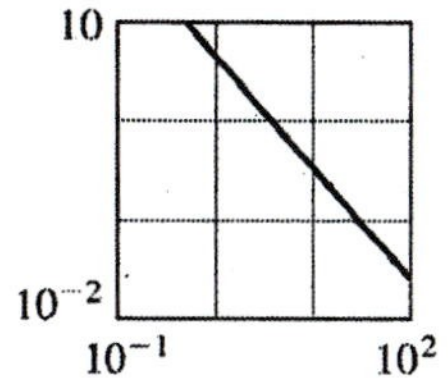

21. $y = 3x^6$, log–log paper

x	1	2	3	4
y	3	192	2187	12,288

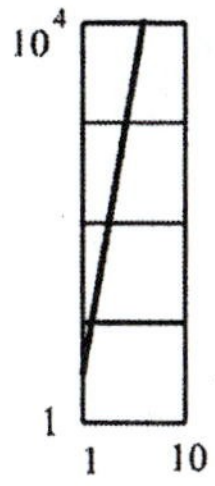

25. $x\sqrt{y} = 4,\ y = \dfrac{16}{x^2}$, log–log paper

x	1	25	50	75	100
y	16	0.0256	0.0064	$0.0028\overline{4}$	0.0016

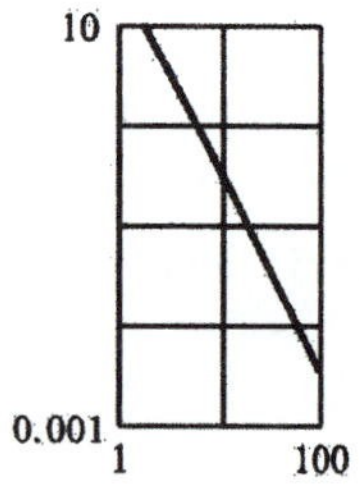

29. $N = N_0 e^{-0.028t},\ N_0 = 1000$

t	0	25	50	75	100
N	1000	496.6	246.6	122.5	60.81

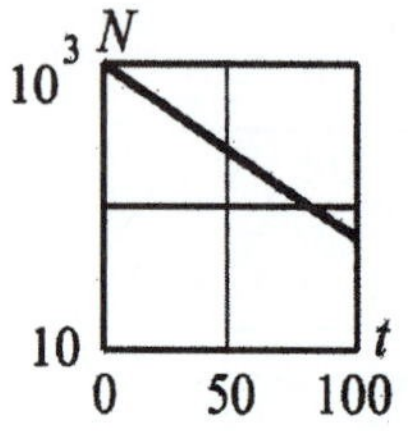

33.

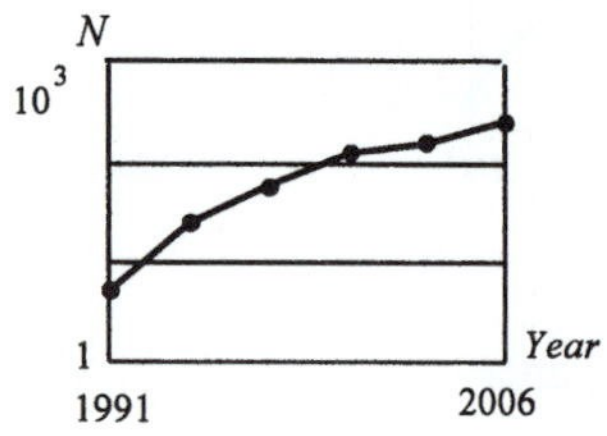

37.

d	0.63	1.3	1.9	2.5	3.8
R	600	190	100	72	46

d	5.0	7.5	10	15
R	29	17	10	6.0

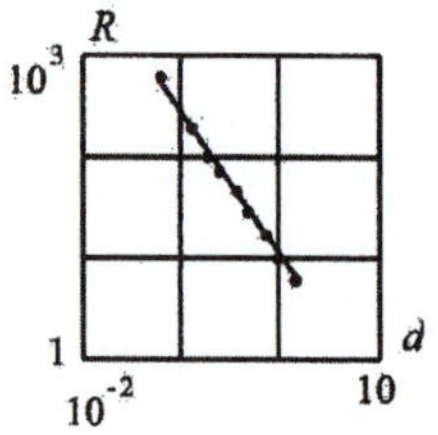

Chapter 13 Review Exercises

1. $\log_{10} x = 4 \Rightarrow x = 10^4 = 10{,}000$

5. $2\log_2 8 = x \Rightarrow 2^x = 64 = 2^6 = 6$

9. $\log_x 36 = 2 \Rightarrow x^2 = 36 = 6^2 = 6$

13. $\log_3 2x = \log_3 2 + \log_3 x$

17.
$$\begin{aligned}\log_2 28 &= \log_2\left(2^2 \cdot 7\right)\\ &= \log_2 2^2 + \log_2 7\\ &= 2\log_2 2 + \log_2 7\\ &= 2\cdot 1 + \log_2 7\\ &= 2 + \log_2 7\end{aligned}$$

21.
$$\begin{aligned}\log_3\left(\frac{9}{x}\right) &= \log_3 9 - \log_3 x\\ &= \log_3 3^2 - \log_3 x\\ &= 2\log_3 3 - \log_3 x\\ &= 2 - \log_3 x\end{aligned}$$

25.
$$\begin{aligned}\log_6 y &= \log_6 4 - \log_{6x}\\ \log_6 y &= \log_6 \frac{4}{x}\\ y &= \frac{4}{x}\end{aligned}$$

29.
$$\begin{aligned}\log_3 y &= \frac{1}{2}\log_3 7 + \frac{1}{2}\log_3 x\\ \log_3 y &= \log_3 \sqrt{7} + \log_3 \sqrt{x}\\ \log_3 y &= \log_3 \sqrt{7x}\\ y &= \sqrt{7x}\end{aligned}$$

33.
$$\begin{aligned}2\left(\log_4 y - 3\log_4 x\right) &= 3\\ \log_4 y - \log_4 x^3 &= \frac{3}{2}\\ \log_4 \frac{y}{x^3} &= \frac{3}{2}\\ \frac{y}{x^3} &= 4^{3/2} = 8\\ y &= 8x^3\end{aligned}$$

37. $y = 0.5(5^x)$

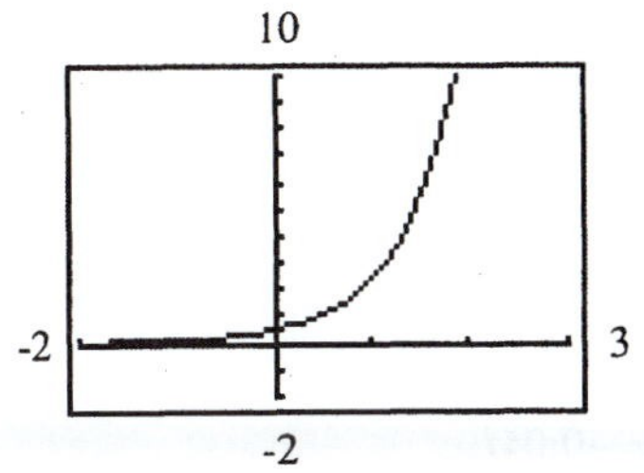

41. $y = \log_{3.15} x$

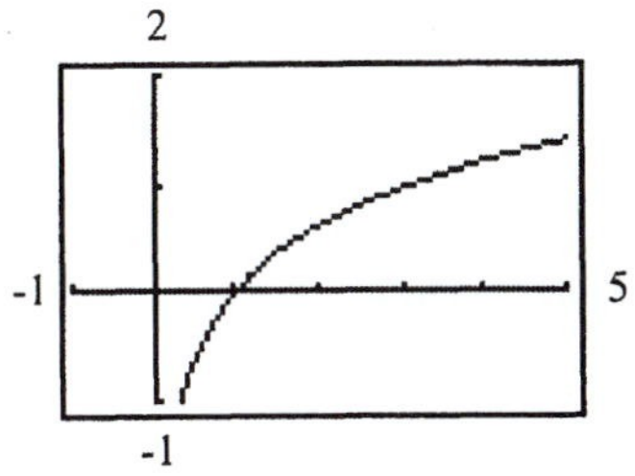

45. $\ln 8.86 = \dfrac{\log_{10} 8.86}{\log_{10} e} = 2.18$

49. $\log_{10} 65.89 = \dfrac{\ln 65.89}{\ln 10} = 1.819$

53.
$$\begin{aligned} e^{2x} &= 5 \\ \ln e^{2x} &= \ln 5 \\ 2x \cdot \ln e &= \ln 5 \\ 2x \cdot 1 &= \ln 5 \\ x &= \frac{\ln 5}{2} \end{aligned}$$

57.
$$\begin{aligned} \log_4 z + \log_4 6 &= \log_4 12 \\ \log(z \cdot 6) &= \log_4 12 \\ 6z &= 12 \\ z &= 2 \end{aligned}$$

61. $y = 8^x$

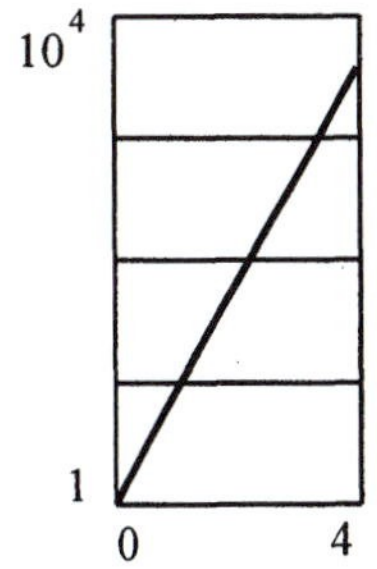

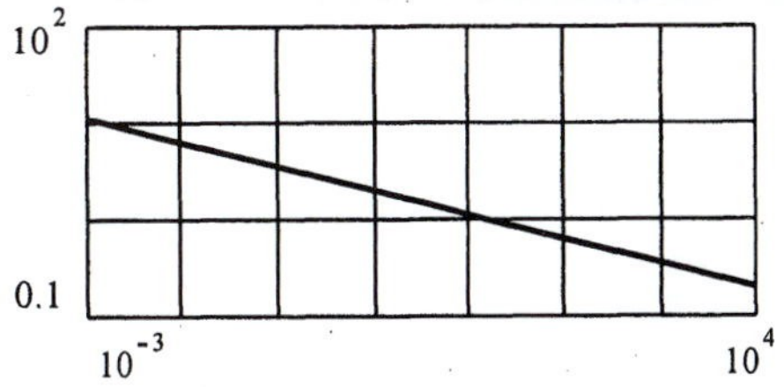

65. $10^{\log 4} = 4$

69. $2\log 3 - \log 6 = \log 1.5$

$0.1760912591 = 0.1760912591$

```
2log(3)-log(6)
         .1760912591
log(1.5)
         .1760912591
```

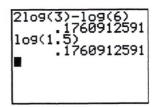

71.
$$\begin{aligned} \sqrt[3]{\ln e^8} - \sqrt{\log 10^4} &= \sqrt[3]{8 \ln e} - \sqrt{4 \log 10} \\ &= \sqrt[3]{8} - \sqrt{4} \\ &= 2 - 2 \\ &= 0 \end{aligned}$$

73. $f(x) = 2\log_b x$

$f(8) = 2\log_b 8 = \log_b 8^2 = \log_b 64 = 3$

$b^3 = 64 = 4^3 \Rightarrow b = 4$

$f(x) = 2\log_4 x$

$f(4) = 2\log_4 4 = 2$

77. Graph $y_1 = \dfrac{\ln x}{\ln 5} - 2x + 7$ and use the zero feature to solve.

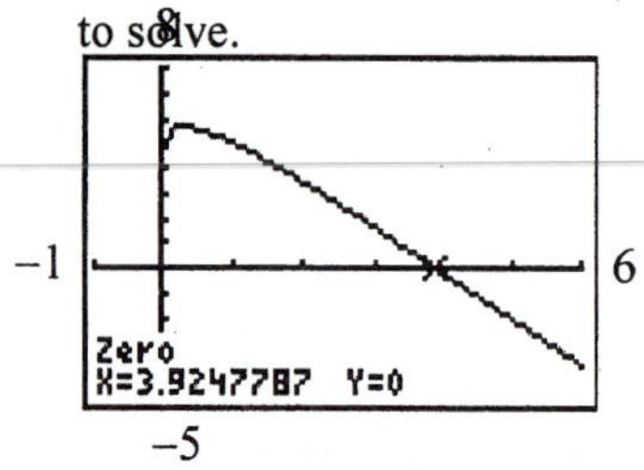

81. $\ln \dfrac{I}{I_0} = -\beta h$

$\dfrac{I}{I_0} = e^{-\beta h}$

$I = I_0 e^{-\beta h}$

85. $P = 2.25e^{0.00486t}$

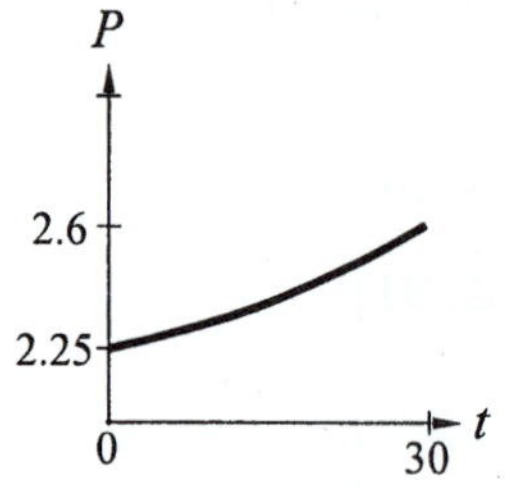

89. $2\ln\omega = \ln 3g + \ln\sin\theta - \ln l$

$\ln\omega^2 = \ln\dfrac{3g\sin\theta}{l}$

$\omega^2 = \dfrac{3g\sin\theta}{l}$

$\sin\theta = \dfrac{\omega^2 l}{3g}$

93. $m_1 - m_2 = 2.5\log\dfrac{b_2}{b_1}$

$-1.4 - 6.0 = 2.5\log\dfrac{b_2}{b_1}$

$-2.9 = \log\dfrac{b_2}{b_1}$

$\dfrac{b_2}{b_1} = 10^{-2.96}$

$\dfrac{b_1}{b_2} = 10^{2.96} = 910$ times brighter

97. $x = k(\ln I_0 - \ln I)$

$x = 5.00(\ln I_0 - \ln 0.850 I_0)$

$x = 5.00\ln\dfrac{I_0}{0.850 I_0}$

$x = 0.813$ cm

101. $\ln n = -0.04t + \ln 20$

$\ln n - \ln 20 = -0.04t$

$\ln\dfrac{n}{20} = -0.04t$

$\dfrac{n}{20} = e^{-0.04t}$

$n = 20e^{-0.04t}$

105. $y = (2\ln x)/3 + \ln 4 - \ln(\ln e^2)$ requires $x > 0$

$y = \dfrac{2}{3}\ln x + \ln 4 - \ln(2\ln e)$

$y = \ln x^{2/3} + \ln 4 - \ln 2$

$y = \ln x^{2/3} + \ln\dfrac{4}{2}$

$y = \ln x^{2/3} + \ln 2$

$y = \ln(2x^{2/3})$ which only requires $x \neq 0$.

(a) The two equations are equivalent for $x > 0$.

(b) The graph of $y = (2\ln x)/3 + \ln 4 - \ln(\ln e^2)$ contains only the right hand branch of the graph of $y = \ln(2x^{2/3})$.

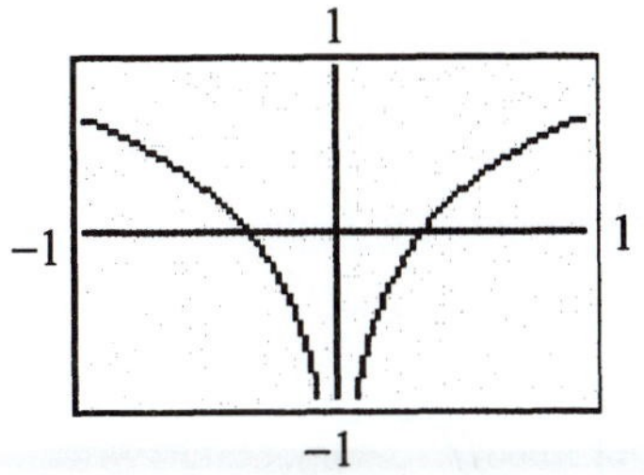

$y = \ln\left(2x^{2/3}\right)$

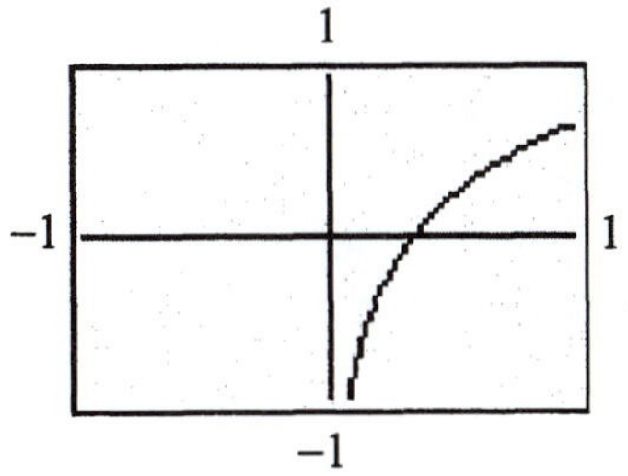

$y = (2\ln x)/3 + \ln 4 - \ln\left(\ln e^{2}\right)$

Chapter 14

ADDITIONAL TYPES OF EQUATIONS AND SYSTEMS OF EQUATIONS

14.1 Graphical Solution of Systems Of Equations

1. Graph $y_1 = 3x^2 + 6x$.

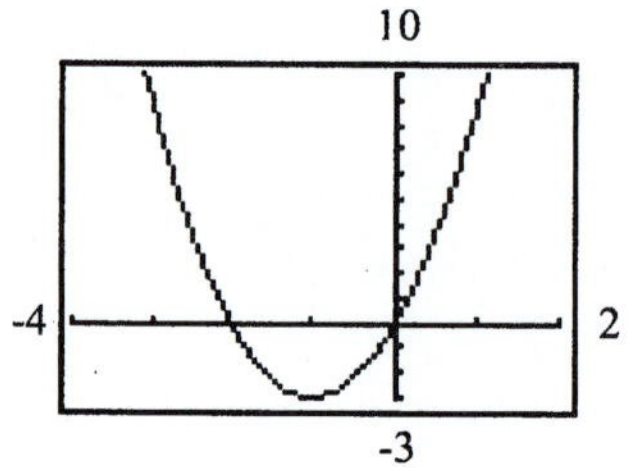

5. $y = 2x$

$x^2 + y^2 = 16 \Rightarrow y = \pm\sqrt{16 - x^2}$.

Graph $y_1 = 2x$, $y_2 = \sqrt{16 - x^2}$, and $y_3 = -\sqrt{16 - x^2}$.

Use the intersect feature to solve.

$x = 1.8,\ y = 3.6;\ x = -1.8,\ y = -3.6$

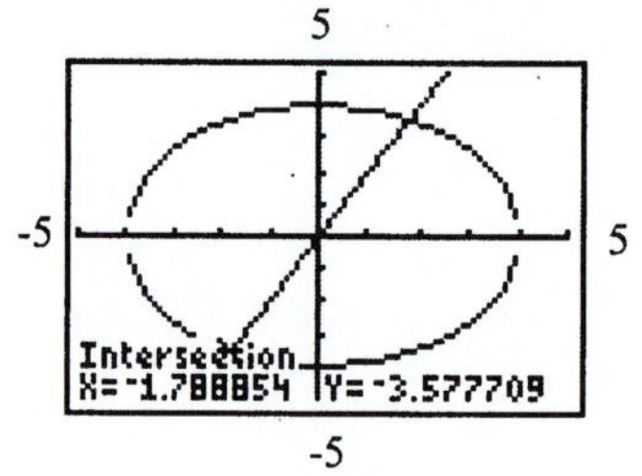

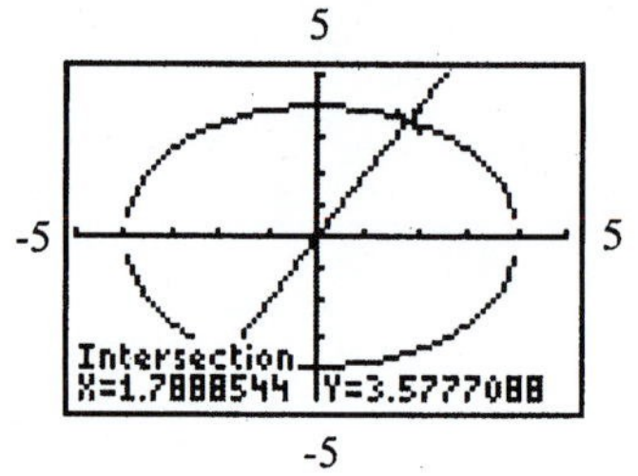

9. $y = x^2 - 2$

$4y = 12x - 7 \Rightarrow y = \dfrac{12x - 7}{4}$

Graph $y_1 = x^2 - 2$ and $y_2 = (12x - 7)/4$.

Use the intersect feature to solve.

$x = 1.5,\ y = 0.2$

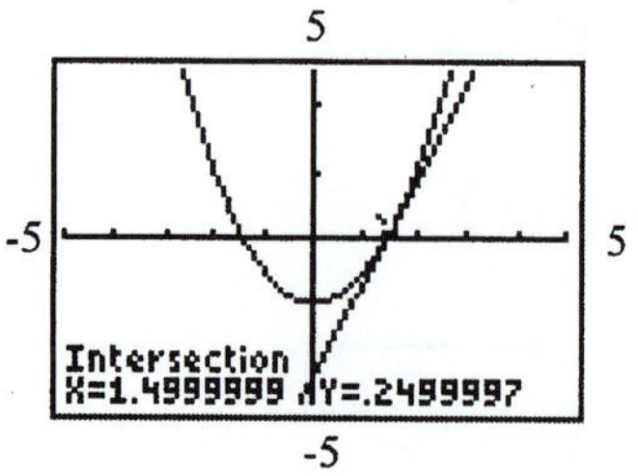

13. $x^2 + y^2 = 9 \Rightarrow 9 = \pm\sqrt{9 - x^2}$

$y = -x^2 + 4$

Graph $y_1 = \sqrt{9 - x^2}$, $y_2 = -\sqrt{9 - x^2}$, $y_3 = -x^2 + 4$.

Use the intersect feature to solve.

$x = 1.2,\ y = 2.8;\ x = -1.1,\ y = 2.8;$

$x = -2.4,\ y = -1.8;\ x = 2.3,\ y = -1.8$

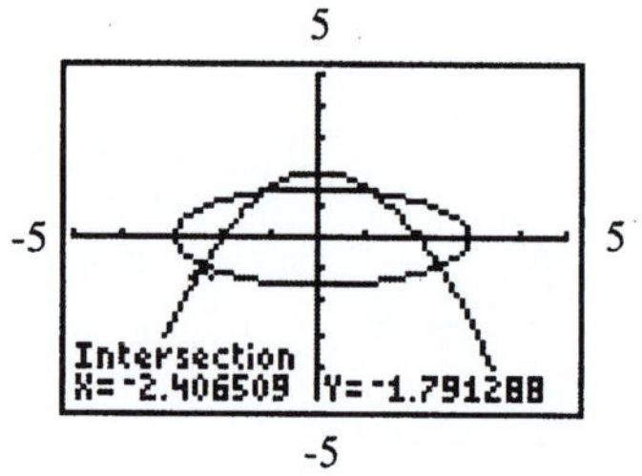

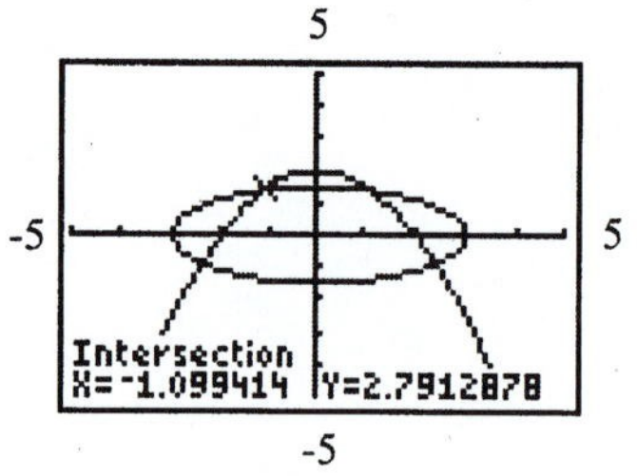

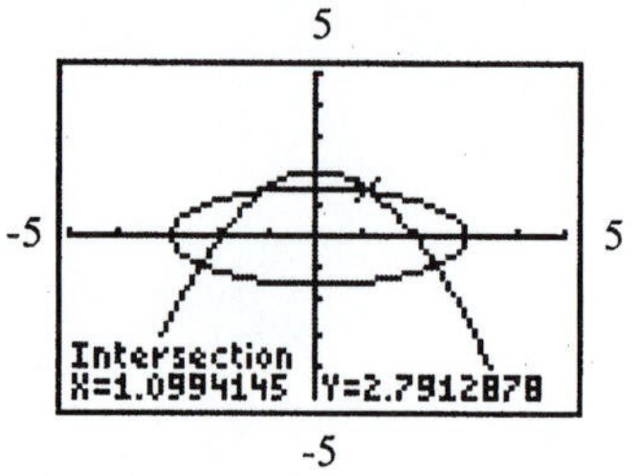

17. $2x^2 + 3y^2 = 19 \Rightarrow y = \pm\sqrt{(19 - 2x^2)/3}$ and $x^2 + y^2 = 9 \Rightarrow y = \pm\sqrt{9 - x^2}$. Graph $y_1 = \sqrt{(19 - 2x^2)/3}$, $y_2 = -\sqrt{(19 - 2x^2)/3}$, $y_3 = \sqrt{9 - x^2}$; $y_4 = -\sqrt{9 - x^2}$ and use the intersect feature to solve. $x = -2.8$, $y = 1.0$; $x = 2.8$, $y = 1.0$; $x = -2.8$, $y = -1.0$; $x = 2.8$, $y = -1.0$

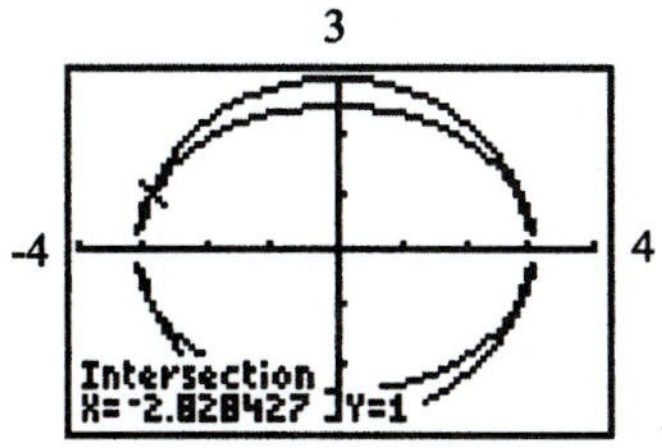

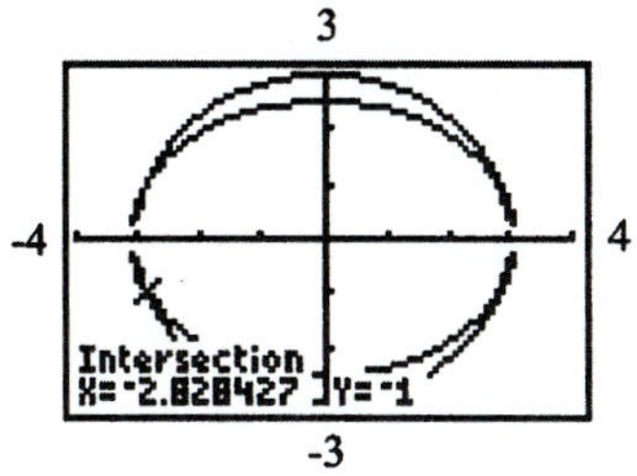

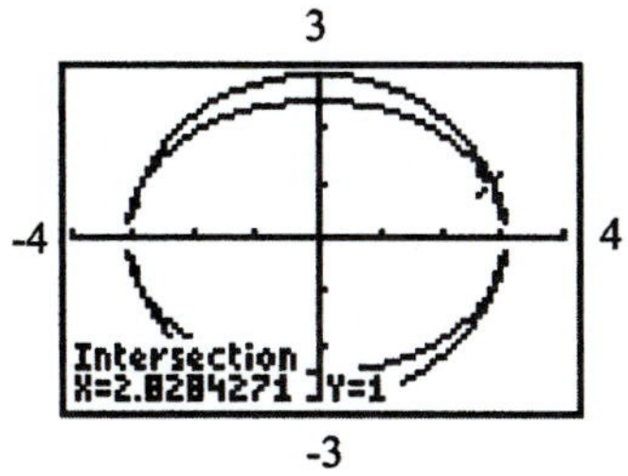

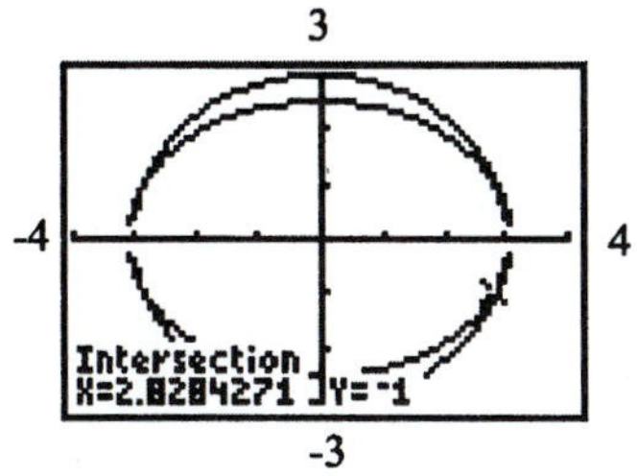

$x = -0.8$, $y = 2.5$; $x = 2.6$, $y = 0.7$

21. $y = x^2$

$y = \sin x$

Graph $y_1 = x^2$ and $y_2 = \sin x$ and use the intersect feature to solve.

$x = 0.0$, $y = 0.0$ and $x = 0.9$, $y = 0.8$

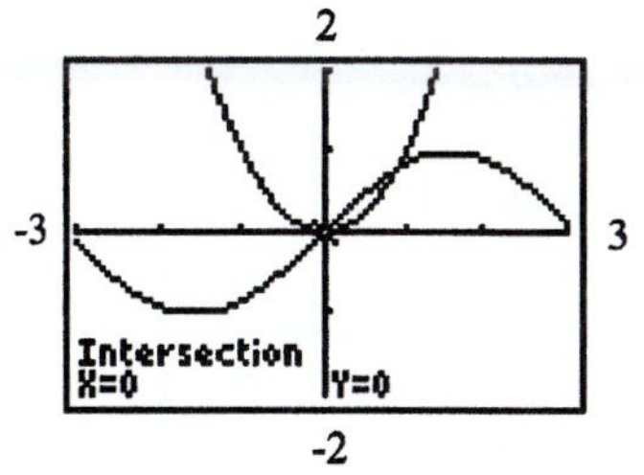

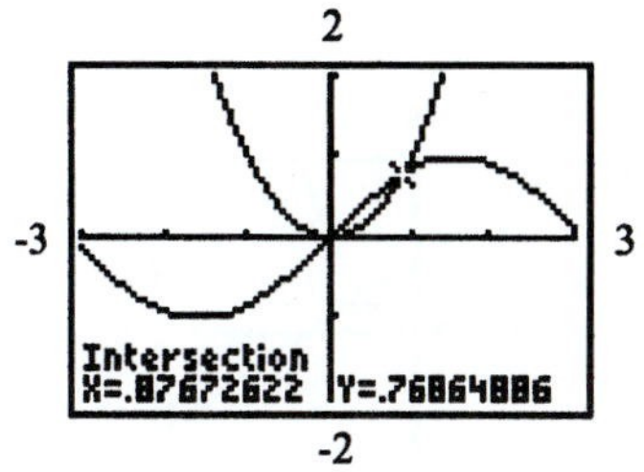

25. $x^2 - y^2 = 7 \Rightarrow y = \pm\sqrt{x^2 - 7}$, $y = 4\log_2 x = \dfrac{4\ln x}{\ln 2}$.

Graph $y_1 = \sqrt{x^2 - 7}$, $y_2 = -\sqrt{x^2 - 7}$, and $y_3 = \dfrac{4\ln x}{\ln 2}$.

Use the intersect feature to solve.

$x = 16.3$, $y = 16.1$

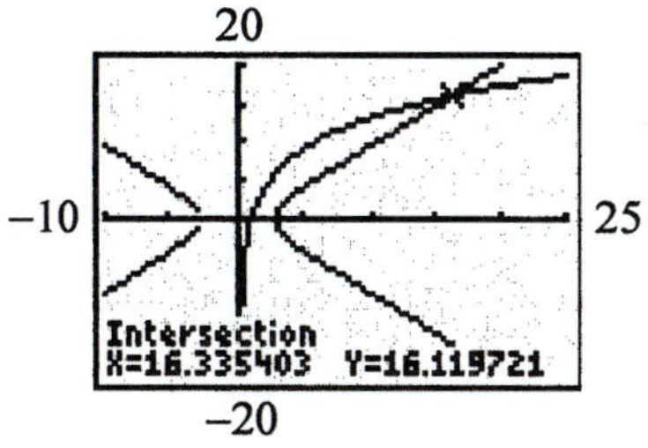

29. $$10^{x+y}=150$$
$$(x+y)\log 10=\log 150$$
$$y=\log 150-x$$
$$y=x^2$$

Graph $y_1=\log 150-x,\ y_2=x^2$ and use the intersect feature to solve.

$x=-2.06,\ y=4.23;\ x=1.06,\ y=1.12$

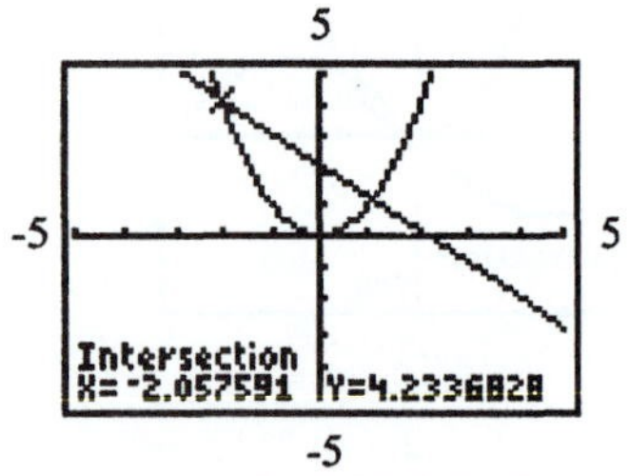

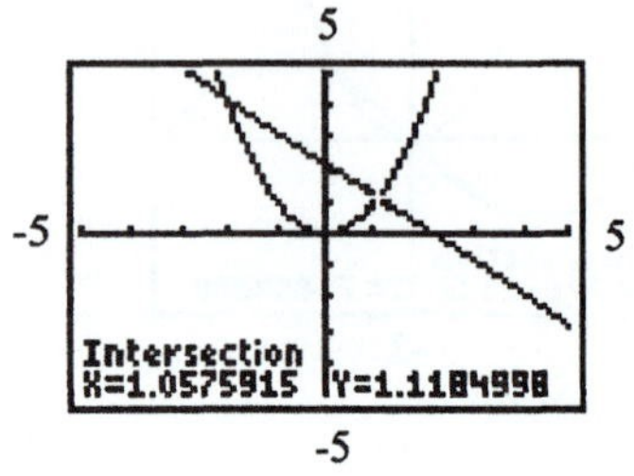

33. $y=3x,\ y>0;\ x^2+y^2=5.2^2=27.04,$ $x>0,\ y>0.$ Graph $y_1=3x,\ x>0$ and $y_2=\sqrt{27.04-x^2},\ x>0$ and use the intersect feature to solve. The solution is 4.9 mi N, 1.6 mi E.

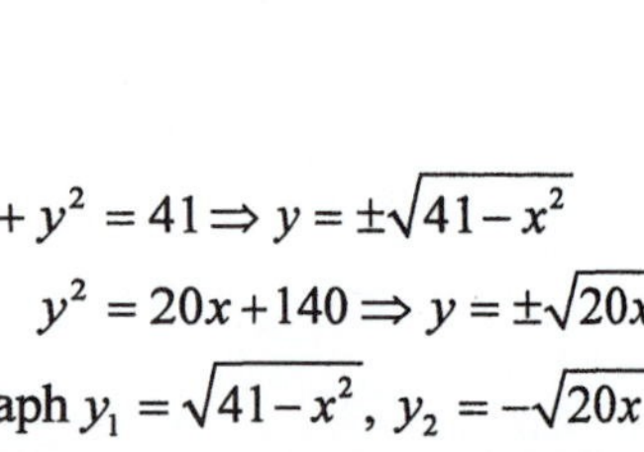

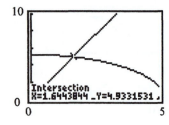

37. $x^2+y^2=41\Rightarrow y=\pm\sqrt{41-x^2}$
$$y^2=20x+140\Rightarrow y=\pm\sqrt{20x+140}$$

Graph $y_1=\sqrt{41-x^2},\ y_2=-\sqrt{20x+140}.$ From the graph, there is no intersection. No, the meteorite will not strike the earth.

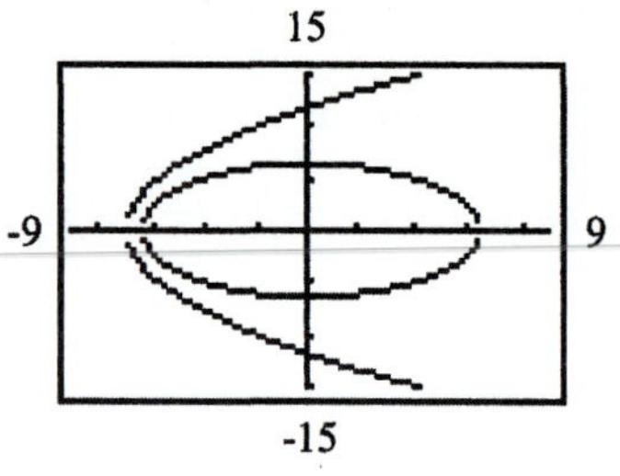

14.2 Algebraic Solution of Systems of Equations

1. $2x+y=4\Rightarrow y=4-2x,$ substitute into second equation $x^2-y^2=4$
$$x^2-(4-2x)^2=4$$
$$x^2-(16-16x+4x^2)=4$$
$$x^2-16+16x-4x^2=4$$
$$3x^2-16x+20=0$$
$$(x-2)(3x-10)=0$$
$$x-2=0 \quad\text{or}\quad 3x-10=0$$
$$x=2 \qquad x=\frac{10}{3}$$
$$y=4-2(2)=0 \qquad y=4-2\left(\frac{10}{3}\right)=-\frac{8}{3}$$

The solutions are $x=2,\ y=0;\ x=\frac{10}{3},\ y=-\frac{8}{3}.$

5. (1) $y=x+1$

(2) $y=x^2+1$
$$y=y$$
$x^2+1=x+1$ with y from (1) and (2)
$$x^2-x=0$$
$$x(x-1)=0$$
$$x=0 \quad\text{or}\quad x-1=0$$
$$y=x+1=1 \qquad x=1$$
$$y=x+1=2$$

The solutions are $x=0,\ y=1;\ x=1,\ y=2.$

9. (1) $x+y=1 \Rightarrow y=1-x$

(2) $x^2-y^2=1$

(2) $x^2-(1-x)^2=1$ with $y=1-x$ from (1)

$$x^2-1+2x-x^2=1$$
$$-1+2x=1$$
$$2x=2$$
$$x=1 \quad y=1-(1)=0$$

The solution is $x=1,\ y=0$.

13. (1) $wh=1$

(2) $w+h=2 \Rightarrow = 2-w$

(1) $w(2-w)=1$ with h from (2)

$$2w-w^2=1$$
$$w^2-2w+1=0$$
$$(w-1)^2=0$$
$$w-1=0$$
$$w=1 \Rightarrow (2)1+h=2,\ h=1$$

The solution is $w=1,\ h=1$.

17. $y=x^2$

$$y=3x^2-50$$
$$y=y \Rightarrow x^2=3x^2-50$$
$$2x^2=50$$
$$x^2=25$$
$$x=\pm 5$$
$$y=25$$

The solutions are $x=5,\ y=25,\ x=-5,\ y=25$.

21. (1) $D^2-1=R \Rightarrow D^2=1+R$

(2) $D^2-2R^2=1$

$$1+R-2R^2=1$$
$$R-2R^2=0$$
$$R(1-2R)=0$$

(2) with D^2 from (1)

$R=0$ or $1-2R=0$

$$2R=1$$
$$R=\frac{1}{2}$$

(1) with $R=0$, $D^2=1+0$

$$D=\pm 1$$

(1) with $R=\frac{1}{2}$, $D^2=1+\frac{1}{2}=\frac{3}{2}=\frac{6}{4}$

$$D=\frac{\pm\sqrt{6}}{2}$$

Solutions:

$R=0,\ D=1$

$R=0,\ D=-1$

$R=\frac{1}{2},\ D=\frac{\sqrt{6}}{2}$

$R=\frac{1}{2},\ D=\frac{-\sqrt{6}}{2}$

25. (1) $x^2+3y^2=37$

(2) $2x^2-9y^2=14$

$3\cdot(1) \Rightarrow 3x^2+9y^2=111$

(2) $2x^2-9y^2=14$

$$5x^2=125$$
$$x^2=25$$
$$x=\pm 5$$

(1) $(\pm 5)^2+3y^2=37$

$$25+3y^2=37$$
$$3y^2=12$$
$$y^2=4$$
$$y=\pm 2$$

Solution: $(5, 2), (5, -2), (-5, 2), (-5, -2)$

29. $x-y=a-b \Rightarrow y=x-a+b=x-(a-b)$

$x^2-y^2=a^2-b^2 \Rightarrow x^2-(x-(a-b))^2=a^2-b^2$

$$x^2-\left(x^2-2(a-b)x+(a-b)^2\right)=a^2-b^2$$
$$x^2-x^2+2(a-b)x=a^2+2ab-b^2=a^2-b^2$$
$$2(a-b)x+2ab-2a^2=0$$
$$(a-b)x-a(a-b)=0$$
$$x=a$$
$$y=x-a+b=a-a+b$$
$$y=b$$

The solution is $x=9,\ y=b$.

33.

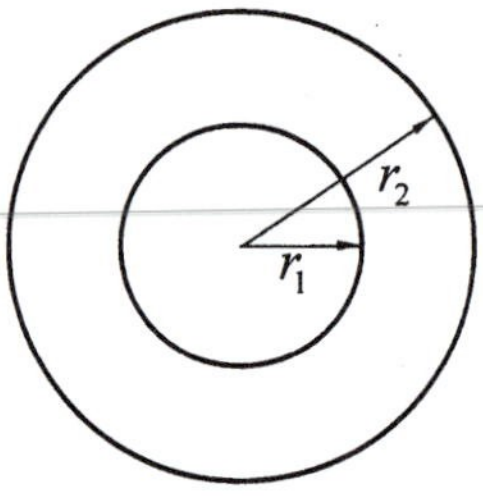

$$r_1 = r_2 - 2.00$$
$$\pi r_2^2 - \pi r_1^2 = 37.7$$
$$\pi r_2^2 - \pi(r_2 - 2.00)^2 = 37.7$$
$$\pi r_2^2 - \pi r_2^2 + 4.00\pi r_2 - 4.00\pi = 37.7$$
$$r_2 = \frac{37.7 + 4.00\pi}{4.00\pi}$$
$$r_2 = 4.00$$
$$r_1 = r_2 - 2.00 = 2.00$$

The radii are 2.00 cm and 4.00 cm.

37.

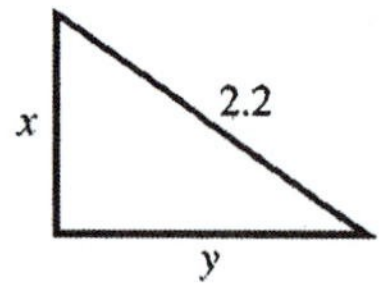

(1) $x + y + 2.2 = 4.6$

$$x + y = 2.4 \Rightarrow y = 2.4 - x$$

(2) $x^2 + y^2 = 2.2^2$

$$x^2 + y^2 = 4.84$$

(2) $x^2 + (2.4 - x)^2 = 4.84$, (2) with y from (1).

$$x^2 + 5.76 - 4.8x + x^2 = 4.84$$
$$2x^2 - 4.8x + 0.92 = 0$$
$$x = \frac{-(-4.8) \pm \sqrt{(-4.8)^2 - 4(2)(0.92)}}{2(2)}$$
$$= \frac{4.8 \pm \sqrt{15.68}}{4} = +2.2, -0.2$$

$x = 2.2$ in (1), $2.2 + y = 2.4$

$$y = 0.2$$

$x = 0.2$ in (1), $0.2 + y = 2.4$

$$y = 2.2$$

The lengths of the sides of the truss are 2.2 m
And 0.2 m.

41.

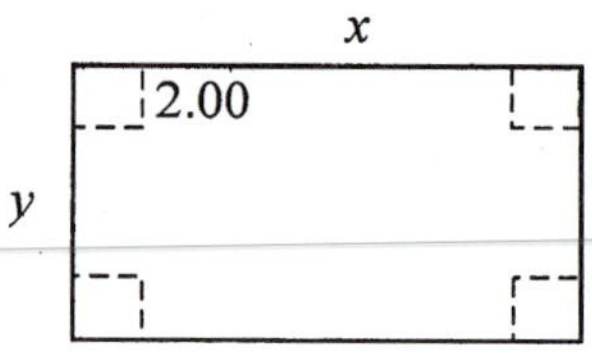

$$xy(2.00) = 224 \Rightarrow y = \frac{112}{x}$$
$$(x + 4.00)(y + 4.00) = 216$$
$$(x + 4.00)\left(\frac{112}{x} + 4.00\right) = 216$$
$$x^2 - 22.0x + 112 = 0$$
$$(x - 8.00)(x - 14.0) = 0$$

$x = 8.00$ or $x = 14.0$

$y = 14.0$ $\quad$ $y = 8.0$

The dimensions of the sheet are
8.00 in. by 12.0 in.

43. Let V_w = wind velocity

V_{vv} = velocity with wind $= 610 + V_w$

V_{aw} = velocity against wind $= 610 - V_w$

$$\frac{3660}{V_{ww}} + 1.6 = \frac{3660}{V_{aw}}$$
$$\frac{3660}{610 + V_w} + 1.6 = \frac{3660}{610 - V_w}$$
$$-1.6_w^2 - 7320v_w + 595{,}360 = 0$$

Solve using quadratic formula,

$V_w = 80$ mi/h

14.3 Equations in Quadratic Form

1. $2x^4 - 7x^2 = 4$

$$2x^4 - 7x^2 - 4 = 0, \text{ let } y = x^2$$
$$2y^2 - 7y - 4 = 0$$
$$(y - 4)(2y + 1) = 0$$

$y-4=0 \quad \text{or} \quad 2y+1=0$

$y=4 \qquad y=-\frac{1}{2}$

$x^2=4 \qquad x^2=-\frac{1}{2}$

$x=\pm 2 \qquad x=\pm\frac{\sqrt{2}}{2}j$

Check:

$2(\pm 2)^4-7(\pm 2)^2=32-28=4$

$2\left(\pm\frac{\sqrt{2}}{2}j\right)^4-7\left(\pm\frac{\sqrt{2}}{2}j\right)^2=\frac{1}{2}+\frac{7}{2}=4$

5. $x^{-2}-2x^{-1}-8=0$

Let $y=x^{-1}$, $y^2=x^{-2}$, then

$y^2-2y-8=0$

$(y-4)(y+2)=0$

$y-4=0 \quad \text{or} \quad y+2=0$

$y=4 \qquad y=-2$

$x^{-1}=4 \qquad x^{-1}=-2$

$x=\frac{1}{4} \qquad x=\frac{-1}{x}$

Check:

$\left(\frac{1}{4}\right)^{-2}-2\left(\frac{1}{4}\right)^{-1}-8=16-8-8=0$

$\left(-\frac{1}{2}\right)^{-2}-2\left(-\frac{1}{2}\right)^{-1}-8=4+4-8=0$

9. $2x-7\sqrt{x}+5=0$

Let $y=\sqrt{x}$, $y^2=x$, then

$2y^2-7y+5=0$

$(2y-5)(y-1)=0$

$2y-5=0 \quad \text{or} \quad y-1=0$

$2y=5 \qquad y=1$

$y=\frac{5}{2} \qquad \sqrt{x}=1$

$\sqrt{x}=\frac{2}{5} \qquad x=1$

$x=\frac{25}{4}$

Check:

$2\left(\frac{25}{4}\right)-7\sqrt{\frac{25}{4}}+5=\frac{25}{2}-\frac{35}{2}+\frac{35}{2}+\frac{10}{2}=0$

$2(1)-7\sqrt{1}+5=2-7+5=0$

13. $x^{2/3}-2x^{1/3}-15=0$

Let $y=x^{1/3}$, $y^2=x^{2/3}$, then

$y^2-2y-15=0$

$(y-5)(y+3)=0$

$y-5=0 \quad \text{or} \quad y+3=0$

$y=5 \qquad y=-3$

$x=125 \qquad x=-27$

Check:

$125^{2/3}-2(125)^{1/3}-15=25-10-15=0$

$(-27)^{2/3}-2(-27)^{1/3}-15=9+6-15=0$

17. $(x-1)-\sqrt{x-1}=20$

Let $y=\sqrt{x-1}$, $y^2=x-1$, then

$y^2-y-20=0$

$(y-5)(y+4)=0$

$y-5=0$

$y=25$

$\sqrt{x-1}=5$

$x-1=25$

$x=26$

or

$y+4=0$

$y=-4$ reject since $y=\sqrt{x-1}$ requires $y\geq 0$

Check:

$(26-1)-\sqrt{26-1}=25-\sqrt{25}$

$=25-5$

$=20$

21. $x-3\sqrt{x-2}=6$

Let $y=\sqrt{x-2}$, $y^2=x-2\Rightarrow y^2+2=x$, then

$y^2+2-3y=6$

$y^2-3y-4=0$

$(y-4)(y+1)=0$

$y-4=0$

$y=4$

$\sqrt{x-2}=4$

$x-2=16$

$x=18$

or

$y+1=0$

$y=-1$

$\sqrt{x-2}=-1$ has no solution since $\sqrt{x-2}\geq 0$

Check:

$18-3\sqrt{18-2}=18-3\sqrt{16}$

$=18-3(4)$

$=18-12$

$=6$

25. $e^{2x}-e^x=0$

$e^x\left(e^x-1\right)=0$

$e^x=0$, no solution

$e^x-1=0$

$e^x=1$

$x=0$

Check: $e^{2(0)}-e^0=1-1=0$

29. $x+2=3\sqrt{x}$

$x-3\sqrt{2}+2=0$, let $y=\sqrt{x}$, $y\geq 0$, $x\geq 0$

$y^2-3y+2=0$

$(y-2)(y-1)=0$

$y-2=0$ or $y-1=0$

$y=2$ $\quad y=1$

$\sqrt{x}=2$ $\quad \sqrt{x}=1$

$x=4$ $\quad x=1$

Check:

$4+2\overset{?}{=}3\sqrt{4}$

$6\overset{?}{=}3(2)$

$6=6$

$1+2\overset{?}{=}3\sqrt{1}$

$3=3$

Both solutions check.

33. $\log\left(x^4+4\right)-\log 5x^2=0$

$\log\dfrac{x^4+4}{5x^2}=0\Rightarrow\dfrac{x^4+4}{5x^2}=1$

$x^4+4=5x^2$

$\left(x^2\right)^2-5x^2+4=0$

$\left(x^2-4\right)\left(x^2-1\right)=0$

$x^2=4$ or $x^2=1$

$x=\pm 2$ $\quad x=\pm 1$

All four solutions check.

37. $\sqrt{F}=2\sqrt{p}/(1-p)$

$\sqrt{16}=\dfrac{2\sqrt{p}}{1-p}$

$4(1-p)=2\sqrt{p}$

$2(1-p)=\sqrt{p}$

$2-2p=\sqrt{p}$

$2p+\sqrt{p}-2=0$, let $x=\sqrt{p}$, $x\geq 0$, $p\geq 0$

$2x^2+x-2=0$

$x=\dfrac{-1\pm\sqrt{1^2-4(2)(-2)}}{2(2)}=\dfrac{-1\pm\sqrt{17}}{4}$

$\sqrt{p}=\dfrac{-1+\sqrt{17}}{4}$

$p=0.610$

14.4 Equations with Radicals

1. $2\sqrt{3x-1}=3$

$4(3x-1)=9$

$12x-4=9$

$12x=13$

$x=\frac{13}{12}$

Check: $2\sqrt{3\left(\frac{13}{12}\right)-1}\overset{?}{=}3$

$3=3$

5. $\sqrt{x-8}=2$; square both sides

$x-8=4$

$x=12$

Check:

$\sqrt{12-8}=\sqrt{4}=2$

9. $\sqrt{3x+2}=3x$

$3x+2=9x^2$

$9x^2-3x-2=0$

$(3x+1)(3x-2)=0$

$3x+1=0$ or $3x-2=0$

$3x=-1$ $\quad$ $3x=2$

$x=\frac{-1}{3}$ $\quad$ $x=\frac{2}{3}$

Check:

$\sqrt{3\cdot\frac{1}{3}+2}$	$3\cdot\frac{-1}{3}$
$\sqrt{-1+2}$	-1
$\sqrt{1}$	
1	

$x=\frac{-1}{3}$ is not a solution.

Check:

$\sqrt{3\cdot\frac{2}{3}+2}$	$3\cdot\frac{2}{3}$
$\sqrt{2+2}$	2
$\sqrt{4}$	
2	

$x=\frac{2}{3}$ is the solution.

13. $\sqrt[3]{y-5}=3$

$y-5=3^3$

$=27$

$y=32$

The solution is $y=32$, which checks.

17. $\sqrt{x^2-9}=4$

$x^2-9=16$

$x^2=25$

$x=\pm5$

Check:

$\sqrt{(\pm5)^2-9}\overset{?}{=}4$

$\sqrt{25-9}\overset{?}{=}4$

$\sqrt{16}\overset{?}{=}4$

$4=4$

$x=\pm5$ is the solution.

21. $\sqrt{5+\sqrt{x}}=\sqrt{x}-1$

$5+\sqrt{x}=\left(\sqrt{x}-1\right)^2$

$5+\sqrt{x}=x-2\sqrt{x}+1$

$x-3\sqrt{x}-4=0$

$\left(\sqrt{x}-4\right)\left(\sqrt{x}+1\right)=0$

$x=16$

$\sqrt{x}=-1$, not possible

The solution is $x=16$ since it checks.

25. $2\sqrt{x+2}-\sqrt{3x+4}=1$

$$2\sqrt{x+2}=1+\sqrt{3x+4}$$
$$4(x+2)=\left(1+\sqrt{3x+4}\right)^2$$
$$4x+8=1+2\sqrt{3x+4}+3x+4$$
$$x+3=2\sqrt{3x+4}$$
$$(x+3)^2=4(3x+4)$$
$$x^2+6x+9=12x+16$$
$$x^2-6x-7=0$$
$$(x-7)(x+1)=0$$

The solutions are $x=-1$ and $x=7$, which check.

29. $\sqrt{6x+5}-\sqrt{x+4}=2 \Rightarrow \sqrt{6x+5}=\sqrt{x+4}+2$

$$6x+5=x+4+4\sqrt{x+4}+4$$
$$5x-3=4\sqrt{x+4}$$
$$25x^2-30x+9=16x+64$$

$25x^2-46x-55=0$ from which, using quadratic formula

$$x=\frac{23\pm4\sqrt{119}}{25}.$$

33. $\sqrt{x-2}=\sqrt[4]{x-2}+12$

$$\left(\sqrt{x-2}-12\right)^2=\left(\sqrt[4]{x-2}\right)^2$$
$$\sqrt{x-2}=x-2-24\sqrt{x-2}+144$$
$$\left(25\sqrt{x-2}\right)^2=(x+142)^2$$
$$625(x-2)=x^2+284x+20,164$$
$$x^2-341x+21,414=0$$
$$(x-258)(x-83)=0$$

The solution is $x=258$ and $x=83$ does not check.

37. $\sqrt{2x+1}+3\sqrt{x}=9$

$$\sqrt{2x+1}=9-3\sqrt{x}$$
$$2x+1=81-54\sqrt{x}+9x$$
$$54\sqrt{x}=7x+80$$
$$2916x=49x^2+1120x+6400$$
$$49x^2-1796x+6400=0$$
$$(x-4)(49x-1600)=0$$

$x-4=0$ or $49x=1600$

$x=4$ $\quad x=\dfrac{1600}{49}$

Check:

$$\sqrt{2(4)+1}+3\sqrt{4}\overset{?}{=}9$$
$$3+6\overset{?}{=}9$$
$$9=9$$
$$\sqrt{2\left(\frac{1600}{49}\right)+1}\overset{?}{=}9$$
$$\frac{57}{7}+\frac{120}{7}\overset{?}{=}9$$
$$\frac{177}{9}\neq 9$$

$x=4$ is the solution

41. $\sqrt{x-1}+x=3 \Rightarrow \sqrt{x-1}=3-x$

$$x-1=9-6x+x^2$$
$$x^2-7x+10=0$$
$$(x-5)(x-2)=0$$

$x=5$ or $x=2$

$x=5$ does not check.

The solution is $x=2$.

The equations, $\sqrt{x-1}=x-3$ and $\sqrt{x-1}=3-x$, are different and have different solutions.

45. $kC=\sqrt{R_1^2-R_2^2}+\sqrt{r_1^2-r_2^2}-A$

$$kC+A-\sqrt{R_1^2-R_2^2}=\sqrt{r_1^2-r_2^2}$$
$$\left(kC+A-\sqrt{R_1^2-R_2^2}\right)=r_1^2-r_2^2$$
$$r_1^2=\left(kC+A-\sqrt{R_1^2-R_2^2}\right)^2+r_2^2$$

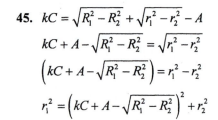

49.

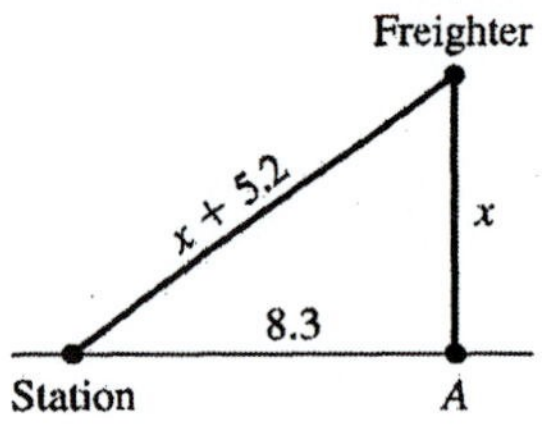

$$(x+5.2)^2 = x^2 + 8.3^2$$
$$x^2 + 10.4x + 27.04 = x^2 + 68.89$$
$$10.4x = 41.85$$
$$x = 4.0$$
$$x + 5.2 = 4.0 + 5.2 = 9.2$$

The freighter is 9.2 km from the station.

Chapter 14 Review Exercises

1. Graph $y = \dfrac{6-x}{2}$, $y_2 = 4x^2$, and use the intersect.

$x = -0.9,\ y = 3.5;\ x = 0.8,\ y = 2.6$

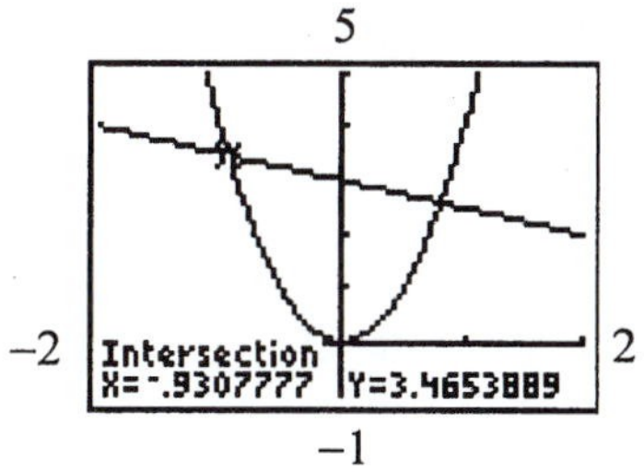

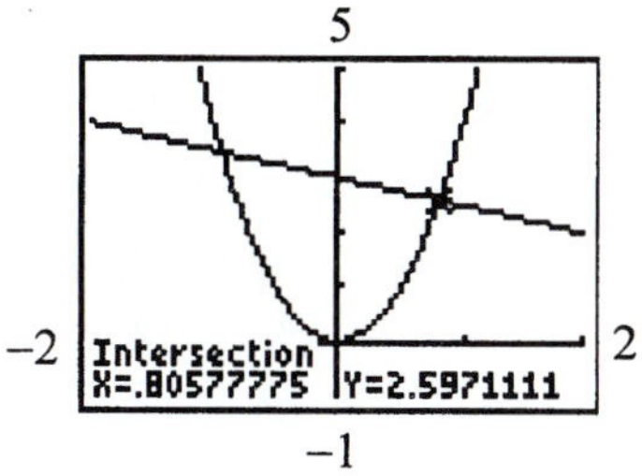

5. Graph $y_1 = x^2 + 1$, $y_2 = \sqrt{\dfrac{29-4x^2}{16}}$, $y_3 = -\sqrt{\dfrac{29-4x^2}{16}}$ and use the intersect. $x = -0.6,\ y = 1.3;\ x = 0.6,$ $y = 1.3$

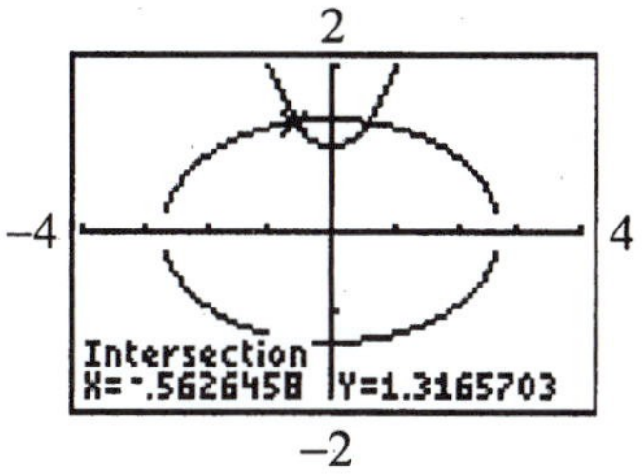

9. Graph $y_1 = x^2 - 2x$, $y_2 = 1 - e^{-x}$, and use intersect.

$x = 2.2,\ y = 0.8$

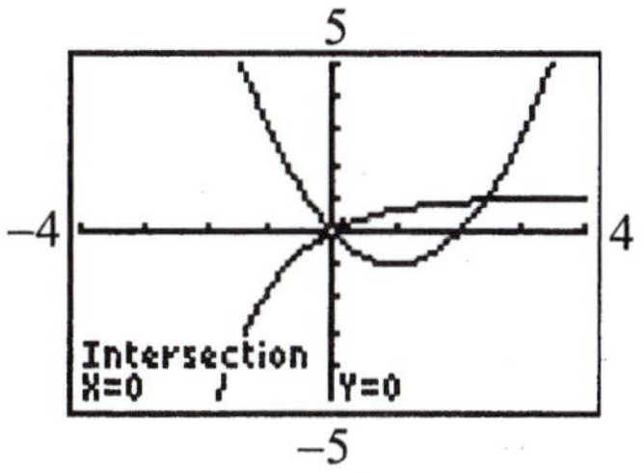

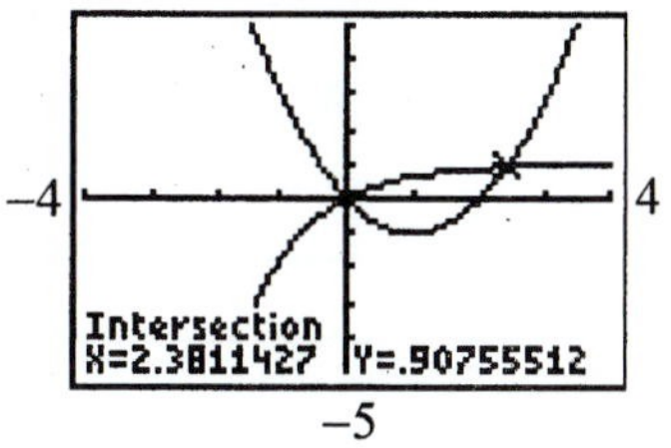

13. (1) $2L = R^2$

(2) $R^2 + L^2 = 3$

$$2L + L^2 = 3$$
$$L^2 + 2L - 3 = 0$$
$$(L+3)(L-1) = 0$$

$L + 3 = 0 \qquad L - 1 = 0$

$L = -3 \qquad L = 1$

$2(-3) = R^2 \qquad 2(1) = R$

no solution $\qquad R = \pm\sqrt{2}$

$R = \pm\sqrt{2},\ L = 1$

17. (1) $4x^2 - 7y^2 = 21$

(2) $x^2 + 2y^2 = 99$

(1) $8x^2 - 14y^2 = 42$

(2) $7x^2 + 14y^2 = 693$

$$15x^2 = 735$$
$$x^2 = 49$$
$$x = \pm 7$$

(2) $(\pm 7)^2 + 2y^2 = 99$

$$49 + 2y^2 = 99$$
$$2y^2 = 50$$
$$y^2 = 25$$
$$y = \pm 5$$

$x = 7,\ y = \pm 5;\ x = -7,\ y = \pm 5$

21. $x^4 - 20x^2 + 64 = 0$

Let $y = x^2$, $y^2 = x^4$

$$y^2 - 20y + 64 = 0$$
$$(y-16)(y-4) = 0$$

$y - 16 = 0$ or $y - 4 = 0$

$y = 16 \qquad y = 4$

$x^2 = 16 \qquad x^2 = 4$

$x = \pm 4 \qquad x = \pm 2$

25. $D^{-2} + 4D^{-1} - 21 = 0$

Let $x = D^{-1}$, $x^2 = D^{-2}$

$$x^2 + 4x - 21 = 0$$
$$(x+7)(x-3) = 0$$

$x+7=0$ or $x-3=0$

$x=-7$ $\quad x=3$

$D^{-1}=-7$ $\quad D^{-1}=3$

$D=\dfrac{-1}{7}$ $\quad D=\dfrac{1}{3}$

29. $\dfrac{4}{r^2+1}+\dfrac{7}{2r^2+1}=2$

$4(2r^2+1)+7(r^2+1)=2(r^2+1)(2r^2+1)$

$8r^2+4+7r^2+7=2(2r^4+3r^2+1)$

$15r^2+11=4r^4+6r^2+2$

$4r^4-9r^2-9=0$. Let $x=r^2$, $x^2=r^4$

$4x^2-9x-9=0$

$(4x+3)(x-3)=0$

$4x+3=0$ or $x-3=0$

$4x=-3$ $\quad x=3$

$x=\dfrac{-3}{4}$ $\quad r^2=3$

$r^2=\dfrac{-3}{4}$ $\quad r=\pm\sqrt{3}$

$r=\dfrac{\pm\sqrt{3}}{2}j$

33. $\sqrt{5x+9}+1=x$

$\sqrt{5x+9}=x-1$

$5x+9=x^2-2x+1$

$x^2-7x-8=0$

$(x-8)(x+1)=0$

$x-8=0$ or $x+1=0$

$x=8$ $\quad x=-1$

Check:

$\sqrt{5\cdot 8+9}+1$	8	$\sqrt{5(-1)+9}+1$	-1
$\sqrt{49}$		$\sqrt{-5+9}+1$	
$7+1$		$\sqrt{4}+1$	
8		$2+1$	
		3	

$x=8$ is a solution. $\quad x=-1$ is not a solution.

37. $\sqrt{n+4}+2\sqrt{n+2}=3$

$\sqrt{n+4}=3-2\sqrt{n+2}$

$n+4=9-12\sqrt{n+2}+4(n+2)$

$n+4=9-12\sqrt{n+2}+4n+8$

$12\sqrt{n+2}=13+3n$

$144(n+2)=169+78n+9n^2$

$144n+288=169+78n+9n^2$

$9n^2-66n-119=0$

From the quadratic formula,

$$n=\frac{-(-66)\pm\sqrt{(-66)^2-4(9)(-119)}}{2(9)}=\frac{66\pm\sqrt{8640}}{18}$$

Check:

$$\sqrt{\frac{66+\sqrt{8640}}{18}+4}+2\sqrt{\frac{66+\sqrt{8640}}{18}+2}=10.1639\cdots$$

$\dfrac{66+\sqrt{8640}}{18}$ does not check

$$\sqrt{\frac{66-\sqrt{8640}}{18}+4}+2\sqrt{\frac{66-\sqrt{8640}}{18}+2}=3$$

$\dfrac{66-\sqrt{8640}}{18}$ is the solution.

41. $x^3-2x^{3/2}-48=0$

$x^3-2(x^3)^{1/2}-48=0$, let $x^3=y$

$y-2y^{1/2}-48=0$, let $z=y^{1/2}$

$z^2-2z-48=0$

$(z-8)(z+6)=0$

$z-8=0$ or $z+6=0$

$z=8$ $\quad z=-6$

$y^{1/2}=8$ $\quad y^{1/2}=-6$, reject

$y=64$

$x^3=64$

$x=4$

```
4→X
                    4
X³-2X^(3/2)-48
                    0
█
```

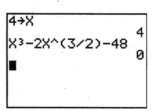

$x=4$ is the solution.

45. $\sqrt[3]{x^3-7}=x-1$

$x^3-7=x^3-3x^2+3x-1$

$3x^2-3x-6=0$

$x^2-x-2=0$

$x-2=0 \qquad x+1=0$

$x=2 \qquad x=-1$

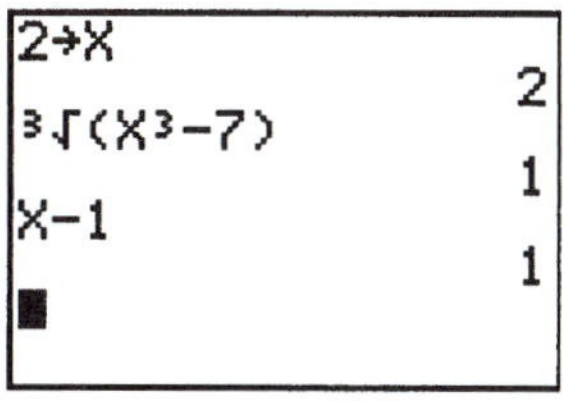

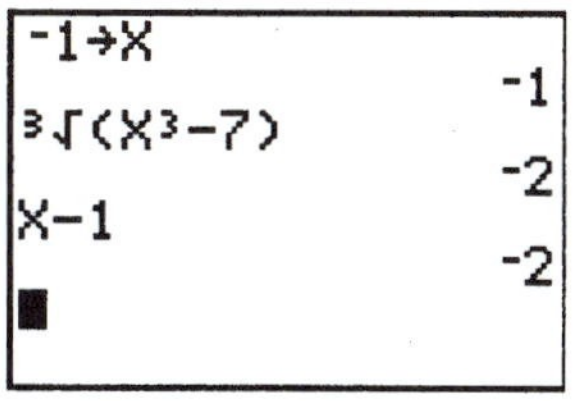

$x=2,\ x=-1$ is the solution.

49. $\sqrt{\sqrt{x}-1}=2$

$\sqrt{x}-1=4$

$\sqrt{x}=5$

$x=25$

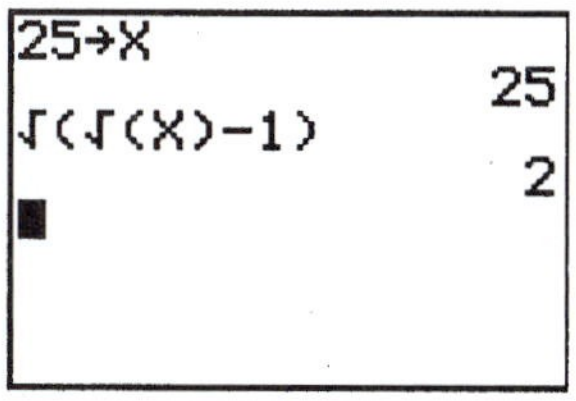

53. $L=\dfrac{h}{2\pi}\sqrt{l(l+1)}$

$L^2=\dfrac{h^2}{4\pi^2}\cdot l(l+1)$

$\dfrac{4\pi^2L^2}{h^2}=l^2+l$

$l^2+l-\dfrac{4\pi^2L^2}{h^2}=0$; from the quadratic formula,

$l=\dfrac{-1\pm\sqrt{1^2-4\cdot1\left(-\frac{4\pi^2L^2}{h^2}\right)}}{2(1)}=\dfrac{-1\pm\sqrt{1+\frac{16\pi^2L^2}{h^2}}}{2}$

$l=\dfrac{-1\pm\sqrt{1+\frac{16\pi^2L^2}{h^2}}}{2}$. The + was chosen because $l>0$

57. (1) $16t_1^2+16t_2^2=45$ where $t_1,\ t_2>0$

(2) $t_2=2t_1$

(1) with t_2 from (2) $\quad 16t_1^2+16(2t_1)^2=45$

$16t_1^2+64t_1^2=45$

$80t_1^2=45$

$t_1^2=\dfrac{45}{80}$

$t_1=0.75$ s

$t_2=2\cdot t_1=2(0.75)$

$t_2=1.5$ s

61.

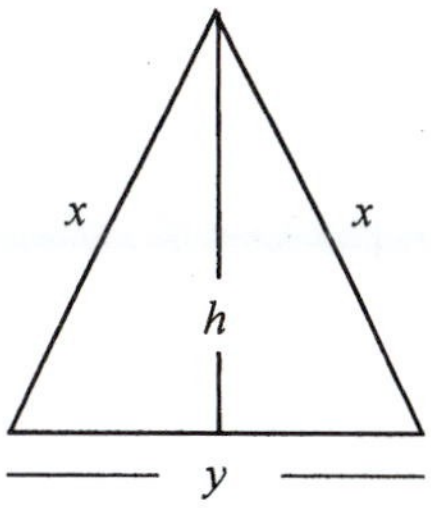

perimeter $=2x+y=72$

area $=\dfrac{1}{2}y\cdot h=240\Rightarrow h=\dfrac{480}{y}$

$x^2=\dfrac{y^2}{4}+h^2\Rightarrow x^2=\dfrac{y^2}{4}+\dfrac{480^2}{y^2}$

$y^4-4x^2y^2+4\cdot480^2=0$

$y=\sqrt{\dfrac{4x^2\pm\sqrt{\left(4x^2\right)^2-4\left(4-480^2\right)}}{2}}$

Graph $y_1=72-2x$

$y_2=\sqrt{\dfrac{4x^2+\sqrt{16x^2-4\left(4\cdot480^2\right)}}{2}}$

$y_3=\sqrt{\dfrac{4x^2-\sqrt{16x^2-4\left(4\cdot480^2\right)}}{2}}$

and use the intersect feature to solve.

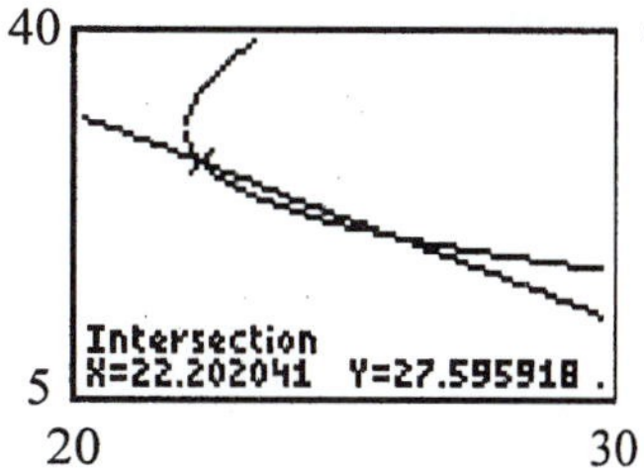

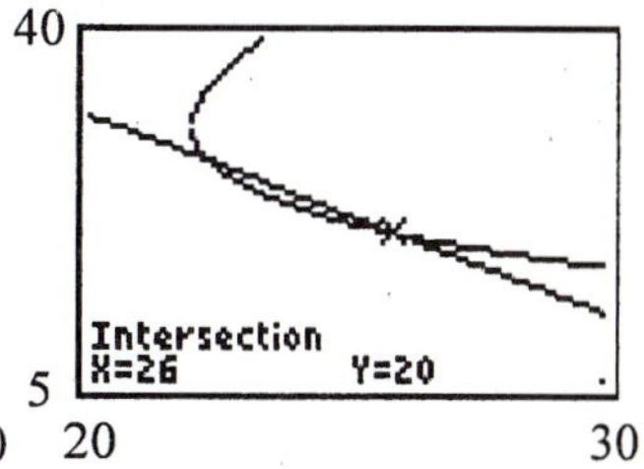

Since both solutions check, pick the one which requires no rounding.

The lengths of sides the banner are 26 in. and 20 in.

65. Area $= lw = 1770 \Rightarrow w = \dfrac{1770}{l}$

$$l^2 + w^2 = 62^2$$

$$l^2 + \frac{1770^2}{l^2} = 62^2$$

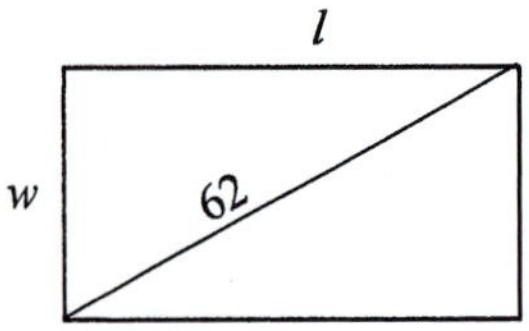

$$l^4 - 62^2 l^2 + 1770^2 = 0$$

$$l = \sqrt{\frac{62^2 \pm \sqrt{\left(-62^2\right)^2 - 4\left(1770^2\right)}}{2}} = \begin{cases} 52 \text{ for } +,\ w = 34 \\ 34 \text{ for } -,\ w = 52 \end{cases}$$

The dimensions of the rectangle are $l = 52$ mm and $w = 34$ mm.

69. Houston to Mobile: $510 = v \cdot t_1 \Rightarrow t_1 = \dfrac{510}{v}$

Mobile to Houston: $510 = (v + 6.0) \cdot t_2 \Rightarrow t_2 = \dfrac{510}{v + 6.0}$

$$t_1 + t_2 = 35$$

$$\frac{510}{v} + \frac{510}{v + 6.0} = 35 \Rightarrow 510(v + 6.0) + 510v = 35v(v + 6.0)$$

$35v^2 - 810v - 3100 = 0$ from which, using the quadratic formula, $v = 26$, $v = -3.3$ (reject, $v > 0$)

$v = 26$ m/h, Houston to Mobile

$v + 6.0 = 32$ m/h, Mobile to Houston

Chapter 15

EQUATIONS OF HIGHER DEGREE

15.1 The Remainder and Factor Theorems; Synthetic Division

1. Using the remainder theorem find the remainder, for $(3x^3 - x^2 - 20x + 5) \div (x+3)$.

$R = f(-3) = 3(-3)^3 - (-3)^2 - 20(-3) + 5 = -25$

5.
$$\begin{array}{r} x^2 - x + 3 \\ x+1 \overline{\smash{)}x^3 + \quad 2x + 3} \\ \underline{x^3 + x^2} \\ -x^2 + 2x \\ \underline{-x^2 - \ \ x} \\ 3x + 3 \\ \underline{3x + 3} \\ 0 \end{array}$$

$f(r) = R;\ r = -1$

$f(-1) = (-1)^3 + 2(-1) + 3$

$= -1 - 2 + 3$

$= 0$

Therefore, $R = 0$

9.
$$\begin{array}{r} 2x^3 - 2x - 18 \\ x - \tfrac{3}{2} \overline{\smash{)}2x^4 - 3x^3 - 2x^2 - 15x - 16} \\ \underline{2x^4 - 3x^3} \\ -2x^2 - 15x \\ \underline{-2x^2 + 3x} \\ -18x - 16 \\ \underline{-18x + 27} \\ -43 \end{array}$$

$f(r) = R;\ r = \frac{3}{2}$

$f\left(\frac{3}{2}\right) = 2\left(\frac{3}{2}\right)^4 - 3\left(\frac{3}{2}\right)^3 - 2\left(\frac{3}{2}\right)^2 - 15\left(\frac{3}{2}\right) - 16 = -43$

Therefore, $R = -43$

13. $f(3) = 2\cdot 3^4 - 7\cdot 3^3 - 3^2 + 8 = -28$, the remainder

17. $4x^3 + x^2 - 16x - 4,\ x - 2;\ r = 2$

$f(2) = 4(2)^3 + 2^2 - 16(2) - 4$

$= 32 + 4 - 32 - 4 = 0$

$x - 2$ is a factor since $f(r) = R = 0$.

21. $x^{61} + 1,\ x + 1;\ r = -1$

$f(-1) = (-1)^{61} + 1 = 0$

$x + 1$ is a factor since $f(r) = R = 0$.

25. $(x^3 + 2x^2 - 3x + 4) \div (x+1) = x^2 + x - 4 + \dfrac{8}{x+1}$

1	2	−3	4
	−1	−1	4
1	1	−4	8

29. $(x^7 - 128) \div (x - 2)$

$= x^6 + 2x^5 + 4x^4 + 8x^3 + 16x^2 + 32x + 64$

$R = 0$

1	0	0	0	0	0	0	−128
	2	4	8	16	32	64	128
1	2	4	8	16	32	64	0

33. $2x^5 - x^3 + 3x^2 - 4;\ x + 1$

2	0	−1	3	0	−4
	−2	2	−1	−2	2
2	−2	1	2	−2	−2

$R = -2$,

$x + 1$ is not a factor.

37. $2Z^4 - Z^3 - 4Z^2 + 1;\ 2Z - 1$

$$\begin{array}{rrrrr} 2 & -1 & -4 & 0 & 1 \\ & 1 & 0 & -2 & -1 \\ \hline 2 & 0 & -4 & -2 & 0 \end{array}$$

$R = 0,$

$Z - \frac{1}{2}$ is a factor $\Rightarrow 2x - 1$ is a factor.

41. $x^4 - 5x^3 - 15x^2 + 5x + 14;\ 7$

$$\begin{array}{rrrrr} 1 & -5 & -15 & 5 & 14 \\ & 7 & 14 & -7 & -14 \\ \hline 1 & 2 & -1 & -2 & 0 \end{array}$$

$R = 0$, 7 is a zero.

45. $f(x) = 2x^3 + 3x^2 - 19x - 4 = (x+4)g(x)$

$\Rightarrow g(x) = \dfrac{2x^3 + 3x^2 - 19x - 4}{x+4}$

$$\begin{array}{rrrr} 2 & 3 & -19 & -4 \\ & -8 & 20 & -4 \\ \hline 2 & -5 & 1 & 0 \end{array}$$

$g(x) = 2x^2 - 5x + 1$

49. $f(x) = 2x^3 + kx^2 - x + 14;\ x - 2$

we want $f(r) = R = 0$

$f(2) = 2(2)^3 + k(2)^2 - 2 + 14$

$= 16 + 4k - 2 + 14$

$= 28 + 4k = 0$

$4k = -28$

$k = -7$, then $x - 2$ will be a factor.

53. Suppose r is a zero of $f(x)$, then $f(r) = 0$. But $f(r) = -g(r) = 0 \Rightarrow g(r) = 0$, so r is also a zero of $g(x)$. Yes, they have the same zeros.

(b) $(s^3 + 5s^2 + 4s + 20) \div (s+5)$

$$\begin{array}{rrrr} 1 & 5 & 4 & 20 \\ & -5 & 0 & -20 \\ \hline 1 & 0 & 4 & 0 \end{array}$$

$R = 0$, $s + 5$ is a factor.

57. $V^3 - 6V^2 + 12V = 8$

$V^3 - 6V^2 + 12V - 8 = 0$

$$\begin{array}{rrrr} 1 & -6 & 12 & -8 \\ & 2 & -8 & 8 \\ \hline 1 & -4 & 4 & 0 \end{array}$$

$R = 0$; Yes, $V = 2\ \text{cm}^3$

15.2 The Roots of an Equation

1. $f(x) = (x-1)^3(x^2 + 2x + 1) = 0$

$(x-1)^3 = 0$ or $x^2 + 2x + 1 = 0$

$x = 1$ $\qquad (x+1)^2 = 0$

A triple root $\qquad x + 1 = 0$

$x = -1$, a double root

the five roots are 1, 1, 1, −1, −1

5. $(x^2 + 6x + 9)(x^2 + 4) = 0$

$(x+3)^2(x^2 + 4) = 0$, by inspection

$x = -3$ double root, $x = \pm 2j$

$R = \dfrac{-(-1) \pm \sqrt{1-4}}{2} = \dfrac{1 \pm \sqrt{-3}}{2} = \dfrac{1 \pm j\sqrt{3}}{2}$

$r_1 = 1,\ r_2 = \frac{1}{2}(1 + j\sqrt{3}),\ r_3 = \frac{1}{2}(1 - j\sqrt{3})$

9.

$$\begin{array}{rrrr} 2 & 11 & 20 & 12 \\ & -3 & -12 & -12 \\ \hline 2 & 8 & 8 & 0 \end{array}$$

$2x^3 + 11x^2 + 20x + 12 = \left(x + \frac{3}{2}\right)(2x^2 + 8x + 8)$

$= 2\left(x + \frac{3}{2}\right)(x^2 + 4x + 4)$

$= 2\left(x + \frac{3}{2}\right)(x+2)(x+2)$

$r_1 = -\frac{3}{2},\ r_2 = -2,\ r_3 = -2$

13.

$$\begin{array}{rrrrr} 1 & 1 & -2 & 4 & -24 \\ & 2 & 6 & 8 & 24 \\ \hline 1 & 3 & 4 & 12 & 0 \end{array}$$

$t^4 + t^3 - 2t^2 + 4t - 24 = (t-2)(t^3 + 3t\ + 4t + 12)$

1	3	4	12
	−3	0	−12
1	0	4	0

$t^4+t^3-2t^2+4t-24=(t-2)(t+3)(t^2+4)$

$=(t-2)(t+3)(t-2j)(t+2j)$

$r_1=2,\ r_2=-3,\ r_3=-2j,\ r_4=2j$

17.

6	5	−15	0	4
	−3	−1	8	−4
6	2	−16	8	0

$6x^4+5x^3-15x^2+4$

$=\left(x+\dfrac{1}{2}\right)(6x^3+2x^2-16x+8)$

$=2\left(x+\dfrac{1}{2}\right)(3x^3+x^2-8x+4)$

3	1	−8	4
	2	2	−4
3	3	−6	0

$6x^4+5x^3-15x^2+4$

$=2\left(x+\dfrac{1}{2}\right)\left(x-\dfrac{2}{3}\right)(x^2+3x-6)$

$=6\left(x+\dfrac{1}{2}\right)\left(x-\dfrac{2}{3}\right)(x^2+x-2)$

$=6\left(x+\dfrac{1}{2}\right)\left(x-\dfrac{2}{3}\right)(x+2)(x-1)$

$r_1=-\dfrac{1}{2},\ r_2=\dfrac{2}{3},\ r_3=-2,\ r_4=1$

21. $x^5-3x^4+4x^3-4x^2+3x-1=0$ (1 is a triple root)

1	−3	4	−4	3	−1
	1	−2	2	−2	1
1	−2	2	−2	1	0
	1	−1	1	−1	
1	−1	1	1	0	
	1	0	1		
1	0	1	0		

$(x-1)(x^2+1)$

The roots are $1, 1, 1, -j, j$.

25. $x^6+2x^5-4x^4-10x^3-41x^2-72x-36=0$

(-1 is a double root; $2j$ is a root)

1	2	−4	−10	−41	−72	−36
	−1	−1	5	5	36	36
1	1	−5	−5	−36	−36	0
	−1	0	5	0	36	0
1	0	−5	0	−36	0	0
	2j	−4	−18j	36	0	0
1	2j	−9	−18j	0	0	0
	−2	0	18j	0	0	0
1	0	−9	0	0	0	0

$(x+1)^2(x-2j)(x+2j)(x^2-9)$

The roots are $-1, -1, 2j, -2j, -3, 3$.

29. $f(x)=(x-(1+j))(x-(1-j))(x)$ is a polynomial, degree 3 with root $1+j$.

$f(x)=(x^2-x(1-j)-x(1+j)+2)(x)$

$=(x^2-x+xj-x-xj+2)(x)$

$=(x^2-2x+2)x$

$=x^3-2x^2+2x$ is the required polynomial.

15.3 Rational and Irrational Roots

1. $f(x)=4x^5+x^4+4x^3-x^2+5x+6=0$ has two sign changes and thus no more than two positive roots.

$f(-x)=-4x^5+x^4-4x^3-x^2-5x+6=0$ has three sign changes and thus no more than three negative roots.

5. $x^3+2x^2-5x-6=0;$ there are 3 roots.

$f(x)=x^3+x^2-5x+3;$ there are at most 2 positive roots.

$f(-x)=-x^3+x^2+5x+3;$ there is one negative root.

Possible rational roots are $\pm1, \pm2, \pm, \pm6$.

Trying -1, we have

1	2	−5	−6
	−1	−1	6
1	1	−6	0

Hence, -1 is a root and the remaining factor is $x^2+x-6=(x+3)(x-2)$.

The remaining roots are -3, 2.

3	8	0	−1
	1	3	1
3	9	3	0

$\frac{1}{3}$ is a root

$$3t^3+8t^2-1=\left(t-\frac{1}{3}\right)\left(3t^2+9t+3\right)$$

$$t=\frac{-9\pm\sqrt{9^2-4(3)(3)}}{2(3)}=\frac{-3\pm\sqrt{5}}{2}$$

roots: $\frac{1}{3}, \frac{-3\pm\sqrt{5}}{2}$

9. $3x^3+11x^2+5x-3=0$; there are three roots.

$f(x)=3x^3+11x^2+5x-3$; there is 1 positive root.

$f(-x)=-3x^3+11x^2-5x-3$; there are at most two negative roots.

Possible rational roots are $\pm\frac{1}{3}, \pm1, \pm3$. We try to find the one positive root first. $\frac{1}{3}$ is the first root with 0 remainder.

3	1	5	−3
	11	4	3
3	12	9	0

Thus, $\frac{1}{3}$ is the positive root. The remaining factor

$3x^2+12x+9=3\left(x^2+4x+3\right)=3(x+1)(x+3)$.

The remaining roots are $-1, -3$.

13. $5n^4-2n^3+40n-17=0$; there are four roots.

$f(n)$ has 3 sign changes, at most 3 positive roots.

$f(-n)=5n^4+2n^3-40n-16$ has one sign change, at most one negative root.

Possible rational roots: $\pm16, \pm\frac{16}{5}, \pm2, \pm\frac{2}{5}, \pm4, \pm\frac{4}{5}, \pm8, \pm\frac{8}{5}$

5	−2	0	40	−16
	−10	24	−48	16
5	−12	24	−8	0

−2 is a root

$$5n^4-2n^3+40n-16=(n+2)\left(5n^3-12n^2+24n-8\right)$$

5	−12	24	−8
	2	−4	8
5	−10	20	0

$\frac{2}{5}$ is a root

$$5n^4-2n^3+40n-16=(n+2)\left(n-\frac{2}{5}\right)\left(5n^2-10n+20\right)$$

$$n=\frac{-(-10)\pm\sqrt{(-10)^2-4(5)(20)}}{2(5)}=1\pm i\sqrt{3}$$

roots: $-2, \frac{2}{5}, 1+i\sqrt{3}$

17. $f(D)=D^5+D^4-9D^3-5D^2+16D+12=0$ has $n=5$ and therefore five roots. $f(D)$ has two sign changes and therefore at most two positive roots.

$f(-D)=-D^5+D^4+9D^3-5D^2-16D+12$ has three sign changes and therefore at most three negative roots.

$$\text{Possible rational roots}=\frac{\text{factors of }12}{\text{factors of }1}=\frac{\pm1, \pm2, \pm3, \pm4, \pm6, \pm12}{\pm1}$$

1	1	−9	−5	16	12
	2	6	−6	−22	−12
1	3	−3	−11	−6	0

2 is a root

$D^5+D^4-9D^3-5D^2+16D+12$

$=(D-2)(D^4+3D^3-3D^2-11D-6)$

1	3	−3	−11	−6
	2	10	14	6
1	5	7	3	0

2 is a root of $D^4+3D^3-3D^2-11D-6$

$D^5+D^4-9D^3-5D^2+16D+12$

$=(D-2)(D-2)(D^3+5D^2+7D+3)$

1	5	7	3
	−1	−4	−3
1	4	3	0

-1 is a root of D^3+5D^2+7D+3

$D^5+D^4-9D^3-5D^2+16D+12$

$=(D-2)(D-2)(D+1)(D^2+4D+3)$

$=(D-2)(D-2)(D+1)(D+1)(D+3)$

roots: $2, 2, -1, -1, -3$

21. $x^3-2x^2-5x+4=0$

Graph $y_1=x^3-2x^2-5x+4$ and use the zero feature to solve. The zeros are $-1.86, 0.68, 3.18$.

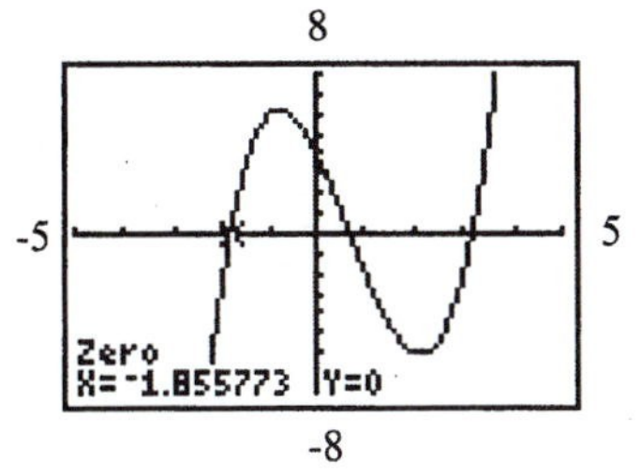

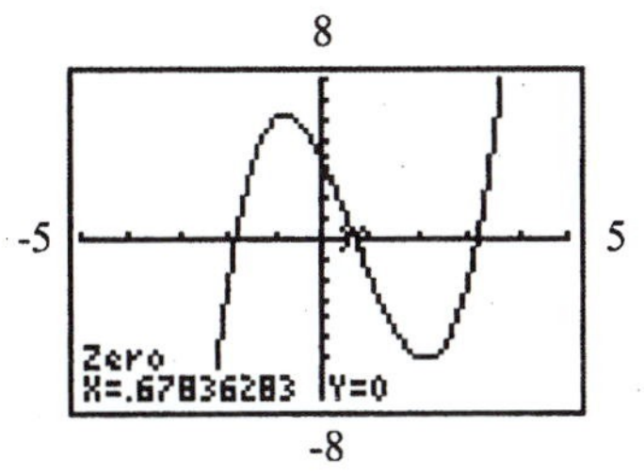

25. $x^3-6x^2+10x-4=0$ (0 and 1)

Graph $y_1=x^3-6x^2+10x-4$ and use the zero feature to solve. $x=0.59$

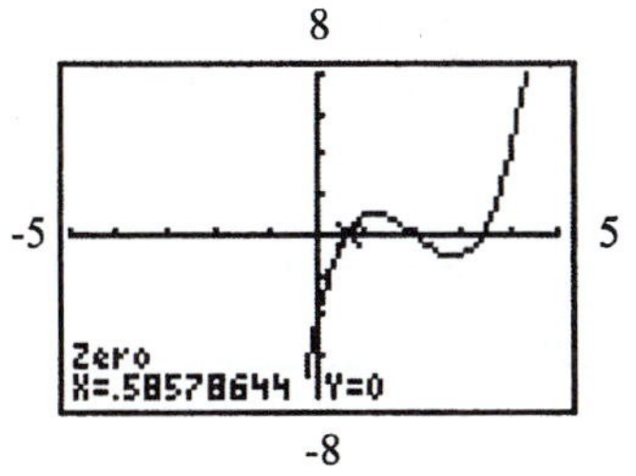

29. $y=x^4-11x^2,\ y=12x-4$

$x^4-11x^2=12x-4$

$x^4-11x^2-12x+4=0$, use rational root theorem and synthetic division to solve.

1	0	−11	−12	−4
	−2	4	14	−4
1	−2	−7	2	0

1	−2	−7	2
	−2	8	−2
1	−4	1	0

Solve $x^2-4x+1=0$ with quadratic formula.

$$x=\frac{-(-4)\pm\sqrt{(-4)^2-4(1)(1)}}{2(1)}=2\pm\sqrt{3}$$

The solutions are $x=-2,\ y=-28j,\ x=2+\sqrt{3},\ y=20+12\sqrt{3};\ x=2-\sqrt{3},\ y=20-12\sqrt{3}$.

33. $\alpha=0.2t^3+t^2,\ \alpha=2.0\text{ rad/s}^2$

$2.0=-0.2t^3+t^2;\ 0.2t^3-t^2+2.0=0$

Graph $y_1=0.2x^3-x^2+2$. Using the zero feature $y=0$ for $x=1.8, 4.5$. Therefore, $\alpha=2.0\text{ rad/s}^2$ for $t=1.8$, s, 4.5 s.

37. $V=0.1t^4-1.0t^3+3.5t^2-5.0t+2.3,\ 0\le t\le 5.0\text{ s}$

Graph $y_1=0.1x^4-1.0x^3+3.5x^2-5.0x+2.3$ and use the zero feature to solve.

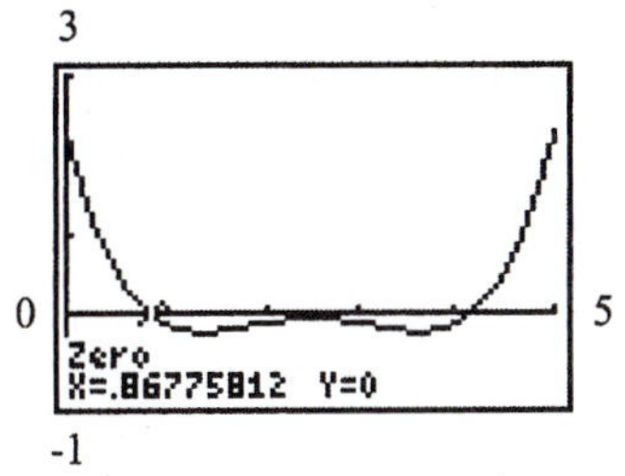

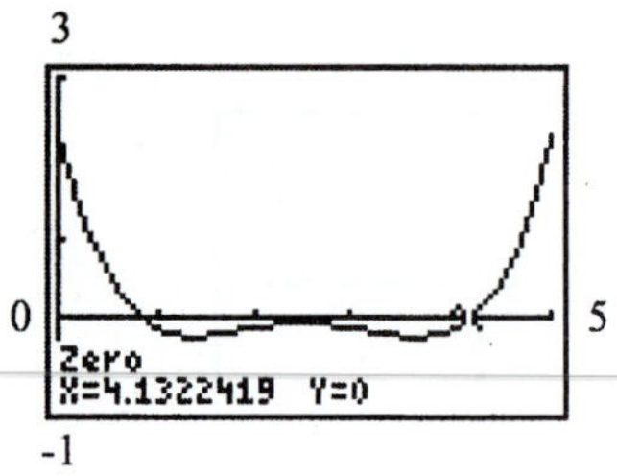

$V=0$ at $t=-0.9$ s and $t=4.1$ s.

41. Let r = smallest radius, the radii are r, $r+1$, $r+2$, $r+3$.

$$\frac{4}{3}\pi(r+3)^3=\frac{4}{3}\pi r^3+\frac{4}{3}\pi(r+1)^3+\frac{4}{3}\pi(r+2)^3$$

$$r^3+3\cdot r^2\cdot 3+3\cdot r\cdot 3^2+3^3=r^3+r^3+3r^2+3r+1+$$
$$r^3+3\cdot r^2\cdot 2+3\cdot r\cdot 2^2+2^3$$

$$r^3+9r^2+27r+27=3r^3+9r^2+15r+9$$

$2r^3-12r-18=0$ has one sign change and therefore one positive root.

$$\text{Possible rational roots}=\frac{\text{factors of }18}{\text{factors of }2}=\frac{\pm1,\ \pm2,\ \pm3,\ \pm6,\ \pm9,\ \pm18}{\pm1,\ \pm2}$$

2	0	−12	−18
	6	18	18
2	6	6	0

3 is a root of $2r^3-12r-18$

$2r^3-12r-18=(x-3)(2x^2+6x+6)$

$2x^2+6x+6=0$

$x^2+3x+3=0$, has no real solutions

The radii are 3.0 mm, 4.0 mm, 5.0 mm, 6.0 mm.

45. $f(x)=ax^3-bx^2+c=0$. There are three roots. Since $f(x)$ has two sign changes, there are at most two positive roots. $f(-x)=-ax^3-bx^2+c=0$ which has one sign change and thus, one negative root. Since the complex roots occur in conjugate pairs, the possible roots are one negative and two positive or one negative and two complex.

Chapter 15 Review Exercises

1. $2(1)^3-4(1)^2-(1)+4=1$

$(2x^3-4x^2-x+4)\div(x-1)$ has remainder 1

5.

1	1	1	−2	−3
	−1	0	−1	3
1	0	1	−3	0

remainder = 0, therefore $x+1$ is a factor of $x^4+x^3+x^2-2x-3$

9.

1	3	6	1
	1	4	10
1	4	10	11

$$(x^3+3x^2+6x+1)\div(x-1)=(x^2+4x+10)+\frac{11}{x-1}$$

13.

1	3	−20	−2	56
	−6	18	12	−60
1	−3	−2	10	−4

$$\frac{x^4+3x^3-20x^2-2x+56}{x+6}=x^3-3x^2-2x+10+\frac{-4}{x+6}$$

17.

1	5	0	−6
	−3	−6	18
1	2	−6	12

remainder = 12, therefore -3 is not a root of $y^3+5y^2-6=0$

21.

1	−4	−7	10
	5	5	−10
1	1	−2	0

$x^2+x-2=0\Rightarrow(x+2)(x-1)=0\Rightarrow x=-2,\ x=1$

$r_1=5,\ r_2=-2,\ r_3=1$

25.

$$\begin{array}{rrrrr} 4 & 0 & -1 & -18 & 9 \\ & 2 & 1 & 0 & -9 \\ \hline 4 & 2 & 0 & -18 & 0 \end{array}$$

$$4p^4 - p^2 - 18p + 9 = \left(p - \frac{1}{2}\right)\left(4p^3 + 2p^2 - 18\right)$$

$$\begin{array}{rrrr} 4 & 2 & 0 & -18 \\ & 6 & 12 & 18 \\ \hline 4 & 8 & 12 & 0 \end{array}$$

$$4p^4 - p^2 - 18p + 9$$
$$= \left(p - \frac{1}{2}\right)\left(p - \frac{3}{2}\right)\left(4p^2 + 8p + 12\right)$$
$$= 4\left(p - \frac{1}{2}\right)\left(p - \frac{3}{2}\right)\left(p^2 + 2p + 3\right)$$

$p^2 + 2p + 3$ has roots $= -1 \pm \sqrt{2}j$

roots $\frac{1}{2}, \frac{3}{2}, -1 \pm \sqrt{2}j$

29.

$$\begin{array}{rrrrrr} & 3 & -1 & -11 & -12 & -4 \\ & -1 & -2 & 3 & 8 & 4 \\ \hline 1 & 2 & -3 & -8 & -4 & 0 \end{array}$$

$$s^5 + 3s^4 - s^3 - 11s^2 - 12s - 4$$
$$= (s+1)\left(s^4 + 2s^3 - 3s^2 - 8s - 4\right)$$

$$\begin{array}{rrrrr} 1 & 2 & -3 & -8 & -4 \\ & -1 & -1 & 4 & 4 \\ \hline 1 & 1 & -4 & -4 & 0 \end{array}$$

$$s^5 + 3s^4 - s^3 - 11s^2 - 12s - 4$$
$$= (s+1)(s+1)\left(s^3 + s^2 - 4s - 4\right)$$

$$\begin{array}{rrrr} 1 & 1 & -4 & -4 \\ & -1 & 0 & 4 \\ \hline 1 & 0 & -4 & 0 \end{array}$$

$$s^5 + 3s^4 - 11s^3 - 11s^2 - 4$$
$$= (s+1)(s+1)(s+1)\left(s^2 - 4\right)$$

roots: $-1, -1, -1, 2, -2$

33. $x^3 + x^2 - 10x + 8 = 0$ has three roots.

Possible rational roots $= \frac{\pm 1, \pm 2, \pm 4, \pm 8}{\pm 1}$

$$\begin{array}{rrrr} 1 & 1 & -10 & 8 \\ & 1 & 2 & -8 \\ \hline 1 & 2 & -8 & 0 \end{array}$$

1 is a root

$$x^3 + x^2 - 10x + 8 = (x-1)\left(x^2 + 2x - 8\right)$$
$$= (x-1)(x+4)(x-2)$$

roots: $1, -4, 2$

37. $6x^3 - x^2 - 12x - 5 = 0$ has three roots.

Possible rational roots $= \frac{\pm 1, \pm 5}{\pm 1, \pm 2, \pm 3, \pm 6}$

$$\begin{array}{rrrr} 6 & -1 & -12 & -5 \\ & 10 & 15 & 5 \\ \hline 6 & 9 & 3 & 0 \end{array}$$

$\frac{5}{3}$ is a root

$$6x^3 - x^2 - 12x - 5 = \left(x - \frac{5}{3}\right)\left(6x^2 + 9x + 3\right)$$
$$= 3\left(x - \frac{5}{3}\right)\left(2x^2 + 3x + 1\right)$$
$$= 3\left(x - \frac{5}{3}\right)(2x+1)(x+1)$$

roots: $\frac{5}{3}, \frac{-1}{2}, -1$

41. A polynomial of degree five with real coefficients has five zeros. Since the complex zeros occur in conjugate pairs, the possibilities are: real 5, complex 0; real 3, complex 2; and real 1, complex 4.

45.

$$\begin{array}{rrrr} 3 & k & -8 & -8 \\ & -6 & 12 - 2k & -8 + 4k \\ \hline 3 & -6 + k & 4 - 2k & -16 + 4k \end{array}$$

remainder $= 4k - 16 = 0$

$$4k = 16$$
$$k = 4$$

49. $x^2 = y + 3;\ \ xy = 2$

$y = x^2 - 3 \quad y = \frac{2}{x}$

$$x^2 - 3 = \frac{2}{x}$$

$$x^3 - 3x - 2 = 0$$

1	0	−3	−2
	−1	1	2
1	−1	−2	0
	−1	2	
1	−2	0	

$x^3-3x-2=(x+1)(x+1)(x-2)\Rightarrow x=-1,\ 2$

The solutions are $x=-1,\ y=2;\ x=2,\ y=1.$

53. $n=x^3-9x^2+15x+600=580$

$x^3-9x^2+15x+20=0$

1	−9	15	20
	4	−20	−20
1	−5	−5	0

$\Rightarrow x=4$

580 crimes were committed in April.

57. $\dfrac{1}{p}+\dfrac{1}{q}=\dfrac{1}{f}$

$\dfrac{1}{p}+\dfrac{1}{p+4}=\dfrac{1}{\frac{p+1}{p}}\Rightarrow p^3+2p^2=6p-4=0$

1	2	−6	−4
	2	8	4
1	4	2	0

$p^3+2p^2-6p-4=0$

$(p-2)(p^2+4p+2)=0$ (both roots of p^2+4p+2 are negative).

$p=2$ in.

61. $\sqrt{(x+0.900)^2-x^2}$

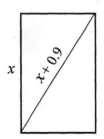

$A=lw\Rightarrow x\sqrt{(x+0.900)^2-x^2}=24.3$

$x^2(x^2+1.80x+0.810-x^2)=24.3^2$

$1.80x^3+0.810x^2-24.3^2=0$

Graph $y_1=1.80x^3+0.810x^2-24.3^2$ and use the zero feature to solve.

$x=6.75$ in., $\sqrt{(x+0.900)^2-x^2}=3.60$ in.

the dimensions are 6.75 in. by 3.60 in.

Chapter 16

MATRICES

16.1 Definitions and Basic Operations

1. $\begin{bmatrix} 8 & 1 & -5 & 9 \\ 0 & -2 & 3 & 7 \end{bmatrix} + \begin{bmatrix} -3 & 6 & 4 & 0 \\ 6 & 6 & -2 & 5 \end{bmatrix}$

$= \begin{bmatrix} 8+(-3) & 1+6 & -5+4 & 9+0 \\ 0+6 & -2+6 & 3+(-2) & 7+5 \end{bmatrix}$

$= \begin{bmatrix} 5 & 7 & -1 & 9 \\ 6 & 4 & 1 & 12 \end{bmatrix}$

5. $\begin{bmatrix} x & 2y & z \\ \frac{r}{4} & -s & -5t \end{bmatrix} = \begin{bmatrix} -2 & 10 & -9 \\ 12 & -4 & 5 \end{bmatrix}$

$x = -2, \quad 2y = 10, \quad x = -9$

$y = 5,$

$\frac{r}{4} = 12, \quad -s = -4, \quad -5t = 5$

$r = 48 \quad s = 4 \quad t = -1$

9. $\begin{bmatrix} x-3 & x+y \\ x-z & y+z \\ x+t & y-t \end{bmatrix} = \begin{bmatrix} 5 & 3 \\ 4 & -1 \end{bmatrix}$, cannot solve, matrices have different dimensions.

13. $\begin{bmatrix} 50+(-55) & -82+82 \\ -34+45 & 57+14 \\ -15+26 & 62+(-67) \end{bmatrix} = \begin{bmatrix} -5 & 0 \\ 11 & 71 \\ 11 & -5 \end{bmatrix}$

17. Since A and C do not have the same number of columns, they cannot be added.

21. $A - 2B = \begin{bmatrix} -1 & 4 & -7 \\ 2 & -6 & 11 \end{bmatrix} - 2\begin{bmatrix} 7 & 9 & -6 \\ 4 & -1 & -8 \end{bmatrix}$

$= \begin{bmatrix} -1 & 4 & -7 \\ 2 & -6 & 11 \end{bmatrix} - \begin{bmatrix} 14 & 18 & -12 \\ 8 & -2 & -16 \end{bmatrix}$

$= \begin{bmatrix} -15 & -14 & 5 \\ -6 & -4 & 27 \end{bmatrix}$

25. Since A and C do not have the same number of columns, they cannot be subtracted.

29.

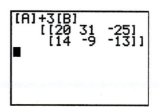

```
[A]+3[B]
  [[20 31  -25]
   [14 -9  -13]]
```

33.

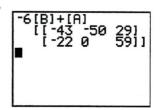

```
-6[B]+[A]
  [[-43 -50 29]
   [-22 0   59]]
```

37. $-(A-B) = -\begin{bmatrix} -1-4 & 2+1 & 3+3 & 7-0 \\ 0-5 & -3-0 & -1+1 & 4-1 \\ 9-1 & -1-11 & 0-8 & -2-2 \end{bmatrix}$

$= \begin{bmatrix} 5 & -3 & -6 & -7 \\ 5 & 3 & 0 & -3 \\ -8 & 12 & 8 & 4 \end{bmatrix}$

$B - A = \begin{bmatrix} 4+1 & -1-2 & -3-3 & 0-7 \\ 5-0 & 0+3 & -1+1 & 1-4 \\ 1-9 & 11+1 & 8-0 & 2+2 \end{bmatrix}$

$= \begin{bmatrix} 5 & -3 & -6 & -7 \\ 5 & 3 & 0 & -3 \\ -8 & 12 & 8 & 4 \end{bmatrix}$

41. $\begin{bmatrix} 96 & 75 & 0 & 0 \\ 62 & 44 & 24 & 0 \\ 0 & 35 & 68 & 78 \end{bmatrix} + 2\begin{bmatrix} 96 & 75 & 0 & 0 \\ 62 & 44 & 24 & 0 \\ 0 & 35 & 68 & 78 \end{bmatrix}$

$= \begin{bmatrix} 288 & 225 & 0 & 0 \\ 186 & 132 & 72 & 0 \\ 0 & 105 & 204 & 234 \end{bmatrix}$

16.2 Multiplication of Matrices

1. $A=\begin{bmatrix}1 & 2\\ 0 & -3\\ 2 & 1\end{bmatrix}, B=\begin{bmatrix}-1 & 6 & 5 & -2\\ 3 & 0 & 1 & -4\end{bmatrix}$

$$AB=\begin{bmatrix}1 & 2\\ 0 & -3\\ 2 & 1\end{bmatrix}\begin{bmatrix}-1 & 6 & 5 & -2\\ 3 & 0 & 1 & -4\end{bmatrix}$$

$$AB=\begin{bmatrix}-1+6 & 6+0 & 5+2 & -2-8\\ 0-9 & 0+0 & 0-3 & 0+12\\ -2+3 & 12+0 & 10+1 & -4-4\end{bmatrix}$$

$$AB=\begin{bmatrix}5 & 6 & 7 & -10\\ -9 & 0 & -3 & 12\\ 1 & 12 & 11 & -8\end{bmatrix}$$

5. $\begin{bmatrix}2 & -3 & 1\\ 0 & 7 & -3\end{bmatrix}\begin{bmatrix}90\\ -25\\ 50\end{bmatrix}=\begin{bmatrix}2(80)+(-3)(-25)+1(50)\\ 0(90)+7(-25)+(-3)(50)\end{bmatrix}$

$$=\begin{bmatrix}305\\ -325\end{bmatrix}$$

9. $\begin{bmatrix}-1 & 7\\ 3 & 5\\ 10 & -1\\ -5 & 12\end{bmatrix}\begin{bmatrix}2 & 1\\ 5 & -3\end{bmatrix}=\begin{bmatrix}-1(2) & -1(1)+7(-3)\\ 3(2) & 3(1)+5(-3)\\ 10(2) & 10(1)+(-1)(-3)\\ -5(2) & -5(1)+12(-3)\end{bmatrix}$

$$=\begin{bmatrix}33 & -22\\ 31 & -12\\ 15 & 13\\ 50 & -41\end{bmatrix}$$

13.

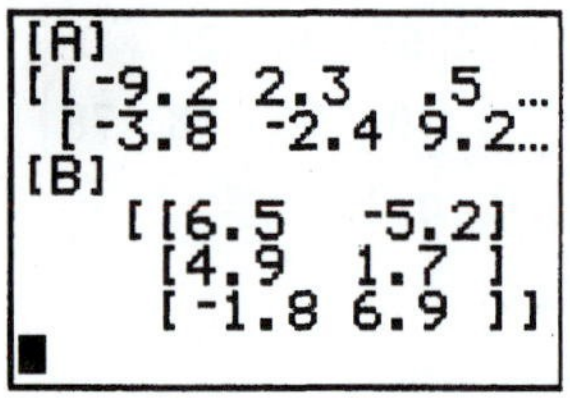

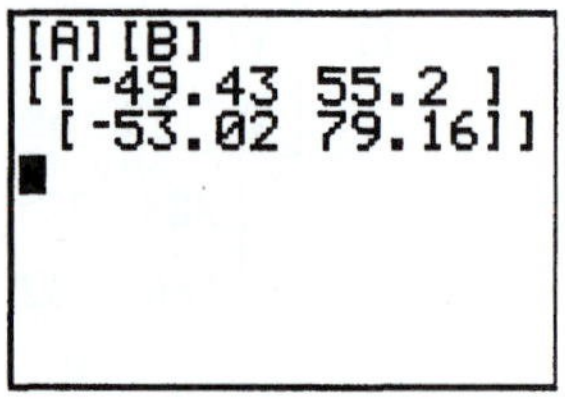

17. $AB=\begin{bmatrix}-10 & 25 & 40\\ 42 & -5 & 0\end{bmatrix}\begin{bmatrix}6\\ -15\\ 12\end{bmatrix}=\begin{bmatrix}45\\ 327\end{bmatrix}$

BA is not possible because the number of columns in *B* is not equal to the number of rows in *A*.

21. $AI=\begin{bmatrix}1 & 3 & -5\\ 2 & 0 & 1\\ -1 & -2 & 4\end{bmatrix}\begin{bmatrix}1 & 0 & 0\\ 0 & 1 & 0\\ 0 & 0 & 1\end{bmatrix}$

$$=\begin{bmatrix}1(1)+3(0)+(-5)(0) & 1(0)+3(1)+(-5)(0)\\ 2(1)+0(0)+1(0) & 2(0)+0(1)+1(0)\\ 1(1)+(-2)(0)+4(0) & 1(0)+(-2)(1)+4(0)\end{bmatrix}$$

$$\begin{bmatrix}1(0)+3(0)+(-5)(1)\\ 2(0)+0(0)+1(1)\\ 1(0)+(-2)(0)+4(1)\end{bmatrix}=\begin{bmatrix}1 & 3 & -5\\ 2 & 0 & 1\\ -1 & -2 & 4\end{bmatrix}$$

$$IA=\begin{bmatrix}1 & 0 & 0\\ 0 & 1 & 0\\ 0 & 0 & 1\end{bmatrix}\begin{bmatrix}1 & 3 & -5\\ 2 & 0 & 1\\ -1 & -2 & 4\end{bmatrix}$$

$$=\begin{bmatrix}1(1)+0(2)+0(1) & 1(3)+0(0)+0(-2)\\ 0(1)+1(2)+0(1) & 0(3)+1(0)+0(-2)\\ 0(1)+0(2)+1(1) & 0(3)+0(0)+1(-2)\end{bmatrix}$$

$$\begin{bmatrix}1(-5)+0(1)+0(4)\\ 0(-5)+1(1)+0(4)\\ 0(-5)+0(1)+1(4)\end{bmatrix}$$

$$=\begin{bmatrix}1 & 3 & -5\\ 2 & 0 & 1\\ -1 & -2 & 4\end{bmatrix}$$

Therefore, $AI = IA = A$

$$=\begin{bmatrix}-1 & 2 & 0\\ 4 & -3 & 1\\ 2 & 1 & 3\end{bmatrix}$$

Therefore, $AI = IA = A$

23. $AB=\begin{bmatrix}5 & -2\\ -2 & 1\end{bmatrix}\begin{bmatrix}1 & 2\\ 2 & 5\end{bmatrix}$

$$=\begin{bmatrix}5(1)+(-2)(2) & 5(2)+(-2)(5)\\ (-2)(1)+1(2) & -2(2)+1(5)\end{bmatrix}$$

$$=\begin{bmatrix}1 & 0\\ 0 & 1\end{bmatrix}$$

25. $AB = \begin{bmatrix} 1 & -2 & 3 \\ 2 & -5 & 7 \\ -1 & 3 & -5 \end{bmatrix}\begin{bmatrix} 4 & -1 & 1 \\ 3 & -2 & -1 \\ 1 & -1 & -1 \end{bmatrix}$

$$= \begin{bmatrix} 1(4)+(-2)(3)+3(1) & 1(-1)+(-2)(-2)+3(-1) & 1(1)+(-2)(-1)+3(-1) \\ 2(4)+(-5)(3)+7(1) & 2(-1)+(-5)(-2)+7(-1) & 2(1)+(-5)(-1)+7(-1) \\ -1(4)+3(3)+(-5)(1) & -1(-1)+3(-2)+(-5)(-1) & -1(1)+3(-1)+(-5)(-1) \end{bmatrix}$$

$$= \begin{bmatrix} 1 & 0 & 0 \\ 0 & 1 & 0 \\ 0 & 0 & 1 \end{bmatrix}$$

Therefore, $B = A^{-1}$ since $AB = I$.

29. $\begin{bmatrix} 3 & 1 & 2 \\ 1 & -3 & 4 \\ 2 & 2 & 1 \end{bmatrix}\begin{bmatrix} -1 \\ 2 \\ 1 \end{bmatrix} = \begin{bmatrix} 3(-1)+1(2)+2(1) \\ 1(-1)+(-3)(2)+4(1) \\ 2(-1)+2(2)+1(1) \end{bmatrix}$

$$\neq \begin{bmatrix} 1 \\ -3 \\ 1 \end{bmatrix};$$

A is not the proper matrix of solution values.

33.

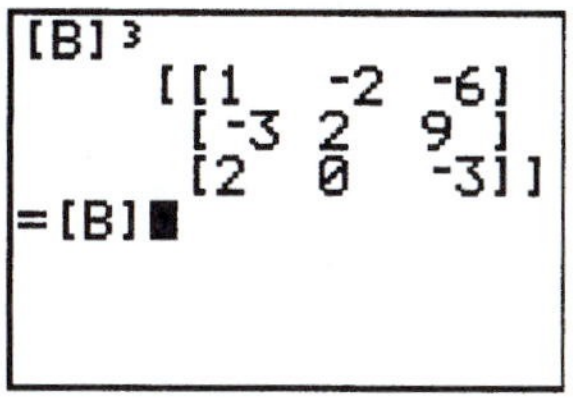

37. $I = \begin{bmatrix} 1 & 0 \\ 0 & 1 \end{bmatrix}, -I = \begin{bmatrix} -1 & 0 \\ 0 & -1 \end{bmatrix}$

$$(-I)^2 = \begin{bmatrix} -1 & 0 \\ 0 & -1 \end{bmatrix}\begin{bmatrix} -1 & 0 \\ 0 & -1 \end{bmatrix}$$

$$= \begin{bmatrix} (-1)(-1)+(0)(0) & (-1)(0)+0(-1) \\ 0(-1)+(-1)(0) & 0(0)+(-1)(-1) \end{bmatrix}$$

$$= \begin{bmatrix} 1 & 0 \\ 0 & 1 \end{bmatrix} = I$$

41. $S_y^2 = \begin{bmatrix} 0 & -j \\ j & 0 \end{bmatrix}\begin{bmatrix} 0 & -j \\ j & 0 \end{bmatrix}$

$$= \begin{bmatrix} 0(0)-j^2 & 0(-j)-j(0) \\ j(0)+0(j) & j(-j)-0(0) \end{bmatrix} = \begin{bmatrix} -j^2 & 0 \\ 0 & -j^2 \end{bmatrix}$$

$$= \begin{bmatrix} -(-1) & 0 \\ 0 & -(-1) \end{bmatrix}$$

$$S_y^2 = \begin{bmatrix} 1 & 0 \\ 0 & 1 \end{bmatrix} = I$$

45. $[x \quad y]\begin{bmatrix} 2.76 & -1 \\ 1 & 2.81 \end{bmatrix}\begin{bmatrix} x \\ y \end{bmatrix} = [7.76\times 10^7]$

$$[2.76x + y - x + 2.81y]\begin{bmatrix} x \\ y \end{bmatrix} = [7.76\times 10^7]$$

$$[2.76x^2 + xy - xy + 2.81y^2] = [7.76\times 10^7]$$

$2.76x^2 + 2.81y^2 = 7.76\times 10^7$, an ellipse

16.3 Finding the Inverse of a Matrix

1. $A = \begin{bmatrix} 2 & -3 \\ 4 & -5 \end{bmatrix}$

$\det A = 2(-5)-(-3)(4) = -10+12 = 2$

$$A^{-1} = \frac{1}{2}\begin{bmatrix} -5 & 3 \\ -4 & 2 \end{bmatrix} = \begin{bmatrix} -\frac{5}{2} & \frac{3}{2} \\ -2 & 1 \end{bmatrix}$$

Check: $AA^{-1} = \begin{bmatrix} 2 & -3 \\ 4 & -5 \end{bmatrix}\begin{bmatrix} -\frac{5}{2} & \frac{3}{2} \\ -2 & 1 \end{bmatrix} = \begin{bmatrix} 1 & 0 \\ 0 & 1 \end{bmatrix}$

5. $\begin{bmatrix} -1 & 5 \\ 4 & 10 \end{bmatrix}$

Interchange the elements of the principal diagonal and change the signs of the off-diagonal elements.

$$\begin{bmatrix} 10 & -5 \\ -4 & -1 \end{bmatrix}$$

Find the determinant of the original matrix.

$$\begin{vmatrix} -1 & 5 \\ 4 & 10 \end{vmatrix} = -30$$

Divide each element of the second matrix by -30.

$$-\frac{1}{30}\begin{bmatrix} 10 & -5 \\ -4 & -1 \end{bmatrix} = \begin{bmatrix} -\frac{1}{3} & \frac{1}{6} \\ \frac{2}{15} & \frac{1}{30} \end{bmatrix}$$

9. $\begin{bmatrix} -50 & -45 \\ 26 & 80 \end{bmatrix}$

Interchange the elements of the principal diagonal and change the signs of the off-diagonal elements.

$\begin{bmatrix} 80 & 45 \\ -26 & -50 \end{bmatrix}$

Find the determinant of the original matrix.

$\begin{bmatrix} -50 & -45 \\ 26 & 80 \end{bmatrix} = -2830$

Divide each element of the second matrix by -2830.

$$-\frac{1}{2830}\begin{bmatrix} 80 & 45 \\ -26 & -50 \end{bmatrix} = \begin{bmatrix} -\frac{8}{283} & -\frac{9}{566} \\ \frac{13}{1415} & \frac{5}{283} \end{bmatrix}$$

13. $\left[\begin{array}{cc|cc} 2 & 4 & 1 & 0 \\ -1 & -1 & 0 & 1 \end{array}\right] R1 \to R1 + R2 \left[\begin{array}{cc|cc} 1 & 3 & 1 & 1 \\ -1 & -1 & 0 & 1 \end{array}\right]$

$R2 \to R2 + R1 \left[\begin{array}{cc|cc} 1 & 3 & 1 & 1 \\ 0 & 2 & 1 & 2 \end{array}\right]$

$R2 \to \frac{1}{2}R2 \left[\begin{array}{cc|cc} 1 & 3 & 1 & 1 \\ 0 & 1 & \frac{1}{2} & 1 \end{array}\right]$

$R1 \to R1 - 3R2 \left[\begin{array}{cc|cc} 1 & 0 & -\frac{1}{2} & -2 \\ 0 & 1 & \frac{1}{2} & 1 \end{array}\right];$

$A^{-1} = \begin{bmatrix} -\frac{1}{2} & -2 \\ \frac{1}{2} & 1 \end{bmatrix}$

17. $\left[\begin{array}{cc|cc} 25 & 30 & 1 & 0 \\ -10 & -14 & 0 & 1 \end{array}\right] R1 \to \frac{1}{25}R1$

$\left[\begin{array}{cc|cc} 1 & 1.2 & 0.04 & 0 \\ -10 & -14 & 0 & 1 \end{array}\right] R2 \to 10R1 + 2$

$\left[\begin{array}{cc|cc} 1 & 1.2 & 0.04 & 0 \\ 0 & -2 & 0.4 & 1 \end{array}\right] R2 \to \frac{1}{-2}R2$

$\left[\begin{array}{cc|cc} 1 & 1.2 & 0.04 & 0 \\ 0 & 1 & -0.2 & -0.5 \end{array}\right] R1 \to -1.2R2 + R1$

$\left[\begin{array}{cc|cc} 1 & 0 & 0.28 & 0.6 \\ 0 & 1 & -0.2 & -0.5 \end{array}\right]; A^{-1} = \begin{bmatrix} 0.28 & 0.6 \\ -0.2 & -0.5 \end{bmatrix}$

21. $\left[\begin{array}{ccc|ccc} 1 & 3 & 2 & 1 & 0 & 0 \\ -2 & -5 & -1 & 0 & 1 & 0 \\ 2 & 4 & 0 & 0 & 0 & 1 \end{array}\right]$ $\begin{array}{l} R2 \to 2R1 + R2 \\ R3 \to -2R1 + R3 \end{array}$

$\left[\begin{array}{ccc|ccc} 1 & 3 & 2 & 1 & 0 & 0 \\ 0 & 1 & 3 & 2 & 1 & 0 \\ 0 & -2 & -4 & -2 & 0 & 1 \end{array}\right]$ $\begin{array}{l} R1 \to -3R2 + R1 \\ R3 \to 2R2 + R3 \end{array}$

$\left[\begin{array}{ccc|ccc} 1 & 0 & -7 & -5 & -3 & 0 \\ 0 & 1 & 3 & 2 & 1 & 0 \\ 0 & 0 & 2 & 2 & 2 & 1 \end{array}\right]$ $R3 \to \frac{1}{2}R3$

$\left[\begin{array}{ccc|ccc} 1 & 0 & -7 & -5 & -3 & 0 \\ 0 & 1 & 3 & 2 & 1 & 0 \\ 0 & 0 & 1 & 1 & 1 & \frac{1}{2} \end{array}\right]$ $\begin{array}{l} R2 \to -3R3 + R2 \\ R1 \to 7R3 + R1 \end{array}$

$\left[\begin{array}{ccc|ccc} 1 & 0 & 0 & 2 & 4 & \frac{7}{2} \\ 0 & 1 & 0 & -1 & -2 & \frac{-3}{2} \\ 0 & 0 & 1 & 1 & 1 & \frac{1}{2} \end{array}\right]; A^{-1} = \begin{bmatrix} 2 & 4 & \frac{7}{2} \\ -1 & -2 & \frac{-3}{2} \\ 1 & 1 & \frac{1}{2} \end{bmatrix}$

25.

```
[A]
          [[2  8]
           [-1 6]]
[A]-1▸Frac
     [[3/10  -2/5]
      [1/20  1/10]]
■
```

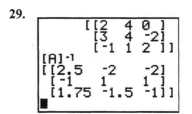

29.

```
     [[2  4  0 ]
      [3  4 -2]
      [-1 1  2 ]]
[A]-1
[[2.5   -2    -2]
 [-1    1     1 ]
 [1.75 -1.5  -1]]
■
```

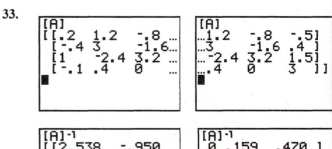

33.

```
[A]
[[.2  1.2   -.8  ...
 [-.4 3     -1.6...
 [1   -2.4  3.2 ...
 [-.1 .4    0   ...
■
```

```
[A]
...1.2   -.8   -.5]
...3     -1.6  .4 ]
...-2.4  3.2   1.5]
....4    0     3  ]]
■
```

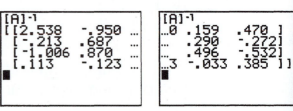

```
[A]-1
[[2.538   -.950 ...
 [-.213   .687  ...
 [-1.006  .870  ...
 [.113    -.123 ...
■
```

```
[A]-1
...0  .159  .470 ]
...   .290  -.272]
...   .496  -.532]
...3  -.033 .385 ]]
■
```

$$A^{-1} = \begin{bmatrix} 2.538 & -0.950 & 0.159 & 0.470 \\ -0.213 & 0.687 & 0.290 & -0.272 \\ -1.006 & 0.870 & 0.496 & -0.532 \\ 0.113 & -0.123 & -0.033 & 0.385 \end{bmatrix}$$

37. $\frac{1}{ad-bc}\cdot\begin{bmatrix} a & b \\ c & d \end{bmatrix}\cdot\begin{bmatrix} d & -b \\ -c & a \end{bmatrix}$

$$= \frac{1}{ad-bc}\cdot\begin{bmatrix} ad-bc & -ab+ab \\ cd-cd & -bc+ad \end{bmatrix}$$

$$\frac{1}{ad-bc}\cdot\begin{bmatrix} ad-bc & 0 \\ 0 & ad-bc \end{bmatrix} = \begin{bmatrix} \frac{ad-bc}{ad-bc} & \frac{0}{ad-bc} \\ \frac{0}{ad-bc} & \frac{ad-bc}{ad-bc} \end{bmatrix}$$

$$= \begin{bmatrix} 1 & 0 \\ 0 & 1 \end{bmatrix}$$

41. $\begin{bmatrix} a_{11} & a_{12} \\ a_{21} & a_{22} \end{bmatrix}$

Interchange the elements of the principal diagonal and change the signs of the off-diagonal elements.

$$\begin{bmatrix} a_{22} & -a_{12} \\ -a_{21} & a_{11} \end{bmatrix}$$

Find the determinant of the original matrix.

$$\begin{vmatrix} a_{11} & a_{12} \\ a_{21} & a_{22} \end{vmatrix} = a_{11}a_{22} - a_{12}a_{21}$$

$$A^{-1} = \frac{1}{a_{11}a_{22} - a_{12}a_{21}}\begin{bmatrix} a_{22} & -a_{12} \\ -a_{21} & a_{11} \end{bmatrix}$$

$$V = A^{-1}I = \frac{1}{a_{11}a_{22} - a_{12}a_{21}}\begin{bmatrix} a_{22} & -a_{12} \\ -a_{21} & a_{11} \end{bmatrix}\begin{bmatrix} i_1 \\ i_2 \end{bmatrix}$$

$$= \frac{1}{a_{11}a_{22} - a_{12}a_{21}}\begin{bmatrix} a_{22}i_1 & -a_{12}i_2 \\ -a_{21}i_1 & a_{11}i_2 \end{bmatrix}$$

$$v_1 = \frac{a_{22}i_1 - a_{12}i_2}{a_{11}a_{22} - a_{12}a_{21}},\ v_2 = \frac{-a_{21}i_1 + a_{11}i_2}{a_{11}a_{22} - a_{12}a_{21}}$$

16.4 Matrices and Linear Equations

1. $2x - y = 7$

$5x - 3y = 19$

$$A = \begin{bmatrix} 2 & -1 \\ 5 & -3 \end{bmatrix},\ C = \begin{bmatrix} 7 \\ 19 \end{bmatrix},\ A^{-1} = \begin{bmatrix} 3 & -1 \\ 5 & -2 \end{bmatrix}$$

$$A^{-1}C = \begin{bmatrix} 3 & -1 \\ 5 & -2 \end{bmatrix}\begin{bmatrix} 7 \\ 19 \end{bmatrix} = \begin{bmatrix} 2 \\ -3 \end{bmatrix} = \begin{bmatrix} x \\ y \end{bmatrix}$$

$x = 2,\ y = -3$ is the solution.

5. $x + 2y = 7$

$2x + 3y = 11$

$$A = \begin{bmatrix} 1 & 2 \\ 2 & 3 \end{bmatrix},\ C = \begin{bmatrix} 7 \\ 11 \end{bmatrix},\ A^{-1} = \begin{bmatrix} -3 & 2 \\ 2 & -1 \end{bmatrix}$$

$$A^{-1}C = \begin{bmatrix} -3 & 2 \\ 2 & -1 \end{bmatrix}\begin{bmatrix} 7 \\ 11 \end{bmatrix} = \begin{bmatrix} 1 \\ 3 \end{bmatrix} = \begin{bmatrix} x \\ y \end{bmatrix}$$

$x = 1,\ y = 3$

9. $C = \begin{bmatrix} 5 \\ -1 \\ -2 \end{bmatrix}$

$$A^{-1} = \begin{bmatrix} 2 & 4 & \frac{7}{2} \\ -1 & -2 & -\frac{3}{2} \\ 1 & 1 & \frac{1}{2} \end{bmatrix};\ A^{-1}C = \begin{bmatrix} 2 & 4 & \frac{7}{2} \\ -1 & -2 & -\frac{3}{2} \\ 1 & 1 & \frac{1}{2} \end{bmatrix}\begin{bmatrix} 5 \\ -1 \\ -2 \end{bmatrix}$$

$$= \begin{bmatrix} 10 & -4 & 7 \\ -5 & 2 & 3 \\ 5 & -1 & -1 \end{bmatrix}\begin{bmatrix} -1 \\ 0 \\ 3 \end{bmatrix} \quad x = -1,\ y = 0,\ z = 3$$

13. $A = \begin{bmatrix} 2.5 & 2.8 \\ 3.5 & -1.6 \end{bmatrix};\ C = \begin{bmatrix} -3.0 \\ 9.6 \end{bmatrix};\ \begin{vmatrix} 2.5 & 2.8 \\ 3.5 & -1.6 \end{vmatrix}$

$$= -4 - 9.8 = -13.8;\ A^{-1} = -\frac{1}{13.8}\begin{bmatrix} -1.6 & -2.8 \\ -3.5 & 2.5 \end{bmatrix}$$

$$A^{-1}C = -\frac{1}{13.8}\begin{bmatrix} -1.6 & -2.8 \\ -3.5 & 2.5 \end{bmatrix}\begin{bmatrix} -3.0 \\ 9.6 \end{bmatrix}$$

$$= -\frac{1}{13.8}\begin{bmatrix} -22.08 \\ 34.5 \end{bmatrix} = \begin{bmatrix} -0.348 + 1.949 \\ -0.762 - 1.74 \end{bmatrix} = \begin{bmatrix} 1.6 \\ -2.5 \end{bmatrix}$$

$x = 1.6,\ y = -2.5$

17. $A=\begin{bmatrix}2&4&1\\-2&-2&-1\\-1&2&1\end{bmatrix}$; $C=\begin{bmatrix}5\\-6\\0\end{bmatrix}$. To find A^{-1}:

$$\left[\begin{array}{ccc|ccc}2&4&1&1&0&0\\-2&-2&-1&0&1&0\\-1&2&1&0&0&1\end{array}\right]\begin{array}{l}R1\to\frac{1}{2}R1\\R2\to R2+R1\\R3\to R3+\frac{1}{2}R1\end{array}$$

$$\left[\begin{array}{ccc|ccc}1&2&\frac{1}{2}&\frac{1}{2}&0&0\\0&2&0&1&1&0\\0&4&\frac{3}{2}&\frac{1}{2}&0&1\end{array}\right]\begin{array}{l}R1\to R1-R2\\R2\to\frac{1}{2}R2\\R3\to R3-2R2\end{array}$$

$$\left[\begin{array}{ccc|ccc}1&0&\frac{1}{2}&\frac{1}{2}&-1&0\\0&1&0&\frac{1}{2}&\frac{1}{2}&0\\0&0&\frac{3}{2}&-\frac{3}{2}&-2&1\end{array}\right]\begin{array}{l}R1\to R1-\frac{1}{3}R3\\ \\R3\to\frac{2}{3}R3\end{array}$$

$$\left[\begin{array}{ccc|ccc}1&0&0&0&-\frac{1}{3}&-\frac{1}{3}\\0&1&0&0.5&0.5&0\\0&0&1&-1&-\frac{4}{3}&\frac{2}{3}\end{array}\right]$$

$$A^{-1}C=\begin{bmatrix}0&-\frac{1}{3}&-\frac{1}{3}\\\frac{1}{2}&\frac{1}{2}&0\\-1&-\frac{4}{3}&\frac{2}{3}\end{bmatrix}\begin{bmatrix}5\\-6\\0\end{bmatrix}=\begin{bmatrix}2\\2.5-3\\-5+8\end{bmatrix}=\begin{bmatrix}2\\-\frac{1}{2}\\3\end{bmatrix};$$

$x=2,\ y=-\dfrac{1}{2},\ z=3$

21. $$\begin{bmatrix}u\\v\\w\end{bmatrix}=\begin{bmatrix}1&-3&-2\\3&2&6\\4&-1&3\end{bmatrix}^{-1}\begin{bmatrix}9\\20\\25\end{bmatrix}=\begin{bmatrix}-\frac{12}{11}&-1&\frac{14}{11}\\-\frac{15}{11}&-1&\frac{12}{11}\\1&1&-1\end{bmatrix}\begin{bmatrix}9\\20\\25\end{bmatrix}$$

$$=\begin{bmatrix}2\\-5\\4\end{bmatrix}$$ (Inverse from calculator)

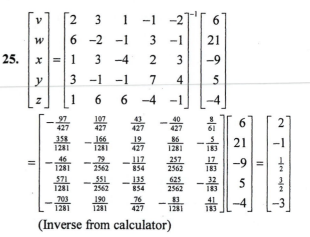

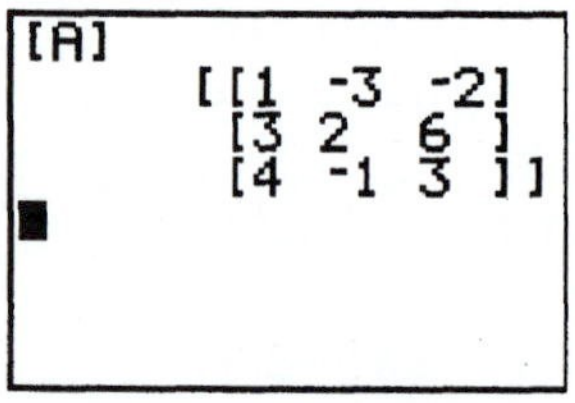

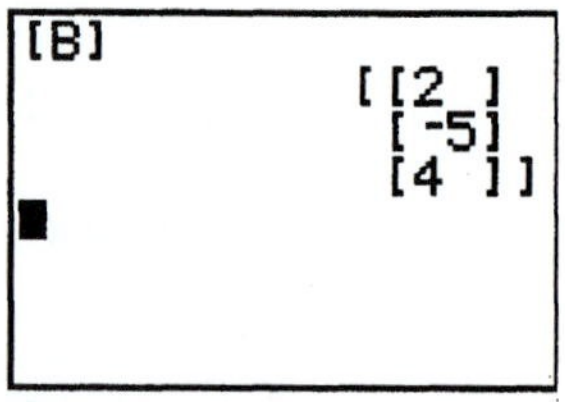

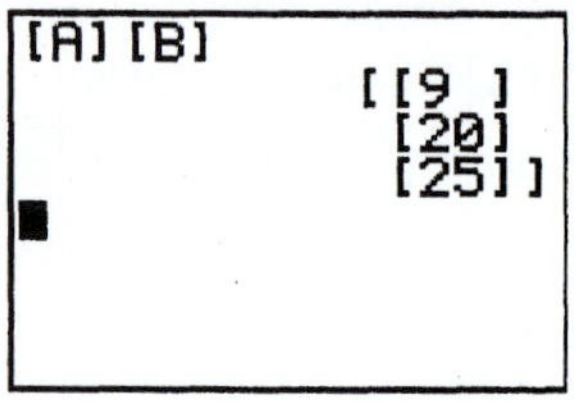

25. $$\begin{bmatrix}v\\w\\x\\y\\z\end{bmatrix}=\begin{bmatrix}2&3&1&-1&-2\\6&-2&-1&3&-1\\1&3&-4&2&3\\3&-1&-1&7&4\\1&6&6&-4&-1\end{bmatrix}^{-1}\begin{bmatrix}6\\21\\-9\\5\\-4\end{bmatrix}$$

$$=\begin{bmatrix}-\frac{97}{427}&\frac{107}{427}&\frac{43}{427}&-\frac{40}{427}&\frac{8}{61}\\\frac{358}{1281}&-\frac{166}{1281}&\frac{19}{427}&\frac{86}{1281}&-\frac{5}{183}\\-\frac{46}{1281}&-\frac{79}{2562}&-\frac{117}{854}&\frac{257}{2562}&\frac{17}{183}\\\frac{571}{1281}&-\frac{551}{2562}&-\frac{135}{854}&\frac{625}{2562}&-\frac{32}{183}\\-\frac{703}{1281}&\frac{190}{1281}&\frac{76}{427}&-\frac{83}{1281}&\frac{41}{183}\end{bmatrix}\begin{bmatrix}6\\21\\-9\\5\\-4\end{bmatrix}=\begin{bmatrix}2\\-1\\\frac{1}{2}\\\frac{3}{2}\\-3\end{bmatrix}$$

(Inverse from calculator)

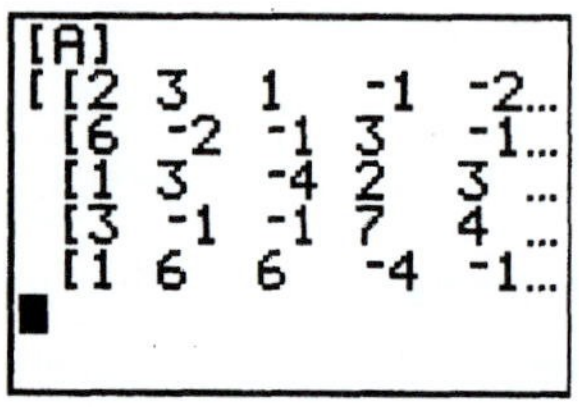

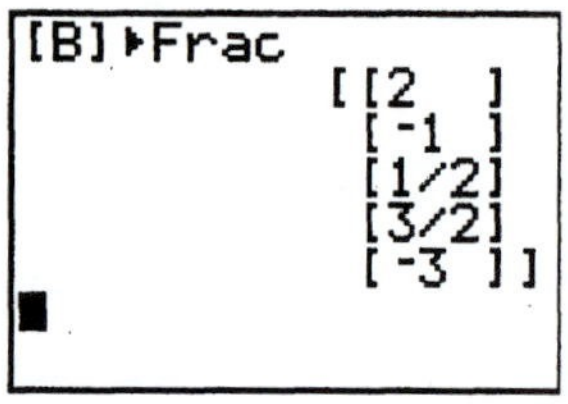

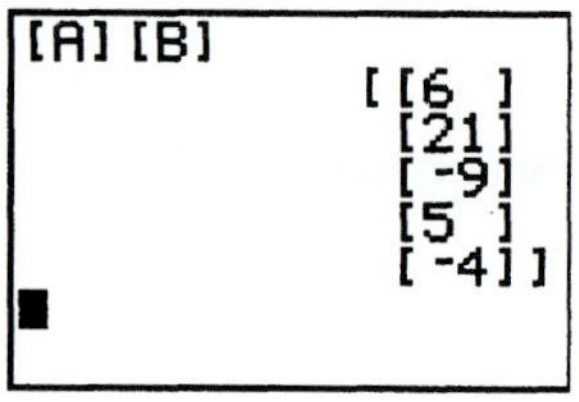

29. $2x-y=4$

$3x+y=1$

$A=\begin{bmatrix}2&-1\\3&1\end{bmatrix}$, $C=\begin{bmatrix}4\\1\end{bmatrix}$, $A^{-1}=\begin{bmatrix}0.2&0.2\\-0.6&0.4\end{bmatrix}$

$A^{-1}C=\begin{bmatrix}0.2&0.2\\-0.6&0.4\end{bmatrix}\begin{bmatrix}4\\1\end{bmatrix}=\begin{bmatrix}1\\-2\end{bmatrix}=\begin{bmatrix}x\\y\end{bmatrix}$

Since $x=1,\ y=-2$ is also a solution of $x-2y=5$, the graphs of all three equations contain the point $(1,-2)$.

33. $x+y=18$

$x-2y=-6$

$$A=\begin{bmatrix}1 & 1\\1 & -2\end{bmatrix},\ C=\begin{bmatrix}18\\-6\end{bmatrix},\ A^{-1}=\begin{bmatrix}\frac{2}{3} & \frac{1}{3}\\\frac{1}{3} & -\frac{1}{3}\end{bmatrix}$$

$$A^{-1}C=\begin{bmatrix}\frac{2}{3} & \frac{1}{3}\\\frac{1}{3} & -\frac{1}{3}\end{bmatrix}\begin{bmatrix}18\\-6\end{bmatrix}=\begin{bmatrix}x\\y\end{bmatrix}=\begin{bmatrix}10\\8\end{bmatrix}$$

The voltages of the batteries are 10 V and 8 V.

37. $x+y+z=10$

$4y=x$

$0.05y+0.06z=10(0.02)$

$$A=\begin{bmatrix}1 & 1 & 1\\-1 & 4 & 0\\0 & 0.05 & 0.06\end{bmatrix};\ C=\begin{bmatrix}10\\0\\0.2\end{bmatrix}$$

To find A^{-1}:

$$\left[\begin{array}{ccc|ccc}1 & 1 & 1 & 1 & 0 & 0\\-1 & 4 & 0 & 0 & 1 & 0\\0 & 0.05 & 0.06 & 0 & 0 & 1\end{array}\right]\ R2\to R2+R1$$

$$\left[\begin{array}{ccc|ccc}1 & 1 & 1 & 1 & 0 & 0\\0 & 5 & 1 & 1 & 1 & 0\\0 & 0.05 & 0.06 & 0 & 0 & 1\end{array}\right]\ \begin{array}{l}R1\to R1-\frac{1}{5}R2\\R2\to\frac{1}{5}R2\\R3\to R3-0.01R2\end{array}$$

$$\left[\begin{array}{ccc|ccc}1 & 0 & \frac{4}{5} & \frac{4}{5} & -\frac{1}{5} & 0\\0 & 1 & \frac{1}{5} & \frac{1}{5} & \frac{1}{5} & 0\\0 & 0 & 0.05 & -0.01 & -0.01 & 1\end{array}\right]\ \begin{array}{l}R1\to R1-16R3\\R2\to R2-4R3\\R3\to 20R3\end{array}$$

$$\left[\begin{array}{ccc|ccc}1 & 0 & 0 & 0.96 & -0.04 & -16\\0 & 1 & 0 & 0.24 & 0.24 & -4\\0 & 0 & 1 & -0.2 & -0.2 & 20\end{array}\right];$$

$$A^{-1}C=\begin{bmatrix}0.96 & -0.04 & -16\\0.24 & 0.24 & -4\\-0.2 & -0.2 & 20\end{bmatrix}\begin{bmatrix}10\\0\\0.2\end{bmatrix}$$

$$=\begin{bmatrix}9.6-3.2\\2.4-0.8\\-2+4\end{bmatrix}=\begin{bmatrix}6.4\\1.6\\2.0\end{bmatrix}=6.4\text{ L, }1.6\text{ L, }2.0\text{ L}$$

16.5 Gaussian Elimination

1. $3x-2y=3 \quad R1\to\frac{1}{3}R1$

$2x+y=4$

$x-\frac{2}{3}y=1 \quad R2\to -2R1+R2$

$2x+\ y=4$

$x-\frac{2}{3}y=1$

$\frac{7}{3}y=2 \quad R\to\frac{3}{7}R2$

$x-\frac{2}{3}y=1$

$y=\frac{6}{7}$

$x-\frac{2}{3}\left(\frac{6}{7}\right)=1$

$x \quad =\frac{11}{7}$

The solution is $x=\frac{11}{7}$, $y=\frac{6}{7}$.

5. $5x-3y=2 \quad R1\to\frac{1}{5}R1$

$-2x+4y=3$

$x-\frac{3}{5}y=\frac{2}{5} \quad R2\to 2R1+R2$

$-2x+4y=3$

$x-\frac{3}{5}y=\frac{2}{5}$

$\frac{14}{5}y=\frac{19}{5} \quad R2\to\frac{5}{14}R2$

$x-\frac{3}{5}y=\frac{2}{5}$

$y=\frac{19}{14}$

$x-\frac{3}{5}\left(\frac{19}{14}\right)=\frac{2}{5}\Rightarrow x=\frac{17}{14}$

The solution is $x=\frac{17}{14}$, $y=\frac{19}{14}$.

9. $x+3y+3z=-3 \quad R2\to -2R1+R2$

$2x+2y+z=-5 \quad R3\to 2R1+R3$

$-2x-y+4z=\ 6$

$x+3y+\ 3z=-3$

$-4y-\ 5z=\ 1 \quad R2\to -\frac{1}{4}R2$

$5y+10z=\ 0$

$x+3y+3z=-3$

$y+\ \frac{5}{4}z=-\frac{1}{4} \quad R3\to -5R2+R3$

$5y + 10z = 0$

$x + 3y + 3z = -3$

$y + \frac{5}{4}z = -\frac{1}{4} \quad R3 \to -5R2 + R3$

$5y + 10z = 0$

$x + 3y + 3z = -3$

$y + \frac{5}{4}z = -\frac{1}{4}$

$\frac{15}{4}z = \frac{5}{4} \Rightarrow z = \frac{1}{3}$

$y + \frac{5}{4}\left(\frac{1}{3}\right) = -\frac{1}{4} \Rightarrow y = -\frac{2}{3}$

$x + 3\left(-\frac{2}{3}\right) + 3\left(\frac{1}{3}\right) = -3 \Rightarrow x = -2$

The solution is $x = -2$, $y = -\frac{2}{3}$, $z = \frac{1}{3}$.

13. $x - 4y + z = 2 \quad R2 \to -3R1 + R2$

$3x - y + 4z = -4$

$x - 4y + z = 2$

$11y + z = -10 \Rightarrow z = -10 - 11y$

There are an unlimited number of solutions.

$x = -3$, $y = -1$, $z = 1$; $x = 12$, $y = 0$, $z = -10$

17. $x + 3y + z = 4$

$2x - 6y - 3z = 10 \quad R2 \to -2R1 + R2$

$4x - 9y + 3z = 4 \quad R3 \to -4R1 + R3$

$x + 3y + z = 4$

$-12y - 5z = 2 \quad R3 \to -\frac{21}{12}R2 + R3$

$-21y - z = -12$

$x + 3y + z = 4$

$-12y - 5z = 2$

$\frac{31}{4}z = -\frac{31}{2} \Rightarrow z = -2$

$-12y - 5(-2) = 2 \Rightarrow y = \frac{2}{3}$

$x + 3\left(\frac{2}{3}\right) + (-2) = 4 \Rightarrow x = 4$

The solution is $x = 4$, $y = \frac{2}{3}$, $z = -2$.

21. $3x + 5y = -2$

$24x - 18y = 13 \quad R2 \to -8R1 + R2$

$15x - 33y = 19 \quad R3 \to -5R1 + R3$

$6x + 68y = -33 \quad R4 \to -2R1 + R4$

$3x + 5y = -2$

$-58y = 29 \Rightarrow y = -\frac{1}{2}$

$-58y = 29 \Rightarrow y = -\frac{1}{2}$

$58y = -29 \Rightarrow y = -\frac{1}{2}$

$3x + 5\left(-\frac{1}{2}\right) = -2 \Rightarrow x = \frac{1}{6}$

The solution is $x = \frac{1}{6}$, $y = -\frac{1}{2}$.

25. $s + 2t - 3u = 2$

$3s + 6t - 9u = 6 \quad R2 \to -3R1 + R2$

$7s + 14t - 21u = 13 \quad R3 \to -7R1 + R3$

$s + 2t - 3u = 2$

$0 = 0$

$0 = -1$

The system is inconsistent.

29. $a_1x + b_1y = c_1$

$a_2x + b_2y = c_2 \quad R2 \to -\frac{a_2}{a_1}R1 + R2$

$a_1x + b_1y = c_1$

$\left(\frac{-a_2b_1}{a_1} + b_2\right)y = \frac{-a_2c_1}{a_1} + c_2$

$$y = \frac{-\frac{a_2c_1}{a_1} + c_2}{-\frac{a_2b_1}{a_1} + b_2} = \frac{a_1c_2 - a_2c_1}{a_1b_2 - a_2b_1} = \frac{\begin{vmatrix} a_1 & c_1 \\ a_2 & c_2 \end{vmatrix}}{\begin{vmatrix} a_1 & b_1 \\ a_2 & b_2 \end{vmatrix}}$$

$$a_1x + b_1 \cdot \frac{a_1c_2 - a_2c_1}{a_1b_2 - a_2b_1} = c_1$$

$$x = \frac{-b_1(a_1c_2 - a_2c_1)}{a_1b_2 - a_2b_1} + \frac{c_1(a_1b_2 - a_2b_1)}{a_1(a_1b_2 - a_2b_1)}$$

$$x = \frac{-a_1b_1c_2 + a_2b_1c_1 + a_1b_2c_1 - a_2b_1c_1}{a_1(a_1b_2 - a_2b_1)}$$

$$x = \frac{a_1(b_2c_1 - b_1c_2)}{a_1(a_1b_2 - a_2b_1)} = \frac{\begin{vmatrix} c_1 & b_1 \\ c_2 & b_2 \end{vmatrix}}{\begin{vmatrix} a_1 & b_1 \\ a_2 & b_2 \end{vmatrix}}$$

33. $x + y + z = 650$

$-x + 2y - z = 10 \quad R2 \to R1 + R2$

$3.00x + 2.00y + 2.00z = 1550 \quad R3 \to -3R1 + R3$

$x + y + z = 650$

$3.00y = 660 \Rightarrow y = 220$

$-y - z = -400$

$-220 - z = -400 \Rightarrow z = 180$

$x + 220 + 180 = 650 \Rightarrow x = 250$

The production rates are 250 parts/h, 220 parts/h, and 180 parts/h.

16.6 Higher Order Determinants

1. $\begin{vmatrix} 3 & 0 & 0 \\ 1 & 1 & 0 \\ 2 & 1 & 3 \end{vmatrix} = 9$, switch first and third column

$\begin{vmatrix} 0 & 0 & 3 \\ 0 & 1 & 1 \\ 3 & 1 & 2 \end{vmatrix} = -9$

5. $\begin{vmatrix} 3 & -2 & 4 & 2 \\ 5 & -1 & 2 & -1 \\ 3 & -2 & 4 & 2 \\ 0 & 3 & -6 & 0 \end{vmatrix} = 0$, Row 1 and R3 are identical.

9. $\begin{vmatrix} 2 & -3 & -1 \\ -4 & 1 & -3 \\ 1 & -3 & 2 \end{vmatrix} = -40$,

Column 3 of given determinant was multiplied by -1.

13. Expand by first row,

$$\begin{vmatrix} 3 & 1 & 0 \\ -2 & 3 & -1 \\ 4 & 2 & 5 \end{vmatrix} = 3\begin{vmatrix} 3 & -1 \\ 2 & 5 \end{vmatrix} - \begin{vmatrix} -2 & -1 \\ 4 & 5 \end{vmatrix} + 0\begin{vmatrix} -2 & 3 \\ 4 & 2 \end{vmatrix}$$

$$= 3(3(5) - 2(-1)) - (-2(5) - 4(-1))$$

$$= 57$$

17. Expand by first column,

$$\begin{vmatrix} 1 & 3 & -3 & 5 \\ 4 & 2 & 1 & 2 \\ 3 & 2 & -2 & 2 \\ 0 & 1 & 2 & -1 \end{vmatrix} = 1\begin{vmatrix} 2 & 1 & 2 \\ 2 & -2 & 2 \\ 1 & 2 & -1 \end{vmatrix} - 4\begin{vmatrix} 3 & -3 & 5 \\ 2 & -2 & 2 \\ 1 & 2 & -1 \end{vmatrix}$$

$$+3\begin{vmatrix} 3 & -3 & 5 \\ 2 & 1 & 2 \\ 1 & 2 & -1 \end{vmatrix} - 0\begin{vmatrix} 3 & -3 & 5 \\ 2 & 1 & 2 \\ 2 & -2 & 2 \end{vmatrix}$$

$$= 1(12) - 4(12) + 3(-12) = -72$$

21. $\begin{vmatrix} 3 & 0 & 0 \\ -2 & 1 & 4 \\ 4 & -2 & 5 \end{vmatrix}$ $R2 \to -\frac{4}{5}R3 + R2$

$$\begin{vmatrix} 3 & 0 & 0 \\ -2 & 1 & 4 \\ 4 & -2 & 5 \end{vmatrix} = \begin{vmatrix} 3 & 0 & 0 \\ -\frac{26}{5} & \frac{13}{5} & 0 \\ 5 & -2 & 5 \end{vmatrix} = 3\left(\frac{13}{5}\right)(5) = 39$$

25. $\begin{vmatrix} 4 & 3 & 6 & 0 \\ 3 & 0 & 0 & 4 \\ 5 & 0 & 1 & 2 \\ 2 & 1 & 1 & 7 \end{vmatrix}$ $R2 \to -\frac{3}{4}R1 + R2$, $R3 \to -\frac{5}{4}R1 + R3$, $R4 \to -\frac{1}{2}R1 + R4$

$$\begin{vmatrix} 4 & 3 & 6 & 0 \\ 3 & 0 & 0 & 4 \\ 5 & 0 & 1 & 2 \\ 2 & 1 & 1 & 7 \end{vmatrix} = \begin{vmatrix} 4 & 3 & 6 & 0 \\ 0 & -\frac{9}{4} & -\frac{9}{2} & 4 \\ 0 & -\frac{15}{4} & -\frac{13}{2} & 2 \\ 0 & -\frac{1}{2} & -2 & 7 \end{vmatrix} \begin{matrix} \\ \\ R3 \to -\frac{15}{9}R2 + R3 \\ R4 \to -\frac{2}{9}R2 + R4 \end{matrix}$$

$$= \begin{vmatrix} 4 & 3 & 6 & 0 \\ 0 & -\frac{9}{4} & -\frac{9}{2} & 4 \\ 0 & 0 & 1 & -\frac{14}{3} \\ 0 & 0 & -1 & \frac{55}{9} \end{vmatrix} \; R4 \to R3 + R4$$

$$= \begin{vmatrix} 4 & 3 & 6 & 0 \\ 0 & -\frac{9}{4} & -\frac{9}{2} & 4 \\ 0 & 0 & 1 & -\frac{14}{3} \\ 0 & 0 & 0 & \frac{13}{9} \end{vmatrix} = 4\left(-\frac{9}{4}\right)(1)\left(\frac{13}{9}\right) = -13$$

29. $\begin{vmatrix} 1 & 2 & 0 & 1 & 0 \\ 0 & 2 & 1 & 0 & 1 \\ 1 & 0 & -1 & 1 & -1 \\ -2 & 0 & -1 & 2 & 1 \\ 1 & 0 & 2 & -1 & -2 \end{vmatrix}$ $R3 \to -R1 + R3$, $R4 \to 2R1 + R4$, $R5 \to -R1 + R5$

$$= \begin{vmatrix} 1 & 2 & 0 & 1 & 0 \\ 0 & 2 & 1 & 0 & 1 \\ 0 & -2 & -1 & 0 & -1 \\ 0 & 4 & -1 & 4 & 1 \\ 0 & -2 & 2 & 2 & -2 \end{vmatrix} \begin{matrix} \\ \\ R3 \to R2 + R3 \\ R4 \to -2R2 + R4 \\ R5 \to R2 + R5 \end{matrix}$$

$$= \begin{vmatrix} 1 & 2 & 0 & 1 & 0 \\ 0 & 2 & 1 & 0 & 1 \\ 0 & 0 & 0 & 0 & 0 \\ 0 & 4 & -1 & 4 & 1 \\ 0 & -2 & 2 & -2 & -2 \end{vmatrix} \; R3 \Leftrightarrow R5$$

$$= -\begin{vmatrix} 1 & 2 & 0 & 1 & 0 \\ 0 & 2 & 1 & 0 & 1 \\ 0 & -2 & 2 & -2 & -2 \\ 0 & 4 & -1 & 4 & 1 \\ 0 & 0 & 0 & 0 & 0 \end{vmatrix} \begin{matrix} \\ \\ R3 \to R2 + R3 \\ R4 \to -2R2 + R4 \\ \\ \end{matrix}$$

$$= -\begin{vmatrix} 1 & 2 & 0 & 1 & 0 \\ 0 & 2 & 1 & 0 & 1 \\ 0 & 0 & 3 & -2 & -1 \\ 0 & 0 & -3 & 4 & -1 \\ 0 & 0 & 0 & 0 & 0 \end{vmatrix} \begin{matrix} \\ \\ \\ R4 \to R3 + R4 \\ \\ \end{matrix}$$

$$= -\begin{vmatrix} 1 & 2 & 0 & 1 & 0 \\ 0 & 2 & 1 & 0 & 1 \\ 0 & 0 & 3 & -2 & -1 \\ 0 & 0 & 0 & 2 & -2 \\ 0 & 0 & 0 & 0 & 0 \end{vmatrix} = -(1)(2)(3)(2)(0) = 0$$

33. $x + 2y - z = 6$
$y - 2z - 3t = -5$
$3x - 2y + t = 2$
$2x + y + z - t = 0$

$$\begin{vmatrix} 1 & 2 & -1 & 0 \\ 0 & 1 & -2 & -3 \\ 3 & -2 & 0 & 1 \\ 2 & 1 & 1 & -1 \end{vmatrix} = 61$$

$$x = \frac{\begin{vmatrix} 6 & 2 & -1 & 0 \\ -5 & 1 & -2 & -3 \\ 2 & -2 & 0 & 1 \\ 0 & 1 & 1 & -1 \end{vmatrix}}{61} = \frac{61}{61} = 1$$

$$y = \frac{\begin{vmatrix} 1 & 6 & -1 & 0 \\ 0 & -5 & -2 & -3 \\ 3 & 2 & 0 & 1 \\ 2 & 0 & 1 & -1 \end{vmatrix}}{61} = \frac{122}{61} = 2$$

$$z = \frac{\begin{vmatrix} 1 & 2 & 6 & 0 \\ 0 & 1 & -5 & -3 \\ 3 & -2 & 2 & 1 \\ 2 & 1 & 0 & -1 \end{vmatrix}}{61} = \frac{-61}{61} = -1$$

$$t = \frac{\begin{vmatrix} 1 & 2 & -1 & 6 \\ 0 & 1 & -2 & -5 \\ 3 & -2 & 0 & 2 \\ 2 & 1 & 1 & 0 \end{vmatrix}}{61} = \frac{183}{61} = 3$$

The solution is $x = 1,\ y = 2,\ z = -1,\ t = 3$.

$$x = \frac{\begin{vmatrix} 2 & 1 & 1 & 0 \\ 4 & 3 & -1 & 2 \\ 0 & 1 & 2 & 1 \\ 4 & 0 & 2 & 0 \end{vmatrix}}{14} = \frac{28}{14} = 2$$

$$y = \frac{\begin{vmatrix} 2 & 2 & 1 & 0 \\ 0 & 4 & -1 & 2 \\ 0 & 0 & 2 & 1 \\ 3 & 4 & 2 & 0 \end{vmatrix}}{14} = \frac{-14}{14} = -1$$

$$z = \frac{\begin{vmatrix} 2 & 1 & 1 & 2 \\ 0 & 3 & -1 & 4 \\ 0 & 1 & 2 & 0 \\ 3 & 0 & 2 & 4 \end{vmatrix}}{14} = \frac{42}{14} = 3$$

$$t = \frac{\begin{vmatrix} 2 & 1 & 1 & 2 \\ 0 & 3 & -1 & 4 \\ 0 & 1 & 2 & 0 \\ 3 & 0 & 2 & 4 \end{vmatrix}}{14} = \frac{42}{14} = 3$$

The solution is $x = 2,\ y = -1,\ z = -1,\ t = 3$.

37. $D + E + 2F = 1$
$2D - E + G = -2$
$D - E - F - 2G = 4$
$2D - E + 2F - G = 0$

$$\begin{vmatrix} 1 & 1 & 2 & 0 \\ 2 & -1 & 0 & 1 \\ 1 & -1 & -1 & -2 \\ 2 & -1 & 2 & -1 \end{vmatrix} = -18$$

$$D=\frac{\begin{vmatrix}1&1&2&0\\-2&-1&0&1\\4&-1&-1&-2\\0&-1&2&-1\end{vmatrix}}{-18}=\frac{-18}{-18}=1$$

$$E=\frac{\begin{vmatrix}1&1&2&0\\2&-2&0&1\\1&4&-1&-2\\2&0&2&-1\end{vmatrix}}{-18}=\frac{-36}{-18}=2$$

$$F=\frac{\begin{vmatrix}1&1&1&0\\2&-1&-2&1\\1&-1&4&-2\\2&-1&0&-1\end{vmatrix}}{-18}=\frac{18}{-18}=-1$$

$$G=\frac{\begin{vmatrix}1&1&2&1\\2&-1&0&-2\\1&-1&-1&4\\2&-1&2&0\end{vmatrix}}{-18}=\frac{36}{-18}=-2$$

The solution is $D=1$, $E=2$, $F=-1$, $G=-2$.

41. $\begin{vmatrix}2a&2b&2c\\2d&2e&2f\\2g&2h&2i\end{vmatrix}=2\begin{vmatrix}a&b&c\\2d&2e&2f\\2g&2h&2i\end{vmatrix}$

$=2(2)\begin{vmatrix}a&b&c\\d&e&f\\2g&2h&2i\end{vmatrix}$

$=2(2)(2)\begin{vmatrix}a&b&c\\d&e&f\\g&h&i\end{vmatrix}$

The value of the determinant is changed by a factor of 8.

45. $\begin{vmatrix}C&-1&0&0\\-1&C&-1&0\\0&-1&C&-1\\0&0&-1&C\end{vmatrix}=C^4-3C^2+1=0,$

solve graphically,

$C=0.618, 1.618$

Chapter 16 Review Exercises

1. $\begin{pmatrix}2a\\a-b\end{pmatrix}=\begin{pmatrix}8\\5\end{pmatrix}$; $2a=8$;

$a=4$; $a-b=5$; $4-b=5$; $b=-1$

5. $\begin{bmatrix}\cos\pi&\sin\frac{\pi}{6}\\x+y&x-y\end{bmatrix}=\begin{bmatrix}x&y\\a&b\end{bmatrix}$

$x=\cos\pi=-1,\ y=\sin\frac{\pi}{6}=\frac{1}{2}$

$a=x+y=-1+\frac{1}{2}=-\frac{1}{2}$

$b=x-y=-1-\frac{1}{2}=-\frac{3}{2}$

9. $B-A=\begin{bmatrix}-1&0\\4&-6\\-3&-2\\1&-7\end{bmatrix}-\begin{bmatrix}2&-3\\4&1\\-5&0\\2&-3\end{bmatrix}=\begin{bmatrix}-3&3\\0&-7\\2&-2\\-1&-4\end{bmatrix}$

13.

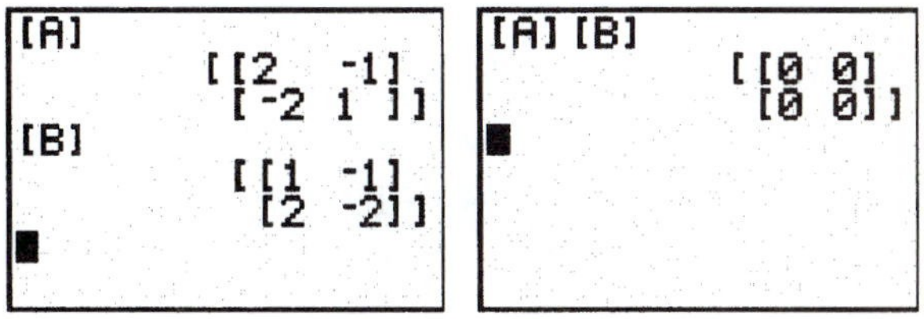

17. $\begin{bmatrix}2&-5\\2&-4\end{bmatrix}$

Interchange elements of principal diagonal and change signs of off-diagonal elements.

$\begin{bmatrix}-4&5\\-2&2\end{bmatrix}$

Find the determinant of original matrix.

$\begin{vmatrix}2&-5\\2&-4\end{vmatrix}=2$

Divide each element of second matrix by 2.

$\frac{1}{2}\begin{bmatrix}-4&5\\-2&2\end{bmatrix}=\begin{bmatrix}-2&\frac{5}{2}\\-1&1\end{bmatrix}$

21. $\left[\begin{array}{ccc|ccc} 1 & 1 & -2 & 1 & 0 & 0 \\ -1 & -2 & 1 & 0 & 1 & 0 \\ 0 & 3 & 4 & 0 & 0 & 1 \end{array}\right]$

$$R1 + R2 \to R2 \left[\begin{array}{ccc|ccc} 1 & 1 & -2 & 1 & 0 & 0 \\ 0 & -1 & -1 & 1 & 1 & 0 \\ 0 & 3 & 4 & 0 & 0 & 1 \end{array}\right]$$

$$3R2 + R3 \to R3 \left[\begin{array}{ccc|ccc} 1 & 1 & -2 & 1 & 0 & 0 \\ 0 & -1 & -1 & 1 & 1 & 0 \\ 0 & 0 & 1 & 3 & 3 & 1 \end{array}\right]$$

$$R2 + R1 \to R1 \left[\begin{array}{ccc|ccc} 1 & 0 & -3 & 2 & 1 & 0 \\ 0 & -1 & -1 & 1 & 1 & 0 \\ 0 & 0 & 1 & 3 & 3 & 1 \end{array}\right]$$

$$R3 + R2 \to R2 \left[\begin{array}{ccc|ccc} 1 & 0 & -3 & 2 & 1 & 0 \\ 0 & -1 & 0 & 4 & 4 & 1 \\ 0 & 0 & 1 & 3 & 3 & 1 \end{array}\right]$$

$$3R3 + R1 \to R1 \left[\begin{array}{ccc|ccc} 1 & 0 & 0 & 11 & 10 & 3 \\ 0 & -1 & 0 & 4 & 4 & 1 \\ 0 & 0 & 1 & 3 & 3 & 1 \end{array}\right]$$

$$-R2 \to R2 \left[\begin{array}{ccc|ccc} 1 & 0 & 0 & 11 & 10 & 3 \\ 0 & 1 & 0 & -4 & -4 & -1 \\ 0 & 0 & 1 & 3 & 3 & 1 \end{array}\right]$$

$$A^{-1} = \begin{bmatrix} 11 & 10 & 3 \\ -4 & -4 & -1 \\ 3 & 3 & 1 \end{bmatrix}$$

25. $A = \begin{bmatrix} 2 & -3 \\ 4 & -1 \end{bmatrix}; C = \begin{bmatrix} -9 \\ -13 \end{bmatrix}; \begin{vmatrix} 2 & -3 \\ 4 & -1 \end{vmatrix} = 10;$

$$A^{-1} = \frac{1}{10}\begin{bmatrix} -1 & 3 \\ -4 & 2 \end{bmatrix} = \begin{bmatrix} -\frac{1}{10} & \frac{3}{10} \\ -\frac{4}{10} & \frac{2}{10} \end{bmatrix}$$

$$A^{-1}C = \begin{bmatrix} -\frac{1}{10} & \frac{3}{10} \\ -\frac{4}{10} & \frac{2}{10} \end{bmatrix}\begin{bmatrix} -9 \\ -13 \end{bmatrix} = \begin{bmatrix} \frac{9}{10} & -\frac{39}{10} \\ \frac{36}{10} & -\frac{26}{10} \end{bmatrix}\begin{bmatrix} -3 \\ 1 \end{bmatrix}$$

$x = -3,\ y = 1$

29. $A = \begin{bmatrix} 2 & -3 & 2 \\ 3 & 1 & -3 \\ 1 & 4 & 1 \end{bmatrix}; C = \begin{bmatrix} 7 \\ -6 \\ -13 \end{bmatrix}$

$$\left[\begin{array}{ccc|ccc} 2 & -3 & 2 & 1 & 0 & 0 \\ 3 & 1 & -3 & 0 & 1 & 0 \\ 1 & 4 & 1 & 0 & 0 & 1 \end{array}\right]$$

$$-3R3 + R2 \to R2 \left[\begin{array}{ccc|ccc} 2 & -3 & -2 & 1 & 0 & 0 \\ 3 & 1 & -3 & 0 & 1 & 0 \\ 0 & -11 & -6 & 0 & 1 & -3 \end{array}\right]$$

$$-3R1 + 2R2 \to R2 \left[\begin{array}{ccc|ccc} 2 & -3 & 2 & 1 & 0 & 0 \\ 0 & 11 & -12 & -3 & 2 & 0 \\ 0 & -11 & -6 & 1 & 0 & -3 \end{array}\right]$$

$$\tfrac{1}{2}R1 \to R1 \left[\begin{array}{ccc|ccc} 1 & -\frac{3}{2} & 1 & \frac{1}{2} & 0 & 0 \\ 0 & -11 & -12 & -3 & 2 & 0 \\ 0 & -11 & -6 & 0 & 1 & -3 \end{array}\right]$$

$$R2 + R3 \to R3 \left[\begin{array}{ccc|ccc} 1 & -\frac{3}{2} & 1 & \frac{1}{2} & 0 & 0 \\ 0 & 11 & -12 & -3 & 2 & 0 \\ 0 & 0 & -18 & -3 & 3 & -3 \end{array}\right]$$

$$\tfrac{1}{11}R2 \to R2 \left[\begin{array}{ccc|ccc} 1 & -\frac{3}{2} & 1 & \frac{1}{2} & 0 & 0 \\ 0 & 1 & -\frac{12}{11} & -\frac{3}{11} & \frac{2}{11} & 0 \\ 0 & 0 & -18 & -3 & 3 & -3 \end{array}\right]$$

$$\tfrac{3}{2}R2 + R1 \to R1 \left[\begin{array}{ccc|ccc} 1 & 0 & -\frac{7}{11} & \frac{1}{11} & \frac{3}{11} & 0 \\ 0 & 1 & -\frac{12}{11} & -\frac{3}{11} & \frac{2}{11} & 0 \\ 0 & 0 & -18 & -3 & 3 & -3 \end{array}\right]$$

$$-\tfrac{1}{18}R3 \to R3 \left[\begin{array}{ccc|ccc} 1 & 0 & -\frac{7}{11} & \frac{1}{11} & \frac{3}{11} & 0 \\ 0 & 1 & -\frac{12}{11} & -\frac{3}{11} & \frac{2}{11} & 0 \\ 0 & 0 & 1 & \frac{1}{6} & -\frac{1}{6} & \frac{1}{6} \end{array}\right]$$

$$\tfrac{12}{11}R3 + R2 \to R2 \left[\begin{array}{ccc|ccc} 1 & 0 & -\frac{7}{11} & \frac{1}{11} & \frac{3}{11} & 0 \\ 0 & 1 & 0 & -\frac{1}{11} & 0 & \frac{2}{11} \\ 0 & 0 & 1 & \frac{1}{6} & -\frac{1}{6} & \frac{1}{6} \end{array}\right]$$

$$\tfrac{7}{11}R3 + R1 \to R1 \left[\begin{array}{ccc|ccc} 1 & 0 & 0 & \frac{13}{66} & \frac{11}{66} & \frac{7}{66} \\ 0 & 1 & 0 & -\frac{1}{11} & 0 & \frac{2}{11} \\ 0 & 0 & 1 & \frac{1}{6} & -\frac{1}{6} & \frac{1}{6} \end{array}\right]$$

$$A^{-1}C = \begin{bmatrix} \frac{13}{66} & \frac{11}{66} & \frac{7}{66} \\ -\frac{1}{11} & 0 & \frac{2}{11} \\ \frac{1}{6} & -\frac{1}{6} & \frac{1}{6} \end{bmatrix}\begin{bmatrix} 7 \\ -6 \\ -13 \end{bmatrix}$$

$$= \begin{bmatrix} \frac{91}{66} & \frac{66}{66} & -\frac{91}{66} \\ -\frac{7}{11} & 0 & -\frac{26}{11} \\ \frac{7}{6} & 1 & -\frac{13}{6} \end{bmatrix} = \begin{bmatrix} -1 \\ -3 \\ 0 \end{bmatrix}$$

33. $2x - 3y = -9$

$4x - y = -13 \quad R2 \to -2R1 + R2$

$2x - 3y = -9$

$5y = 5 \Rightarrow y = 1$

$2x - 3(1) = -9 \Rightarrow x = -3$

The solution is $x = 3,\ y = 1$.

37. $x + 2y + 3z = 1$

$3x - 4y - 3z = 2 \quad R2 \to -3R1 + R2$

$7x - 6y + 6z = 2 \quad R3 \to -7R1 + R3$

$x + 2y + 3z = 1$

$-10y - 12z = -1$

$-20y - 15z = -5 \quad R3 \to -2R2 + R3$

$x + 2y + 3z = 1$

$-10y - 12z = -1$

$9z = -3 \Rightarrow z = -\frac{1}{3}$

$-10y - 12\left(-\frac{1}{3}\right) = -1 \Rightarrow y = \frac{1}{2}$

$x + 2\left(\frac{1}{2}\right) + 3\left(-\frac{1}{3}\right) = 1 \Rightarrow x = 1$

The solution is $x = 1,\ y = \frac{1}{2},\ z = -\frac{1}{3}$

41. $2u - 3v + 2w = 7$

$3u + v - 3w = -6$

$u + 4v + w = -13$

$$\begin{vmatrix} 2 & -3 & 2 \\ 3 & 1 & -3 \\ 1 & 4 & 1 \end{vmatrix} = 66$$

$$u = \frac{\begin{vmatrix} 7 & -3 & 2 \\ -6 & 1 & -3 \\ -13 & 4 & 1 \end{vmatrix}}{66} = \frac{-66}{66} = -1$$

$$v = \frac{\begin{vmatrix} 2 & 7 & 2 \\ 3 & -6 & -3 \\ 1 & -13 & 1 \end{vmatrix}}{66} = \frac{-198}{66} = -3$$

$$w = \frac{\begin{vmatrix} 2 & -3 & 7 \\ 3 & 1 & -6 \\ 1 & 4 & -13 \end{vmatrix}}{66} = \frac{0}{66} = 0$$

The solution is u $= -1,\ v = -3,\ w = 0$.

$$x = \frac{\begin{vmatrix} 2 & 2 & 1 \\ 3 & 3 & -6 \\ 1 & 3 & 3 \end{vmatrix}}{60} = \frac{30}{60} = \frac{1}{2}$$

$$y = \frac{\begin{vmatrix} 3 & 2 & 1 \\ 2 & 3 & -6 \\ 1 & 1 & 3 \end{vmatrix}}{60} = \frac{20}{60} = \frac{1}{3}$$

$$z = \frac{\begin{vmatrix} 3 & 2 & 2 \\ 2 & 3 & 3 \\ 1 & 3 & 1 \end{vmatrix}}{60} = \frac{-10}{60} = -\frac{1}{6}$$

45. $3x - 2y + z = 6$

$2x + 0y + 3z = 3$

$4x - y + 5z = 6$

$$A = \begin{bmatrix} 3 & -2 & 1 \\ 2 & 0 & 3 \\ 4 & -1 & 5 \end{bmatrix},\ X = \begin{bmatrix} x \\ y \\ z \end{bmatrix},\ C = \begin{bmatrix} 6 \\ 3 \\ 6 \end{bmatrix}$$

$$X = A^{-1}C = \begin{bmatrix} 3 \\ 1 \\ -1 \end{bmatrix} \quad \text{check: } AX = \begin{bmatrix} 6 \\ 3 \\ 6 \end{bmatrix}$$

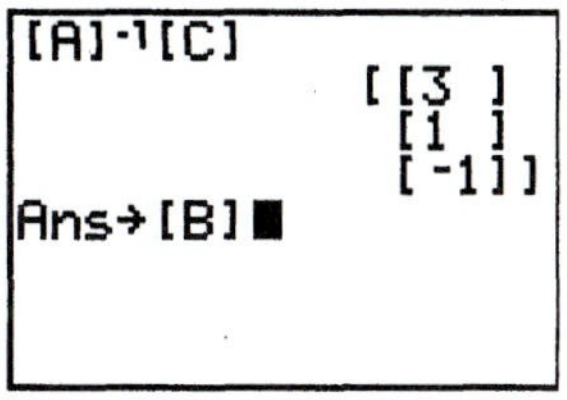

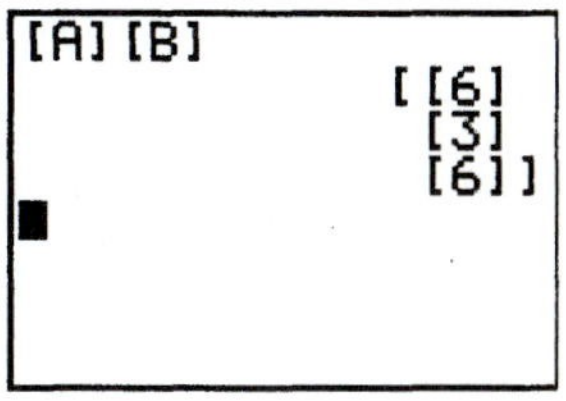

49. $3x - y + 6z - 2t = 8$

$2x + 5y + z + 2t = 7$

$4x - 3y + 8z + 3t = -17$

$3x + 5y - 3z + t = 8$

$$A = \begin{bmatrix} 3 & -1 & 6 & -2 \\ 2 & 5 & 1 & 2 \\ 4 & -3 & 8 & 3 \\ 3 & 5 & -3 & 1 \end{bmatrix},\ X = \begin{bmatrix} x \\ y \\ z \\ t \end{bmatrix},\ C = \begin{bmatrix} 8 \\ 7 \\ -17 \\ 8 \end{bmatrix}$$

$$X = A^{-1}C = \begin{bmatrix} -\frac{1}{3} \\ 3 \\ \frac{2}{3} \\ 4 \end{bmatrix} \quad \text{check: } AX = \begin{bmatrix} 8 \\ 7 \\ -17 \\ 8 \end{bmatrix}$$

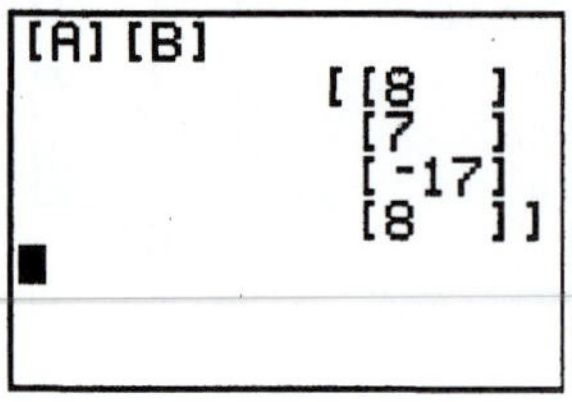

53. $A^2 = \begin{bmatrix} 1 & 0 \\ 3 & 4 \end{bmatrix}\begin{bmatrix} 1 & 0 \\ 3 & 4 \end{bmatrix} = \begin{bmatrix} 1 & 0 \\ 15 & 16 \end{bmatrix}$ from calculator

$A^3 = A^2A = \begin{bmatrix} 1 & 0 \\ 15 & 16 \end{bmatrix}\begin{bmatrix} 1 & 0 \\ 3 & 4 \end{bmatrix} = \begin{bmatrix} 1 & 0 \\ 63 & 64 \end{bmatrix}$

from calculator

$A^4 = A^3A = \begin{bmatrix} 1 & 0 \\ 63 & 64 \end{bmatrix}\begin{bmatrix} 1 & 0 \\ 3 & 4 \end{bmatrix} = \begin{bmatrix} 1 & 0 \\ 255 & 256 \end{bmatrix}$

from calculator

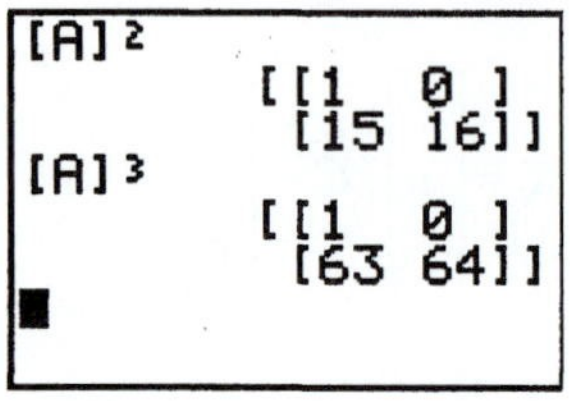

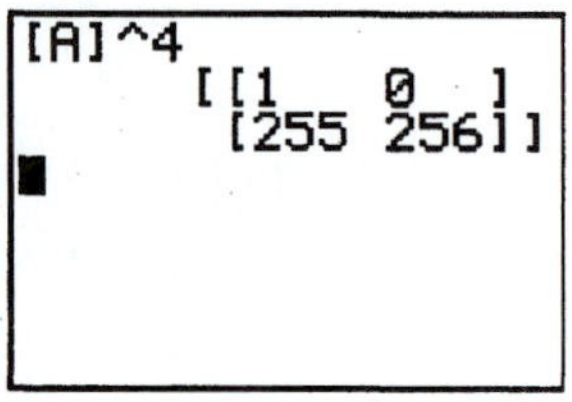

57. $\begin{vmatrix} 4 & 2 & 3 \\ 1 & -5 & -2 \\ -3 & 4 & -3 \end{vmatrix}$

$= 4(-5(-3) - 4(-2)) - 2(1(-3) - (-3)(-2))$
$+3(1(4) - (-3)(-5)) = 77$

61. $\begin{vmatrix} 4 & 2 & 3 \\ 1 & -5 & -2 \\ -3 & 4 & -1 \end{vmatrix}$ $\begin{matrix} -\frac{1}{4}R1 + r2 \to R2 \\ \frac{3}{4}R1 + R3 \to R3 \end{matrix}$

$\begin{vmatrix} 4 & 2 & 3 \\ 0 & -\frac{11}{2} & -\frac{11}{4} \\ 0 & \frac{11}{2} & -\frac{3}{4} \end{vmatrix}$ $R2 + R3 \to R3$

$\begin{vmatrix} 4 & 2 & 3 \\ 0 & -\frac{11}{2} & -\frac{11}{4} \\ 0 & 0 & -\frac{7}{2} \end{vmatrix} = 4\left(-\frac{11}{2}\right)\left(-\frac{7}{2}\right) = 77$

65. $N = \begin{bmatrix} 0 & -1 \\ 1 & 0 \end{bmatrix}; N^{-1}$

$= \frac{1}{0-(-1)}\begin{bmatrix} 0 & 1 \\ -1 & 0 \end{bmatrix} = 1\begin{bmatrix} 0 & -1 \\ -1 & 0 \end{bmatrix} = -N$

69. $A = \begin{bmatrix} 1 & -2 \\ 0 & 3 \end{bmatrix}, B = \begin{bmatrix} -3 & 1 \\ 2 & -1 \end{bmatrix}$

$(A+B)(A-B)$

$= \left(\begin{bmatrix} 1 & -2 \\ 0 & 3 \end{bmatrix} + \begin{bmatrix} -3 & 1 \\ 2 & -1 \end{bmatrix}\right)\left(\begin{bmatrix} 1 & -2 \\ 0 & 3 \end{bmatrix} - \begin{bmatrix} -3 & 1 \\ 2 & -1 \end{bmatrix}\right)$

$= \begin{bmatrix} -2 & -1 \\ 2 & 2 \end{bmatrix}\begin{bmatrix} 4 & -3 \\ -2 & 4 \end{bmatrix} = \begin{bmatrix} -6 & 2 \\ 4 & 2 \end{bmatrix}$

$A^2 - B^2 = \begin{bmatrix} 1 & -2 \\ 0 & 3 \end{bmatrix}^2 - \begin{bmatrix} -3 & 1 \\ 2 & -1 \end{bmatrix}^2$

$= \begin{bmatrix} 1 & -8 \\ 0 & 9 \end{bmatrix} - \begin{bmatrix} 11 & -4 \\ -8 & 3 \end{bmatrix}$

$= \begin{bmatrix} -10 & -4 \\ 8 & 6 \end{bmatrix}$

$(A+B)(A-B) \neq A^2 - B^2$

73. $A = \begin{bmatrix} 2 & 3 \\ 3 & 2 \end{bmatrix}, C = \begin{bmatrix} 26 \\ 24 \end{bmatrix}, A^{-1} = \begin{bmatrix} -\frac{2}{5} & \frac{3}{5} \\ \frac{3}{5} & -\frac{2}{5} \end{bmatrix}$

$A^{-1}C = \begin{bmatrix} -\frac{2}{5} & \frac{3}{5} \\ \frac{3}{5} & -\frac{2}{5} \end{bmatrix}\begin{bmatrix} 26 \\ 24 \end{bmatrix} = \begin{bmatrix} 4 \\ 6 \end{bmatrix} = \begin{bmatrix} R_1 \\ R_2 \end{bmatrix}$

$R_1 = 4\,\Omega,\ R_2 = 6\,\Omega$

77. $2R_1 + 3R_2 = 26$

$3R_1 + 2R_2 = 24 \quad -\frac{3}{2}R1 + R2 \to R2$

$2R_1 + 3R_2 = 26$

$-\frac{5}{2}R_2 = -15 \Rightarrow R_2 = 6$

$2R_1 + 3(6) = 26 \Rightarrow R_1 = 4$

The solution is $R_1 = 4\,\Omega,\ R_2 = 6\,\Omega.$

81. $110t - d = 0$, suspect

$$135t - d = \frac{135(3.0)}{60}, \text{ police}$$

$$A = \begin{bmatrix} 110 & -1 \\ 135 & -1 \end{bmatrix}, C = \begin{bmatrix} 0 \\ \frac{135(3.0)}{60} \end{bmatrix}, A^{-1} = \begin{bmatrix} -\frac{1}{25} & \frac{1}{25} \\ -\frac{27}{25} & \frac{22}{25} \end{bmatrix}$$

$$A^{-1}C = \begin{bmatrix} -\frac{1}{25} & \frac{1}{25} \\ -\frac{27}{25} & \frac{22}{25} \end{bmatrix} \begin{bmatrix} 0 \\ \frac{135(3.0)}{60} \end{bmatrix} = \begin{bmatrix} 0.27 \\ 5.94 \end{bmatrix} = \begin{bmatrix} t \\ d \end{bmatrix}$$

$t = 0.27, \quad t - \frac{3.0}{60} = 0.22$

The police overtake the suspect 0.22 s after passing the intersection.

85.

	Standard Transmission	Automatic Transmission
4 cylinders	12,000	15,000
6 cylinders	24,000	8,000
8 cylinders	4,000	30,000

$$A = \begin{pmatrix} 12{,}000 & 15{,}000 \\ 24{,}000 & 8{,}000 \\ 4{,}000 & 30{,}000 \end{pmatrix}$$

$$B = \begin{pmatrix} 15{,}000 & 20{,}000 \\ 12{,}000 & 3{,}000 \\ 2{,}000 & 22{,}000 \end{pmatrix}$$

$$A + B = \begin{pmatrix} 27{,}000 & 35{,}000 \\ 36{,}000 & 11{,}000 \\ 6{,}000 & 52{,}000 \end{pmatrix}$$

89. Answers may vary, but the basic idea is that the matrix entries show the inventory of a particular product in a particular store in a compact way.

Chapter 17

INEQUALITIES

17.1 Properties of Inequalities

1. $x+1<0$ is true for all values of x less than -1. Therefore, the values of x that satisfy this inequality are written as $x<-1$.

5. $4+3<9+3$; $7<12$; property 1

9. $\frac{4}{-1}>\frac{9}{-1}$; $-4>-9$; property 3

13. $x>-2$

17. $1<x<7$

21. $x<1$ or $3<x\le 5$

25. x is greater than 0 and less than or equal to 2.

29. $x<3$

33. $0\le x<5$

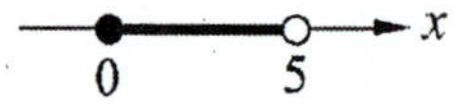

37. $x<-1$ or $1\le x<4$

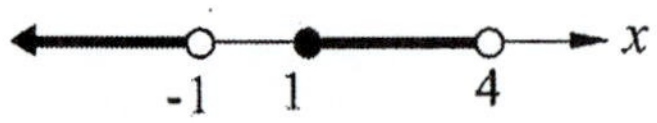

41. $t<-0.3$ or $t>-0.3$

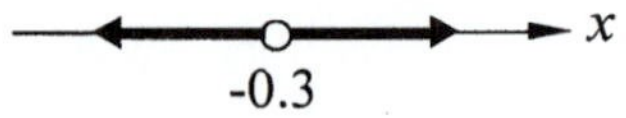

$(x\leqslant 5 \text{ or } x\geqslant 8)$

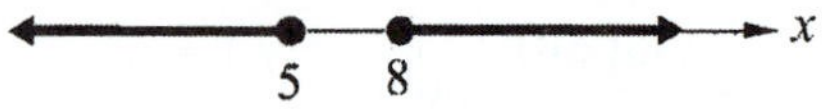

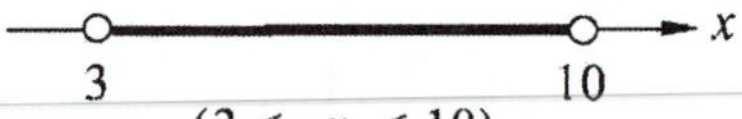

$(3< x <10)$

$(x\leqslant 5 \text{ or } x\geqslant 8)$ and $(3< x <10)$

45.

$0<a<b$	given
$0<a^2<ab$	since $a>0$
$0<ab<b^2$	since $b>0$
$a^2<ab<b^2$	
$a^2<b^2$	is an absolute inequality

49. Note: $x^2=|x|^2$.

For $x>0$, $y<0\Rightarrow xy<0<|x||y|$

$$2|x||y|>2xy$$
$$x^2+2|x||y|+y^2>x^2+2xy+y^2$$
$$|x|^2+2|x||y|+|y|^2>(x+y)^2$$
$$(|x|+|y|)^2>|x+y|^2$$
$$|x|+|y|>|x+y|$$

53. $2000\le M\le 1{,}000{,}000$

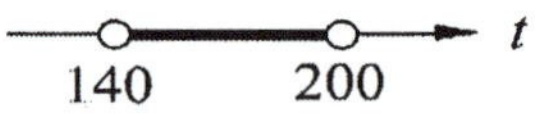

57. $0<n\le 2565$ steps

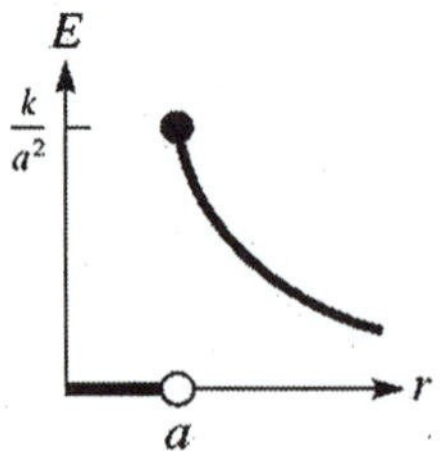

17.2 Solving Linear Inequalities

1. $21-2x\geq 15$

$-2x\geq -6$

$x\leq 3$

5. $x-3>-4$

$x>-4+3$

$x>-1$

9. $3x-5\leq -11$

$3x\leq -11+5$

$3x\leq -6$

$x\leq -2$

13. $4x-5\leq 2x$

$4x-2x\leq 5$

$2x\leq 5$

$x\leq \frac{5}{2}$

17. $2.50(1.50-3.40x)<3.84-8.45x$

$3.75-8.50x<3.84-8.45x$

$-0.09<0.05x$

$x>-1.80$

21. $-1<2x+1<3$

$-2<2x<2$

$-1<x<1$

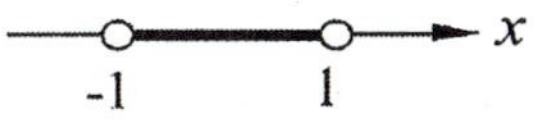

25. $2x<x-1\leq 3x+5$

$0<-x-1\leq x+5$

$0<-x-1$ and $-x-1\leq x+5$

$x<-1$ and $-6\leq 2x$

$x<-1$ and $x\geq -3$

$-3\leq x<-1$

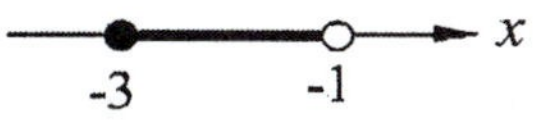

29. $3x-2<8-x$

$4x<10$

$x<\frac{5}{2}$

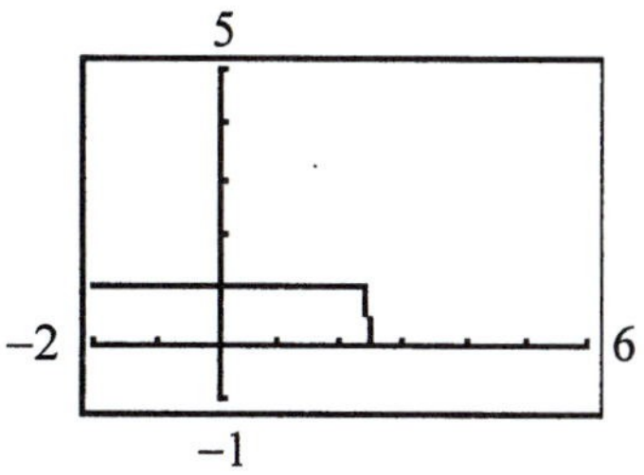

33. $0.1<0.5-2t<0.9$

$-0.4<-2t<0.4$

$0.2>t>-0.2$

$-0.2<t<0.2$

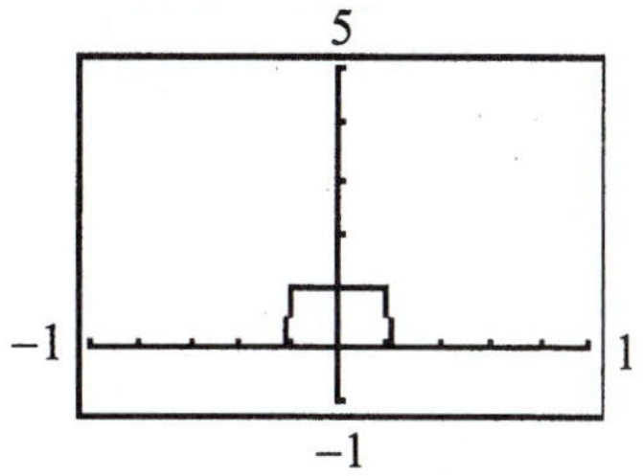

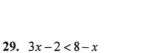

37. $f(x)=\sqrt{2x-10}$

$2x-10\geq 0$

$2x\geq 10$

$x\geq 5$

41. $25n>350+15n$

$10n>350$

for $n>35$ h, the second position pays more.

45. $100<130(1.42)w<150$

$0.54\text{ m}<w<0.81\text{ m}$

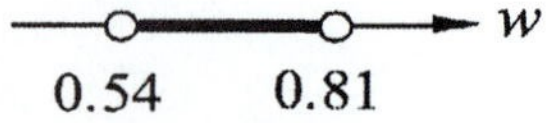

49. $0\leq x\leq 800-300$

$0\leq x\leq 500$

$y=x+400-200=x+200$

$x=y-200$

$0\leq y-200\leq 500$

$200\leq y\leq 700$

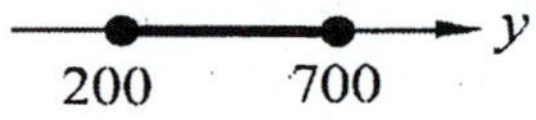

17.3 Solving Nonlinear Inequalities

1. $x^2+3>4x$

$x^2-4x+3>0$

$(x-3)(x-1)>0$

The critical values are 1, 3.

	$(x-3)$	$(x-1)$	Sign
$x<1$	$-$	$-$	$+$
$1<x<3$	$-$	$+$	$-$
$x>3$	$+$	$+$	$+$

$x^2-4x+3>0$ when $x<1$ or $x>3$

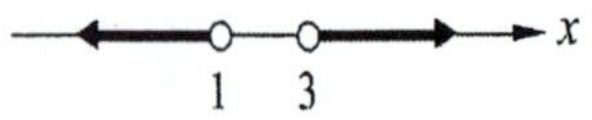

5. $x^2-16<0$

$(x+4)(x-4)<0$

The critical values are $x=-4$ and $x=4$.

	$(x+4)$	$(x-4)$	Sign
$x<-4$	$-$	$-$	$+$
$x-4<x<4$	$+$	$-$	$-$
$x>4x$	$+$	$+$	$+$

$x^2-16<0$ for $-4<x<4$

9. $2x^2-12\leq -5x$

$2x^2+5x-12<0$

$(2x-3)(x+4)\leq 0$

The critical values are $x=\dfrac{3}{2}$, $x=-4$.

	$(2x-3)$	$(x+4)$	Sign
$x<-4$	$-$	$-$	$+$
$-4<x<3/2$	$-$	$+$	$-$
$0<x<3/2$	$+$	$+$	$+$

$(2x-3)(x+4)\leq 0$ for $-4\leq x\leq \dfrac{3}{2}$

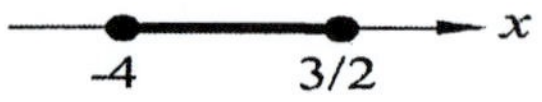

13. $R^2+4>0$

R^2+4 is never less than 4,

so all values of R are solutions.

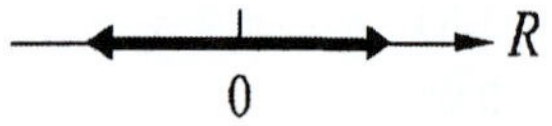

17. $s^3+2s^2-s\geq 2$

$s^2(s+2)-1(s+2)\geq 0$

$(s^2-1)(s+2)\geq 0$

$(s+1)(s-1)(s+2)\geq 0$

The critical values are $s=-1$, $s=1$, $s=-2$.

	$(s-1)(s+1)(s+2)$	Sign
$s<-2$	$-\quad-\quad-$	$-$
$-2<s<-1$	$-\quad-\quad+$	$+$
$-1<s<1$	$-\quad+\quad+$	$-$
$s>1$	$+\quad+\quad+$	$+$

$(s+1)(s-1)(s+2)\ge 0$ for
$-2\le s\le -1$ or $s\ge 1$

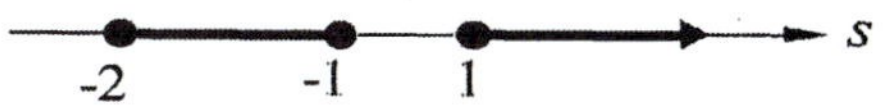

21. $\dfrac{x^2-6x-7}{x+5}>0$

$\dfrac{(x-7)(x+1)}{x+5}>0$

The critical values are $x=7$, $x=-1$, $x=-5$.

	$(x-7)(x+1)(x+5)$	Sign
$x<-5$	$-\quad-\quad-$	$-$
$-5<x<-1$	$-\quad-\quad+$	$+$
$-1<x<7$	$-\quad+\quad+$	$-$
$x>7$	$+\quad+\quad+$	$+$

$\dfrac{(x-7)(x+1)}{x+5}>0$ for $-5<x<-1$ or $x>7$

25. $3x^2+5x\ge 2$

$3x^2+5x-2\ge 0$

$(3x-1)(x+2)\ge 0$

The critical values are $x=\dfrac{1}{3}$, $x=-2$.

	$(3x-1)(x+2)$	Sign
$x<-2$	$-\quad-$	$+$
$-2<x<1/3$	$-\quad+$	$-$
$x>1/3$	$+\quad+$	$+$

$(3x-1)(x+2)\ge 0$ for $x\le -2$ or $x\ge \dfrac{1}{3}$

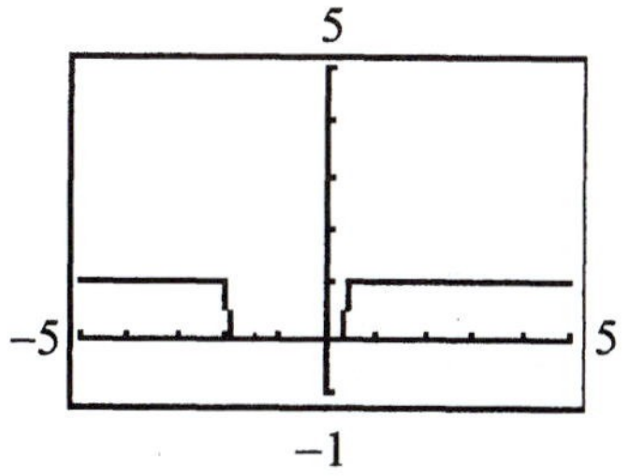

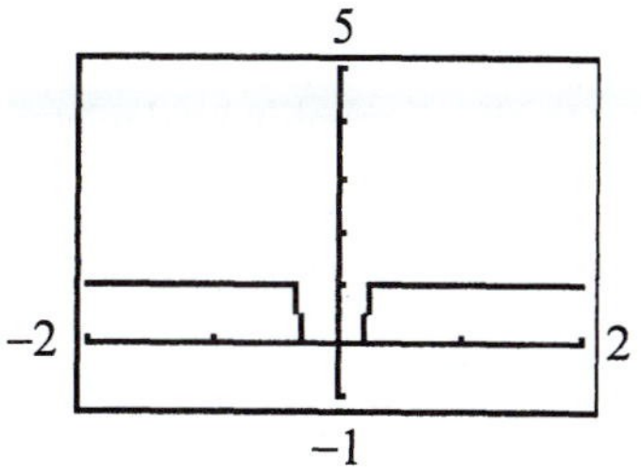

29. $\dfrac{6-x}{3-x-4x^2}\ge 0$

$\dfrac{6-x}{(1+x)(3-4x)}\ge 0;\ \left(x\ne -1,\ x\ne \dfrac{3}{4}\right)$

The critical values are $x=6$, $x=-1$ and $x=\dfrac{3}{4}$.

	$(6-x)/(1+x)(3-4x)$	Sign
$x<-1$	$+\quad-\quad+$	$-$
$-1<x<3/4$	$+\quad+\quad+$	$+$
$3/4<x<6$	$+\quad+\quad-$	$-$
$x>6$	$-\quad+\quad-$	$+$

$\dfrac{6-x}{(1+x)(3-4x)}\ge 0$ for $-1<x<\dfrac{3}{4}$ or $x\ge 6$

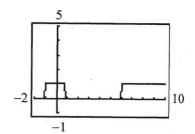

33. $\sqrt{(x-1)(x+2)}$ is real if $(x-1)(x+2) \ge 0$

The critical values are $x = 1$ and $x = -2$.

	$(x-1)$	$(x+2)$	Sign
$x < -2$	$-$	$-$	$+$
$-2 < x < 1$	$-$	$+$	$-$
$x > 1$	$+$	$+$	$+$

$(x-1)(x+2) > 0$ for $x \le -2$ or $x \ge 1$

37. To solve $x^3 - x > 2$ using a graphing calculator, let $y_1 = x^3 - x - 2$.

$y > 0$ for $x > 1.52$

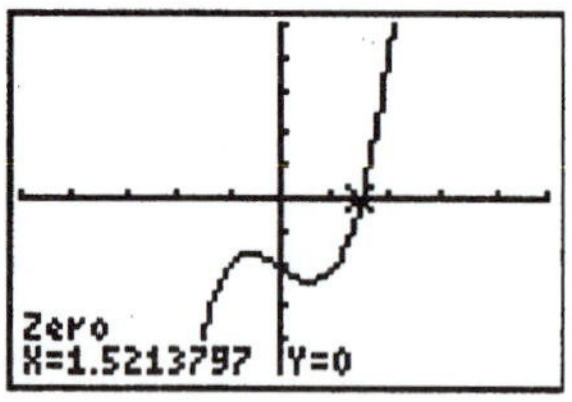

41. To solve $2^x > x + 2$ using a graphing calculator, let $y_1 = 2^x - x - 2$.

$y > 0$ for $x < -1.69$, $x > 2.00$

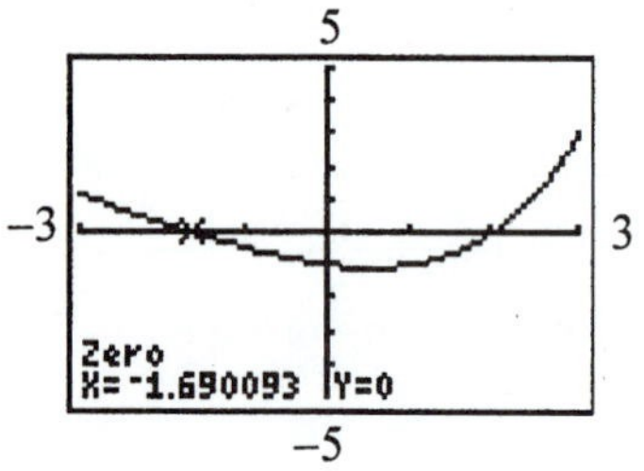

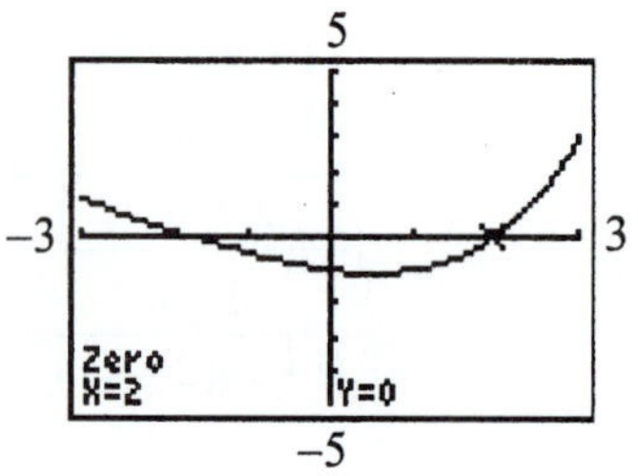

45. $x^2 > x$

$x^2 - x > 0$

$x(x-1) > 0$

The critical values are $x = 0$, $x = 1$.

	x	$(x-1)$	Sign
$x < 0$	$-$	$-$	$+$
$0 < x < 1$	$+$	$-$	$-$
$x > 1$	$+$	$+$	$+$

$x(x-1) > 0$ for $x < 0$ or $x > 1$

Is $x^2 > x$ for all x? No.

$x^2 > x$ for $x < 0$ or $x > 1$.

$x^2 > x$ is not true for $0 \le x \le 1$.

49.
$$2^{x+2} > 3^{2x-3}$$
$$\log 2^{x+2} > \log 3^{2x-3}$$
$$(x+2)\log 2 > (2x-3)\log 3$$
$$x\log 2 + 2\log 2 > 2x\log 3 - 3\log 3$$
$$x\log 2 + \log 4 > 2x\log 3 - \log 27$$
$$\log 4 + \log 27 > x(2\log 3 - \log 2)$$
$$\log 108 > x(\log 9 - \log 2)$$
$$\log 108 > x\log\frac{9}{2}$$
$$x < \frac{\log 108}{\log\frac{9}{2}}$$

53. $p = 6i - 4i^2$, $6i - 4i^2 > 2$ and $i = 1$.

$4i^2 - 6i + 2 < 0$

$2i^2 - 3i + 1 < 0$

$(2i-1)(i-1) < 0$

	$(2i-1)$	$(i-1)$	Sign
$i < 0.5$	$-$	$-$	$+$
$0.5 < i < 1$	$+$	$-$	$-$
$i > 1$	$+$	$+$	$+$

$(2i-1)(i-1) < 0$ for $0.5 < i < 1$ A

57. $C > 1.00 \Rightarrow 1 > 1.00C^{-1} \Leftrightarrow 1.00C^{-1} < 1$

$C^{-1} = C_1^{-1} + C_2^{-1} < 1$

$C_1^{-1} + 4.00^{-1} < 1 \Rightarrow C_1^{-1} < 0.750$

$C_1 < 1.33\ \mu\text{F}$

61. $l = w + 2.0$; $w(w+2.0) < 35$; $w \ge 3.0$ mm

$w^2 + 2.0w - 35 < 0$

$(w+7.0)(w-5.0) < 0$

The critical values are $w = -7.0$ and $w = 5.0$.

	$(w+7)$	$(w-5)$	Sign
$w < -7$	$-$	$-$	$+$
$-7 < w < 5$	$+$	$-$	$-$
$w > 5$	$+$	$+$	$+$

$(w+7.0)(w-5.0) < 0$ for $-7.0 < w < 5.0$; $w \ge 3.0$, so $3.0 \le w < 5.0$ mm

17.4 Inequalities Involving Absolute Values

1. $|2x-1|<5$

$-5<2x-1<5$

$-4<2x<6$

$-2<x<3$

5. $|5x+4|>6$

$5x+4<-6$ or $5x+4>6$

$5x<-10$ or $5x>2$

$x<-2$ or $x>\frac{2}{5}$

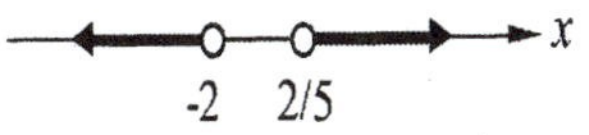

9. $|3-4x|>3$

$3-4x<-3$ or $3-4x>3$

$-4x<-6$ or $-4x>0$

$x<\frac{6}{4}=\frac{3}{2}$ or $x<0$

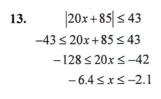

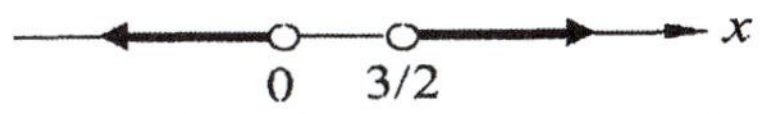

13. $|20x+85|\le 43$

$-43\le 20x+85\le 43$

$-128\le 20x\le -42$

$-6.4\le x\le -2.1$

17. $8+3|3-2x|<11$

$3|3-2x|<3$

$|3-2x|<1$

$-1<3-2x<1$

$-4<-2x<-2$

$2>x>1$

$1<x<2$

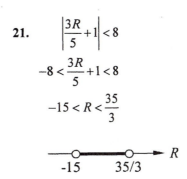

21. $\left|\frac{3R}{5}+1\right|<8$

$-8<\frac{3R}{5}+1<8$

$-15<R<\frac{35}{3}$

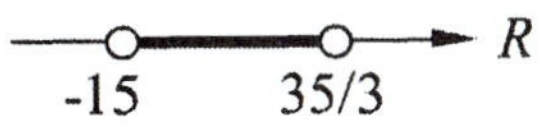

25. $|2x-5|<3 \Leftrightarrow 1<x<4$

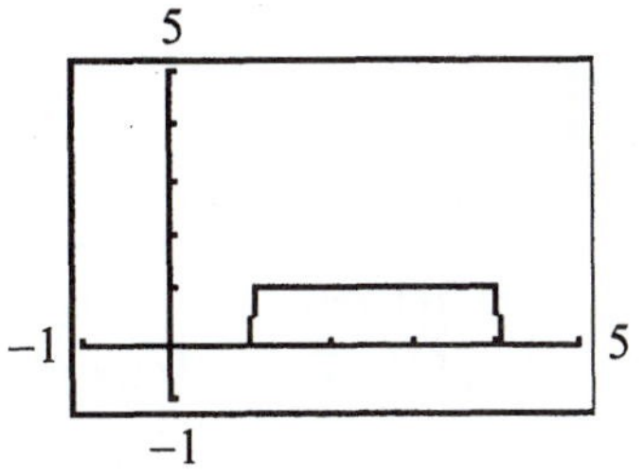

29. $|x^2+x-4|>2$

$x^2+x-4>2$ or $x^2+x-4<-2$

$x^2+x-6>0$ or $x^2+x-2<0$

(A) $(x+3)(x-2)>0$

(B) $(x-1)(x+2)<0$

(A) Critical values are $x=-3$, $x=2$.

	$(x+3)$	$(x-2)$	Sign
$x<-3$	−	−	+
$-3<x<2$	+	−	−
$x>2$	+	+	+

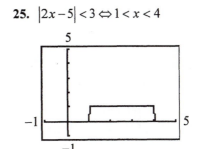

$(x+3)(x-2)>0$ for $x<-3$ or $x>2$

(B) Critical values are $x = 1$, $x = -2$.

	$(x-1)$	$(x+2)$	Sign
$x < -2$	$-$	$-$	$+$
$-2 < x < 1$	$-$	$+$	$-$
$x > 1$	$+$	$+$	$+$

$(x-1)(x+2) < 0$ for $-2 < x < 1$

The solution consists of values of x that are in (A) or (B): $x < -3$, $-2 < x < 1$, $x > 2$

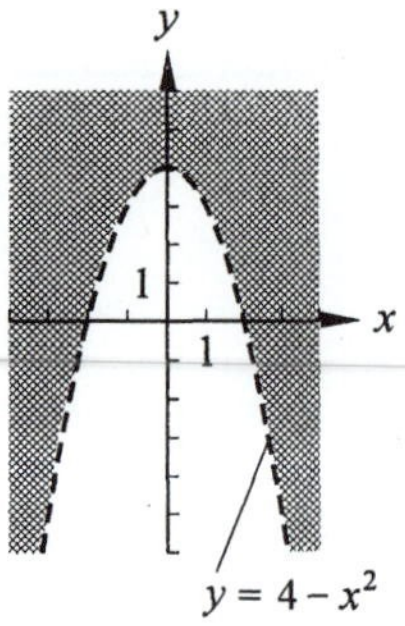

33. Solve for x if $|x| < a$ and $a \le 0$.

$|x| < a \le 0$

$|x| < 0$, no values since $|x| \ge 0$.

37. $|t - 27| \le 23$

$-23 \le t - 27 \le 23$

$4 \le t \le 50$

4 km, minimum thickness of earth's crust.

50 km, maximum thickness of earth's crust.

41. $3.675 - 0.002 \le d \le 3.675 + 0.002$

$-0.002 \le d - 3.675 \le 0.002$

$|d - 3.675| \le 0.002$ cm

17.5 Graphical Solution of Inequalities with Two Variables

1. $y < 3 - x$

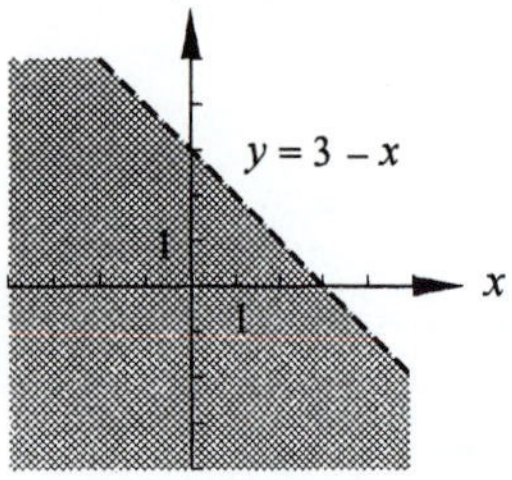

5. $y \ge 2x + 5$; graph $y = 2x + 5$. Use a solid line to indicate that points on it satisfy the inequality. Shade the region above the line.

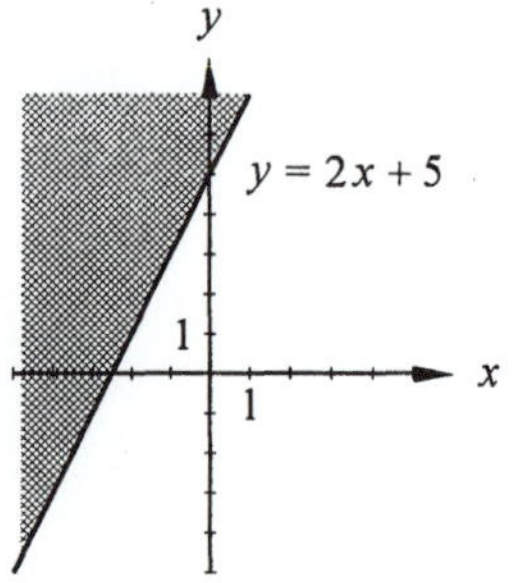

9. $y < x^2$; graph $y = x^2$. Use a dashed curve to indicate that points on it do not satisfy the inequality. Shade the region below the curve.

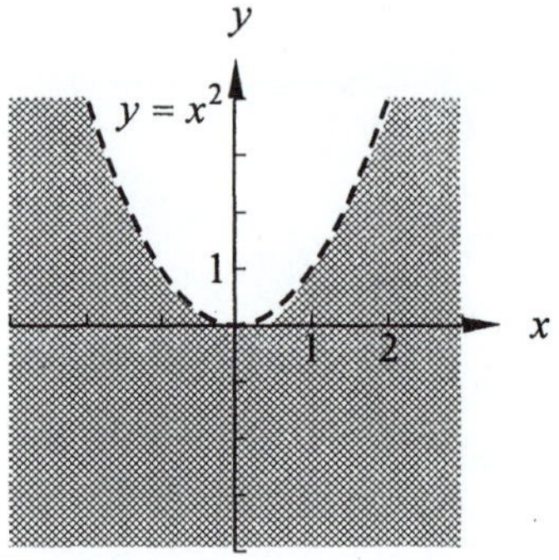

11. $2x^2 - 4x - y > 0$; $-y > 4x - 2x^2$; $y < 2x^2 - 4x$; graph $y = 2x^2 - 4x$. Use a dashed curve to indicate that points on it do not satisfy the inequality. Shade the region below the curve.

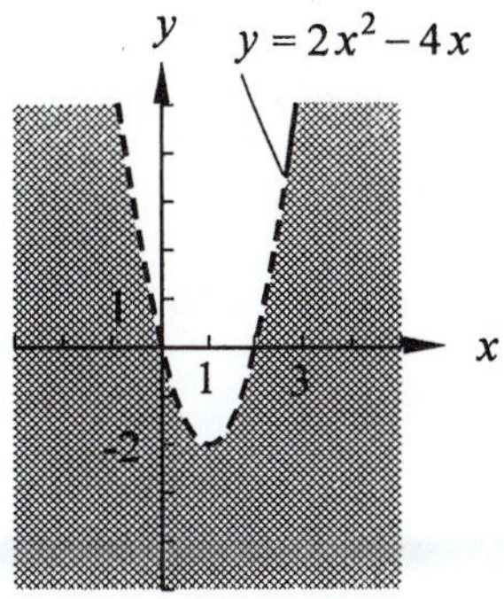

13. $y < 32x - x^4$; graph $y = 32 - x^4$. Use a dashed curve to indicate that points on it do not satisfy the inequality. Shade the region below the curve.

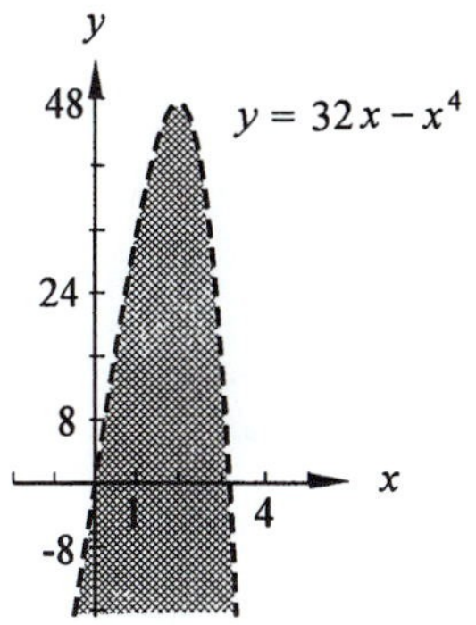

17. $y > 1 + \sin 2x$; graph $y = 1 + \sin 2x$. Use a dashed curve to indicate that the points on it do not satisfy the inequality. Shade the region above the curve.

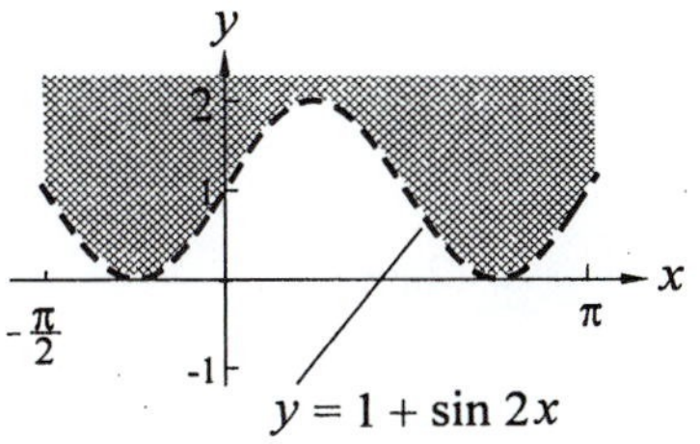

21. $|y| > |x|$. For $y > 0$, $|y| > |x|$ becomes $y = |x|$. Graph $y = |x|$ with dashed line and shade region above graph.
For $y < 0$, $|y| > |x|$ becomes $-y > |x| \Rightarrow$ $y < -|x|$. Graph $y = -|x|$ with dashed line and shade region below graph. The solution is both regions.

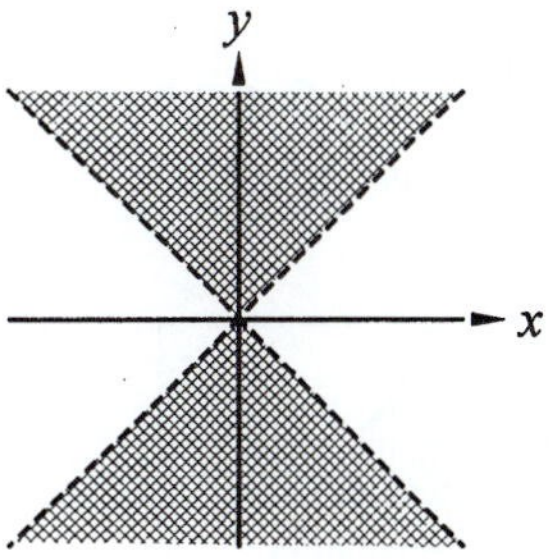

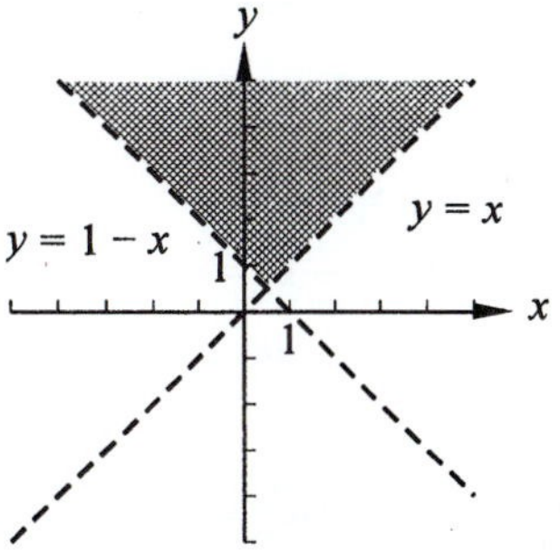

25. $y \le 2x^2$ and $y > x - 2$. Graph $y = 2x^2$ using a solid curve. Shade the region below the curve. Graph $y = x - 2$ using a dashed line. Shade the region above the line. The region where the shadings overlap satisfies both inequalities.

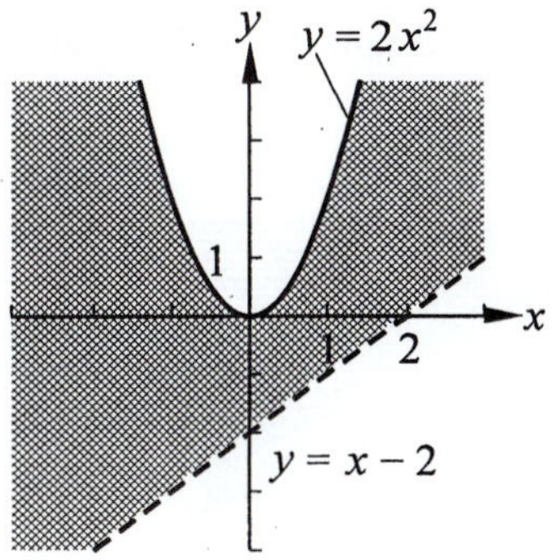

29. $0y \geq$ and $y \leq \sin x$; $0 \leq x \leq 3\pi$. Graph $y = \sin x$ using a solid curve. Shade the region below the curve and above the x-axis for $0 \leq x \leq 3\pi$.

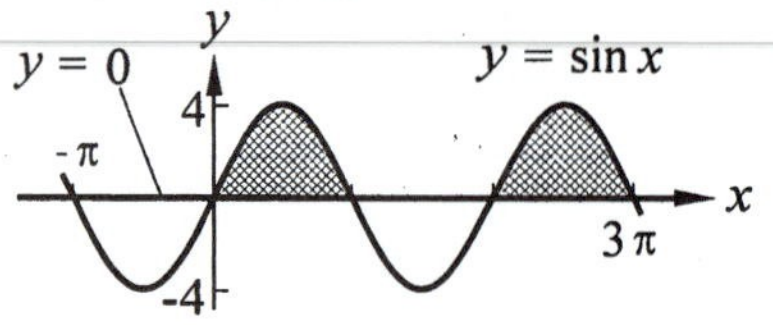

33. $2x + y < 5 \Rightarrow y < -2x + 5$ graph $y_1 = -2x + 5$. The boundary line is dashed.

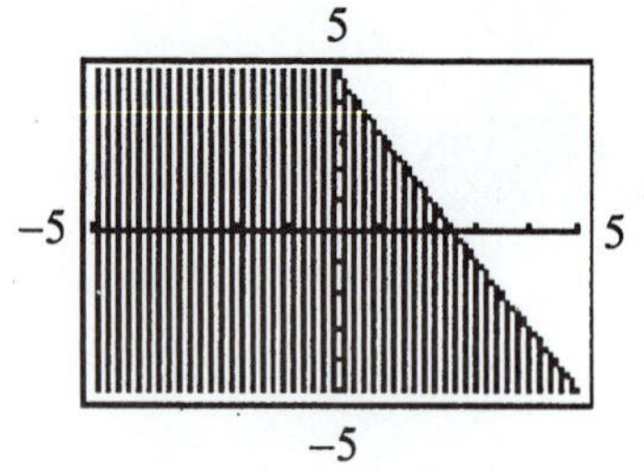

37. $y > 2x - 1$. Graph $y_1 = 2x - 1$.
$y < x^4 - 8$. Graph $y_2 = x^4 - 8$.
The boundary lines are dashed.

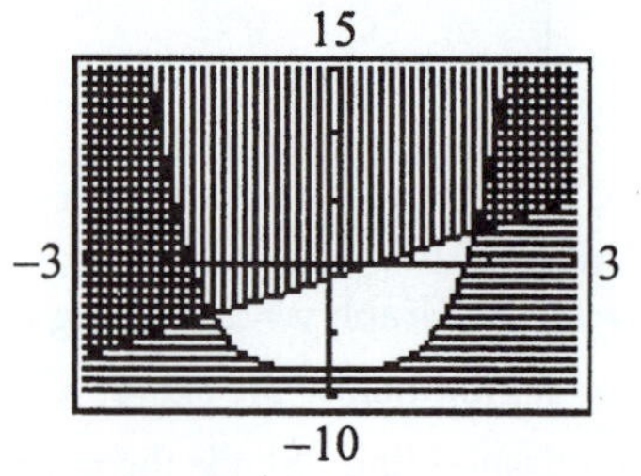

41. $y \leq |2x - 3|$. Graph $y_1 = |2x - 3|$. The boundary line is solid.

$y > 1 - 2x^2$. Graph $y_2 = 1 - 2x^2$. The boundary line is dashed.

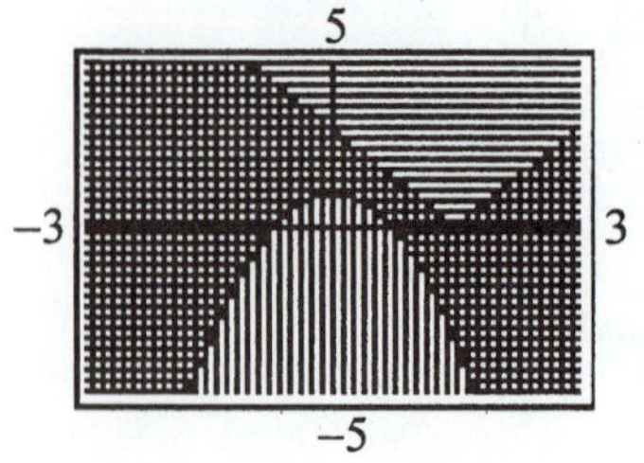

45. $Ax + By > C \Rightarrow By > -Ax + C$
Since $B < 0$ division of both sides by B reverses the inequality.

49. $A \leq 300$ m; 200 m $\leq B \leq 400$ m

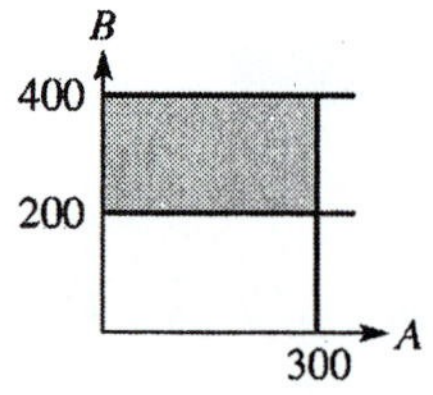

53. Let x and y = pumping rates.
$75x + 45y > 4500$

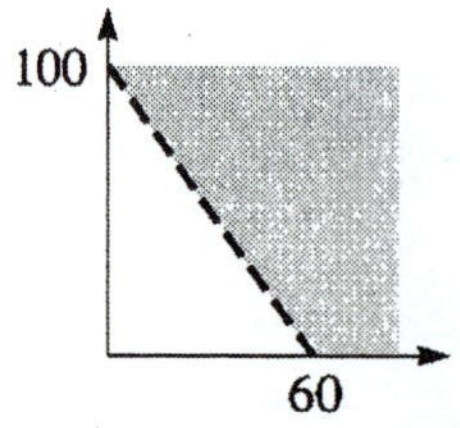

17.6 Linear Programming

1. Maximize $F = 2x + 3y$ subject to $x \geq 0$, $y \geq 0$, $x + y \leq 6$, $2x + y \leq 8$.

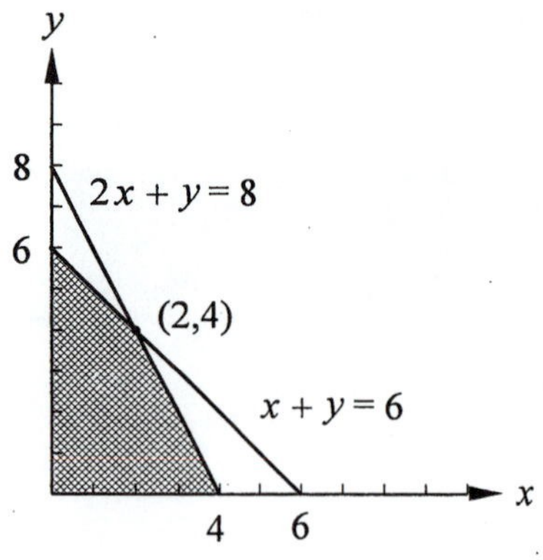

point	value of F
(0, 0)	0
(0, 6)	18
(2, 4)	16
(4, 0)	8

5. Maximum P: $P = 3x + 5y$ subject to

$x \geq 0,\ y \geq 0$

$2x + y \leq 6$

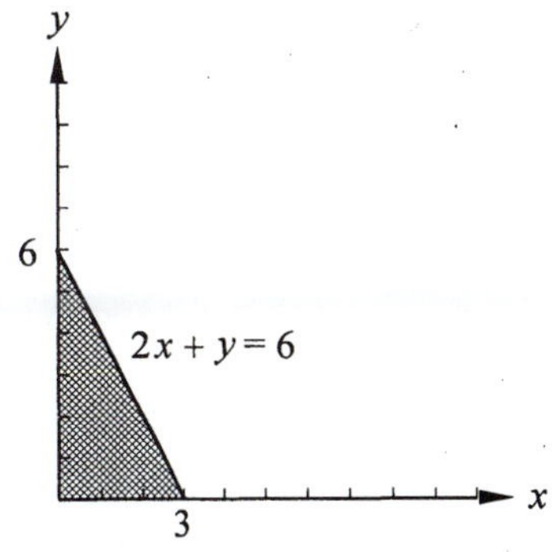

Vertex	$P = 3x + 5y$
$(0, 0)$	0
$(0, 6)$	30
$(3, 0)$	9

max $P = 30$ at $(0, 6)$

9. Minimum C: $C = 4x + 6y$ subject to

$x \geq 0,\ y \geq 0$

$x + y \geq 5$

$x + 2y \geq 7$

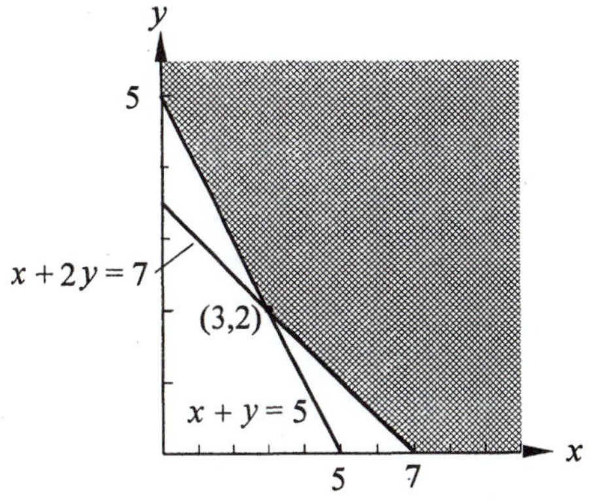

Vertex	$C = 4x + 6y$
$(0, 5)$	30
$(3, 2)$	24
$(7, 0)$	28

min $C = 24$ at $(3, 2)$

13. Maximum P: $P = 9x + 2y$ subject to

$x \geq 0,\ y \geq 0$

$2x + 5y \leq 10$

$4x + 3y \leq 12$

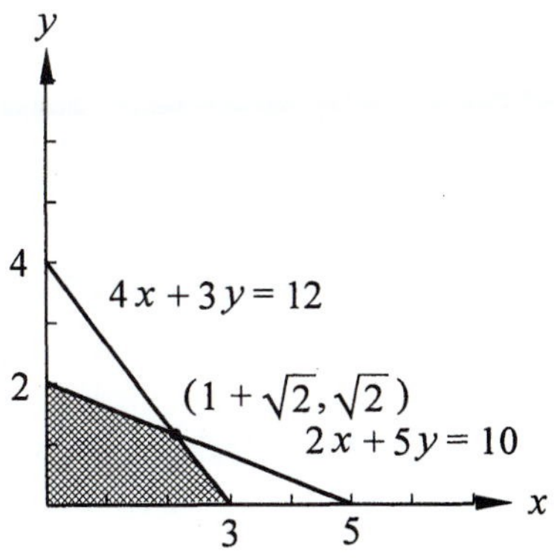

Vertex	$P = 9x + 2y$
$(0, 0)$	0
$(0, 2)$	4
$(1+\sqrt{2},\ \sqrt{2})$	$9 + 11\sqrt{2} = 15.5$
$(3, 0)$	27

max $P = 27$ at $(3, 0)$

17. $x =$ amount invested at 6%

$y =$ amount invested at 5%

$I = 0.06x + 0.05y$

$x \geq 0,\ y \geq 0$

$x + y \leq 9000$

$x \leq 2y$

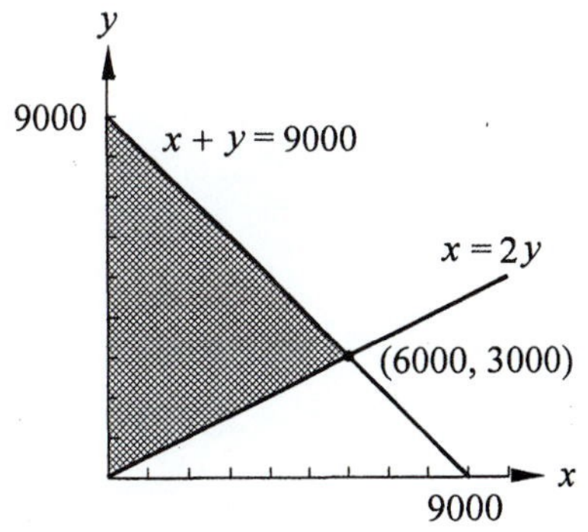

Vertex	$I = 0.06x + 0.05y$
$(0, 0)$	0
$(0, 9000)$	450
$(6000, 3000)$	510

max I = \$510 with \$6000 at 6% and \$3000 at 5%

21. x = cereal A, y = cereal B, c = cost

$c = 12x + 18y$

$x + 2y \ge 10$; $y = -\frac{1}{2}x + 5$; graph: shade region above the graph.

$5x + 3y \ge 30$; $y = -\frac{5}{3}x + 10$; graph: shade regiobn above graph.

Minimum c is the intersection of $y = -\frac{1}{2}x + 5$ and $y = -\frac{5}{3}x + 10$.

Solve simultaneously by substitution:

$-\frac{1}{2}x + 5 = -\frac{5}{3}x + 10$; $x = \frac{30}{7}$ oz, $y = \frac{20}{7}$ oz

OR

Use coordinates of vertices:

$c = 12(10) + 18(0) = 120$ cents

$c = 12(0) + 18(10) = 180$ cents

$c = 12\left(\frac{30}{7}\right) + 18\left(\frac{20}{7}\right) = \frac{720}{7} = 103$ cents

The minimum cost occurs when $x = \frac{30}{7}$, $y = \frac{20}{7}$; i.e., $4\frac{2}{7}$ oz of A, $2\frac{6}{7}$ oz of B.

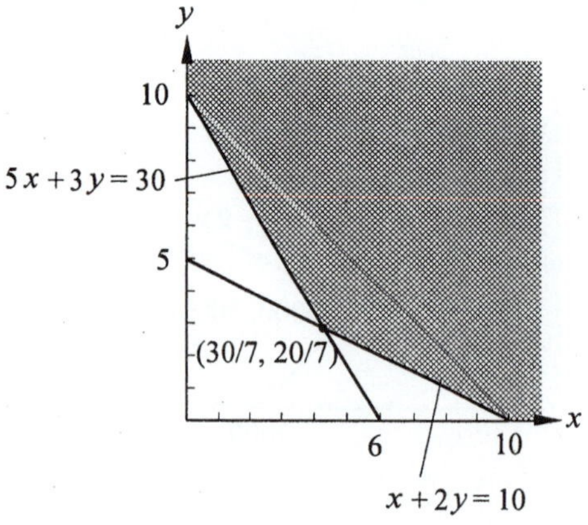

Chapter 17 Review Exercises

1. $2x - 12 > 0 \Rightarrow 2x > 12 \Rightarrow x > 6$

5. $5x^2 + 9x < 2$

$(x+2)(5x-1) < 0$

	$(x+2)(5x-1)$		Sign
$x < -2$	−	−	+
$-2 < x < \frac{1}{5}$	+	−	−
$x > \frac{1}{5}$	+	+	+

$-2 < x < \frac{1}{5}$

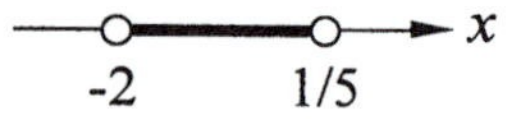

9. $\frac{(2x-1)(3-x)}{(x+4)} > 0$

Critical values are $-4, \frac{1}{2}, 3$.

	$2x-1$	$3-x$	$x+4$	$\frac{(2x-1)(3-x)}{(x+4)}$
$x < -4$	−	+	−	+
$-4 < x < \frac{1}{2}$	−	+	+	−
$\frac{1}{2} < x < 3$	+	+	+	+
$x > 3$	+	−	+	−

Solution: $x < -4$ or $\frac{1}{2} < x < 3$

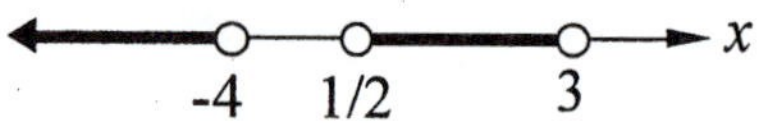

13. $|3x + 2| \le 4$

$-4 \le 3x + 2 \le 4$

$-6 \le 3x \le 2$

$-2 \le x \le \frac{2}{3}$

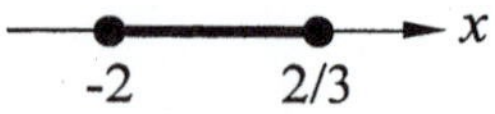

17. $5-3x<0 \Leftrightarrow 3x>5 \Leftrightarrow x>\frac{5}{3}$

Graph $y_1=5-3x<0$.

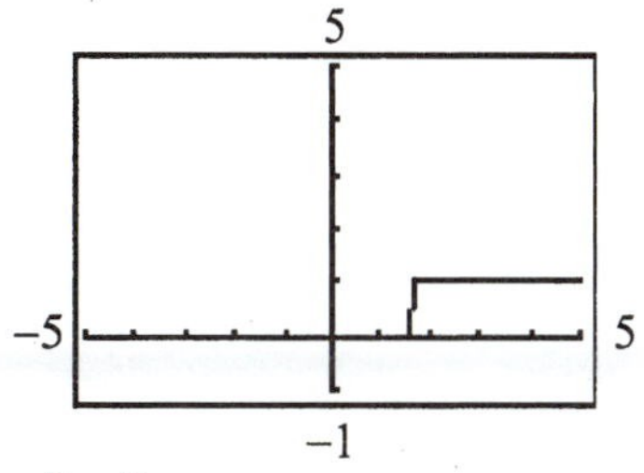

21. $\frac{8-R}{2R+1}\le 0$

Critical values are $-\frac{1}{2}$, 8.

	$8-R$	$2R+1$	$\frac{8-R}{2R+1}$
$R<-\frac{1}{2}$	+	+	−
$-\frac{1}{2}<R<8$	−	+	+
$R\ge 8$	−	+	−

$R<\frac{1}{2}$ or $R\ge 8$

Graph $y_1=(8-x)(2x+1)\le 0$

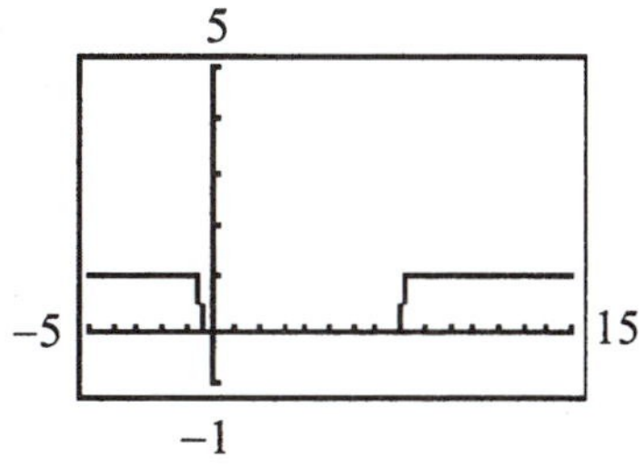

25. $x^3+x+1<0$; Graph $y_1=x^3+x+1$ and use the zero feature. $(x<-0.68)$

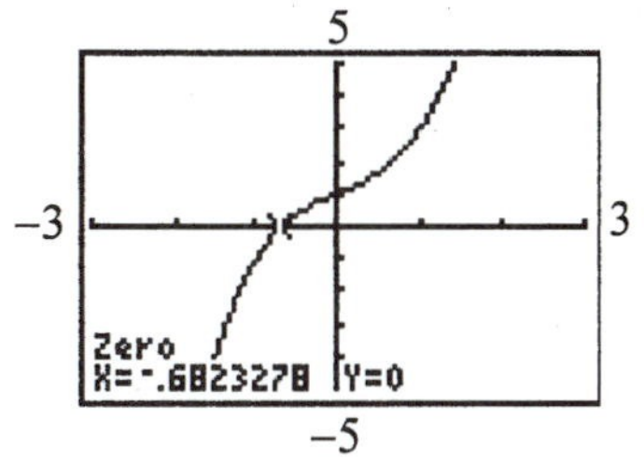

29. $y>12-3x$. Graph $y=12-3x$. Use a dashed line to indicate that points on it do not satisfy the inequality. Shade the region above the line.

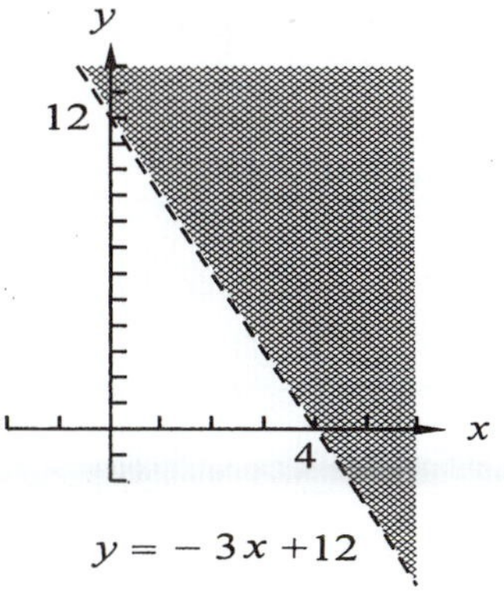

33. $y>6-2x^2$. Graph $y=6-2x^2$. Use a dashed line to indicate points on boundary line are not part of solution. Shade region above line.

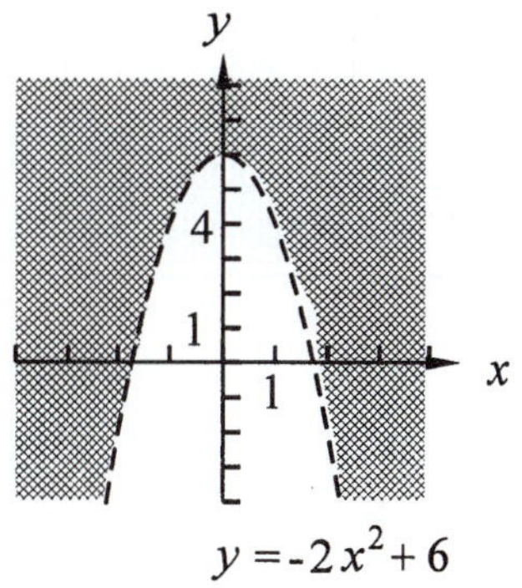

37. $y>x+1$, $y<4-x^2$. Graph $y=x+1$ and $y=4-x^2$. Use a dashed line to indicate boundary line is not part of solution. Shade region below parabola and above line.

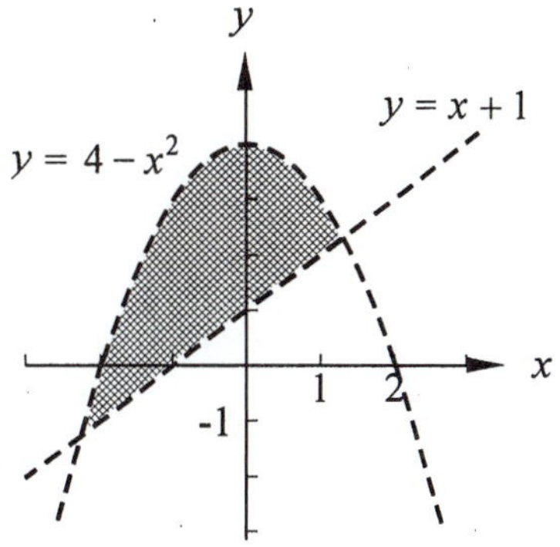

41. $y < 3x + 5$. Graph $y_1 = 3x + 5$ and shade below the line. Boundary line is not part of solution.

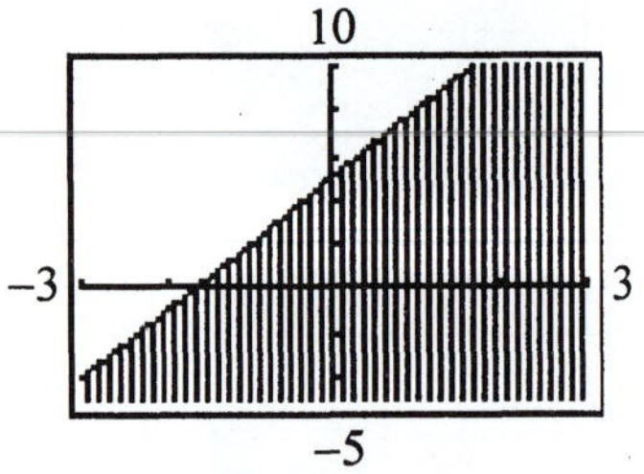

45. $y < 32x - x^4$. Graph $y_1 = 32x - x^4$ and shade below curve. Boundary line is not part of solution.

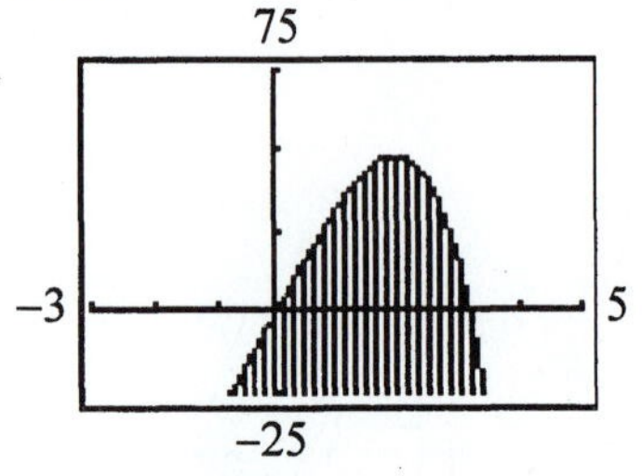

49. $\sqrt{3-x}$ is a real number for $3 - x \geq 0$ $\Rightarrow 3 \geq x \Leftrightarrow x \leq 3$.

53. Maximize P: $P = 2x + 9y$ subject to

$x \geq 0,\ y \geq 0$

$x + 4y \leq 13$

$3y - x \leq 8$

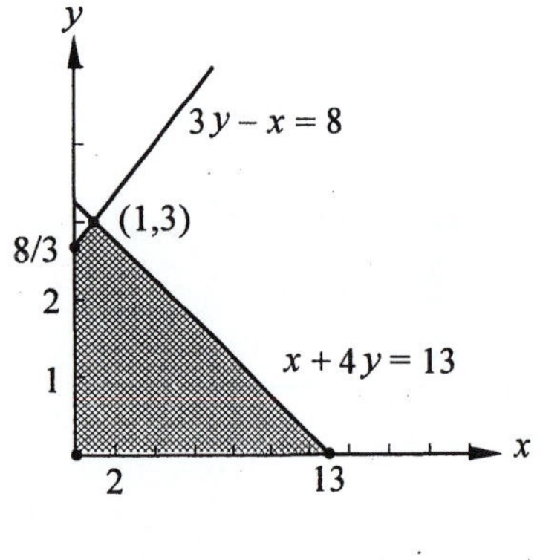

Vertex	$P = 2x + 9y$
$(0, 0)$	0
$(0, \frac{8}{3})$	24
$(1, 3)$	29
$(13, 0)$	26

max $P = 29$ at $(1, 3)$

57. When is $|a+b| < |a| + |b|$. There are 4 cases.

(1) $a \geq 0,\ b \geq 0 \Rightarrow a + b \geq 0$, the given inequality is

$$a + b < a + b$$
$$0 < 0,\ F$$

(2) $a < 0,\ b < 0 \Rightarrow a + b < 0$, the given inequality is

$$-(a+b) < -a + (-b)$$
$$-(a+b) < -(a+b)$$
$$0 < 0,\ F$$

(3) $a < 0,\ b > 0,\ |a| > b \Rightarrow a + b < 0$, the given inequality is

$$-(a+b) < -a + b$$
$$-a - b < -a + b$$
$$-b < b,\ T$$

(4) $a < 0,\ b > 0,\ |a| < b \Rightarrow a + b > 0$, the given inequality is

$$a + b < -a + b$$
$$a < -a,\ T$$

Note: in cases (3) and (4) a and b can be reversed without loss of generality.

$|a+b| < |a| + |b|$ when a and b have opposite signs.

61. $(x-5)(x+2) < 0 \Rightarrow x^2 - 3x - 10 < 0$

65.

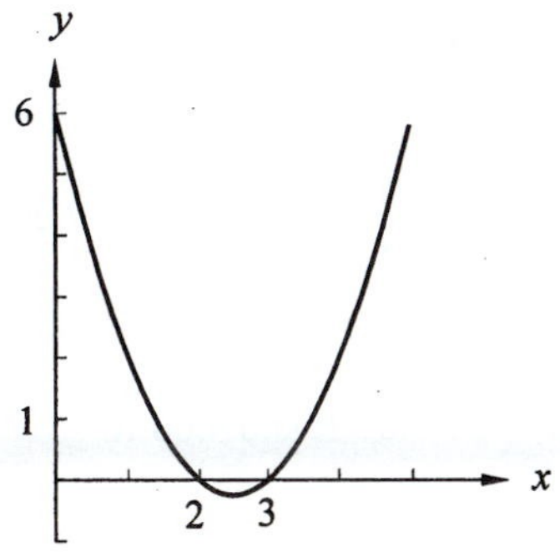

69. $2x + 5y = 50 \Rightarrow y = \dfrac{50 - 2x}{5}$

$5 < y < 8 \Rightarrow 5 < \dfrac{50 - 2x}{5} < 8 \Rightarrow 5 < x < 12.5$

The cost of production for the first type is between \$5 and \$12.

73. $0.8x + 0.9y = 360{,}000, \quad 0 < x \le 261{,}000$

$$x = \frac{360{,}000 - 0.9y}{0.8}$$

$$0 < \frac{360{,}000 - 0.9y}{0.8} \le 261{,}000 \Rightarrow$$

$168{,}000 \le y < 400{,}000$ BTU

77. $e = 100\left(1 - r^{-0.4}\right) > 50 \Rightarrow r^{0.4} > 2$

$\ln r^{0.4} > \ln 2$

$0.4 \ln r > \ln 2$

$\ln r > \ln / 0.4$

$r > e^{\ln 2/0.4} = 5.7$

$r > 5.7$

81. x = the number of regular models

y = the number of deluxe models

$x + y \le 450$

$P = 8x + 15y$ is the profit

t = time spent on one regular model

$2t$ = time spent on one deluxe model

xt = time spent on regular models

$2yt$ = time spent on deluxe models

$xt + 2yt$ = total time = $600t \Rightarrow$

$x + 2y \le 600$

Maximize $P = 8x + 15y$ subject to

$x \ge 0,\ y \ge 0$

$x + y \le 450,\ x + 2y \le 600$

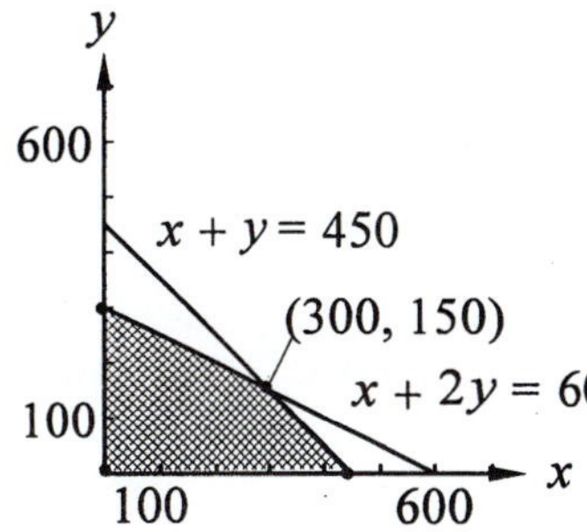

Chapter 18

VARIATION

18.1 Ratio and Proportion

1. $\dfrac{3.5\text{ in.}}{x}=\dfrac{1\text{ in.}}{16\text{ mi}}$

$x=56\text{ mi}$

5. $\dfrac{96\text{ h}}{3\text{ days}}=\dfrac{96\text{ h}}{72\text{ h}}=\dfrac{4}{3}$

9. $\dfrac{0.14\text{ kg}}{3500\text{ mg}}=\dfrac{0.14\text{ kg}}{0.0035\text{ kg}}=40$

13. $\dfrac{45\text{ N}}{110\text{ N}}=0.41$

17. $R=\dfrac{s^2}{A_w}=\dfrac{32.0^2}{195}=5.25$

21. $\dfrac{2540-2450}{2540}=\dfrac{90}{2540}=0.035=3.5\%$

25. $\dfrac{36.6}{84.4}=\dfrac{0.0447}{V_1};\ 36.6V_1=3.77;\ V_1=0.103\text{ m}^3$

29. $\dfrac{1.00\text{ in.}^2}{6.45\text{ cm}^2}=\dfrac{x}{36.3\text{ cm}^2};\ x=\dfrac{1.00(36.3)}{6.45}=5.63\text{ in.}^2$

33. $750\text{ mi}=750\text{ mi}\times\dfrac{2.00\text{ km}}{1.24\text{ mi}}=1210\text{ km}$

37. $\dfrac{62{,}500}{2.00}=\dfrac{x}{0.75};\ x=\dfrac{62{,}500(0.75)}{2.00}=23{,}400\text{ cm}^3$

41. $x+y=15.0$

$\dfrac{x}{y}=\dfrac{2}{3}\Rightarrow x=\dfrac{2}{3}y$

$\dfrac{2}{3}y+y=15.0\Rightarrow\dfrac{5}{3}y=15.0\Rightarrow y=9.00$

$x=\dfrac{2}{3}(9.00)=6.00$

The lengths are 6.00 ft and 9.00 ft.

45. $\dfrac{17}{595}=\dfrac{500}{x};\ 17x=297{,}500;\ x=17{,}500$ chips

18.2 Variation

1. $C=kd=\pi d$

$k=\pi$

5. $v=kr$

9. $P=\dfrac{k}{\sqrt{A}}$

13. The area varfies directly as the square of the radius.

17. $V=kH^2;\ 2=k\cdot 64^2$

$k=\dfrac{2}{64^2};\ V=\dfrac{2H^2}{64^2}=\dfrac{H^2}{2048}$

21. $y=kx;\ 20=k(8);\ k=2.5;\ y=2.5x$

$y=2.5(10)=25$

25. $y=\dfrac{kx}{z};\ 60=\dfrac{k(4)}{10};\ k=150;\ y=\dfrac{150x}{z}$

$y=\dfrac{150(6)}{5}=180$

29. $A=k_1x,\ B=k_2x$

$A+B=k_1x+k_2x=(k_1+k_2)x$ which shows $A+B$ varies directly as x.

33. $H=km;\ 2.93\times10^5=k(875);\ k=335\text{ J/g}$

$H=335\text{ m};\ H=335(625)=2.09\times10^5\text{ J}$

37. $E = kp$; $1200 = k(0.75)$; $k = \dfrac{1200}{0.75} = 1600$ kJ/%

$E = 1600(0.35) = 560$ kJ

41. (a) a varies inversely with mass.

(b) $a = \dfrac{k}{m}$; $30 = \dfrac{k}{2}$; $k = 60\ \text{g}\cdot\text{cm/s}^2$

$a = \dfrac{60}{m}$

45. $F = kAv^2$; $19.2 = k(3.72)(31.4)^2$;

$k = 5.23\times10^{-3}\ \text{lb}\cdot\text{s/ft}^3$

$F = 5.23\times10^{-3}Av^2$

49. $s = k\sqrt{T}$; $460 = k\sqrt{273}$; $k = \dfrac{460}{16.5} = 27.9\ \text{m/}\left(\text{s}\cdot\text{K}^{1/2}\right)$

$s = 27.9\sqrt{300} = 480$ m/s

53. $P = kRi^2$; $10.0 = k(40.0)(0.500)^2$; $k = \dfrac{10.0}{10.0} = 1.00$

$P = 20.0\left(2.00^2\right) = 80.0$ W

57. Note: Make sure calculator is in rad mode.

$x = k\omega^2(\cos\omega t)$

$-11.4 = k\left(0.524^2\right)\cos\left[(1.00)(0.524)\right]$

$k = \dfrac{-11.4}{0.238} = -48.0$ ft

$x = -48.0(0.524)^2\cos\left[(2.00)(0.524)\right]$

$= -6.58\ \text{ft/s}^2$

Chapter 18 Review Exercises

1. $\dfrac{4\text{ Mg}}{20\text{ kg}} = \dfrac{4000\text{ kg}}{20\text{ kg}} = 200$

5. $\pi = \dfrac{c}{d}$

$\dfrac{4.2736}{1.3603} = 3.1417$

9. $p = \dfrac{F}{A} = \dfrac{37.4}{2.25^2}$

$p = 7.39\ \text{lb/in.}^2$

13. $\text{Commission rate} = \dfrac{\text{Commission}}{\text{Selling price}} = \dfrac{20{,}900}{380{,}000}$

$= 0.055$

17. $\dfrac{1.3\text{ in.}}{20.0\text{ mi}} = \dfrac{6.0\text{ in.}}{x}$

$x = 92$ mi

21. 2 min 5 s = 125 s

10.0 min = 600 s

$\dfrac{50\text{ pages}}{125\text{ s}} = \dfrac{p}{600\text{ s}} \Rightarrow p = 240$ pages

25. $\dfrac{25.0\text{ ft}}{2.00\text{ in.}} = \dfrac{x}{5.75\text{ in.}}$; $x = 71.9$ ft

29. $\dfrac{80.0}{98.0} = \dfrac{x}{37.0}$

33. $y = kx^2$; $27 = k\left(3^2\right)$; $k = 3$; $y = 3x^2$

37. $\dfrac{F_1}{F_2} = \dfrac{L_2}{L_1}$; $F_1 = 4.50$ lb, $F_2 = 6.75$ lb, $L_1 = 17.5$ in.

$\dfrac{4.50}{6.75} = \dfrac{L_2}{17.5}$

$(6.75)L_2 = (4.50)(17.5)$

$L_2 = 11.7$ in.

41. $R = kA$

$850 = k(900)$

$k = 0.94$

$R = 0.94$ A

45. $\dfrac{F-550}{L-10} = \dfrac{550-250}{10-22}$

$F = -25L + 800$

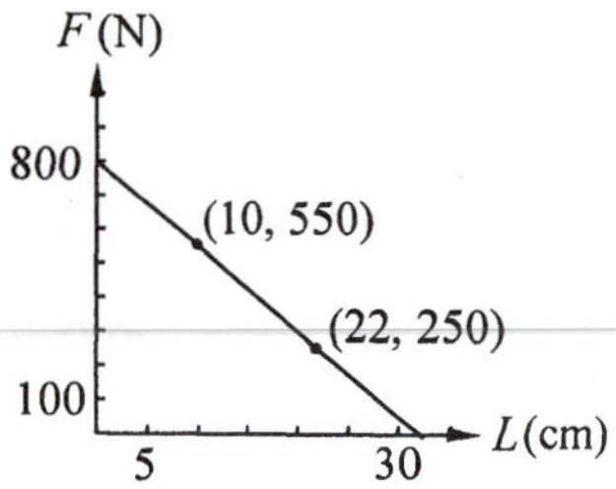

49. $C = kV$

$$6.3 = k(220) \Rightarrow k = \frac{6.3}{220}$$

$$C = \frac{6.3}{220}(150) = 4.3\ \mu\text{C}$$

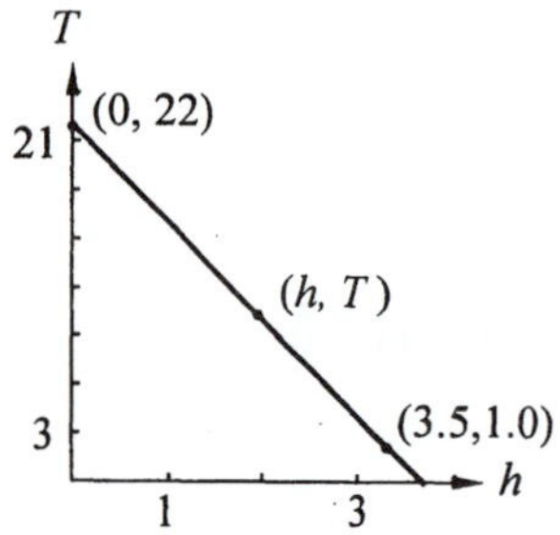

53. $d = kt^2$

$$64.0 = k(2.00)^2 \Rightarrow k = \frac{64.0}{(2.00)^2}$$

$$d = \frac{64.0}{(2.00)^2}(3.00)^2 = 144\text{ ft}$$

57. $f = \dfrac{k}{\sqrt{C}}$

$$25.0 = \frac{k}{\sqrt{95.0}} \Rightarrow k = 25.0\sqrt{95.0}$$

$$f = \frac{25.0\sqrt{95.0}}{\sqrt{25.0}} = 48.7\text{ Hz}$$

61. $15\text{ m} = 49.2\text{ ft}$

$$w = k \cdot L^3$$

$$3500 = k \cdot 42^3 \Rightarrow k = \frac{3500}{42^3}$$

$$w = \frac{3500}{42^3} \cdot (49.2)^3 = 5600\text{ lb}$$

65.

$$R = kv_0^2 \sin 2\theta$$

$$5.12\times10^4 = k(850)^2 \sin\left(2\left(22.0^\circ\right)\right)$$

$$k = \frac{5.12\times10^4}{850^2 \sin\left(44.0^\circ\right)}$$

$$R = \frac{5.12\times10^4}{850^2 \sin 44.0^\circ} 750^2\left(2\left(43.2^\circ\right)\right)$$

$$R = 5.73\times10^4\text{ m}$$

69. $V = \dfrac{kr^4}{d};\ V_1 = \dfrac{k(1.25r)^4}{0.98d}$

$$= \frac{2.44kr^4}{0.98d}$$

$$= 2.49\left(\frac{kr^4}{d}\right)$$

$$= 2.49V$$

73. $L = \dfrac{kt}{d};\ 1200 = \dfrac{k\cdot 30}{8.0};\ k = 320\text{ Btu}\cdot\text{in./min}$

$$L = \frac{320t}{d} = \frac{320(90)}{6.0} = 4800\text{ Btu}$$

77. Let $V_1 = \pi r_1^2 h_1$ be the original volume then the new volume is

$V_2 = \pi r_2^2 h_2 = 0.9\pi r_1^2 h_1$ from which

$\left(\dfrac{r_2}{r_1}\right)^2 \cdot \dfrac{h_2}{h_1} = 0.9$ and since $\dfrac{r_2}{r_1} = \dfrac{h_2}{h_1}$

$$\frac{r_2^3}{r_1^3} = 0.9$$

$$r_2 = \sqrt[3]{0.9}r_1 = 0.97r_1$$

Reducing the radius and height by 3% will reduce the volume by 10%.

Chapter 19

SEQUENCES AND THE BINOMIAL THEOREM

19.1 Arithmetic Sequences

1. $a_1 = 5,\ a_{32} = -88,\ n = 32$

$-88 = 5 + (32-1)d$

$31d = -93$

$d = -3$

5. 4, 6, 8, 10, 12

9. $a_8 = 1 + (8-1) = 22$

13. $a_{80} = -0.7 + (80-1)0.4 = 30.9$

17. $S_{20} = \frac{20}{2}(4+40) = 440$

21. $45 = 5 + (n-1)8$

$45 = 8n - 3$

$n = 6$

$S_6 = \frac{6}{2}(5+45) = 150$

25. $a_{30} = a_1 + (29)(3) = a_1 + 87$

$1875 = \frac{30}{2}(a_1 + a_1 + 87)$

$125 = 2a_1 + 87$

$a_1 = 19;\ a_{30} = 106$

29. $a_n = 5k + (n-1)(0.5k)$

$S_n = \frac{n}{2}\left[5k + (5k + (n-1)(0.5k))\right]$

$104k = \frac{n}{2}(5k + 5k + 0.5kn - 0.5k)$

$208k = n(9.5k + 0.5kn)$

$n^2 + 18n - 416 = 0$

$(n+32)(n-13) = 0$

$n = -32$ (not valid)

$n = 13;\ a_{13} = 5k + (12)(0.5k) = 5k + 6k = 11k$

33. $d = \frac{720-560}{10-6} = 40,\ a_6 = 560 = a_1 + (5)(40)$

$a_1 = 360$

$S_{10} = 5(360 + 720) = 5400$

37. $a_{n+1} = a_n + 2 \Rightarrow d = 2$

$a_n = a_1 + (n-1)d = 3 + (n-1)2 = 2n + 1$

41. $a_1,\ b,\ c,\ a_4,\ a_5$

$b + d = c \Rightarrow d = c - b$

$a_1 = b - d = b - (c-b) = 2b - c$

$a_4 = c + d = c + (c-b) = 2c - b$

$a_5 = c + 2d = c + 2(c-b) = 3c - 2b$

45. $3-x,\ -x,\ \sqrt{9-2x};\ 9-2x \ge 0 \Rightarrow x \le \frac{9}{2}$

$3 - x + d = -x$

$d = -3$

$-x + (-3) = \sqrt{9-2x}$

$x^2 + 6x + 9 = 9 - 2x$

$x^2 + 8x + 9 = 0$

$x(x+8) = 0 \Rightarrow x = -8;\ x = 0$, reject

Check: $x = -8$

$3-(-8),\ -(-8),\ \sqrt{9-2(-8)}$

11, 8, 5, A.S. with $d = -3$

Check: $x = 0$

$3-0,\ -0,\ \sqrt{9-2(0)}$

3, 0, 3 is not an A.S.

49. $a_1 = 20,\ d = -1,\ n = 15$

$a_n = a_1 + (n-1)d = 20 + (15-1)(-1)$

$a_{15} = 20 + (15-1)(-1) = 6$

$S_n = \frac{n}{2}(a_1 + a_n)$

$S_{15} = \frac{15}{2}(20+6)$

$S_{15} = 195$ logs in pile

53. $a_1 = 1800,\ d = -150,\ a_n = 0$

$0 = 1800 + (n-1)(-150) = 1800 - 150n + 150$

$150n = 1800 + 150 = 1950;\ n = 13$ (12 more years)

$S_{13} = \frac{13}{2}(1800+0) = \$11,700$, the sum of all depreciations, which is the cost of the car.

57. $S_n = \frac{n}{2}(a_1 + a_n) = \frac{n}{2}\left[a_1 + (a_1 + (n-1))d\right]$

$= \frac{n}{2}\left[2a_1 + (n-1)d\right]$

19.2 Geometric Sequences

1. Find a_7 for $a_1 = \frac{8}{625},\ r = -\frac{5}{2}$

$a_n = a_1 r^{n-1}$

$a_7 = a_1 r^{7-1} = \frac{8}{625}\left(-\frac{5}{2}\right)^6$

$a_7 = \frac{25}{8}$

5. $\frac{1}{6}, \frac{1}{6}\cdot 3, \frac{1}{6}\cdot 3^2, \frac{1}{6}\cdot 3^3, \frac{1}{6}\cdot 3^4$

$\frac{1}{6}, \frac{1}{2}, \frac{3}{2}, \frac{9}{2}, \frac{27}{2}$

9. $r = -25 \div 125 = -0.2,\ a_1 = 125,\ n = 7$

$a_7 = 125(-0.2)^{7-1} = \frac{1}{125}$

13. $10^{100}, -10^{98}, 10^{96}, \ldots n = 51$

$r = \frac{-10^{98}}{10^{100}} = -10^{-2}$

$a_{51} = 10^{100}\cdot\left(-10^{-2}\right)^{50} = 1$

17. $384, 192, 96, \cdots$

$384\cdot r = 192 \Rightarrow r = \frac{1}{2}$

$S_7 = \frac{384\left(1-\left(\frac{1}{2}\right)^7\right)}{1-\frac{1}{2}} = 762$

21. $a_6 = \left(\frac{1}{16}\right)(4)^{6-1} = \left(\frac{1}{16}\right)(4)^5 = 64$

$S_6 = \frac{\frac{1}{16}\left(1-4^6\right)}{1-4} = \frac{\frac{1}{16}(1-4096)}{-3} = \frac{4095}{48} = \frac{1365}{16}$

25. $27 = a_1 r^{4-1};\ a_1 = \frac{27}{r^3}$

$40 = a_1\frac{\left(1-r^4\right)}{1-r}$

$= a_1\frac{\left(1+r^2\right)(1+r)(1-r)}{1-r}$

$= a_1\left(1+r^2\right)(1+r)$

Substitute a from first equation in second equation:

$40 = \frac{27}{r^3}\left(1+r^2\right)(1+r);\ 40r^3 = 27 + 27r + 27r^2 + 27r^3;$

$13r^3 - 27r^2 - 27r - 27 = 0$

Using synthetic division, 3 gives a remainder of zero. Therefore, $r = 3,\ a_1\left(3^{4-1}\right);\ 27a_1 = 27;\ a_1 = 1$

29. $3, 3^{x+1}, 3^{2x+1}, \cdots$ is a G.S. since

$\frac{3^{x+1}}{3} = 3^x$

$\frac{3^{2x+1}}{3^{x+1}} = 3^x.$

$a_1 = 3,\ r = 3^x$

$a_{20} = 3\cdot\left(3^x\right)^{20-1} = 3^{19x+1}$

33. G.S: $2, 6, 2x+8, \cdots$

$6 = 2r \Rightarrow r = 3$

$6\cdot r = 2x + 8 \Rightarrow 6\cdot 3 = 2x + 8 \Rightarrow x = 5$

37. $S_n = \dfrac{a_1(1-r^n)}{1-r},\ r \ne 1$

$S_3 = 7a_1 = \dfrac{a_1(1-r^3)}{1-r} \Rightarrow 7 - 7r = 1 - r^3$

$r^3 - 7r + 6 = 0$

$(r+3)(r-2)(r-1) = 0$

$r = -3,\ r = 2,\ r = 1$ reject since $r \ne 1$

41. $r = 1 - 0.125 = 0.875,\ a_1 = 3.27\text{ mA},\ n = 9.2$

$a_{9.2} = 3.27(0.875)^{8.2} = 1.09\text{ mA}$

45. $0.9^5(9800) = 5800^\circ\text{C}$

49. $a_1 = 80,\ r = 0.65$

$a_{11} = 80(0.65)^{10} = 1.1$

$20^\circ + 1.1^\circ = 21.1^\circ\text{ C}$

53. $a_n = a_1 r^{n-1} = a_1 r^n r^{-1} = a_1 r^{n/r}$

$a_1 r^n = a_n r$

$S_n = \dfrac{a_1(1-r^n)}{1-r} = \dfrac{a_1 - a_1 r^n}{1-r} = \dfrac{a_1 - r a_n}{1-r}$

57. A.S.: $8, x, y, \cdots \Rightarrow \left.\begin{matrix} 8 + d = x \\ x + d = y \end{matrix}\right\} \Rightarrow y = 2x - 8$

G.S: $x, y, 36, \cdots \Rightarrow \left.\begin{matrix} rx = y \\ ry = 36 \end{matrix}\right\} \Rightarrow y^2 = 36x$

$\left.\begin{matrix} y^2 = 36x \\ y = 2x - 8 \end{matrix}\right\} \Rightarrow x = 16,\ y = 24;\ x = 1,\ y = -6$

A.S: 8, 16, 24, $\cdots$

G.S: 16, 24, 36, $\cdots$

A.S: 8, 1, $-6, \cdots$

G.S: 1, -6, 36, $\cdots$

19.3 Infinite Geometric Series

1. Given the G.S. $4 + \frac{1}{2} + \frac{1}{16} + \frac{1}{128} + \cdots$ find the sum.

$a_1 = 4,\ r = \dfrac{1}{8}$

$S = \dfrac{a}{1-r} = \dfrac{4}{1-\frac{1}{8}}$

$S = \dfrac{32}{7}$

5. $S = \dfrac{a_1}{1-r}$

$\dfrac{25}{0.6} = \dfrac{0.5}{1-r}$

$r = \dfrac{1}{5}$

9. $a_1 = 1,\ r = \dfrac{7}{8},\ S = \dfrac{1}{1-\frac{7}{8}} = 8$

13. $a_1 = 2 + \sqrt{3},\ r = \dfrac{1}{2+\sqrt{3}}$

$S = \dfrac{2+\sqrt{3}}{1 - \frac{1}{2+\sqrt{3}}} = \dfrac{2+\sqrt{3}}{\frac{2+\sqrt{3}-1}{2+\sqrt{3}}}$

$= \dfrac{(2+\sqrt{3})(2+\sqrt{3})}{1+\sqrt{3}} = \dfrac{7+4\sqrt{3}}{1+\sqrt{3}} \times \dfrac{1-\sqrt{3}}{1-\sqrt{3}}$

$= \dfrac{7 - 7\sqrt{3} + 4\sqrt{3} - 12}{1-3}$

$= \dfrac{-5-3\sqrt{3}}{-2} = \dfrac{1}{2}(5 + 3\sqrt{3})$

17. $0.49999\ldots = 0.4 + 0.09 + 0.009 + 0.0009 + \cdots$

$= 0.4 + \dfrac{0.09}{1 - \frac{1}{10}}$

$= 0.5$

21. $0.181818\ldots = 0.18 + 0.0018 + 0.000018 + \cdots$

$a_1 = 0.18,\ r = 0.01$

$S = \dfrac{0.18}{1 - 0.01} = \dfrac{2}{11}$

25. $0.36666... = 0.3 + 0.06666...$

For the G.S. $0.066...$, $a = 0.06$, $r = 0.1$

$$S = \frac{0.06}{1-0.1} = \frac{0.06}{0.9} = \frac{1}{15}$$

Therefore,

$$0.36666... = \frac{3}{10} + \frac{1}{15} = \frac{11}{30}$$

29. $50, a_2, 2, \cdots \Rightarrow 50 \cdot r = a_2 r = 2 \Rightarrow$

$$\frac{a_2}{50} = \frac{2}{a_2} \Rightarrow a_2 = \pm 10$$

$$50, 10, 2, \cdots \text{ has } r = \frac{1}{5} \Rightarrow S = \frac{50}{1+\frac{1}{5}} = 62.5$$

$$50, -10, 2, \cdots \text{ has } r = -\frac{1}{5} \Rightarrow S = \frac{50}{1+\frac{1}{5}} = 41.\overline{6}$$

33. $a_1 = 5.882 \text{ g}, \; r = \frac{5.782}{5.882} = 0.9830$

$$S = \frac{5.882}{1-0.9830} = \frac{5.882}{0.0170} = 346 \text{ g}$$

37. $1 + 2x + 4x^2 + \cdots = \frac{2}{3} = \frac{1}{1-2x}$

$$2 - 4x = 3$$

$$4x = -1$$

$$x = -\frac{1}{4}$$

19.4 The Binomial Theorem

1. $(2x+3)^5 = (2x)^5 + 5(2)^4(3) + \frac{5(4)}{2!}(2x)^3(3)^2$

$$+\frac{5(4)(3)}{3!}(2x)^2(3)^3$$

$$+\frac{5(4)(3)(2)}{4!}(2x)^1(3)^4 + (3)^5$$

$$(2x+3)^5 = 32x^5 + 240x^4 + 720x^3 + 1080x^2 + 810x + 243$$

5. $(2x-1)^4 = (2x)^4 + (2x)^3(-1) + \frac{4(3)}{2}(2x)^2(-1)^2$

$$+\frac{4(3)(2)}{6}(2x)(-1)^3 + \frac{4(3)(2)(1)}{24}(-1)^4$$

$$= 16x^4 - 32x^3 + 24x^2 - 8x + 1$$

9. $(n+2\pi)^5 = n^5 + 5n^4(2\pi) + \frac{5(4)}{2!}n^3(2\pi)^2$

$$+\frac{5(4)(3)}{3!}n^2(2\pi)^3$$

$$+\frac{5(4)(3)(2)}{4!}n(2\pi)^4 + (2\pi)^5$$

$$(n+2\pi)^5 = n^5 + 10\pi n^4 + 40\pi^2 n^3 + 80\pi^3 n^2 + 80\pi^4 n + 32\pi^5$$

13. From Pascal's triangle, the coefficients for $n = 4$ are 1, 4, 6, 4, 1.

$$(5x-3)^4 = [5x + (-3)]^4$$

$$= 1(5x)^4 + 4(5x)^3(-3) + 6(5x)^2(-3)^2 + 4(5x)(-3)^3 + (-3)^4$$

$$= 625x^4 - 1500x^3 + 1350x^2 - 540x + 81$$

17. $(x+2)^{10} = x^{10} + 10x^9(2) + \frac{(10)(9)}{2}x^8(2)^2$

$$+\frac{(10)(9)(8)}{6}x^7(2)^3 + \cdots$$

$$= x^{10} + 20x^9 + 180x^8 + 960x^7 + \cdots$$

21. $(x^{1/2} - 4y)^{12} = (x^{1/2})^{12} - 12(x^{1/2})^{11}(4y)$

$$+\frac{12 \cdot 11}{2!}(x^{1/2})^{10}(4y)^2$$

$$-\frac{12 \cdot 11 \cdot 10}{3!}(x^{1/2})^9(4y)^3$$

$$= x^6 - 48x^{11/2}y + 1056x^5y^2 - 14{,}080x^{9/2}y^3 + \cdots$$

25. $(1.05)^6 = (1 + 0.05)^6$

$$= 1^6 + 6(1)^5(0.05) + \frac{6(5)}{2!}(1)^4(0.05)^2 = 1.3375$$

$= 1.338$ to 3 decimal places using three terms

29. $(1+x)^8 = 1 + 8x + \frac{8(7)}{2}x^2 + \frac{8(7)(6)}{6}x^3 + \cdots$

$$= 1 + 8x + 28x^2 + 56x^3 + \cdots$$

33. $\sqrt{1+x}=(1+x)^{1/2}=1+\frac{1}{2}x+\frac{\frac{1}{2}\left(-\frac{1}{2}\right)}{2}x^2$

$+\frac{\frac{1}{2}\left(-\frac{1}{2}\right)\left(-\frac{3}{2}\right)}{6}x^3+\cdots$

$=1+\frac{1}{2}x-\frac{1}{8}x^2+\frac{1}{16}x^3\cdots$

37. (a) $17!+4!=3.557\times10^{14}$

(b) $21!=5.109\times10^{19}$

(c) $17!\times4!=8.536\times10^{15}$

(d) $68!=2.480\times10^{96}$

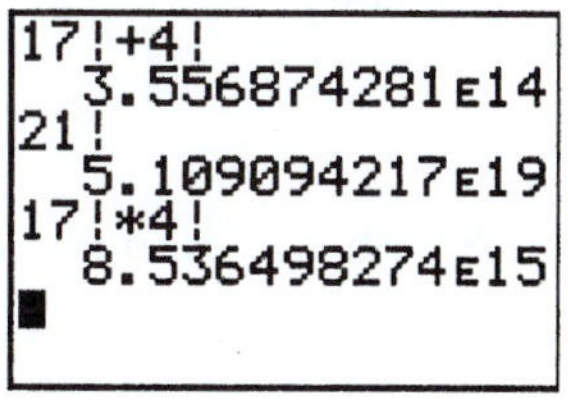

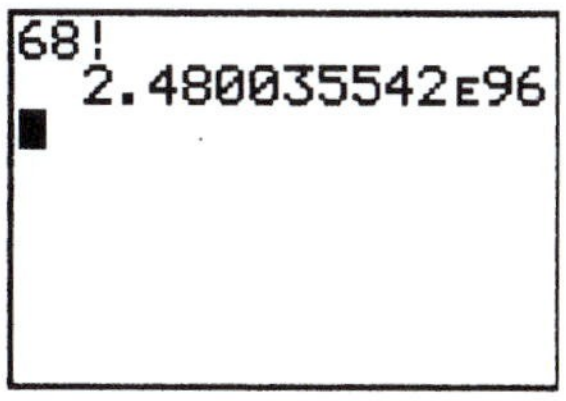

41. The term involving b^5 will be the sixth term.

$r=5,\ n=8$

The sixth term is $\frac{8(7)(6)(5)(4)}{5(4)(3)(2)}a^3b^5=56a^3b^5$.

45. if $n>4$, $n!=n(n-1)(n-2)\cdots5\cdot4\cdot3\cdot2\cdot1$

$=n(n-1)(n-2)\cdots4\cdot3\cdot(5\cdot2)$

$=n(n-1)(n-2)\cdots4\cdot3\cdot(10)$

which will always end in a zero

49. $\sqrt{6}=\sqrt{4(1.5)}=2\sqrt{1+0.5}$

$=2(1+0.5)^{1/2}$

$=2\left[1+\frac{1}{2}(0.5)+\frac{\frac{1}{2}\left(\frac{1}{2}-1\right)}{2!}(0.5)^2\right.$

$\left.+\frac{\frac{1}{2}\left(\frac{1}{2}-1\right)\left(\frac{1}{2}-2\right)}{3!}(0.5)^3\right]$

$=2.453125$

$\sqrt{6}=2.45$ to hundredths using four terms

53. $\left(a^2+x^2\right)^{-1/2}=\frac{1}{a}\left[1+\left(\frac{x}{a}\right)^2\right]^{-1/2}$

$=\frac{1}{a}\left[1-\frac{1}{2}\left(\frac{x^2}{a^2}\right)+\frac{3}{8}\left(\frac{x^4}{a^4}\right)+\cdots\right]$

$=1-x\left(\frac{1}{a}\right)\left[1-\frac{x^2}{2a^2}+\frac{3x^4}{8a^4}+\cdots\right]$

$=1-\frac{x}{a}+\frac{x^3}{2a^3}-\cdots$

Chapter 19 Review Exercises

1. $d=5;\ a_n=1+(17-1)5=1+80=81$

5. $d=8-3.5=4.5;\ a_{16}=-1+(16-1)(4.5)=66.5$

9. $S_n=\frac{15}{2}(-4+17)=\frac{15}{2}(13)=\frac{195}{2}$

13. $a_n=17+(9-1)(-2)=17-16=1$

$S_n=\frac{9}{2}(1+17)=81$

17. $S_n=\frac{n}{2}(a_1+a_n);\ a_1=80,\ a_n=-25,\ S_n=220$

$220=\frac{n}{2}(80-25)$

$440=n(80-25)=55n$

$n=\frac{440}{55}=8$

$a_n=a_1+(n-1)d$

$-25=80+(8-1)d$

$=80+7d$

$d=\frac{-105}{7}=-15$

21. $S_{12}=\frac{12}{2}(-1+32)=6(31)=186$

25. $r=\frac{6}{9}=\frac{2}{3};\ S=\frac{a_1}{1-r}=\frac{0.9}{1-\frac{2}{3}}=2.7$

29. $0.030303...=0.03+0.0003+0.000003...$

$a_a=0.03;\ r=0.01$

$S=\frac{0.03}{1-0.01}=\frac{0.03}{0.99}=\frac{3}{99}=\frac{1}{33}$

33. $(x-2)^4=[x+(-2)]^4$

$$=x^4+4x^3(-2)+\frac{4(3)x^2(-2)^2}{2}+\frac{4(3)(2)}{3(2)}x(-2)^3+(-2)^4$$
$$=x^4-8x^3+24x^2-32x+16$$

37. $\left(a+2e^2\right)^{10}=a^{10}+10a^{10-1}\left(2e^2\right)$

$$+\frac{10(10-1)}{2!}a^{10-2}\left(2e^2\right)^2+\frac{10(10-1)(10-2)}{3!}a^{10-3}\left(2e^2\right)^3+\cdots$$
$$=a^{10}+10a^9\left(2e^2\right)+45a^8\left(4e^4\right)+120a^7\left(8e^6\right)+\cdots$$
$$=a^{10}+20a^9e^2+180a^8e^4+960a^7e^6+\cdots$$

41. $(1+x)^{12}=1+12x+\frac{12(12-1)}{2!}x^2$

$$+\frac{12(12-2)}{3!}x^3+\cdots$$
$$=1+12x+66x^2+220x^3+\cdots$$

45. $\left[1+\left(-a^2\right)\right]^{1/2}=1+\frac{1}{2}\left(-a^2\right)+\frac{\frac{1}{2}\left(-\frac{1}{2}\right)\left(-a^2\right)^2}{2}$

$$+\frac{\frac{1}{2}\left(-\frac{1}{2}\right)\left(-\frac{3}{2}\right)\left(-a^2\right)}{3(2)}$$
$$=1-\frac{1}{2}a^2-\frac{1}{8}a^4-\frac{1}{16}a^6-\cdots$$

49. $a=2,\ d=2,\ n=1000$

$a_n=2+(1000-1)2=2+999(2)=2000$

$S=\frac{1000}{2}(2+2000)=1,001,000$

53. Let $a_3,\ a_4,\ a_5,\ a_6,\ a_7$ be the GS 6, a_4, 9, a_6, 12

Suppose r = common ratio, the GS is 6, $6r$, $6r^2$, $6r^3$, $6r^4$ which gives

$9=6r^2$ and $6r^4=12$

$r^2=\frac{9}{6}$ $\qquad r^4=2$

$r^2=\sqrt{2}$

Since $\frac{9}{6}\neq\sqrt{2}$, 6, a_4, 9, a_6, 12 cannot be a GS.

57. $(1.06)^{-6}=(1+0.06)^{-6}=1^{-6}+\frac{-6}{1}1^{-7}(0.06)^1$

$$+\frac{-6(-7)}{1\cdot 2}1^{-8}(0.06)^2$$
$$=0.7156$$

61. 24, 22, 20, ⋯ is an AS with $a_1=24$, $d=-=2$

$a_n=a_1+(n-1)d\Rightarrow 24+(n-1)(n-2)=4\Rightarrow$

$n=11$

65. There are 15 pieces with a distance between $d=\frac{224.0}{14}$.

Total length of pieces $=10.0(15)+\frac{d}{\tan 84.8^\circ}+\frac{2d}{\tan 84.8^\circ}+\cdots+\frac{14d}{\tan 84.8^\circ}$

$$=10.0(15)+\frac{d}{\tan 84.8^\circ}(1+2+3+\cdots+14)$$

69. $0.0040,\ 0.0040(2),\ 0.0040(2)^2,\cdots,\ 0.0040(2)^{40}$

$0.0040(2)^{40}=4.4\times10^9$ in. = 69,000 mi

73. $10+10(0.9)+10(0.9)^2+\cdots=\frac{10}{1-0.9}=100$ cm

77. $250(0.4)^4=\$6.40$

81. Let $x=\frac{a-1}{2}m^2$ and $y=\frac{a}{a-1}$

$$(1+x)^y = 1 + yx + \frac{y(y-1)}{2}x^2 \ldots \text{ (3 terms)}$$

$$= 1 + \left(\frac{a}{a-1}\right)\left(\frac{a-1}{2}m^2\right) + \frac{\left(\frac{a}{a-1}\right)\left(\frac{a}{a-1}-1\right)}{2}\left(\frac{a-1}{2}m^2\right)^2$$

$$= 1 + \frac{a}{2}m^2 + \frac{\left(\frac{a}{a-1}\right)\left(\frac{1}{a-1}\right)}{2}\left(\frac{(a-1)^2}{2^2}m^4\right)$$

$$= 1 + \frac{a}{2}m^2 + \frac{a}{2^3}m^4 = 1 + \frac{1}{2}am^2 + \frac{1}{8}am^4$$

85. Let $a_1 = 1000$ units. If 75% are killed, 25% remain after the first application.

$$r = \frac{a_2}{a_1} = \frac{250}{1000} = 0.25$$

If 99% are destroyed, 0.1% remain.

$0.001 \times 1000 = 1$ insect remains.

$$1 = 1000(0.25)^n$$

$$0.001 = 0.25^n$$

$$\log 0.001 = \log 0.25^n$$

$$\log 0.001 = n \log 0.25$$

$$n = \frac{\log 0.001}{\log 0.25} = 5 \text{ applications}$$

89. For odd n, the middle term of an arithmetic sequence is $a + \frac{n-1}{2}d$.

$$S_n = a + (a+d) + (a+2d) + \cdots + (a+(n-1)d),$$

n odd

$$S_n = \frac{a + (a+(n-1)d)}{2}n$$

$$\frac{S_n}{n} = \frac{a + a + (n-1)d}{2} = \frac{2a + (n-1)d}{2}$$

$$= a + \frac{(n-1)d}{2}$$

$\frac{S_n}{n}$ = middle term.

93. Let A = initial deposit,

t = time of a compounding period.

$V = A + Art = A(1+rt)$, after one compounding period

$V = A(1+rt) + A(1+rt)rt = A(1+rt)^2$, after two compounding periods

$V = A(1+rt)^n$, at the end of one year.

For $A = \$1000$ and $r = 0.1 = 10\%$

$V = 1000(1+0.1t)^n$.

As n, the number of compounding periods, increases t, the length of a compounding period decreases. The product $nt = 1$.

For example, if the compounding is done monthly $n = 12$ and $t = \frac{1}{12}$, so that $nt = 12 \cdot \frac{1}{2} = 1$.

Thus, $V = 1000\left(1 + 0.1 \cdot \frac{1}{n}\right)^n$ which will increase as n increases the more compounding periods the more interest. Write V as

$$V = 1000\left(1 + \frac{0.1}{n}\right)^{\frac{0.1}{0.1}n} = 1000\left[\left(1 + \frac{0.1}{n}\right)^{\frac{1}{0.1}n}\right]^{0.1}$$

As n increases $\left(1 + \frac{0.1}{n}\right)^{\frac{1}{0.1}n}$ approaches e. Hence, the maximum value is $V_{\text{max}} = 1000 \cdot e^{0.1} = \$1105.17$ as compared with $V = 1000\left(1 + 0.1 \cdot \frac{1}{12}\right)^{12}$ $= \$1104.71$ for monthly compounding.

Chapter 20

ADDITIONAL TOPICS IN TRIGONOMETRY

20.1 Fundamental Trigonometric Identities

1. $\sin x = \dfrac{\tan x}{\sec x}$

$= \dfrac{\frac{\sin s}{\cos x}}{\frac{1}{\cos x}}$

$= \dfrac{\sin x}{\cos x} = \cdot \dfrac{\cos x}{21}$

$= \sin x$

5. Verify $\sin^2\theta + \cos^2\theta = 1$ for $\theta = \dfrac{4\pi}{3}$

$\left(\sin\dfrac{4\pi}{3}\right)^2 = \left(-\dfrac{1}{2}\sqrt{3}\right)^2 = \dfrac{3}{4}$

$\left(\cos\dfrac{4\pi}{3}\right)^2 = \left(-\dfrac{1}{2}\right)^2 = \dfrac{1}{4}$

$\dfrac{3}{4} + \dfrac{1}{4} = 1$

9. $\cos\theta\cot\theta(\sec\theta - 2\tan\theta)$

$= \cos\theta\dfrac{\cos\theta}{\sin\theta}\left(\dfrac{1}{\cos\theta} - 2\dfrac{\sin\theta}{\cos\theta}\right)$

$= \cot\theta - 2\cos\theta$

13. $\sin x + \sin x\tan^2 x = \sin x\left(1 + \tan^2 x\right) = \sin x\sec^2 x$

17. $\csc^4 y - 1 = \left(\csc^2 y + 1\right)\left(\csc^2 y - 1\right)$

$= \left(\csc^2 y + 1\right)\left(\cot^2 y\right)$

21. $\sin x\sec x = \sin x \times \dfrac{1}{\cos x} = \dfrac{\sin x}{\cos x} = \tan x$

25. $\sin x\left(1 + \cot^2 x\right) = \sin x\left(\csc^2 x\right) = \sin x\left(\dfrac{1}{\sin^2 x}\right)$

$= \dfrac{1}{\sin x} = \csc x$

29. $\cot\theta\sec^2\theta - \cot\theta = \dfrac{\cos\theta}{\sin\theta} \times \dfrac{1}{\cos^2\theta} - \dfrac{\cos\theta}{\sin\theta}$

$= \dfrac{1}{\sin\theta\cos\theta} - \dfrac{\cos\theta}{\sin\theta}$

$= \dfrac{1 - \cos^2\theta}{\sin\theta\cos\theta} = \dfrac{\sin^2\theta}{\sin\theta\cos\theta}$

$= \dfrac{\sin\theta}{\cos\theta}$

33. $\cos^2 x - \sin^2 x = 1 - \sin^2 x - \sin^2 x = 1 - 2\sin^2 x$

37. $2\sin^4 x - 3\sin^2 x + 1 = \left(2\sin^2 x - 1\right)\left(\sin^2 x - 1\right)$

$= \left(2\sin^2 x - 1\right)\left(-\cos^2 x\right)$

$= \cos^2 x\left(1 - 2\sin^2 x\right)$

41. $1 + \sin^2 x + \sin^4 x \cdots + =$ infinite series

$a_1 = 1,\ r = \sin^2 x$

$S = \dfrac{1}{1 - \sin^2 x} = \dfrac{1}{\cos^2 x} = \sec^2 x$

45. $\cot x(\sec x - \cos x) = \dfrac{\cos x}{\sin x} \cdot \dfrac{1}{\cos x} - \dfrac{\cos x}{\sin x} \cdot \cos x$

$= \dfrac{1}{\sin x} - \dfrac{\cos^2 x}{\sin x}$

$= \dfrac{1 - \cos^2 x}{\sin x}$

$= \dfrac{\sin^2 x}{\sin x} = \sin x$

49. $\dfrac{\cos x + \sin x}{1 + \tan x} = \dfrac{\cos x + \sin x}{1 + \frac{\sin x}{\cos x}} \cdot \dfrac{\cos x}{\cos x}$

$= \dfrac{(\cos x + \sin x) \cdot \cos x}{\cos x + \sin x} = \cos x$

53.

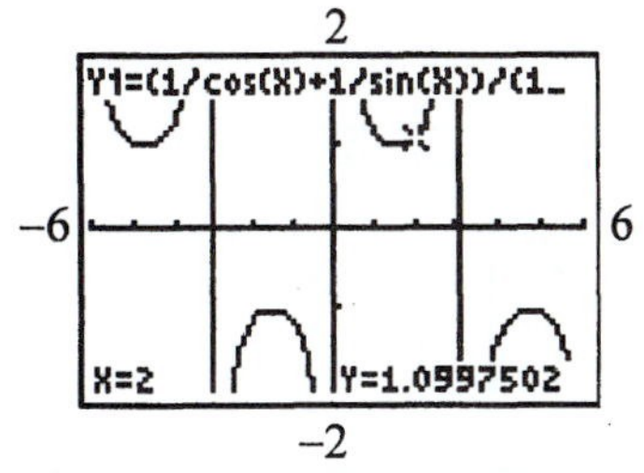

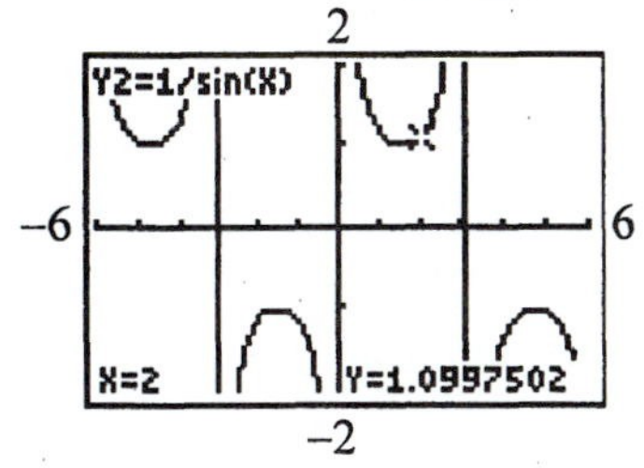

57. No. $\frac{2\cos^2 x-1}{\sin x\cos x} \neq \tan x-\cot x$

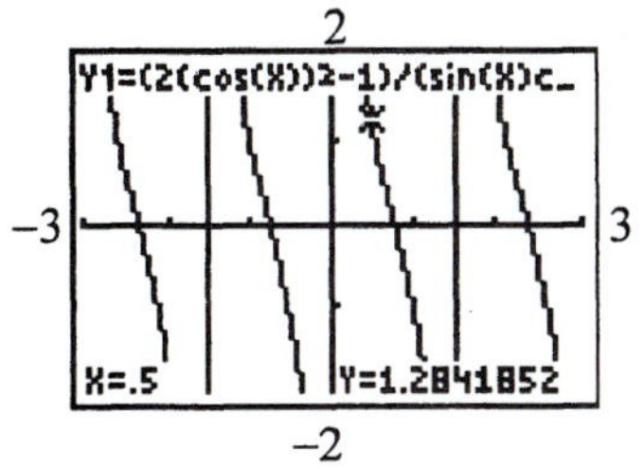

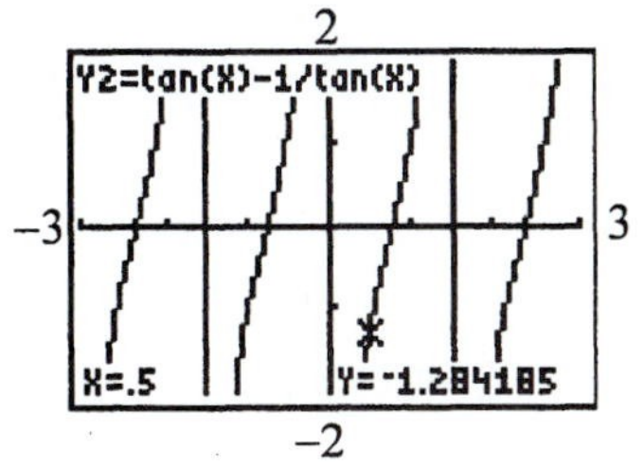

61. $l = a\csc\theta + a\sec\theta$

$$= a(\csc\theta + \sec\theta)$$

$$= a\left(\frac{1}{\sin\theta} + \frac{1}{\cos\theta}\right)$$

$$= a\left(\frac{1}{\sin\theta} + \frac{\sin\theta}{\cos\theta}\cdot\frac{1}{\sin\theta}\right)$$

$$= a\left(\frac{1}{\sin\theta} + \frac{\tan\theta}{\sin\theta}\right)$$

65. $\sin^2 x\left(1-\sec^2 s\right)+\cos^2 x\left(1+\sec^4 x\right)$

$$= \sin^2 x - \sin^2 x\sec^2 x + \cos^2 x + \cos^2 x\sec^4 x$$

$$= \sin^2 x - \frac{\sin^2 x}{\cos^2 x} + \cos^2 x + \frac{\cos^2 x}{\cos^4 x}$$

$$= \sin^2 x - \tan^2 x + \cos^2 x + \sec^2 x$$

$$= 1 - \tan^2 x + \sec^2 x$$

$$= 1 - \left(\sec^2 x - 1\right) + \sec^2 x$$

$$= 1 - \sec^2 x + 1 + \sec^2 x = 2$$

69. $x = \cos\theta$;

$$\sqrt{1-x^2} = \sqrt{1-\cos^2\theta} = \sqrt{\sin^2\theta} = \sin\theta$$

20.2 The Sum and Difference Formulas

1. $\sin\alpha = \frac{12}{13}$ (α in first quadrant) and $\sin\beta = -\frac{3}{5}$ for β in third quadrant.

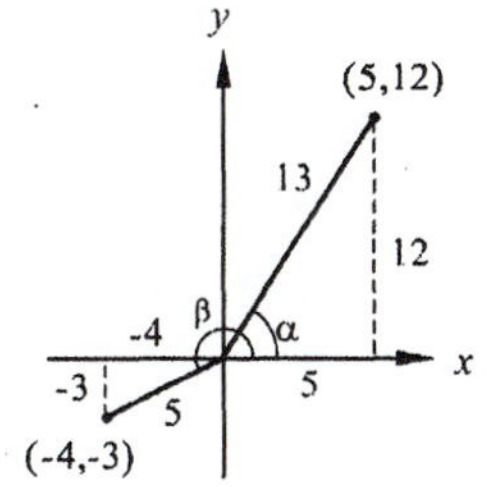

$$\cos(\alpha+\beta) = \cos\alpha\cos\beta - \sin\alpha\sin\beta$$

$$= \frac{15}{13}\cdot\frac{-4}{5} - \frac{12}{13}\cdot\frac{-3}{5} = \frac{16}{65}$$

5. Given: $15^\circ = 60^\circ - 45^\circ$

$$\cos(\alpha-\beta) = \cos\alpha\cos\beta + \sin\alpha\sin\beta$$

$$\cos 15^\circ = \cos\left(60^\circ - 45^\circ\right)$$

$$= \cos 60^\circ\cos 45^\circ + \sin 60^\circ\sin 45^\circ$$

$$= \frac{1}{2}\times\frac{\sqrt{2}}{2} + \frac{\sqrt{3}}{2}\times\frac{\sqrt{2}}{2}$$

$$= \frac{\sqrt{2}}{4} + \frac{\sqrt{6}}{4} = \frac{\sqrt{2}+\sqrt{6}}{4} = 0.9659$$

9. Using the results of exercise 7:

$\cos\alpha = \frac{3}{5}$

$\cos\beta = -\frac{12}{13}$

$\sin\alpha = \frac{4}{5}$

$\sin\beta = \frac{5}{13}$

$$\begin{aligned}\cos(\alpha+\beta) &= \cos\alpha\cos\beta - \sin\alpha\sin\beta\\ &= \frac{3}{5}\left(-\frac{12}{13}\right) - \frac{4}{5}\left(\frac{5}{13}\right)\\ &= \frac{-36-20}{65} = -\frac{56}{65}\end{aligned}$$

13. $\cos\pi\cos x + \sin\pi\sin x = (-1)\cos x + (0)\sin x$

$= -\cos x$

17. $\tan(x-\pi) = \dfrac{\tan x + \tan\pi}{1-\tan x \times \tan\pi} = \tan x$

21. $\sin 122^\circ \cos 32^\circ - \cos 122^\circ \sin 32^\circ$ is of the form $\sin\alpha\cos\beta - \cos\alpha\sin\beta$, where $\alpha = 122^\circ$ and $\beta = 32^\circ$

$\sin\alpha\cos\beta - \cos\alpha\sin\beta = \sin(\alpha-\beta)$ so

$$\begin{aligned}&\sin 122^\circ \cos 32^\circ - \cos 122^\circ \sin 32^\circ\\ &= \sin\left(122^\circ - 32^\circ\right)\\ &= \sin 90^\circ = 1\end{aligned}$$

25. $\sin(x+y)\sin(x-y)$

$$\begin{aligned}&= (\sin x\cos y + \cos x\sin y)(\sin x\cos y - \cos x\sin y)\\ &= \sin^2 x\cos^2 y - \cos^2 x\sin^2 y\\ &= \sin^2 x\left(1-\sin^2 y\right) - \left(1-\sin^2 x\right)\left(\sin^2 y\right)\\ &= \sin^2 x - \sin^2 x\sin^2 y - \sin^2 y + \sin^2 x\sin^2 y\\ &= \sin^2 x - \sin^2 y\end{aligned}$$

29.

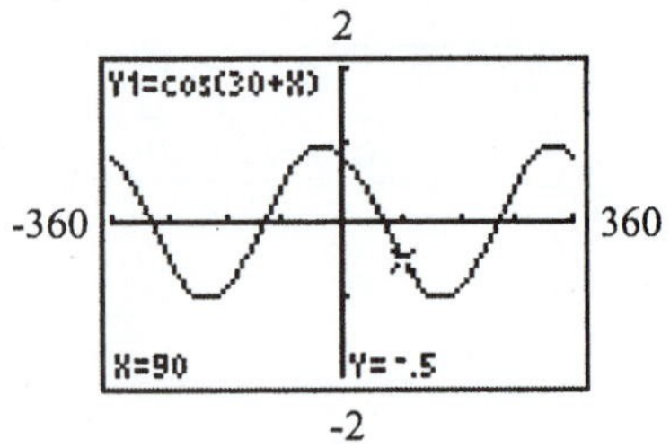

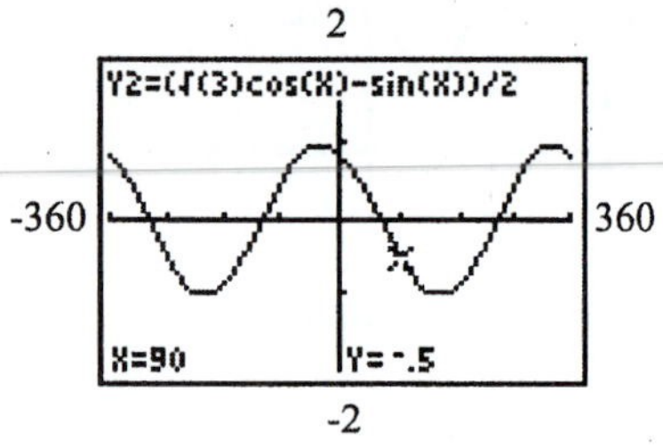

33. $\tan(\alpha\pm\beta)$

$$= \frac{\sin(\alpha\pm\beta)}{\cos(\alpha\pm\beta)} = \frac{\sin\alpha\cos\beta \pm \cos\alpha\sin\beta}{\cos\alpha\cos\beta \mp \sin\alpha\sin\beta}$$

(divide numerator and denominator by $\cos\alpha\cos\beta$)

$$= \frac{\dfrac{\sin\alpha\cos\beta}{\cos\alpha\cos\beta} \pm \dfrac{\cos\alpha\sin\beta}{\cos\alpha\cos\beta}}{\dfrac{\cos\alpha os\beta}{\cos\alpha\cos\beta} \mp \dfrac{\sin\alpha\sin\beta}{\cos\alpha\cos\beta}}$$

$$= \frac{\dfrac{\sin\alpha}{\cos\alpha} \pm \dfrac{\sin\beta}{\cos\beta}}{1 \mp \dfrac{\sin\alpha}{\cos\alpha} \times \dfrac{\sin\beta}{\cos\beta}}$$

$$= \frac{\tan\alpha \pm \tan\beta}{1 \mp \tan\alpha\tan\beta}$$

37. $\alpha+\beta = x;\ \alpha-\beta = y;\ \alpha = \frac{1}{2}(x+y);\ \beta = \frac{1}{2}(x-y)$

$$\begin{aligned}&\sin x + \sin y\\ &= \sin(\alpha+\beta) + \sin(\alpha-\beta)\\ &= 2\left[\frac{1}{2}\sin(\alpha+\beta) + \frac{1}{2}\sin(\alpha-\beta)\right]\\ &= 2\sin\alpha\cos\beta = 2\sin\frac{1}{2}(x+y)\cos\frac{1}{2}(x-y)\end{aligned}$$

41. $\dfrac{\sin 2x}{\sin x} = \dfrac{\sin(x+x)}{\sin x} = \dfrac{\sin x\cos x + \sin x\cos x}{\sin x}$

$= 2\cos x$

45. $I\cos(\theta+30^\circ)+I\cos(\theta+150^\circ)+I\cos(\theta+270^\circ)$

$$= I\left[\cos\theta\cos 30^\circ - \sin\theta\sin 30^\circ + \cos\theta\cos 150^\circ - \sin\theta\sin 150^\circ + \cos\theta\cos 270^\circ - \sin\theta\sin 270^\circ\right]$$

$$= I\left[\frac{\sqrt{3}}{2}\cos\theta - \frac{1}{2}\sin\theta - \frac{\sqrt{3}}{2}\cos\theta - \frac{1}{2}\sin\theta + (0)\cos 270^\circ - (-1)\sin\theta\right]$$

49. $i_0\sin(\omega t+\alpha) = i_0[\sin\omega t\cos\alpha + \sin\alpha\cos\omega t]$

$$= i_0\cos\alpha\sin\omega t + i_0\sin\alpha\cos\omega t$$
$$= i_1\sin\omega t + i_2\cos\omega t$$

20.3 Double-Angle Formulas

1. If $\alpha = \frac{\pi}{3}$,

$$\tan\frac{2\pi}{3} = \tan\left(2\cdot\frac{\pi}{3}\right) = \frac{2\tan\frac{\pi}{3}}{1-\tan^2\frac{\pi}{3}} = \frac{2(\sqrt{3})}{1-(\sqrt{3})^2} = -\sqrt{3}$$

5. $60^\circ = 2(30^\circ)$; $\sin 2\alpha = 2\sin\alpha\cos\alpha$

$$\sin 2(30^\circ) = 2\sin 30^\circ\cos 30^\circ = 2\left(\frac{1}{2}\right)\left(\frac{\sqrt{3}}{2}\right) = \frac{\sqrt{3}}{2} = \frac{1}{2}\sqrt{2}$$

9. $\sin 258^\circ = -0.9781476$

$$\sin 258^\circ = \sin 2(129^\circ) = 2\sin 129^\circ\cos 129^\circ = -0.9781476$$

13. $\tan\frac{2\pi}{5} = \frac{2\tan\frac{\pi}{5}}{1-\tan^2\frac{\pi}{5}} = 3.0777$

17. $\sin x = 0.5$ (QII) $\Rightarrow x = \frac{5\pi}{6}$

$$\tan 2x = \frac{2\tan x}{1-\tan^2 x} = \frac{2\tan\frac{5\pi}{6}}{1-\tan^2\frac{5\pi}{6}} = -\sqrt{3}$$

21. $1-2\sin^2 4x = \cos 2(4x) = \cos 8x$

25. $4\sin^2 2x - 2 = 2(2\sin^2 2x - 1)$

$$= -2(1-2\sin^2 2x) = -2\cos 2(2x) = -2\cos 4x$$

29. $\frac{\sin 3x}{\sin x} - \frac{\cos 3x}{\cos x} = \frac{\sin 3x\cos x - \sin x\cos 3x}{\sin x\cos x}$

$$= \frac{\sin(3x-x)}{\sin x\cos x} = \frac{\sin 2x}{\sin x\cos x} = \frac{2\sin x\cos x}{\sin x\cos x} = 2$$

33. $\frac{\cos x - \tan x\sin x}{\sec x} = \frac{\cos x - \frac{\sin x\sin x}{\cos x}}{\frac{1}{\cos x}}$

$$= \frac{\cos^2 x - \sin^2 x}{\cos x}\times\frac{\cos x}{1} = \cos^2 x - \sin^2 x = \cos 2x$$

37. $1-\cos 2\theta = 1-(1-2\sin^2\theta)$

$$2\sin^2\theta = \frac{2}{\csc^2\theta} = \frac{2}{1+\cot^2\theta}$$

41. Both graphs are the same.

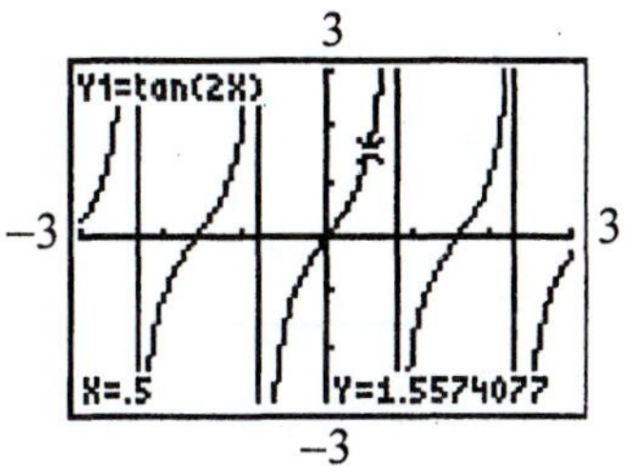

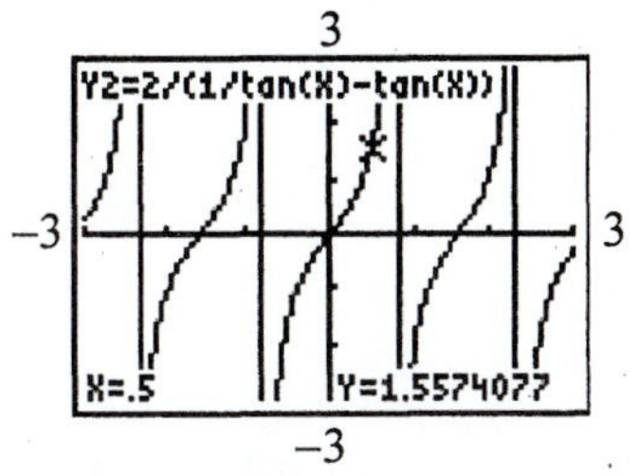

45. $\sin 3x$

$$= \sin(2x + x) = \sin 2x \cos x + \cos 2x \sin x$$
$$= (2\sin x \cos x)(\cos x) + (\cos^2 x - \sin^2 x)(\sin x)$$
$$= 2\sin x \cos^2 x + \sin x \cos^2 x - \sin^3 x$$
$$= 3\sin x \cos^2 x - \sin^3 x$$
$$= 3\sin x(1 - \sin^2 x) - \sin^3 x$$
$$= 3\sin x - 4\sin^3 x$$

49. $\cos 2x + \sin 2x \tan x = \cos^2 x - \sin^2 x + 2\sin x \cos x \cdot \dfrac{\sin x}{\cos x}$

$$= \cos^2 x - \sin^2 x + 2\sin^2 x$$
$$= \cos^2 x + \sin^2 x = 1$$

53. $y = \sqrt{(\sin x + \cos x)^2}$

$$= \sqrt{\sin^2 x + 2\sin x \cos x + \cos^2 x}$$

$y = \sqrt{1 + 2\sin x \cos x}$ and since $\sqrt{x^2} = |x|$

$y = \sqrt{(\sin x + \cos x)^2} = |\sin x + \cos x|$

Graph $y_1 = \sqrt{(\sin x + \cos x)^2}$, $y_2 = \sqrt{1 + 2\sin x \cos x}$, $y_3 = |\sin x + \cos x|$

All three graphs are the same.

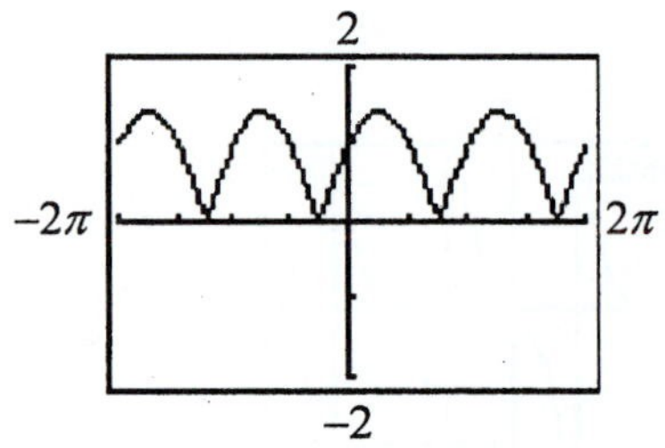

57. $R = vt\cos\alpha;\ t = \dfrac{(2v\sin\alpha)}{g}$

$$R = v\left(\frac{2v\sin\alpha}{g}\right)\cos\alpha = \frac{v^2(2\sin\alpha\cos\alpha)}{g}$$
$$= \frac{v^2\sin 2\alpha}{g}$$

20.4 Half-Angle Formulas

1. $\sqrt{\dfrac{1+\cos 114^\circ}{2}} = \cos\dfrac{1}{2}(114^\circ) = \cos 57^\circ$

$\sqrt{\dfrac{1+\cos 114^\circ}{2}} = 0.544639035$, calculator

$\cos 57^\circ = 0.544639035$, calculator

5. $\sin 105^\circ = \sin\dfrac{210^\circ}{2} = \sqrt{\dfrac{1-\cos 210^\circ}{2}} = \sqrt{\dfrac{1-(-\sqrt{3}/2)}{2}}$

$$= 0.9659$$

9. $\sqrt{\dfrac{1-\cos 236^\circ}{2}} = \sin\dfrac{1}{2}(236^\circ) = \sin 118^\circ$

$$= 0.8829476$$

13. $\sin\dfrac{\alpha}{2} = \sqrt{\dfrac{1-\cos\alpha}{2}}$

$$\sqrt{\frac{1-\cos 6x}{2}} = \sin\frac{6x}{2} = \sin 3x$$

17. $\sqrt{4 - 4\cos 10\theta} = \sqrt{4(1-\cos(2(5\theta)))}$

$$= 2\sqrt{2}\sqrt{\frac{1-\cos 5\theta}{2}}$$
$$= 2\sqrt{2}\sin 5\theta$$

21. $\sin\dfrac{\alpha}{2} = \sqrt{\dfrac{1-\cos\alpha}{2}} = \sqrt{\dfrac{1-\frac{12}{13}}{2}}$

$$= \sqrt{\frac{1}{13}\times\frac{1}{2}} = \sqrt{\frac{1}{26}}$$
$$= \sqrt{\frac{1}{26}\times\frac{26}{26}} = \frac{1}{26}\sqrt{26}$$

25. $\csc\frac{\alpha}{2} = \frac{1}{\sin\frac{\alpha}{2}} = \frac{1}{\pm\sqrt{\frac{1-\cos\alpha}{2}}} = \frac{\sqrt{2}}{\pm\sqrt{1-\cos\alpha}}$

$= \pm\sqrt{\frac{2}{1-\cos\alpha}} = \pm\sqrt{\frac{2}{1-\frac{1}{\sec\alpha}}}$

$= \pm\sqrt{\frac{2\sec\alpha}{\sec\alpha - 1}}$

29. $\frac{1-\cos\alpha}{2\sin\frac{\alpha}{2}} = \frac{1-\cos\alpha}{2\sqrt{\frac{1-\cos\alpha}{2}}} \times \frac{\sqrt{\frac{1-\cos\alpha}{2}}}{\sqrt{\frac{1-\cos\alpha}{2}}}$

$= \frac{(1-\cos\alpha)\sqrt{\frac{1-\cos\alpha}{2}}}{2\left(\frac{1-\cos\alpha}{2}\right)}$

$= \sqrt{\frac{1-\cos\alpha}{2}} = \sin\frac{\alpha}{2}$

33. $2\sin^2\frac{\alpha}{2} - \cos^2\frac{\alpha}{2} = \frac{1-3\cos\alpha}{2}$

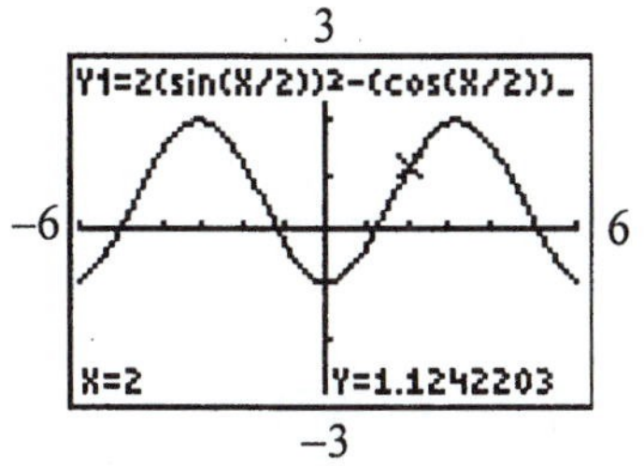

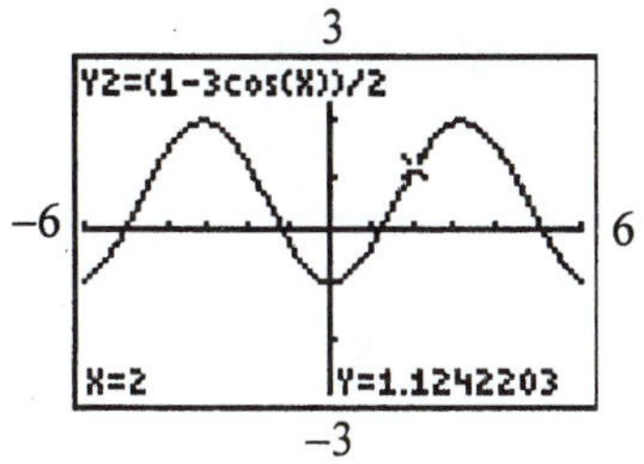

37. Find $\tan\theta$ if $\sin\frac{\theta}{2} = \frac{3}{5}$.

$\sin\frac{\theta}{2} = \frac{3}{5} \Rightarrow \frac{\theta}{2}$ in QI or $\frac{\theta}{2}$ in QII

For $\frac{\theta}{2}$ in QI, $\sin\frac{\theta}{2} = \frac{3}{5} \Rightarrow \cos\frac{\theta}{2} = \frac{4}{5}$

$\sin 2\cdot\frac{\theta}{2} = \sin\theta = 2\sin\frac{\theta}{2}\cos\frac{\theta}{2} = 2\cdot\frac{3}{5}\cdot\frac{4}{5} = \frac{24}{25}$

41. $\sin^2\omega t = \sin^2\left[\left(\frac{1}{2}\right)(2\omega t)\right]$

$= \left(\sqrt{\frac{1-\cos 2\omega t}{2}}\right)^2$

$= \frac{1-\cos 2\omega t}{2}$

20.5 Solving Trigonometric Equations

1. $\tan\theta - 1 = 0,\ 0 \le \theta < 2\pi$

$\tan\theta = 1$

$\theta = \tan^{-1} 1 = \frac{\pi}{4}, \frac{5\pi}{4}$

Check: $\sin\left(\frac{\frac{\pi}{3}}{2}\right) \overset{?}{=} 1 - \cos\frac{\pi}{3}$

$\frac{1}{2} = \frac{1}{2}$

Check: $\sin\left(\frac{\frac{5\pi}{3}}{2}\right) \overset{?}{=} 1 - \cos\frac{5\pi}{3}$

$\frac{1}{2} = \frac{1}{2}$

Check: $\sin 0 \overset{?}{=} 1 - \cos 0$

$0 = 0$

$x = 0, \frac{\pi}{3}, \frac{5\pi}{3}$

5. $\sin x - 1 = 0,\ 0 \le x < 2\pi;\ \sin x = 1;\ x = \frac{\pi}{2}$

9. $4\cos^2 x - 1 = 0;\ 0 \le x < 2\pi$

$4\cos^2 x = 1;\ \cos^2 x = \frac{1}{4};\ \cos x = \pm\frac{1}{2}$

$x = \frac{\pi}{3}, \frac{2\pi}{3}, \frac{4\pi}{3}, \frac{5\pi}{3}$

13. $\sin 2x \sin x + \cos x = 0,\ 0 \le x < 2\pi$

$(2\sin x \cos x)(\sin x) + \cos x = 0$

$2\sin^2 x \cos x + \cos x = 0;\ \cos x(2\sin^2 x + 1) = 0$

$\cos x = 0;\ x = \dfrac{\pi}{2}, \dfrac{3\pi}{2};\ 2\sin x + 1 = 0;\ 2\sin^2 x = -1$

$\sin^2 x = -\dfrac{1}{2}$ which has no real solution; thus,

$x = \dfrac{\pi}{2}, \dfrac{3\pi}{2}$

17. $4\tan x - \sec^2 x = 0;\ 4\tan x - (1 + \tan^2 x) = 0$

$4\tan x - 1 - \tan^2 x = 0;\ \tan^2 x - 4\tan x + 1 = 0$

$\tan^2 x - 4\tan x = -1;\ \tan^2 x - 4\tan x + 4 = -1 + 4$

(completing the square)

$(\tan x - 2)^2 = 3;\ \tan x - 2 = \pm\sqrt{3}$

$\tan x = 2 \pm \sqrt{3} = 3.732,\ 0.2679$

$x = \tan^{-1} 0.2679 = 0.2618,\ \pi + 0.2618 = 3.403$

$x = \tan^{-1} 3.732 = 1.309,\ \pi + 1.309 = 4.451$

$x = 0.2618,\ 1.309,\ 3.403,\ 4.451$

21. $\tan x + 1 = 0;\ \tan x = -1;\ x_{\text{ref}} = \dfrac{\pi}{4}$

(tan negative QII, QIV)

$x = \pi - \dfrac{\pi}{4} = \dfrac{3\pi}{4} \approx 2.36$ or

$x = 2\pi - \dfrac{\pi}{4} = \dfrac{7\pi}{4} \approx 5.50$

Graph $y_1 = \tan x + 1$. Use zero feature to solve.

$x = 2.36,\ 5.50.$

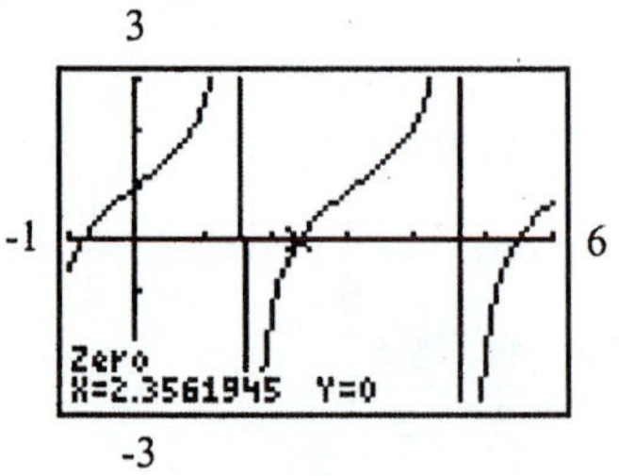

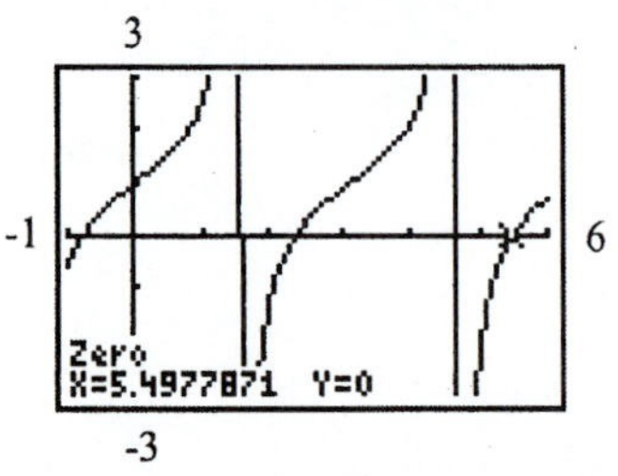

25. $4\sin^2 x - 3 = 0;\ 4\sin^2 x = 3;\ \sin^2 x = \dfrac{3}{4};$

$\sin x = \pm\sqrt{\dfrac{3}{4}} = \pm\dfrac{\sqrt{3}}{2};\ x_{\text{ref}} = \dfrac{\pi}{3}$

(sin positive or negative-all quadrants)

$x = \dfrac{\pi}{3} = 1.05;\ \pi - \dfrac{\pi}{3} = \dfrac{2\pi}{3} = 2.09$

$x = \pi + \dfrac{\pi}{3} = \dfrac{4\pi}{3} = 4.19;\ x = 2\pi - \dfrac{\pi}{3} = \dfrac{5\pi}{3} = 5.24$

Graph $y_1 = 4\sin^2 xx - 3$. Use zero feature to solve.

$x = 1.05,\ 2.09,\ 4.19,$ and $5.24.$

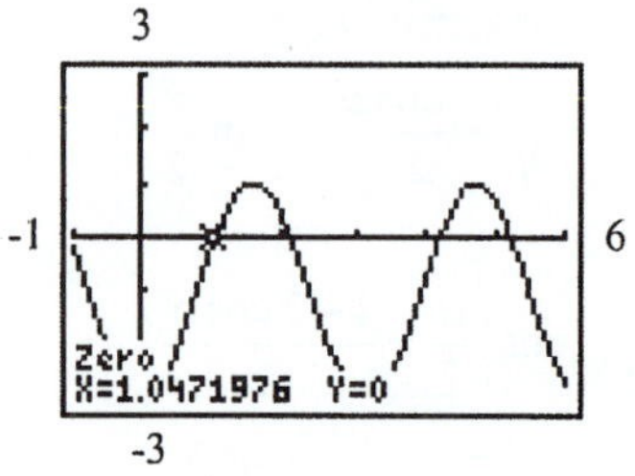

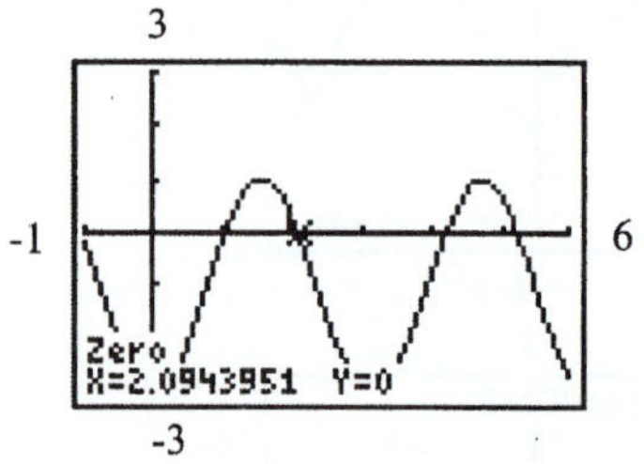

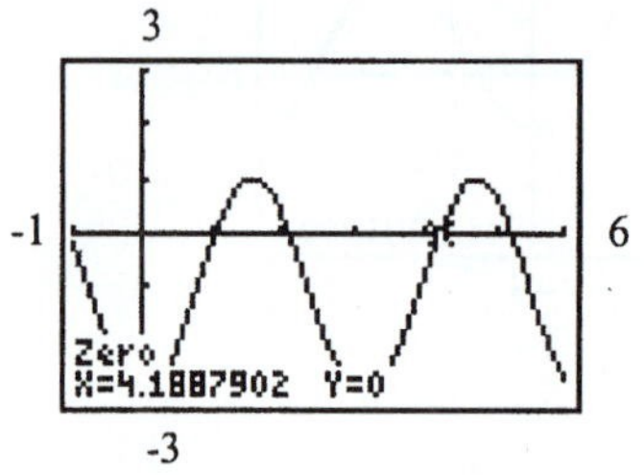

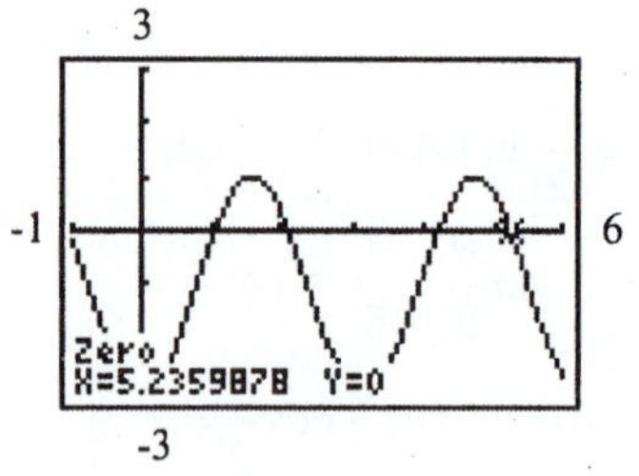

29. $2\sin x - \tan x = 0;\ 2\sin x - \dfrac{\sin x}{\cos x} = 0;$

$\sin x\left(2 - \dfrac{1}{\cos x}\right) = 0;\ \sin x = 0;\ x = 0.00;$

$x = \pi = 3.14$

$2 - \dfrac{1}{\cos x} = 0;\ \dfrac{1}{\cos x} = 2;\ \cos x = \dfrac{1}{2};\ x_{\text{ref}} = \dfrac{\pi}{3};$

$x = \dfrac{\pi}{3} = 1.05;\ x = 2\pi - \dfrac{\pi}{3} = \dfrac{5\pi}{3} = 5.24$

Graph $y_1 = 2\sin x - \tan x$. Use zero feature to solve. $x = 0.00, 1.05, 3.14, 5.24.$

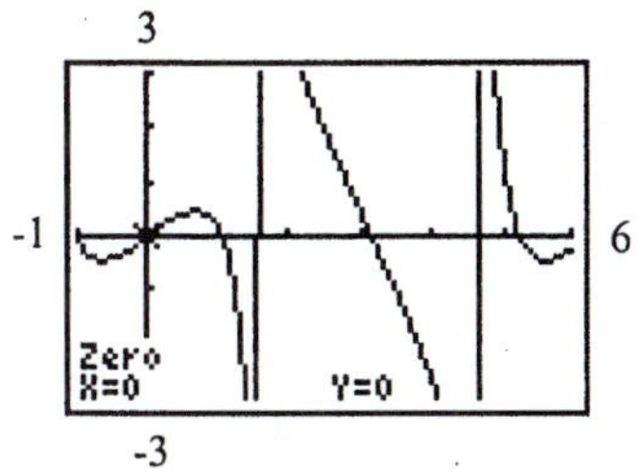

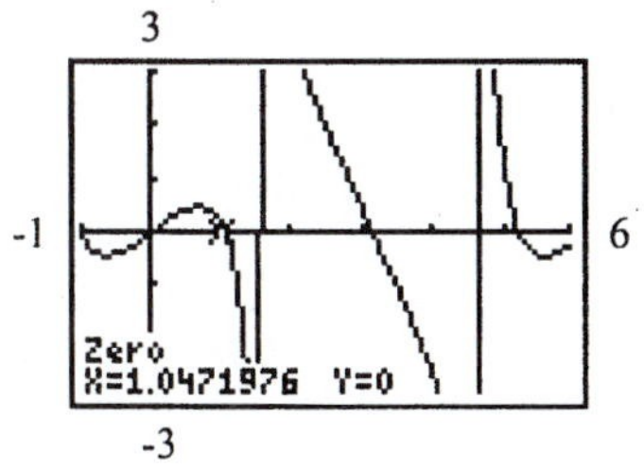

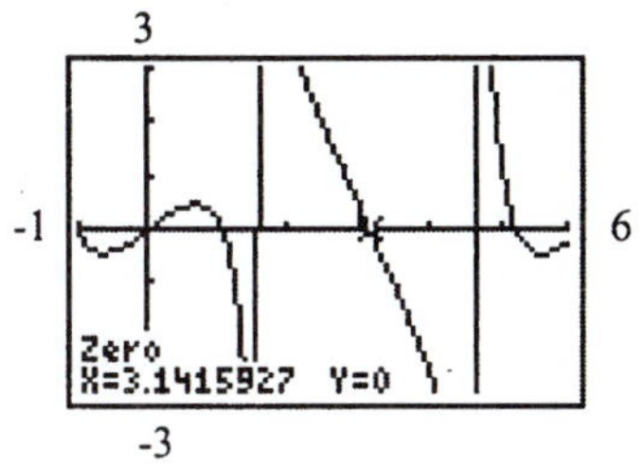

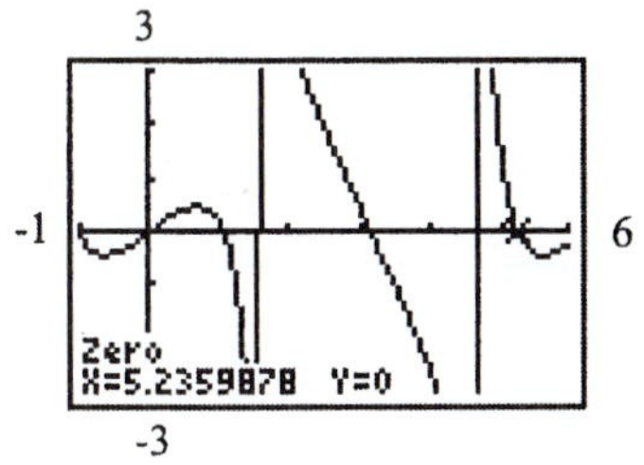

33. $\tan x + 3\cot x = 4;\ \tan x + \dfrac{3}{\tan x} = 4$

$\tan^2 x + 3 = 4\tan x;\ \tan^2 x - 4\tan x + 3 = 0;$

$(\tan x - 1)(\tan x - 3) = 0$

$\tan x = 1;\ x = \dfrac{\pi}{4} = 0.7854$ or

$x = \pi + \dfrac{\pi}{4} = \dfrac{5\pi}{4} = 3.927;$

$\tan x - 3 = 0;\ \tan x = 3;\ x = 1.249$ or

$x = \pi + 1.249 = 4.391$

Graph $y_1 = \tan x + 3/\tan x - 4$. Use zero feature to solve. $x = 0.79, 1.25, 3.93, 4.39.$

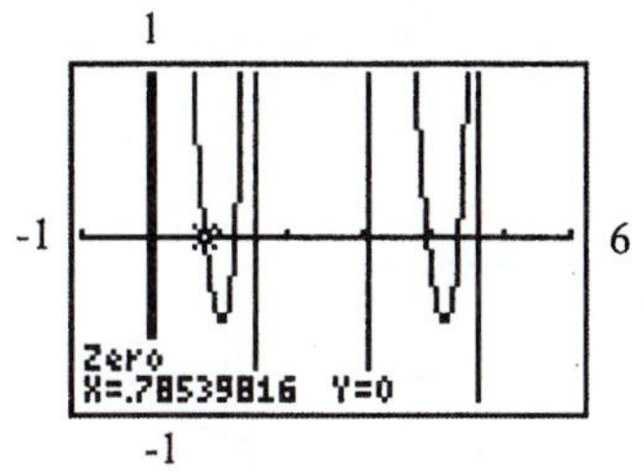

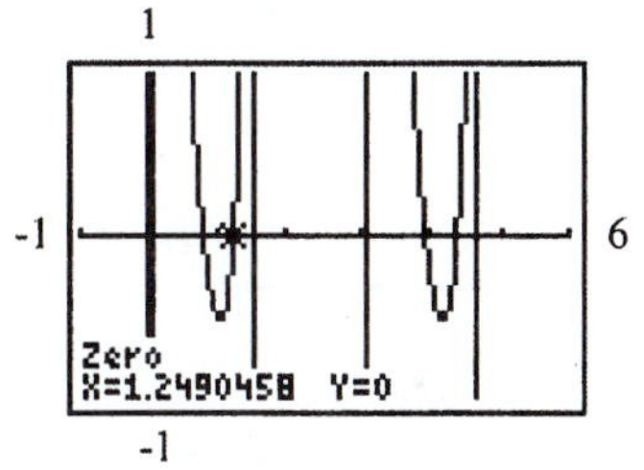

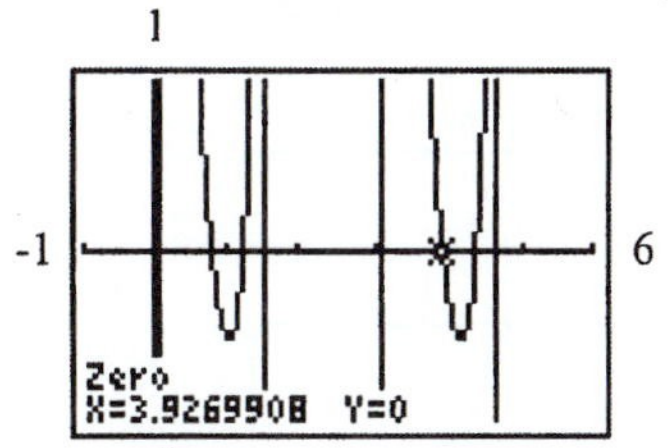

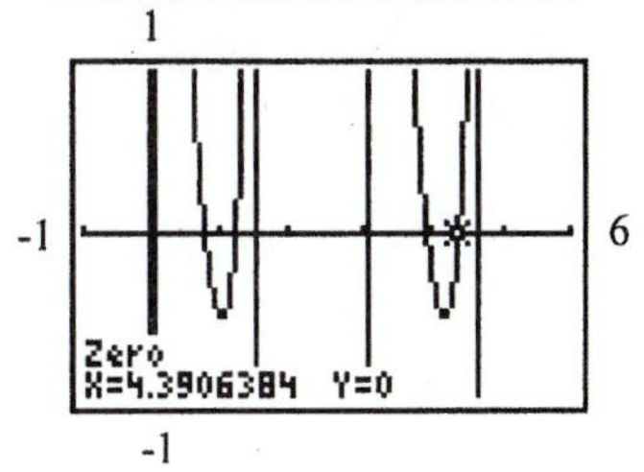

37. $$2\sin 2x - \cos x \sin^3 x = 0$$
$$2(2\sin x\cos x) = \cos x \sin^3 x = 0$$
$$\sin x \cos x\left(4 - \sin^2 x\right) = 0$$

$\sin x \cos x = 0$ $\qquad\qquad\qquad$ $4 - \sin^2 x = 0$

$\sin x = 0 \quad \cos x = 0$ $\qquad\qquad$ $\sin x = \pm 2$

$x = 0, \pi \qquad x = \dfrac{\pi}{2}, \dfrac{3\pi}{2}$ $\qquad$ no solution

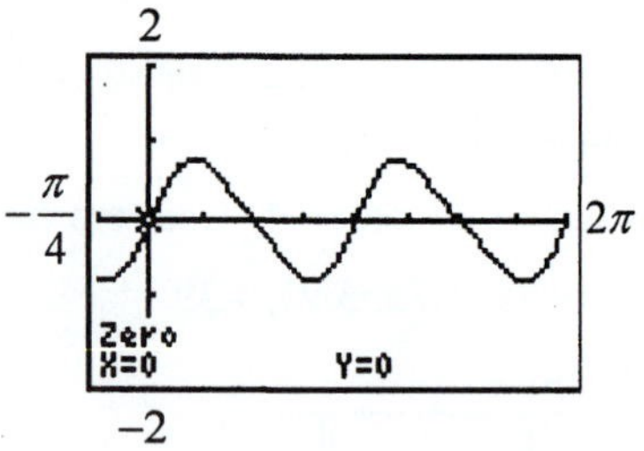

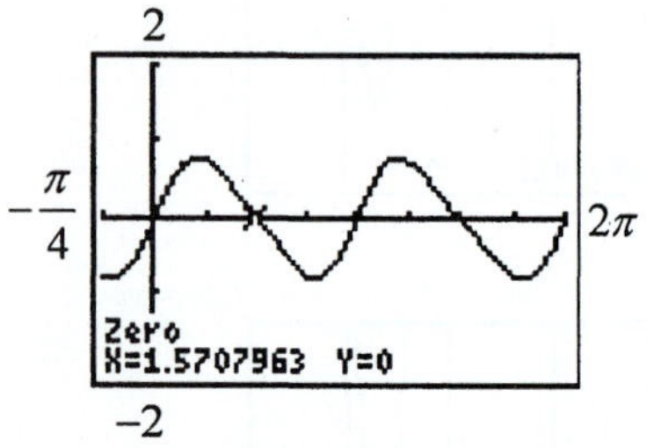

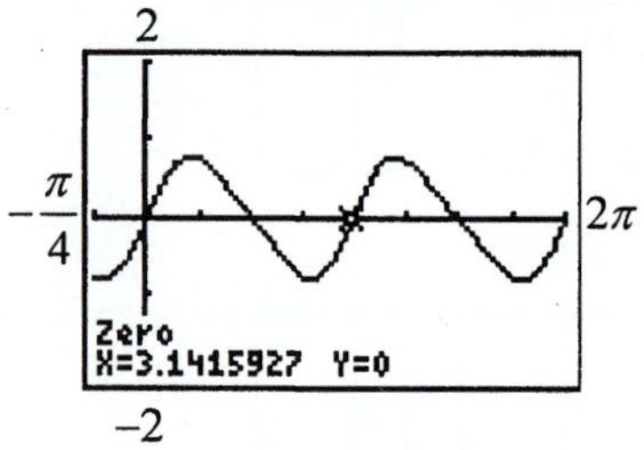

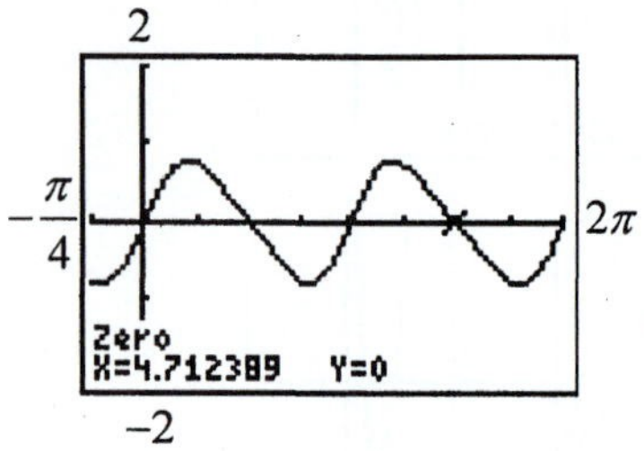

41. $\sin\theta + \cos\theta + \tan\theta + \cot\theta + \sec\theta + \csc\theta = f(\theta) = 1$

θ is a positive acute angle $\Rightarrow 0 < \theta < \dfrac{\pi}{2}$ for which

$0 < \sin\theta < 1,\ 0 < \cos\theta < 1,\ \tan\theta > 0,\ \cot\theta > 0,$

$\sec\theta > 1,\ \csc\theta > 1$ which implies $f(\theta) = \sin\theta$

$+\cos\theta + \tan\theta + \cot\theta + \sec\theta + \csc\theta > 1 \neq 1.$

$f(\theta) = 1$ has no solution.

45. $g = 9.8000 = 9.7805\left(1 + 0.0053\sin^2\theta\right)$

$$\sin^2\theta = \frac{\frac{9.8000}{9.7805} - 1}{0.0053}$$

$$\sin\theta = \sqrt{\frac{\frac{9.8000}{9.7805} - 1}{0.0053}}$$

$$\theta = 37.8^\circ$$

49. $\dfrac{p^2\tan\theta}{0.0063 + p\tan\theta} = 1.6;\ p = 4.8$

$$\frac{4.8^2\tan\theta}{0.0063 + 4.8\tan\theta} = 1.6$$

$$4.8^2\tan\theta = 1.6(0.0063 + 4.8\tan\theta)$$
$$= 0.01008 + 7.68\tan\theta$$
$$4.8^2\tan\theta - 7.68\tan\theta = 0.01008$$
$$15.36\tan\theta = 0.01008$$
$$\tan\theta = \frac{0.01008}{15.36} = 6.5625\times 10^{-4}$$
$$\theta = 6.56\times 10^{-4}$$

53. $3\sin x - x = 0$. Graph $y_1 = 3\sin x - x$. Use zero feature to solve. $x = -2.28, 0.00, 2.28$. Using the zoom feature, more accurate values can be found.

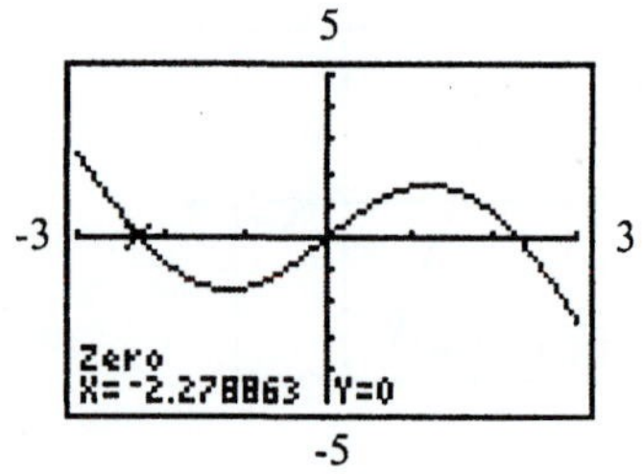

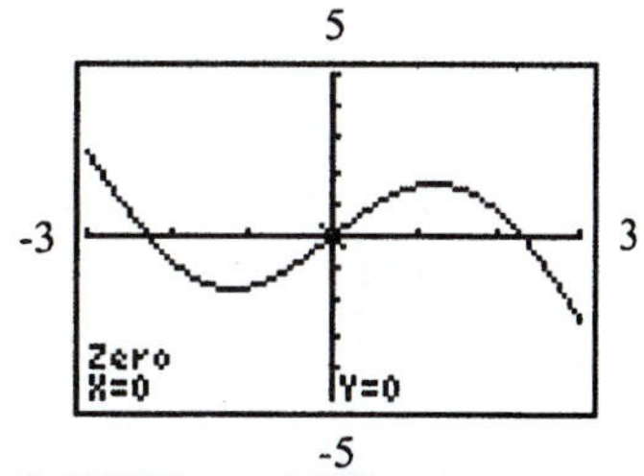

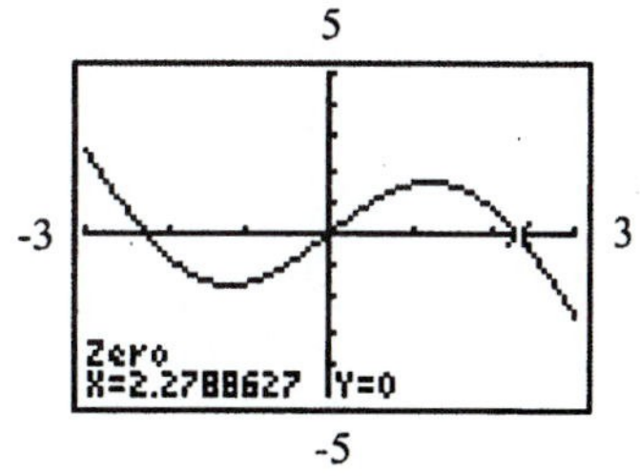

57. $2\ln x = 1 - \cos 2x$; $2\ln x - 1 + \cos 2x = 0$
Graph $y_1 = 2\ln x - 1 + \cos 2x$. Use zero feature to solve. $x = 2.10$.

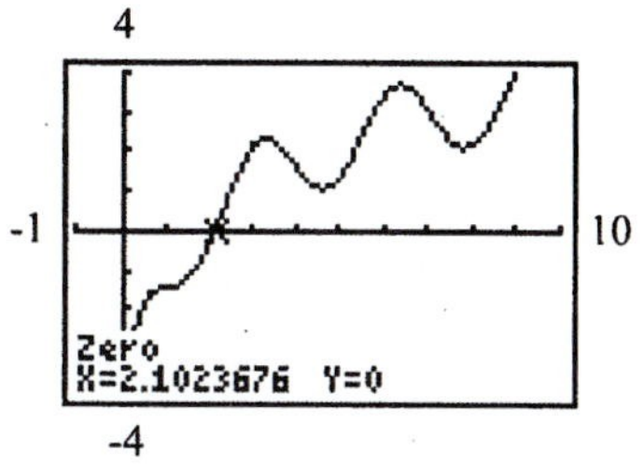

20.6 The Inverse Trigonometric Functions

1. $y = \tan^{-1} 3A$ is read as "y is the angle whose tangent is $3A$." In this case, $3A = \tan y$.

5. y is an angle whose cotangent is $3x$.

9. y is five times the angle whose cosine is $2x - 1$.

13. $\tan^{-1} 1 = \frac{\pi}{4}$ since $\tan \frac{\pi}{4} = 1$ and $-\frac{\pi}{2} < \frac{\pi}{4} < \frac{\pi}{2}$

17. Let $\sec^{-1} 0.5 = x$, then

$$\sec x = 0.5$$
$$\frac{1}{\cos x} = 0.5$$
$$\cos x = 2$$

and since $-1 \le \cos x \le 1$ there is no value for x.

21. $\sin\left(\tan^{-1}\sqrt{3}\right) = \sin\frac{\pi}{3} = \frac{1}{2}\sqrt{3}$

25. $\cos\left[\tan^{-1}(-5)\right] = \frac{1}{\sqrt{26}}$

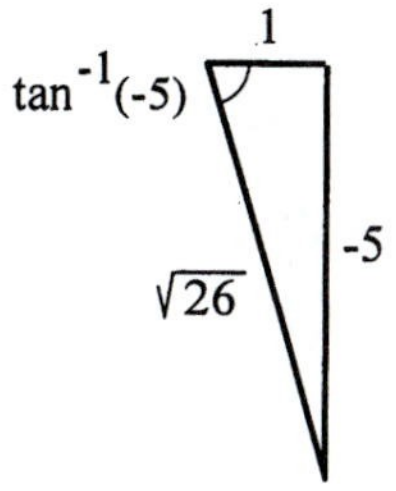

29. $\tan^{-1} x = \sin^{-1}\frac{2}{5}$

$$x = \tan\left(\sin^{-1}\frac{2}{5}\right)$$
$$x = \frac{2}{\sqrt{21}}$$

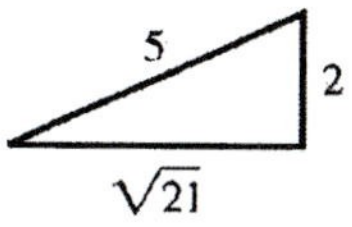

33. $\tan^{-1}(-3.7321) = -1.3090$

37. $\tan\left[\cos^{-1}(-0.6281)\right] = \tan 2.250 = -1.2389$

41. $y = \sin 3x$; $3x = \sin^{-1} y$; $x = \frac{1}{3}\sin^{-1} y$

45. $1 - y = \cos^{-1}(1 - x)$

$$\cos(1 - y) = 1 - x$$
$$x = 1 - \cos(1 - y)$$

49. Let $\alpha - \sin^{-1}$, $\beta = \cos^{-1} y$

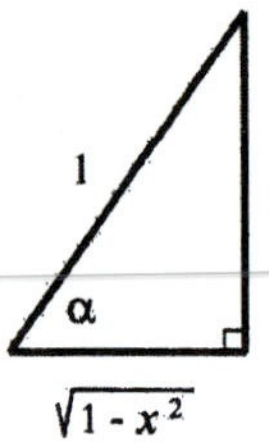

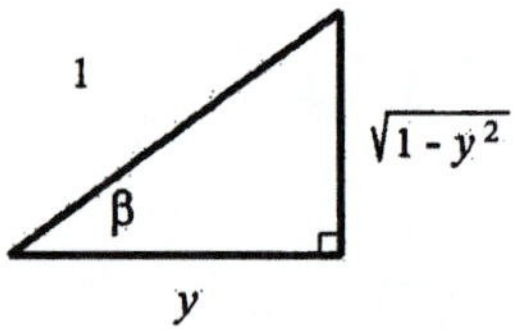

$$\sin\left(\sin^{-1}x+\cos^{-1}y\right)=\sin(\alpha+\beta)$$
$$=\sin\alpha\cos\beta+\sin\beta\cos\alpha$$
$$=xy+\sqrt{1-y^2}\sqrt{1-x^2}$$
$$=xy+\sqrt{1-y^2-x^2+x^2y^2}$$

53. $\sin\left(2\sin^{-1}x\right)=\sin 2\theta=2\sin\theta\cos\theta$

$$=2\left(\frac{x}{1}\right)\left(\frac{\sqrt{1-x^2}}{1}\right)=2x\sqrt{1-x^2}$$

In triangle, θ is set up such that its sine is $2x$. This gives an opposite side of $2x$, hypotenuse 1, and adjacent side $\sqrt{1-4x^2}$.

57. $y=A\cos 2(\omega t+\phi)$

$$\frac{y}{A}=\cos 2(\omega t+\phi)$$
$$\cos^{-1}\frac{y}{A}=2(\omega t+\phi)=2\omega t+2\phi$$
$$\cos^{-1}\frac{y}{A}-2\phi=2\omega t$$
$$\frac{\cos^{-1}\frac{y}{A}-2\phi}{2\omega}=t$$
$$t=\frac{1}{2\omega}\cos^{-1}\frac{y}{A}-\frac{\phi}{\omega}$$

61. Let $\alpha=\sin^{-1}\frac{3}{5}$ and $\beta=\sin^{-1}\frac{5}{13}$; $\sin\alpha=\frac{3}{5}$

$$\cos\alpha=\sqrt{1-\tfrac{9}{25}}=\sqrt{\frac{16}{25}}=\frac{4}{5};\ \sin\beta=\frac{5}{13}$$
$$\cos\beta=\sqrt{1-\tfrac{25}{169}}=\sqrt{\frac{144}{169}}=\frac{12}{13}$$
$$\sin^{-1}\frac{3}{5}+\sin^{-1}\frac{5}{13}=\alpha+\beta$$
$$\sin(\alpha+\beta)=\sin\alpha\cos\beta+\cos\alpha\sin\beta$$
$$=\frac{3}{5}\left(\frac{12}{13}\right)+\frac{4}{5}\left(\frac{5}{13}\right)$$
$$=\frac{36}{65}+\frac{20}{65}=\frac{56}{65}$$

65. $y=\sin^{-1}x+\sin^{-1}(-x),\ -1\le x\le 1,$

$$-\frac{\pi}{2}\le y\le\frac{\pi}{2}$$
$$\sin y=\sin\left(\sin^{-1}x\right)+\sin\left(\sin^{-1}(x)\right)$$
$$\sin y=x+(-x)=0$$
$$y=0$$
$$\sin^{-1}x+\sin^{-1}(-x)=0$$

69. $\tan B=\frac{y}{b}\Rightarrow y=b\tan B$

$$\tan A=\frac{y}{a}=\frac{b\tan B}{a}$$
$$A=\tan^{-1}\frac{b\tan B}{a}$$

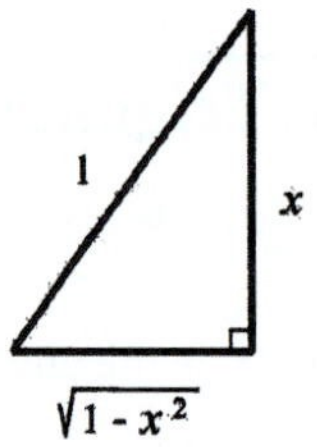

73. $\tan\alpha=\frac{y}{x}\Rightarrow a=\tan^{-1}\left(\frac{y}{x}\right)$

$$\tan(\alpha+\theta)=\frac{y+50}{x}\Rightarrow\alpha+\theta=\tan^{-1}\left(\frac{y+50}{x}\right)$$
$$\tan^{-1}\left(\frac{y}{x}\right)+\theta=\tan^{-1}\left(\frac{y+50}{x}\right)$$
$$\theta=\tan^{-1}\left(\frac{y+50}{x}\right)-\tan^{-1}\left(\frac{y}{x}\right)$$

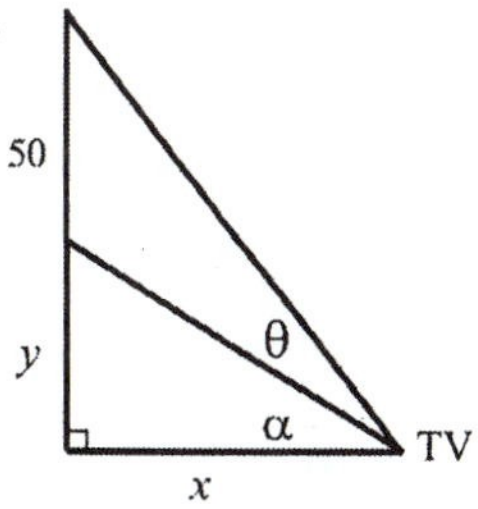

Chapter 20 Review Exercises

1. $\sin 120^\circ = \sin(90^\circ + 30^\circ)$

$= \sin 90^\circ \cos 30^\circ + \cos 90^\circ \sin 30^\circ$

$= 1\left(\frac{\sqrt{3}}{2}\right) + 0\left(\frac{1}{2}\right) = \frac{\sqrt{3}}{2} = \frac{1}{2}\sqrt{3}$

5. $\cos\pi = \cos 2\left(\frac{\pi}{2}\right) = \cos^2\frac{\pi}{2} - \sin^2\frac{\pi}{2} = 0 - (1)^2 = -1$

9. $\sin 14^\circ \cos 38^\circ + \cos 14^\circ \sin 38^\circ$

$= \sin(14^\circ + 38^\circ)$

$= \sin 52^\circ = 0.7880108$

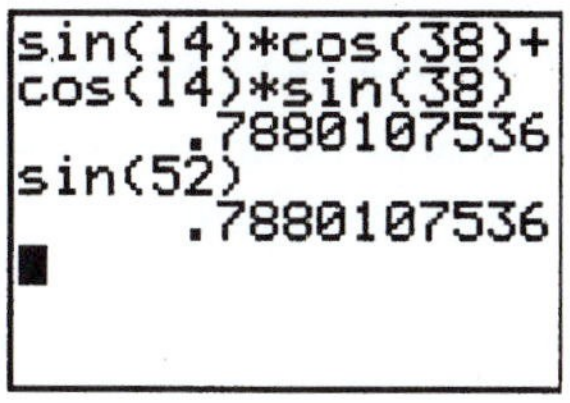

13. $\cos 73^\circ \cos(-142^\circ) + \sin 73^\circ \sin(-142^\circ)$

$= \cos 73^\circ \cos 142^\circ - \sin 73^\circ \sin 142^\circ$

$= \cos(73^\circ + 142^\circ) = \cos 215^\circ$

$= -0.8191520443$

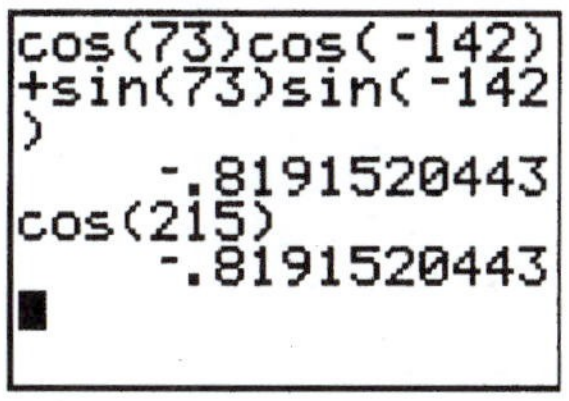

17. $\sin 2x \cos 3x + \cos 2x \sin 3x$

$= \sin\alpha\cos\beta + \cos\alpha\sin\beta$

$= \sin(\alpha + \beta)$ where $\alpha = 2x,\ \beta = 3x$

$\sin(\alpha+\beta) = \sin(2x + 3x) = \sin 5x$

21. $2 - 4\sin^2 6x = 2(1 - 2\sin^2 6x)$

$= 2(-2\sin^2\alpha) = 2(\cos 2\alpha)$

$= 2\cos 12x$ where $\alpha = 6x$

25. $\sin^{-1}(-1) = -\frac{\pi}{2}$ since $\sin\left(-\frac{\pi}{2}\right) = -1$ and

$-\frac{\pi}{2} \le -\frac{\pi}{2} \le \frac{\pi}{2}$

29. $\tan\left[\sin^{-1}(-0.5)\right] = \tan\left(-\frac{\pi}{6}\right) = -\frac{\sqrt{3}}{3}$

33. $\frac{\sec y}{\cos y} - \frac{\tan y}{\cot y} = \sec^2 y - \tan^2 y$

$= 1 + \tan^2 y - \tan^2 y$

$= 1$

37. $\frac{\sec^4 x - 1}{\tan^2 x} = \frac{(\sec^2 x - 1)(\sec^2 x + 1)}{\tan^2 x}$

$\frac{(\sec^2 x + 1)(\tan^2 x)}{\tan^2 x} = \sec^2 x + 1$

$1 + \tan^2 x + 1 = 2 + \tan^2 x$

41. $\frac{1 - \sin^2\theta}{1 - \cos^2\theta} = \frac{\cos^2\theta}{\sin^2\theta}$ since $\sin^2\theta + \cos^2\theta = 1$

$= \left(\frac{\cos\theta}{\sin\theta}\right)^2$

$= (\cot\theta)^2 = \cot^2\theta$

45. $\frac{\sec x}{\sin x} - \sec x \sin x = \frac{1}{\cos x \sin x} - \frac{\sin x}{\cos x}\cdot\frac{\sin x}{\sin x}$

$= \frac{1 - \sin^2 x}{\sin x \cos x} = \frac{\cos^2 x}{\sin x \cos x}$

$= \frac{\cos x}{\sin x} = \cot x$

49. $\dfrac{\sin x\cot x+\cos x}{2\cot x}=\dfrac{\sin x\cdot\frac{\cos x}{\sin x}+\cos x}{2\frac{\cos x}{\sin x}}\cdot\dfrac{\sin x}{\sin x}$

$=\dfrac{\sin x\cos x+\sin x\cos x}{2\cos x}$

$=\dfrac{2\sin x\cos x}{2\cos x}=\sin x$

53.

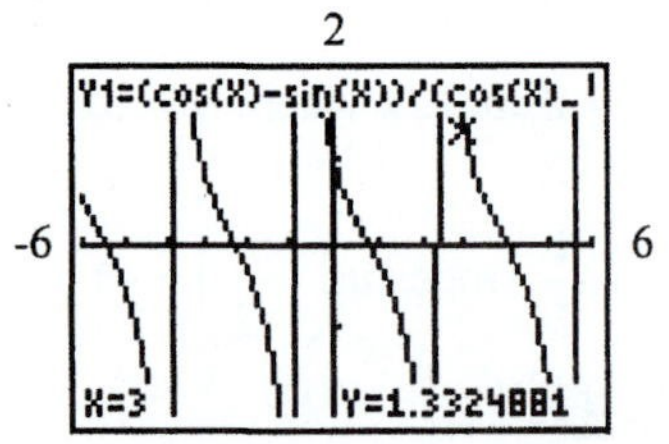

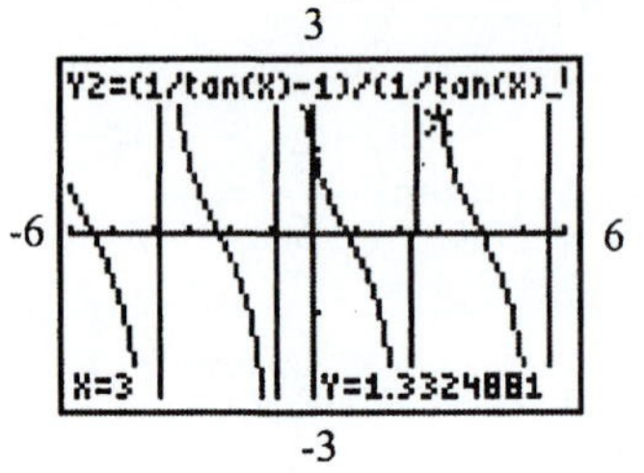

57.

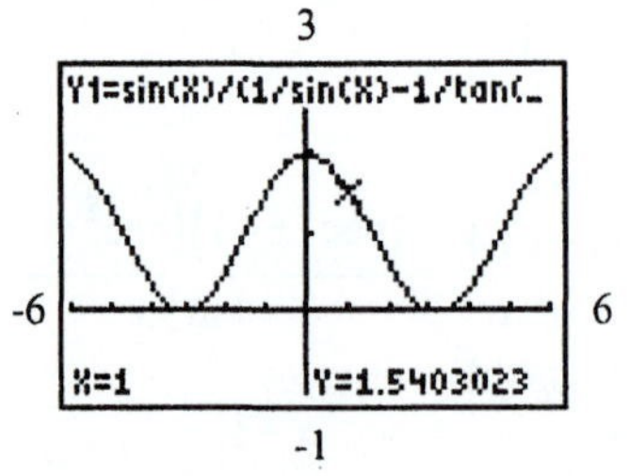

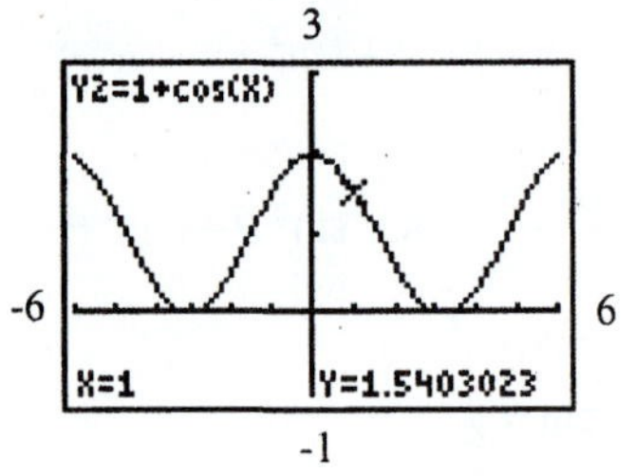

61. $y=2\cos 2x;\ \dfrac{y}{2}=\cos 2x$

$\cos^{-1}\dfrac{y}{2}=2x;\ \dfrac{1}{2}\cos^{-1}\dfrac{1}{2}y=x$

65. $3(\tan x-2)=1+\tan x$

$3\tan x-1=1+\tan x$

$2\tan x=7$

$\tan x=\dfrac{7}{2}$

$x=\tan^{-1}\dfrac{7}{2}=1.2925$

Since $\tan x$ is positive in QIII also, $\pi+1.2925=4.4341$ is also a value for x that is within the specified range of values for x.

69. $2\sin^2\theta+3\cos\theta-3=0,\ 0\le\theta<2\pi$

$2(1-\cos^2\theta)+3\cos\theta-3=0$

$2-2\cos^2\theta+3\cos\theta-3=0$

$2\cos^2\theta-3\cos\theta+1=0$

$(\cos\theta-1)(2\cos\theta-1)=0$

$\cos\theta-1=0$ or $2\cos\theta-1=0$

$\theta=0$ $\qquad\cos\theta=\dfrac{1}{2}$

$\theta=\dfrac{\pi}{3},\dfrac{5\pi}{3}$

73. $\sin 2x=\cos 3x$

$\sin 2x=\cos(2x+x)$

$2\sin x\cos x=\cos 2x\cos x-\sin 2x\sin x$

$2\sin x\cos x=\cos 2x\cos x-2\sin x\cos x\sin x$

$2\sin x\cos x-\cos 2x\cos x+2\sin^2 x\cos x=0$

$\cos x\left(2\sin x-\cos 2x+\sin^2 x\right)=0$

$\cos x=0\Rightarrow x=\dfrac{\pi}{2},\dfrac{3\pi}{2}$ or

$2\sin x-\left(1-2\sin^2 x\right)+2\sin^2 x=0$

$4\sin^2 x+2\sin x-1=0$

$\sin x=\dfrac{-2\pm\sqrt{2^2-4(4)(-1)}}{2(4)}=\dfrac{-1\pm\sqrt{5}}{4}$

$\sin x=\dfrac{-1+\sqrt{5}}{4}$ or $\sin x=\dfrac{-1-\sqrt{5}}{4}$

$x=\dfrac{\pi}{10},\dfrac{9\pi}{10}$ $\qquad x=\dfrac{13\pi}{10},\dfrac{17\pi}{10}$

$\sin 2x=\cos 3x$ has solutions

$\left\{\dfrac{\pi}{10},\dfrac{\pi}{2},\dfrac{9\pi}{10},\dfrac{13\pi}{10},\dfrac{3\pi}{2},\dfrac{17\pi}{10}\right\}$

77. $\tan x + \cot x = \dfrac{\sin x}{\cos x} + \dfrac{\cos x}{\sin x}$

$= \dfrac{\sin^2 x + \cos^2 x}{\sin x \cos x}$

$= \dfrac{1}{\sin x}\dfrac{1}{\cos x}$

$= \csc x \sec x$, identity

81. $x + \ln x - 3\cos^2 x = 2$

$x + \ln x - 3\cos^2 x - 2 = 0$

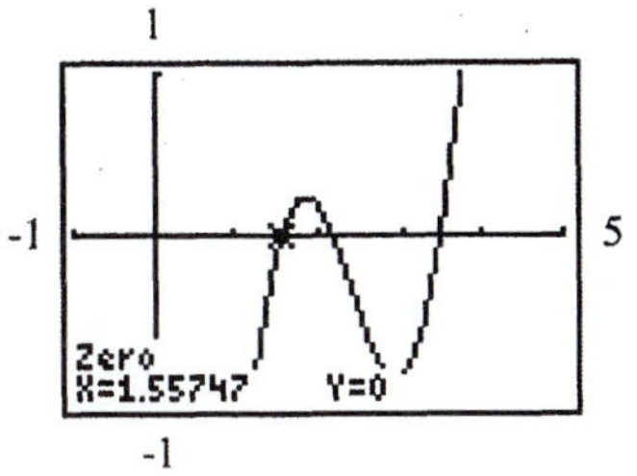

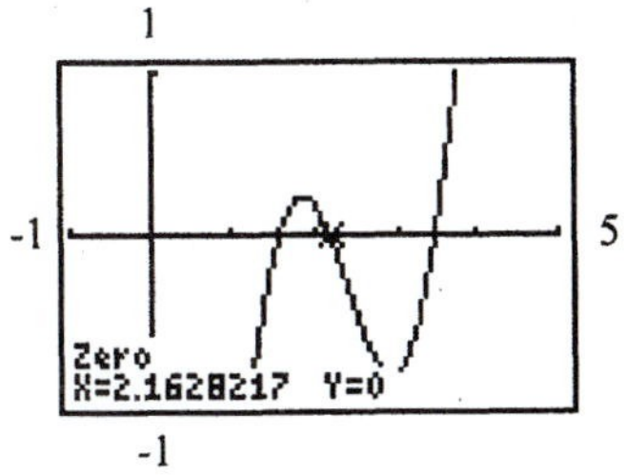

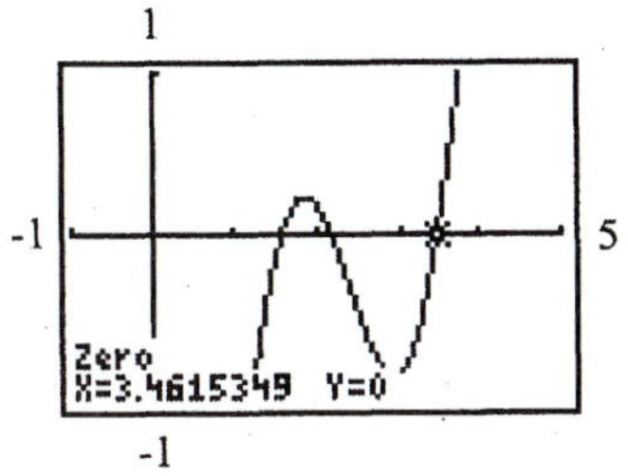

85. $\tan\left(\cot^{-1} x\right) = \tan\theta = \dfrac{1}{x}$

$\theta = \cot^{-1} x$

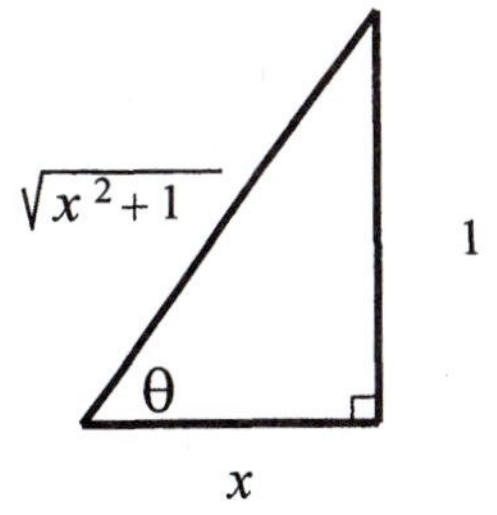

89.

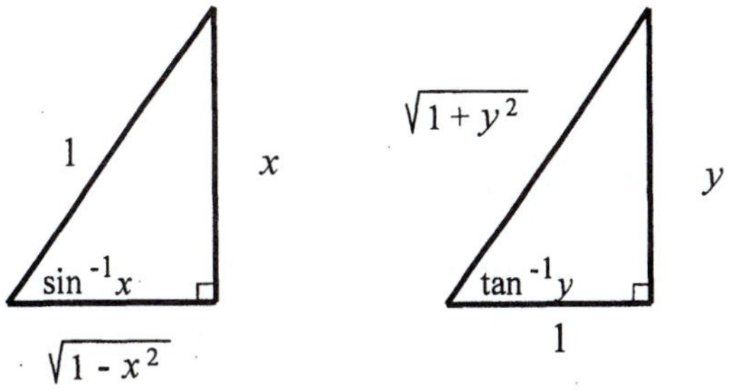

$\cos\left(\sin^{-1} x + \tan^{-1} y\right) = \cos\left(\sin^{-1} x\right)\cos\left(\tan^{-1} y\right)$

$- \sin\left(\sin^{-1} x\right)\sin\left(\tan^{-1} y\right)$

$= \sqrt{1+x^2}\dfrac{1}{\sqrt{1+y^2}} - x\dfrac{1}{\sqrt{+y^2}}$

$= \dfrac{\sqrt{1+x^2} - xy}{\sqrt{1+y^2}}$

93. $\dfrac{x}{\sqrt{1+x^2}} = \dfrac{\tan\theta}{\sqrt{1+\tan^2\theta}} = \dfrac{\tan\theta}{\sqrt{\sec^2\theta}} = \dfrac{\tan\theta}{\sec\theta}$

$= \dfrac{\sin\theta}{\cos\theta}\cdot\cos\theta = \sin\theta$

97. $\sin 2x > 2\cos x,\quad 0 \le x < 2\pi$

$2\sin x\cos x - 2\cos x > 0$

$\cos x(\sin x - 1) > 0$, there are two cases

I $\cos x > 0$ and $\sin x - 1 > 0$

$0 < x < \frac{\pi}{2}$ $\sin x > 1$, no solution

$\frac{3\pi}{2} < x < 2\pi$

II $\cos x < 0$ and $\sin x - 1 < 0$

$\frac{\pi}{2} < x < \frac{3\pi}{2}$ $\sin x - 1 < 0$

$0 < x < \frac{\pi}{2}$ or $\frac{\pi}{2} < x < 2\pi$

101. $R = \sqrt{Rx^2 + Ry^2}$

$$= \sqrt{(A\cos\theta - B\sin\theta)^2 + (A\sin\theta + B\cos\theta)^2}$$
$$= \sqrt{A^2\cos^2\theta + A^2\sin^2\theta + B^2\cos^2\theta + B^2\sin^2\theta}$$
$$= \sqrt{A^2(\cos^2\theta + \sin^2\theta) + B^2(\cos^2\theta + \sin^2\theta)}$$
$$= \sqrt{(\cos^2\theta + \sin^2\theta)(A^2 + B^2)} = \sqrt{1(A^2 + B^2)}$$

105.
$$\omega t = \sin^{-1}\frac{\theta - \alpha}{R}$$
$$\sin(\omega t) = \frac{\theta - \alpha}{R}$$
$$R\sin(\omega t) = \theta - \alpha$$
$$\theta = R\sin(\omega t) + \alpha$$

109. $p = VI\cos\phi\cos^2\omega t - VI\sin\phi\cos\omega t\sin\omega t$

$$p = VI\cos\omega t(\cos\phi\cos\omega t - \sin\phi\sin\omega t)$$
$$p = VI\cos\omega t\cos(\omega t + \phi)$$

113.
$$2\sin\theta\cos^2\theta - \sin^3\theta = 0, \quad 0 < \theta < 90^\circ$$
$$\sin\theta(2\cos^2\theta - \sin^2\theta) = 0$$
$$\sin\theta(2\cos^2\theta - (1 - \cos^2\theta)) = 0$$
$$\sin\theta(3\cos^2\theta - 1) = 0$$

$\sin\theta = 0$ or $\cos^2\theta = \frac{1}{3}$

no solution $\cos\theta = \sqrt{\frac{1}{3}}$

$\theta = 54.7^\circ$

Chapter 21

PLANE ANALYTIC GEOMETRY

21.1 Basic Definitions

1. The distance between $(3, -1)$ and $(-2, 5)$ is

$$d=\sqrt{(3-(-2))^2+(-1-5)^2}$$
$$=\sqrt{61}$$

5. Given: $(x_1, y_1)=(3, 8); (x_2, y_2)=(-1, -2)$

$$d=\sqrt{(x_2-x_1)^2+(y_2-y_1)^2}$$
$$=\sqrt{(-1-3)^2+(-2-8)^2}$$
$$=\sqrt{(-4)^2+(-10)^2}$$
$$=\sqrt{16+100}=\sqrt{116}$$
$$=\sqrt{4\times 29}=2\sqrt{29}$$

9. Given: $(x_1, y_1)=(-12, 20); (x_2, y_2)=(32, -13)$

$$d=\sqrt{(x_2-x_1)^2+(y_2-y_1)^2}$$
$$=\sqrt{(32+12)^2+(-13-20)^2}$$
$$=\sqrt{(44)^2+(-33)^2}$$
$$=\sqrt{1936+1089}=\sqrt{3025}$$
$$=55$$

13. Given: $(x_1, y_1)=(1.22, -3.45);$

$(x_2, y_2)=(-1.07, -5.16)$

$$d=\sqrt{(x_2-x_1)^2+(y_2-y_1)^2}$$
$$=\sqrt{(-1.07-1.22)^2+(-5.16-(-3.45))^2}$$
$$=\sqrt{(-2.29)^2+(-5.16+3.45)^2}$$
$$=\sqrt{(-2.29)^2+(-1.71)^2}=\sqrt{8.1682}$$
$$=2.86$$

17. Given: $(x_1, y_1)=(4, -5); (x_2, y_2)=(4, -8)$

$$m=\frac{y_2-y_1}{x_2-x_1}=\frac{-8-(-5)}{4-4}$$

Since $x_2-x_1=4-4=0$, the slope is undefined.

21. Given: $(x_1, y_1)=\left(\sqrt{32}, -\sqrt{18}\right);$

$(x_2, y_2)=\left(-\sqrt{50}, \sqrt{8}\right)$

$$m=\frac{y_2-y_1}{x_2-x_1}=\frac{\sqrt{8}-\left(-\sqrt{18}\right)}{-\sqrt{50}-\sqrt{32}}=\frac{-5}{9}$$

25. Given: $\alpha=30^\circ$; $m=\tan\alpha$, $0^\circ<\alpha<180^\circ$

$$\tan 30^\circ=\frac{\sqrt{3}}{3}$$

29. Given: $m=0.364$; $m=\tan\alpha$; $0.364=\tan\alpha$;

$\alpha=20.0^\circ$

33. Given: $(x, y)=(6, -1); (x_1, y_1)=(4, 3)$

$(x_2, y_2)=(-5, 2); (x_3, y_3)=(-7, 6)$

$$m_1=\frac{y-y_1}{x-x_1}=\frac{-1-3}{6-4}=\frac{-4}{2}=-2$$

$$m_2=\frac{y_2-y_3}{x_2-x_3}=\frac{2-6}{-5-(-7)}=\frac{-4}{-5+7}=\frac{-4}{2}=-2$$

$m_1=m_2$ for all parallel lines.

37. Given: distance between $(-1, 3)$ and $(11, k)$ is 13.

$$d=\sqrt{(x_1-x_2)^2+(y_1-y_2)^2}$$
$$13=\sqrt{(-1-11)^2+(3-k)^2}$$
$$=\sqrt{(-12)^2+(3-k)^2}$$
$$169=144+(3-k)^2;$$
$$(3-k)^2=25;\ 3-k=\pm 5$$
$$-k=-3\pm 5;\ k=-2, 8$$

41. $d_1 = \sqrt{(9-7)^2 + [4-(-2)]^2}$
$= \sqrt{2^2 + 6^2} = \sqrt{40} = 2\sqrt{10}$
$d_2 = \sqrt{(9-3)^2 + (4-2)^2} = \sqrt{6^2 + 2^2}$
$= \sqrt{40} = 2\sqrt{10}$
$d_1 = d_2$ so the triangle is isosceles.

45. $d_1 = \sqrt{(3-5)^2 + (-1-3)^2} = \sqrt{(-2)^2 + (-4)^2}$
$= \sqrt{4+16} = \sqrt{20}$
$m_1 = \frac{y-y_1}{x-x_1} = \frac{5-3}{3-(-1)} = \frac{5-3}{3+1} = \frac{2}{4} = \frac{1}{2}$
$d_2 = \sqrt{(5-1)^2 + (3-5)^2} = \sqrt{(4)^2 + (-2)^2}$
$= \sqrt{16+4} = \sqrt{20}$
$m_2 = \frac{y-y_1}{x-x_1} = \frac{5-1}{3-5} = \frac{4}{-2} = -2$
$m_1 = \frac{-1}{m_2}$, $m_1 \perp m_2$
$A = \frac{1}{2}d_1 d_2 = \frac{1}{2}\sqrt{20}\sqrt{20} = \frac{1}{2}(20) = 10$

49. $\left(\frac{-4+6}{2}, \frac{9+1}{2}\right) = \left(\frac{2}{2}, \frac{10}{2}\right) = (1, 5)$

53. The distance between (x, y) and $(0, 0) = 3$.
$\sqrt{(x-0)^2 + (y-0)^2} = 3$
$x^2 + y^2 = 9$

57. $m = \frac{y_2 - y_1}{x_2 - x_1} = \frac{5-0}{-2-x} = 3$
$x = -\frac{11}{3}$

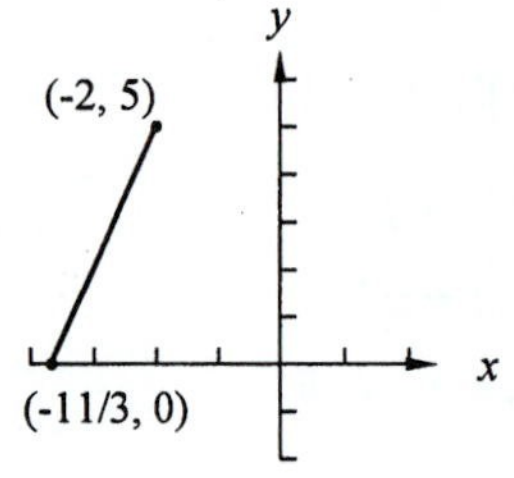

61. $a^2 + b^2 = x^2 = (a-b)^2 + b^2$
$a^2 + b^2 = a^2 - 2ax + x^2 + b^2$
$x = 2a$
$a^2 + b^2 = 4a^2 \Rightarrow a = \frac{b}{\sqrt{3}}$
$m_1 = \frac{b-0}{\frac{b}{\sqrt{3}} - 0} = \sqrt{3}$
$m_2 = \frac{b-0}{\frac{b}{\sqrt{3}} - \frac{2b}{\sqrt{3}}} = -\sqrt{3}$

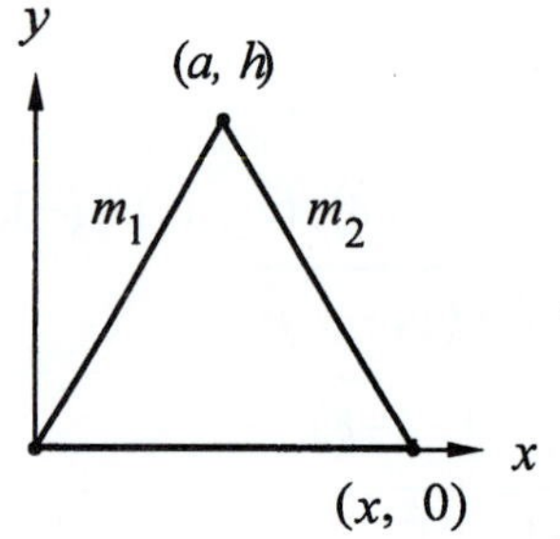

21.2 The Straight Line

1. $m = -2, (x_1, y_1) = (4, -1)$
$y - y_1 = m(x - x_1)$
$y - (-1) = -2(x-4)$
$y + 1 = -2x + 8$
$y + 2x - 7 = 0$

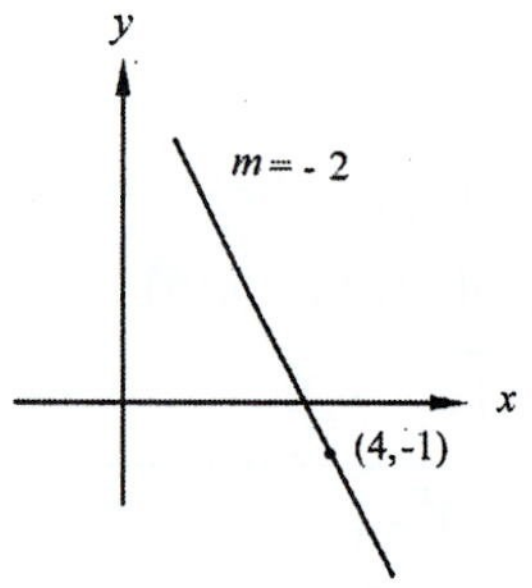

5. Given: $m = 4$; $(x_1, y_1) = (-3, 8)$

$y - y_1 = m(x - x_1)$

$y - 8 = 4[x - (-3)] = 4(x + 3) = 4x + 12$

$y = 4x + 20$ or $4x - y + 20 = 0$

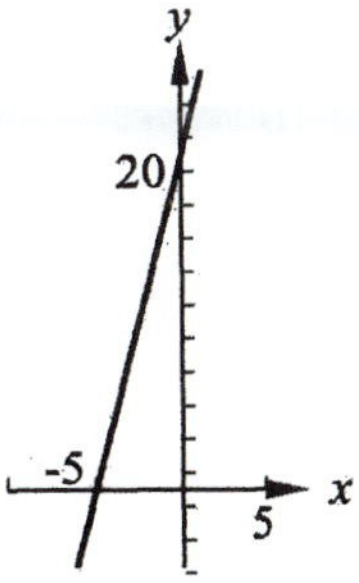

9. Given: $(x_1, y_1) = (-7, 12)$ $\alpha = 45^\circ$

$m = \tan\alpha = \tan 45^\circ = 1$

$y - y_1 = m(x - x_1)$

$y - 12 = 1(x + 7) = x + 7$; $y = x + 19$

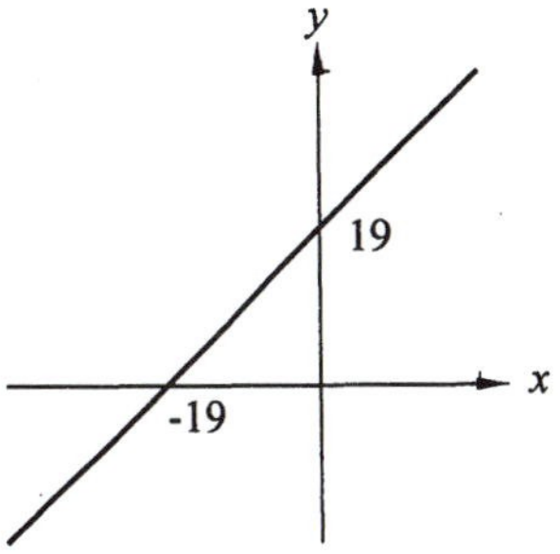

13. Parallel to y-axis and 3 units left of y-axis.

$x = -3$

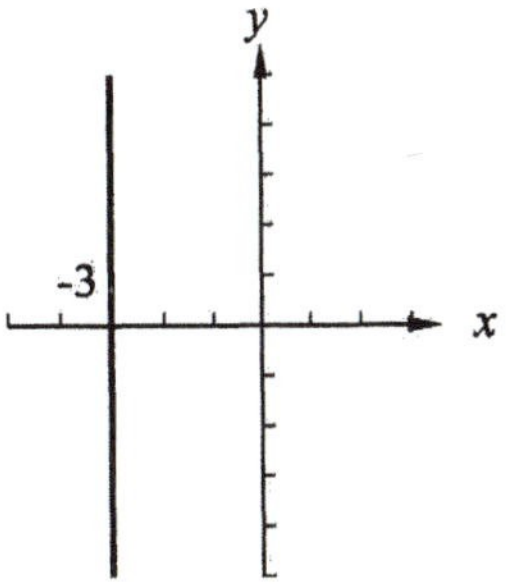

17. Parallel to a line through $(7, -1)$ and $(4, 3)$;

$$m = \frac{y_2 - y_1}{x_2 - x_1} = \frac{3-(-1)}{4-7} = \frac{4}{-3} = -\frac{4}{3}$$

y-intercept is $b = -2$

$y = mx + b$; $y = -\frac{4}{3}x - 2$; $4x + 3y + 6 = 0$

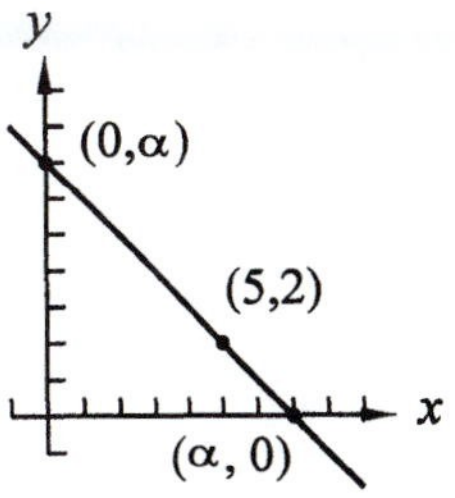

21. Given: $4x - y = 8$, $y = 4x - 8$, $m = 4$, $b = -8$

When $x = 0$, $y = -8$

$y = 0$, $x = 2$

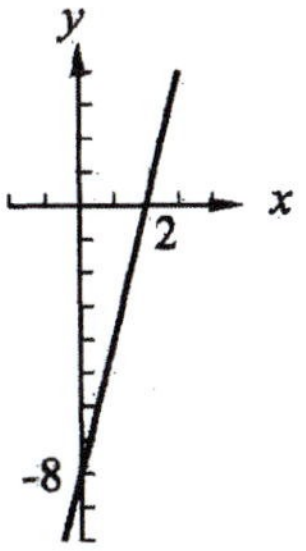

25. Given: $3x - 2y - 1 = 0$

$3x - 2y - 1 = 0$; $-2y = -3x + 1$

$$y = \frac{-3}{-2}x + \frac{1}{-2};\ y = \frac{3}{2}x - \frac{1}{2}$$

$\text{Slope} = \frac{3}{2} = m$;

$y\text{-intercept} = -\frac{1}{2} = b$

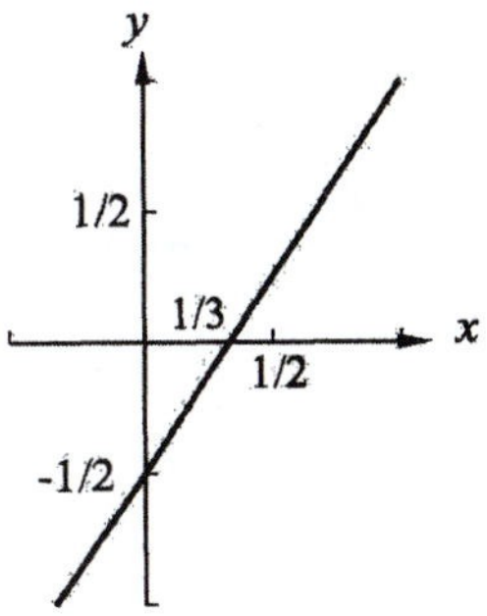

29. $3x-2y+5=0;\ -2y=-3x-5;$

$y=\frac{-3}{-2}x+\frac{-5}{-2};\ y=\frac{3}{2}x+\frac{5}{2};$

$\text{slope}=\frac{3}{2}=m_1$

$4y=6x-1;\ y=\frac{6}{4}x-\frac{1}{4};$

$y=\frac{3}{2}x-\frac{1}{4};\ \text{slope}=\frac{3}{2}=m_2$

$m_1=m_2$ for all parallel lines.

33. $5x+2y=3\Rightarrow y=\frac{-5}{2}\cdot x+\frac{3}{2}$

$10y=7-4x\Rightarrow y=\frac{-4}{10}x+\frac{7}{10}$

$m_1\cdot m_2=\frac{-5}{2}\cdot\frac{-4}{10}=1\neq -1$

$m_1\neq m_2$

Lines are neither perpendicular nor parallel.

37. Given: $4x-ky=6\,\|\,6x+3y+2=0$

$6x+3y+2=0;\ 3y=-6x-2$

$y=\frac{-6}{3}x-\frac{2}{3};\ y=-2x-\frac{2}{3};$ slope is -2

$4x-ky=6;\ -ky=-4x+6$

$y=\frac{-4}{-k}x+\frac{6}{-k};\ y=\frac{4}{k}x-\frac{6}{k};$ slope is $\frac{4}{k}$

Since the lines are parallel, the slopes are equal.

$\frac{4}{k}=-2;\ 4=-2k;\ k=-2$

41.

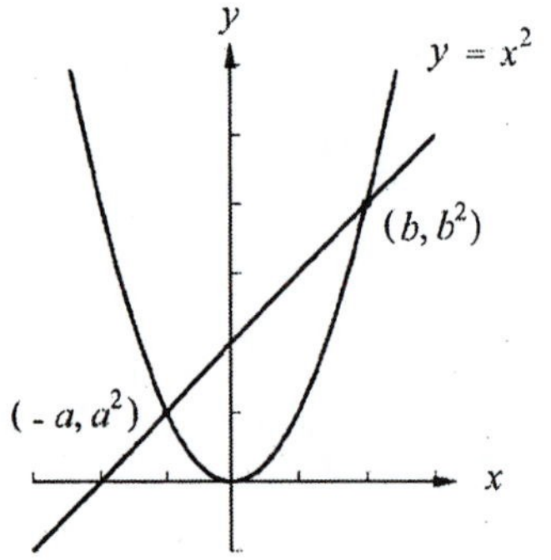

$m=\frac{b^2-a^2}{b-(-a)}$

$m=\frac{(b+a)(b-a)}{\cancel{b+a}}$

$m=b-a$

45. $8x+10y=3\Rightarrow y_1=-\frac{4}{5}x+\frac{3}{10}$

$2x-3y=5\Rightarrow y_2=\frac{2}{3}x-\frac{5}{3}$

$4x-6y=-3\Rightarrow y_3=\frac{2}{3}x+\frac{1}{2}$

$5y+4x=1\Rightarrow y_4=-\frac{4}{5}x+\frac{1}{5}$

$m_1=m_4=-\frac{4}{5}$ and $m_2=m_3=\frac{2}{3}$

showing the lines form a parallelogram.

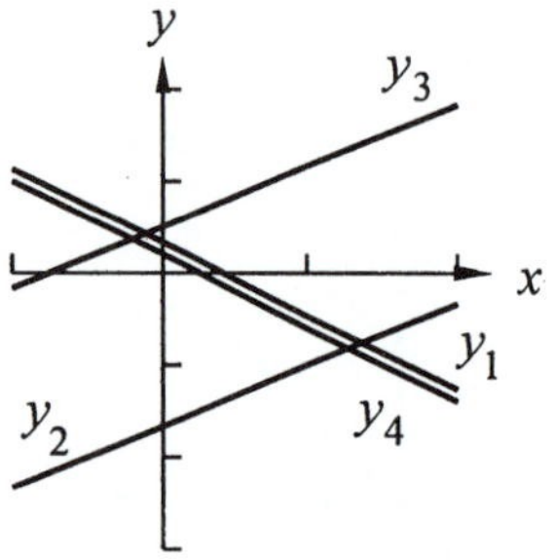

49. $\frac{F-32}{212-32}=\frac{R-0}{80-0}$

$F=\frac{9}{4}R+32$

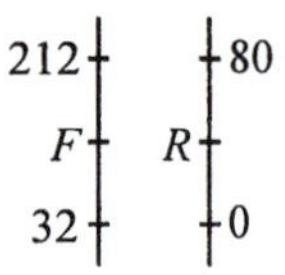

53. $x=0;\ T=3^\circ\text{C}$

$T=kx+T_1$

$3=0+T_1$

$T=kx+3$

$x=15\text{ cm};\ T=23^\circ\text{ C}$

$23^\circ=k(15)+3$

$k = \frac{20}{15} = \frac{4^{\circ}}{3}$ C/cm, slope = change in temperature in $^{\circ}$C per cm. Therefore, $T = \frac{4}{3}x + 3.$

57. $m = \tan\left(180^{\circ} - 0.0032^{\circ}\right)$

$b = 24\mu\text{m} = 24\times10^{-6}\text{m} = 2.4\times10^{-5}$ m

$m = -5.6\times10^{-5}$

$y = mx + b = -5.6\times10^{-5}x + 2.4\times10^{-5}$

$y = (-5.6x + 2.4)10^{-5}$

61. $n = 1200\sqrt{t} + 0$

$m = 1200$

$b = 0$

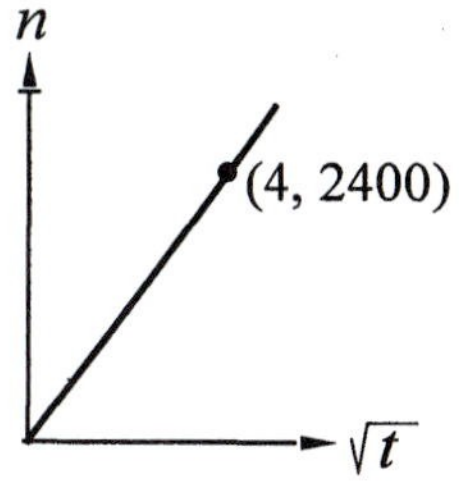

t	$\sqrt{t}$	h
0	0	0
1	1	1200
4	2	2400

65. Slope is found by measuring between points. The vertical displacement and the horizontal displacement between the extreme points is in a 1 to 2 ratio; $m = \frac{1}{2}$.

Since the graph is linear, the log equation is of the form $\log y = m\log x + \log a$, where a is the intercept $(1, a)$.

$y = ax^n$; $y = 3x^6$

$a = 3$, $n = 4$

x	y
1.0	3.0
1.1	4.4
1.2	6.2
1.3	8.6

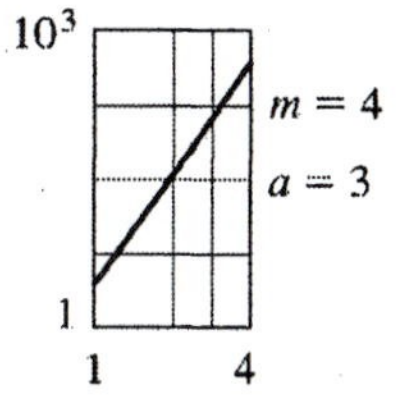

$\log y = \log a + n\log x$

$\log y = \log 3 + 3\log x$

Verify

(1) Slope is $\frac{\log y - \log a}{\log x} = 4.$

Vertical and horizontal measures in millimeters between points are shown. Each slope is 4.

(2) The intercept is $a = 3$.

The line crosses the vertical axis at $x = 1.0$, $y = 3.0$.

21.3 The Circle

1. $(x-1)^2 + (y+1)^2 = 16$ has center at $(1, -1)$ and $r = 4$

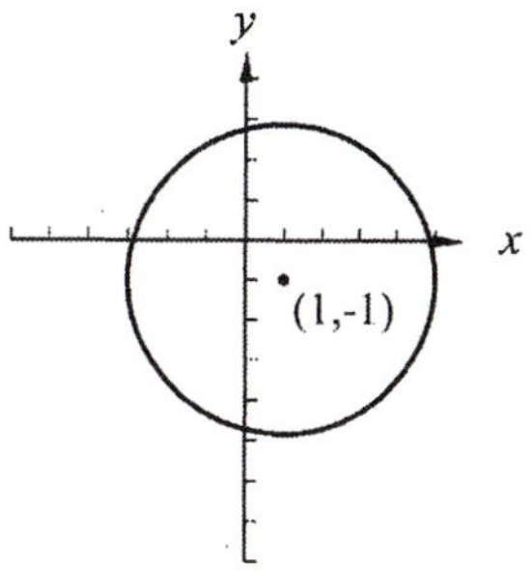

5. $(x-2)^2 + (y-1)^2 = 25$

$C(2, 1)$, radius is 5.

9. $(x-h)^2 + (y-k)^2 = r^2$; $C(0, 0)$, $r = 3$

$(x-0)^2 + (y-0)^2 = 3^2$, $x^2 + y^2 = 9$

13. $(x-12)^2 + (y-(-15))^2 = 18^2$; $C(12, -15)$, $r = 18$

$(x-12)^2 + (y+15)^2 = 324;$

$x^2 + y^2 - 24x + 30y + 45 = 0$

17. Concentric with $(x-2)^2 + (y-1)^2 = 4$ gives center at $(2, 1)$.

$r = \sqrt{(2-4)^2 + (1-(-1))^2} = 2\sqrt{2}$

The equation is $(x-2)^2 + (y-1)^2 = 8$

21. The center is $(-2, 2)$ and radius is 2.

$(x-h)^2+(y-k)^2=r^2$

$(x+2)^2+(y-2)^2=2^2$

$x^2+4x+4+y^2-4y+4=4$

$x^2+y^2+4x-4y+4=0$

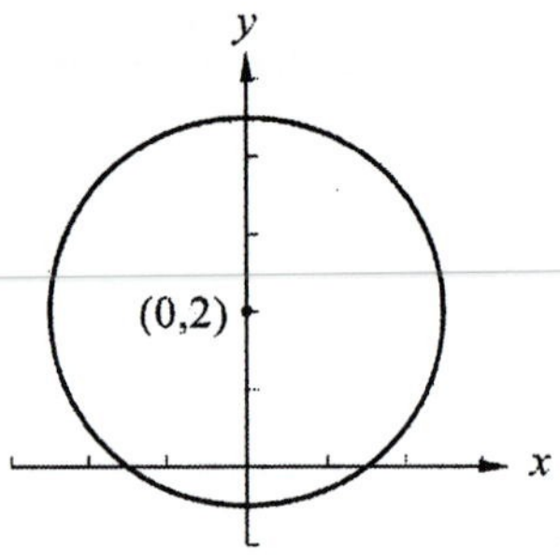

25. $x^2+(y-3)^2=4$ is the same as

$(x-0)^2+(y-3)^2=2^2$, so

Therefore, $h=0$, $k=3$, $r=2$

$C(0, 3)$

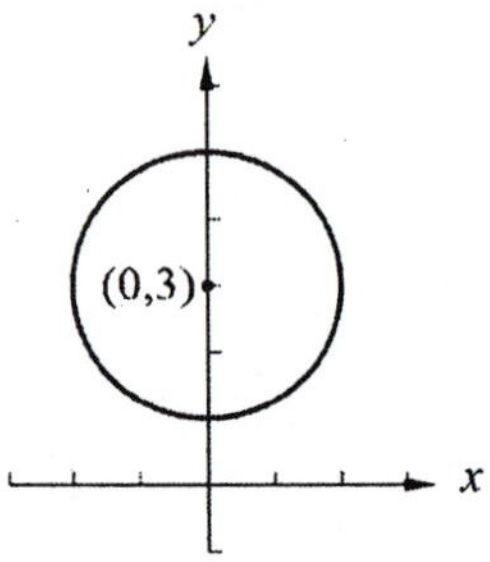

29. $x^2+y^2-2x-8=0$

$x^2-2x+1+y^2=9$

$(x-1)^2+(y-0)^2=9$

$h=1$, $k=0$, $r=3$

$C(1, 0)$

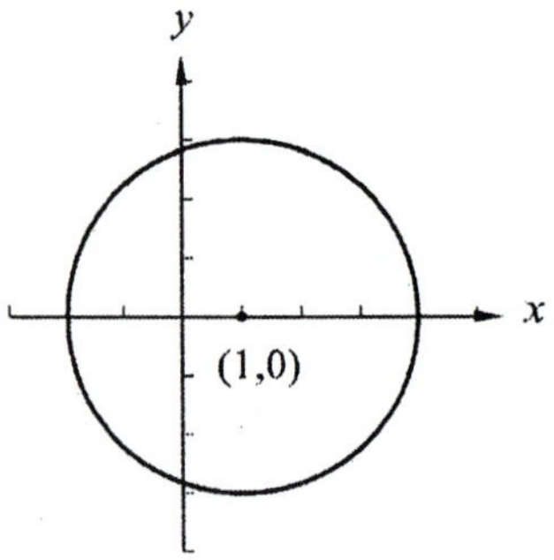

33. $4x^2+4y^2-16y=9$

$4(x-0)^2+4(y^2-4y+4)=9+16$

$(x-0)^2+(y-2)=\frac{25}{4}$

$h=0$, $k=2$, $r=\frac{5}{2}$

$C(0, 2)$

37. $(-x)^2+y^2=100;\ x^2+y^2=100$

Symmetrical to y-axis

$(-x)^2+(-y)^2=100;\ x^2+y^2=100$

Symmetrical to origin

41. Find all points for which $y=0$.

$x^2-6x+(0)^2-7=0;\ x^2-6x-7=0\Rightarrow$

$(x+1)(x-7)=0;\ x=-1$ or $x=7;$

$(-1, 0)$ and $(7, 0)$

45. $x^2+y^2+5y-4=0$

$y^2+5y+(x^2-4)=0$; solve for y

$$y=\frac{-5\pm\sqrt{5^2-4(x^2-4)}}{2}=\frac{-5\pm\sqrt{41-4x^2}}{2}$$

$$=-2.5\pm\sqrt{10.25-x^2}$$

Set the range for

$x_{min}=-6,\ x_{max}=6,\ y_{min}=-6,\ y_{max}=2$

$y_1=-2.5+\sqrt{10.25-x^2}$

$y_2=-2.5-\sqrt{10.25-x^2}$

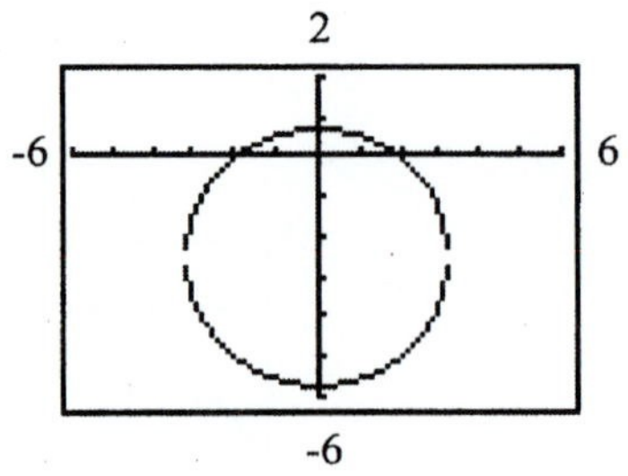

49. $x^2 + y^2 - 2x - 4y + 3 = 0$

$x^2 - 2x + 1 + y^2 - 4y + 4 = -3 + 1 + 4 = 2$

$(x-1)^2 + (y-2)^2 = 2 \Rightarrow C(1, 2),\ r = \sqrt{2}$

$\sqrt{(0.1-1)^2 + (3.1-2)^2} = 1.42126704 > \sqrt{2}$

$= 1.414213562 \Rightarrow$ the point is outside the circle.

53. $2.00x^2 + 2.00y^2 = 5.73 \Rightarrow x^2 + y^2 = \dfrac{5.73}{2.00}$

$2.80x^2 + 2.80y^2 = 8.91 \Rightarrow x^2 + y^2 = \dfrac{8.91}{2.80}$

thickness $= \sqrt{\dfrac{8.91}{2.80}} - \sqrt{\dfrac{5.73}{2.00}} = 0.0912$

57. 60 Hz = 60 cycles/s = 37.7 m/s; $(h, k) = (0, 0)$

60 cycles = 37.7 m; 1 cycle = 0.628 m

$r = 0.628 \text{ m} \div 2\pi$; r=0.10

$x^2 + y^2 = (0.10)^2$; $x^2 + y^2 = 0.0100$

61.

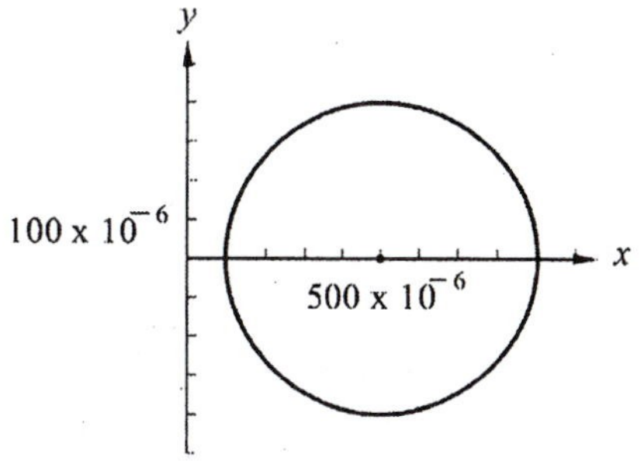

Center of circle is at $(500, 0)$.

Radius of circle is 300.

$(x-h)^2 + (y-k)^2 = r^2$

$(x-500)^2 + (y-0)^2 = 400^2$

$x^2 - 1000x + 25\times10^4 + y^2 = 16\times10^4$

$x^2 + y^2 - 1000x + 9\times10^4 = 0$

$x^2 + y^2 - 1000\times10^{-6} + 9\times10^{-8} = 0$

or $(x - 500\times10^{-6})^2 + y^2 = (400\times10^{-6})^2$

$= 0.16\times10^{-6}$

21.4 The Parabola

1. $y^2 = 20x \Rightarrow 4p = 20 \Rightarrow p = 5.\ F(5, 0)$;

directrix $x = -5$

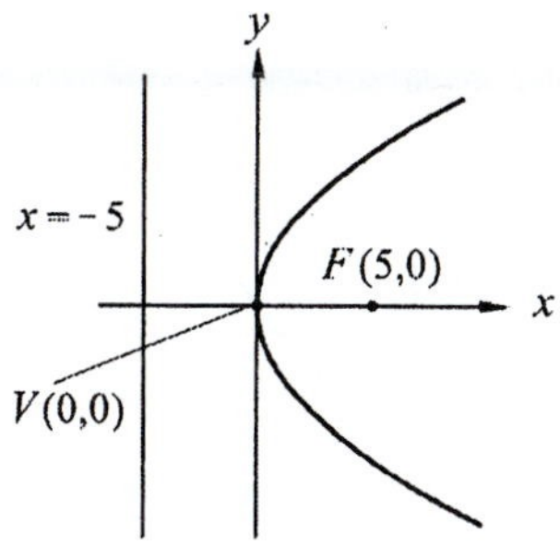

5. $y^2 = 4x$

$y^2 = 4px$

$y^2 = 4x = 4(1)x;\ p = 1$

$F(1, 0)$; directrix $x = -1$

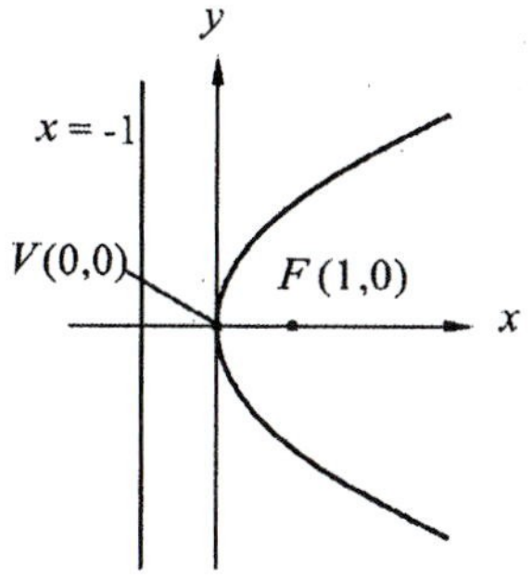

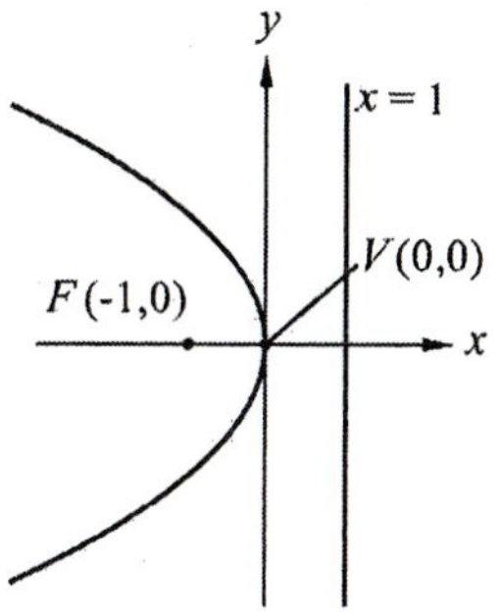

9. $x^2 = 72y$

$x^2 = 4py$

$x^2 = 72y = 4(18)y;\ p = 18$

$F(0, 18)$; directrix $y = -18$

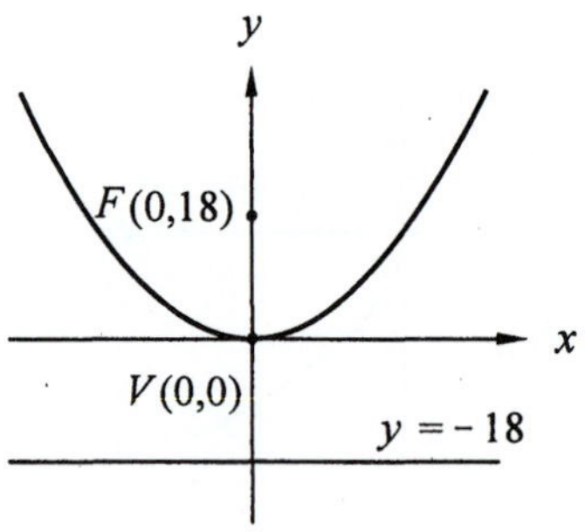

13. $2y^2 - 5x = 0$

$2y^2 = 5x$

$y^2 = \frac{5}{2}x = 4px$

$p = \frac{5}{8}$

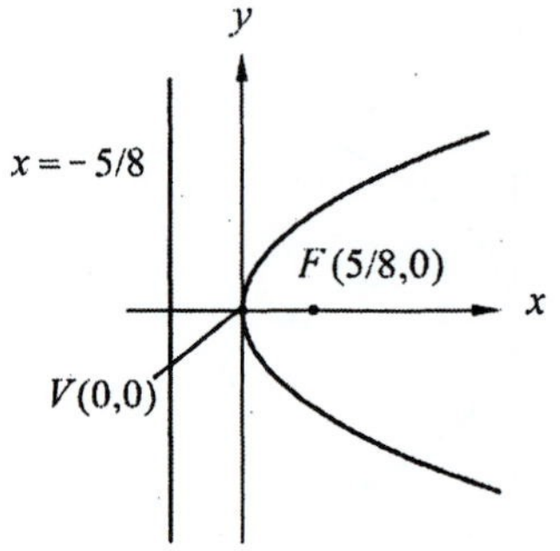

17. $F(3, 0)$; directrix $x = -3,\ p = 3$

$y^2 = 4px$

$y^2 = 4(3)x;$

$y^2 = 12x$

21. $V(0, 0)$, directrix $y = -0.16$

$F(0, 0.16),\ p = 0.16$

$x^2 = 4py = 4(0.16)y$

$x^2 = 0.64y$

25. $V(0, 0)$

Therefore, $x^2 = 4py$

$(-1)^2 = 4p(8);\ 1 = 32p;\ p = \frac{1}{32}$

Therefore, $x^2 = \frac{1}{8}y.$

29. $y^2 = 4px \Rightarrow 3^2 = 4p(3) \Rightarrow p = \frac{3}{4}$

$y^2 = 4\left(\frac{3}{4}\right)x$

$y^2 = 3x$

33. $y^2 = 2x \Rightarrow x = \frac{y^2}{2},\ x^2 = -16y$

$\left(\frac{y^2}{2}\right)^2 = -16y$

$y^4 + 64y = 0$

$y(y^3 + 64) = 0$

$y = 0$ or $y^3 + 64 = 0 \Rightarrow y^3 = -64$

$2x = 0^2,\ x = 0$ $\qquad y = -4$

$(-4)^2 = 2x \Rightarrow x = 8$

The points of intersection are $(0, 0)$ and $(8, -4)$.

37. $y^2 + 2x + 8y + 13 = 0$; solve for y

$y^2 + 8y + (2x + 13) = 0$

$y = \frac{-8 \pm \sqrt{8^2 - 4(2x + 13)}}{2}$

$y = \frac{-8 \pm \sqrt{12 - 8x}}{2}$

$y_1 = -4 + \sqrt{3 - 2x},\ y_2 = -4 - \sqrt{3 - 2x}$

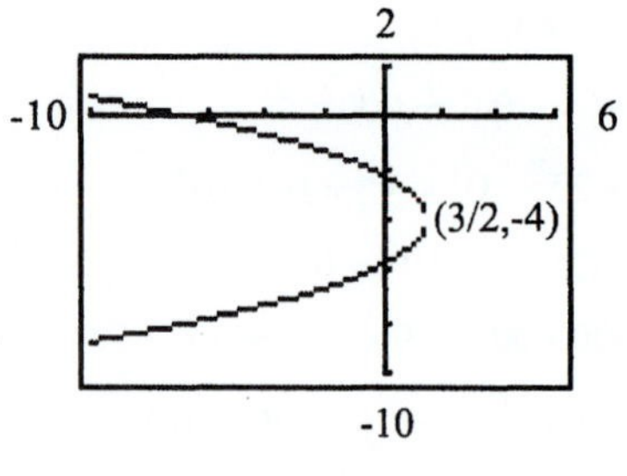

41. $y^2 = 4px$

When $x = p$; $y^2 = 4p^2$; $y = 2p$

Therefore, latus rectum intersects parabola at $(p, 2p)$. Therefore, length of latus rectum is $2(2p) = 4p$.

45. Let the vertex of the parabola be at the origin. $x^2 = 4py$. A point on the parabola will be $(2100, 300)$. Substitute this into the equation and solve for p.

$2100^2 = 4p(300)$; $1200p = 4,410,000$; $p = 3675$

$x^2 = 4(3675)y = 14,700y$

49. $y^2 = 4px$

$(1.20)^2 = 4p(0.00625)$

$p = 57.6$ m

53. $y^2 = 4px$

$15.0^2 = 4p(6.5)$

$p = 8.65$

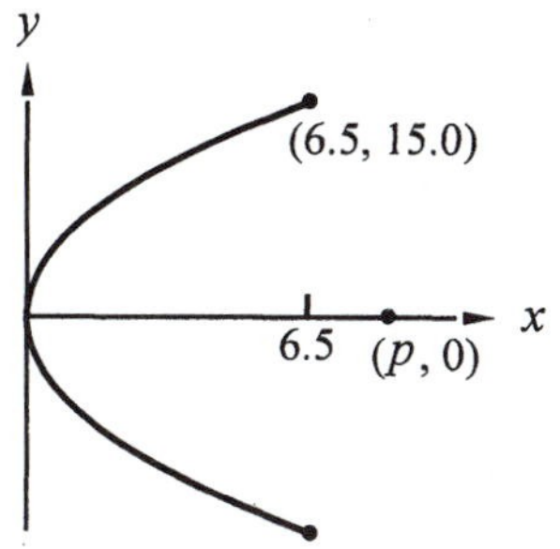

57. Path of ship channel is a parabola with focus at $(0, -2)$ and vertex $(0, 0)$ and directrix $y = 2$. Therefore, $p = -2$. Parabola of the type $x^2 = -4py$; therefore, $x^2 = -8y$. If positions of island and shoreline are interchanged, the equation would be $x^2 = 8y$. If a parabola of the form $y^2 = 4px$ is assumed, the path would be $y^2 = 8x$ or $y^2 = -8x$.

21.5 The Ellipse

1. $\frac{x^2}{25} + \frac{y^2}{36} = 1$,

$a^2 = 36$, $a = 6$,

$b^2 = 25$, $b = 5$

$V(0, \pm 6)$, minor axis: $(\pm 5, 0)$

$a^2 = b^2 + c^2$

$36 = 25 + c^2$

$c = \sqrt{11}$

$F\left(0, \pm\sqrt{11}\right)$

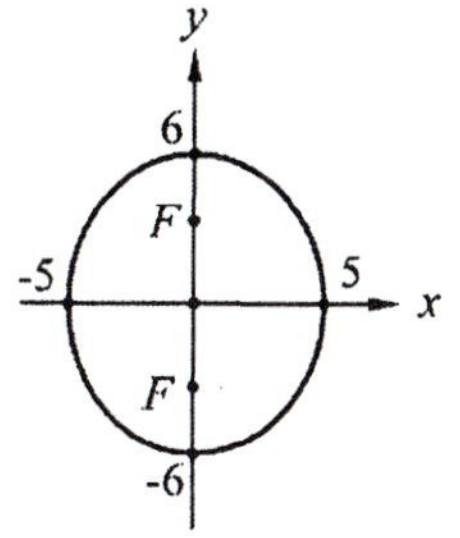

5. $\frac{x^2}{25} + \frac{y^2}{144} = 1$

$a^2 = 144$, $a = 12$,

$b^2 = 25$, $b = 5$

$V(0, \pm 12)$, minor axis: $(\pm 5, 0)$

$a^2 = b^2 + c^2$

$144 = 25 + c^2$

$c = \sqrt{119} \approx 10.9$

$F\left(0, \pm\sqrt{119}\right)$

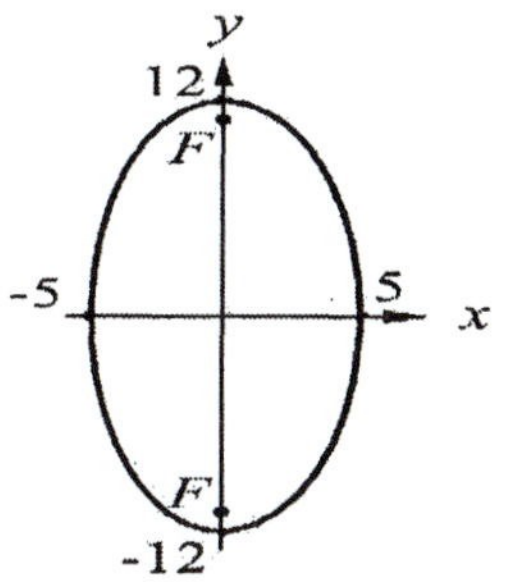

$V(0, \pm 6)$, $F\left(0, \pm\sqrt{11}\right)$, x-intercepts $(\pm 5, 0)$

9. $4x^2 + 9y^2 = 36$

$\dfrac{x^2}{36} + \dfrac{y^2}{36} = 1$

$\dfrac{x^2}{9} + \dfrac{y^2}{4} = 1$

$a^2 = 9,\ b^2 = 4$

$c^2 = 9 - 4 = 5,\ c = \sqrt{5}$

$V(\pm 3,\ 0),\ F\left(\pm\sqrt{5},\ 0\right),$

y-intercepts $(0,\ \pm 2)$

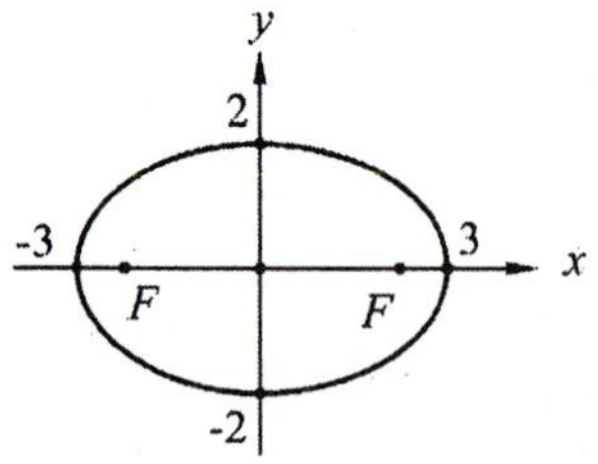

$V(\pm 12,\ 0),\ F\left(\pm 2\sqrt{35},\ 0\right),$ y-intercepts $(0,\ \pm 2)$

13. $y^2 = 8\left(2 - x^2\right)$

$8x^2 + y^2 = -16$

$\dfrac{8x^2}{16} + \dfrac{y^2}{16} = 1$

$\dfrac{x^2}{2} + \dfrac{y^2}{16} = 1$

$\dfrac{y^2}{16} + \dfrac{x^2}{2} = 1$

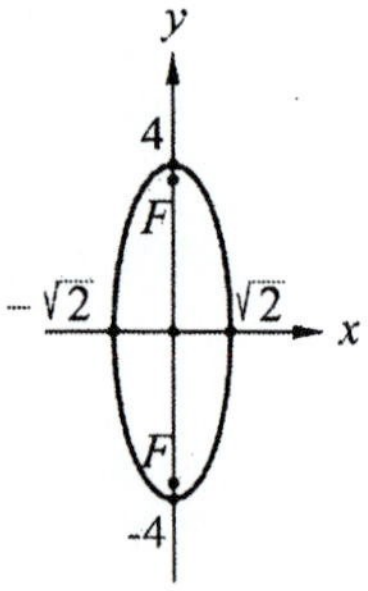

$a^2 = 16,\ b^2 = 2,\ c^2 = 16 - 2 = 14$

$V(0,\ \pm 4),\ F\left(0,\ \pm\sqrt{14}\right),$ x-intercepts $\left(\pm\sqrt{2},\ 0\right)$

17. $V(15,\ 0);\ F(9,\ 0)$

$a = 15,\ a^2 = 225;$

$c = 9,\ c^2 = 81;\ a^2 - c^2 = b^2$

$b^2 = 144;\ \dfrac{x^2}{a^2} + \dfrac{y^2}{b^2} = 1;$

$\dfrac{x^2}{225} + \dfrac{y^2}{144} = 1$

$144x^2 + 225y^2 = 32{,}400$

21. $F(8,\ 0) \Rightarrow c = 8$

end of minor axis: $(0,\ 12) \Rightarrow b = 12$

$a^2 = b^2 + c^2 \Rightarrow a^2 = 12^2 + 8^2 \Rightarrow a = \sqrt{208}$

$\dfrac{x^2}{\sqrt{208}^2} + \dfrac{y^2}{12^2} = 1$

$\dfrac{x^2}{208} + \dfrac{y^2}{144} = 1$

$\dfrac{x^2}{64} + \dfrac{15y^2}{144} = 1;\ 144x^2 + 960y^2 = 9216$

$3x^2 + 20y^2 = 192$

25. $(x_1,\ y_1) = (2,\ 2),\ (x_2,\ y_2) = (1,\ 4)$

$\dfrac{x^2}{b^2} + \dfrac{y^2}{a^2} = 1$

Substitute: $\dfrac{4}{b^2} + \dfrac{4}{a^2} = 1$

Therefore, $4a^2 + 4b^2 = a^2b^2$

$\dfrac{1}{b^2} + \dfrac{16}{a^2} = 1$

Therefore, $a^2 + 16b^2 = a^2b^2$

$a^2 + 16b^2 = a^2b^2$

$16b^2 = a^2b^2 - a^2 = a^2\left(b^2 - 1\right)$

Therefore, $a^2 = \dfrac{16b^2}{b^2 - 1}$

Substitute:

$4a^2 + 4b^2 = a^2b^2;\ \dfrac{64b^2}{b^2 - 1} + 4b^2 = \dfrac{16b^4}{b^2 - 1}$

$64b^2 + 4b^4 - 4b^2 = 16b^4;\ -12b^4 + 60b^2 = 0$

$12b^2\left(-b^2 + 5\right) = 0$

$b^2 = 5$

Therefore, $a^2 = \dfrac{16(5)}{4} = 20$

Therefore, $\dfrac{y^2}{20} + \dfrac{x^2}{5} = 1$

or: $5y^2 + 20x^2 = 100;\ 4x^2 + y^2 = 20$

29. $4x^2 + 9y^2 = 40, \quad y^2 = 4x$

$4x^2 + 9(4x) = 40$

$x^2 + 9x - 10 = 0$

$(x+10)(x-1) = 0$

$x = -10$ or $x = 1$

$y^2 = 4(-10)$ $\quad$ $y^2 = 4(1)$

$y^2 = -40$, no solution $\quad$ $y = \pm 2$

Graphs intersect at $(1, 2), (1, -2)$

33. $4x^2 + 3y^2 + 16x - 18y + 31 = 0$; solve for y

$3y^2 - 18y + \left(4x^2 + 16x + 31\right) = 0$

$$y = \frac{18 \pm \sqrt{(-18)^2 - 4(3)\left(4x^2 + 16x + 31\right)}}{2(3)}$$

$$= \frac{18 + \sqrt{-48x^2 - 192x - 48}}{2}$$

$$y_1 = 3 + \frac{\sqrt{-12x^2 - 48x - 12}}{3}$$

$$y_2 = 3 - \frac{\sqrt{-12x^2 - 48x - 12}}{3}$$

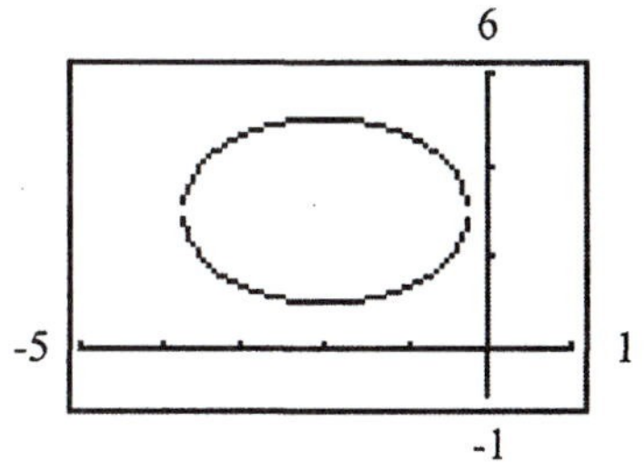

37. $x^2 + y^2 = 1,\ k > 0$

Therefore, $\dfrac{x^2}{1} + \dfrac{y^2}{\frac{1}{k}} = 1$

$\dfrac{1}{k} = a^2 = 1$

Therefore, $\sqrt{\dfrac{1}{k}} = a > 1 \Rightarrow \dfrac{1}{k} > 1$

Therefore, $k < 1$

41. $100x^2 + 49y^2 < 4900$

$100x^2 + 49y^2 = 4900$

$\dfrac{x^2}{49} + \dfrac{y^2}{100} = 1$ is the elliptical dashed boundary

Test point $(0, 0)$

$100(0)^2 + 49(0)^2 < 4900$

$0 < 4900,\ T$

Shade inside ellipse

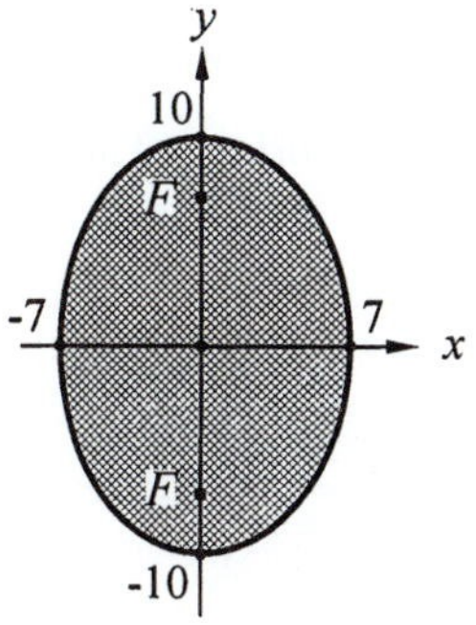

45. $P = Ri^2$

$P_T = R_1 i_1^2 + R_2 i_2^2$

$64 = 2i_1^2 + 8i_2^2$

$\dfrac{2}{64} i_1^2 + \dfrac{8}{64} i_2^2 = 1$

$\dfrac{i_1^2}{32} + \dfrac{i_2^2}{8} = 1$

$i_1^2 + 4i_2^2 = 32$

$a^2 = 32;\ a = \sqrt{32} \approx 5.7$

$b^2 = 8;\ b = \sqrt{8} \approx 2.8$

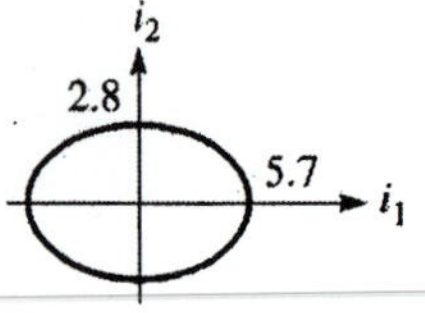

49. If the two vertices of each base are fixed at $(-3, 0)$ and $(3, 0)$, and the sum of the two leg lengths is also fixed, the third vertex lies on an ellipse. The base is 6 cm, so

$d_1 + d_2 = 14 \text{ cm} - 6 \text{ cm} = 8 \text{ cm}$

$(-3, 0)$ and $(3, 0)$ are foci $(-c, 0)$ and $(c, 0)$

$d_1 + d_2 = 2a = 8;\ a = 4$

$a^2 - c^2 = b^2$

$4^2 - 3^2 = b^2$

$b^2 = 7,\ a^2 = 16$

The equation is $\frac{x^2}{16} + \frac{y^2}{7} = 1,\ 7x^2 + 16y^2 = 112$

53. $a = \frac{64}{2} = 32,\ b = 18$

Let the center be at the origin. The equation of the ellipse is

$\frac{x^2}{32^2} + \frac{y^2}{18^2} = 1;\ \frac{x^2}{1024} + \frac{y^2}{324} = 1$

If $x = 22,\ \frac{22^2}{1024} + \frac{y^2}{324} = 1$

$y = 13$ ft

21.6 The Hyperbola

1. $\frac{y^2}{16} - \frac{x^2}{4} = 1$

$a^2 = 16,\ a = 4$

$b^2 = 4,\ b = 2$

$c^2 = a^2 + b^2 = 20$

$c = 2\sqrt{5}$

$V(0, \pm 4)$

conjugate axis: $(\pm 2, 0)$

$F(0, \pm 2\sqrt{5})$

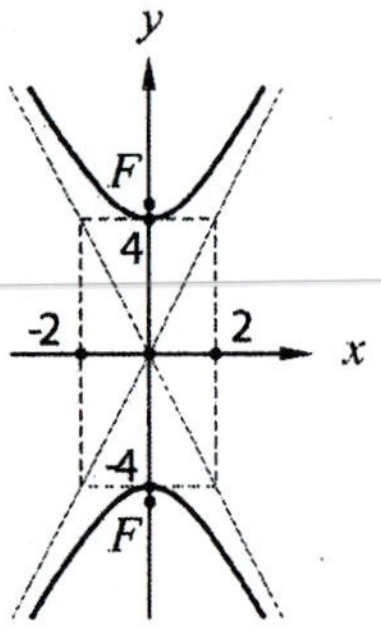

5. $\frac{y^2}{9} - \frac{x^2}{1} = 1$

$a^2 = 9,\ a = 3$

$b^2 = 1,\ b = 1$

$c^2 = 10;\ c = \sqrt{10}$

$V(0, \pm 3),\ F(0, \pm\sqrt{10})$

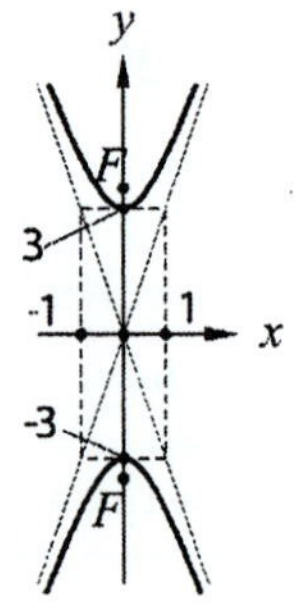

9. $4x^2 - y^2 = 4$

$\frac{x^2}{4} - \frac{y^2}{4} = 1;$

$\frac{x^2}{1} - \frac{y^2}{4} = 1$

$a^2 = 1;\ b^2 = 4;\ c^2 = 5$

$V(\pm 1, 0),\ F(\pm\sqrt{5}, 0)$

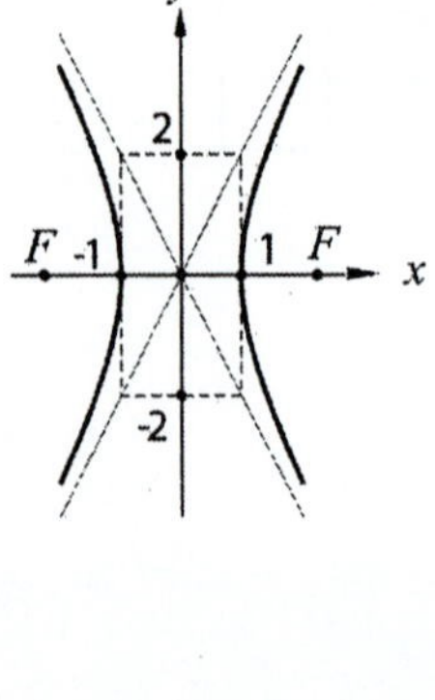

13. $y^2 = 4(x^2 + 1)$

$4x^2 - y^2 + 4 = 0$

$4x^2 - y^2 = -4$

$\frac{4x^2}{-4} - \frac{y^2}{-4} = \frac{-4}{-4}$

$-x^2 + \frac{y^2}{4} = 1$

$\frac{y^2}{4} - \frac{x^2}{1} = 1$

$a^2 = 4;\ b^2 = 1;\ c^2 = 5$

$V(0, \pm 2),\ F(0, \pm\sqrt{5})$

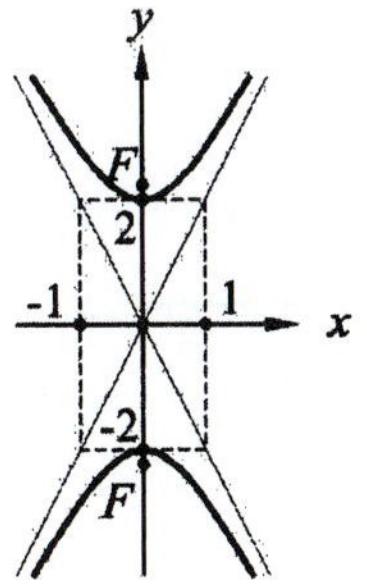

17. $V(3, 0);\ F(5, 0)$

$a = 3;\ c = 5;\ a^2 = 9;\ c^2 = 25$

$b^2 = c^2 - a^2 = 25 - 9 = 16$

$\frac{x^2}{a^2} - \frac{y^2}{b^2} = 1;\ \frac{x^2}{9} - \frac{y^2}{16} = 1;$

$16x^2 - 9y^2 = 144$

21. (x, y) is $(2, 3)$; $F(2, 0), (-2, 0)$; $c = \pm 2,\ c^2 = 4$

$d_1 = \sqrt{(2-(-2))^2 + (3-0)^2}$

$= \sqrt{4^2 + 3^2} = \sqrt{16+9}$

$= \sqrt{25} = 5$

$d_2 = \sqrt{(2-2)^2 + (3-0)^2}$

$= \sqrt{0+9} = \sqrt{9} = 3$

$d_1 - d_2 = 2a;\ 5 - 3 = 2a;\ 2 = 2a;\ 1 = a;\ a^2 = 1$

$c^2 = 4;\ b^2 = c^2 - a^2 = 3$

$\frac{x^2}{a^2} - \frac{y^2}{b^2} = 1;\ \frac{x^2}{1} - \frac{y^2}{3} = 1$

$3x^2 - y^2 = 3$

25. $V(1, 0) \Rightarrow a = 1,\ a^2 = 1$

Asymptote $y = \frac{b}{a}x = \frac{b}{1}x = 2x \Rightarrow b = 2,\ b^2 = 4$

$\frac{x^2}{1} - \frac{y^2}{4} = 1$

29. $xy = 2;\ y = \frac{2}{x}$

x	y
$\pm\frac{1}{2}$	± 4
± 1	± 2
± 2	± 1
± 4	$\pm\frac{1}{2}$
± 8	$\pm\frac{1}{4}$

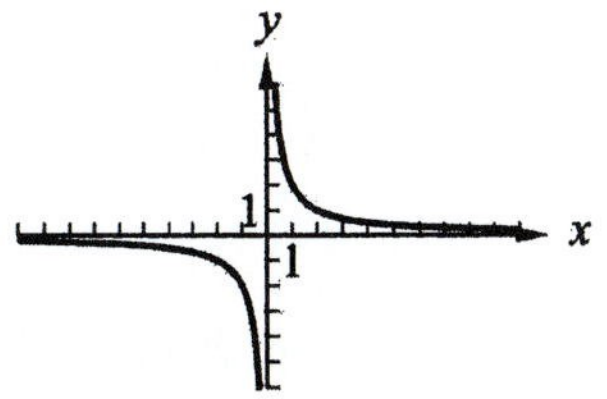

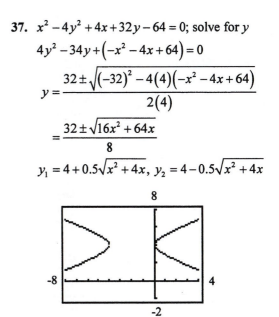

33. $2x^2 + y^2 = 17,\ y^2 - x^2 = 5 \Rightarrow y^2 = x^2 + 5$

$2x^2 + x^2 + 5 = 17$

$3x^2 = 12$

$x^2 = 4$

$x = \pm 2$

$y^3 = x^2 + 5 = 4 + 5 = 9 \Rightarrow 9 = \pm 3$

points of intersection $(2, \pm 3), (-2, \pm 3)$

37. $x^2 - 4y^2 + 4x + 32y - 64 = 0$; solve for y

$4y^2 - 34y + (-x^2 - 4x + 64) = 0$

$y = \frac{32 \pm \sqrt{(-32)^2 - 4(4)(-x^2 - 4x + 64)}}{2(4)}$

$= \frac{32 \pm \sqrt{16x^2 + 64x}}{8}$

$y_1 = 4 + 0.5\sqrt{x^2 + 4x},\ y_2 = 4 - 0.5\sqrt{x^2 + 4x}$

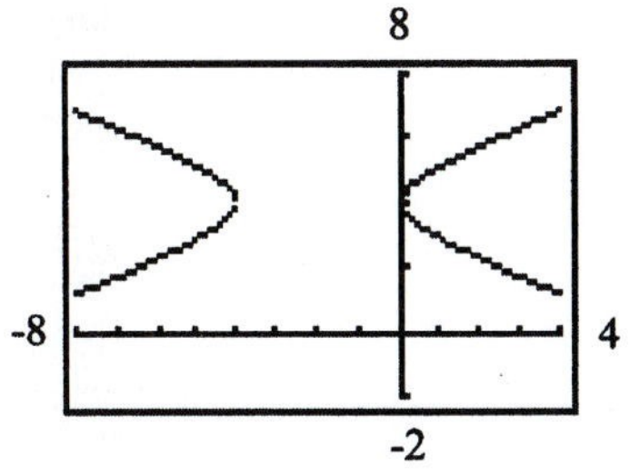

41. $V(0, 1)$, $F(0, \sqrt{3})$; $c^2 = a^2 + b^2$ where $c = \sqrt{3}$ and

$a = 1$; $b^2 = \sqrt{3}^2 - 1^2 = 2$

$$\frac{y^2}{1^2} - \frac{x^2}{\sqrt{2}^2} = 1$$

The transverse axis of the first equation is length $2a = 2\sqrt{1}$ along the y-axis. Its conjugate axis is length $2b = 2\sqrt{2}$ along the x-axis.

The transverse axis of the conjugate hyperbola is length $2\sqrt{2}$ along the x-axis, and its conjugate axis is length $2\sqrt{1}$ along the y-axis.

The equation, then, is $\frac{x^2}{\sqrt{2}^2} - \frac{y^2}{\sqrt{1}^2} = 1$

$\frac{x^2}{2} - \frac{y^2}{1} = 1$ or $x^2 - 2y^2 = 2$

45. $\frac{x^2}{169} + \frac{y^2}{144} = 1 \Rightarrow \frac{x^2}{13^3} + \frac{y^2}{12^2} = 1$

$13^3 = 12^2 + c^2 \Rightarrow c = 5$

$\frac{x^2}{a^2} - \frac{y^2}{b^2} = 1,$

$c^2 = a^2 + b^2 \Rightarrow 5^2 = a^2 + b^2, 0 < a < 5$

$b^2 = 25 - a^2$

$$\frac{(4\sqrt{2})^2}{a^2} - \frac{9}{25 - a^2} = 1$$

$a^4 - 66a^2 + 800 = 1$

$(a^2 - 16)(a^2 - 50) = 0$

$a^4 - 16 = 0$ or $a^2 - 50 = 0$

$a = 4$ $\quad a = \sqrt{50} > 5$, reject

$b^2 = 25 - 4^2 = 9 \Rightarrow b = 3$

$\frac{x^2}{16} - \frac{y^2}{9} = 1$

49. $600 = vy$

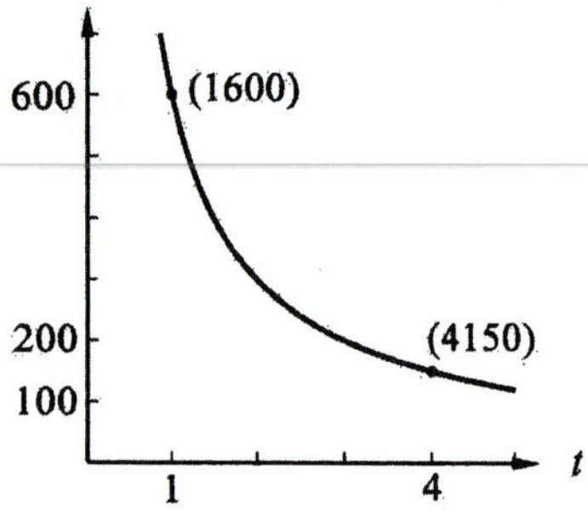

53.

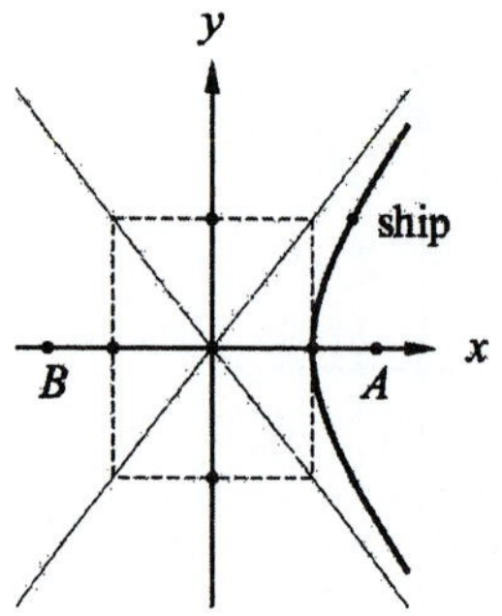

$d_1 - d_2 =$ constant

Let t_1 be time for signal to go from B to ship.

The t_2 is time for signal to go from A to ship.

Where $t_2 = t_1 - 1.20$ ms

$v = \frac{s}{t}$, therefore, $s = vt$

Therefore, $d_1 = 300t_1$ and $d_2 = 300(t_1 - 1.20)$

$d_1 - d_2 = 2a$

$300t_1 - 300(t_1 - 1.20) = \text{constant} = 360 \text{ km} = 2a$

Therefore, the ship could lie anywhere on the hyperbolic arc sketched.

Foci at $(\pm 300, 0)$, therefore $c = 300$

Vertices at $(\pm 180, 0)$, therefore $a = 180$; therefore $b = 240$

21.7 Translation of Axes

1. $\dfrac{(x-3)^2}{25}-\dfrac{(y-2)^2}{9}=1$, hyperbola: $a=5$, $b=3$

Center: $(3, 2)$. Transverse axis parallel to x-axis.

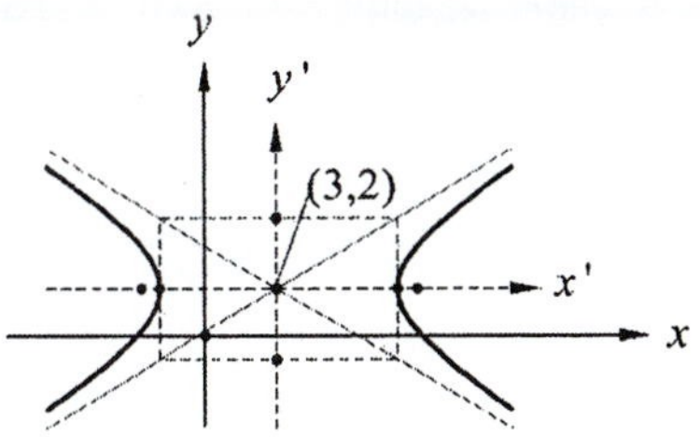

5. $\dfrac{(x-1)^2}{4}-\dfrac{(y-2)^2}{9}=1$; eq. (21-32), hyperbola

Center: $(1, 2)$; $a=2$; $b=3$

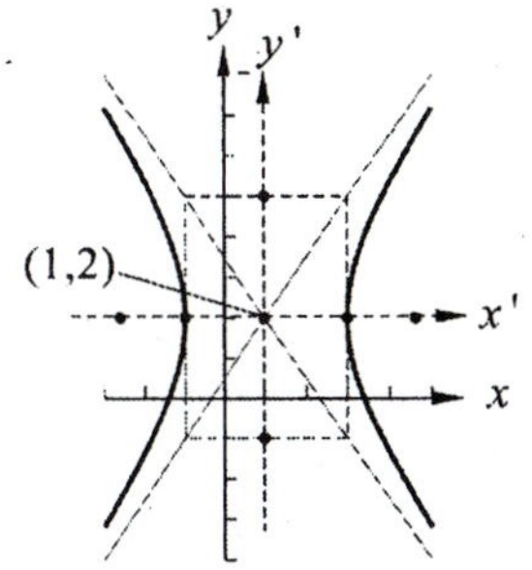

9. $(x+3)^2=-12(y-1)$; eq. (21-29), parabola

$x'=x+3$; $y'=y-1$

$x'^2=-12y'$

Origin O' at $(h, k)=(-3, 1)$

$x'^2=4(3y')$; therefore $p=3$

Vertex $(-3, 1)$, focus $(-3, -2)$, directrix $y=4$

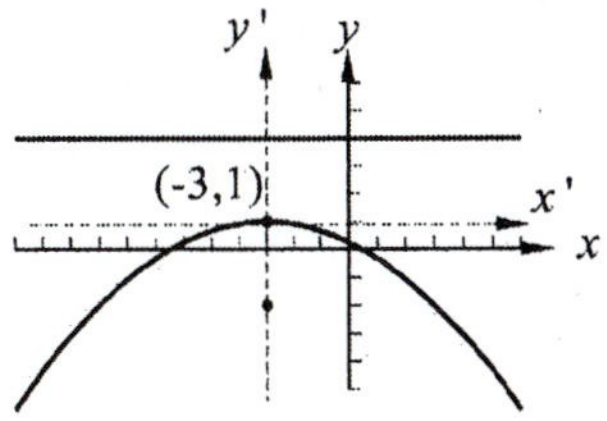

13. $F(12, 0)$, $V(6, 0)$, $p=6$

$(y-k)^2=4p(x-h)^2$

$(y-0)^2=4\cdot 6(x-6)$

$y^2=24(x-6)$

17. Ellipse: center $(-2, 1)$, vertex $(-2, 5)$, passes through $(0, 1)$.

$\dfrac{(y-1)^2}{4^2}+\dfrac{(x+2)^2}{b^2}=1$

$\dfrac{(1-1)^2}{4^2}+\dfrac{(0+2)^2}{b^2}=1$

$b^2=4$

$\dfrac{(y-1)^2}{16}+\dfrac{(x+2)^2}{4}=1$

21. Hyperbola: $V(2,1)$, $V(-4, 1)$, $F(-6, 1)$

Center: $(h, k)=(-1, 1)$

$2a=6$; $a=3$; $c=5$

Therefore, $b^2=c^2-a^2=25-9=16$

Transverse axis parallel to x-axis.

Therefore, $\dfrac{(x-h)^2}{a^2}-\dfrac{(y-k)^2}{b^2}=1$

$\dfrac{(x+1)^2}{9}-\dfrac{(y-1)^2}{16}=1$

or $16x^2-9y^2+32x+18y-137=0$

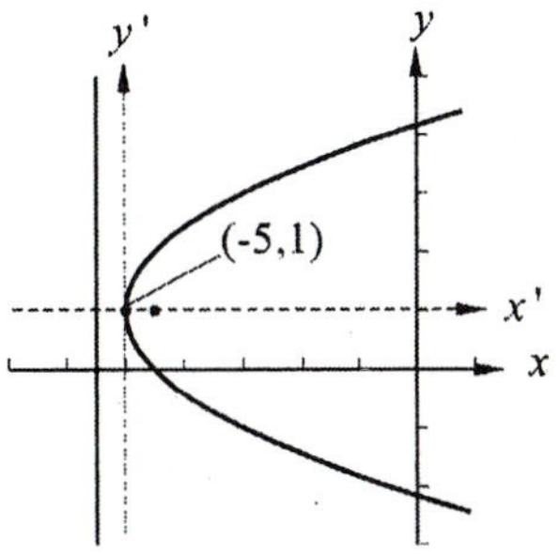

25. $x^2+4y=24$

$x^2=-4y+24$

$x^2=-4(y-6)$ is a parabola with $V(0, 6)$

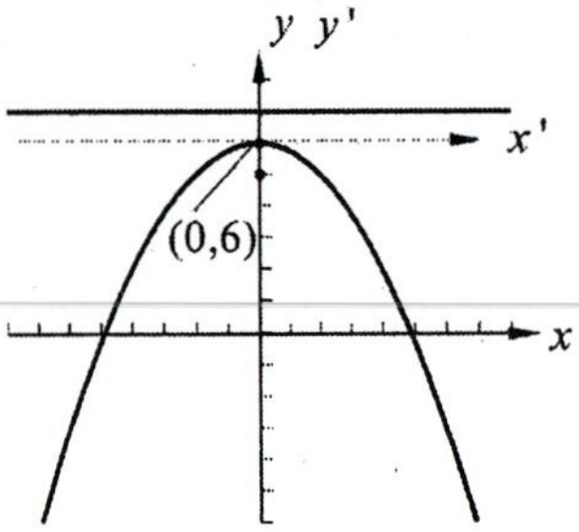

29. $9x^2 - y^2 + 8y = 7$

$9x^2 - y^2 + 8y - 7 = 0$

$9x^2 - (y^2 - 8y + 7 + 9) = -9$

$9x^2 - (y-4)^2 = -9$

$\dfrac{9x^2}{-9} + \dfrac{(y-4)^2}{-9} = 1$

$-x^2 + \dfrac{(y-4)^2}{9} = 1$

$\dfrac{(y-4)^2}{9} - \dfrac{x^2}{1} = 1$

Hyperbola, (h, k) is $(0, 4)$; $a = 3$; $b = 1$

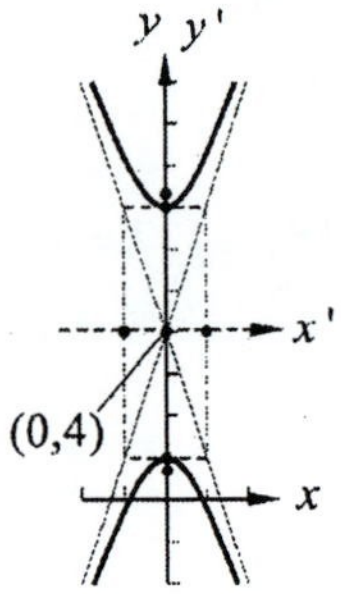

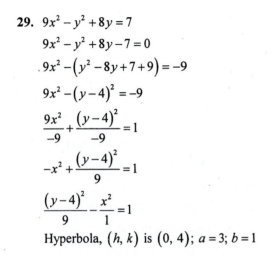

33. $4x^2 - y^2 + 32x + 10y + 35 = 0$

$4(x^2 + 8x) - (y^2 - 10y) = -35$

$4(x^2 + 8x + 16) - (y^2 - 10y + 25) = -35 + 64 - 25$

$\dfrac{(x+4)^2}{1^2} - \dfrac{(y-5)^2}{2^2} = 1$, hyperbola

$C(-4, 5)$

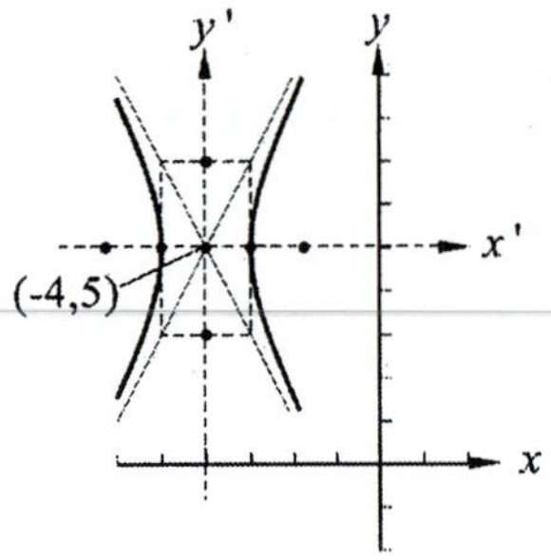

37. $5x^2 - 3y^2 + 95 = 40x$

$5(x^2 - 8x + 16) - 3y^2 = -95 + 80 = -15$

$\dfrac{y^2}{5} - \dfrac{(x-4)^2}{3} = 1$, hyperbola, center $(4, 0)$

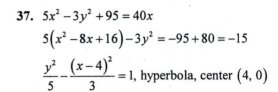

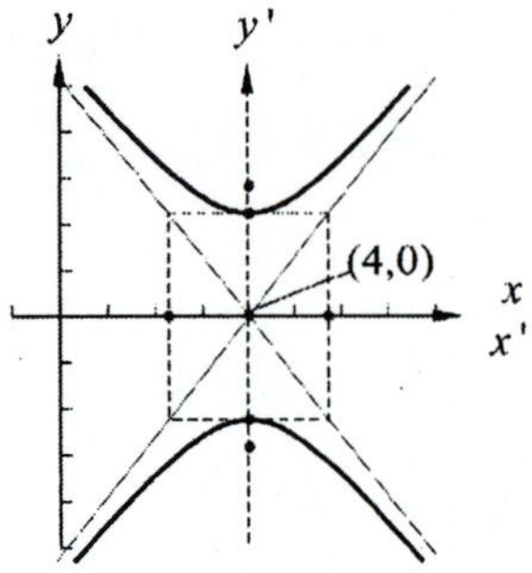

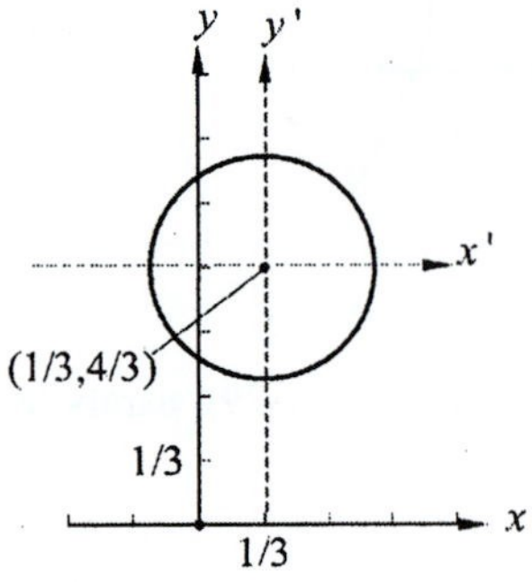

41. Hyperbola: asymptotes: $x - y = -1$ or $x + 1 = y$, and $x + y = -3$ or $y = -x - 3$; vertices $(3, -1)$ and $(-7, -1)$. The center is at the point of interaction of the asymptotes. The equation for the asymptotes are solved simultaneously by adding, $2y = -2$; $y = -1$; $-1 = x + 1$; $x = -2$. Therefore, the coordinates of the center are $(-2, -1)$. Since the slopes are 1 and -1, $a=b$, where a is the distance from the center $(-2,-1)$ to the vertex $(3,-1)$; $a=5$, $b=5$.

$$\frac{(x-h)^2}{a^2}-\frac{(y-k)^2}{b^2}=1;$$

$$\frac{[x-(-2)]^2}{25}-\frac{[y-(-1)]^2}{25}=1$$

$$\frac{(x+2)^2}{25}-\frac{(y+1)^2}{25}=1$$

$$x^2+4x+4-(y^2+2y+1)=25;$$

$$x^2+4x+4-2y-1=25$$

$$x^2-y^2+4x-2y-22=0$$

45. Parabola: vertex and focus on x-axis.

$$y^2=4p(x-h)$$

49. $(x-h)^2=4p(y-k)$

$$(x-95)^2=4p(y-60)$$

Solve for $4p$ using $(x, y)=(0, 0)$

$$(-95)^2=4p(-60)$$

$$4p=\frac{95^2}{-60}$$

$$(x-95)^2=\frac{95^2}{-60}(y-60)$$

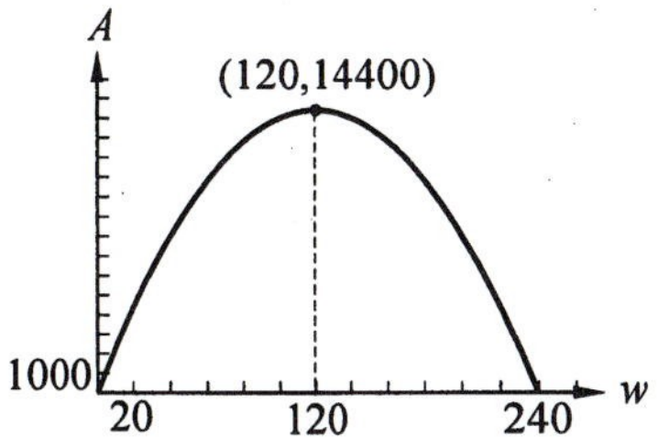

21.8 The Second-Degree Equation

1. $2x^2=3+2y^2$

$2x^2-2y^2-3=0$, A, C have different signs, $B=0$, hyperbola

5. $2x^2-y^2-1=0$

A and C have different signs, $B=0$; hyperbola

9. $2.2x^2-x-y=1.6$

$A\neq 0$; $C=0$; $B=0$; parabola

13. $36x^2=12y(1-3y)+1$

$$36x^2=12y-36y^2+1$$

$$36x^2+0\cdot xy+36y^2-12x-0\cdot y-1=0,$$

$A=C$, $B=0$, circle

17. $2xy+x-3y=6$

$A=0$; $B\neq 0$; $C=0$; hyperbola

21. $(x+1)^2+(y+1)^2=2(x+y+1)$

$$x^2+2x+1+y^2+2y+1=2x+2y+2$$

$x^2+y^2=0$, point $(0, 0)$ is the only solution

25. $x^2=8(y-x-2)$

$$x^2=8y-8x-16$$

$$x^2+8x-8y+16=0$$

$A\neq 0$; $B=0$;

$C=0$; parabola

$$x^2+8x-8y+16=0$$

$$x^2+8x+16=8y$$

$(x+4)^2=4(2)y$; $p=2$

Vertex $(-4, 0)$, focus $(-4, 2)$

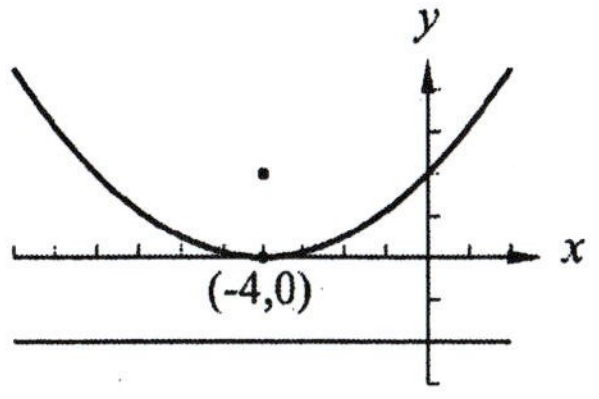

29. $y^2+42=2x(10-x)$

$$y^2+42=20x-2x^2$$

$y^2+2x^2-20x+42=0$; ellipse

$$\frac{y^2}{2}+x^2-10x=-21$$

$$\frac{y^2}{2}+x^2-10x+25=-21+25$$
$$\frac{y^2}{2}+(x-5)^2=4$$
$$\frac{y^2}{8}+\frac{(x-5)^2}{4}=1$$
(h, k) at $(5, 0)$, $V\left(5, \pm 2\sqrt{2}\right)$

$a=\sqrt{8}=2\sqrt{2};\ b=2$

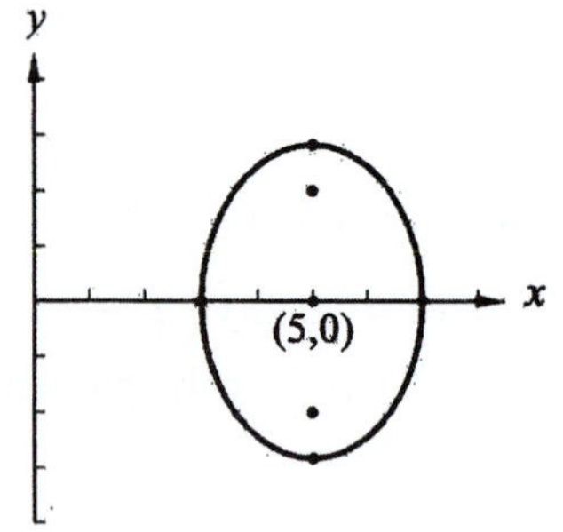

33. $4(y^2-4x-2)=5(4y-5)$

$4y^2-16x-8=20y-25$

$4y^2-20y-16x+17=0$

$A=0;\ C=4;\ B=0;$ parabola

$$y^2-5y-4x+\frac{17}{4}=0$$
$$y_1=\frac{5+\sqrt{25-4\left(-4x+\frac{17}{4}\right)}}{2}=\frac{5+\sqrt{16x+8}}{2}$$
$$y_2=\frac{5-\sqrt{16x+8}}{2}$$

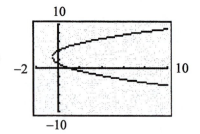

37. (a) If $k=1$, $x^2+ky^2=a^2;\ x^2+(1)y^2=a^2$

$x^2+y^2=a^2$ (circle)

(b) If $k<0$, $x^2+ky^2=a^2;\ x^2-|k|y^2=a^2$

$$\frac{x^2}{a^2}-\frac{y^2}{a^2/|k|}=1 \text{ (hyperbola)}$$

(c) If $k>0\ (k\neq 1)$, $x^2+ky^2=a^2$

$$\frac{x^2}{a^2}+\frac{y^2}{a^2/k}=1 \text{ (ellipse)}$$

41. In $Ax^2+Bxy+Cy^2+Dx+Ey+F=0$, $A=B=C=0,\ D\neq 0,\ E\neq 0,\ F\neq 0$, then the equation is $Dx+Ey+F=0$ whose locus is a straight line.

45. (a) Beam is perpendicular to floor. We have a circle.

(b) Beam is not perpendicular to floor. We have an ellipse.

*See conic section diagrams, Fig. 21.94 pg. 594 of text.

21.9 Rotation of Axes

1. $x^2-y^2=25,\ \theta=45^\circ;\ x=x'\cos 45^\circ-y'\sin 45^\circ$

$$=\frac{x'}{\sqrt{2}}-\frac{y'}{\sqrt{2}};\ y=x'\sin 45^\circ+y'\cos 45^\circ=\frac{x'}{\sqrt{2}}+\frac{y'}{\sqrt{2}}$$
$$x^2-y^2=\left(\frac{x'}{\sqrt{2}}-\frac{y'}{\sqrt{2}}\right)^2-\left(\frac{x'}{\sqrt{2}}+\frac{y'}{\sqrt{2}}\right)^2=25;$$
$$\frac{x'^2}{2}-\frac{2x'y'}{2}+\frac{y'^2}{2}-\frac{x'^2}{2}-\frac{2x'y'}{2}-\frac{y'^2}{2}=25$$

$2x'y'+25=0$, hyperbola

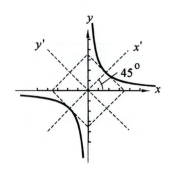

5. $x^2+2xy+x-y-3=0$

$B^2-4AC=2^2-4(1)(0)=4>0$, hyperbola

9. $13x^2+10xy+13y^2+6x-42y-27=0$

$B^2-4AC=10^2-4(13)(13)=-576<0$, ellipse

13. $3x^2+4xy=4$

$$\tan 2\theta=\frac{B}{A-C}=\frac{4}{3-0}=\frac{4}{3}$$

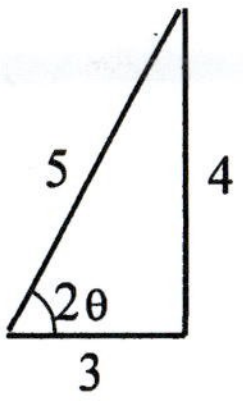

$$\cos 2\theta=\frac{3}{5}$$

$$\sin\theta=\sqrt{\frac{1-\cos 2\theta}{2}}=\sqrt{\frac{1-\frac{3}{5}}{2}}=\frac{1}{\sqrt{5}};$$

$$\cos\theta=\sqrt{\frac{1+\cos 2\theta}{2}}=\sqrt{\frac{1+\frac{3}{5}}{2}}=\frac{2}{\sqrt{5}};\ x=\frac{2x'-y'}{\sqrt{5}},$$

$$y=\frac{x'+2y'}{\sqrt{5}}$$

$$3\left(\frac{2x'-y'}{\sqrt{5}}\right)^2+4\left(\frac{2x'-y'}{\sqrt{5}}\right)\left(\frac{x'+2y'}{\sqrt{5}}\right)=4;$$

$$3\left(4x'^2-4x'y'+y'^2\right)+4\left(2x'^2+3x'y'-2y'^2\right)=20$$

$$20x'^2-5y'^2=20$$

$$4x'^2-y'^2=4,\ \text{hyperbola}$$

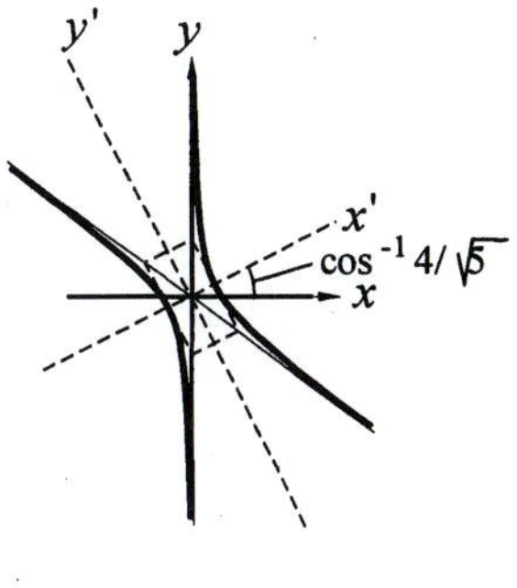

17. $16x^2-24xy+9y^2-60x-80y+400=0$

$$\tan 2\theta=\frac{B}{A-C}=\frac{-24}{16-9}=\frac{-24}{7}$$

$$\cos 2\theta=-\frac{7}{25}$$

$$\sin\theta=\sqrt{\frac{1-\cos 2\theta}{2}}=\sqrt{\frac{1-\frac{-7}{25}}{2}}=\frac{4}{5}$$

$$\cos\theta=\sqrt{\frac{1+\cos 2\theta}{2}}=\sqrt{\frac{1+\frac{-7}{25}}{2}}=\frac{3}{5}$$

$$x=\frac{3x'-4y'}{5},\ y=\frac{4x'+3y'}{5}$$

$$16\left(\frac{3x'-4y'}{5}\right)^2-24\left(\frac{3x'-4y'}{5}\right)\left(\frac{4x'+3y'}{5}\right)$$

$$+9\left(\frac{4x'+3y'}{5}\right)^2-60\left(\frac{3x'-4y'}{5}\right)-80\left(\frac{4x'+3y'}{5}\right)$$

$$+400=0$$

$$16\left(9x'^2-24x'y'+16y'^2\right)$$

$$-24\left(12x'^2+9x'y'-16x'y'-12y'^2\right)$$

$$+9\left(16x'^2+24x'y'+9y'^2\right)-300(3x'-4y')$$

$$-400(4x'+3y')+10{,}000=0$$

$$625y'^2-2500x'+10{,}000=0$$

$$y'^2-4x'+16=0$$

$$(y'-0)^2-4(x'-4)=0$$

$$y'^2-4x''=0$$

$$y''^2=4x'',\ \text{parabola}$$

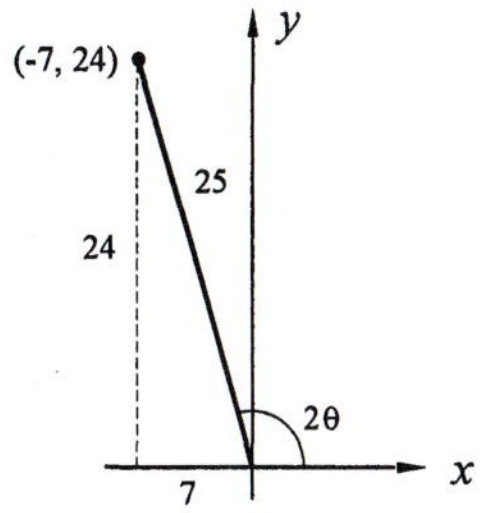

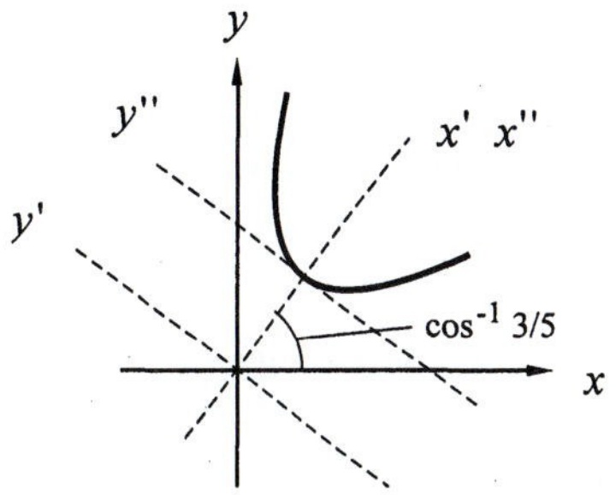

21.10 Polar Coordinates

1. $\left(3, \frac{\pi}{3}\right)$ and $\left(3, -\frac{5\pi}{3}\right)$ represent the same point.

$\left(-3, \frac{\pi}{3}\right)$ and $\left(3, -\frac{2\pi}{3}\right)$ are on the opposite side of the pole.

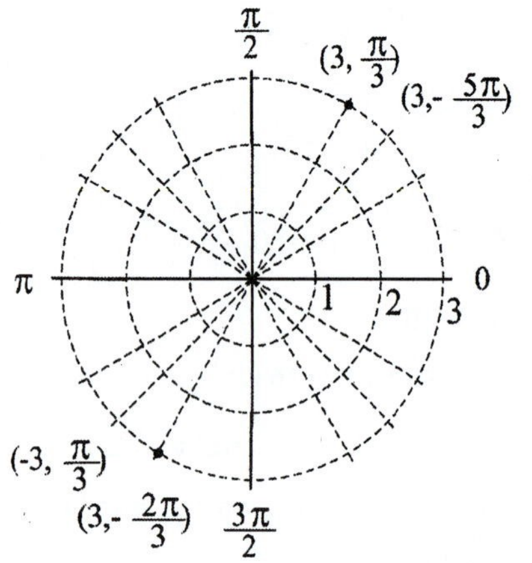

5. $\left(3, \frac{\pi}{6}\right)$; $r = 3$, $\theta = \frac{\pi}{6}$

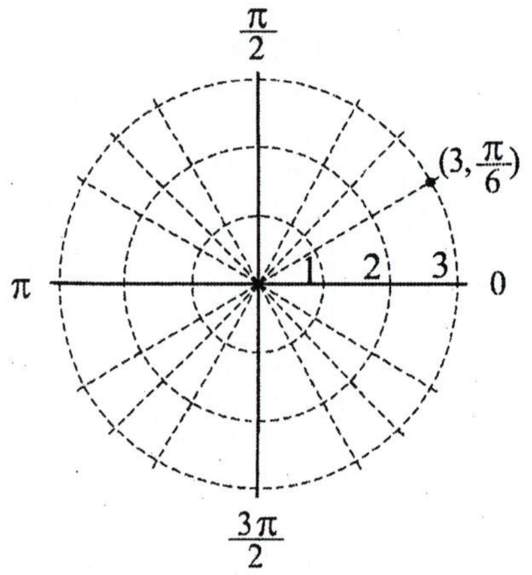

9. $\left(-8, \frac{7\pi}{6}\right)$; negative r is reversed in direction from positive r.

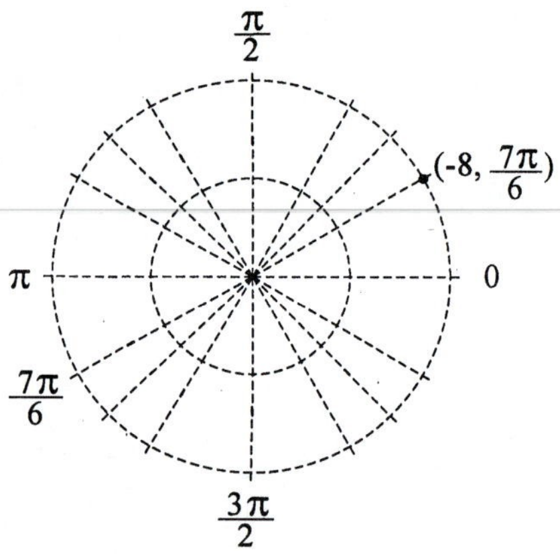

13. $(2, 2)$; $\frac{\pi}{180^\circ} = \frac{2}{\theta}$; $\theta = \frac{360}{\pi} = 114.6^\circ$

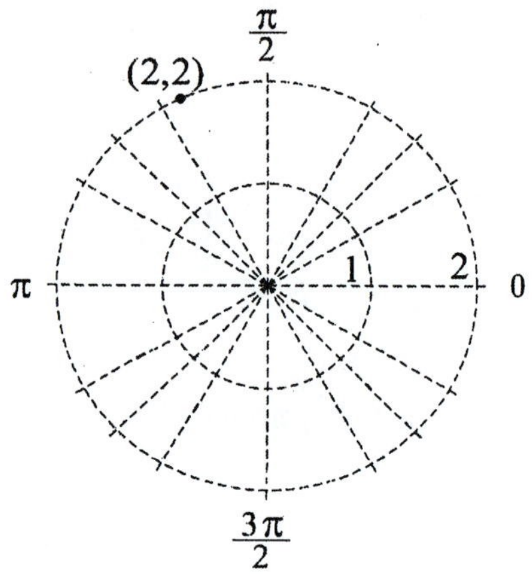

17. $(\sqrt{3}, 1)$ is (x, y), QI

$\tan\theta = \frac{y}{x}$

$\theta = \tan^{-1}\frac{y}{x} = \tan^{-1}\frac{1}{\sqrt{3}} = \tan^{-1}\frac{\sqrt{3}}{3}$;

$\theta = 30^\circ = \frac{\pi}{6}$

$r = \sqrt{x^2 + y^2} = \sqrt{(\sqrt{3})^2 + 1^2}$

$= \sqrt{3+1} = \sqrt{4} = 2$

(r, θ) is $\left(2, \frac{\pi}{6}\right)$

21. (r, θ) is $\left(8, \frac{4\pi}{3}\right)$, QIII

$x = r\cos\theta = 8\cos\frac{4\pi}{3} = 8\left(-\frac{1}{2}\right) = -4$

$y = r\sin\theta = 8\left(-\frac{\sqrt{3}}{2}\right) = -4\sqrt{3}$

(x, y) is $\left(-4, -4\sqrt{3}\right)$

25. $x = 3$

$r\cos\theta = x = 3;\ r = \frac{3}{\cos\theta} = 3\sec\theta$

29. $x^2 + (y-2)^2 = 4$

$x^2 + y^2 - 4y + 4 = 4$

$r^2 - 4 \cdot r\sin\theta = 0$

$r = 4\sin\theta$

33. $x^2 + y^2 = 6y$

$r^2 = 6 \cdot r\sin\theta$

$r = 6\sin\theta$

37. $r = \sin\theta;\ r^2 = r\sin\theta;\ r^2 = x^2 + y^2$

$x^2 + y^2 = r^2 = r\sin\theta = y;\ x^2 + y^2 - y = 0,$

circle

41. $r = \frac{2}{\cos\theta - 3\sin\theta}$

$r\cos\theta - 3r\sin\theta = 2$

$x - 3y = 2,$ line

45. $r = 2(1+\cos\theta);\ x = r\cos\theta;\ \frac{x}{r} = \cos\theta$

$r^2 = x^2 + y^2;\ r = \sqrt{x^2 + y^2}$

$r = 2(1+\cos\theta) = 2\left(1 + \frac{x}{r}\right) = 2 + \frac{2x}{r};\ r^2 = 2r + 2x$

Multiply through by r.

$x^2 + y^2 = 2\sqrt{x^2+y^2} + 2x;\ x^2 + y^2 - 2x = 2\sqrt{x^2+y^2}$

$\left(x^2 + y^2 - 2x\right)^2 = 4\left(x^2 + y^2\right)$

$x^4 + y^4 - 4x^3 + 2x^2y^2 - 4xy^2 + 4x^2 = 4x^2 + 4y^2$

$x^4 + y^4 - 4x^3 + 2x^2y^2 - 4xy^2 + 4x^2 - 4x^2 - 4y^2 = 0$

$x^4 + y^4 - 4x^3 + 2x^2y^2 - 4xy^2 - 4y^2 = 0$

49. As the graph shows the point $(2, 3\pi/4)$ is on the curve $r = 2\sin 2\theta$ even though $(2, 3\pi/4)$ is not a solution to $r = 2\sin 2\theta$. $(2, 3\pi/4)$ and $(-2, 7\pi/4)$ are the same point and $(-2, 7\pi/4)$ is a solution to $r = 2\sin 2\theta$.

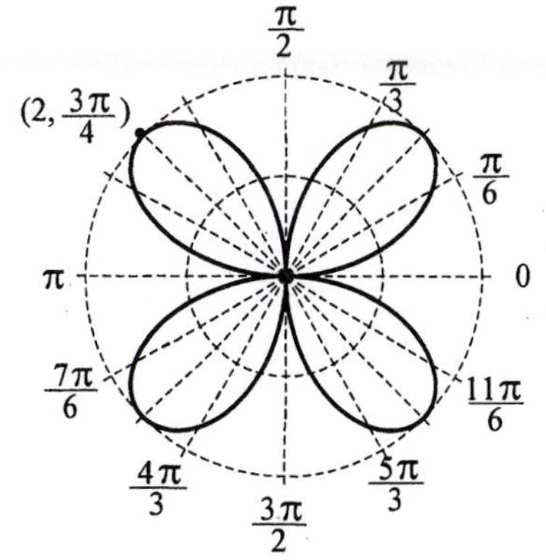

53. Each pair of vertices subtends a central angle of $\frac{\pi}{3}$ at the pole. The coordinatges of the other vertices are $(2, 0), \left(2, \frac{\pi}{3}\right), \left(2, \frac{2\pi}{3}\right), \left(2, \frac{4\pi}{3}\right), \left(, \frac{5\pi}{3}\right)$

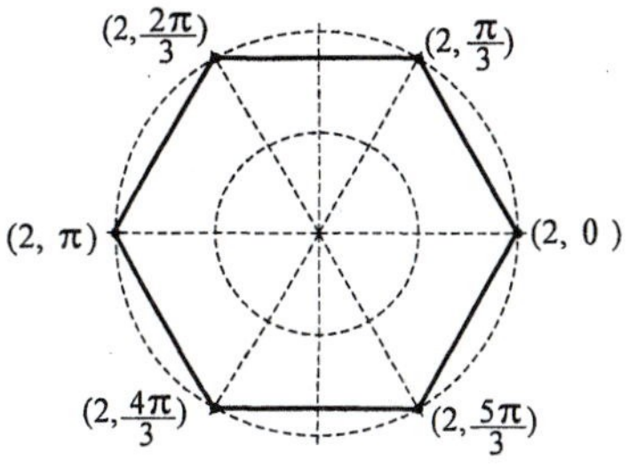

57. $r = 3 - \sin\theta;\ y = r\sin\theta$

$\sqrt{x^2+y^2} = 3 - \frac{y}{r} = 3 - \frac{y}{\sqrt{x^2+y^2}}$

$\sqrt{x^2+y^2} = \frac{3\sqrt{x^2+y^2} - y}{\sqrt{x^2+y^2}}$

$x^2 + y^2 = 3\sqrt{x^2+y^2} - y;\ x^2 + y^2 + y = 3\sqrt{x^2+y^2}$

Square both sides.

$$x^4 + 2x^2y^2 + 2x^2y + y^4 + 2y^3 + y^2 = 9\left(x^2 + y^2\right)$$
$$= 9x^2 + 9y^2$$

Therefore,

$$x^4 + 2x^2y^2 + 2x^2y + y^4 + 2y^3 - 9x^2 - 8y^2 = 0$$

21.11 Curves in Polar Coordinates

1. The graph of $\theta = \frac{5\pi}{6}$ is a straight line through the pole. $\theta = \frac{5\pi}{6}$ for all possible values of r.

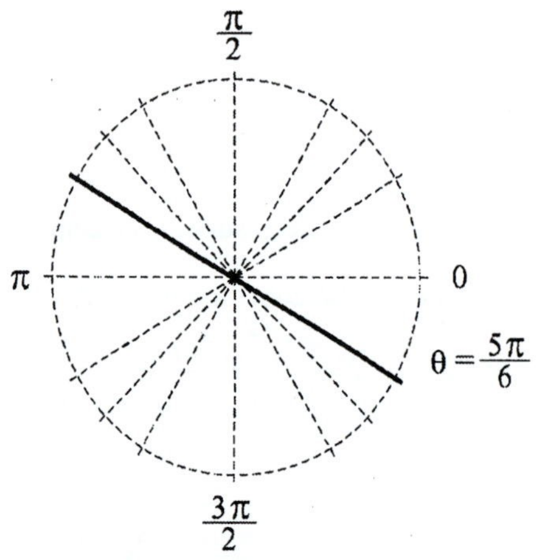

5. $r = 5$ for all θ. Graph is a circle with radius 5.

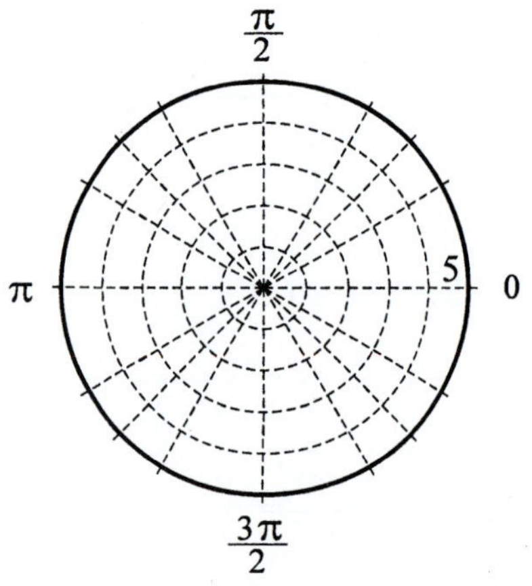

9. $r = 4\sec\theta = \frac{4}{\cos\theta}$; vertical line

θ	r
0	4
$\frac{\pi}{6}$	4.6
$\frac{\pi}{4}$	5.7
$\frac{\pi}{3}$	8
$\frac{\pi}{2}$	*
$\frac{2\pi}{3}$	−8
$\frac{3\pi}{4}$	−5.7
$\frac{5\pi}{6}$	4.6
π	−4
$\frac{5\pi}{4}$	−5.7
$\frac{3\pi}{2}$	*
$\frac{7\pi}{4}$	5.7
2π	4

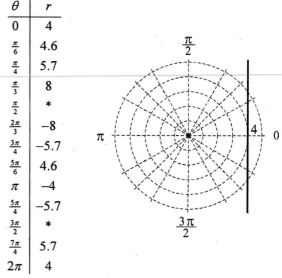

* denotes undefined

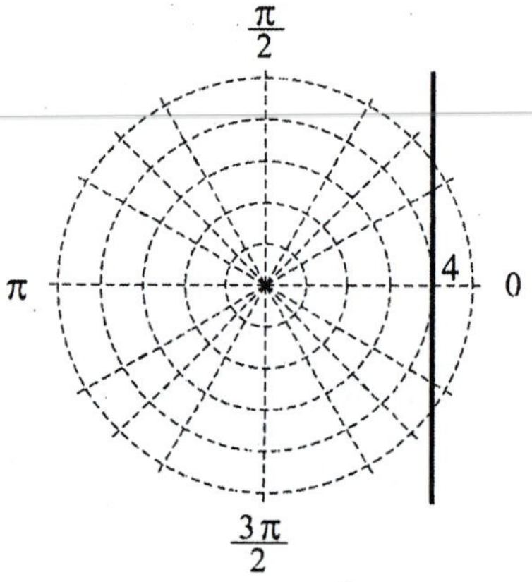

13. $r = 1 - \cos\theta$; cardioid

θ	r
0	0
$\frac{\pi}{4}$	0.3
$\frac{\pi}{2}$	1
$\frac{3\pi}{4}$	1.7
π	2
$\frac{5\pi}{4}$	1.7
$\frac{3\pi}{2}$	1
$\frac{7\pi}{4}$	0.3

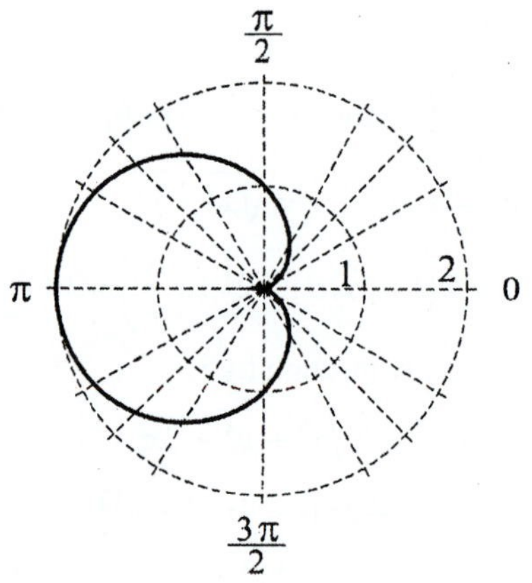

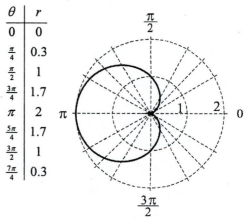

17. $r = 4\sin 2\theta$; rose (4 petals)

θ	r
0	0
$\frac{\pi}{8}$	2.8
$\frac{\pi}{4}$	4
$\frac{3\pi}{8}$	−2.8
$\frac{\pi}{2}$	0
$\frac{5\pi}{8}$	2.8
$\frac{3\pi}{4}$	−4
$\frac{7\pi}{8}$	−2.8
π	0
$\frac{9\pi}{8}$	2.8
$\frac{5\pi}{4}$	4
$\frac{11\pi}{8}$	2.8
$\frac{3\pi}{2}$	0
$\frac{13\pi}{8}$	−2.8
$\frac{7\pi}{4}$	−4
2π	−2.8

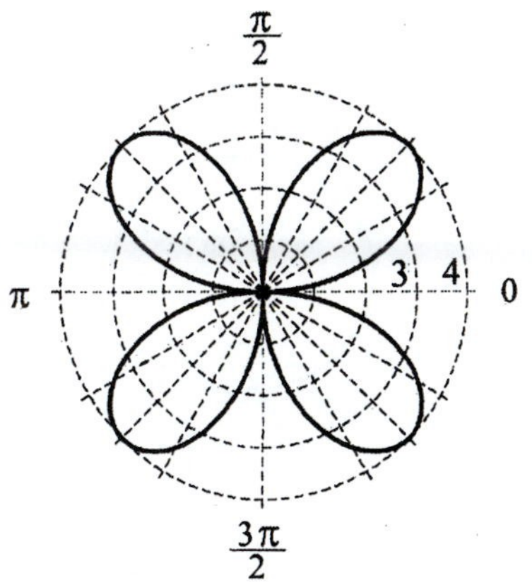

21. $r = 2^{\theta}$; spiral

θ	r
0	1
$\frac{\pi}{4}$	1.7
$\frac{\pi}{2}$	3.0
$\frac{3\pi}{4}$	5.1
π	8.8
$\frac{5\pi}{4}$	15.2
$\frac{3\pi}{2}$	26.2
$\frac{7\pi}{4}$	45.2
2π	77.9

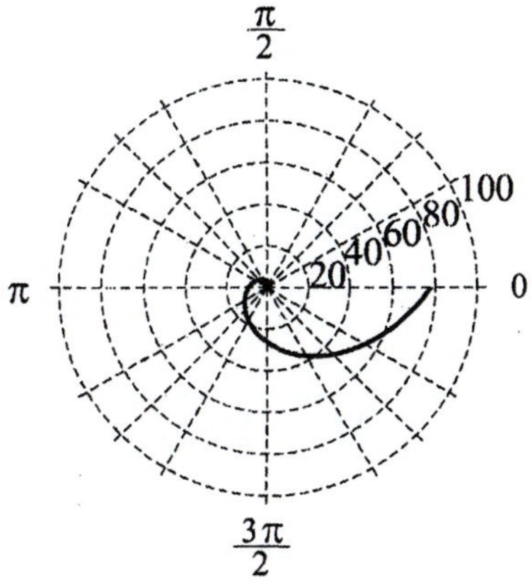

25. $r = \dfrac{3}{2 - \cos\theta}$; ellipse

θ	r
0	3
$\frac{\pi}{4}$	2.32
$\frac{\pi}{2}$	1.5
$\frac{3\pi}{4}$	1.11
π	1
$\frac{5\pi}{4}$	1.11
$\frac{3\pi}{2}$	1.5
$\frac{7\pi}{4}$	2.32
2π	3

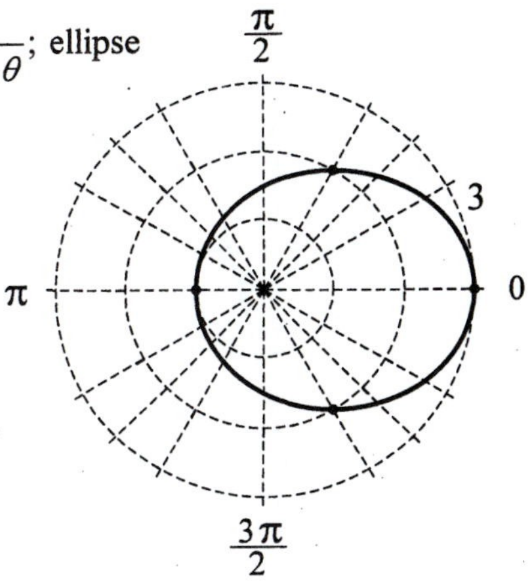

29. $r = 4\cos\frac{1}{2}\theta$

θ	r	θ	r
0	4.0	$\frac{13\pi}{6}$	−3.9
$\frac{\pi}{6}$	3.9	$\frac{9\pi}{4}$	−3.7
$\frac{\pi}{4}$	3.7	$\frac{7\pi}{3}$	−3.5
$\frac{\pi}{3}$	3.5	$\frac{5\pi}{2}$	−2.8
$\frac{\pi}{2}$	2.8	$\frac{8\pi}{3}$	−2.0
$\frac{2\pi}{3}$	2.0	$\frac{11\pi}{4}$	−1.5
$\frac{3\pi}{4}$	1.5	$\frac{17\pi}{6}$	−1.0
$\frac{5\pi}{6}$	1.0	3π	0
π	0	$\frac{19\pi}{6}$	1.0
$\frac{7\pi}{6}$	−1.0	$\frac{13\pi}{4}$	1.5
$\frac{5\pi}{4}$	−1.5	$\frac{10\pi}{3}$	2.0
$\frac{4\pi}{3}$	−2.0	$\frac{7\pi}{2}$	2.8
$\frac{3\pi}{2}$	−2.8	$\frac{11\pi}{3}$	3.5
$\frac{5\pi}{3}$	−3.5	$\frac{15\pi}{4}$	3.7
$\frac{7\pi}{4}$	−3.7	$\frac{23\pi}{6}$	3.9
$\frac{11\pi}{6}$	−3.9	4π	4.0
2π	−4.0		

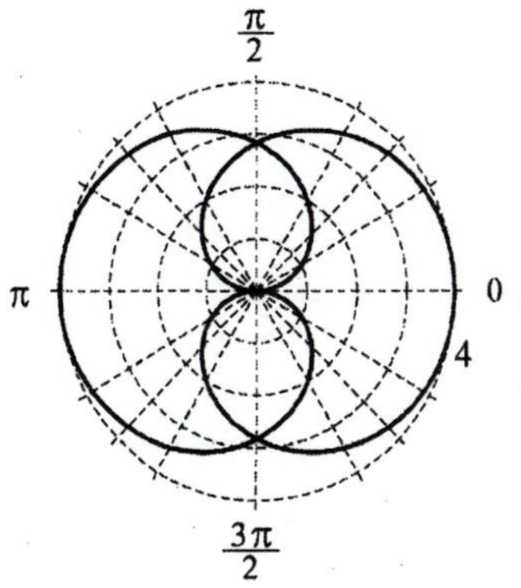

33. $r = \theta \quad (-20 \le \theta \le 20)$

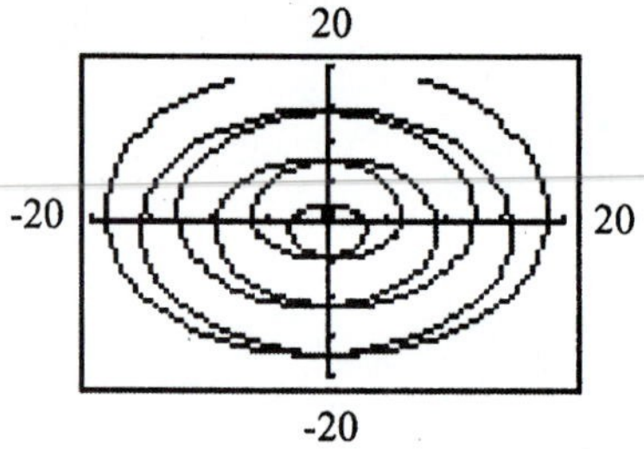

37. $r = 3\cos 4\theta$

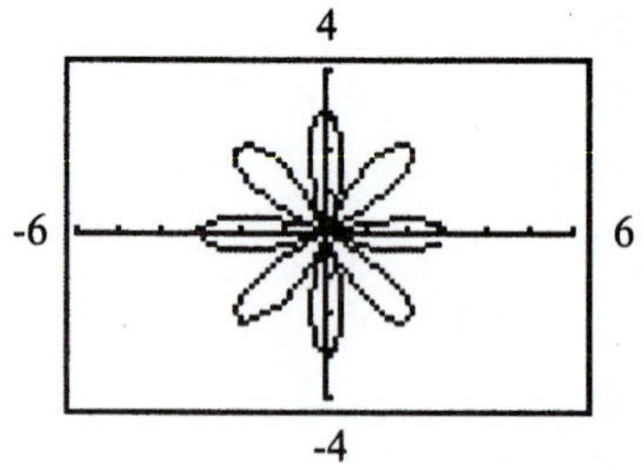

41. $r + 2 = \cos 2\theta \Rightarrow r = \cos 2\theta - 2$

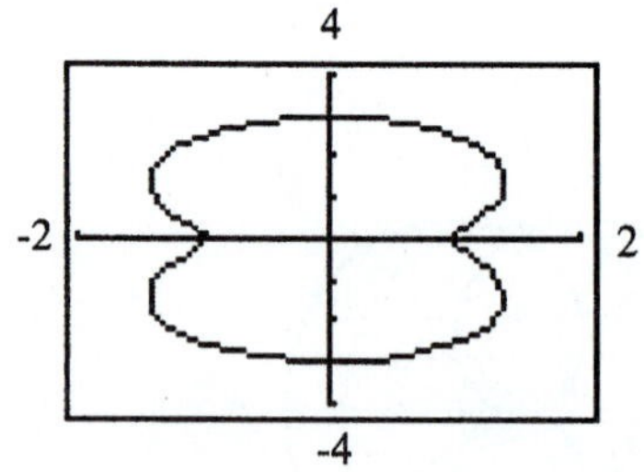

45. From the calculator screen the curves intersect at $(0, 0)$ and $(1, 1)$ where the tangent lines are horizontal and vertical showing the curves intersect at right angles.

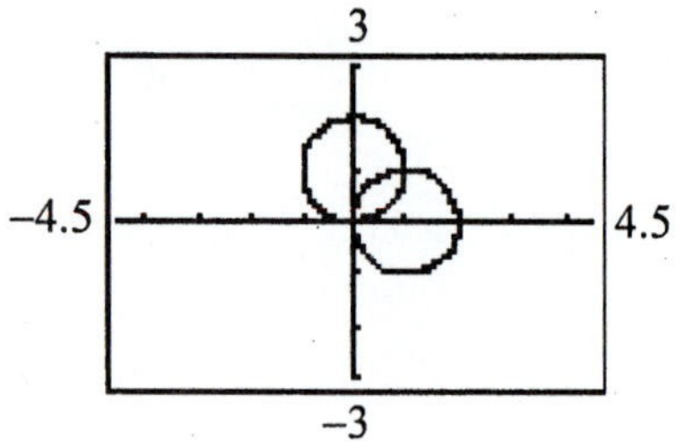

49. $r = \dfrac{25}{10 + 4\cos\theta}$

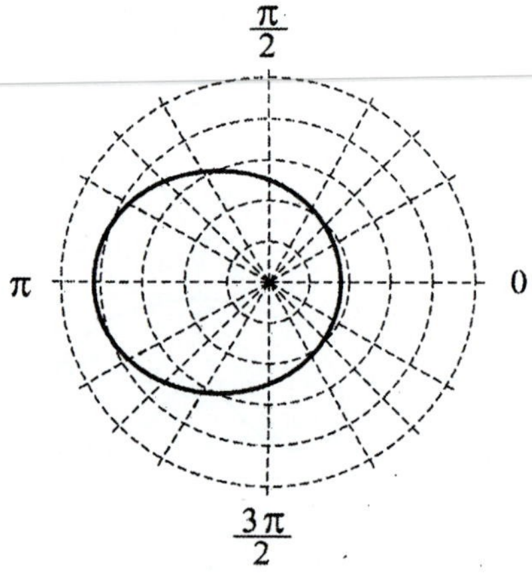

53.

exercise	equation	graph
17	$r = 4\sin 2\theta$	rose with 4 petals
18	$r = 2\sin 3\theta$	rose with 3 petals
37	$r = 3\cos 4\theta$	rose with 8 petals
38	$r = 3\sin 5\theta$	rose with 5 petals

The graph of $r = a\sin n\theta$ or $r = b\cos n\theta$ is a rose with n loops for n odd and a rose with $2n$ loops for n even.

Chapter 21 Review Exercises

1. Given straight line; (x_1, y_1) is $(1, -7)$; $m = 4$

$y - y_1 = m(x - x_1)$; $y - (-7) = 4(x - 1)$

$y + y = 4x - 4$; $y = 4x - 4 - 7$

$y = 4x - 11$ or $4x - y - 11 = 0$

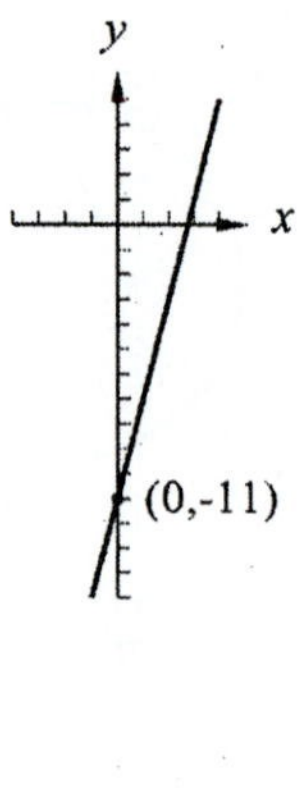

5. $x^2+y^2=6x$

$x^2-6x+9+y^2=9$

$(x-3)^2+y^2=3^2$; circle, center $(3, 0)$, $r=3$

The concentric circle has equation

$(x-3)^2+y^2=r^2$ and passes through $(4, -3)$

$(4-3)^2+(-3)^2=r^2$

$r^2=10$

$(x-3)^2+y^2=10$

$x^2-6x+9+y^2=10$

$x^2-6x+y^3-1=0$

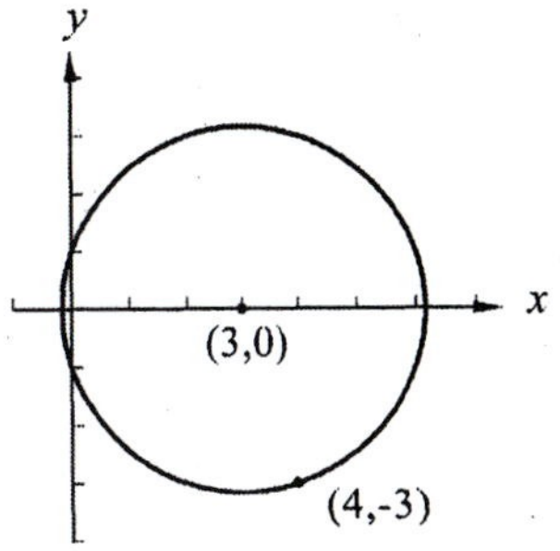

9. $a=10$, $c=8$

$a^2=b^2+c^2$

$100=b^2+8^2 \Rightarrow b^2=36$

$\frac{x^2}{100}+\frac{y^2}{36}=1$ or $9x^2+25y^2=900$

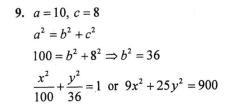

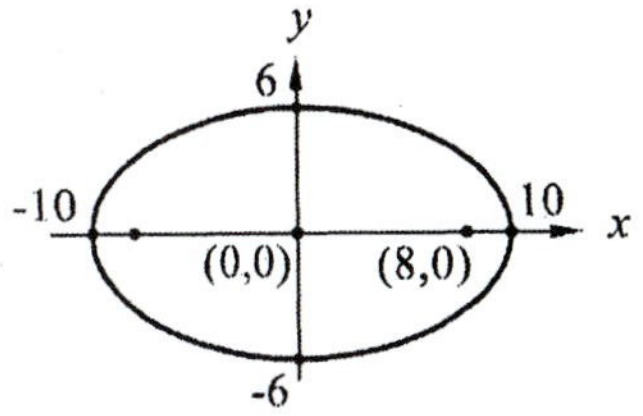

13. Given: $x^2+y^2+6x-7=0$

$(x^2+6x)+(y^2)=7$; $(x^2+6x+9)+y^2=7+9$

$(x+3)^2+(y+0)^2=16$

$[x-(-3)]^2+(y-0)^2=4^2$

$C(h, k)=(-3, 0)$; $r=4$

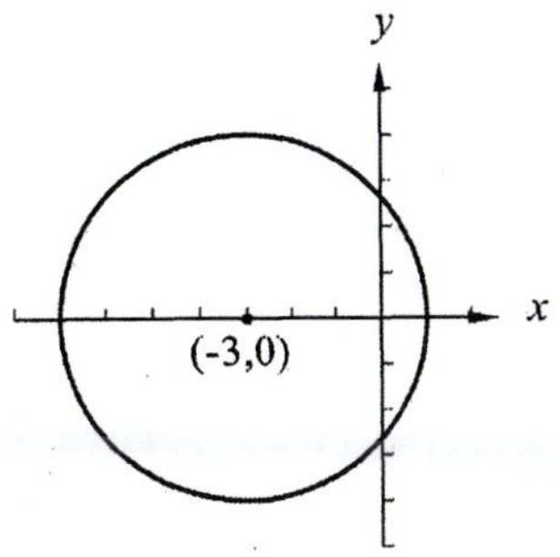

17. Given: $4x^2+y^2=1$

$\frac{x^2}{\frac{1}{4}}+\frac{y^2}{1}=1$

$a=1$, $b=\frac{1}{4}$, $c=\sqrt{1-\frac{1}{4}}=\frac{\sqrt{3}}{2}$

vertices $(0, 1)$, $(0, -1)$

foci: $\left(0, \frac{\sqrt{3}}{2}\right), \left(0, \frac{-\sqrt{3}}{2}\right)$

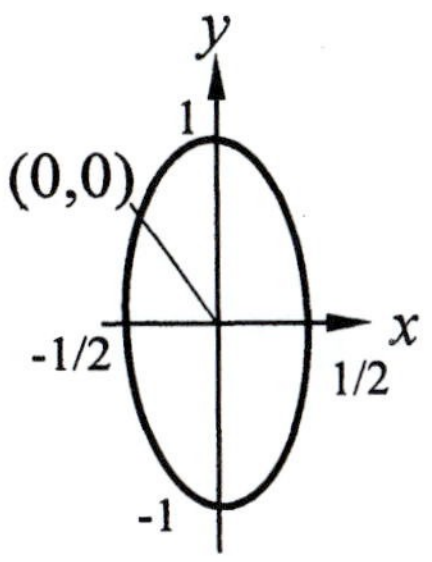

21. Given: $x^2-8x-4y-16=0$

$x^2-8x=4y+16$; $x^2-8x+16=4y+16+16$

$(x-4)^2=4y+32$; $(x-4)^2=4(+8)$

$(x-4)^2=4(1)(y+8)$; $p=1$

vertex (h, k) is $(4, -8)$; focus is $(4, -7)$

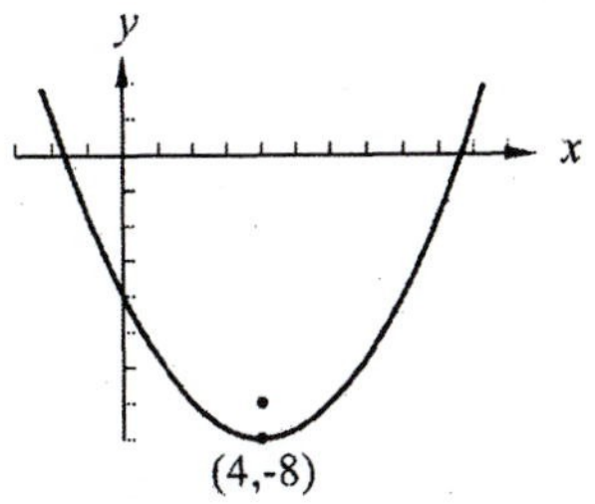

25. $x^2 - 2xy + y^2 + 4x + 4y = 0$

$B^2 - 4AC = (-2)^2 - 4(1)(1) = 0$, parabola

$\tan 2\theta = \dfrac{B}{A-C} = \dfrac{-2}{1-0} \Rightarrow \theta = 45^\circ$

$$\left(x'\cdot\frac{1}{\sqrt{2}} - y'\cdot\frac{1}{\sqrt{2}}\right)^2 - 2\left(x'\cdot\frac{1}{\sqrt{2}} - y'\cdot\frac{1}{\sqrt{2}}\right)$$

$$\left(x'\cdot\frac{1}{\sqrt{2}} + y'\cdot\frac{1}{\sqrt{2}}\right) + \left(x'\cdot\frac{1}{\sqrt{2}} + y'\cdot\frac{1}{\sqrt{2}}\right)^2$$

$$+4\left(x'\cdot\frac{1}{\sqrt{2}} - y'\cdot\frac{1}{\sqrt{2}}\right) + 4\left(x'\cdot\frac{1}{\sqrt{2}} + y'\cdot\frac{1}{\sqrt{2}}\right) = 0$$

$y'^2 = -2\sqrt{2}x' \Rightarrow V(0, 0)$

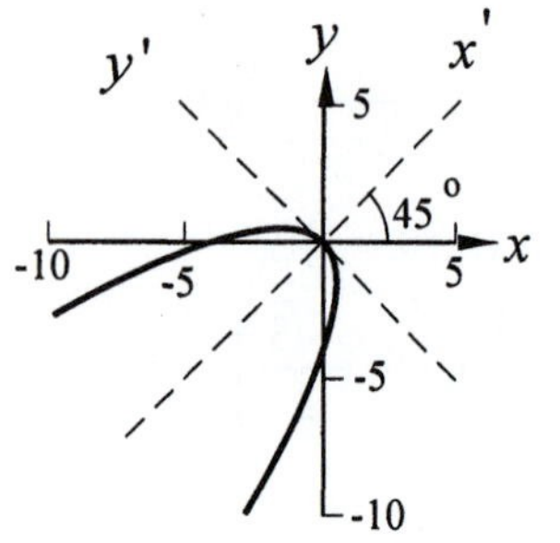

29. $r = 4\cos 3\theta$

Let $\theta = 0$ to π in steps of $\dfrac{\pi}{48}$.

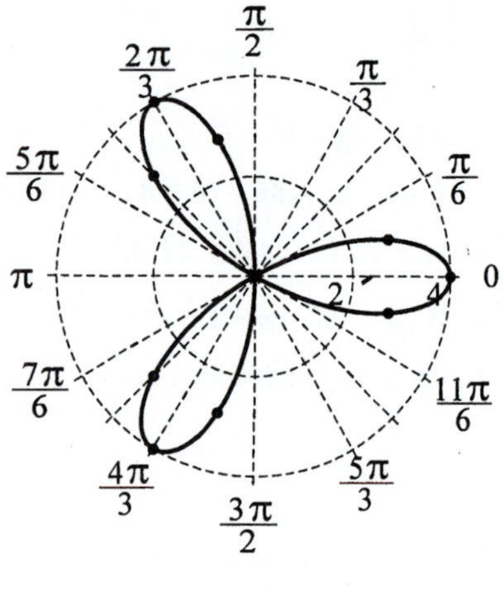

33. $r = 2\sin\dfrac{\theta}{2}$

Let $\theta = 0$ to 4π in steps of $\dfrac{\pi}{24}$.

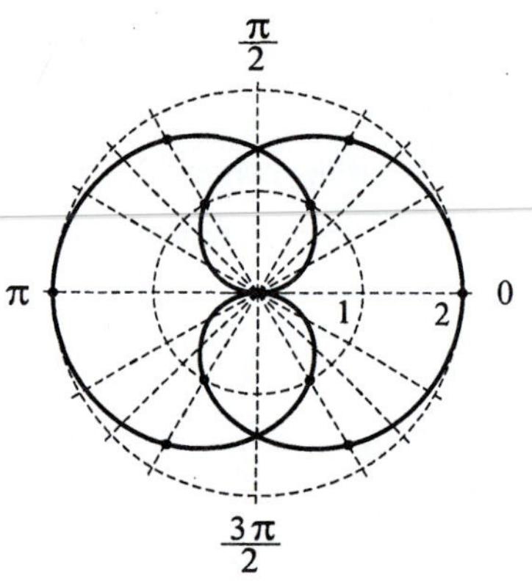

37. $x^2 + xy + y^2 = 2$

$x^2 + y^2 + xy = 2$

$r^2 + (r\cos\theta)(r\sin\theta) = 2$

$r^2 = \dfrac{2}{1 + \sin\theta\cos\theta}$

41. $r = \dfrac{4}{2 - \cos\theta}$

$2r - r\cos\theta = 4$

$2r = 4 + r\cos\theta = 4 + x$

$4r^2 = 16 + 8x + x^2$

$4(x^2 + y^2) = 4x^2 + 4y^2 = x^2 + 8x + 16$

$3x^2 + 4y^2 - 8x - 16 = 0$

45. $x^2 + y^2 - 4y - 5 = 0$

$y^2 - 4x^2 - 4 = 0$

From the graph, two real solutions.

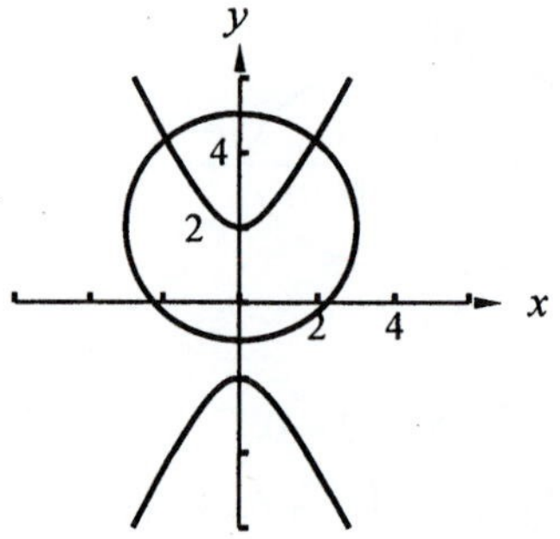

49. $x^2+3y+2-(1+x)^2=0$

$3y=(1+x)^2-x^2-2$

$y=\frac{(1+x)^2-x^2-2}{3}=\frac{1+2x+x^2-x^2-2}{3}=\frac{2x-1}{3}$

$=\frac{2}{3}x-\frac{1}{3}$

Graph $y_1=\frac{2}{3}x-\frac{1}{3}$

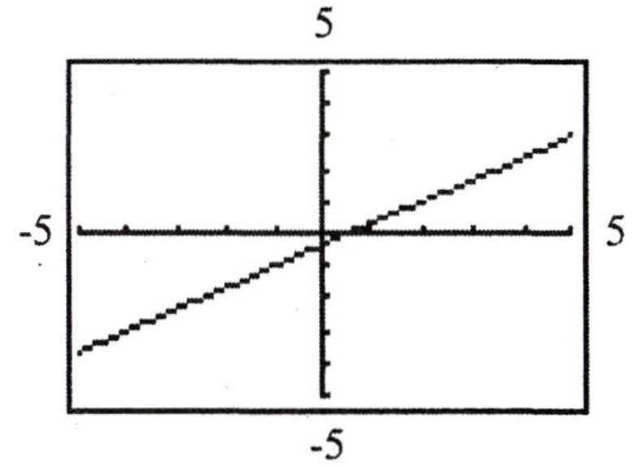

53. $x^2-4y^2+4x+24y-48=0$. Solve for y by completing the square.

$y^2-6y=0.25x^2+x-12$

$y^2-6y+9=0.25x^2+x-3$

$(y-3)^2=0.25x^2+x-3$

$y=\pm\sqrt{0.25x^2+x-3}+3$

Graph $y_1=\sqrt{0.25x^2+x-3}+3$

$y_2=-\sqrt{0.25x^2+x-3}+3$

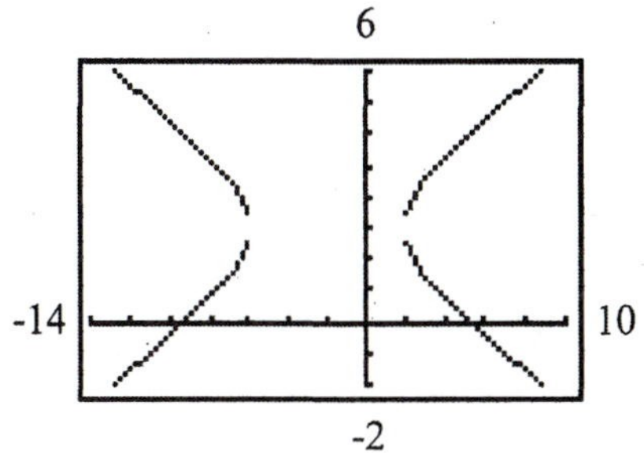

57. $r=2-3\csc\theta=2-\frac{3}{\sin\theta}$

Graph $r_1=2-\frac{3}{\sin\theta}$

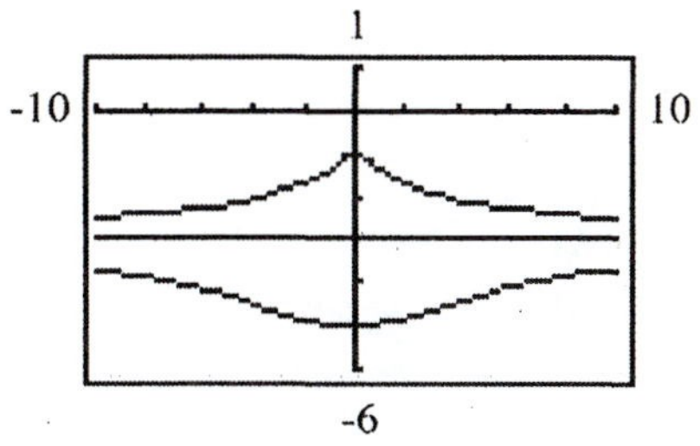

61.

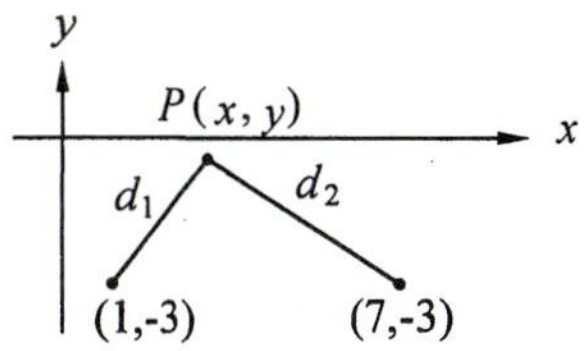

$d_1+d_2=8$ describes an ellipse with center $(4,-3)$,

$a=4\Rightarrow a^2=16$, $c=3\Rightarrow c^2=9$.

$a^2=b^2+c^2\Rightarrow 16=b^2+9\Rightarrow b^2=7$

$\frac{(x-4)^2}{16}+\frac{(y+3)^2}{7}=1$

65. $\frac{x-(-3)}{13-3}=\frac{-3-(-5)}{3-(-2)}$

$x=1$

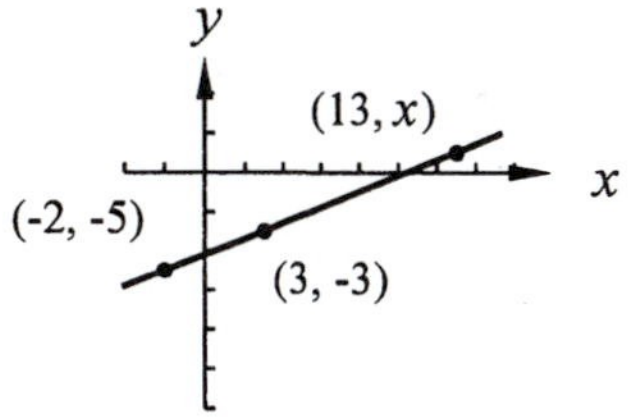

69. $a^2+b^2=(-3-2)^2+(11+1)^2+(14-2)^2+(4+1)^2$

$a^2+b^2=338$

$c^2=(14+3)^2+(4-11)^2=338$,

points form a right triangle

$m_a=\frac{11+1}{-3-2}=\frac{12}{-5}$, $m_b=\frac{4+1}{14-2}=\frac{5}{12}$

$m_a\cdot m_b=\frac{12}{-5}\cdot\frac{5}{12}=-1\Rightarrow a\perp b$,

points form a right triangle

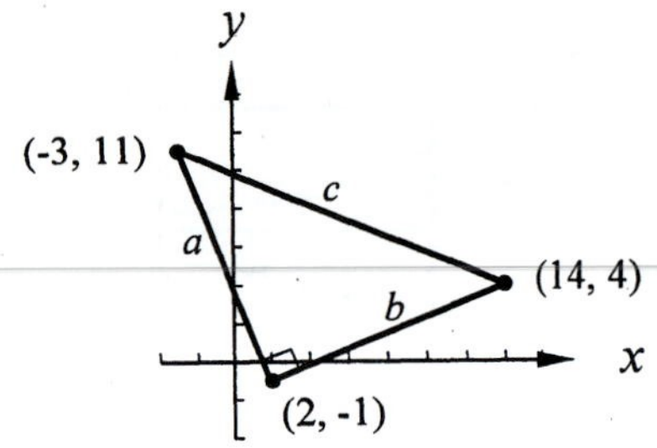

73. $(x+jy)^2+(x-jy)^2=2$

$x^2+2jy-y^2+x^2-2jy-y^2=2$

$x^2-y^2-2=0$, hyperbola

77. From definition,

$$(x-3)^2+(y-1)^2=(y+3)^2$$
$$x^2-6x+9+y^2-2y+1=y^2+6y+9$$
$$x^2-6x-8y+1=0$$

From translation of axes,

$(h, k)=(3, -1)$, $p=2 \Rightarrow 4p=8$

$$(x-h)^2=4p(y-k)$$
$$(x-3)^2=8(y+1)$$
$$x^2-6x+9=8y+8$$
$$x^2-6x-8y+1=0$$

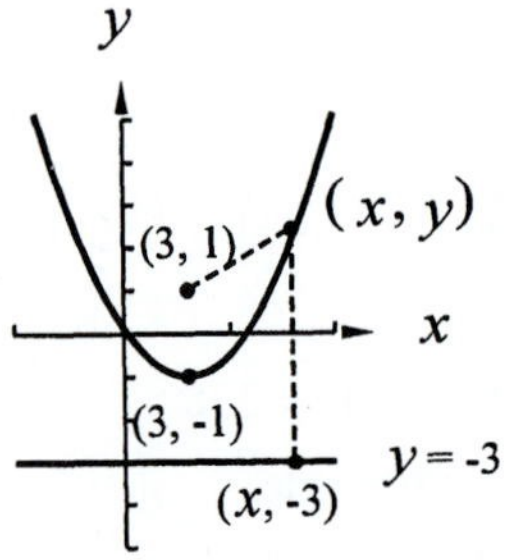

81. $v=v_0+at$

$18.5=5.75+a(5.50)$

$a=2.32$

$v=5.75+2.32t$

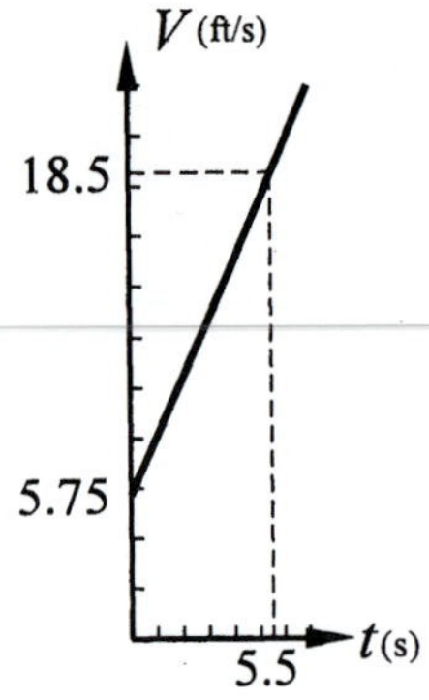

85. $A=\pi r^2=\pi\left(490 \tan 7^\circ\right)$

$A=11,000 \text{ ft}^2$

89. $y^2=4px$

$40^2=4p(50) \Rightarrow 4p=32$

$y^2=32x$

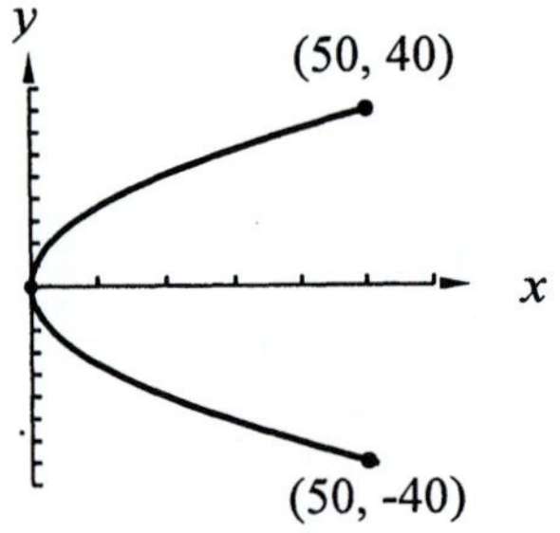

93. $P=12.0i-0.500i^2$

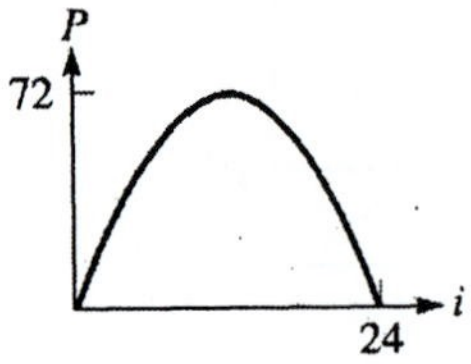

97. If the pins are on the x-axis and centered, the equation of the ellipse is

$$\frac{x^2}{5^2}+\frac{y^2}{3^2}=1$$

$$a^2=b^2+c^2 \Rightarrow 5^2=3^2+c^2 \Rightarrow c=4$$

The coordinates of the pins are $(\pm 4,\ 0)$ so they are 8 cm apart. The length of the string is

$$l=2a+2c=2(3)+2(4)$$

$$l=18 \text{ cm}$$

101. Let $P(x,\ y)$ be the coordinates of the recorder in a coordinate system with origin at the target. Let rilfe be at $(0,\ r)$, then

$\sqrt{x^2+(y-r)^2}=v_s(t_0+t_1)$ where t_0 is the time for the bullet to reach the target and t_1 is the time for sound to reach the detector from the target.

$$\sqrt{x^2+y^2}=v_s t_1$$

$$\sqrt{x^2+(y-r)^2}-\sqrt{x^2+y^2}=v_s t_0=\text{constant}=2a$$

$$\sqrt{x^2+(y-r)^2}=2a+\sqrt{x^2+y^2}$$

$$x^2+y^2-2ry+r^2=4a^2+4a\sqrt{x^2+y^2}+x^2+y^2$$

$$-2ry+\left(r^2-4a^2\right)=4a\sqrt{x^2+y^2}$$

$$4r^2y^2-4r\left(r^2-4a^2\right)y+\left(r^2-4a^2\right)=16a^2\left(x^2+y^2\right)$$

$$\left(4r^2-16a^2\right)y^2-4r\left(r^2-4a^2\right)y-16a^2x^2=4a^2-r^2$$

which has the form $Ay^2-By-dx^2=e$

$$y^2-\frac{B}{A}y+\frac{B^2}{4A^2}-\frac{d}{A}x^2=\frac{e}{A}+\frac{B^2}{4A^2}$$

$$\left(y-\frac{B}{2A}\right)^2-Dx^2=F$$

$$\frac{\left(y-\frac{B}{2A}\right)^2}{F}-\frac{x^2}{\frac{F}{D}}=1,$$

which is a hyperbola.

Summary: The distance from the rifle to P minus the distance from the target to P = constant which is related to the distance from the rifle to the target.

105. $r=200(\sec\theta+\tan\theta)^{-5}/\cos\theta,\quad 0<\theta<\pi/2$

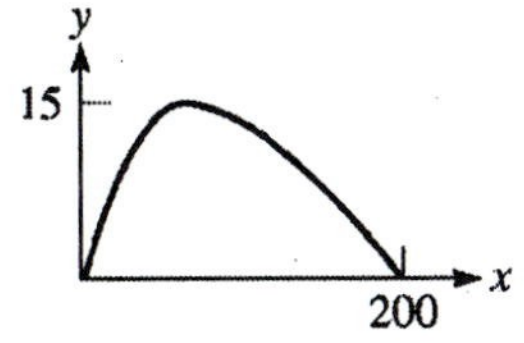

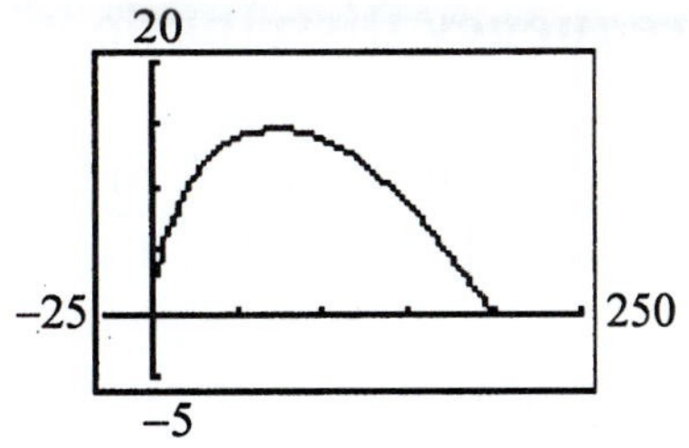

Chapter 22

INTRODUCTION TO STATISTICS

22.1 Frequency Distributions

1.

Est. (hrs)	0 – 5	6 – 11	12 – 17	18 – 23	24 – 29
Freq.	5	12	19	9	5

5.

Number	Frequency
53	1
54	3
55	1
56	3
57	2
58	4
59	3
60	1
61	1
62	0
63	1

9.

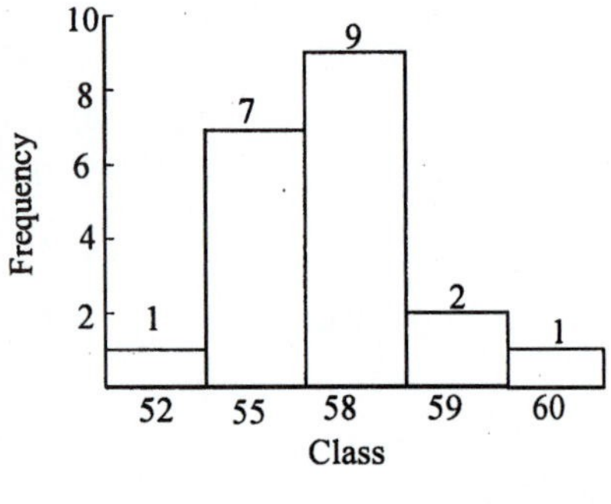

13.

Number	18	19	20	21	22	23	24	25
Frequency	1	3	2	4	3	1	0	1

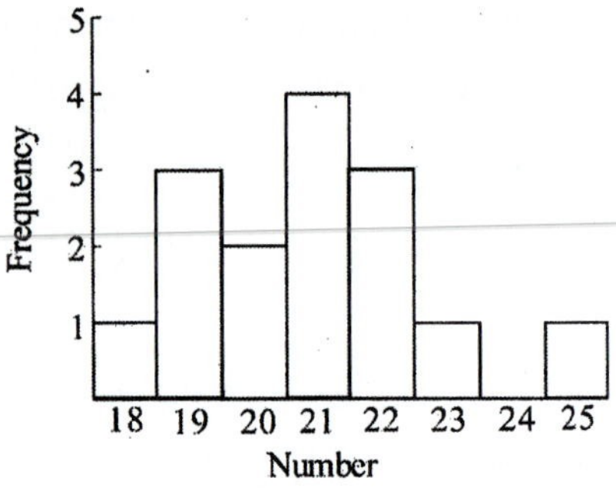

17.

Time(s)	Frequency
2.21	2
2.22	7
2.23	18
2.24	41
2.25	56
2.26	32
2.27	8
2.28	3
2.29	3

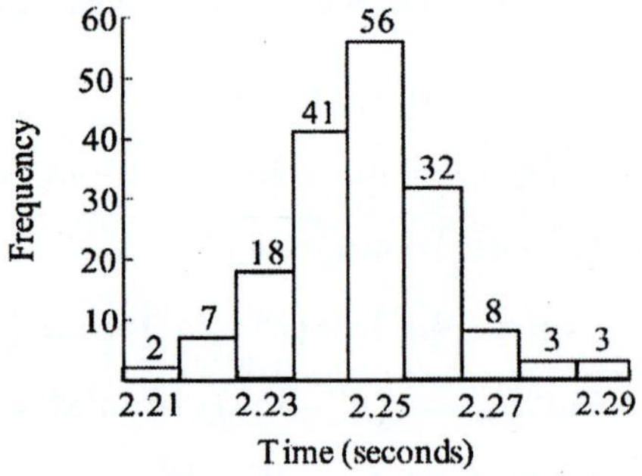

21.

Dist. (ft)	f, (%)
155 – 159	1.7
160 – 164	12.5
165 – 169	26.7
170 – 174	30.0
175 – 179	20.0
180 – 184	8.3
185 – 189	0.8

25.

Dosage (mR)	Frequency
3.73 − 3.87	1
3.88 − 4.02	2
4.03 − 4.17	2
4.18 − 4.32	7
4.33 − 4.47	7
4.48 − 4.62	1

29.

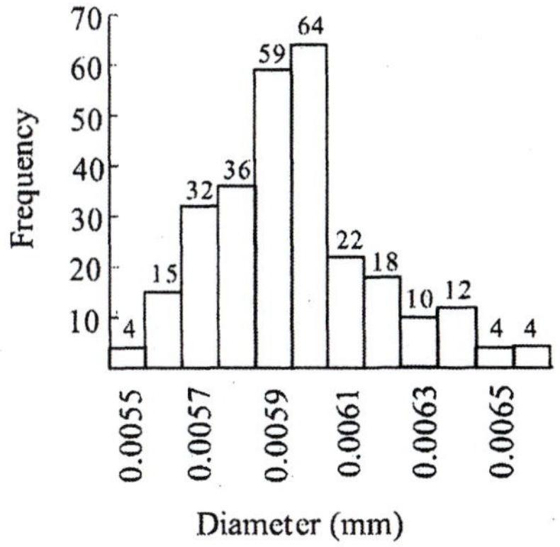

22.2 Measures of Central Tendency

1. 1, 2, 2, 3, 4, 4, 4, 6, 7, 7, 8, 9, 9, 11

There are 14 numbers. The median is halfway between the seventh and eighth numbers.

$$\text{median} = \frac{4+6}{2} = 5$$

5. Arrange the numbers in numerical order:

2, 3, 3, 3, 4, 4, 4, 4, 5, 5, 6, 6, 6, 7, 7

There are 15 numbers. The middle number is eighth. Since the eighth number is 4, the median is 4.

9. The arithmetic mean is:

$$\bar{x} = \frac{2+3+3+3+4+4+4+4+5+5+6+6+6+7+7}{15}$$

$$= \frac{69}{15} = 4.6$$

13. The mode is the number that occurs most frequently, which is 4 since it occurs 4 times.

17. Arrange in ascending order; $n = 20\left(10^{\text{th}}, 11^{\text{th}}\right)$, $M = 57.5$

21. $\bar{x} = \frac{\sum xf}{\sum f} = \frac{313}{15} = 20.9$

25. 60th-61st, $M = 172$ ft

29. $M = 4.36$ mR; $f = 2$

33. $\bar{x} = \frac{\sum xf}{\sum f} = \frac{1.6673}{280} = 0.00595$ mm

37. $\bar{x} = \frac{\sum xf}{\sum f} = \frac{(862)(200)}{1000} = 862$ kW·h

41. Lowest = 375; highest = 600; midrange = $\frac{975}{2}$ = \$487.50

45. $\bar{x} = \frac{\sum xf}{\sum f} = \frac{12{,}375}{14} = 883$

The mean increased by 235 from 648 to 883.

$\bar{x} = 40/10 = 4$

$$\frac{\sum\left(x-\bar{x}\right)^2}{n-1} = \frac{42}{10-1} = \frac{42}{9}$$

$$s = \sqrt{\frac{42}{9}} = 2.2$$

22.3 Standard Deviation

1.

x	$x-\bar{x}$	$(x-\bar{x})^2$
6	2	4
5	1	1
4	0	0
7	3	9
6	2	4
2	−2	4
1	−3	9
1	−3	9
5	1	1
3	−1	1
40		42

5.

x	$x-\bar{x}$	$(x-\bar{x})^2$	x^2
0.45	−0.055	0.00303	0.2025
0.46	−0.045	0.00203	0.2116
0.47	−0.035	0.00123	0.2209
0.48	−0.025	6.3×10^{-4}	0.2304
0.48	−0.025	6.3×10^{-4}	0.2304
0.49	−0.015	2.3×10^{-4}	0.2401
0.49	−0.015	2.3×10^{-4}	0.2401
0.49	−0.015	2.3×10^{-4}	0.2401
0.50	−0.005	2.5×10^{-5}	0.25
0.51	0.005	2.5×10^{-5}	0.2601
0.53	0.025	6.3×10^{-4}	0.2809
0.53	0.025	6.3×10^{-4}	0.2809
0.53	0.025	6.3×10^{-4}	0.2809
0.55	0.045	0.00203	0.3025
0.55	0.045	0.00203	0.3025
0.57	0.045	0.00203	0.3249

$\sum x = 8.08,\ \sum(x-\bar{x})^2 = 0.0184,\ \sum x^2 = 4.0988$

$\bar{x} = \frac{\sum x}{n} = \frac{8.08}{16} = 0.505$

$s = \sqrt{\frac{\sum(x-\bar{x})^2}{n-1}} = \sqrt{\frac{0.0184}{15}} = 0.035$

x	$x-\bar{x}$	$(x-\bar{x})^2$	x^2
103	−4.182	17.488	10,609
104	−3.182	10.124	10,816
104	−3.182	10.124	10816
104	−3.182	10.124	10,816
105	−2.182	4.7603	11,025
105	−2.182	4.7603	11,025
106	−1.182	1.3967	11,236
106	−1.182	1.3967	11,236
106	−1.182	1.3967	11,236
107	−0.1818	0.03306	11,449
107	−0.1818	0.03306	11,449
108	0.81818	0.66942	11,664
108	0.81818	0.66942	11,664
108	0.81818	0.66942	11,664
108	0.81818	0.66942	11,664
108	0.81818	0.66942	11,664
109	1.8182	3.3058	11,881
109	1.8182	3.3058	11,881
109	1.8182	3.3058	11,881
110	2.8182	14.579	12,321
111	3.8182	7.9421	12,100
113	5.8182	33.851	12,769

$\sum x = 2358,\ \sum(x-\bar{x})^2 = 131.\overline{27},\ \sum x^2 = 252{,}866$

$\bar{x} = \frac{\sum x}{n} = \frac{2358}{22} = 107.\overline{18}$

$s = \sqrt{\frac{\sum(x-\bar{x})^2}{n-1}} = \sqrt{\frac{131.\overline{27}}{22-1}} = 2.50$, equation 22.2

9. Using equation 22.3

$$s = \sqrt{\frac{n\sum x^2 - (\sum x)^2}{n(n-1)}} = \sqrt{\frac{16(4.0988)-(8.08)^2}{16(15)}}$$

$= 0.035$

13. $\bar{x} = 0.505,\ s = 0.035$

```
1-Var Stats
 x̄=.505
 Σx=8.08
 Σx²=4.0988
 Sx=.0350238014
 σx=.0339116499
↓n=16
```

17. Computer Instruction, exercise 13, section 22.1

x	x^2
18	324
19	361
19	361
19	361
20	400
20	400
21	441
21	441
21	441
21	441
22	484
22	484
22	484
23	529
25	625
313	6577

$$s=\sqrt{\frac{15(6577)-(313)^2}{15(14)}}$$

$s=1.8$

21. Strobe light times, Exercise 17, Section 22.1

x	f	fx	fx^2
2.21	2	4.42	9.7682
2.22	7	15.54	34.4988
2.23	18	40.14	89.5122
2.24	41	91.84	205.7216
2.25	56	126	283.5
2.26	32	72.32	163.4432
2.27	8	18.16	41.2232
2.28	3	6.84	15.5952
2.29	3	6.87	15.7323
	170	382.13	858.9947

$\bar{x}=382.13/170=2.2478$

$$s=\sqrt{\frac{170(858.9947)-(382.13)^2}{170(169)}}$$

$s=0.014$ s

25. Fiber-optic cable dismeters, Exercise 29, Section 22.1

$s=0.00022$ mm

```
1-Var Stats
 x̄=.0059546429
 Σx=1.6673
 Σx²=.00994211
 Sx=2.234782E-4
 σx=2.230788E-4
↓n=280
```

22.4 Normal Distributions

1. $\mu=10$, $\alpha=5$ and $\mu=20$, $\alpha=5$ are the same height with the first centered at $x=10$ and the second centered at $x=20$.

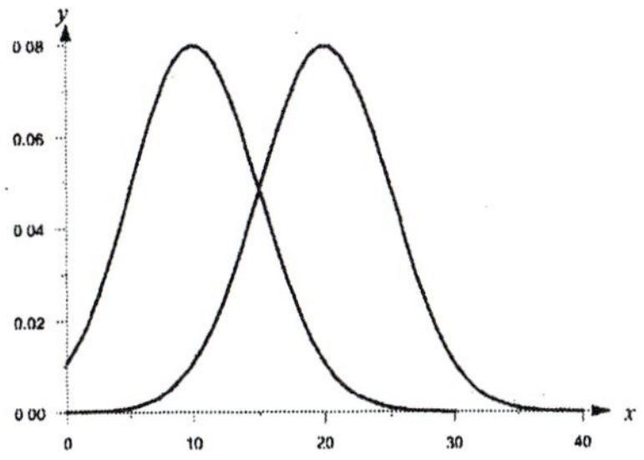

5. $\mu=10$, $\sigma=5$ is the graph in Fig. 22.9.

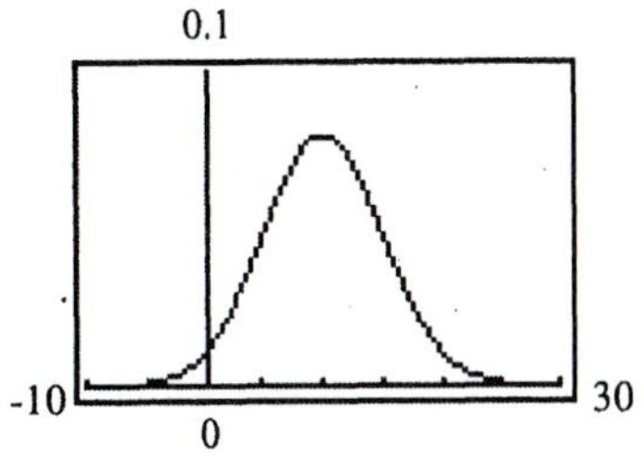

9. $200(0.68)=136$ bags

13. $\mu = 1.50,\ \sigma = 0.05.$ x between 1.45 and 1.55

$$z = \frac{1.45 - 1.50}{0.05} = -1,\ z = \frac{1.55 - 1.50}{0.05} = 1$$

$$(0.3413 + 0.3413)(500) = 341.3$$

341 batteries will have a voltage between 1.45 V and 1.55 V.

17. $\mu = 100,000,\ \sigma = 10,000.$ x between 85,000 and 100,000

$$z = \frac{85,000 - 100,000}{10,000} = -1.5$$

$$(0.4332)(5000) = 2166$$

2166 tires will last between 85,000 km and 100,000 km

21. $\sigma_{\bar{x}} = \frac{\sigma}{\sqrt{n}} = \frac{10,000}{\sqrt{5000}} = 141.$ About 68% have a mean lifetime from 99,859 km to 100,141 km.

25. Since 50% of area is to right of $z = 0$, 25.8% of area will be to the left of $z = 0$ which gives a $z -$ value of 0.7.

29. $\bar{x} + s = 2.262$ and $\bar{x} - s = 2.234$. The readings within these bounds are 2.24, 2.25, and 2.26 with frequencies of $32 + 56 + 41 = 129$. Thus 129 of the 170 readings or 76% fall within these bounds as compared to a normal distribution with 68%.

22.5 Statistical Process Control

1.

Subgroup	Amount of Drug (mg) of five capsules				
1	497	499	502	493	498
2	497	499	500	495	502
3	496	500	507	503	502
4	512	503	488	500	497
5	504	505	500	508	502
6	495	495	501	497	497
7	503	500	507	499	498
8	494	498	497	501	496
9	502	504	505	500	502
10	500	502	500	496	497
11	502	498	510	503	497
12	497	498	496	502	500
13	504	500	495	498	501
14	500	499	498	501	494
15	498	496	502	501	505
16	500	503	504	499	505
17	487	496	499	498	494
18	498	497	497	502	497
19	503	501	500	498	504
20	496	494	503	502	501

Subgroup	Mean x	Range R
1	497.8	9
2	498.6	7
3	501.6	11
4	500.0	24
5	503.8	8
6	497.0	6
7	501.4	9
8	497.2	7
9	502.6	5
10	499.0	6
11	502.0	13
12	498.6	6
13	499.6	9
14	498.4	7
15	500.4	9
16	502.2	6
17	494.8	12
18	498.2	5
19	501.2	6
20	499.2	9
Sums	9993.6	174
Means	499.7	8.7

$$UCL\left(\bar{x}\right) = \bar{\bar{x}} + A_2 R = 499.7 + 0.577(8.7) = 504.7 \text{ mg}$$

$$\text{LCL}\left(\bar{x}\right) = \bar{\bar{x}} - A_2 R = 499.7 - 0.577(8.7) = 494.7 \text{ mg}$$

5.

Hour	Torques	(N·m)	of five	engines	
1	366	352	354	360	362
2	370	374	362	366	356
3	358	357	365	372	361
4	360	368	367	359	363
5	352	356	354	348	350
6	366	361	372	370	363
7	365	366	361	370	362
8	354	363	360	361	364
9	361	358	356	364	364
10	368	366	368	358	360
11	355	360	359	362	353
12	365	364	357	367	370
13	360	364	372	358	365
14	348	360	352	360	354
15	358	364	362	372	361
16	360	361	371	366	346
17	354	359	358	366	366
18	362	366	367	361	357
19	363	373	364	360	358
20	372	362	360	365	367

Subgroup	Mean x	Range R
1	358.8	14
2	365.6	18
3	362.6	15
4	363.4	9
5	352.0	8
6	366.4	11
7	364.8	9
8	360.4	10
9	360.6	8
10	364.0	10
11	357.8	9
12	364.6	13
13	363.8	14
14	354.8	12
15	363.4	14
16	360.8	25
17	360.6	12
18	362.6	10
19	363.6	15
20	365.2	12
Sum	7235.8	248
Mean	361.79	12.4

$CL: \ \bar{x} = 361.79 \text{ N}\cdot\text{m}$

$$UCL(\bar{x}) = \bar{x} + A_2\bar{R} = 361.79 + 0.577(12.4)$$
$$= 368.9 \text{ N}\cdot\text{m}$$
$$LCL(\bar{x}) = \bar{x} - A_2\bar{R} = 361.79 - 0.577(12.4)$$
$$= 354.6 \text{ N}\cdot\text{m}$$

9.

Subgrp	Output	voltages	five	adapters	
1	9.03	9.08	8.85	8.92	8.90
2	9.05	8.98	9.20	9.04	9.12
3	8.93	8.96	9.14	9.06	9.00
4	9.16	9.08	9.04	9.07	8.97
5	9.03	9.08	8.93	8.88	8.95
6	8.92	9.07	8.86	8.96	9.04
7	9.00	9.05	8.90	8.94	8.93
8	8.87	8.99	8.96	9.02	9.03
9	8.89	8.92	9.05	9.10	8.93
10	9.01	9.00	9.09	8.96	8.98
11	8.90	8.97	8.92	8.98	9.03
12	9.04	9.06	8.94	8.93	8.92
13	8.94	8.99	8.93	9.05	9.10
14	9.07	9.01	9.05	8.96	9.02
15	9.01	8.82	8.95	8.99	9.04
16	8.93	8.91	9.04	9.05	8.90
17	9.08	9.03	8.91	8.92	8.96
18	8.94	8.90	9.05	8.93	9.01
19	8.88	8.82	8.89	8.94	8.88
20	9.04	9.00	8.98	8.93	9.05
21	9.00	9.03	8.94	8.92	9.05
22	8.95	8.95	8.91	8.90	9.03
23	9.12	9.04	9.01	8.94	9.02
24	8.94	8.99	8.93	9.05	9.07
	123				

Subgroup	Mean x	Range R
1	8.956	0.23
2	9.078	0.22
3	9.018	0.21
4	9.064	0.19
5	8.974	0.20
6	8.970	0.21
7	8.964	0.15
8	8.974	0.16
9	8.978	0.21
10	9.008	0.13
11	8.960	0.13
12	8.978	0.14
13	9.002	0.17
14	9.022	0.11
15	8.962	0.22
16	8.966	0.15
17	8.980	0.17
18	8.966	0.15
19	8.882	0.12
20	9.000	0.12
21	8.988	0.13
22	8.948	0.13
23	9.026	0.18
24	8.996	0.14
Sum	215.66	3.97
Mean	8.986	0.1654

$CL: \ \overline{\overline{x}} = 8.986 \text{ V}$

$UCL(\overline{x}) = \overline{\overline{x}} + A_2\overline{R} = 8.986 + 0.577(0.1654)$
$= 9.081 \text{ V}$

$LCL(\overline{x}) = \overline{\overline{x}} - A_2\overline{R} = 8.986 - 0.577(0.1654)$
$= 8.891 \text{ V}$

17.

Week	Accounts with errors	Proportion with errors
1	52	0.052
2	36	0.036
3	27	0.027
4	58	0.058
5	44	0.044
6	21	0.021
7	48	0.048
8	63	0.063
9	32	0.032
10	38	0.038
11	27	0.027
12	43	0.043
13	22	0.022
14	35	0.035
15	41	0.041
16	20	0.020
17	28	0.028
18	37	0.037
19	24	0.024
20	42	0.042
Total	738	

$$CL: \ \overline{p} = \frac{738}{1000(20)} = 0.0369$$

$$\sigma_p = \sqrt{\frac{\overline{p}(1-\overline{p})}{n}}$$
$$= \sqrt{\frac{0.0369(1-0.0369)}{1000}} = 0.00596$$

$$UCL(p) = 0.0369 + 3(0.00596) = 0.0548$$
$$LCL(p) = 0.0369 - 3(0.00596) = 0.0190$$

22.6 Linear Regression

1.

x	y	xy	x^2
1	3	3	1
2	7	14	4
3	9	27	9
4	9	36	16
5	12	60	25
15	40	140	55

$n = 5$

$$m = \frac{n\sum xy - \sum x \sum y}{n\sum x^2 - \left(\sum x\right)^2} = \frac{5(140) - 15(40)}{5(55) - 15^2}$$

$$b = \frac{\sum x^2 \sum y - \sum xy \sum x}{n\sum x^2 - \left(\sum x\right)^2} = \frac{55(40) - 140(15)}{5(55) - 15^2}$$

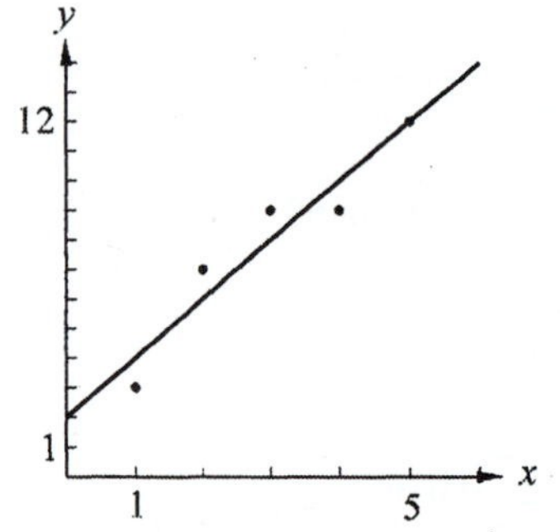

5.

$t(h)$	1.0	2.0	4.0	8.0	10.0	12.0
$y(mg/dL)$	8.7	8.4	7.7	7.3	5.7	5.2

t	y	ty	t^2
1.0	8.7	8.7	1.0
2.0	8.4	16.8	4.0
4.0	7.7	30.8	16.0
8.0	7.3	58.4	64.0
10.0	5.7	57	100
12.0	5.2	62.4	144
37.0	43	234.1	329

$n = 6$

$$m = \frac{6(234.1) - 37.0(43)}{6(329) - 37^2} = -0.308$$

$$b = \frac{329(43) - 234.1(37)}{6(329) - 37^2} = 9.07$$

$y = -0.308t + 9.07$

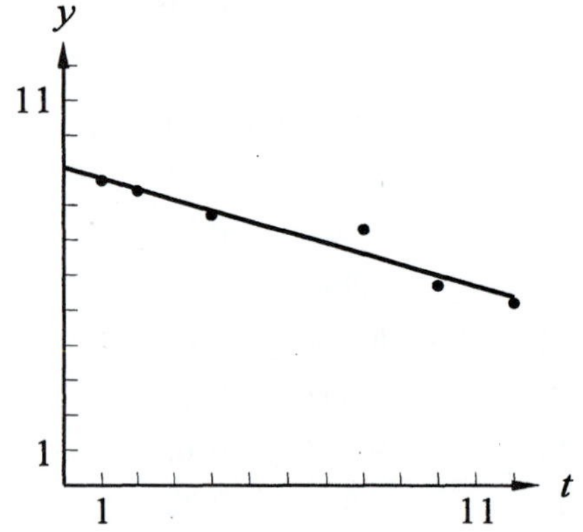

9.

x	h	$xh \cdot (10)^3$	$x^2 \cdot (10)^3$
0	0	0	0
500	1130	565	250
1000	2250	2250	1000
1500	3360	5040	2250
2000	4500	9000	4000
2500	5600	14,000	6250
7500	16,840	30,855	13,750

$n = 6$

$$m = \frac{6(30{,}855 \times 10^3) - (7500)(16{,}840)}{6(13{,}750 \times 10^3) - (7500)^2} = 2.24$$

$$b = \frac{(13{,}750 \times 10^3)(16{,}840) - (30{,}855 \times 10^3)(7500)}{6(13{,}750 \times 10^3) - (7500)^2}$$

$= 5.24$

$h = mx + b;\ h = 2.24x + 5.24$

Plot points:

x	y
750	1690
2250	5045

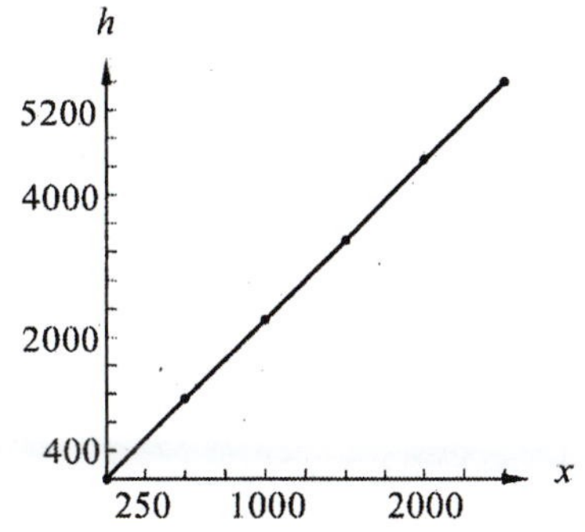

13.

f	V	fV	f^2
0.550	0.350	0.19250	0.302500
0.605	0.600	0.36300	0.366025
0.660	0.850	0.56100	0.435600
0.735	1.10	0.80850	0.540225
0.805	1.45	1.16725	0.648025
0.880	1.80	1.58400	0.774400
4.235	6.15	4.67625	3.066775

$n = 6$

$$m = \frac{6(4.67625)-(4.235)(6.15)}{6(3.066775)-(4.235)^2} = 1.432$$

$$b = \frac{(3.066775)(6.15)-(4.67625)(4.235)}{6(3.066775)-(4.235)^2} = -2.03$$

Therefore, since f is in PHz,

$V = 4.32\times10^{-15} f - 2.03$

$$f_0 = \frac{2.03}{4.32\times10^{-15}} = 0.470\times10^{15}\text{ Hz} = 0.470\text{ PHz}$$

Plot points:

f (PHz)	V
0.600	0.562
0.800	1.426

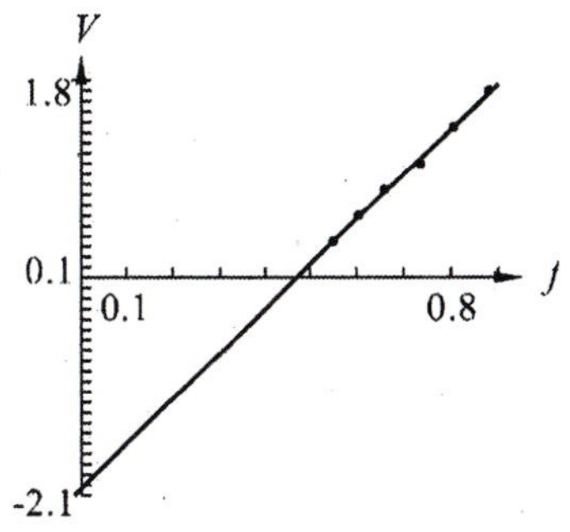

17.

x	1	3	6	5	8	10	4	7	3	8
y	15	12	10	8	9	2	11	9	11	7

$n = 10$

$$s = \sqrt{\overline{x^2} - \left(\overline{x}\right)^2}$$

$$s_x = \sqrt{\frac{373}{10} - \left(\frac{55}{10}\right)^2} = 2.655$$

$$s_y = \sqrt{\frac{990}{10} - \left(\frac{94}{10}\right)^2} = 3.262$$

$m = -1.1064$

$$r = -1.1064\left(\frac{2.655}{3.262}\right) = -0.901$$

22.7 Nonlinear Regression

***1.**

x	$f(x) = x^2$	y	x^2y	$(x^2)^2$
0	0	2	0	0
1	1	3	3	1
2	4	10	40	16
3	9	25	225	81
4	16	44	704	256
5	25	65	1625	625
	55	149	2597	979

$n = 6$

$$m = \frac{6(2597)-55(149)}{6(979)-55^2} = 2.59$$

$$b = \frac{979(149)-2597(55)}{6(979)-55^2} = 1.07$$

$y = 2.59x^2 + 1.07$

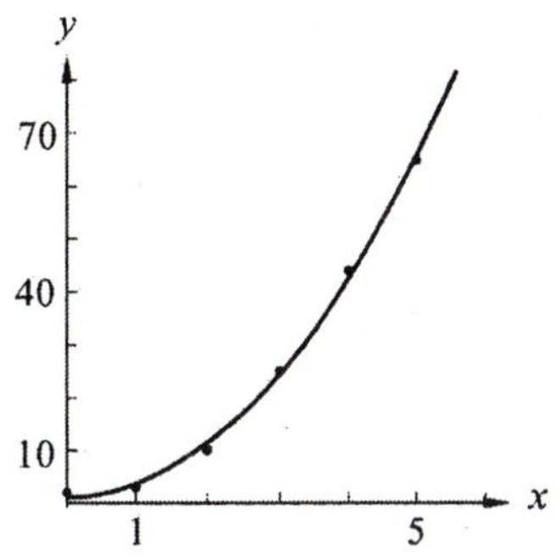

5. $y = mt^2 + b$

t	y	t^2	yt^2	$(t^2)^2$
1.0	6.0	1.0	6.0	1.0
2.0	23	4.0	92	16.0
3.0	55	9.0	495	81.0
4.0	98	16.0	1568	256
5.0	148	25.0	3700	625
	330	55.0	5861.0	979.0

$n = 5$

$$m = \frac{5(5861)-(55.0)(330)}{5(979.0)-(55.0)^2} = 5.97$$

$$b = \frac{(979.0)(330)-(5861)(55.0)}{5(979.0)-(55.0)^2} = 0.38$$

$y = 5.97t^2 + 0.38$

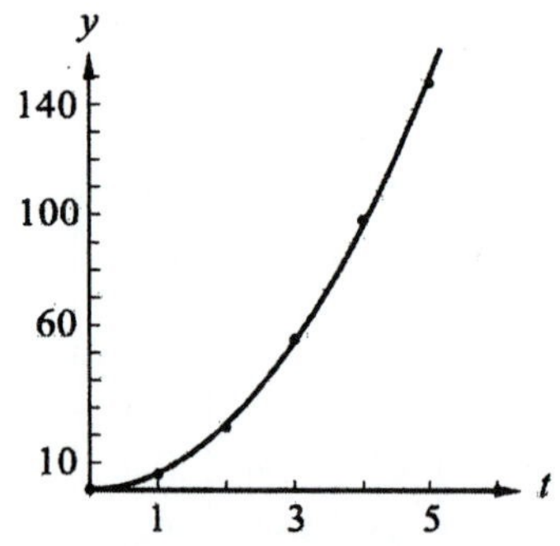

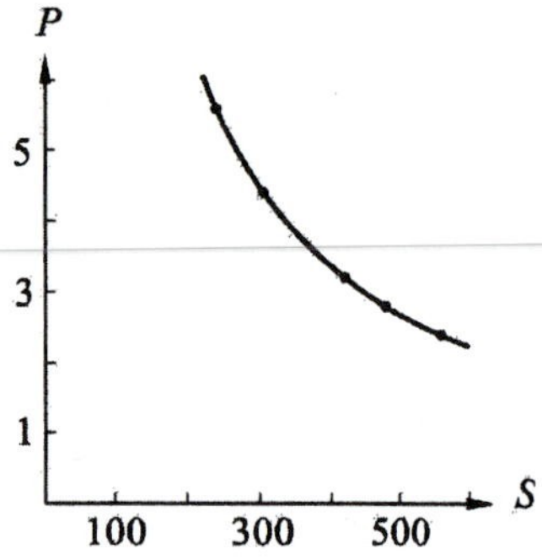

9.

S	$\frac{1}{S}$	P	$\left(\frac{1}{S}\right)P$	$\left(\frac{1}{S}\right)^2$
240	0.004166	5.60	0.023333	1.7361×10^{-5}
305	0.003278	4.40	0.014426	1.0749×10^{-5}
420	0.002380	3.20	0.007619	5.6689×10^{-5}
480	0.002083	2.80	0.005833	4.3402×10^{-5}
560	0.001785	2.40	0.004285	3.1887×10^{-5}
2005	0.013695	18.40	0.055497	4.1308×10^{-5}

$$m = \frac{5(0.554976)-(0.013695)(18.40)}{5(4.13089\times10^{-5})-(0.013695)^2} = 1343$$

$$b = \frac{(4.13089\times10^{-5})(18.40)-(0.05549)(0.01369)}{5(4.13089\times10^{-5})-(0.013695)^2}$$

$$= 1.226612\times10^{-3} \text{ or } 0$$

$$y = 1343\left(\frac{1}{S}\right) + 0 = \frac{1343}{S}$$

Chapter 22 Review Exercises

1. Enter the 20 numbers as list L_1 in the calculator. Then 2nd LIST ← to obtain.

From which

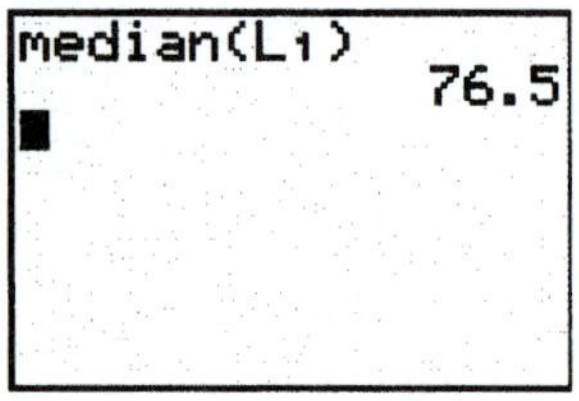

5.

Percent of on-time flights	Frequency	Relative Frequency (%)
67 – 70	3	15
71 – 74	4	20
75 – 78	8	40
79 – 82	3	15
83 – 86	2	10
Total	20	100

9.

Percent of on-time flights	Cumulative Frequency
less than 71	3
less than 75	7
less than 79	15
less than 83	18
less than 87	20

13. See 22.R.11.

17. Enter the 9 power numbers in the calculator as list L_1 and the corresponding frequencies as list L_2.
700 W is the median

```
median(L1,L2)
             700
```

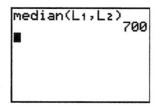

21. Using L_1 and L_2 from problem 17,
17.3 W is the standard deviation

```
stdDev(L1,L2)
     17.32050808
```

25. Enter counts as L_1 and intervals as L_2.
The median is 4.

Counts	0	1	2	3	4	5	6	7	8	9	10
Intervals	3	10	25	45	29	39	26	11	7	2	3

```
median(L1,L2)
               4
```

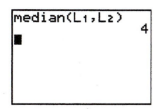

29. Enter speeds in L_1 and number cars in L_2, then

```
mean(L1,L2)
     66.22727273
```

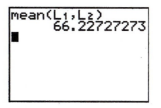

66.2 mi/h is the mean.

33. $CL:\ \bar{p} = \dfrac{540}{500 \cdot 20} = 0.0540$

$$\sigma_p = \sqrt{\frac{\bar{p}\left(1-\bar{p}\right)}{n}} = \sqrt{\frac{0.054\left(1-0.054\right)}{500}} = 0.01011$$

$$UCL\left(p\right) = 0.054 + 3\left(0.01011\right) = 0.0843$$

$$LCL\left(p\right) = 0.054 - 3\left(0.01011\right) = 0.0237$$

37.

```
normalcdf(1.50,2
.50,2.20,0.50)
     .6449902243
Ans*500
     322.4951121
```

There are about 322 readings between 1.5 and 2.5 μg/m^3.

41.

T	R	TR	T^2
0.0	25.0	0	0
20.0	26.8	536	400
40.0	28.9	1156	1600
60.0	31.2	1872	3600
80.0	32.8	2624	6400
100	34.7	3470	10,000
300	179.4	9658	22,000

$\bar{T} = \dfrac{300}{6} = 50.0$ $\qquad \left(\bar{T}\right)^2 = 2500$

$\bar{R} = \dfrac{179.4}{6} = 29.90$ $\qquad \bar{T}\bar{R} = 50.0\left(29.9\right) = 1495$

$\overline{TR} = \dfrac{9658}{6} = 1610$ $\qquad \overline{T^2} = \dfrac{22,000}{6} = 3667$

$s_T^2 = \overline{T^2} - \left(\overline{T}\right)^2 = 3667 - 2500 = 1167$

$m = \dfrac{\overline{TR} - \overline{T}\,\overline{R}}{s_T^2} = \dfrac{1610 - 1495}{1167} = 0.0985$

$b = \overline{R} - m\overline{T} = 29.90 - 0.0977(50.0) = 25.02$

$R = mT + b;\ R = 0.0985T + 25.0$

(Answers may vary due to rounding.)

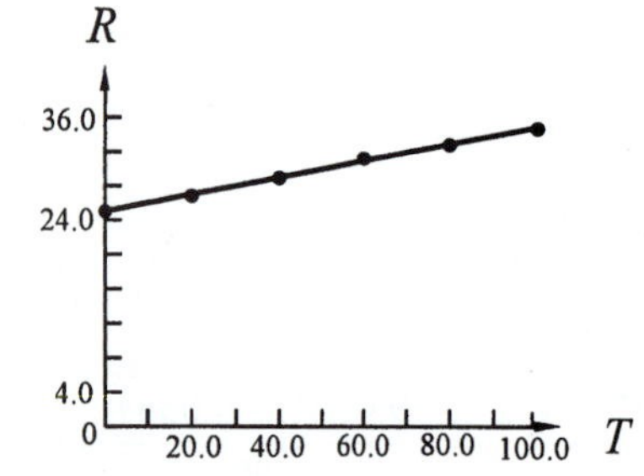

45.

	x	y	xy	x^2
t	t^2	s	$(t^2)(s)$	$(t^2)^2$
0.0	0.0	3000	0	0
3.0	9.0	2960	26,640	81
6.0	36.0	2820	101,520	1296
9.0	81.0	2600	210,600	6561
12.0	144.0	2290	329,760	20,736
15.0	225.0	1900	427,500	50,625
18.0	324.0	1410	456,840	104,976
63.0	819.0	16,980	1,552,860	184,275

$$m = \frac{n\sum xy - \left(\sum x\right)\left(\sum y\right)}{n\sum x^2 - \left(\sum x\right)^2}$$

$$= \frac{7(1,552,860) - (819)(16,980)}{7(184,275) - (819)^2} = -4.90$$

$$b = \frac{\left(\sum x^2\right)\left(\sum y\right) - \left(\sum xy\right)\left(\sum x\right)}{n\sum x^2 - \left(\sum x\right)^2}$$

$$= \frac{(184,275)(16,980) - 1,552,860(819)}{7(184,275) - (819)^2} = 3000$$

$s = -4.90t^2 + 3000$

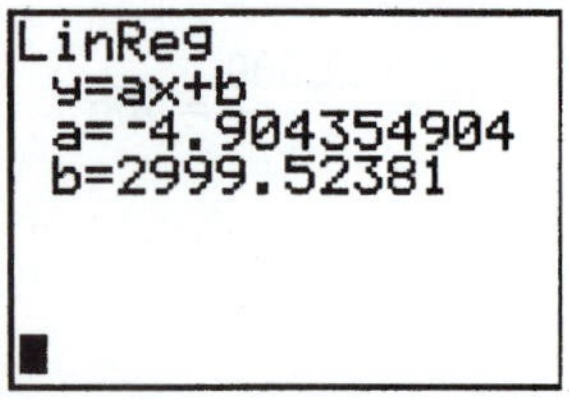

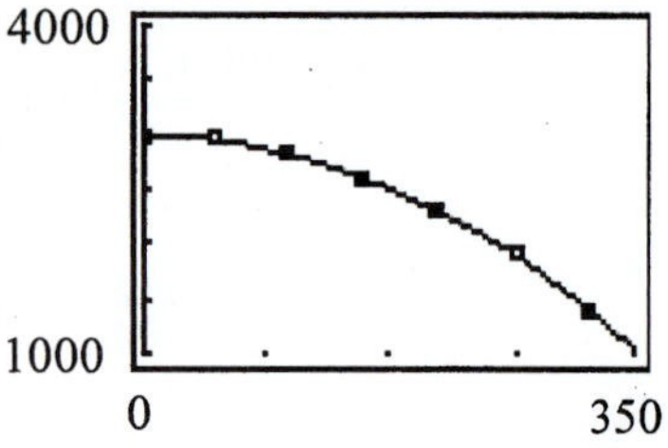

49.

$x(m)$	x^2	$y(m)$
0	0	15
100	10,000	17
200	40,000	23
300	90,000	33
400	160,000	47
500	250,000	65

Use LinReg for x^2 and y.

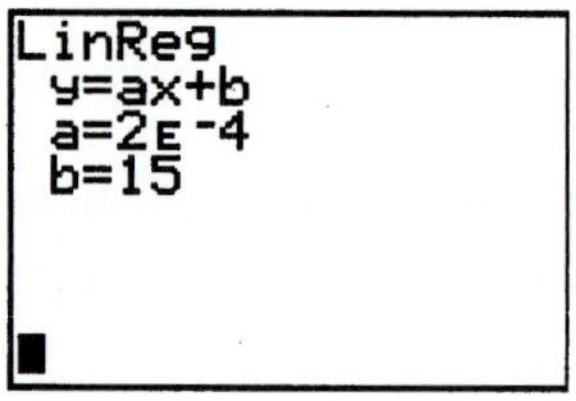

$y = 0.0002x^2 + 15$

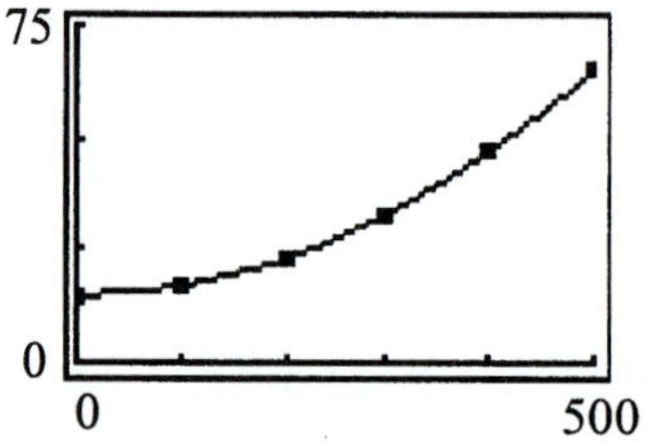

53. Enter t in L_1 as x and A in L_2 as y and use PwrReg.

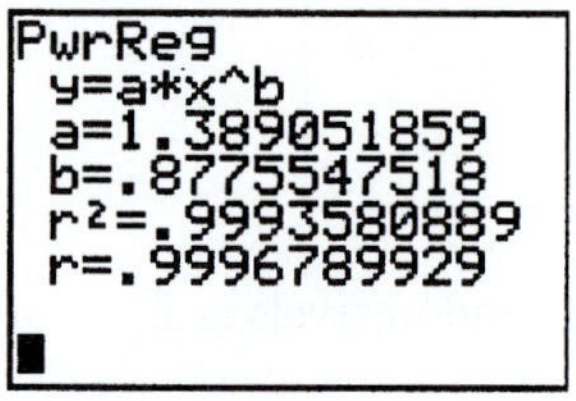

$y = 0.997x + 0.581$

57. There are 7 numbers. The geometric mean is

$\sqrt[7]{(8.0)(8.2)(8.8)(9.5)(9.7)(10.0)(10.7)}$

$= \sqrt[7]{5692009.7} = 5692009.7^{1/7} = 9.2$ ppm

61. Collect data in a chart of income versus education. Find equation of the least squares curve and use it to predict income based on education.

Chapter 23

THE DERIVATIVE

23.1 Limits

1. $f(x)=\dfrac{1}{x+2}$ is not continuous at $x=-2$ because $f(-2)=\dfrac{1}{-2+2}$ is a division by zero and the function is not defined. The condition that the function must exist is not satisfied.

5. $f(x)=3x-2$ is continuous for all real x since it is defined for all x, and any small change in x will produce only a small change in $f(x)$.

9. $f(x)=\sqrt{\dfrac{x}{x-2}}$ is continuous for $x\le 0$ and $x>2$. The function is not defined for $0<x\le 2$. $0<x<2$ gives the square root of a negative and $x=2$ gives division by zero.

13. The graph is not continuous at $x=2$. A small change in x does not produce a small change in y at $x=2$. The function is continuous for $x\le 2$, and continuous for $x>2$.

17. (a) $f(2)=-1$

(b) $\lim\limits_{x\to 2} f(x)$ does not exist

21. $f(x)=\begin{cases} x^2 & \text{for } x<2 \\ 2 & \text{for } x\ge 2\end{cases}$

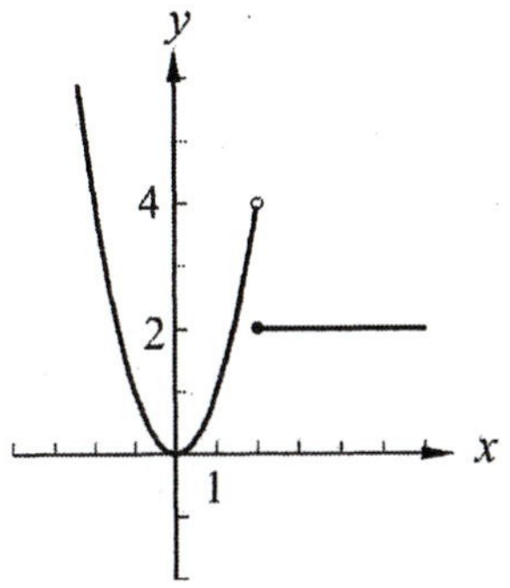

Not continuous at $x=2$. Small change in x around $x=2$ produces a large change in $f(x)$.

25. $f(x)=\dfrac{x^3-x}{x-1}$

x	0.900	0.990	0.999
$f(x)$	1.7100	1.9701	1.9970

x	1.001	1.010	1.100
$f(x)$	2.0030	2.0301	2.3100

Therefore, $\lim\limits_{x\to 1} f(x)=2$

29. $f(x)=\dfrac{2x+1}{5x-3}$

x	10	100	1000
$f(x)$	0.4468	0.4044	0.4004

Therefore, $\lim\limits_{x\to\infty} f(x)=0.4$

33.
$$\lim_{x\to 0}\frac{x^2+x}{x}=\lim_{x\to 0}\frac{x(x+1)}{x}=\lim_{x\to 0}(x+1)=0+1=1$$

37.
$$\lim_{h\to 3}\frac{h^3-27}{h-3}=\lim_{h\to 3}\frac{(h-3)(h^2+3h+9)}{h-3}=\lim_{h\to 3}(h^2+3h+9)=3^2+3(3)+9=27$$

41. For $p=-1$, $\sqrt{p}=\sqrt{-1}=i$. $\lim\limits_{p\to -1}\sqrt{p}(p+1.3)$ does not exist.

45.
$$\lim_{x\to\infty}\frac{3x^2+4.5}{x^2-1.5}=\lim_{x\to\infty}\frac{3+\frac{4.5}{x^2}}{1-\frac{1.5}{x^2}}=3$$

49. $\lim\limits_{x\to 0}\dfrac{x^2-3x}{x}$

x	–0.1	–0.01	–0.001
$f(x)$	–3.10	–2.99	–2.9

x	0.001	0.01	0.1
$f(x)$	–2.999	–2.99	–2.9

We see that $f(x)\to -3$ as $x\to 0$. We now find the limit by changing the algebraic form.

53.

t	T
0	100
1	90
2	81

This is a geometric progression

$a_n = a_1 r^{n-1}$, $r = 0.9$, $a_1 = 100$

Therefore, $T = 100(0.9)^t$

$\lim_{t\to 10} 100(0.9)^t = 34.9^\circ\text{C}$

$\lim_{t\to\infty} 100(0.9)^t = 100\lim_{t\to\infty} 100(0.9)^t = 100(0) = 0^\circ\text{C}$

57. Let $y_1 = \dfrac{2^x - 4}{x-2}$

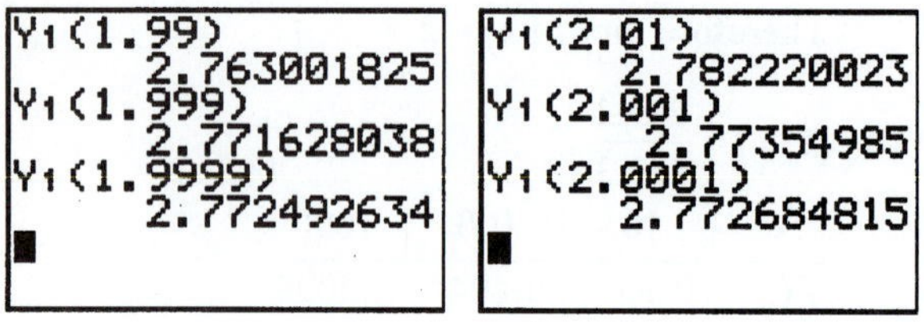

$\lim_{x\to 2} \dfrac{2^x - 4}{x-2} = 2.77$

61. (a) $\lim_{x\to 2^-} f(x) = -1$

(b) $\lim_{x\to 2^+} f(x) = 2$

(c) $\lim_{x\to 2} f(x)$ does not exist

65. For $x > 0$, $f(x) = \dfrac{x}{|x|} = \dfrac{x}{x} = 1$ from which

$\lim_{x\to 0^+} f(x) = 1$

For $x < 0$, $f(x) = \dfrac{x}{|x|} = \dfrac{x}{-x} = -1$ from which

$\lim_{x\to 0^-} f(x) = 1$

Since $RHL \neq LHL$, limit D.N.E. which means $f(x)$ is not continuous.

23.2 The Slope of a Tangent to a Curve

1. Find the slope of a tangent line to $y = x^2 + 3x$ at $(3, 18)$

$$m_{PQ} = \frac{f(3+h) - f(3)}{h}$$

$$= \frac{\left[(3+h)^2 + 3(3+h)\right] - \left[3^2 + 3(3)\right]}{h}$$

$$m_{PQ} = \frac{9 + 6h + h^2 + 9 + 3h - 9 - 9}{h} = \frac{h^2 + 9h}{h}$$

$$m_{PQ} = h + 9$$

$$m_{\tan} = \lim_{h\to 0} m_{PQ} = \lim_{h\to 0}(h+9) = 9$$

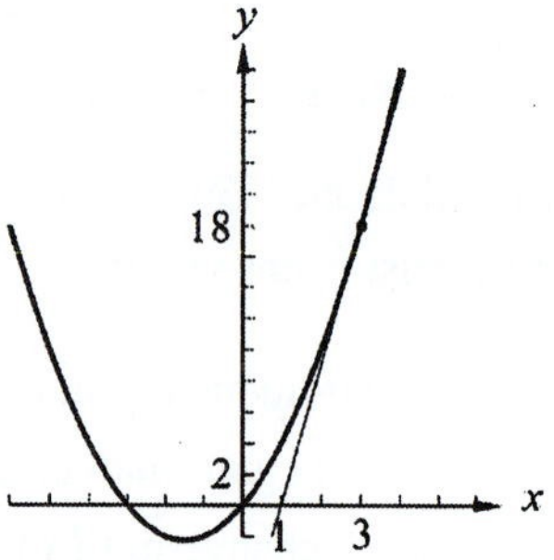

5. $y = 2x^2 + 5x$; $P = (-2, -2)$

	Q_1	Q_2	Q_3	Q_4	P
x_2	−1.5	1.9	1.99	1.999	−2
y_2	−3	−2.28	−2.0298	−2.00299	−2
$y_2 - (-2)$	−1	−0.28	−0.0298	−0.00299	
$x_2 - (-2)$	0.5	0.1	0.01	0.001	
$m = \frac{y_2-(-2)}{x_2-(-2)}$	−2	−2.8	−2.98	−2.998	
$m_{\tan} = -3$					

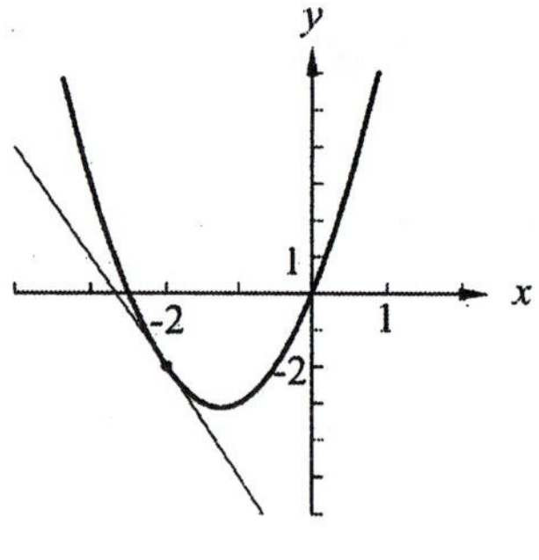

9. $y = 2x^2 + 5x;\ P(-2, 2)$

$$m_{PQ} = \frac{f(-2+h) - f(-2)}{h}$$

$$= \frac{2(-2+h)^2 + 5(-2+h) - \left[2(-2)^2 + 5(-2)\right]}{h}$$

$$m_{PQ} = \frac{8 - 8h + 2h^2 - 10 + 5h - 8 + 10}{h} = \frac{-3h + 2h^2}{h}$$

$$m_{PQ} = -3 + 2h$$

$$m_{\tan} = \lim_{h\to 0}(-3 + 2h) = -3$$

13. $y = 2x^2 + 5x;\ x = -2,\ x = 0.5$

$$m_{PQ} = \frac{f(x_1 + h) - f(x_1)}{h}$$

$$= \frac{2(x_1 + h)^2 + 5(x_1 + h) - (2x_1^2 + 5x_1)}{h}$$

$$m_{PQ} = \frac{2x_1^2 + 4x_1h + 2h^2 + 5x_1 + 5h - 2x_1^2 - 5x_1}{h}$$

$$m_{PQ} = \frac{4x_1h + 2h^2 + 5h}{h} = 4x_1 + 2h + 5$$

$$m_{\tan} = \lim_{h\to 0}(4x_1 + 2h + 5) = 4x_1 + 5$$

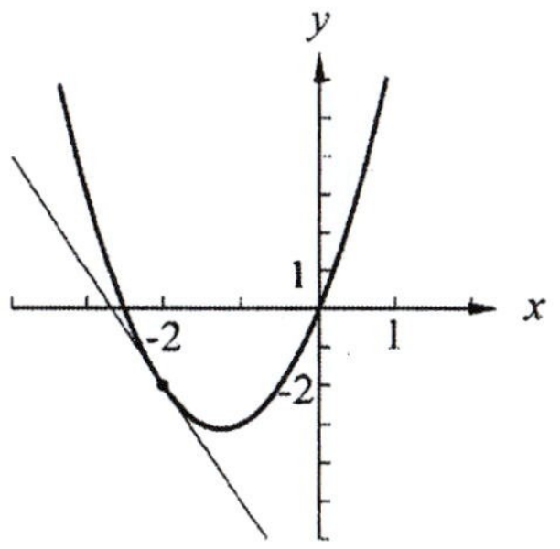

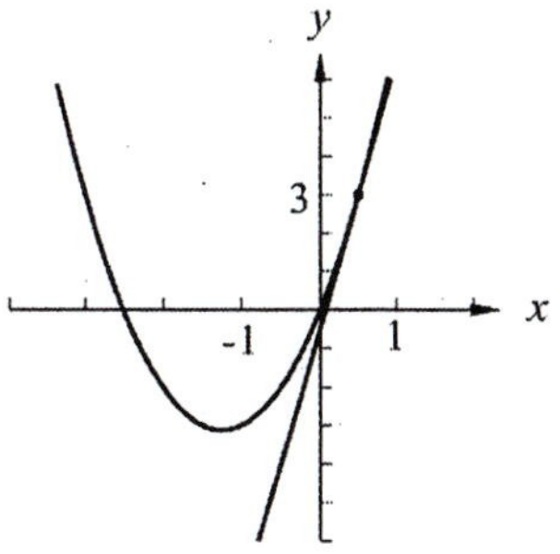

For $x_1 = -2,\ m_{\tan} = 4(-2) + 5 = -3$

For $x_1 = 0.5,\ m_{\tan} = 4(0.5) + 5 = 7$

17. $y = 6x - x^2;\ x = -2,\ x = 3$

$$m_{PQ} = \frac{f(x_1 + h) - f(x_1)}{h}$$

$$= \frac{6(x_1 + h)^2 - (x_1 + h)^2 - (6x_1 - x_1^2)}{h}$$

$$= \frac{6x_1 + 6h - x_1^2 - 2x_1h + h^2 - 6x_1 + x_1^2}{h}$$

$$m_{PQ} = \frac{6h - 2x_1h + h^2}{h} = 6 - 2x_1 + h$$

$$m_{\tan} = \lim_{h\to 0}(6 - 2x_1 + h) = 6 - 2x_1$$

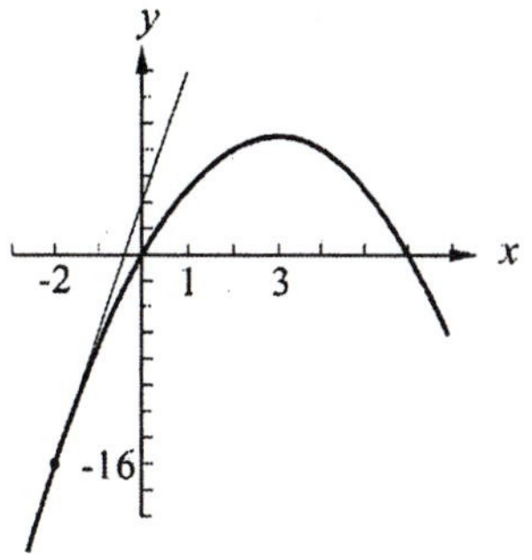

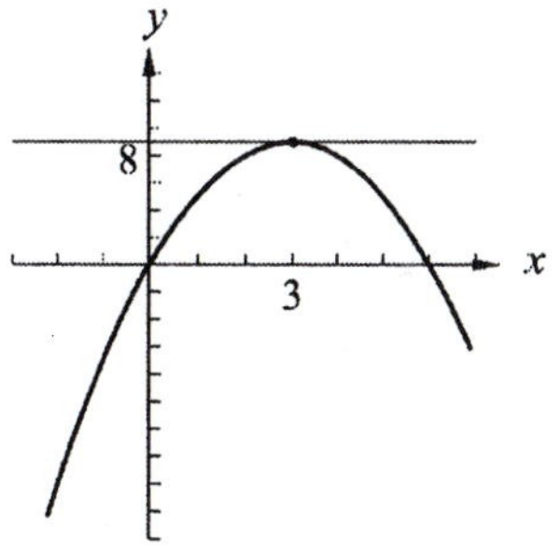

For $x_1 = -2,\ m_{\tan} = 6 - 2(-2) = 10$

For $x_1 = 3,\ m_{\tan} = 6 - 2(3) = 0$

20. $y = 4 - x^4$; $x = 0$, $x = 1$, $x = 2$

$$m_{PQ} = \frac{f(x_1 + h) - f(x_1)}{h} = \frac{4 - (x_1 + h)^4 - (4 - x_1^4)}{h}$$

$$m_{PQ} = \frac{4 - x_1^4 - 4x_1^3 h - 6x_1^2 h^2 - 4x_1 h^3 - h^4 - 4 + x_1^4}{h}$$

$$= \frac{-4x_1^3 h - 6x_1^2 h^2 - 4x_1 h^3 - h^4}{h}$$

$$m_{PQ} = -4x_1^3 - 6x_1^2 h - 4x_1 h^2 - h^3$$

$$m_{\tan} = \lim_{h \to 0}\left(-4x_1^3 - 6x_1^2 h - 4x_1 h^2 - h^3\right) = -4x_1^3$$

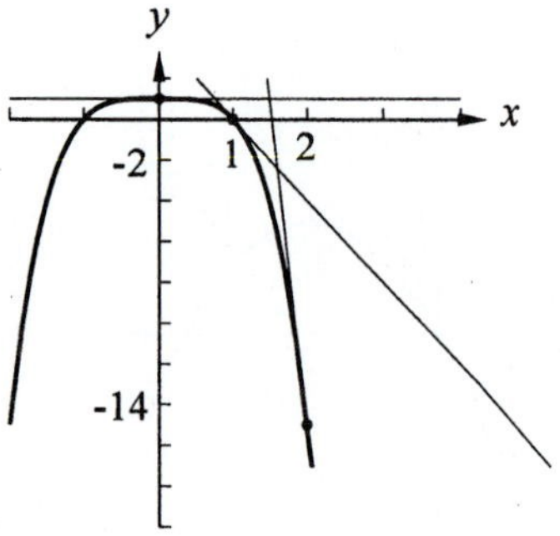

For $x_1 = 0$, $m_{\tan} = -4(0)^3 = 0$

For $x_1 = 1$, $m_{\tan} = -4(1)^3 = -4$

For $x_1 = 2$, $m_{\tan} = -4(2)^3 = -32$

21. $y = x^5$; $x = 0$, $x = 0.5$, $x = 1$

$$m_{PQ} = \frac{f(x_1 + h) - f(x_1)}{h} = \frac{(x_1 + h)^5 - x_1^5}{h}$$

$$= \frac{\left(x_1^5 + 5x_1^4 h + 10x_1^3 h^2 + 10x_1^2 h^3 + 5x_1 h^4 + h^5 - x_1^5\right)}{h}$$

$$m_{PQ} = 5x_1^4 + 10x_1^3 h + 10x_1^2 h^2 + h^4$$

$$m_{\tan} = \lim_{h \to 0}\left(5x_1^4 + 10x_1^3 h + 10x_1^2 h^2 + 5x_1 h^3 + h^4\right) = 5x_1^4$$

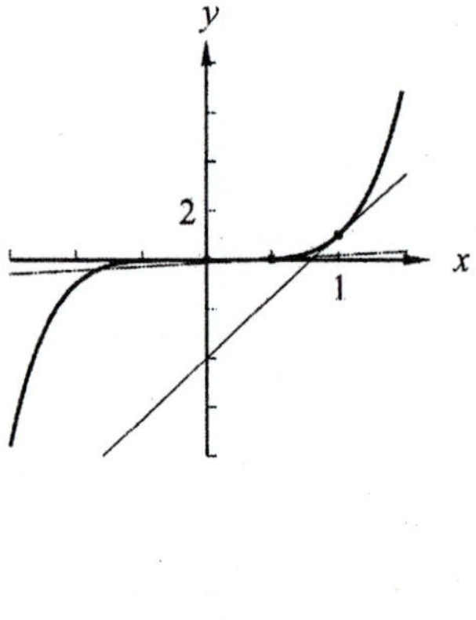

For $x_1 = 0$, $m_{\tan} = 5(0)^4 = 0$

For $x_1 = 0.5$, $m_{\tan} = 5(0.5)^4 = \frac{5}{16}$

For $x_1 = 1$, $m_{\tan} = 5(1)^4 = 5$

25. $y = \frac{1}{3}x^6$; $x = 0$, $x = 0.5$, $x = 1$

Graph $y_1 = \frac{1}{3}x^6$. Use DRAW feature to draw tangent line and *dy* / *dx* feature to find $m_{\tan}$.

$x_1 = 0$, $m_{\tan} = 0$
$x_1 = 0.5$, $m_{\tan} = 0.0625$
$x_1 = 1$, $m_{\tan} = 2$

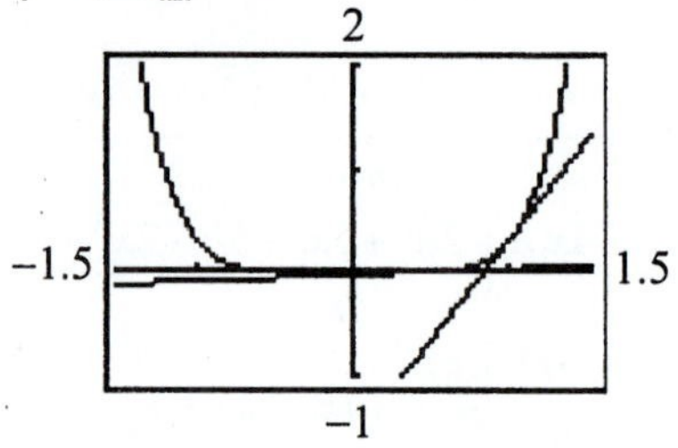

29. $y = 9 - x^3$; $P(2, 1)$, $Q(2.1, -0.261)$

From P to Q, x changes by 0.1 unit and $f(x)$ by -1.261 units. The average change in $f(x)$ for one unit change in x is $\frac{-1.261}{0.1} = -12.61$.

$$m_{PQ} = \frac{f(x_1 + h) - f(x_1)}{h} = \frac{9 - (x_1 + h)^3 - (9 - x_1^3)}{h}$$

$$m_{PQ} = -3x_1^2 - 3x_1 h - h^2$$

$$m_{\tan} = \lim_{h \to 0}\left(-3x_1^2 - 3x_1 h - h^2\right) = -3x_1^2$$

$x = 2$, $m_{\tan} = -12$ (Instantaneous rate of change)

$m_{PQ} = -12.61$ (Average rate of change)

33. $m_{\tan} = -6x_1\Big|_{x_1 = -1} = 6 \Rightarrow m_{\perp} = -\frac{1}{6}$

23.3 The Derivative

1. $y = 4x^2 + 3x$

$$f(x+h) = 4(x+h)^2 + 3(x+h)$$
$$= 4(x^2 + 2xh + h^2) + 3x + 3h$$
$$f(x+h) = 4x^2 + 8xh + 4h^2 + 3x + 3h$$
$$f(x+h) - f(x)$$
$$= 4x^2 + 8xh + 4h^2 + 3x + 3h - 4x^2 - 3x$$
$$= 8xh + 4h^2 + 3h$$
$$\frac{f(x+h)-f(x)}{h} = \frac{8xh + 4h^2 + 3h}{h} = 8x + 4h + 3$$
$$\lim_{h\to 0}\frac{f(x+h)-f(x)}{h} = \lim_{h\to 0}(8x + 4h + 3) = 8x + 3$$
$$f'(x) = 8x + 3$$

5. $y = 1 - 2x$

$$f'(x) = \lim_{h\to 0}\frac{f(x+h)-f(x)}{h}$$
$$= \lim_{h\to 0}\frac{1 - 2(x+h) - (1 - 2x)}{h}$$
$$f'(x) = \lim_{h\to 0}\frac{1 - 2x - 2h - 1 + 2x}{h} = \lim_{h\to 0}\frac{-2h}{h}$$
$$= \lim_{h\to 0} -2 = -2$$

9. $y = \pi x^2$

$$f'(x) = \lim_{h\to 0}\frac{f(x+h)-f(x)}{h}$$
$$= \lim_{h\to 0}\frac{\pi(x+h)^2 - \pi x^2}{h}$$
$$= \lim_{h\to 0}\frac{\pi x^2 + 2\pi xh + h^2 - \pi x^2}{h}$$
$$f'(x) = \lim_{h\to 0}\frac{2\pi xh + h^2}{h} = \lim_{h\to 0}(2\pi x + h)$$
$$f'(x) = 2\pi x$$

13. $y = 8x - 2x^2$

$$f'(x) = \lim_{h\to 0}\frac{f(x+h)-f(x)}{h}$$
$$= \lim_{h\to 0}\frac{8(x+h)^2 - 2(x+h)^2 - (8x - 2x^2)}{h}$$
$$= \lim_{h\to 0}\frac{8x + 8h - 2x^2 - 4xh - 2h^2 - 8x + 2x^2}{h}$$
$$f'(x) = \lim_{h\to 0}\frac{8h - 4xh - 2h^2}{h} = \lim_{h\to 0}(8 - 4x - 2h)$$
$$f'(x) = 8 - 4x$$

17. $y = \dfrac{\sqrt{3}}{x+2}$

$$f'(x) = \lim_{h\to 0}\frac{f(x+h)-f(x)}{h} = \lim_{h\to 0}\frac{\frac{\sqrt{3}}{x+h+2} - \left(\frac{\sqrt{3}}{x+2}\right)}{h}$$
$$f'(x) = \lim_{h\to 0}\frac{\frac{\sqrt{3}}{x+h+2} - \left(\frac{\sqrt{3}}{x+2}\right)}{h} = \lim_{h\to 0}\frac{\frac{\sqrt{3}(x+2) - \sqrt{3}(x+h+2)}{(x+h+2)(x+2)}}{h}$$
$$f'(x) = \lim_{h\to 0}\frac{\frac{\sqrt{3}(x+2) - \sqrt{3}(x+h+2)}{(x+h+2)(x+2)}}{h} = \lim_{h\to 0}\frac{\frac{\sqrt{3}x + 2\sqrt{3} - \sqrt{3}x - \sqrt{3}h - 2\sqrt{3}}{(x+h+2)(x+2)}}{h}$$
$$f'(x) = \lim_{h\to 0}\frac{\frac{-\sqrt{3}h}{(x+h+2)(x+2)}}{h} = \lim_{h\to 0}\frac{-\sqrt{3}}{(x+h+2)(x+2)}$$
$$f'(x) = \frac{-\sqrt{3}}{(x+2)^2}$$

$$f'(x) = \lim_{h\to 0}\frac{f(x+h)-f(x)}{h} = \lim_{h\to 0}\frac{\frac{x+h}{x+h-1} - \frac{x}{x-1}}{h}$$
$$f'(x) = \lim_{h\to 0}\frac{\frac{(x+h)(x-1) - x(x+h-1)}{(x+h-1)(x-1)}}{h}$$
$$f'(x) = \lim_{h\to 0}\frac{-h}{(x+h-1)(x-1)}$$
$$f'(x) = \lim_{h\to 0}\frac{-1}{(x+h-1)(x-1)}$$
$$f'(x) = \frac{-1}{(x-1)^2}$$

21. $y = \frac{2}{x^2}$

$$f'(x) = \lim_{h\to 0}\frac{f(x+h)-f(x)}{h} = \lim_{h\to 0}\frac{\frac{2}{(x+h)^2}-\frac{2}{x^2}}{h}$$

$$f'(x) = \lim_{h\to 0}\frac{2x^2-2(x^2+2xh+h^2)}{hx^2(x+h)^2}$$

$$= \lim_{h\to 0}\frac{2x^2-2x^2-4xh-2h^2}{hx^2(x+h)^2}$$

$$f'(x) = \lim_{h\to 0}\frac{-4x-2h}{x^2(x+h)^2} = \frac{-4x}{x^4}$$

$$f'(x) = \frac{-4}{x^3}$$

25. $y = x^4 - \frac{2}{x}$

$$f(x+h) = (x+h)^4 - \frac{2}{x+h}$$

$$= x^4 + 4x^3h + 6x^2h^2 + 4xh^2 + 4xh^3 + h^4 - \frac{2}{x+h}$$

$$f(x+h) - f(x)$$

$$= 4x^3h + 6x^2h^2 + 4xh^3 + h^4 - \frac{2x-2(x+h)}{x(x+h)}$$

$$f(x+h) - f(x)$$

$$= 4x^3h + 6x^2h^2 + 4xh^3 + h^4 - \frac{-2h}{x(x+h)}$$

$$\frac{f(x+h)-f(x)}{h}$$

$$= 4x^3 + 6x^2h + 4xh^2 + h^3 + \frac{2}{x(x+h)}$$

$$f'(x) = \lim_{h\to 0}\frac{f(x+h)-f(x)}{h} = 4x^3 + \frac{2}{x^2}$$

$$f'(x) = 4x^3 + \frac{2}{x^2}$$

29. $y = \frac{11}{3x+2}$; $(3, 1)$

$$f(x+h) - f(x) = \frac{11}{3(x+h)} - \frac{11}{3x+2}$$

$$= \frac{-33h}{(3x+2)(3(x+h)+2)}$$

$$\frac{f(x+h)-f(x)}{h} = \frac{-33}{(3x+2)(3(x+h)+2)}$$

$$f'(x) = \lim_{h\to 0}\frac{-33}{(3x+2)(3(x+h)+2)} = \frac{-33}{(3x+2)^2}$$

$$\left.\frac{dy}{dx}\right|_{(3,\,1)} = \frac{-33}{(3(3)+2)^2} = \frac{-3}{11}$$

33. $y = \frac{3}{x^2-1}$

$$\frac{f(x+h)-f(x)}{h} = \frac{\frac{3}{(x+h)^2-1}-\frac{3}{x^2-1}}{h}$$

$$= \frac{-3(2x+h)}{(x^2-1)(x^2+2xh+h^2-1)}$$

$$f'(x) = \lim_{h\to 0}\frac{-3(2x+h)}{(x^2-1)(x^2+2xh+h^2-1)}$$

$$= \frac{-6x}{(x^2-1)^2},\ x \neq \pm 1$$

Function is differentiable for all $x \neq \pm 1$.

37. $y = f(x) = 2x^2 - 16x$

$f'(x) = 4x - 16 = 0$ for $x = 4$

$f(4) = 2(4)^2 - 16(4) = -32$

At $(4, -32)$ on curve $y = 2x^2 - 16x$ the tangent line is horizontal.

41. For $y = x^4 + x^3 + x^2 + x$, $\frac{dy}{dx} = 4x^3 + 3x^2 + 2x + 1$ from which, as a guess, $y = x^n$, $n > 0$, would have $\frac{dy}{dx} = nx^{n-1}$.

23.4 The Derivative as an Instantaneous Rate of Change

1. $s = 48t - 16t^2$

$$v = \lim_{h \to 0} \frac{48(t+h) - 16(t+h)^2 - 48t + 16t^2}{h}$$

$$= \lim_{h \to 0}\left(-16(h + 2t - 3)\right) = -32t + 48$$

$$\left.\frac{ds}{dt}\right|_{t=2} = -32(2) + 48 = -16 \text{ ft/s}$$

$$\left.\frac{ds}{dt}\right|_{t=4} = -32(4) + 48 = -80 \text{ ft/s}$$

5. $y = \frac{6}{3x+1}; (-3, -2)$

$$m_{\tan} = \lim_{h \to 0} \frac{\frac{16}{3(x+h)+1} - \frac{16}{3x+1}}{h}$$

$$m_{\tan} = \lim_{h \to 0} \frac{\frac{48x+16-48x-48h-16}{(3(x+h)+1)(3x+1)}}{h}$$

$$m_{\tan} = \lim_{h \to 0} \frac{-48}{(3(x+h)+1)(3x+1)} = \frac{-48}{(3x+1)^2}$$

$$m_{\tan} = \left.\frac{dy}{dx}\right|_{(-3,\,-2)} = \frac{-48}{(3(-3)+1)^2} = -\frac{3}{4}$$

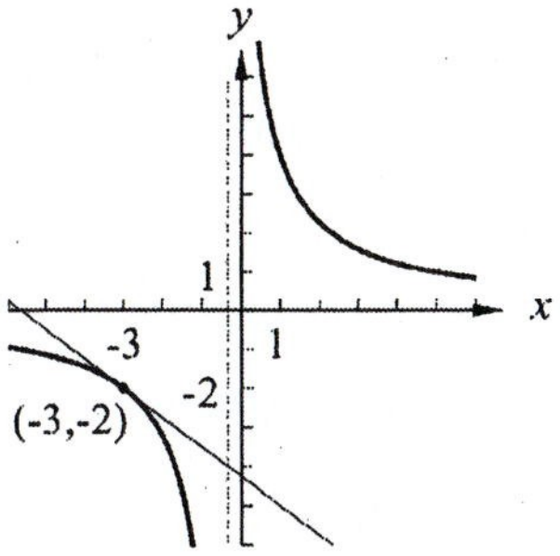

9. $s = 3t^2 - 4t;\ t = 2$

$$s = 3(2)^2 - 4(2) = 4$$

t(s)	1.0	1.5	1.9	1.99	1.999
s(ft)	−1.0	0.75	3.23	3.9203	3.9092003
$4 - s$(ft)	5.0	3.25	0.77	0.0797	0.007997
$h = 2 - t$(s)	1.0	0.5	0.1	0.01	0.001
$v = \frac{4-s}{h}$(ft/s)	5.0	6.5	7.7	7.97	7.997

$v = 8$ ft/s when $t = 2$ s

13. $s = 3t^2 - 4t;\ t = 2$

$$v = \lim_{h \to 0} \frac{3(t+h)^2 - 4(t+h) - 3t^2 + 4t}{h}$$

$$= \lim_{h \to 0} \frac{3t^2 + 6th + 3h^2 - 4t - 4h - 3t^2 + 4t}{h}$$

$$v = \lim_{h \to 0} \frac{6th + 3h^2 - 4h}{h} = \lim_{h \to 0}(6t + 3 - 4) = 6t - 4$$

$$v = \left.\frac{ds}{dt}\right|_{t=2} = 6(2) - 4 = 8 \text{ ft/s}$$

17. $s = 12t^2 - t^4$

$$v = \frac{ds}{dt} = \lim_{h \to 0} \frac{12(t+h)^2 - (t+h)^4 - 12t^2 + t^4}{h}$$

$$v = \lim_{h \to 0} \frac{12(t^2 + 2th + h^2) - t^4 - 4t^3h - 6t^2h^2 - 4th^3 - t^4 - 12t^2 + t^4}{h}$$

$$v = \lim_{h \to 0} \frac{24th + 12h^2 - 4t^3h - 6t^2h^2 - 4th^3}{h}$$

$$v = \lim_{h \to 0}\left(24t + 12h - 4t^3 - 6t^2h - 4th^2\right) = 24 - 4t^3$$

21. $v = 6t^2 - 4t + 2$

$$a = \frac{dv}{dt} = \lim_{h \to 0} \frac{6(t+h)^2 - 4(t+h) + 2 - 6t^2 + 4t - 2}{h}$$

$$a = \lim_{h \to 0} \frac{6t^2 + 12th + 6t^2 - 4t - 4h + 2 - 6t^2 + 4t - 2}{h}$$

$$a = \lim_{h \to 0}(12t - 4)$$

$$a = 12t - 4$$

25. $c = 2\pi r$

$$\frac{dc}{dt} = 2\pi\frac{dr}{dt} = 2\pi(-0.0015)$$

$$\frac{dc}{dt} = -0.0094 \text{ cm/min}$$

29. $q = 30 - 2t$

$$i = \frac{dq}{dt} = \lim_{h \to 0} \frac{30 - 2(t+h) - 30 + 2t}{h} = \lim_{h \to 0} \frac{-2h}{h}$$

$i = -2$

33. $P = 500 + 250\text{ m}^2$

$$\frac{dP}{dm} = \lim_{h \to 0} \frac{500 + 250(m+h)^2 - 500 - 250m^2}{h}$$

$$\frac{dP}{dm} = \lim_{h \to 0} \frac{250m^2 + 500mh + 250h^2 - 250m^2}{h}$$

$$\frac{dP}{dm} = \lim_{h \to 0} (500 + 250h) = 500m$$

$$\left.\frac{dP}{dm}\right|_{m=0.92} = 500(0.92) = 460\text{ W}$$

37. $V = \dfrac{48}{t+3}$

$$\frac{dV}{dt} = \lim_{h \to 0} \frac{\frac{48}{t+h+3} - \frac{48}{t+3}}{h} = \frac{-48}{(t+3)^2}$$

$$\left.\frac{dV}{dt}\right|_{t=3} = \frac{-48}{(3+3)^2} = -\$1300/\text{year}$$

41. $r = k\sqrt{\lambda}$, $r = 3.72 \times 10^{-2}$ m, $\lambda = 59.2 \times 10^{-8}$ m

$$k = \frac{3.72 \times 10^{-2}}{\sqrt{59.2 \times 10^{-8}}} = 48.35$$

$$\frac{dr}{d\lambda} = \lim_{h \to 0} \frac{k\sqrt{\lambda + h} - k\sqrt{\lambda}}{h} = \frac{k}{2\sqrt{\lambda}} = \frac{48.35}{2\sqrt{\lambda}}$$

$$\frac{dr}{d\lambda} = \frac{24.2}{\sqrt{\lambda}}$$

23.5 Derivatives of Polynomials

1. $v = r^9$

$$\frac{dv}{dr} = 9r^{9-1}$$

$$\frac{dv}{dr} = 9r^8$$

5. $y = x^5;\ \dfrac{dy}{dx} = 5x^{5-1} = 5x^4$

9. $y = 5x^4 - 3\pi;\ \dfrac{dy}{dx} = 5(4x^{4-1}) - 0 = 20x^3$

13. $p = 5r^3 - 2r + 1;\ \dfrac{dp}{dr} = 5(3r^2) - 2 + 0$

$= 15r^2 - 2$

17. $f(x) = -6x^7 + 5x^3 + \pi^2$

$$\frac{f(x)}{dx} = -6(7x^6) + 5(3x^2) + 0 = -42x^6 + 15x^2$$

21. $y = 6x^2 - 8x + 1;$

$$\frac{dy}{dx} = \frac{d(6x^2)}{dx} - \frac{d(8x)}{dx} + \frac{d(1)}{dx} = 12x - 8 + 0$$

Since the derivative is a function of only x, we now evaluate it for $x = 2$.

$$\left.\frac{dy}{dx}\right|_{x=2} = 12(2) - 8 = 24 - 8 = 16$$

25. $y = 2x^6 - 4x^2;\ m_{\tan} = \dfrac{dy}{dx} = 12x^5 - 8x$

$$\left.\frac{dy}{dx}\right|_{x=-1} = m_{\tan} = 12(-1)^5 - 8(-1) = -12 + 8 = -4$$

Move the trace to $x = -1$ and observe that the function is decreasing and that the slope is negative.

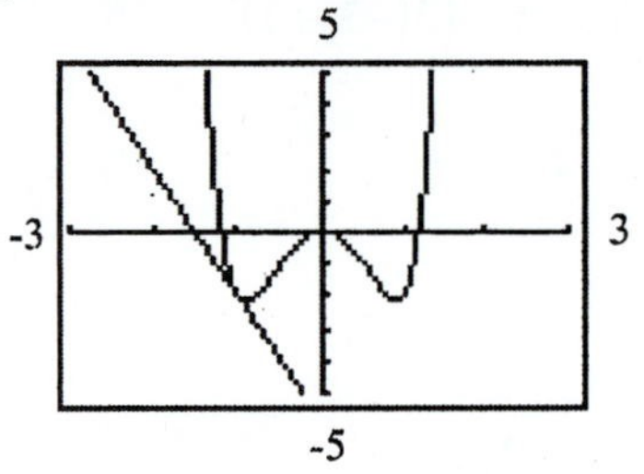

29. $s = 6t^5 - 5t + 2;\ v = \dfrac{ds}{dt} = 30t^4 - 5$

33. $s = 2t^3 - 4t^2;\ t = 4$

$$v = \frac{ds}{dt} = 2(3t^2) - 4(2t) = 6t^2 - 8t$$

$$v\big|_{t=4} = 6(4^2) - 8(4) = 64$$

37. $y = 3x^2 - 6x;\ m_{\tan} = \dfrac{dy}{dx} = 6x - 6$

Tangent is parallel where slope is zero.
Therefore, $6x - 6 = 0;\ x = 1$

41. $x-3y=16 \Rightarrow y=\frac{1}{3}x-\frac{16}{3};\ m_{\perp}=-3$

$y=2x^2-7x$

$\frac{dy}{dx}=4x-7=-3 \Rightarrow x=1$

$y=2(1)^2-7(1)=-5$

The tangent line to $y=2x-7x$ at $(1,-5)$ is perpendicular to the line $x-3y=16$.

45. $V=\pi r^2 \cdot h$

$V=\pi r^2 \cdot r=r^3$

$\frac{dV}{dr}=3\pi r^2$

49. $R=16.0+0.450T+0.0125T^2$

$\frac{dR}{dT}=0.450+0.0250T\big|_{T=115}=3.33\ \Omega/^\circ\text{C}$

53. $h=0.000104x^4-0.0417x^3+4.21x^2-8.33x$

$\frac{dh}{dx}=0.000416x^3-0.1251x^2+8.42x-8.33\big|_{x=120}$

$\frac{dh}{dx}=-80.5\ \text{m/km}$

23.6 Derivatives of Products and Quotients of Functions

1. $p(x)=(5-3x^2)(3-2x),\ u=5-3x^2,\ v=3-2x$

$\frac{d(uv)}{dx}=u\frac{dv}{dx}+v\frac{du}{dx}$

$p'(x)=(5-3x^2)(-2)+(3-2x)(-6x)$

$p'(x)=18x^2-18x-10$

5. $s=(3t+2)(2t-5)$

$u=3t+2,\ v=2t-5$

$\frac{du}{dt}=3 \quad \frac{dv}{dt}=2$

$\frac{ds}{dt}=(3t+2)(2)+(2t-5)(3)$

$=6t+4+6t-15=12t-11$

9. $y=(2x-7)(5-2x);\ u=(2x-7);\ v=(5-2x)$

$\frac{dy}{dx}=(2x-7)(-2)+(5-2x)(2)$

$=-4x+14+10-4x=-8x+24$

$y=(2x-7)(5-2x)$

$=10x-4x^2-35+14x=-4x^2+24x-35$

$\frac{dy}{dx}=-8x+24$

13. $y=\frac{x}{2x+3};\ u=x;\ \frac{du}{dx}=1;\ v=2x+3;\ \frac{dv}{dx}=2$

$\frac{dy}{dx}=\frac{(2x+3)(1)-x(2)}{(2x+3)^2}$

$=\frac{2x+3-2x}{(2x+3)^2}=\frac{3}{(2x+3)^2}$

17. $y=\frac{6x^2}{3-2x};\ u=6x^2;$

$\frac{du}{dx}=12x;\ v=3-2x;\ \frac{dv}{dx}=-2$

$\frac{dy}{dx}=\frac{(3-2x)(12x)-(6x^2)(-2)}{(3-2x)^2}$

$=\frac{36x-24x^2+12x^2}{(3-2x)^2}=\frac{36x-12x^2}{(3-2x)^2}$

21. $f(x)=\frac{3x+8}{x^2+4x+2}$

$\frac{df(x)}{dx}=\frac{(x^2+4x+2)(3)-(3x+8)(2x+4)}{(x^2+4x+2)^2}$

$=\frac{-3x^2-16x-26}{(x^2+4x+2)^2}$

25. $y=(3x-1)(4-7x)$

$\frac{dy}{dx}=(3x-1)(-7)+(4-7x)(3)$

$=-21x+7+12-21x=-42x+19$

$\frac{dy}{dx}\big|_{x=3}=-42(3)+19=-126+19=-107$

29. $y=\dfrac{3x-5}{2x+3}$; $y=3x-5$; $v=2x+3$; $du=3dx$;

$dv=2dx$

$$\frac{dy}{dx}=\frac{(2x+3)(3)-(3x-5)(2)}{(2x+3)^2}=\frac{6x+9-6x+10}{(2x+3)^2}$$

$$=\frac{19}{(2x+3)^2}$$

$$\left.\frac{dy}{dx}\right|_{x=-2}=\frac{19}{(2(-2)+3)^2}=\frac{19}{1}=19$$

33. For $u=c$ the product rule $\dfrac{d(u\cdot v)}{dx}=u\cdot\dfrac{dv}{dx}+v\cdot\dfrac{du}{dx}$

becomes $\dfrac{d(cv)}{dx}=c\dfrac{dv}{dx}+v\dfrac{dc}{dx}=c\dfrac{dv}{dx}+v(0)$

$\dfrac{d(cv)}{dx}=c\dfrac{dv}{dx}$, Equation 23.10

37. $\dfrac{d}{dx}\left(x^2f(x)\right)=xf'(x)+2xf(x)$

41.

$$y=(4x+1)(x^4-1);\ (-1,0)$$

$$\frac{dy}{dx}=(4x+1)(4x^3)+(x^4-1)4$$

$$=16x^4+4x^3+4x^4-4$$

$$\left.\frac{dy}{dx}\right|_{x=-1}=20x^4+4x^3-4\Big|_{x=-1}=12$$

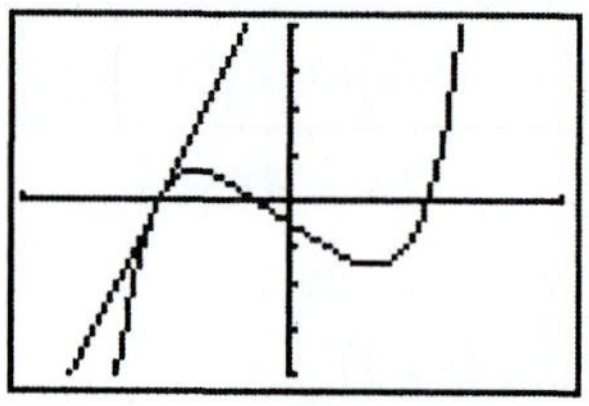

45. $P=VI=\left(0.048-1.20t^2\right)(2.00-0.800t),\ 0\le t\le 2$

$$\frac{dP}{dt}=2.88t^2-4.80t-0.0384\Big|_{t=0.15}=-0.694\text{ W/s}$$

49. $s=\left(t^2-8t\right)\left(2t^2+t+1\right)$

$$\frac{ds}{dt}=\left(t^2-8t\right)(4t+1)+\left(2t^2+t+1\right)(2t-8)$$

$$v=4t^3+t^2-32t^2-8t+4t^3+2t^2+2t-16t^2-8t-8$$

$$v=8t^3-45t^2-14t-8$$

53. $r_f=\dfrac{2\left(R^2+Rr+r^2\right)}{3(R+r)}$

$$\frac{dr_f}{dR}=\frac{2(2R+r)(3)(R+r)-2\left(R^2+Rr+r^2\right)(3)}{9(R+r)^2}$$

$$=\frac{6(2R+r)(R+r)-6\left(R^2+Rr+r^2\right)}{9(R+r)^2}$$

$$=\frac{12R^2+18Rr+6r^2-6R^2-6Rr-6r^2}{9(R+r)^2}$$

$$=\frac{6R^2+12Rr}{9(R+r)^2}=\frac{6R(R+2)}{9(R+r)^2}=\frac{2R(R+2r)}{3(R+r)^2}$$

23.7 The Derivative of a Power of a Function

1. $p(x)=\left(2+3x^3\right)^4$

$$p'(x)=4\left(2+3x^3\right)^3\left(9x^2\right)$$

$$p'(x)=36x^2\left(2+3x^3\right)^3$$

5. $y=\sqrt{x}=x^{1/2}$

$$\frac{dy}{dx}=\frac{1}{2}x^{1/2-1}=\frac{1}{2}x^{-1/2}=\frac{1}{2}\left(\frac{1}{x^{1/2}}\right)=\frac{1}{2x^{1/2}}$$

9. $y=\dfrac{3}{\sqrt[3]{x}}=\dfrac{3}{x^{1/3}}=3x^{-1/3}$; $\dfrac{dy}{dx}=3\left(-\dfrac{1}{3}x^{-1/3-1}\right)$

$$=-1x^{-4/3}=-1\left(\frac{1}{x^{4/3}}\right)=-\frac{1}{x^{4/3}}$$

13. $y=5\left(x^2+1\right)^4(2x)=10x\left(x^2+1\right)^4$

$$\frac{dy}{dx}=5\left(x^2+1\right)^4(2x)=10x\left(x^2+1\right)^4$$

17. $y=\left(2x^3-3\right)^{1/3}$;

$$\frac{dy}{dx}=\frac{1}{3}\left(2x^3-3\right)^{-2/3}\left(6x^2\right)=\frac{2x^2}{\left(2x^3=3\right)^{2/3}}$$

21. $y=4\left(2x^4-5\right)^{0.75}$; $u=2x^4-5$; $\frac{du}{dx}=8x^3$

$$\frac{dy}{dx}=4\left[0.75\left(2x^4-5\right)^{-0.25}\left(8x^3\right)\right]=\frac{24x^3}{\left(2x^4-5\right)^{0.25}}$$

25. $u=v\sqrt{8v+5}=v(8v+5)^{1/2}$

$$\frac{du}{dv}=v\left(\frac{1}{2}\right)(8v+5)^{-1/2}(8)+(8v+5)^{1/2}(1)$$

$$=4v(8v+5)^{-1/2}+(8v+5)^{1/2}(1)$$

$$=\frac{4v}{(8v+5)^{1/2}}+\frac{(8v+5)}{(8v+5)^{1/2}}=\frac{12v+5}{(8v+5)^{1/2}}$$

29. $y=\frac{2x\sqrt{x+2}}{x+4}$

$$\frac{dy}{dx}=\frac{(x+4)\left[2x\cdot\frac{1}{2\sqrt{x+2}}+2\sqrt{x+2}\right]-2x\sqrt{x+2}(1)}{(x+4)^2}$$

$$\frac{dy}{dx}=\frac{x(x+4)+2(x+4)(x+2)-2x(x+2)}{(x+4)^2\sqrt{x+2}}$$

$$\frac{dy}{dx}=\frac{x^2+4x+2x^2+12x+16-2x^2-4x}{(x+4)^2\sqrt{x+2}}$$

$$\frac{dy}{dx}=\frac{x^2+12x+16}{(x+4)^2\sqrt{x+2}}$$

33. $y=\sqrt{3x+4}$; $x=7$

$y=(3x+4)^{1/2}$; $u=3x+4$; $n=\frac{1}{2}$; $\frac{du}{dx}=3$

$$\frac{dy}{dx}=\frac{1}{2}(3x+4)^{-1/2}(3)=\frac{3}{2}(3x+4)^{-1/2}$$

$$=\frac{3}{2\sqrt{3x+4}}$$

$$\left.\frac{dy}{dx}\right|_{x=7}=\frac{3}{2\sqrt{3(7)+4}}=\frac{3}{2\sqrt{25}}=\frac{3}{2(5)}=\frac{3}{10}$$

37. $\frac{d}{dx}\left(x^{3/2}\right)=\frac{d}{dx}\left(x\left(x^{1/2}\right)\right)=x\left(\frac{1}{2}x^{-1/2}\right)+x^{1/2}(1)$

$$=\frac{1}{2}x^{1/2}+x^{1/2}=\frac{3}{2}x^{1/2}$$

41. $y=\frac{x^2}{\sqrt{x^2+1}}=\frac{x^2}{\left(x^2+1\right)^{1/2}}$

$$\frac{dy}{dx}=\frac{\left(x^2+1\right)^{1/2}2x-x^2\frac{1}{2}\left(x^2+1\right)^{-1/2}(2x)}{\left(x^2+1\right)}$$

$$\frac{dy}{dx}=\frac{2x\left(x^2+1\right)^{1/2}-\frac{x^3}{\left(x^2+1\right)^{1/2}}}{\left(x^2+1\right)}$$

$$=\frac{\frac{2x\left(x^2+1\right)-x^3}{\left(x^2+1\right)^{1/2}}}{\left(x^2+1\right)}$$

$$\frac{dy}{dx}=\frac{x^3+2x}{\left(x^2+1\right)^{3/2}}=0$$

$x^3+2x=0$; $x\left(x^2+2\right)=0$

$x=0$ or $x=\pm\sqrt{-2}$ (imaginary). Therefore,

$\frac{dy}{d}=0$ for $x=0$

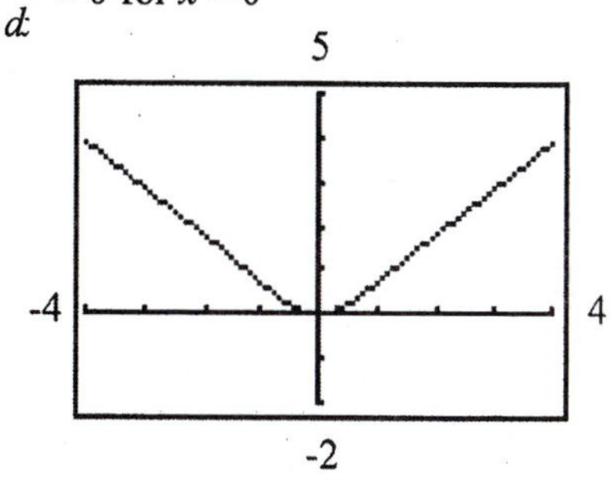

45. $y^2=4x$; $y=\sqrt{4x}=2\sqrt{x}=2x^{1/2}$

$$m_{\tan}=\frac{dy}{dx}=\frac{d\left(2x^{1/2}\right)}{dx}=2\left(\frac{1}{2}\right)x^{-1/2}=\frac{1}{\sqrt{x}}$$

$$m_{\tan}\Big|_{x=1}=\frac{1}{\sqrt{1}}=\frac{1}{1}=1$$

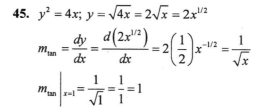

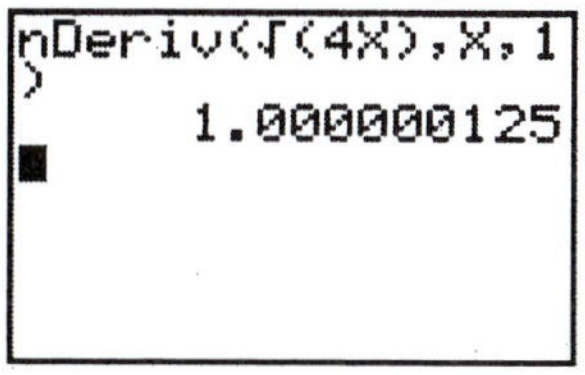

49. $s=\left(8t-t^2\right)^{2/3}$; $t=6.25s$

$$\frac{ds}{dt}=\frac{2}{3}\left(8t-t^2\right)^{-1/3}(8-2t)$$

$$v=\frac{2(8-2t)}{3\sqrt[3]{8t-t^2}};\ v\Big|_{t=6.25}=-1.35\text{ cm/s}$$

53. $v=\sqrt{\frac{l}{a}+\frac{a}{l}}=\left(\frac{l}{a}+\frac{a}{l}\right)^{1/2}$

$$\frac{dv}{dl}=\frac{1}{2}\left(\frac{1}{a}+\frac{a}{l}\right)^{-1/2}\left(\frac{1}{a}-\frac{a}{l^2}\right)=0$$

$$\frac{1}{a}-\frac{a}{l^2}=0$$

$$\frac{1}{a}=\frac{a}{l^2}$$

$$l^2=a^2$$

$$l=a$$

57. $\lambda_r=\frac{2a\lambda}{\sqrt{4a^2-\lambda^2}}=\frac{2a\lambda}{\left(4a^2-\lambda^2\right)^{1/2}}$

$$\frac{d\lambda_r}{d\lambda}$$

$$=\frac{\left(4a^2-\lambda^2\right)^{1/2}(2a)-2a\lambda\left(\tfrac{1}{2}\right)\left(4a^2-\lambda^2\right)^{-1/2}(-2\lambda)}{\left(4a^2-\lambda^2\right)}$$

$$=\frac{2a\left(4a^2-\lambda^2\right)^{1/2}+2a\lambda^2\left(4a^2-\lambda^2\right)^{-1/2}}{\left(4a^2-\lambda^2\right)}$$

$$=\frac{\left(4a^2-\lambda^2\right)^{-1/2}\left[(2a)\left(4a^2-\lambda^2\right)+2a\lambda^2\right]}{\left(4a^2-\lambda^2\right)}$$

$$=\frac{8a^3}{\left(4a^2-\lambda^2\right)^{3/2}}$$

$$\frac{dl}{dx}=\frac{1}{2}\left[25+(x-3)^2\right]^{-1/2}\cdot 2(x-3)^1$$

$$\frac{dl}{dx}=\frac{x-3}{\left[25+(x-3)^2\right]^{1/2}}=\frac{x-3}{\sqrt{x^2-6x+34}}$$

23.8 Differentiation of Implicit Functions

1. $y^3+2x^2=5$

$$3y^2\cdot\frac{dy}{dx}+4x=0$$

$$\frac{dy}{dx}=\frac{-4x}{3y^2}$$

5. $\frac{d}{dx}\left(\frac{1}{xy}\right)=\frac{xy(0)-\left(x\frac{dy}{dx}+y(1)\right)}{x^2y^2}=\frac{\frac{dy}{dx}}{xy^2}+\frac{1}{x^2y}$

9. $4y-3x^2=x;\ \frac{d}{dx}(4y)-\frac{d}{dx}\left(3x^2\right)=\frac{d}{dx}(x)$

$$4\frac{dy}{dx}-6x=1;\ 4\frac{dy}{dx}=1+6x;\ \frac{dy}{dx}=\frac{1+6x}{4}$$

13. $y^5=x^2-1;\ \frac{d}{dx}\left(y^5\right)=2x;\ 5x\frac{dy}{dx}=2x;\ \frac{dy}{dx}=\frac{2x}{5y^4}$

17. $y+3xy-4=0$

$$\frac{dy}{dx}+3\frac{d}{dx}(xy)=0;\ \frac{dy}{dx}+3\left(x\frac{dy}{dx}+y(1)\right)=0$$

$$\frac{dy}{dx}+3x\frac{dy}{dx}+3y=0;\ \frac{dy}{dx}(1+3x)=-3y$$

$$\frac{dy}{dx}=\frac{-3y}{1+3x}$$

21. $\frac{3x^2}{y^2+1}+y=3x+1$

$$\frac{\left(y^2+1\right)6x-3x^2\left(2\frac{dy}{dx}\right)}{\left(y^2+1\right)^2}+\frac{dy}{dx}=3$$

$$\frac{6x\left(y^2+1\right)-6x^2y\frac{dy}{dx}+\left(y^2+1\right)^2\frac{dy}{dx}}{\left(y^2+1\right)^2}=3$$

$$\left(y^2+1\right)^2\frac{dy}{dx}-6x^2y\frac{dy}{dx}=3\left(y^2+1\right)^2-6x\left(y^2+1\right)$$

$$=3\left(y^2+1\right)\left(y^2+1-2x\right)$$

$$\frac{dy}{dx}\left[\left(y^2+1\right)^2-6x^2y\right]=3\left(y^2+1\right)\left(y^2-2x+1\right)$$

$$\frac{dy}{dx}=\frac{3\left(y^2+1\right)\left(y^2-2x+1\right)}{\left(y^2+1\right)^2-6x^2y}$$

25. $2(x^2+1)^3+(y^2+1)^2=17$

$$6(x^2+1)^2(2x)+2(y^2+1)^1 2y\frac{dy}{dx}=0$$

$$4y(y^2+1)\frac{dy}{dx}=-12x(x^2+1)^2$$

$$\frac{dy}{dx}=\frac{-12(x^2+1)^2}{4y(y^2+1)}=\frac{3x(x^2+1)^2}{y(y^2+1)}$$

29. $5y^4+7=x^4-3y;\ (3,-2)$

$$20y^3\frac{dy}{dx}=4x^3-3\frac{dy}{dx};\ 20y^3\frac{dy}{dx}+3\frac{dy}{dx}=4x^3$$

$$\frac{dy}{dx}(20y^3+3)=4x^3;\ \frac{dy}{dx}=\frac{4x^3}{20y^3+3};$$

$$\left.\frac{dy}{dx}\right|_{(3,-2)}=-\frac{108}{157}$$

33. $x^2+y^2=4x$

$$2x+2y\frac{dy}{dx}=4$$

$$\frac{dy}{dx}=\frac{2-x}{y}=0$$

$$x=2$$

$$2^2+y^2=4(2)$$

$$y^2=4$$

$$y=\pm 2$$

The graph of $x^2+y^2=4x$ has a horizontal tangent line at $(2, 2)$ and $(2, -2)$.

37. $xy+y^2+2=0;\ \frac{d(xy)}{dx}+\frac{d(y^2)}{dx}+\frac{d(2)}{dx}=\frac{d(0)}{dx}$

$$\frac{xdy}{dx}+\frac{ydx}{dx}+\frac{2ydy}{dx}+0=0$$

$$\frac{xdy}{dx}+y(1)+\frac{2ydy}{dx}=0;\ \frac{xdy}{dx}+\frac{2ydy}{dx}=-y$$

$$\frac{dy}{dx}(x+2y)=-y;\ \frac{dy}{dx}=\frac{-y}{x+2y}$$

$$\left.\frac{dy}{dx}\right|_{(-3,1)}=\frac{-1}{-3+2(1)}=\frac{-1}{-1}=1$$

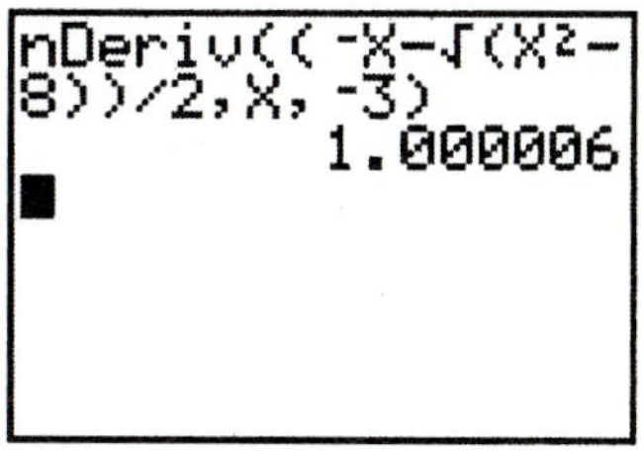

41. $PV=n\left(RT+aP-\frac{bP}{T}\right)$

$$PV=nRT+naP-\frac{nbP}{T}$$

$$V\frac{dP}{dT}=nR+na\frac{dP}{dT}-\frac{nbT\frac{dP}{dT}-nbP}{T^2}$$

$$VT^2\frac{dP}{dT}=nRT^2+naT^2\frac{dP}{dT}-nbT\frac{dP}{dT}+nbP$$

$$\frac{dP}{dT}(VT^2-naT^2+nbT)=nRT^2+nbP$$

$$\frac{dP}{dT}=\frac{nRT^2+nbP}{VT^2-naT^2+nbT}$$

45. $r^2=2rR+2R-2r;\ 2r=2R+\frac{dR}{dr}(2r)+2\frac{dR}{dr}-2$

$$2r-2R+2=2r\frac{dR}{dr}+2\frac{dR}{dr}=\frac{dR}{dr}(2r+2)$$

$$\frac{dR}{dr}=\frac{2(r-R+1)}{2(r+1)}=\frac{r-R+1}{r+1}$$

23.9 Higher Derivatives

1. $y=5x^3-2x^2$

$$y'=15x^2-4x$$

$$y''=30x-4$$

$$y'''=30$$

$$y^{(4)}=0$$

$$y^{(n)}(x)=0,\ n\geq 4$$

5. $f(x) = x^3 - 6x^4;\ f'(x) = 3x^2 - 24x^3;$

$f''(x) = 6x - 72x^2;\ f'''(x) = 6 - 144x;$

$f^{(4)}(x) = -144;\ f^{(n)}(x) = 0,\ n \ge 5$

9. $f(r) = r(4r+9)^3 = r(64r^3 + 432r^2 + 972r + 729)$

$f(r) = 64r^4 + 432r^3 + 972r^2 + 729r$

$f'(r) = 256r^3 + 1296r^2 + 1944r + 729$

$f''(r) = 768r^2 + 2592r + 1944$

$f'''(r) = 1536r + 2592$

$f^{(4)}(r) = 1536$

$f^{(n)}(r)0,\ n \ge 5$

13. $y = 2x + \sqrt{x} = 2x + x^{1/2};\ y' = 2 + \frac{1}{2}x^{-1/2}$

$y'' = \frac{1}{2}\left(-\frac{1}{2}\right)x^{-3/2} = \frac{-1}{4x^{3/2}}$

17. $f(p) = \frac{4.8\pi}{\sqrt{1+2p}} = 4.8\pi(1+2p)^{-1/2}$

$f'(p) = -2.4\pi(1+2p)^{-3/2}(2) = -4.8\pi(1+2p)^{-3/2}$

$f''(p) = 7.2\pi(1+2p)^{-5/2}(2) = \frac{14.4\pi}{(1+2p)^{5/2}}$

21. $y = (3x^2-1)^5;\ y' = 5(3x^2-1)^4(6x)$

$= 30x(3x^2-1)^4;$

$y'' = 30x(4)(3x^2-1)^3(6x) + (3x^2-1)^4(30)$

$= 30(3x^2-1)^3(27x^2-1)$

25. $u = \frac{v^2}{v+15}$

$u' = \frac{(v+15)2v - v^2}{(v+15)^2} = \frac{v^2+30v}{(v+15)^2}$

$u'' = \frac{(v+15)^2(2v+30) - (v^2+30v)(2)(v+15)}{(v+15)^4}$

$u'' = \frac{450}{(v+15)^3}$

29. $x^2 - xy = 1 - y^2;\ 2x - (xy' + y) = -2yy'$

$2x - xy' - y = -2yy';\ 2yy' - xy = y - 2x$

$y' = \frac{y-2x}{2y-x};$

$y'' = \frac{(2y-x)(y'-2) = (y-2x)(2y'-1)}{(2y-x)^2}$

$y'' = \left[(2y-x)\left(\frac{y-2x}{2y-x} - 2\right) - (y-2x)\left(2\frac{y-2x}{2y-x} - 1\right)\right]$

$\div (2y-x)^2$

$y'' = \left[y - 2x - 2(2-y) - \frac{2(y-2x)^2}{2y-x} + y - 2x\right]$

$\div (2y-x)^2$

$y'' = \left[-2(y+x) - \frac{2(y-2x)^2}{(2y-x)}\right] \div (2y-x)^2$

$y'' = (-4y^2 - 2xy + 2x^2 - 2y^2 + 8xy - 8x^2) \div (2y-x)^3$

$y'' = \frac{-6(y^2 - xy + x^2)}{(2y-x)^3}$

33. $y = 3x^{2/3} - \frac{2}{x} = 3x^{2/3} - 2x^{-1}$

$y' = 2x^{-1/3} + 2x^{-2}$

$y'' = -\frac{2}{3}x^{-4/3} - 4x^{-3} = \frac{-2}{3x^{4/3}} - \frac{4}{x^3};$

$y''\Big|_{x=-8} = \frac{-2}{48} + \frac{4}{512}$

$y''\big|_{x=-8} = \frac{-2}{6 \times 8} + \frac{4}{8 \times 64} = -\frac{13}{384}$

37. $s = 26t - 4.9t^2,\ t = 3.0$ s

$v = 26 - 9.8t$

$a = -9.8\ \text{m/s}^2$

41. $\frac{d(uv)}{dx} = u\frac{dv}{dx} + v\frac{du}{dx}$

$\frac{d^2(uv)}{dx^2} = u\frac{d^2v}{dx^2} + \frac{du}{dx}\frac{dv}{dx} + v\frac{d^2u}{dx^2} + \frac{du}{dx}\frac{dv}{dx}$

$= u\frac{d^2v}{dx^2} + 2\frac{du}{dx}\frac{dv}{dx} + v\frac{d^2u}{dx^2}$

$y = (1-2x)^4$

$y' = 4(1-2x)^3(-2) = -8(1-2x)^2$

$y'' = -24(1-2x)^2(-2) = 48(1-2x)^2$

$y''\big|_{x=1} = 48$

45. $P(t)=8000\left(1+0.02t+0.005t^2\right)$

$P'(t)=8000(0.02+0.01t)$

$P''(t)=8000(0.01)=80$

49. $V=L\left(\dfrac{d^2q}{dt^2}\right);\ L=1.60\text{ H}$

$q=\sqrt{2t+1}-1=(2t+1)^{1/2}-1$

$\dfrac{dq}{dt}=\dfrac{1}{2}(2t+1)^{-1/2}(2)=(2t+1)^{-1/2}$

$\dfrac{d^2q}{dt^2}=-\dfrac{1}{2}(2t+1)^{-3/2}(2)=\dfrac{-1}{(2t+1)^{3/2}}$

$V=\dfrac{-1.60}{(2t+1)^{3/2}}$

Chapter 23 Review Exercises

1. $\lim\limits_{x\to 4}(8-3x)=8-3(4)=-4$

5. $\lim\limits_{x\to 2}\dfrac{4x-8}{x^2-4}=\lim\limits_{x\to 2}\dfrac{4(x-2)}{(x-2)(x+2)}$

$=\lim\limits_{x\to 2}\dfrac{4}{x+2}=\dfrac{4}{2+2}=1$

9. $\lim\limits_{x\to\infty}\dfrac{2+\frac{1}{x+4}}{3-\frac{1}{x^2}}=\dfrac{2+0}{3-0}=\dfrac{2}{3}$

13. $y=7+5x$

$\dfrac{dy}{dx}=\lim\limits_{h\to 0}\dfrac{f(x+h)-f(x)}{h}$

$=\lim\limits_{h\to 0}\dfrac{7+5(x+h)-7-5x}{h}$

$\dfrac{dy}{dx}=\lim\limits_{h\to 0}\dfrac{5h}{h}=\lim\limits_{h\to 0}5=5$

17. $y=\dfrac{2}{x^2}$

$\dfrac{dy}{dx}=\lim\limits_{h\to 0}\dfrac{\frac{2}{(x+h)^2}-\frac{2}{x^2}}{h}=\lim\limits_{h\to 0}\dfrac{2x^2-2(x+h)^2}{hx^2(x+h)^2}$

$=\lim\limits_{h\to 0}\dfrac{2x^2-2x^2-4xh-2h^2}{hx^2(x+h)^2}$

$=\lim\limits_{h\to 0}\dfrac{-4xh-2h^2}{hx^2(x+h)^2}=\lim\limits_{h\to 0}\dfrac{-4x-2h}{x^2(x+h)^2}$

$=\dfrac{-4x}{x^4}=\dfrac{-4}{x^3}$

21. $y=2x^7-3x^2+5$

$\dfrac{dy}{dx}=2\left(7x^6\right)-3(2x)+0=14x^6-6x$

25. $f(y)=\dfrac{3y}{1-5y}$

$\dfrac{df(y)}{dy}=\dfrac{(1-5y)(3)-3y(-5)}{(1-5y)^2}=\dfrac{3-15y+15y}{(1-5y)^2}$

$\dfrac{df(y)}{dy}=\dfrac{3}{(1-5y)^2}$

29. $y=\dfrac{3\pi}{\left(5-2x^2\right)^{3/4}}$

$\dfrac{dy}{dx}=\dfrac{\left(5-2x^2\right)^{3/4}(0)-3\pi\left[-3x\left(5-2x^2\right)^{-1/4}\right]}{\left(5-2x^2\right)^{3/2}}$

$=\dfrac{9\pi x\left(5-2x^2\right)^{-1/4}}{\left(5-2x^2\right)^{3/2}}$

$=\dfrac{9\pi x}{\left(5-2x^2\right)^{3/2}\left(5-2x^2\right)^{1/4}}=\dfrac{9\pi x}{\left(5-2x^2\right)^{7/4}}$

33. $y=\dfrac{\sqrt{4x+3}}{2x}=2(4x+3)^{-1/2}$

$\dfrac{dy}{dx}=\dfrac{2x(2)(4x+3)^{-1/2}-(4x+3)^{1/2}(2)}{(2x)^2}$

$=\dfrac{(4x+3)^{-1/2}\left[4x-(4x+3)(2)\right]}{4x^2}$

$=\dfrac{-4x-6}{(4x+3)^{1/2}\left(4x^2\right)}=\dfrac{2(-2x-3)}{2\left(2x^2\right)(4x+3)^{1/2}}$

$=\dfrac{-2x-3}{2x^2(4x+3)^{1/2}}$

37. $y=\frac{4}{x}+2\sqrt[3]{x},\ x=8$

$y=4x^{-1}+2x^{1/3};\ \frac{dy}{dx}=4(-1)x^{-2}+2\left(\frac{1}{3}x^{-2/3}\right)$

$=\frac{-4}{x^2}+\frac{2}{3x^{2/3}}$

$\left.\frac{dy}{dx}\right|_{x=8}=\frac{-4}{8^2}+\frac{2}{3(8)^{2/3}}=\frac{-4}{64}+\frac{2}{3(4)}=\frac{-1}{16}+\frac{1}{6}$

$=\frac{-3}{48}+\frac{8}{48}=\frac{5}{48}$

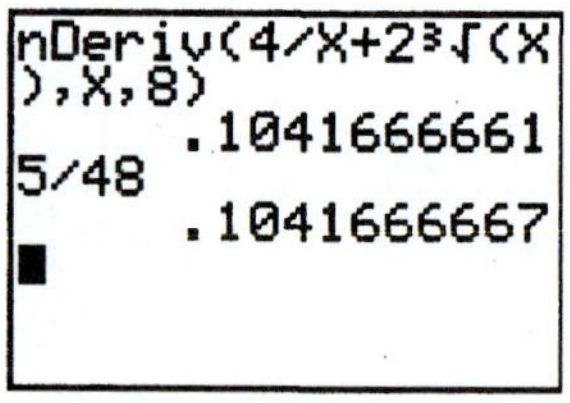

41. $y=3x^4-\frac{1}{x}=3x^4-x^{-1}$

$y'=12x^3+x^{-2}$

$y''=36x^2-2x^{-3}$

45. From graph, as $x\to 0^+$

$1/\sqrt{x}$ increases most rapidly.

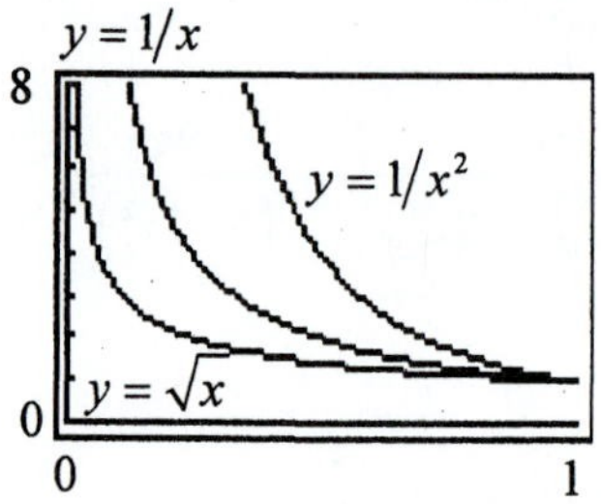

49. $y=\frac{2(x^2-4)}{x-2},\ x\neq 2$

$y=\frac{2(x+2)(x-2)}{(x-2)}$

$=2(x+2)|_{x=2}$

$=2(2+2)=8$

Just to the left of $x=2$, the trace feature gives $x=1.9787234,\ y=7.9574468$ and just to the right of $x=2$, the trace feature gives $x=2.0319149,\ y=8.0638298$ which would appear to give $y=8$ for $x=2$. However, using the value feature shows there is no y-value for $x=2$.

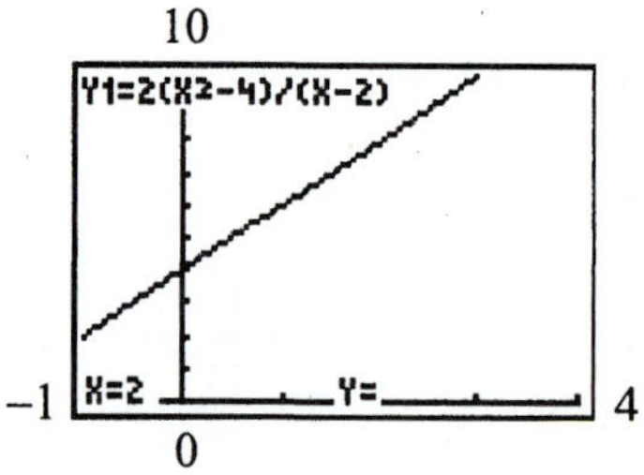

53. $y=7x^4-x^3$

$\frac{dy}{dx}=28x^3-3x^2\Big|_{(-1,\,8)}$

$=28(-1)^3-3(-1)^2$

$=-31=m_{\tan}$ at $(-1,\ 8)$

The nDeriv feature of calculator gives -31.000029

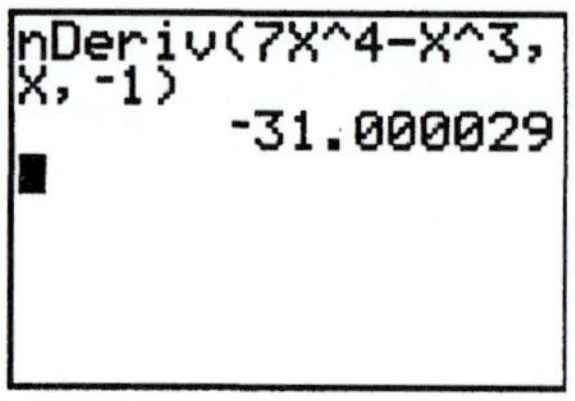

57. $A=5000(1+0.25i)^8$

$\frac{dA}{di}=40,000(1+0.25i)^7(0.25)$

$=10,000(1+0.25i)^7$

61. $y=0.0015x^2+C$

$\frac{dy}{dx}=2(0.0015)x=0.3$

$x=\frac{0.3}{2(0.0015)}$

$y=0.3\left(\frac{0.3}{2(0.0015)}\right)-10$

$$= 0.0015\left(\frac{0.3}{2(0.0015)}\right)^2 + C$$

$$C = 5$$

65. $E = \dfrac{k}{r^2}$

$$\frac{dE}{dr} = \frac{-2k}{r^3}$$

69. $r_f = \dfrac{2(R^3 - r^3)}{3(R^2 - r^2)}$

$$\frac{dr_f}{dR} = \frac{3(R^2 - r^2)(6R^2) - 2(R^3 - r^3)(6R)}{9(R^2 - r^2)^2}$$

$$\frac{dr_f}{dR} = \frac{18R^2(R+r)(R-r) - 12R(R-r)(R^2 + Rr + r^2)}{9(R+r)^2(R-r)^2}$$

$$\frac{dr_f}{dR} = \frac{6R^2(R+r) - 4R(R^2 + Rr + r^2)}{3(R+r)^2(R-r)}$$

$$\frac{dr_f}{dR} = \frac{6R^3 + 6R^2r - 4R^3 - 4R^2r - 4Rr^2}{3(R+r)^2(R-r)}$$

$$\frac{dr_f}{dR} = \frac{2R^3 + 2R^2r - 4Rr^2}{3(R+r)^2(R-r)}$$

$$= \frac{2R(R^2 + Rr - 2r^2)}{3(R+r)^2(R-r)}$$

$$\frac{dr_f}{dR} = \frac{2R(R+2r)(R-r)}{3(R+r)^2(R-r)}$$

$$= \frac{2R(R+2r)}{3(R+r)^2}$$

73. $y = kx(x^4 + 450x^2 - 950) = kx^5 + 450kx^3 - 950kx$

$$\frac{dy}{dx} = 5kx^4 + 1350kx^2 - 950k$$

77. $T = \dfrac{10(1-t)}{0.5t+1}$

$$\frac{dT}{dt} = \frac{(0.5t+1)(-10) - 10(1-t)(0.5)}{(0.5t+1)^2}$$

$$= \frac{-5t - 10 - 5 + 5t}{(0.5t+1)^2}$$

$$\frac{dT}{dt} = \frac{-15}{(0.5t+1)^2}$$

81. $A = lw = 75 \Rightarrow l = \dfrac{75}{w}$

$$p = 2l + 2w = \frac{150}{w} + 2w$$

$$\frac{dp}{dw} = \frac{-150}{w^2} + 2$$

85. $x^2 = h^2 + (vt)^2,\ h = \dfrac{2640}{5280},\ v = 400$

$$2x\frac{dx}{dt} = 2v^2t$$

$$\frac{dx}{dt} = \frac{v^2t}{x}\Bigg|_{v=400,\ x=\sqrt{\left(\frac{2640}{5280}\right)^2 + \left(400\left(\frac{0.600}{60}\right)\right)^2},\ t=\frac{0.600}{60}}$$

$$\frac{dx}{dt} = 397 \text{ mi/h}$$

Chapter 24

APPLICATIONS OF THE DERIVATIVE

24.1 Tangents and Normals

1. $x^2+4y^2=17, (1, 2)$

$2x+8yy'=0$

$y'=\left.\dfrac{-x}{4y}\right|_{(1,\,2)}=\dfrac{-1}{4(2)}=-\dfrac{1}{8}$

$y-2=-\dfrac{1}{8}(x-1)$

$8y-16=-x+1$

$x+8y-17=0$

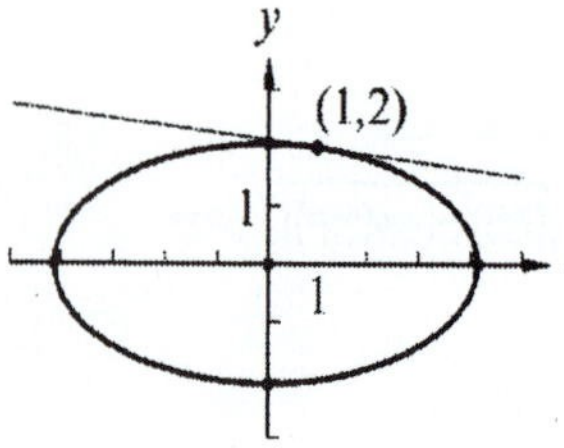

5. $y=\dfrac{1}{x^2+1}; \left(1, \dfrac{1}{2}\right); y=\left(x^2+1\right)^{-1}$

$\dfrac{dy}{dx}=m_{\tan}=-\left(x^2+1\right)^{-2}(2x);\ m_{\tan}=\dfrac{-2}{\left(x^2+1\right)^2};$

$m_{\tan_{x=1}}=-\dfrac{1}{2}$

Eq. T.L.: $y-\dfrac{1}{2}=-\dfrac{1}{2}(x-1);\ 2y-1=-x+1$

$y=-\dfrac{1}{2}x+1$

Therefore, $x+2y-2=0$

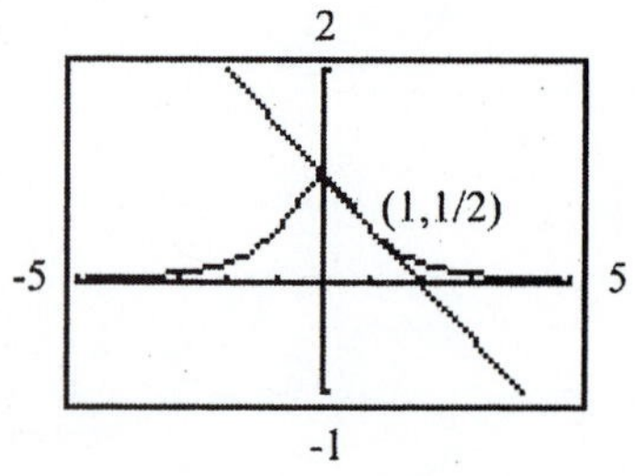

9. $y=\dfrac{6}{\left(x^2+1\right)^2}; \left(1, \dfrac{3}{2}\right); y=6\left(x^2+1\right)^{-2}$

$\dfrac{dy}{dx}=m_{\tan}=-12\left(x^2+1\right)^{-3}(2x)$

$m_{\tan}=\dfrac{-24x}{\left(x^2+1\right)^3};\ m_{\tan(1,\,3/2)}\Big|_{x=1}=-3;$

$m_{\text{normal}}=\dfrac{1}{3}$

Eq. of normal: $y-\dfrac{3}{2}=\dfrac{1}{3}(x-1)$

Therefore, $2x-6y+7=0$

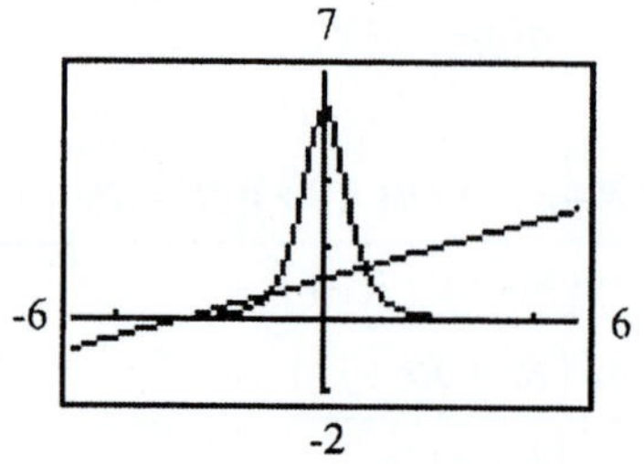

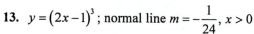

13. $y=(2x-1)^3$; normal line $m=-\dfrac{1}{24}, x>0$

Therefore, $m_{\tan}=24$

$m_{\tan}=3(2x-1)^2(2)=6(x-1)^2$

$6(2x-1)^2=24;\ (2x-1)^2=4;\ 2x-1=\pm 2$

$2x=\pm 2+1=3$ and -1

$x=\dfrac{3}{2},\ y=8$

Eq. of N.L.: $y-8=-\dfrac{1}{24}\left(x-\dfrac{3}{2}\right);$

$24y-192=-x+\dfrac{3}{2}$

$48y-384=-2x+3;\ 2x+48y-387=0$

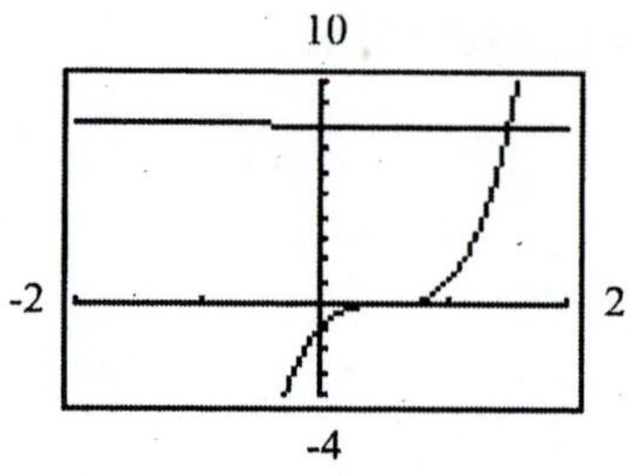

17. $y = x + 2x^2 - x^4$

$y' = 1 + 4x - 4x^3$

At $(1, 2)$, $y' = 1 + 4(1) - 4(1)^3 = 1$

Eq. of T.L.: at $(1, 2)$: $y - 2 = 1(x - 1)$

$y = x + 1$

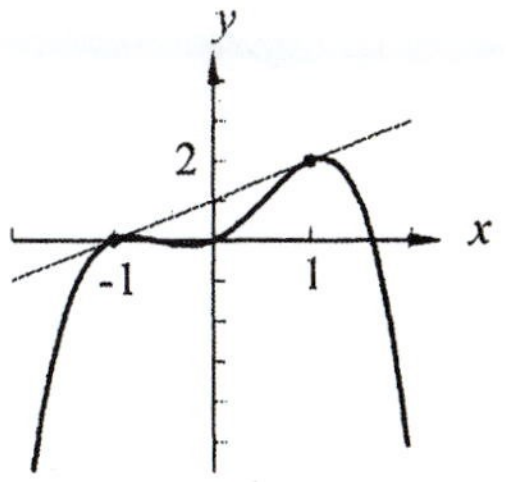

At $(-1, 0)$, $y' = 1 + 4(1) - 4(-1)^3 = 1$

Eq. of TL at $(-1, 0)$:

$y - 0 = 1(x - (-1))$

$y = x + 1$

The tangent lines are the same: $y = x + 1$

21. $x^2 + y^2 = a^2$

$2x + 2yy' = 0$

$$y' = \frac{-x}{y}\bigg|_{(x_1,\, y_1)} = \frac{-x_1}{y_1}$$

$$y - y_1 = \frac{-x_1}{y_1}(x - x_1)$$

$y_1 y - y_1^2 = -x_1 x + x_1^2$

$x_1 x + y_1 y = x_1^2 + y_1^2 = a^2$

25. Equation of a curve:

$x^2 = 4py$

$$p = \frac{x^2}{4y} = \frac{(100)^2}{4(30)}$$

$$p = \frac{250}{3}$$

Therefore, $x^2 = \frac{1000}{3} y$

Eq. of T.L.: at $(100, 30)$

$$2x = \frac{1000}{3}\frac{dy}{dx};\ \frac{dy}{dx} = \frac{6x}{1000};\ m_{\text{TL}}\big|_{x=100} = \frac{6}{10} = \frac{3}{5}$$

$$y - 30 = \frac{3}{5}(x - 100);\ 5y - 150 = 3x - 300$$

$3x - 5y - 150 = 0$

29. $y = \frac{4}{x^2 + 1};\ (-2 < x < 2)$

Supports at $(-1, 2)$, $(0, 4)$, $(1, 2)$

$y = 4(x^2 + 1)^{-1}$

$$\frac{dy}{dx} = m_{\tan} = -4(x^2 + 1)^{-2}(2x)$$

$$m_{\tan} = \frac{-8x}{(x^2 + 1)^2}$$

$m_{\tan}\big|_{x=-1} = 2;\ m_{\tan}\big|_{x=0} = 0;$

$m_{\tan}\big|_{x=1} = -2$

Eq. of N.L.: at $(-1, 2)$:

$$m_{\text{NL}} = -\frac{1}{2}$$

$$y - 2 = -\frac{1}{2}(x + 1);\ 2y - 4 = -x + 1;$$

$x + 2y - 3 = 0$

Eq. of N.L.: at $(0, 4)$

$x = 0$; m_{NL} is undefined (vertical).

Eq. of N.L.: at $(1, 2)$:

$$m_{\text{NL}} = \frac{1}{2}$$

$$y - 2 = \frac{1}{2}(x - 1);\ 2y - 4 = x - 1;$$

$x - 2y + 3 = 0$

24.2 Newton's Method for Solving Equations

1. $x^2-5x+1=0,\ 0<x<1,\ f(x)=x^2-5x+1$

$f(0)=1,\ f(1)=-3$, choose $x_1=0.5$

$f'(x)=2x-5$

$$x_2=x_1-\frac{f(x_1)}{f'(x_1)}=0.5-\frac{0.5^2-5(0.5)+1}{2(0.5)-5}$$

$x_3=0.2086148649$

$x_4=0.2087121505$

$x_5=0.2087121525$

$x_6=0.2087121525$

Using the quadratic formula,

$$x=\frac{(-5)-\sqrt{(-5)^2-4(1)(1)}}{2(1)}$$

$=0.2087121525$

5. $x^3-6x^2+10x-4=0$ (between 0 and 1)

$f(x)=x^3-6x^2+10x-4;\ f'(x)=3x^2-12x+10;$

$f(0)=-4;\ f(1)=1$

Let $x_1=0.7$

n	x_n	$f(x_n)$	$f'(x_n)$	$x_n-\frac{f(x_n)}{f'(x_n)}$
1	0.7	0.403	3.07	0.5687296
2	0.5687296	−0.0694666	4.1456049	0.5854863
3	0.5854863	−0.0012009	4.002547	0.5857863
4	0.5857863	−0.0000005	4.0000012	0.5857864

$x_4=x_3=0.5857864$ to seven decimal places;

$x=0.58578644$, calculator

9. $x^4 - x^3 - 3x^2 - x - 4 = 0$; (between 2 and 3)
$f(x) = x^4 - x^3 - 3x^2 - x - 4$; $f'(x) = 4x^3 - 3x^2 - 6x - 1$
$f(2) = -10$; $f(3) = 20$.
Let $x_1 = 2.3$

n	x_n	$f(x_n)$	$f'(x_n)$	$x_n - \frac{f(x_n)}{f'(x_n)}$
1	2.3	−6.3529	17.998	2.6529781
2	2.6529781	3.0972725	36.657001	2.5684848
3	2.5684848	0.2175007	31.576097	2.5615967
4	2.5615967	0.0013683	−31.179599	2.5615528

$x_3 = x_4 = 2.5615528$ to seven decimal places; $x = 2.5615528$, calculator

13. $2x^2 = \sqrt{2x+1}$ or $2x^2 - \sqrt{2x+1} = 0$, (the positive real solution)
$4x^4 - 2x - 1 = 0$ (Square both sides.)
$f(x) = 4x^4 - 2x - 1$; $f'(x) = 16x^3 - 2$

Let $x_1 = 0.8$.

n	x_n	$f(x_n)$	$f'(x_n)$	$x_n - \frac{f(x_n)}{f'(x_n)}$
1	0.8	−0.9616	6.192	0.9552972
2	0.9552972	0.4207071	11.948757	0.9200879
3	0.9200879	0.0264914	10.462579	0.9175559
4	0.9175559	1.301096×10^{-4}	10.359975	0.9175433
5	0.9175433	3.1869×10^{-9}	10.3594671	0.9175433

The positive root is approximately 0.9175433; to seven decimal places. $x = 0.91754334$, calculator

17. $f(x) = x^3 - 2x^2 - 5x + 4$. From graph, one root lies between −1 and −2, a second between 0 and 1, and a third between 3 and 4

$f'(x) = 3x^2 - 4x - 5$

Let $x_1 = -1.7$

n	x_n	$f(x_n)$	$f'(x_n)$	$x_n - \frac{f(x_n)}{f'(x_n)}$
1	−1.7	1.807	10.47	−1.8725883
2	−1.8725883	−0.2166267	13.0101148	−1.8559377
3	−1.8559377	−0.0021074	12.7572652	−1.8557725
4	−1.8557725	$-2.0650050 \times 10^{-7}$	12.7547651	−1.8557725

Let $x_1 = 0.7$

1	0.7	−0.137	−6.33	0.6783570
2	0.6783570	3.6703855×10^{-5}	−6.3329233	0.6783628

Let $x_1 = 3.1$

1	3.1	−0.929	11.43	3.1812773
2	3.11812773	0.0487608	12.6364672	3.1774186
3	3.1774186	1.1226889×10^{-4}	12.5782926	3.1774097

The roots are −1.8557725, 0.6783628, and 3.1774097.

21. $f(x) = x^2 - a$

$f'(x) = 2x$

$$x_2 = x_1 - \frac{f(x_1)}{f'(x_1)} = x_1 - \frac{x_1^2}{2x_1} = x_1 - \frac{x_1}{2} + \frac{a}{2x_1}$$

$x_2 = \frac{x_1}{2} + \frac{a}{2x_1}$. Similarly, $x_3 = \frac{x^2}{2} + \frac{a}{2x_2}$ which generalizes to

$x_{n+1} = \frac{x_n}{2} + \frac{a}{2x_n}$.

25. $h = -2t^3 + 84t^2 + 480t + 10$

$h' = -6t^2 + 168t + 480$

From the graph $h = 0$ near 50. Let $x_1 = 50$. Newton's method gives

$x_2 = 47.3872549$

$x_3 = 47.10133148$

$x_4 = 47.09801113$

$x_5 = 47.09801068$

$x_6 = 47.09801068$

The rocket strikes the ground when $t = 47$ s.

29. Volume = 2(half spheres) + cylinder

$V = \frac{4}{3}\pi r^3 + \pi r^2 h$; $1500 = \frac{4}{3}\pi r^3 + 12.0\pi r^2$

$f(r) = \frac{4}{3}\pi r^3 + 12.0\pi r^2 - 1500$; $f'(r) = 4\pi r^2 + 24.0\pi r$

$f(5) < 0$; $f(6) > 0$; let $r_1 = 5.1$

n	x_n	$f(x_n)$	$f'(x_n)$	$x_n - \frac{f(x_n)}{f'(x_n)}$
1	5.1	36.2011085	711.382405	5.0491116
2	5.0491116	0.2630481	701.0551673	5.0487364
3	5.0487364	ϕ		

$r_3 = 5.0487364$; $r = 5.05$ ft

24.3 Curvilinear Motion

1. $x = 4t^2 \qquad y = 1 - t^2$

$v_x = \frac{dx}{dt} = 8t\big|_{t=2} = 16 \qquad \frac{dy}{dt} = -2t\big|_{t=2} = -4$

$v = \sqrt{16^2 + (-4)^2} = 16.5$

$\tan\theta = \frac{-4}{16},\ \theta = -14.0^\circ$

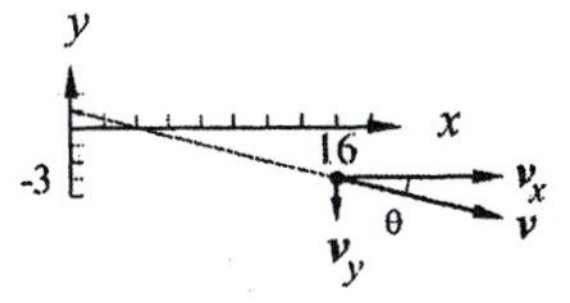

5. $x = t(2t+1)^2$

$\frac{dx}{dt} = t(2)(2t+1)(2) + (2t+1)^2(1)$

$= 12t^2 + 8t + 1 = v_x$

$y = 6(4t+3)^{-1/2}$

$\frac{dy}{dt} = 6\left(-\frac{1}{2}\right)(4t+3)^{-3/2}(4) = \frac{-12}{(4t+3)^{3/2}} = v_y$

$v_x\big|_{t=0.5} = 8$

$v_y\big|_{t=0.5} = -1.0733$

$v = \sqrt{8^2 + (-1.0773)^2}$

$v = 8.07$

$\alpha = \tan^{-1}\frac{1.0733}{8} = 7.641^\circ$

t	x	y
0	0	3.464
0.5	2	2.683
1	9	2.268

Therefore, $\theta = 360^\circ - \alpha = 352.4^\circ$

Therefore, v is 8.07 at $\theta = 352.34^\circ$

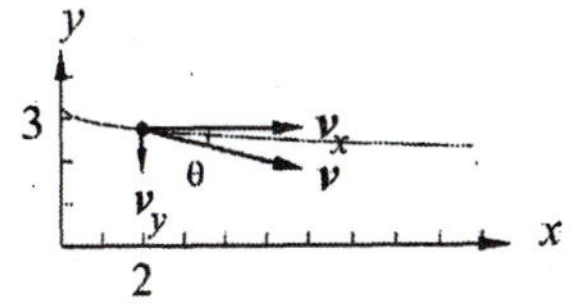

9. $x = t(2t+1)^2;\ v_x = 12t^2 + 8t + 1;\ a_x = 24t + 8;$

$a_x\big|_{t=0.5} = 20.0$

$y = \frac{6}{\sqrt{4t+3}};\ v_y = -12(4t+3)^{-3/2}$

$a_y = \frac{72}{(4t+3)^{5/2}};\ a_y\big|_{t=0.5} = 1.288$

$a = \sqrt{20.0^2 + 1.288^2}$

$a = 20.041$

$\theta = \tan^{-1}\frac{1.288}{20.0}$

$\theta = 3.68^\circ$

Therefore, a is 20.0 at $\theta = 3.7^\circ$.

13. $x = 0.2t^2;\ v_x = 0.4t;\ a_x = 0.4;\ a_x\big|_{t=2.0} = 0.4$

$y = -0.1t^3;\ v_y = -0.3t^2;\ a_y = -0.6t;$

$a_y\big|_{t=2.0} = -1.2$

$a = \sqrt{(0.4)^2 + (1.2)^2} = 1.3$

$\alpha = \tan^{-1}\left(\frac{1.2}{0.4}\right) = 71.6^\circ$

$\theta = 288.4^\circ$

Therefore, a is 1.3 ft/min^2 at $\theta = 288^\circ$.

17. $x = 10\left(\sqrt{1+t^4} - 1\right);\ v_x = \frac{20t^3}{\sqrt{1+t^4}}$

$y = 40t^{3/2};\ v_y = 60t^{1/2};\ (0 \le t \le 100\text{ s})$

For $t = 10$ s:

$v_x = \frac{20t^3}{\sqrt{1+t^4}}\bigg|_{t=10.0\text{ s}} = 200.0\text{ m/s}$

$v_y = 60t^{1/2}\big|_{t=10.0\text{ s}} = 189.7\text{ m/s}$

$v = \sqrt{(200)^2 + (189.7)^2}$

$v = 276\text{ m/s}$

$\theta = \tan^{-1}\frac{189.7}{200} = 43.5^\circ$

For $t = 100$ s:

$v_x\big|_{t=100\text{ s}} = 2000.0\text{ m/s}$

$v_y\big|_{t=100\text{ s}} = 600\text{ m/s}$

$v = \sqrt{(2000)^2 + (600)^2}$

$v = 2088\text{ m/s}; \theta = \tan^{-1}\dfrac{600}{2000} = 16.7^\circ$

At 10 s v is 276 m/s at $\theta = 43.5^\circ$, and at 100 s v is 2090 m/s at $\theta = 16.7^\circ$.

21. $x = 10\left(\sqrt{1+t^4} - 1\right) = 10\left(1+t^4\right)^{1/2} - 10;$

$y = 40t^{3/2}$

$$\frac{dx}{dt} = 5\left(1+t^4\right)^{-1/2}\left(4t^3\right) = 20t^3\left(1+t^4\right)^{-1/2} = \frac{20t^3}{\sqrt{1+t^4}}$$

$$\frac{dy}{dt} = 60t^{1/2} = 60\sqrt{t}$$

$$ax = \frac{\left(1+t^4\right)^{1/2}\left(60t^2\right) - \left(20t^3\right)\left(\frac{1}{2}\right)\left(1+t^4\right)^{-1/2}\left(4t^3\right)}{\left(1+t^4\right)}$$

$$= \frac{\left(1+t^4\right)^{-1/2}\left[60t^2\left(1+t^4\right) - 40t^6\right]}{\left(1+t^4\right)}$$

$$= \frac{60t^2\left(1+t^4\right) - 40t^6}{\left(1+t^4\right)^{3/2}}$$

$ay = 30t^{-1/2}$

$$a_x\big|_{t=10.0} = \frac{6000(10,000) - 40,000,000}{1,000,000} = 20.0$$

$$a_y\big|_{t=10.0} = 30(10)^{-1/2} = \frac{30}{\sqrt{10}} = 9.5$$

$$a = \sqrt{(20)^2 + (9.5)^2} = 22.1\text{ m/s}^2$$

$$\tan\theta = \frac{9.5}{20.0} = 0.475;\ \theta = 25.4^\circ$$

$$a_x\big|_{t=100} = \frac{60\left(10^4\right)\left(10^8\right) - 40\left(10^{12}\right)}{\left(10^8\right)^{3/2}}$$

$$= \frac{6\times10^{13} - 4\times10^{13}}{10^{12}}$$

$$= \frac{2\times10^{13}}{10^{12}} = 20.0$$

$$a_y\big|_{t=100} = 30\left(10^2\right)^{-1/2} = 30\left(10^{-1}\right) = 3.0$$

$$a = \sqrt{(20.0)^2 + (3.0)^2} = 20.2\text{ m/s}^2$$

$$\tan\theta = \frac{3.0}{20.0} = 0.150;\ \theta = 8.5^\circ$$

25. $d = 3.50$ in.; $r = 1.75$ in.; $x^2 + y^2 = 1.75^2$; $\dfrac{dy}{dx} = -\dfrac{x}{y}$

3600 r/min $= 7200\pi$ rad/min $= \omega$

$v = \omega r = 7200\pi(1.75) = 12,600\pi$ in./min

$x^2 + y^2 = 1.75^2$

$y^2 = 1.75^2 - x^2 = 1.75^2 - 1.20^2$

$= 3.062 - 1.44 = 1.622$

$y = 1.274$ in.

$$\frac{dy}{dx} = -\frac{x}{y} = -\frac{1.20}{1.274} = -0.942 = \frac{v_y}{v_x};$$

$v_y = -0.942v_x$

$v = 12,600\pi = \sqrt{v_x^2 + v_y^2}$

$= \sqrt{(-0.942v_x)^2 + v_x^2} = \sqrt{1.893v_x^2} = 1.376v_x$

$v_x = 9158\pi = 28,800$ in./min

$v_y = -0.942v_x = 8654\pi = -27,100$ in./min

24.4 Related Rates

1. $E = 2.800T + -0.012T^2$

$$\frac{dE}{dt} = 2.800\frac{dT}{dt} + 0.024T\frac{dT}{dt}$$

$$\frac{dE}{dt}\bigg|_{T=100^\circ\text{C}} = 2.800(1.00) + 0.024(100)(1.00)$$

$$\frac{dE}{dt}\bigg|_{T=100^\circ\text{C}} = 5.20\text{ V/min}$$

5. $x^2 + 3y^2 + 2y = 10$

$$2x\frac{dx}{dt} + 6y\frac{dy}{dt} + 2\frac{dy}{dt} = 0$$

$$2(3)(2) + 6(-1)\frac{dy}{dt} + 2\frac{dy}{dt} = 0$$

$$\frac{dy}{dt} = -3$$

9. $v = 18\sqrt{T}$

$$\frac{dv}{dt} = \frac{18}{2\sqrt{T}}\frac{dT}{dt} = \frac{9}{\sqrt{25}}(0.2)$$

$= 0.36$ ft/s^2

13. $T = \pi\sqrt{\frac{L}{96}} = \frac{\pi}{\sqrt{96}}L^{1/2}$

$$\frac{dT}{dt} = \frac{\pi}{\sqrt{96}}L^{-1/2}\frac{dL}{dT}$$

$$= \frac{\pi}{2\sqrt{96}}(16.0)^{-1/2}(0.100)$$

$$\frac{dT}{dt} = 0.00401$$

17. $r = \sqrt{0.4\lambda};\ \frac{d\lambda}{dt} = 0.10\times10^{-7};\ r = (0.4\lambda)^{1/2}$

$$\frac{dr}{dt} = \frac{1}{2}(0.4\lambda)^{-1/2}(0.4)\frac{d\lambda}{dt}$$

$$= 0.2(0.4\lambda)^{-1/2}\frac{d\lambda}{dt}$$

$$\frac{dr}{dt}\Big|_{\lambda=6.0\times10^{-7}} = 0.2\left[0.4\left(6.0\times10^{-7}\right)\right]^{-1/2}\left(0.10\times10^{-7}\right)$$

$$= \frac{2\times10^{-9}}{\sqrt{24\times10^{-4}}} - 4.1\times10^{-6}\ \text{m/s}$$

21. $y = \frac{1.5}{18}x \Rightarrow 12 = x$

$v = \frac{1}{2}xy(12) = 6xy, \quad 0 < x < 18,$

$\qquad 0 < y < 1.5$

$$v = 72y^2$$

$$\frac{dv}{dt} = 144y\frac{dy}{dt}$$

$$0.80 = 144(1.0)\frac{dy}{dt}$$

$$\frac{dy}{dt} = 0.0056\ \text{m/min}$$

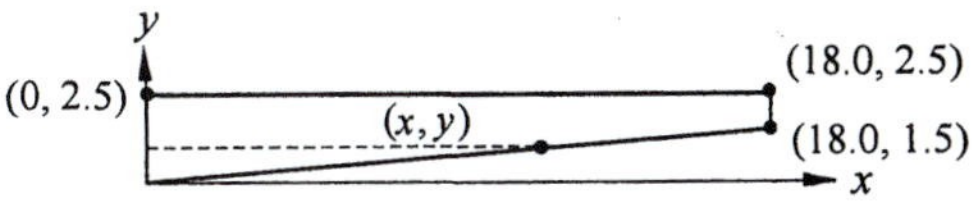

25. $V = x^3;\ \frac{dx}{dt} = -0.50$ mm/min

$$\frac{dV}{dt} = 3x^2\frac{dx}{dt};\ \frac{dV}{dt}\Big|_{x=8.20} = 3(8.20)^2(-0.50)$$

$$= -101\ \text{mm}^3/\text{min}$$

29. $p = \frac{k}{v};\ \frac{dv}{dt} = 20\ \text{cm}^3/\text{min};\ v = 810\ \text{cm}^3$

$$230 = \frac{k}{650};\ k = 1.495\times10^5\ \text{kPa}\times\text{cm}^3$$

$$p = \frac{149{,}500}{v} = 149{,}500v^{-1};\ \frac{dp}{dt} = -149{,}500v^{-2}\frac{dv}{dt}$$

$$\frac{dp}{dt}\Big|_{v=810} = -149{,}500(810)^{-2}(20) = -4.6\ \text{kPa/min}$$

33.

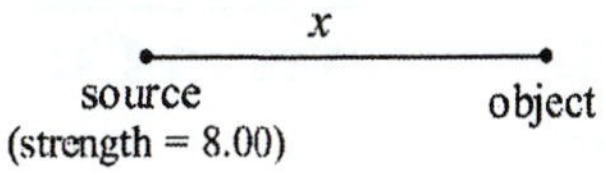

$$I = \frac{8.00k}{x^2}$$

$$\frac{dI}{dt} = \frac{-2(8.00)k}{x^3}\frac{dx}{dt} = \frac{-2(8.00)k}{100^3}(-50.0)$$

$$\frac{dI}{dt} = 0.000800k\ \text{units/s}$$

37.

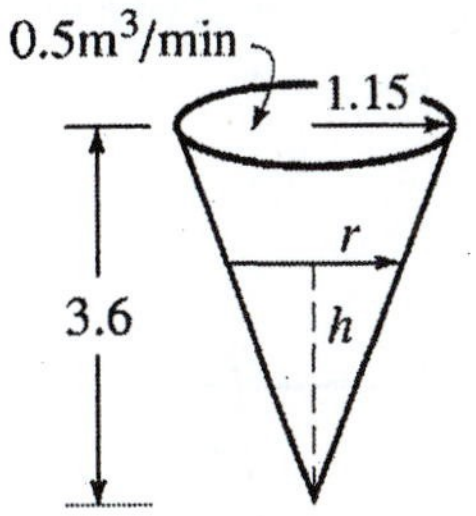

$$V = \frac{1}{3}\pi r^2 h = \frac{1}{3}\pi\left(\frac{1.15}{3.6}\right)^2 h^3$$

$$\frac{dV}{dt} = \frac{dV}{dh}\cdot\frac{dh}{dt}$$

$$0.50 = \pi\cdot\left(\frac{1.15}{3.6}\right)^2\cdot h^2\cdot\frac{dh}{dt}$$

$$0.50 = \pi\cdot\left(\frac{1.15}{3.6}\right)^2(1.8)^2\cdot\frac{dh}{dt}$$

$$\frac{dh}{dt} = 0.48\ \text{m/min}$$

41. $z^2 = 20^2 + x^2$; $\frac{dz}{dt} = -10.0$ ft/s; $z = 36.0$ ft,

$x = 29.9$ ft

$$2z\frac{dz}{dt} = 2x\frac{dx}{dt}$$

$$\frac{dx}{dt} = \frac{z}{x}\frac{dz}{dt}$$

$$\frac{dx}{dt}\Big|_{x=36.0} = \frac{36.0}{29.9}(-10.0)$$

$$\frac{dx}{dt} = -12.0 \text{ ft/s}$$

The negative sign indicates the boat is approaching the wharf.

24.5 Using Derivatives in Curve Sketching

1. $f(x) = x^3 - 6x^2$

$f'(x) = 3x^2 - 12x = 3(x-4)$ with $x = 0$, $x = 4$ as critical values.

If $x < 0$, $f'(x) = 3x(x-4) > 0$. $f(x)$ increasing

If $0 < x < 4$, $f'(x) = 3x(x-4) > 0$. $f(x)$ decreasing

If $x > 4$, $f'(x) = 3x(x-4) > 0$. $f(x)$ increasing

inc. $x < 0$, $x > 4$; dec. $0 < x < 4$

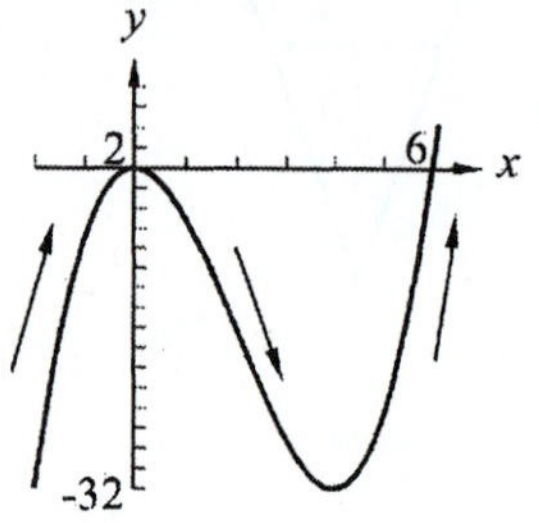

5. $y = x^2 + 2x$; $y' = 2x + 2$; $2x + 2 > 0$

$2x > -2$; $x > -1$; $f(x)$ increases.

$2x + 2 < 0$; $2x < -2$; $x < -1$; $f(x)$ decreases.

9. $y = x^2 + 2x$; $y' = 2x + 2$; $y' = 0$ at $x = 1$

$y'' = 2 > 0$ at $x = -1$ and $(-1, -1)$ is a relative minimum.

13. $y = x^2 + 2x$; $y' = 2x + 2$; $y'' = 2$

Thus, $y'' > 0$ for all x. The graph is concave up for all x and has no points of inflection.

17. $y = x^2 + 2x$

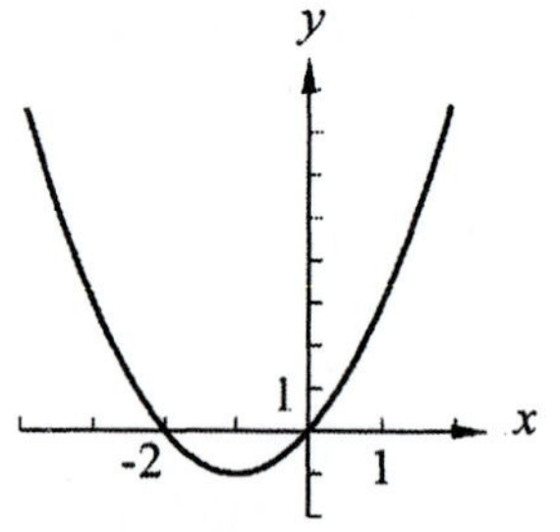

21. $y = 12x - 2x^2$; $y' = 12 - 4x$

$y' = 0$ at $x = 3$; for $x = 3$,

$y = 12(3) - 2(3)^2 = 18$ and $(3, 18)$ is a critical point. $12 - 4x < 0$ for $x > 3$ and the function decreases; $12 - 4x > 0$ for $x < 3$ and the function increases; $y'' = -4$; thus $y'' < 0$ for all x. There are no inflections; the graph is concave down for all x, and $(3, 18)$ is a maximum point.

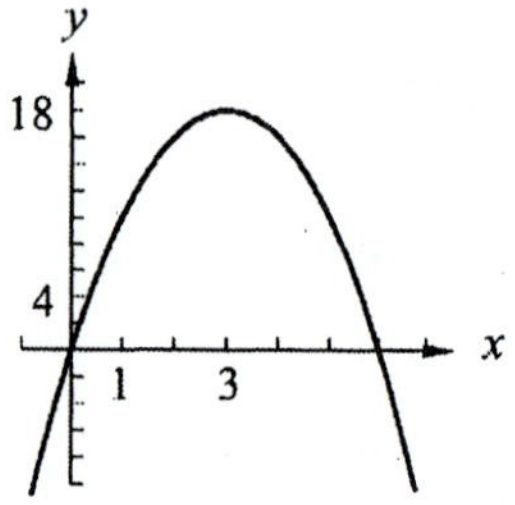

25. $y = x^3 + 3x^2 + 3x + 2$

$$y' = 3x^2 + 6x = 3(x^2 + 2x + 1)$$
$$= 3(x+1)(x+1)$$

$3(x+1)(x+1)=0$ for $x=-1$

$(-1, 1)$ is a critical point.

$3(x+1)(x+1)>0$ for $x<-1$ and the slope is positive.

$3(x+1)(x+1)>0$ for $x>-1$ and the slope is positive.

$y''=6x+6$; $6x+6=0$ for $x=1$, and $(-1, 1)$ is an inflection point.

$6x+6<0$ for $x<-1$ and the graph is concave down.

$6x+6>0$ for $x>-1$ and the graph is concave up.
Since there is no change in slope from positive to negative or vice versa, there are no maximum or minimum points.

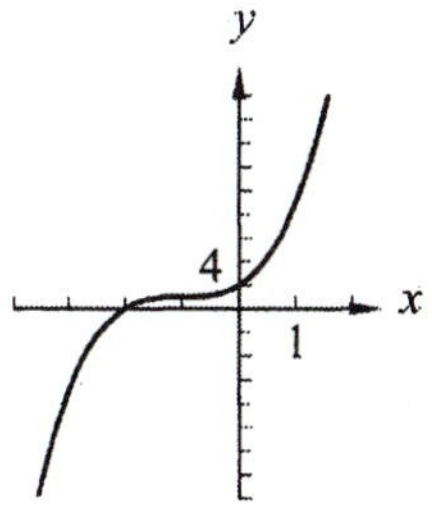

29. $y=4x^3-3x^4+6$

$y'=12x^2-12x^3=12x^2(1-x)=0$

$12x^2(1-x)=0$ for $x=0$ and $x=1$

$(0, 6)$ and $(1, 4)$ are critical points.

$12x^2-12x^3>0$ for $x<0$ and the slope is positive.

$12x-12x^3>0$ for $0<x<1$ and the slope is positive.

$12x-12x^3<0$ for $x>1$ and the slope is negative.

$y''=24x-36x^2$; $24x-36x^2=12x(2-3x)=0$ for $x=0$, $x=\frac{2}{3}$

$(0, 6)$ and $\left(\frac{2}{3}, \frac{178}{27}\right)$ are possible inflection points.

$24x-36x^2<0$ for $x<0$ and the graph is concave down.

$24x-36x^2>0$ for $0<x<\frac{2}{3}$ and the graph is concave up.

$24x-36x^2<0$ for $x>\frac{2}{3}$ and the graph is concave down.

$(1, 7)$ is a relative maximum point since $y'=0$ at $(1, 7)$ and the slope is positive for $x<1$ and negative for $x>1$. $(0, 6)$ and $\left(\frac{2}{3}, \frac{178}{27}\right)$ are inflection pts since there is a concavity change.

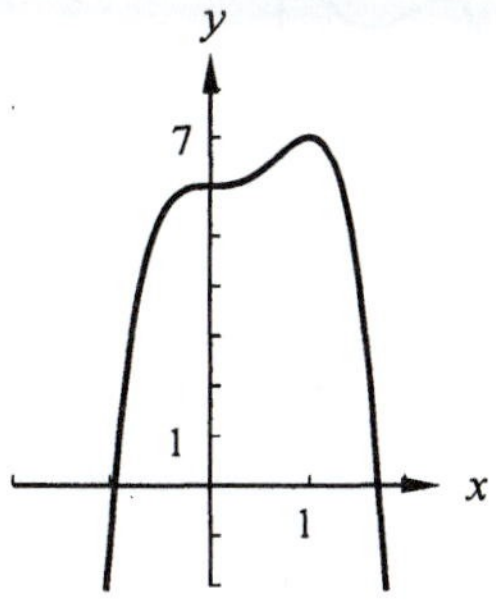

33. $y=x^3=12x$; $y'=3x^2-12$; $y''=6x$. On graphing calculator with $x_{\text{min}}=-5$, $x_{\text{max}}=5$, $y_{\text{min}}=-20$, $y_{\text{max}}=20$, enter $y_1=x^3-12x$; $y_2=3x^2-12$; $y_3=6x$. From the graph is observed that the maximum minimum values of y occur when y' is zero. A maximum value for y occurs when $x=-2$, and a minimum value occurs when $x=2$. An inflection point (change in curvature) occurs when y'' is zero. x is also zero at this point. Where $y'>0$, y inc.; $y'<0$, y dec. $y''>0$, y conc up; $y''<0$, y conc. down, $y''=0$, y has inflection.

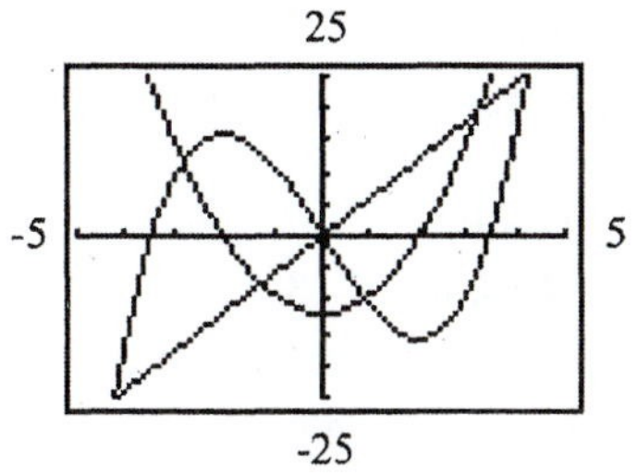

37. The left relative max is above the left relative min and below the right relative min.

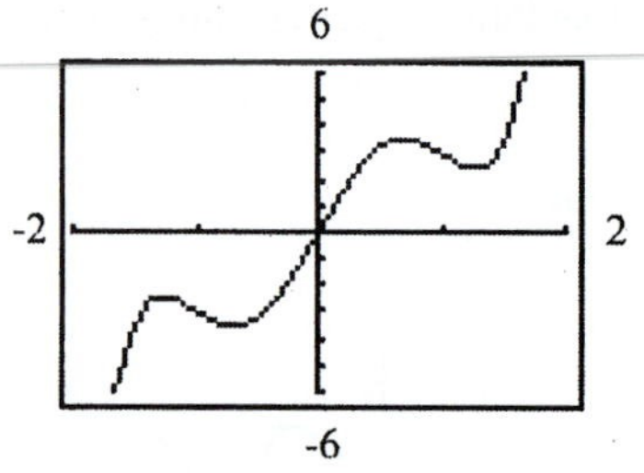

41. $P = 4i - 0.5i^2$

$P' = 4 - 1.0i = 0$ for $i = 4.0$

$P'' = -1.0 < 0$, conc. down everywhere

$P' > 0$ for $x < 4$, P inc.

$P' < 0$ for $x > 4$, P dec.

$P(4) = 8$, $(4, 8)$ max.

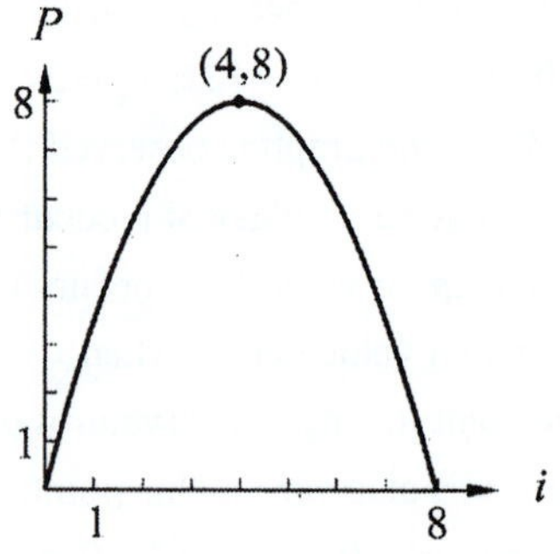

45. $R = 75 - 18i^2 + 8i^3 - i^4$

$R' = -36i + 24i^2 - 4i^3 = -4i(i^2 - 6i + 9)$

$= -4i(i-3)^2$

$R' = 0$ for $i = 0$ and $i = 3$

$(0, 75)$ and $(3, 48)$ are critical points.

$R' > 0$ for $i < 0$, $R' < 0$ for $0 < i < 3$

$R' < 0$ for $i > 3$

Max. at (0, 75), no max. or min. at (3, 48)

$R'' = -36 + 48i - 12i^2 = -12(i-1)(i-3)$

$(1, 64)$ and $(3, 48)$ are possible inflection points.

$R'' < 0$ for $i < 1$, concave down

$R'' > 0$ for $1 < i < 3$, concave up

$R'' < 0$ for $i > 3$, concave down

$(1, 64)$ and $(3, 48)$ are inflection points.

(From calculator graph, $R = 0$ for $i = -1.5$ and $i = 5.0$)

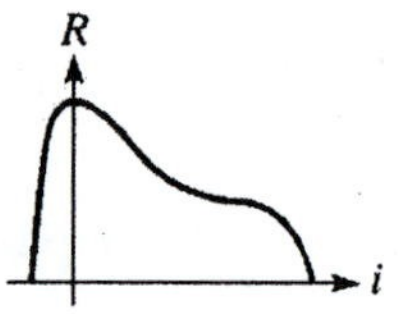

49. $V = x(8-2x)(12-2x)$

$V = 4(24x - 10x^2 + x^3)$

$= 4x^3 - 40x^2 + 96x$

$f'(x) = 4(24 - 20x + 3x^2) = 0$

By quadratic solution:

$x = 1.57$ and $x = 5.10$ (reject)

$f''(x) = 4(-20 + 6x)$

$f''(1.57) < 0$, rel max. $(1.57, 67.6)$

$f''(5.10) > 0$, rel min. $(5.10, -20.2)$

$f''(x) = 0$; $x = \dfrac{20}{6}$, infl. $\left(\dfrac{10}{3}, 23.7\right)$

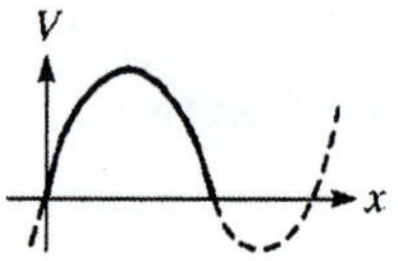

53. $f(-1) = 0$, root at $(-1, 0)$

$f(2) = 2$, point on curve $(2, 2)$

$f'(x) =< 0$ for $x < -1$

Therefore, $f(x)$ decreasing for $x < -1$.

$f'(x) > 0$ for $x > -1$

Therefore, $f(x)$ increasing for $x > -1$.

$f''(x) < 0$ for $0 < x < 2$

Therefore, $f(x)$ concave down for $0 < x < 2$.
$f''(x) > 0$ for $x < 0$ or $x > 2$
Therefore, $f(x)$ concave up for $x < 0$.
Therefore, $f(x)$ concave up for $x > 2$.
Summary: $(-1, 0)$ min. point
Inflection point at y-intercept
$(2, 2)$ inflection point

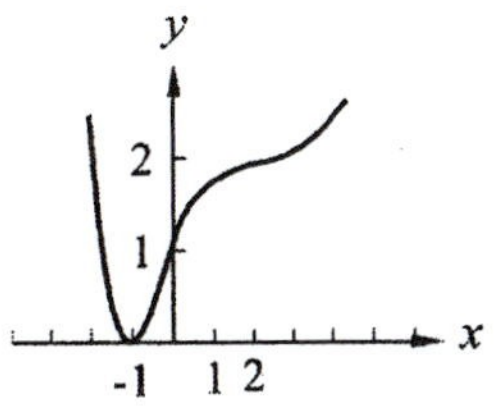

24.6 More on Curve Sketching

1. $y = x - \frac{4}{x}$

(1) Intercepts:
For $x = 0$, y is undefined which means curve is not continuous at $x = 0$ and there are no y-intercepts.

For $y = 0$, $0 = x - \frac{4}{x}$, $x^2 - 4 = 0$, $x = \pm 2$ are the x-intercepts.

(2) Symmetry: none

(3) Behavior as x becomes large:

As $x \to \pm\infty$, $\frac{4}{x} \to 0$ and $y \to x$. $y = x$ is a slant asymptote.

(4) Vertical asymptotes:
y is undefined for $x = 0$. As $x \to 0^-$, $y \to \infty$ and as $x \to 0^+$, $y \to -\infty$. The x-axis is a vertical asymptote.

(5) Domain and range:
domain: $x \neq 0$, range: $-\infty < y < \infty$

(6) Derivatives:

$y' = 1 + \frac{4}{x^2} > 0$, y inc. for $x \neq 0$

$y'' = -\frac{8}{x^3} > 0$ for $x < 0$, y conc. up

$y'' = -\frac{8}{x^3} < 0$ for $x > 0$, y conc. down

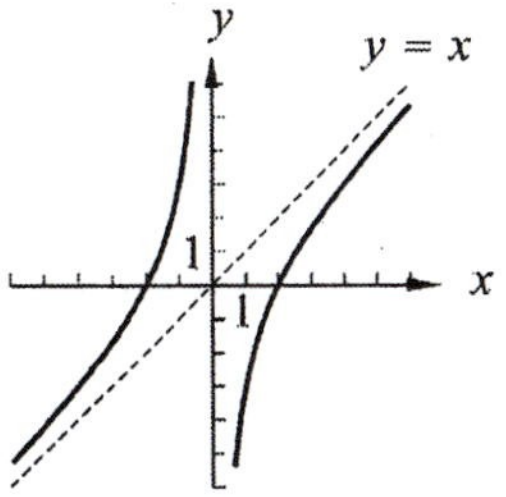

5. $y = x^2 + \frac{2}{x} = \frac{x^3 + 2}{x}$

(1) $\frac{2}{x}$ is undefined for $x = 0$, so the graph is not continuous at the y-axis; i.e., no y-intercept exists.

(2) $\frac{x^3 + 2}{x} = 0$ at $x = \sqrt[3]{-2} = -\sqrt[3]{2}$. There is an x-intercept at $\left(-\sqrt[3]{2}, 0\right)$.

(3) As $x \to \infty$, $x^2 \to \infty$ and $\frac{2}{x} \to 0$, so $x^2 + \frac{2}{x} \to \infty$.

(4) As $x \to 0$ through positive x, $x^2 \to 0$ and $\frac{2}{x} \to \infty$, so $x^2 + \frac{2}{x} \to \infty$.

(5) As $x \to -\infty$, $x^2 \to \infty$ and $\frac{2}{x} \to 0$ so $x^2 + \frac{2}{x} \to \infty$.
$x = 0$ is a vertical asymptote

(6) As $x \to 0$ through negative numbers, $x^2 \to 0$ and $\frac{2}{x} \to -\infty$, so $x^2 + \frac{2}{x} \to -\infty$.

(7) $y' = 2x - 2x^{-2} = 0$ at $x = 1$ and the slope is zero at $(1, 3)$.

(8) $y'' = 2 + 4x^{-3} = 0$ at $x = -\sqrt[3]{2}$ and $\left(-\sqrt[3]{2}, 0\right)$ is an inflection point.

(9) $y'' > 0$ at $x = 1$, so the graph is concave up and $(1, 3)$ is a relative minimum.

(10) Since $\left(-\sqrt[3]{2}, 0\right)$ is an inflection pt, $f''(-1) < 0$ and the graph is concave down. $f''(-2) > 0$ and the graph is concave up.

(11) Not symmetrical about the x or y-axis.

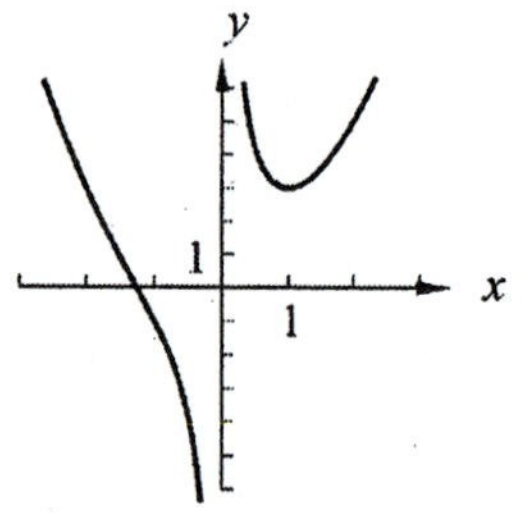

9. $y = \dfrac{x^2}{x+1} = x - 1 + \dfrac{1}{x+1}$

Intercepts:

(1) Function undefined at $x = -1$; not continuous at $x = -1$.

(2) At $x = 0$, $y = 0$. The origin is the only intercept.

(3) Behavior as x becomes large: As $x \to \pm\infty$, $y \to x - 1$, so $y = x - 1$ is a slant asymptote.

Vertical asymptotes:

(4) As $x \to -1$ from the left, $x + 1 \to 0$ through negative values and $\frac{x^2}{x+1} \to -\infty$ since $x^2 > 0$ for all x. As $x \to -1$ from the right, $x + 1 \to 0$ through positive values and $\frac{x^2}{x+1} \to +\infty$. $x = -1$ is an asymptote.

Symmetry:

(5) The graph is not symmetrical about the y-axis or the x-axis.

Derivatives:

(6) $y' = \frac{x^2+2x}{(x+1)^2}$; $y' = 0$ at $x = -2$, $x = 0$.

$(-2, -4)$ and $(0, 0)$ are critical points. Checking the derivative at $x = -3$, the slope is positive, and at $x = -1.5$ the slope is negative. $(-2, -4)$ is a relative maximum point. Checking the derivative at $x = -0.5$, the slope is negative and at $x = 1$ the slope is positive, so $(0, 0)$ is a relative minimum point.

Int. $(0, 0)$, max. $(-2, -4)$, min $(0, 0)$, asym. $x = -1$

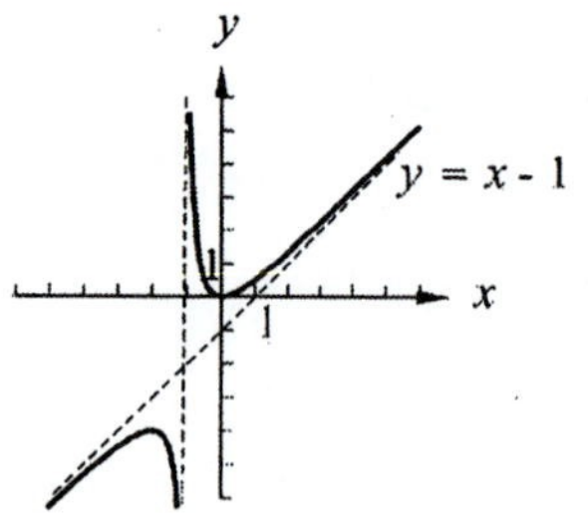

13. $y = \dfrac{4}{x} - \dfrac{4}{x^2}$

Intercepts:

(1) There are no y intercepts since $x = 0$ is undefined.

(2) $y = 0$ when $x = 1$ so $(1, 0)$ is an x-intercept,

Asymptotes:

(3) $x = 0$ is an asymptote; the denominator is 0.
$y = 0$ is an asymptote

Symmetry:

(4) Not symmetrical about the y-axis since $\frac{4}{x} - \frac{4}{x^2}$ is different from $\frac{4}{(-x)} - \frac{4}{(-x)^2}$

(5) Not symmetrical about the x-axis since $y = \frac{4}{x} - \frac{4}{x^2}$ is different from $-y = \frac{4}{x} - \frac{4}{x^2}$

(6) Not symmetrical about the origin since $y = \frac{4}{x} - \frac{4}{x^2}$ is different from $-y = \frac{4}{-x} - \frac{4}{(-x)^2}$

Derivatives:

(7) $y' = -4x^{-2} + 8x^{-3} = 0$ at $x = 2$; $(2, 1)$ is a relative maximum.

(8) $y'' = 8x^{-3} - 24x^{-4} = 0$ at $x = 3$ so $\left(3, \frac{8}{9}\right)$ is a possible inflection.

$y'' < 0$ (concave down) for $x < 3$ and 0 (concave up) for $x > 3$ so $\left(3, \frac{8}{9}\right)$ is an inflection.

Behavior as x becomes large:

(9) As $x \to \infty$ or $-\infty$, $\frac{4}{x}$ and $-\frac{4}{x^2}$ each approach 0.

As $x \to 0$, $\frac{4}{x} - \frac{4}{x^2} = \frac{4x-4}{x^2}$ approaches $-\infty$, through positive or negative values of x.

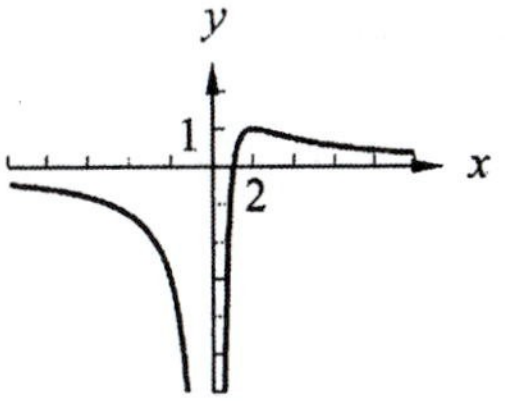

$f'(x) = 0$; $x^2 - 2x + 2 = 0$, no real solution

Consider $x = 0$, $x = 2$

$f'(x) < 0$ for $x < 0$

Therefore, $f(x)$ decreasing for $x < 0$.

$f'(x) < 0$ for $0 < x < 2$

Therefore, $f(x)$ decreasing for $0 < x < 2$.

$f'(x) < 0$ for $x > 2$

Therefore, $f(x)$ decreasing for $x > 2$.

Therefore, range is all y.

$f''(x) = 0 = 2x - 2$, $x = 1$; $(1, 0)$ infl. point

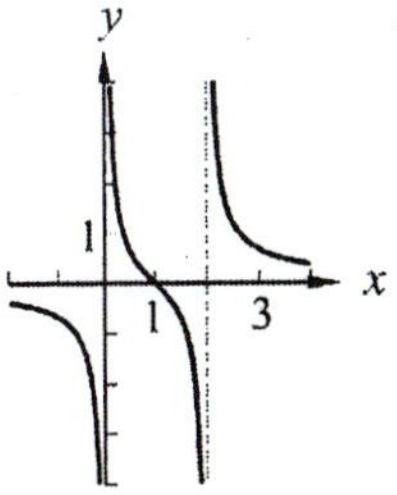

17. $y = \dfrac{9x}{9 - x^2}$

Intercept:

(1) Intercept, $(0, 0)$

(2) Asymptotes at $x = -3$, $x = 3$

Derivatives:

(3) $y' = \dfrac{(9 - x^2)(9) - (9)(-2x)}{(9 - x^2)^2}$

$81 + 9x^2 = 0$; $9x^2 = -81$; $x^2 = \sqrt{-9}$ (imaginary)

No real value max. or min., ± 3 are critical values.

(4) y''

$$= \frac{(9 - x^2)^2 (18x) - (81 + 9x^2)(2)(9 - x^2)(-2x)}{(9 - x^2)^2}$$

$$= \frac{-18x^5 - 324x^3 + 4374x}{(9 - x^2)^4}$$

$-18x^5 - 324x^3 + 4374x = 0$;

$-18x(x^4 + 18x^2 - 243) = 0$

$-18x = 0$; $x = 0$

$x^4 + 18x^2 - 243 = 0$; $(x^2 + 27)(x^2 - 9) = 0$

(imaginary) $x^2 = 9$; $x = \pm 3$ (these are asymptotes)

$y|_{x=0} = 0$; a possible inflection is $(0, 0)$

at $\left(-1, -\frac{9}{8}\right)$, $y'' = -4032$; concave down at $\left(1, \frac{9}{8}\right)$, $y'' = 4032$; concave up, and $(0, 0)$ is an inflection point.

Symmetry:

(5) There is symmetry to the origin.

(6) As $x \to +\infty$ and as $x \to -\infty$, $y \to 0$. Therefore, $y = 0$ is an asymptote.

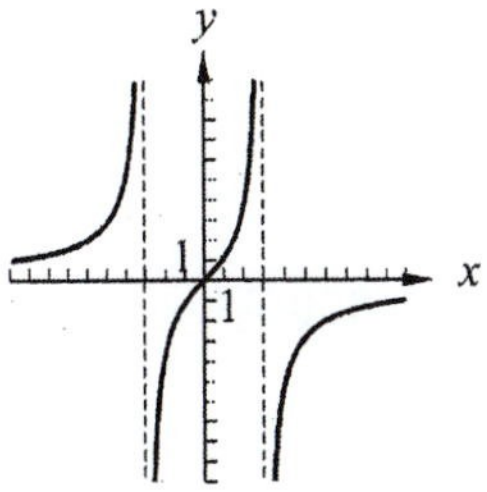

21. $\frac{9x^2}{x^2+9}=9-\frac{81}{x^2+9}$ which simplifies the graphing.

25. $R=\frac{200}{\sqrt{t^2+40{,}000}}$

Intercepts:

(1) Not continuous at $t=0$ (not defined at $t<0$).

(2) No t-intercept; R-intercept at $(0, 1)$

Symmetry:

(3) No symmetry about either axis (R is undefined for $t<0$).

Derivatives:

(4) $R=200\left(t^2+40{,}000\right)^{-1/2}$

$$R'=-100\left(t^2+40{,}000\right)^{-3/2}(2t)$$

$$=-200t\left(t^2+40{,}000\right)^{-3/2};$$

$\frac{-200t}{\left(t^2+40{,}000\right)^{3/2}}=0;\ -200t=0;\ t=0$ is a max, (0, 1).

since

$R\big|_{t=0}=1$ and $R\big|_{t=1}<1$ (R is undefined for $t<0$)

(5)

$$R''=\frac{\left(t^2+40{,}000\right)^{3/2}(-200)-(-200t)\left(\frac{3}{2}\right)\left(t^2+40{,}000\right)^{1/2}(2t)}{\left[\left(t^2+40{,}000\right)^{3/2}\right]^2}$$

$$=-200\left(t^2+40{,}000\right)^{3/2}+600t^2\left(t^2+40{,}000\right)^{1/2}=0$$

$$\left(t^2+40{,}000\right)^{1/2}\left[-200\left(t^2+40{,}000\right)+600t^2\right]=0$$

$$\left(t^2+40{,}000\right)^{1/2}=0;\ t^2+40{,}000=0$$

$t^2=-40{,}000$ (imaginary)

$$-200\left(t^2+40{,}000\right)+600t^2=0$$

$$-200\left[\left(t^2+40{,}000\right)-3t^2\right]=0;\ t^2+40{,}000-3t^2=0$$

$$-2t^2=-40{,}000;\ t^2=20{,}000$$

$t=141$ possible inflection

$R''\big|_{t=140}\le 0;\ R''\big|_{t=142}>0;\ R\big|_{t=141}=0.82$ is an inflection, $(141, 0.82)$

As x becomes large:

(6) As $x\to\infty$, $\sqrt{t^2+40{,}000}$ becomes infinitely large and $\frac{200}{\sqrt{t^2+40{,}000}}$ is a positive value that becomes infinitely small but never zero.

$R=0$ is a horizontal asymptote

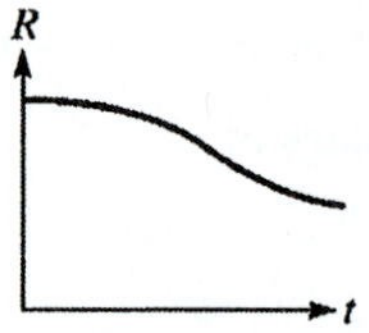

29. $V=\pi r^2h=20$

Therefore, $h=\frac{20}{\pi r^2}$

$A_T=$ top + bottom + wraparound

$$=\pi r^2+\pi r^2+2\pi rh$$

$$=2\pi r^2+2\pi r\left(\frac{20}{\pi r^2}\right)=2\pi r^2+\frac{40}{r}$$

Vertical asymptote at $r=0$.

$$\frac{dA_T}{dr}=4\pi r-\frac{40}{r^2}$$

$$\frac{d^2A}{dr^2}=4\pi+\frac{80}{r^3}$$

$$\frac{dA_T}{dr}=0;\ r=\sqrt[3]{\frac{10}{\pi}}=1.47$$

$f''(1.47)>0$ min. $(1.47, 40.8)$

$$\frac{d^2A_T}{dr^2}=0;\ r=-\sqrt[3]{\frac{20}{\pi}}\ \text{(reject)}$$

r must be >0

$\frac{d^2A}{dr^2}>0$ for $r>0$

Therefore, A_T is concave up for $r>0$.

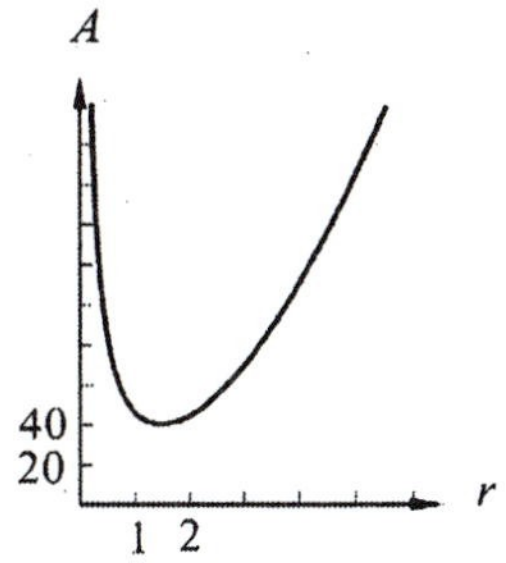

24.7 Applied Maximum and Minimum Problems

5. $P = EI - RI^2; \frac{dP}{dI} = E - 2RI = 0; I = \frac{E}{2R}$

$\frac{d^2P}{dI^2} = -2R < 0$ for all I, therefore max. power

at $I = \frac{E}{2R}$

9. $S = 360A - 0.1A^2$, find maximum S.

$S' = 360 - 0.3A^2;\ A^2 = 1200;\ A = 35\text{ m}^2$

$S'' = -0.6A < 0$ for all valid (positive) A so the graph is concave down and $A = 35\text{ m}^2$ is a max.

Maximum savings are

$S = 360(35) - 0.1(35)^2 = \$8300.$

13. $y = x^2 - 4$

$l =$ distance to origin $= \sqrt{x^2 + y^2}$

$l = \sqrt{x^2 + (x^2 - 4)^2}$

$L = l^2 = x^2 + x^4 - 8x^2 + 16$

$= x^4 - 7x^2 + 16$

$L' = 4x^3 - 14x = 0$

$x = \sqrt{\frac{14}{4}} = 1.87$

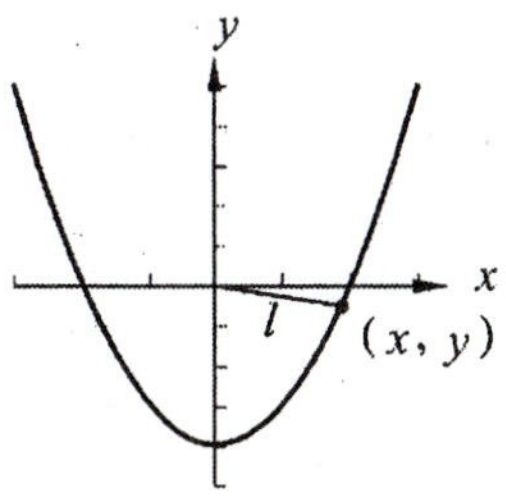

$L'' = 12x^2 - 14\big|_{x=1.87} > 0$, $x = 1.87$ is a minimum

$y = 1.87^2 - 4 = \frac{1}{2}$

$l = \sqrt{1.87^2 + \left(\frac{1}{2}\right)^2} = 1.94$ units is closest particle comes to origin.

17.

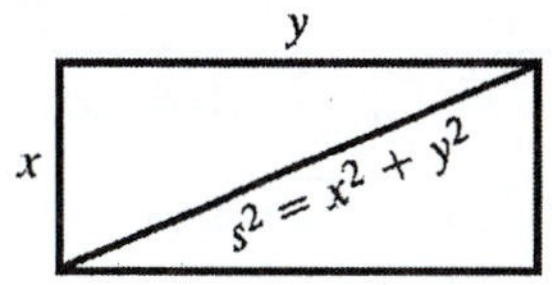

$P = 2x + 2y = 48$

$x + y = 24$

Diagonal will be a minimum if $l = s^2$ is a minimum.

$l = x^2 + y^2$

$l = x^2 + (24 - x)^2$

$l = x^2 + 24x^2 - 48x + x^2$

$l = 2x^2 - 48x + 24x^2$

$\frac{dl}{dx} = 4x - 48 = 0$

$x = 12$

from which $y = 12$

Dimensions are 12 in. by 12 in., a square will minimize the diagonal.

21. $A = xy,\ P = 2x + 2y$

$y = \frac{A}{x}$

$P = 2x + \frac{2A}{x}$

$P' = 2 - \frac{2A}{x^2} = 0 \Rightarrow x = \sqrt{A}$

$A = xy = \sqrt{A}y \Rightarrow y = \sqrt{A}$

$P'' = \frac{4A}{x^3}\Big|_{x=\sqrt{A}} > 0 \Rightarrow P$ is a maximum for

$x = y = \sqrt{A}$, a square.

25. $A = \frac{1}{2}xy;\ 12.0^2 = x^2 + y^2$

$y^2 = 144 - x^2;\ y = \sqrt{144 - x^2}$

$A = \frac{1}{2}x\sqrt{144 - x^2}$

$\frac{dA}{dx} = \frac{1}{2}\left[x \cdot \frac{1}{2}(144 - x^2)^{-1/2}(-2x) + \sqrt{144 - x^2}\right]$

$= \frac{1}{2}\left[\frac{-x^2}{\sqrt{144 - x^2}} + \sqrt{144 - x^2}\right]$

$= \frac{1}{2}\left[\frac{-x^2 + 144 - x^2}{\sqrt{144 - x^2}}\right] = \frac{1}{2}\frac{(144 - 2x^2)}{\sqrt{144 - x^2}}$

$\frac{dA}{dx} = 0 = 144 - 2x^2;\ x = \sqrt{72} = 8.49$

Test $\sqrt{72}$: $f'\sqrt{71} > 0; f'\sqrt{73} < 0$

Therefore, max. $(\sqrt{72}, \sqrt{72})$.

Therefore, legs of triangle will be equal at 8.49 cm for max. area.

29. $lw = 384$

$A = 384 + 4(24) + 2(6.00w) + 2(4.00l)$

$A = 408 + 12.0w + 8.00\frac{384}{w}$

$\frac{dA}{dw} = 12.0 - \frac{8.00(384)}{w^2} = 0$

$w = 16.0,\quad l = \frac{384}{16.0} = 24.0$

Dimensions: $w + 8.00 = 24.0$ cm

$l + 12.0 = 36.0$ cm

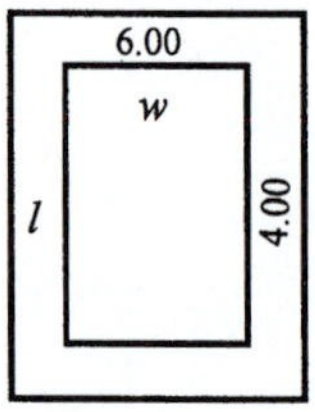

33. $V = (10 - 2x)(15 - 2x),\ 0 < x < 5$

$V = 4x^3 - 50x^2 + 150x$

$V' = 8x^2 - 100x = 150 = 0$

$x = 1.96,\ 6.37 > 5$, reject

$V'' = 12x^2 - 100x + 150\big|_{x=1.96} > 0,$

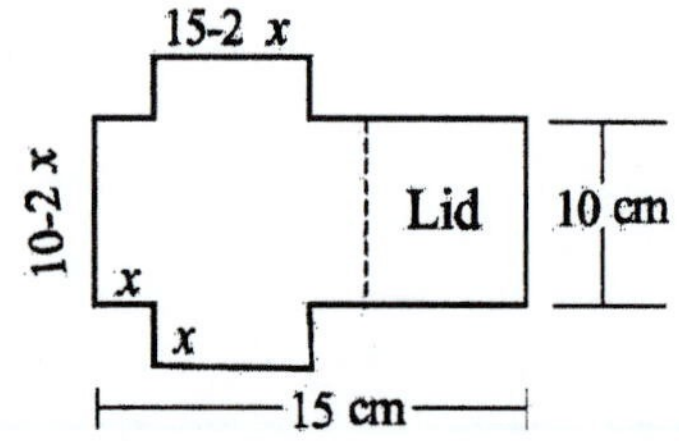

V is a maximum for $x = 2$ cm.

37. $y = k(x^4 - 5Lx^3 + 3L^2x^2)$

$= 2kx^4 - 5kLx^3 + 3kL^2x^2$

$y' = 8kx^3 - 15kLx^2 + 6kL^2x = 0$

$kx(8x^2 - 15Lx + 6L^2) = 0$

$kx = 0;\ x = 0$

$8x^2 - 15Lx + 6L^2 = 0$

$x = \frac{-(-15) \pm \sqrt{(-15)^2 - 4(8)(6)}}{2(8)}$

$= \frac{5 \pm \sqrt{33}}{16}$

$= \frac{15 \pm 5.75}{16}$

$x = 0.58L,\ 1.30L$ (not valid - this distance is greater than L, the length of the beam)

41. $n(x) = \frac{k}{x^2} + \frac{8k}{(8 - x)^2}$

$n'(x) = \frac{-2k}{x^3} + \frac{8k \cdot (-2)(-1)}{(8 - x)^3} = 0 \Rightarrow 8x^3 = (8 - x)^3$

$2x = 8 - x$

$3x = 8$

$x = \frac{8}{3}$

$$n''(x)=\frac{6k}{x^4}+\frac{48k}{(8-x)^4}$$

$$n''\left(\frac{8}{3}\right)>0\Rightarrow n(x)\text{ is min at } x=\frac{8}{3}\text{ km from A.}$$

45. $2x+\pi d=400;\ \pi d=400-2x;\ d=\dfrac{400-2x}{\pi}$

$$A=x(d)=x\left(\frac{400-2x}{\pi}\right)=\frac{400x-2x^2}{\pi}$$

$$A'=\frac{400-4x}{\pi}=0;\ 400-4x=0;\ x=100\text{ m}$$

$$\frac{dP_T}{dx}=-10.00-0.04x=0$$

$x=250$ units in excess of 1000

$\dfrac{d^2P_T}{dx^2}=-0.04<0$ for all x, therefore max. at $x=250$.

Therefore, company should produce 1250 units per week to maximize profit.

49. Let C = total cost

$$C=50,000(10-x)+80,000\sqrt{x^2+2.5^2}$$
$$=500,000-50,000x+80,000\left(x^2+6.25\right)^{1/2}$$
$$C'=-50,000+40,000\left(x^2+6.25\right)^{-1/2}(2x)$$
$$=-50,000+80,000x\left(x^2+6.25\right)^{-1/2}$$
$$=-50,000+\frac{80,000x}{\sqrt{x^2+6.25}}$$
$$=\frac{-50,000\sqrt{x^2+6.25}+80,000x}{\sqrt{x^2+6.25}}=0$$
$$-50,000\sqrt{x^2+6.25}+80,000x=0$$
$$\sqrt{x^2+6.25}=\frac{-80,000x}{-50,000}=\frac{8x}{5};\ x^2+6.25=\frac{64}{25}x^2$$
$$6.25=\frac{64}{25}x^2=\frac{39}{25}x^2;\ x^2=6.25\left(\frac{25}{39}\right)=4.00$$

$x=2.00$ mi; $10-x=8.00$ mi

$$\frac{dA_L}{dx}=(x+20.0)\left(\frac{-7000}{x^2}+40.0\right)+\left(\frac{7000}{x}+40.0\right)1=0$$
$$=\frac{-7000(x+20.0)}{x^2}+\frac{7000+40.0x}{x}=0$$

24.8 Differentials and Linear Approximations

1. $y=\dfrac{4x}{x^3+4}$

$$dy=\frac{\left(x^3+4\right)(4)-4x\left(3x^2\right)}{\left(x^3+4\right)^2}$$
$$=\frac{4x^3+16-12x^3}{\left(x^3+4\right)^2}dx$$
$$dy=\frac{-8x^3+16}{\left(x^3+4\right)^2}dx=\frac{-8\left(x^3-2\right)}{\left(x^3+4\right)^2}$$
$$dy=\frac{8\left(2-x^3\right)}{\left(x^3+4\right)^2}dx$$

5. $y=x^5+x;$

$$\frac{dy}{dx}=5x^4+1$$
$$dy=\left(5x^4+1\right)dx$$

9. $s=2\left(3t^2-5\right)^4;$

$$\frac{ds}{dt}=8\left(3t^2-5\right)^3(6t)$$
$$ds=8\left(3t^2-5\right)^3(6t)\,dt$$
$$ds=48t\left(3t^2-5\right)^3dt$$

13. $y=x^2(1-x)^3$

$$dy=\left[x^2\cdot3(1-x)^2(-1)+(1-x)^3\cdot2x\right]dx$$
$$dy=\left(-3x^2(1-x)^2+2x(1-x)^3\right)dx$$
$$dy=(1-x)^2\left(-3x^2+2x(1-x)\right)dx$$
$$dy=(1-x)^2\left(-5x^2+2x\right)dx$$
$$dy=(1-x)^2(-5x+2)\,dx$$

17. $y = f(x) = 7x^2 + 4x,\ dy = f'(x)\,dx$

$= (14x+4)\,dx$

$\Delta y = f(x+\Delta x) - f(x)$

$= 7(4.2)^2 + 4(4.2) - (7\cdot 4^2 + 4\cdot 4) = 12.28$

$dy = (14\cdot 4 + 4)(0.2) = 12$

21. $f(x) = x^2 + 2x;\ f'(x) = 2x + 2$

$L(x) = f(a) + f'(a)(x-a)$

$= f(0) + f'(0)(x-0)$

$L(x) = 0^2 + 2\cdot 0 + (2\cdot 0 + 2)(x-0)$

$L(x) = 2x$

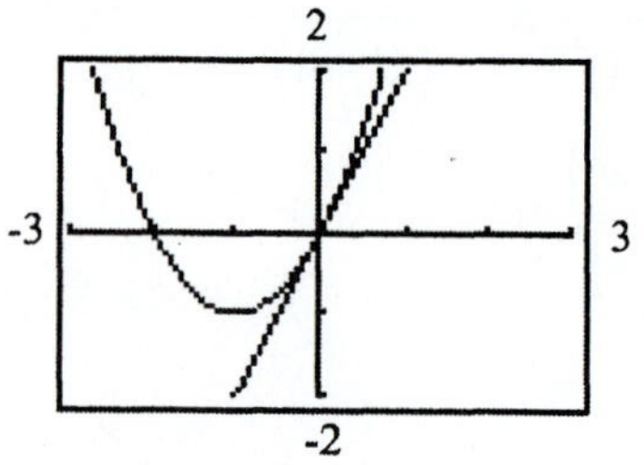

25. $C = 2\pi r,\ r = 6370,\ dr = 250$

$dC = 2\pi\,dr = 2\pi(250) = 1570$ km

29. $\lambda = \dfrac{k}{f},\ 685 = \dfrac{k}{4.38\times 10^{14}};$

$k = 3.00\times 10^{17}$ mm - H_z

$\dfrac{d\lambda}{df} = \dfrac{-k}{f^2}$

$d\lambda = \dfrac{-k}{f^2}\,df$

$d\lambda = \dfrac{-3.00\times 10^{17}}{(4.38\times 10^{14})^2}\cdot(0.20\times 10^{14})$

$d\lambda = -31$ nm

33. $A = \pi r^2$

$dA = 2\pi r\,dr$

$\dfrac{dA}{A} = \dfrac{2\pi r\,dr}{\pi r^2} = 2\dfrac{dr}{r} = 2(2\%) = 4\%$

37. $L(x) = f'(0)(x-0) + f(0)$

$f(x) = (1+x)^k,\qquad f'(x) = k(1+x)^{k-1}$

$f(0) = 1\qquad f'(0) = k$

$L(x) = kx + 1$

41. $C(V) = \dfrac{3.6}{(1+2V)},\ C(4) = 1.2$

$C'(V) = \dfrac{-3.6}{(1+2V)^{3/2}},\ C''(4) = \dfrac{-2}{15}$

$L(V) = -\dfrac{2}{15}(V-4) + 1.2 = -0.13V + 1.73$

Chapter 24 Review Exercises

1. $y = 3x - x^2$ at $(-1, -4)$; $y' = 3 - 2x$

$y'|_{x=-1} = 3 - 2(-1) = 5$

$m = 5$ for tangent line

$y - y_1 = 5(x - x_1)$

$y - (-4) = 5[x - (-1)]$; $y + 4 = 5x + 5$

$5x - y + 1 = 0$

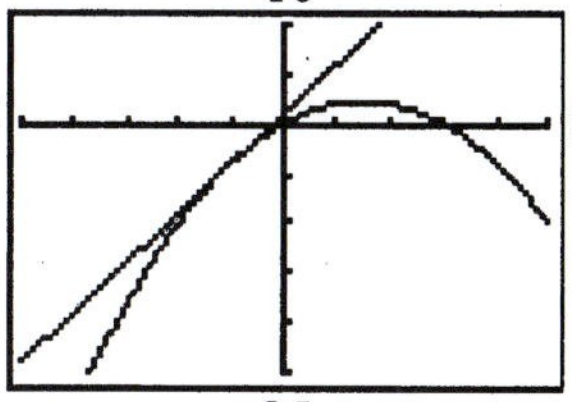

5. $y = \sqrt{x^2 + 3}$; $m = \frac{1}{2}$

$y = (x^2 + 3)^{1/2}$;

$\frac{dy}{dx} = \frac{1}{2}(x^2 + 3)^{-1/2}(2x)$

$= \frac{x}{\sqrt{x^2 + 3}}$

$m_{\tan} = \frac{dy}{dx}$

$2x = \frac{x}{\sqrt{x^2 + 3}} = \frac{1}{2}$

Squaring both sides, $4x^2 = x^2 + 3$; $3x^2 = 3$; $x^2 = 1$; $x = 1$ and $x = -1$. Therefore, the abscissa of the point at which $m = \frac{1}{2}$ is 1 or -1. If $x = 1$, $y = \sqrt{1^2 + 3} = \sqrt{4} = 2$ or -2. If $x = -1$, $y = \sqrt{(-1)^2 + 3} = 2$ or -2. The possible points where the slope of the tangent line is $\frac{1}{2}$ are $(1, 2)$, $(1, -2)$, $(-1, 2)$, $(-1, -2)$. A sketch of the curve shows that the only relative maximum or minimum point is at $m(0, 1.7)$. Therefore, the point is $(1, 2)$.

$y - 2 = \frac{1}{2}(x - 1)$; $y = \frac{1}{2}x + \frac{3}{2}$ is the equation of the tangent line.

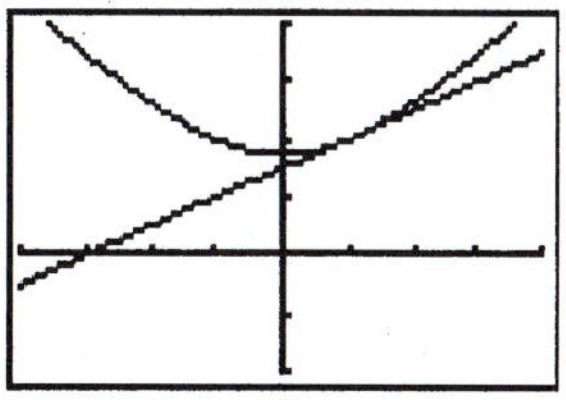

9. $y = 0.5x^2 + x$; $v_y = \frac{dy}{dt} = \frac{x\,dx}{dt} = +\frac{dx}{dt}$; $x = 0.5\sqrt{x}$

Substituting, $v_y = x(0.5\sqrt{x}) + 0.5\sqrt{x}$

Find v_y at $(2, 4)$:

$v_y|_{x=2} = 2(0.5\sqrt{2}) + 0.5\sqrt{2} = \sqrt{2} + 0.5\sqrt{2} = 1.5\sqrt{2} = 2.12$

13. $x^3 - 3x^2 - x + 2 = 0$ (between 0 and 1)

$f(x) = x^3 - 3x^2 - x + 2$; $f'(x) = 3x^2 - 6x - 1$

$f(0) = 0^3 - 3(0^2) - 0 + 2 = 2$; $f(1)$

$= 1^3 - 3(1^2) - 1 + 2 = -1$

The root is possibly closer to 1 than 0. Let $x_1 = 0.6$:

n	x_n	$f(x_n)$	$f'(x_n)$	$x_n - \frac{f(x_n)}{f'(x_n)}$
1	0.6	0.536	-3.52	0.7522727
2	0.7522727	-0.0242935	-3.8158936	0.7459063
3	0.7459063	-0.0000304	-3.8063092	0.7458983

$x_4 = x_3 = 0.7458983$

17. $y = 4x^2 + 16x$

(1) The graph is continuous for all x.

(2) The intercepts are $(0, 0)$ and $(-4, 0)$.

(3) As $x \to +\infty$ and $-\infty$, $y \to +\infty$.

(4) The graph is not symmetrical about either axis or the origin.

(5) $y' = 8x + 16$; $y' = 0$ at $x = -2$. $(-2, -16)$ is a critical point.

(6) $y'' = 8 > 0$ for all x; the graph is concave up and $(-2, -16)$ is a minimum.

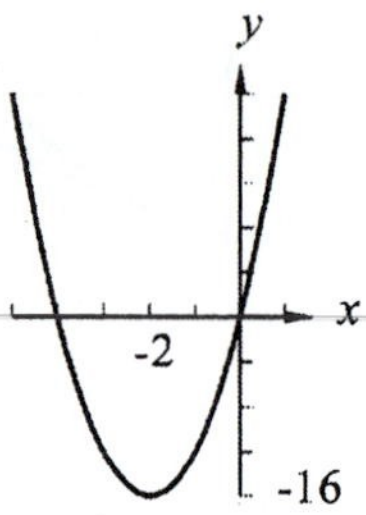

21. $y = x^4 - 32x$

(1) The graph is continuous for all x.

(2) The intercepts are $(0, 0)$ and $(2\sqrt[3]{4}, 0)$.

(3) As $x \to -\infty$, $y \to +\infty$; as $x \to +\infty$, $y \to +\infty$.

(4) The graph is not symmetrical about either axis or the origin.

(5) $y' = 4x^3 - 32 = 0$ for $x = 2$

(6) $y'' = 12x^2$; $y'' = 0$ at $x = 0$; $(0, 0)$ is a possible point of inflection. Since $f''(x) > 0$; the graph is concave up everywhere (except 0) and (0, 0) is not an inflection point. $(2, -48)$ is a minimum

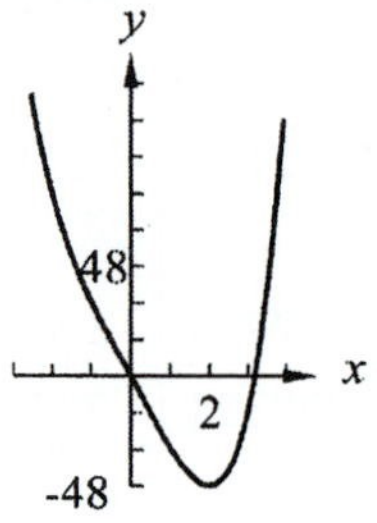

25. $y = f(x)$

$$= 4x^3 + \frac{1}{x}$$

$$dy = f'(x)\,dx$$

$$= \left(2x^2 - \frac{1}{x^2}\right)dx$$

29. $y = f(x) = 4x^3 - 12,\ x = 2,\ \Delta x = 0.1$

$$\Delta y - dy = f(x + \Delta x) - f(x) - f'(x)\,dx$$

$$= 4(2.1)^3 - 12 - \left(4(2)^3 - 12\right) - 12(2)^2(0.1)$$

$$= 0.244$$

33. $V = f(r) = \frac{4}{3}\pi r^3,\ r = 3.500,\ \Delta r = 0.012$

$$dV = f'(r)\,dr = 4\pi r^2 dr = 4\pi(3.500)^2(0.012)$$

$$dV = 1.85 \text{ m}^3$$

37. $Z = \sqrt{R^2 + X^2} = \left(R^2 + X^2\right)^{1/2}$

$$dZ = \frac{1}{2}\left(R^2 + X^2\right)^{-1/2}(2R)\,dR$$

$$dZ = \frac{RdR}{\sqrt{R^2 + X^2}}$$

$$\text{relative error} = \frac{dZ}{Z} = \frac{RdR}{R^2 + X^2}$$

41. $y = x^2 + 2$ and $y = 4x - x^2$

$y' = 2x;\ y' = 4 - 2x$

$2x = 4 - 2x;\ 4x = 4;\ x = 1$

The point (1, 3) belongs to both graphs; the slope of the tangent line is 2.

$y - y_1 = 2(x - x_1);\ y - 3 = 2(x - 1);\ y - 3 = 2x - 2$

$2x - y + 1 = 0$ is the equation of the tangent line.

45. $x = 8t \qquad y = -0.15t^2 \qquad v = \sqrt{8^2 + (-3.6)^2}$

$$\frac{dx}{dt} = 8 \qquad \frac{dy}{dt} = -0.30t \qquad v = \sqrt{64 + 12.96}$$

$$v_x\big|_{t=12} = 8 \qquad v_y\big|_{t=12} = -3.6 \qquad v\big|_{t=12} = \sqrt{76.96}$$

$$= 8.8 \text{ m/s}$$

$$\tan\theta = \frac{-3.6}{8} = -0.45;\ \theta = 336^\circ$$

49. $d = \dfrac{1000}{\sqrt{T^2 - 400}}$

$$\frac{dd}{dt} = \frac{dd}{dT}\frac{dT}{dt} = -\frac{-1000T}{\left(T^2 - 400\right)^{3/2}}\frac{dT}{dt}\Bigg|_{T=28.0,\ \frac{dT}{dt}=2.00}$$

$$\frac{dd}{dt} = \frac{-1000(28.0)}{\left(28.0^2 - 400\right)^{3/2}}(2.00) = -7.44 \text{ cm/s}$$

53. $P = 0.030r^3 - 2.6r^2 + 71r - 200,\ 6 \le r \le 30\ \text{m}^3/\text{s}$

$$\frac{dP}{dr} = 0.09r^2 - 5.2r + 71 = 0$$

$r = 22.1,\ r = 35.6$ (reject), $6 \le r \le 30)$

$$\frac{d^2P}{dr^2} = 0.18 - 5.2\Big|_{22.1} = -1.222 < 0 \text{ maximum}$$

P is a maximum when rate is $22.1\ \text{m}^3/\text{s}$

57.

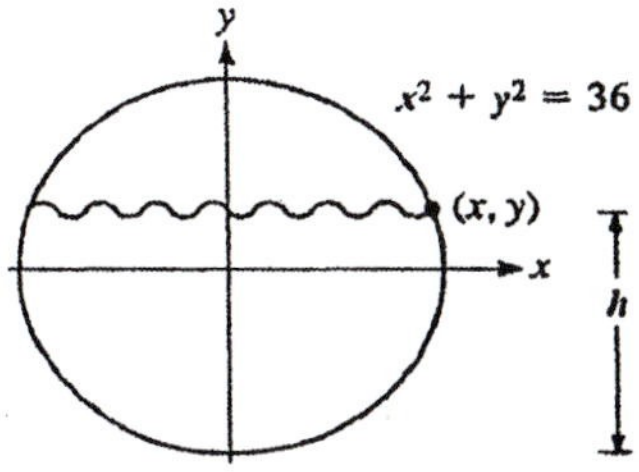

$w = 2x,\ h = 6 + y$

$$w = 2\sqrt{36 - y^2} = 2\cdot\sqrt{36 - (h-6)^2}$$

$$\frac{dw}{dt} = \frac{dw}{dh}\cdot\frac{dh}{dt} = \frac{-2(h-6)}{\sqrt{36-(h-6)^2}}\cdot\frac{dh}{dt}$$

$$\frac{dw}{dt} = \frac{-2(1.5-6)}{\sqrt{36-(1.5-6)^2}}\cdot(-0.250)$$

$= -0.567$ ft/s

61. Let x = distance from home plate to player, then

$$x^2 = 90.0^2 + (18.0t)^2 = 90.0^2 + 18.0^2t^2$$

$$2x\frac{dx}{dt} = 2(18.0)^2 t$$

$$\frac{dx}{dt} = \frac{18.0^2 t}{x}\Bigg|_{x=\sqrt{90^2+(18.0t)^2},\ t=\frac{40.0}{18.0}}$$

$$= \frac{18.0^2\left(\frac{40.0}{18.0}\right)}{\sqrt{90^2 + 18.0^2\left(\frac{40.0}{18.0}\right)^2}} = 7.31 \text{ ft/s}$$

65. Let y = altitude of the plane

z = distance from plane to radar station

x = distance from a point on the ground directly below the plane to the radar station

$$x^2 = x^2 + y^2$$

$$2z\frac{dz}{dt} = 2x\frac{dx}{dt} + 2y\frac{dy}{dt}$$

$$z\frac{dz}{dt} = x\frac{dx}{dt} + y(0)$$

$$\sqrt{x^2+y^2}\,\frac{dz}{dt} = x\frac{dx}{dt}$$

$$\sqrt{5.00^2 + \left(\frac{8000}{5280}\right)^2}\,(-680) = 5.00\frac{dx}{dt}$$

$$\frac{dx}{dt} = -710$$

Actual speed of plane $= |-710| = 710$ mi/h

69. $V = x^2y;\ x^2 + 4xy = 27,\ y = \dfrac{27 - x^2}{4x}$

$$V = x^2\cdot\frac{27-x^2}{4x} = \frac{27x - x^3}{4}$$

$$V' = \frac{27}{4} - \frac{3x^2}{4} = 0 \text{ for } x = 3$$

$$V'' = \frac{-64}{4} < 0 \text{ for } x > 0,\ V \text{ is concave down}$$

$$V = \frac{27(3) - 3^2}{4} = 13.5 \text{ ft}^3 \text{ is the maximum volume.}$$

73. $V = (12 - x)(10 - 2x)(x),\ 0 < x < 5$

$V(x) = 120x - 34x^2 + 2x^3$

$V'(x) = 120 - 68x + 6x^2 = 0$ for $x = 2.2,\ 9.2$ using the quadratic formula. 9.2 is rejected since $0 < x < 5$. Since $V''(x) = -68 + 12x^2\big|_{x=2.2} < 0$ the volume of the drawer is a maximum for $x = 2.2$ in.

77. Let s = cost per cm^2 of stainless steel

$10s$ = cost per cm^2 of silver

$V = \pi r^2 h = 314$

$r = \frac{314}{\pi h}$ and $h = \frac{314}{\pi r^2}$

$\text{Cost} = c = s(\pi r^2 + 2\pi rh) + 10s(\pi r^2)$

$c = 11s\pi r^2 + 2\pi srh,\ r > 0$

$c = 11s\pi r^2 + 2\pi sr\left(\frac{314}{\pi r^2}\right)$

$c(r) = 11s\pi r^2 + \frac{2s(314)}{r} > 0$

$c'(r) = 22s\pi r - \frac{2s(314)}{r^2} = 0$ for $r = 2.09$

which is a minimum since

$c''(r) = 22s\pi + \frac{4s(314)}{r^3} > 0$

$h = \frac{314}{\pi r^2} = 23.0$

The most economical dimensios of the container are $r = 2.09$ cm, $h = 23.0$ cm

81.

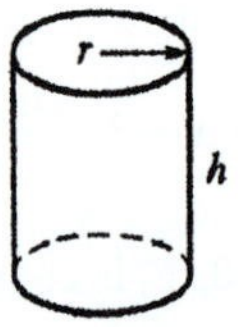

The amount of plastic used is determined by the surface area $S = 2 \cdot \pi r^2 + 2\pi r \cdot h$. $V = \pi r^2 h =$ constant from which $h = \frac{\text{constant}}{\pi r^2}$.

$S = 2\pi r^2 + 2\pi r \cdot \frac{\text{constant}}{\pi r^2}$

$S(r) = 2\pi r^2 + \frac{2 \cdot \text{constant}}{r}$

$\frac{dS}{dr} = 4\pi r - \frac{2 \cdot \text{constant}}{r^2} = 0$ for minimum

$4\pi r^3 = 2 \cdot \text{constant}$

$r^3 = \frac{\text{constant}}{2\pi}$

$\frac{h}{r} = \frac{\text{constant}}{\pi \cdot r^3} = \frac{\text{constant}}{\pi \frac{\text{constant}}{2\pi}} = 2$

The height should be twice the radius to minimize the surface area. In finding the $\frac{h}{r}$ ratio, the constant volume divides out so it is not necessary to specify the volume.

Chapter 25

INTEGRATION

25.1 Antiderivatives

1. $f(x)=12x^3$; power of x required is 4, $F(x)=ax^4$
$F'(x)=4ax^3=12x^3$; $4a=12$, $a=3$
$F(x)=3x^4$

5. $3x^2$; the power of x required in the antiderivative is 3. Therefore, we must multiply by $\frac{1}{3}$. The antiderivative of $3x^2$ is $\frac{1}{3}(3x^3)=x^3$. $a=1$.

9. The power of x required in the antiderivative of $f(x)=9\sqrt{x}$ is $\frac{3}{2}$. Multiply by $\frac{2}{3}$. The antiderivative of $9\sqrt{x}$ is $\frac{2}{3}\cdot 9x^{3/2}$. $a=6$.

13. The power of x required in the antiderivative of $\frac{5}{2}x^{3/2}$ is $\frac{5}{2}$. Multiply by $\frac{2}{5}$. The antiderivative of $\frac{5}{2}x^{3/2}$ is $\frac{2}{5}\left(\frac{5}{2}\right)x^{5/2}=x^{5/2}$.

17. $f(x)=2x^2-x$; $2x^2\to ax^3$; $\frac{d}{dx}\left(ax^3\right)=a3x^2$
$3a=2$; $a=\frac{2}{3}$, therefore, $\frac{2}{3}x^3$; $x\to ax^2$;
$\frac{d}{dx}\left(ax^2\right)=a2x$
$2a=1$; $a=\frac{1}{2}$, therefore, $\frac{1}{2}x^2$.
antiderivative of $2x^2-x$ is $\frac{2}{3}x^3-\frac{1}{2}x^2$.

21. $f(x)=\frac{-7}{x^6}$; $f(x)=-7x^{-6}$; power required is -5 therefore, x^{-5};
$\frac{d}{dx}ax^{-5}=-5ax^{-6}$; $-5a=-7$; $a=\frac{7}{5}$ therefore, $\frac{7}{5}x^{-5}$.
Therefore, antiderivative is $\frac{7}{5x^5}$.

25. The power of x required for the antiderivative of x^2 is 3, so it will be multiplied by $\frac{1}{3}$, the power of x required for $4=4x^0$ is 1, and it will be multiplied by 1. The power of x required for x^{-2} is -1, and it will be multiplied by -1. The antiderivative of x^2-4+x^{-2} is $\frac{1}{3}x^3-4x-\frac{1}{x}$.

29. The antiderivative requires $\left(p^2-1\right)^4$. We multiply by $\frac{1}{4}$. Thus, we have $\frac{1}{4}\left[4\left(p^2-1\right)^4\right]$. The derivative of $\left(p^2-1\right)^4$ is $\left(p^2-1\right)^3(2p)$. The antiderivative of $4\left(p^2-1\right)^3(2p)$ is $\left(p^2-1\right)^4$.

33. The antiderivative requires $(6x+1)^{3/2}$. We multiply by $\frac{2}{3}$. Thus we have $\frac{2}{3}\left(\frac{3}{2}\right)(6x+1)^{3/2}$. The derivative of $(6x+1)^{3/2}$ is $\frac{3}{2}(6x+1)^{1/2}(6)$. The antiderivative of $\frac{3}{2}(6x+1)^{1/2}(6)$ is $(6x+1)^{3/2}$.

37. The antiderivative of $f(x)=\dfrac{-2}{(2x+1)^2}$
$=-2(2x+1)^{-2}$ is
$F(x)=\dfrac{1}{(2x+1)}=(2x+1)^{-1}$ since
$F'(x)=(-1)(2x+1)^{-1-1}(2)=\dfrac{-2}{(2x+1)^2}$.

25.2 The Indefinite Integral

1. $\int 8x\,dx=8\int x^1\,dx=8\dfrac{x^{1+1}}{1+1}+C=4x^2+C$

5. $\int 2x\,dx=2\int x\,dx$; $u=x$; $du=dx$; $n=1$
$2\int x\,dx=2\left(\dfrac{x^{1+1}}{1+1}\right)+C=x^2+C$

9. $\int 8x^{3/2}dx;\ u = x;\ du = dx;\ n = \frac{3}{2}$

$$\int 8x^{3/2}dx = \frac{8x^{(3/2)+1}}{\frac{3}{2}+1} + C = \frac{8x^{5/2}}{\frac{5}{2}} + C = \frac{16}{5}x^{5/2} + C$$

13. $\int(x^2 - x^5)dx = \int x^2 d - x\int x^5 dx$

$$= \frac{x^3}{3} - \frac{x^6}{6} + C = \frac{1}{3}x^3 - \frac{1}{6}x^6 + C$$

17. $\int\left(\frac{t^2}{2} - \frac{2}{t^2}\right)dt = \frac{t^{2+1}}{2(2+1)} - \frac{2t^{-2+1}}{-2+1} + C$

$$= \frac{t^3}{6} + \frac{2}{t} + C$$

21. $\int(2x^{-2.3} + 3^{-2})dx = \int 2x^{-2/3}dx + \int 3^{-2}x^0 dx$

$$= 2\int x^{-2/3}dx + 3^{-2}\int x^0 dx$$

$$= 2 + \frac{1}{3}(3x^{1/3}) + 3^{-2}(x^1)$$

$$= 6x^{1/3} + \frac{1}{9}x + C$$

25. $\int(x^2-1)^5(2xdx);\ u = x^2 - 1;\ du = 2xdx;\ n = 5$

$$\int(x^2-1)^5(2xdx) = \frac{(x^2-1)^6}{6} + C = \frac{1}{6}(x^2-1)^6 + C$$

29. $\int(2\theta^5+5)^7\theta^4 d\theta = \frac{1}{10}\int(2\theta^5+5)^7\cdot(10\theta^4)d\theta$

$$= \frac{1}{10}\cdot\frac{(2\theta^5+5)^8}{8} + C = \frac{(2\theta^5+5)^8}{80} + C$$

33. $\int\frac{xdx}{\sqrt{6x^2+1}} = \int(6x^2+1)^{-1/2}xdx$

$u = 6x^2 + 1;\ du = 12x;\ n = -\frac{1}{2}$

$$\int(6x^2+1)^{-1/2}xdx = \frac{1}{12}\int(6x^2+1)^{-1/2}(12xdx) = \frac{1}{6}\sqrt{6x^2+1} + C$$

37. $\frac{dy}{dx} = 6x^2;\ dy = 6x^2dx$

$$y = \int 6x^2dx = 6\int x^2dx = \frac{6x^3}{3} + C = 2x^3 + C$$

The curve passes through $(0, 2)$. $2 = 2(0^3) + C$,

$C = 2;\ y = 2x^3 + 2$

41. $\int 3x^2dx = x^3 + C$

$\int 3x^2dx \neq x^3$ constant of integration is missing

45. $\int 3(2x+1)^2dx = \frac{3}{2}\int(2x+1)^2(2dx)$

$$= \frac{3}{2}\frac{(2x+1)^3}{3} + C = \frac{(2x+1)^3}{3} + C$$

$\int 3(2x+1)^2dx \neq (2x+1)^3 + C$ the factor of $\frac{1}{3}$ is missing.

49. $f''(x) = \sqrt{x} = x^{1/2}$

$$f'(x) = \frac{x^{1/2+1}}{\frac{1}{2}+1} + C = \frac{2}{3}x^{3/2} + C_1$$

$$f(x) = \frac{2}{3}\frac{x^{3/2+1}}{\frac{3}{2}+1} + C_1x + C_2$$

$$f(x) = \frac{5}{3}x^{5/2} + C_1x + C_2$$

53. $\frac{di}{dt} = 4t - 0.6t^2;\ di = (4t - 0.6t^2)dt$

$$i = \int(4t - 0.6t^2)dt = 2t^2 - 0.2t^3 + C$$

$i = 2A$ when $t = 0$ s; $2 = 2(0)^2 - 0.2(0)^3 + C$;

$C = 2$

$i = 2t^2 - 0.2t^3 + 2$

57. $\frac{df}{dA} = \frac{0.005}{\sqrt{0.01A+1}} = 0.005(0.01A+1)^{-1/2}$

$$f(A) = \int 0.005(0.01A+1)^{-1/2}dA + C$$

$$= \frac{1}{2}\int(0.01A+1)^{-1/2}(0.01dA)$$

$$= (0.01A+1)^{1/2} + C$$

$f = 0$ for $A = 0\text{ m}^2 \Rightarrow f(0) = 0;\ C = -1$

$$f(A) = (0.01A+1)^{1/2} - 1 = \sqrt{0.01A+1} - 1$$

25.3 The Area Under a Curve

1. (a)

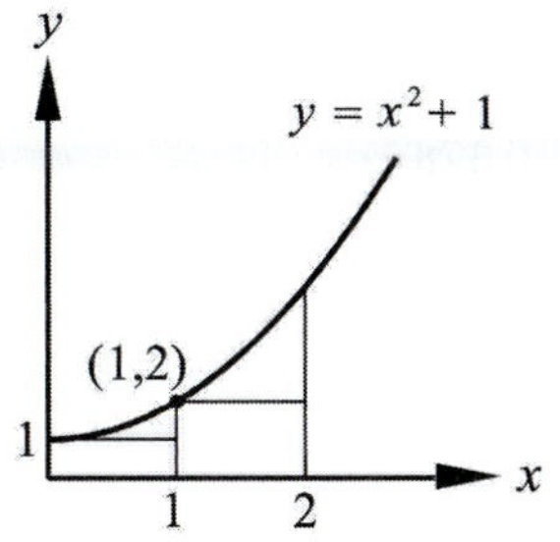

$A = 1(1) + 1(2)$

$A = 3$

(b)

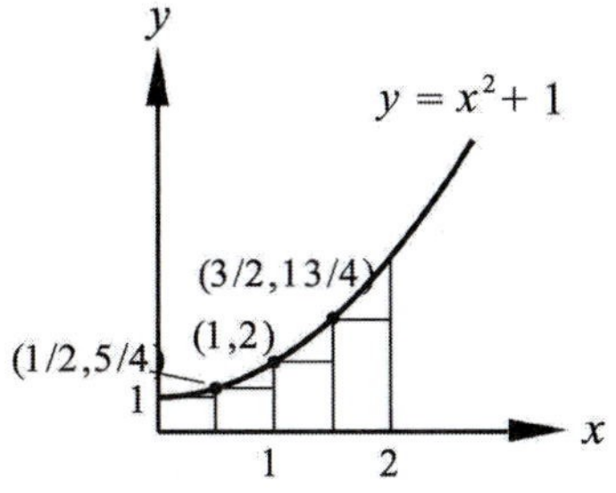

$$A = \frac{1}{2}\left[1 + \frac{5}{4} + 2 + \frac{13}{4}\right]$$

$$A = \frac{15}{4}$$

5. $y = 3x$, between $x = 0$ and $x = 3$

(a)

x	y
1	3
2	6
3	9

$n = 3$; $\Delta x = 1$

$A = 1(0 + 3 + 6) = 9$; (first rectangle has 0 height)

(b)

x	y
0	0
0.3	0.9
0.6	1.8
0.9	2.7
1.2	3.6
1.5	4.5
1.8	5.4
2.1	6.3
2.4	7.2
2.7	8.1
3.0	9.0

$n = 10$; $\Delta x = 0.3$

$$A = 0.3(0 + 0.9 + 1.8 + 2.7 + 3.6 + 4.5 + 5.4 + 6.3 + 7.2 + 8.1)$$
$$= 0.3(40.5) = 12.15$$

(a) $n = 4(\Delta x = 0.5)$

$A = (A_1 + A_2 + A_3 + A_4)$

$A = (y_1 + y_2 + y_3 + y_4)x$

$A = (1 + 2 + 3 + 4)0.5$

$A = (10)(0.5) = 5$

x	y
0	1
$\frac{1}{2}$	2
1	3
$\frac{3}{2}$	4
2	5

9. $y = 4x - x^2$, between $x = 1$ and $x = 4$

(a) $n = 6$, $\Delta x = 0.5$

$A = 0.5(3.00 + 3.75 + 3.75 + 3.00 + 1.75 + 0.00)$

$A = 7.625$

x	y
1.0	3.00
1.5	3.75
2.0	4.00
2.5	3.75
3.0	3.00
3.5	1.75
4.0	0.00

($y = 4.00$ is not the height of any inscribed rectangle)

(b) $n = 10$, $\Delta x = 0.3$

$$A = 0.3(3.00 + 3.51 + 3.84 + 3.96 + 3.75 + 3.36 + 2.79 + 2.04 + 1.11)$$

$A = 8.208$

x	y
1.0	3.00
1.3	3.51
1.6	3.84
1.9	3.99
2.2	3.96
2.5	3.75
2.8	3.36
3.1	2.79
3.4	2.04
3.7	1.11
4.0	0.00

($y = 3.99$ is not the height of any inscribed rectangle)

13. $y = \dfrac{1}{\sqrt{x+1}}$, between $x = 3$ and $x = 8$

(a) $n = 5$, $\Delta x = \dfrac{8-3}{5} = 1$

$$A = \sum_{i=1}^{5} A_i = \sum_{i=1}^{5} y_i \Delta x$$

$$y_1 = f(4)$$

$$A = (0.447 + 0.408 + \cdots + 0.354 + 0.333)(1)$$

$$A = 1.92$$

x	y
3	0.5
4	0.447
5	0.408
6	0.378
7	0.355
8	0.333

(b) $n = 10$, $\Delta x = \dfrac{8-3}{10} = 0.5$

$$A = \sum_{i=1}^{10} A_i = \sum_{i=1}^{10} y_i \Delta x$$

$$y_1 = f(3.5)$$

$$A = (0.471 + 0.447 + \cdots + 0.343 + 0.333)(0.5)$$

$$A = 1.96$$

x	y
3	0.5
3.5	0.471
4	0.447
4.5	0.426
5	0.408
5.5	0.392
6	0.378
6.5	0.365
7	0.354
7.5	0.343
8	0.333

17. $y = x^2$, between $x = 0$ and $x = 2$

$$A_{0,\,2} = \left[\int x^2 dx\right]_0^2 = \frac{x^3}{3}\bigg|_0^2 = \frac{8}{3} - 0 = \frac{8}{3}$$

21. $y = \dfrac{1}{x^2} = x^{-2}$, between $x = 1$ and $x = 5$

$$A_{1,5} = \left[\int x^{-2} dx\right]_1^5 = \frac{x^{-1}}{-1}\bigg|_1^5 = \frac{-1}{x}\bigg|_1^5$$

$$= -\frac{1}{5} - (-1) = \frac{4}{5} = 0.8$$

25. $y = 3x$, $x = 0$ to $x = 3$, $n = 10$, $\Delta x = 0.3$

Using the table in 5(b)

$$A = 0.3(0.9 + 1.8 + 2.7 + 3.6 + 4.5 + 5.4 + 6.3 + 7.2 + 8.1 + 9.0)$$

$$A = 14.85$$

$A_{\text{inscribed}}$	<	A_{exact}	<	$A_{\text{circumscribed}}$
12.15	<	13.5	<	14.85

$\frac{12.15+14.85}{2} = 13.5$ because the extra area above $y = 3x$ using circumscribed rectangles is the same as the omitted area under $y = 3x$ using inscribed rectangles.

25.4 The Definite Integral

1. $\int_1^4 (x^{-2} - 1)\,dx = -\frac{1}{x} - x\Big|_1^4 = -\frac{1}{4} - 4 - \left(-\frac{1}{1} - 1\right)$

$$\int_1^4 (x^{-2} - 1)\,dx = -\frac{9}{4}$$

5. $\int_1^4 x^{5/2}dx = \frac{2}{7}x^{7/2}\Big|_1^4 = \frac{256}{7} - \frac{2}{7} = \frac{254}{7}$

9. $u = 1 - x;\ du = -dx$

$$\int_{-1.6}^{0.7}(1-x)^{1/3}dx = -\int_{-1.6}^{0.7}(1-x)^{1/3}(-dx)$$
$$= -\frac{3}{4}(1-x)^{4/3}\Big|_{-1.6}^{0.7}$$
$$= -\frac{3}{4}(0.2008 - 3.5752)$$

13. $\int_{0.5}^{2.2}\left(\sqrt[3]{x} - 2\right)dx = \int_{0.5}^{2.2}x^{1/3}dx - 2\int_{0.5}^{2.2}dx$

$$= \left(\frac{3}{4}x^{4/3} - 2x\right)\Big|_{0.5}^{2.2}$$
$$= (2.1460 - 4.4) - (0.2976 - 1)$$
$$= -1.5516 = -1.552$$

17. $u = 4 - x^2;\ du = -2xdx$

$$\int_{-2}^{-1}2x\left(4-x^2\right)^3dx = -\int_{-2}^{-1}\left(11-x^2\right)^3(-2xdx)$$
$$= -\frac{\left(4-x^2\right)^4}{4}\Big|_{-2}^{-1} = -\left(\frac{81}{4} - 0\right)$$
$$= -\frac{81}{4}$$

21. $u = 6x + 1;\ du = 6dx$

$$\int_{2.75}^{3.25}\frac{dx}{\sqrt[3]{6x+1}} = \int(6x+1)^{-1/3}dx$$
$$= \frac{1}{6}\int(6x+1)^{-1/3}(6dx)$$
$$= \frac{1}{6}\cdot\frac{3}{2}(6x+1)^{2/3}\Big|_{2.75}^{3.25} = \frac{1}{4}(6x+1)^{2/3}\Big|_{2.75}^{3.25}$$
$$= \frac{1}{4}(7.4904 - 6.7405) = 0.1875$$

25. $u = 4t + 1;\ du = 4dt$

$$\int_3^7\sqrt{16t^2 + 8t + 1}dt = \int_3^7\sqrt{(4t+1)^2}dt$$
$$= \int_3^7\sqrt{(4t+1)}dt$$
$$= \frac{1}{4}\int_3^7(4t+1)^1 4dt = \frac{1}{4}\frac{(4t+1)^2}{2}$$
$$= \frac{1}{8}(4t+1)^2\Big|_3^7 = 84$$

29. $u = x^3 + 9x + 6;\ du = \left(3x^2 + 9\right)dx = 3\left(x^2 + 3\right)dx$

$$\int_0^1\left(x^2+3\right)\left(x^3+9x+6\right)^2dx$$
$$= \frac{1}{3}\int_0^1\left(x^3+9x+6\right)^2\left(x^2+3\right)dx$$
$$= \frac{1}{3}\frac{\left(x^3+9x+6\right)^3}{3} = \frac{1}{9}\left(x^3+9x+6\right)^3\Big|_0^1$$
$$= \frac{1}{9}\left(16^3 - 6^3\right) = \frac{3880}{9}$$

33. $\int_{\sqrt{5}}^3 2z\sqrt{z^4 + 8z^2 + 16}dz$

$$= \int_{\sqrt{5}}^3 2x\left(\left(z^2+4\right)^2\right)^{1/4}dz = \int_{\sqrt{5}}^3 2z\left(z^2+4\right)^{1/2}dz$$
$$= \frac{2}{3}\left(z^2+4\right)^{3/2}\Big|_{\sqrt{5}}^3 = 13.25$$

37. $y^2 = 4x, \quad y > 0$

For $x = 1, \quad y^2 = 4(1) \Rightarrow y = 2$

For $x = 4, \quad y^2 = 4(4) \Rightarrow y = 4$

$$\int_{x=1}^{x=4} ydx = \int_1^4 2\sqrt{x}dx = 2\cdot\frac{2}{3}x^{3/2}\Big|_1^4 = \frac{4}{3}\left(4^{3/2} - 1^{3/2}\right) = \frac{28}{3}$$

41. $\int_{-1}^1 t^{2k}dt = \frac{t^{2k+1}}{2k+1}\Big|_{-1}^1 = \frac{1^{2k+1}}{2k+1} - \frac{(-1)^{2k+1}}{2k+1}$

$$\int_{-1}^1 t^{2k}dt = \frac{1}{2k+1} - \frac{(-1)^{2k}(-1)^1}{2k+1}$$
$$= \frac{1}{2k+1} + \frac{1}{2k+4}$$
$$\int_{-1}^1 t^{2k}dt = \frac{2}{2k+1}$$

45. $W = \int_0^{80}(1000 - 5x)dx$

$$= \left(1000x - \frac{5}{2}x^2\right)\Big|_0^{80} = 1000(80) - \frac{5}{2}(80)^2$$
$$-[0-0]$$
$$= 80{,}000 - 16{,}000 = 64{,}000 \text{ ft}\cdot\text{lb}$$

49. $\dfrac{3N}{2E^{3/2}}\int_0^E E^{3/2}\,dE = \dfrac{3N}{2E^{3/2}}\cdot\dfrac{E^{5/2}}{\frac{2}{5}}\Bigg|_0^E$

$\dfrac{3N}{2E^{3/2}}\int_0^E E^{3/2}\,dE = \dfrac{3N}{2E^{3/2}}\cdot\dfrac{5E^{5/2}}{2} = \dfrac{15NE}{4}$

25.5 Numerical Integration: The Trapezoidal Rule

1. $\int_1^3 \frac{1}{x}dx,\ n = 2,\ h = \dfrac{b-a}{n} = \dfrac{3-1}{2} = 1$

x	y
1	1
2	$\frac{1}{2}$
3	$\frac{1}{3}$

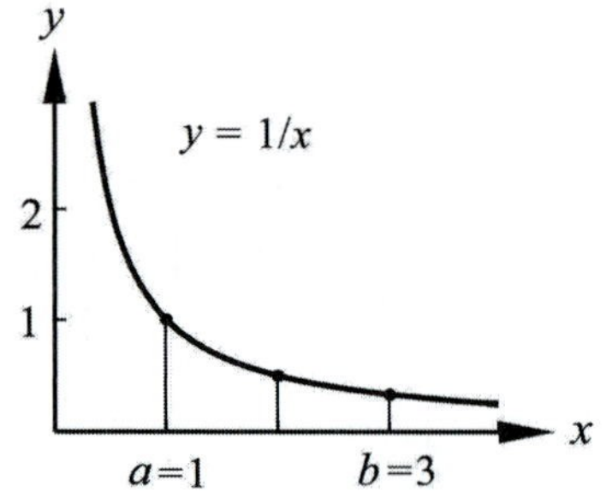

5. $\int_1^4 \left(1+\sqrt{x}\right)dx;\ n = 6;\ h = \dfrac{4-1}{6} = \dfrac{1}{2};\ \dfrac{h}{2} = \dfrac{1}{4}$

n	x_n	y_n
0	1	2
1	1.5	2.22
2	2	2.41
3	2.5	2.58
4	3	2.73
5	3.5	2.87
6	4	3

$A_T = \frac{1}{4}\left[2+2(2.22)+2(2.41)+2(2.58)+2(2.73)\right.$

$\left.+2(2.87)+3\right]$

$A = \frac{1}{4}(30.646) = 7.661$

9. $\int_0^5 \sqrt{25-x^2}\,dx;\ n = 5;\ x = \dfrac{5}{5} = 1;\ \dfrac{h}{2} = \dfrac{1}{2}$

n	x_n	y_n
0	0	5
1	1	4.90
2	2	4.58
3	3	4
4	4	3
5	5	0

$A_T = \frac{1}{2}\left[5+2(4.90)+2(4.58)+2(4)+2(3)+0\right]$

$= 18.98$

13. $\int_0^4 2^x\,dx;\ n = 12;\ h = \dfrac{4}{12} = \dfrac{1}{3};\ \dfrac{h}{2} = \dfrac{1}{6}$

x	y
0	1
$\frac{1}{3}$	1.260
$\frac{2}{3}$	1.587
1	2
$1\frac{1}{3}$	2.520
$1\frac{2}{3}$	3.175
2	4
$2\frac{1}{3}$	5.040
$2\frac{2}{3}$	6.350
3	8
$3\frac{1}{3}$	10.080
$3\frac{2}{3}$	12.699
4	16

$A = \frac{1}{6}\left[1+2(1.260)+2(1.587)+2(2)+2(.520)\right.$

$+2(3.175)+2(4)+2(5.040)+2(6.350)$

$\left.+2(8)+2(10.080)+2(12.699)+16\right]$

$A = 21.74$

17. The approximate value is less than the exact value because the tops of all the trapezoids are below the curve.

21. $L = 2\int_0^{100} \sqrt{1.6\times10^{-7}x^2+1}\,dx$; $n = 10$; $h = \frac{100}{10} = 10$;

$\frac{h}{2} = 5$

x	y
0	2
10	2.000016
20	2.000064
30	2.000144
40	2.000256
50	2.000400
60	2.000576
70	2.000784
80	2.001024
90	2.001296
100	2.001600

$$L = 5[2 + 2(2.000016 + 2.000064 + 2.000144 + 2.000256 + 2.000400 + 2.000576 + 2.000784 + 2.001024 + 2.001296) + 2.001600]$$

$L = 200.54$ ft

25.6 Simpson's Rule

1. $\int_0^1 \frac{dx}{x+2}$, $n = 2$, $h = \frac{1-0}{2} = \frac{1}{2}$

x	y
0	$\frac{1}{2}$
$\frac{1}{2}$	$\frac{2}{5}$
1	$\frac{1}{3}$

$$\int_0^1 \frac{dx}{x+2} = \frac{\frac{1}{2}}{3}\left(\frac{1}{2} + 4\left(\frac{2}{5}\right) + \frac{1}{3}\right) = 0.40\bar{5}$$

5. $\int_1^4 \left(2x + \sqrt{x}\right)dx$; $n = 6$; $\Delta x = \frac{4-1}{6} = \frac{1}{2}$; $\frac{\Delta x}{3} = \frac{1}{6}$

$$A_S = \frac{1}{6}[3 + 4(4.22) + 2(5.41) + 4(6.58) + 2(7.73) + 4(8.87) + 10]$$

$$A_S = \frac{1}{6}(117.999) = 19.67$$

n	x_n	y_n
1	1	3
2	1.5	4.22
3	2	5.41
4	2.5	6.58
5	3	7.73
6	3.5	8.87
7	4	10

$$A = \int_1^4 \left(2x + \sqrt{x}\right)dx = 2\int_1^4 x\,dx + \int x^{1/2}dx$$

$$= \left(x^2 + \frac{2}{3}x^{3/2}\right)\Bigg| = 16 + \frac{16}{3} + - \left(1 + \frac{2}{3}\right)$$

$$= \frac{59}{3} = 19.67$$

9. $\int_1^5 \frac{dx}{x^2+x}$; $n = 10$; $\Delta x = 0.4$; $\frac{\Delta x}{3} = \frac{0.4}{3}$

$$A_S = \frac{0.4}{3}[0.5000 + 4(0.2976) + 2(0.1984) + 4(0.1420) + 2(0.1068) + 4(0.0833) + 4(0.0388) + 0.0333]$$

$$= \frac{0.4}{3}(3.8349) = 0.5114$$

13. $\Delta x = 2$; $\frac{\Delta x}{3} = \frac{2}{3}$

x	y
2	0.67
4	2.34
6	4.56
8	3.67
10	3.56
12	4.78
14	6.87

$$\int_2^{14} y\,dx = \frac{2}{3}[0.67 + 4(2.34) + 2(4.56) + 4(3.67) + 2(3.56) + 4(4.78) + 6.87] = 44.63$$

17. $\bar{x} = 0.9129\int_0^3 x\sqrt{0.3 - 0.1}\,dx$; $n = 12$; $\Delta x = \frac{3}{12} = \frac{1}{4}$;

$\frac{\Delta x}{3} = \frac{1}{12}$

n	x_n	y_n
1	0	0
2	0.25	0.131
3	0.50	0.25
4	0.75	0.356
5	1	0.447
6	1.25	0.523
7	1.50	0.581

n	x_n	y_n
8	1.75	0.619
9	2	0.632
10	2.25	0.616
11	2.5	0.559
12	2.75	0.435
13	3	0

$$\bar{x} = A_S = \frac{1}{12}\big[0+4(0.131)+2(0.25)+4(0.356)$$
$$+2(0.447)+4(0.523)+2(0.581)$$
$$+4(0.619)+2(0.632)+4(0.616)$$
$$+2(0.559)+4(0.435)+(0)\big](0.9129)$$
$$=\frac{1}{12}(15.6572)(0.9129)=1.191$$
$$\bar{x}=1.200 \text{ in.}$$

Chapter 25 Review Exercises

1. $\int(4x^3-x)\,dx=\int 4x^3dx-\int x\,dx=\frac{4x^4}{4}-\frac{x^2}{2}+C$

$$=x^4-\frac{1}{2}x^2+C$$

5. $\int_1^4\left(\frac{\sqrt{x}}{2}+\frac{2}{\sqrt{x}}\right)dx=\frac{1}{2}\int_1^4 x^{1/2}dx+2\int_1^4 x^{-1/2}dx$

$$=\frac{1}{2}\frac{x^{3/2}}{\frac{3}{2}}+\frac{2x^{1/2}}{\frac{1}{2}}\Bigg|_1^4=\frac{1}{3}x^{3/2}+4x^{1/2}\Bigg|_1^4$$
$$=\left[\frac{1}{3}(4)^{3/2}+4(4)^{1/2}\right]$$
$$-\left[\frac{1}{3}(1)^{3/2}+4(1)^{1/2}\right]=\frac{19}{3}$$

9. $\int\left(5+\frac{6}{x^3}\right)dx=\int 5dx+\int\frac{6}{x^3}dx=\int 5dx+\int 6x^{-3}dx$

$$=(5x)+\left(-\frac{6}{2}x^{-2}\right)$$
$$=5x-3x^{-2}=5x-\frac{3}{x^2}+C$$

13. $\int\frac{dn}{(9-5n)^3}=\int(9-5n)^{-3}\,dn$

$$-\frac{1}{5}\int(9-5n)^{-3}(-5\,dn)=-\frac{1}{5}\cdot\frac{(9-5n)^{-2}}{-2}+C$$
$$=\frac{1}{10}\cdot\frac{1}{(9-5n)^2}+C=\frac{1}{10(9-5n)^2}+C$$

17. $\int_0^2\frac{3x\,dx}{\sqrt[3]{1+2x^2}}=\int_0^2(1+2x^2)^{-1/3}(3x)\,dx$

$$u=1+2x^2;\ du=4x\,dx;\ n=-\frac{1}{3}$$
$$\frac{3}{4}\int_0^2(1+2x^2)^{-1/3}(4x)\,dx=\frac{3}{4}\cdot\frac{(1+2x^2)^{2/3}}{\frac{2}{3}}\Bigg|_0^2$$
$$=\frac{9}{8}(1+2x^2)^{2/3}\Big|_0^2$$
$$=\frac{9}{8}\left[1+2(2^2)\right]^{2/3}$$
$$-\frac{9}{8}\left[1+2(0)^2\right]^{2/3}$$
$$=\frac{9}{8}(9)^{2/3}-\frac{9}{8}(1)^{2/3}$$
$$=\frac{9}{8}\left(\sqrt[3]{81}-1\right)=\frac{9}{8}\left(3\sqrt[3]{3}-1\right)$$

21. $\int\frac{(2-3x^2)\,dx}{(2x-x^3)^2}=\int(2x-x^3)^{-2}(2-3x^2)dx;$

$$u=2x-x^3;\ du=2-3x^2;\ n=-2$$
$$\int(2x-x^3)^{-2}(2-3x^2)dx=\frac{(2x-x^3)^{-1}}{-1}+C$$
$$=-\frac{1}{(2x-x^3)}+C$$

25. $\frac{dy}{dx}=3-x^2$

$$y=3x-\frac{x^3}{3}+C$$
$$3=3(-1)-\frac{(-1)^3}{3}+C$$
$$C=\frac{17}{3}$$
$$y=3x-\frac{x^3}{3}+\frac{17}{3}$$

29. $\int_3^8 F(v)\,dv - \int_4^8 F(v)\,dv = \int_3^8 F(v)\,dv + \int_8^4 F(v)\,dv$

$= \int_3^4 F(v)\,dv$

33. $\int_0^1 x^3\,dx = \left.\frac{x^4}{4}\right|_0^1 = \frac{1}{4}$

$\int_1^2 (x-1)^3\,dx = \left.\frac{(x-1)^4}{4}\right|_1^2 = \frac{(2-1)^4}{4} - \frac{(1-1)^4}{4} = \frac{1}{4}$

which shows $\int_0^1 x^3\,dx = \int_1^2 (x-1)^3\,dx$

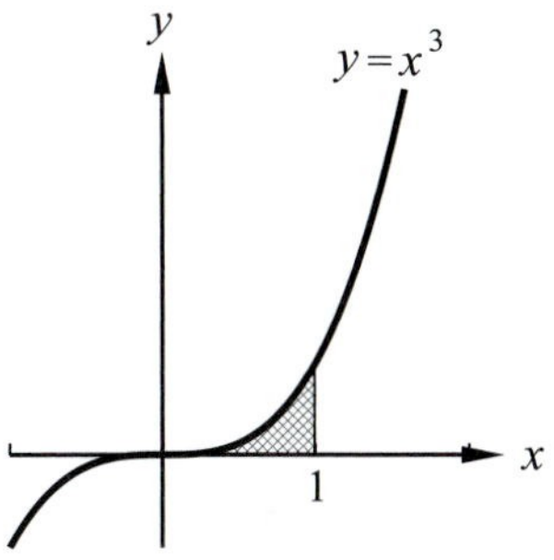

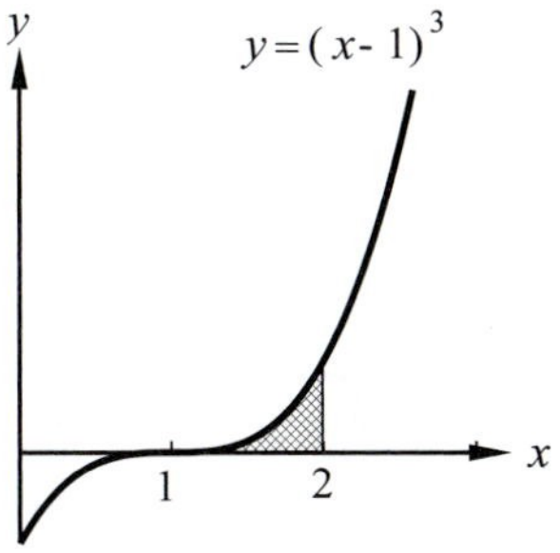

$y=(x-1)^3$ is $y=x^3$ shifted right one unit, so the areas are the same.

37. Since $f(x)>0$, the graph is above the x-axis. Since $f''(x)<0$, for $a \le x < b$, f is concave down for $a \le x \le b$.

$\int_a^b f(x)\,dx > A_{\text{trapezoid}}$

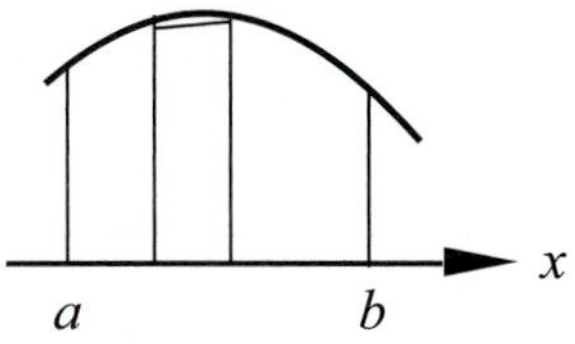

because the tops of the trapezoids are all below the curve.

41. $h = \frac{3-1}{4} = \frac{1}{2}; x_0 = 1, x_1 = 1.5, x_2 = 2, x_3 = 2.5, x_4 = 3$

$\int_1^3 \frac{dx}{2x-1} \approx \frac{h}{3}[y_0 + 4y_1 + 2y_2 + 4y_3 + y_4] \approx \frac{\frac{1}{2}}{3}$

$\left[\frac{1}{2.1-1} + \frac{4}{2(1.5)-1} + \frac{2}{2(2)-1} + \frac{4}{2(2.5)-1} + \frac{1}{2(3)-1}\right]$

$\approx \frac{73}{90} = 0.811$

45. $y = x\sqrt[3]{2x^2+1}$, $a=1$, $b=4$, $n=3$.

$\Delta x = \frac{b-a}{n} = \frac{4-1}{3} = 1$

$x_0 = 1$, $y_0 = 1\sqrt[3]{2\cdot 1^2+1} = \sqrt[3]{3}$, $x_1 = 2$, y_1

$= 2\sqrt[3]{2\cdot 2^2+1} = 2\sqrt[3]{9}$, $x_2 = 3$, $y_2 = 3\sqrt[3]{2\cdot 3^2+1}$

$= 3\sqrt[3]{19}$, $x_3 = 4$, $y_3 = 4\sqrt[3]{2\cdot 4^2+1} = 4\sqrt[3]{33}$

$\int_1^4 x\sqrt[3]{2x^2+1}\,dx \approx \frac{\Delta x}{2}[y_0 + 2y_1 + 2y_2 + y_3] \approx \frac{1}{2}$

$\left[\sqrt[3]{3} + 4\sqrt[3]{9} + 6\sqrt[3]{19} + 4\sqrt[3]{33}\right] \approx 19.3016$

49. $y(x) = 4 + \sqrt{1+8x-2x^2}$.

$A = \frac{1}{2}\left[y(0) + y\left(\frac{1}{2}\right) + y(1) + y\left(\frac{3}{2}\right) + y(2)\left(\frac{5}{2}\right)\right.$

$\left. + y(3) + y\left(\frac{7}{2}\right) + y(4)\right]$

$A = 24.68\text{ m}^2$

53. $Q = \int_0^3 6t^2\,dt = 2t^3\Big|_0^3 = 2(3)^3 = 54\text{ C}$

57. $A = 2\int_0^5 \sqrt{5-y}\,dy = -2\int_5^0 \sqrt{u}\,du$

$= -2\left(\frac{2}{3}\right)u^{3/2}\Big|_5^0$

$u = 5-y$, $du = -dy$

$A = -\frac{4}{3}\left[0 - 5^{3/2}\right]$

$A = 14.9\text{ m}^2$

Chapter 26

Applications of Integration

26.1 Applications of the Indefinite Integral

1. $s = \int (v_0 - 32t)\, dt = v_0 t - 16t^2 + C$

$200 = v_0(0) - 16(0)^2 + C$

$200 = C$

$s = v_0 t - 16t^2 + 200$

$0 = v_0(2.5) - 16(2.5)^2 + 200$

$v_0 = -40$ ft/s

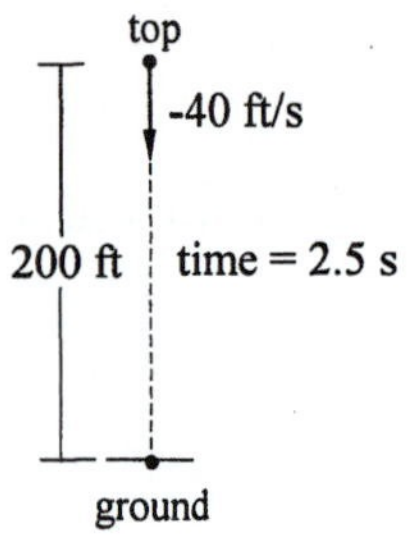

5. $\frac{ds}{dt} = -0.25$ m/s

$ds = -0.25dt$

$s = -0.25 \int dt = -0.25t + C_1$

$t = 0,\ s = 8;$ therefore, $C_1 = 8$

$s = -0.25t + 8.00 = 8.00 - 0.25t$

9. At impact: $t = 0,\ a = -250$ m/s^2,

$v_{\text{impact}} = 96$ km/h $= 26.\overline{6}$ m/s, $s_{\text{impact}} = 0$

At stop: $t = t_{\text{stop}},\ v = 0,\ s = 1.4$ m

$a = -250$

$v = -250t + v_{\text{impact}} = -250t + 26.\overline{6}$

$s = -125t^2 + 26.\overline{6}t + s_{\text{impact}},\ s_{\text{impact}} = 0$

$s = -125t^2 + 26.\overline{6}$

$v = 0 = -250t_{\text{stop}} + 26.\overline{6},\ t_{\text{stop}} = 0.10\overline{6}$

$s = -125\left(0.10\overline{6}\right)^2 + 26.\overline{6} = 1.4$ m

13. $v = \int -32dt = -32t + C;\ v = v_0;\ t = 0,$

$C = v_0;\ v = -32t + v_0;$

$s = \int(-32t + v_0)dt = -16t^2 + v_0 t + C_1$

$s = 0,\ t = 0;\ C_1 = 0;\ s = -16t^2 + v_0;$

$s = 90$ ft when $v = 0$

$90 = 16t^2 + v_0 t;\ 0 = -32t + v_0;\ t = \frac{v_0}{32};$

$90 = -16\left(\frac{v_0}{32}\right)^2 + v_0\left(\frac{v_0}{32}\right);\ \frac{v_0^2}{64} = 90$

$v_0 = \sqrt{64(90)} = 76$ ft/s

17. $q = \int i\, dt = \int 0.230 \times 10^{-6}\, dt = 0.230 \times 10^{-6} t + C;$

$q = 0,\ t = 0;\ C = 0;\ q = 0.230 \times 10^{-6} t$

Find q for $t = 1.50 \times 10^{-3}\,s$:

$q = 0.230 \times 10^{-6}\left(1.50 \times 10^{-3}\right)$

$= 0.345 \times 10^{-9} = 0.345$ nC

21. $V_c = \frac{1}{C}\int i\, dt = \frac{1}{2.5 \times 10^{-6}}\int 0.025dt$

$= \frac{1}{2.5 \times 10^{-6}}(0.025)t = 1.0 \times 10^4 t + C$

$v_c = 0,\ t = 0;\ C = 0;\ v_c = 1.0 \times 10^4 t$

Find v_c for $t = 0.012$ s: $= 1.0 \times 10^4 t(0.012)$

$= 120$ V

25. $\omega = \frac{d\theta}{dt} = 16t + 0.5t^2;\ d\theta = \left(6t + 0.50t^2\right)dt$

$\theta = 8t^2 + \frac{0.50t^3}{3} + C;\ \theta = 0,\ t = 0;\ C = 0;$

$\theta = 8t^2 + \frac{0.50t^3}{3};$ find θ for $t = 10.0$ s

$\theta = 8(10.0)^2 + \frac{0.50}{3}(10.0)^3 = 970$ rad

29. $\frac{dV}{dx}=\frac{-k}{x^2};\ V=-\int\frac{k}{x^2}dx=kx^{-1}+C=\frac{k}{x}+C$

$\lim_{V\to 0}V=\lim_{x\to\infty}\frac{k}{x}+C;\ 0=0+C;\ C=0$; therefore,

$V\big|_{x=x_1}=\frac{k}{x_1}$

26.2 Areas by Integration

1. $A=\int_1^3 y\,dx=\int_1^3 2x^2\,dx$

$=\frac{2x^3}{3}\Big|_1^3$

$A=\frac{2(3)^3}{3}-\frac{2(1)^3}{3}$

$A=\frac{52}{3}$

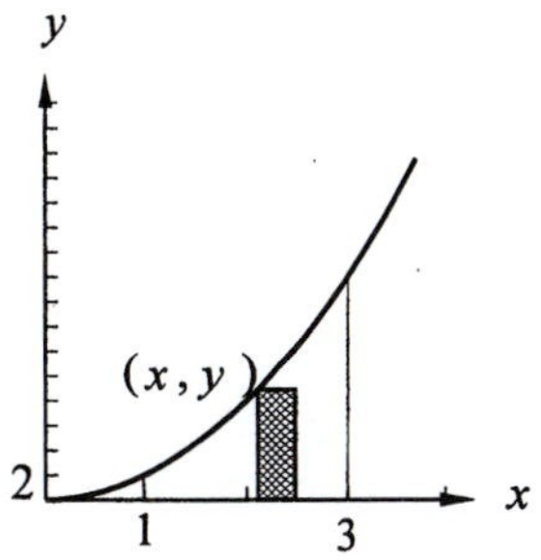

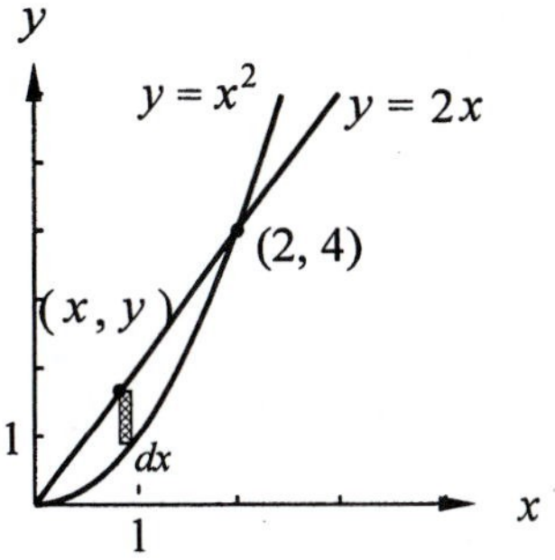

5. $y=6-4x;\ x=0,\ y=0,\ y=3$

$y-6=-4x,\ x=-\frac{1}{4}y+\frac{3}{2}$

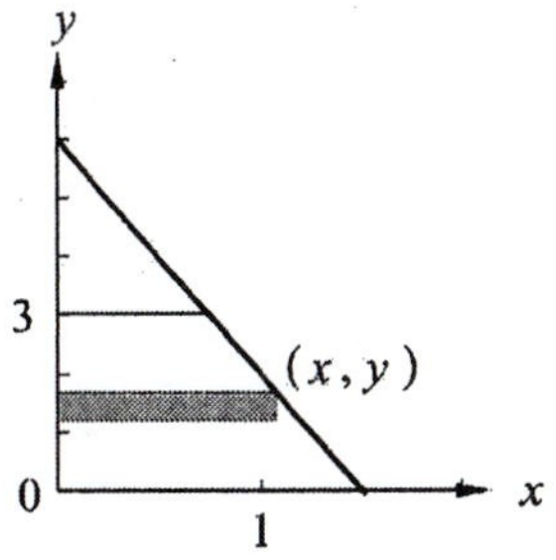

9. $y=x^{-2};\ y=0,\ x=2,\ x=3$

$A=\int_2^3 x^{-2}dy=-x^{-1}\Big|_2^3=-\frac{1}{x}\Big|_2^3$

$=-\frac{1}{3}-\left(-\frac{1}{2}\right)=\frac{1}{6}$

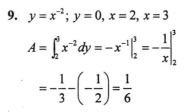

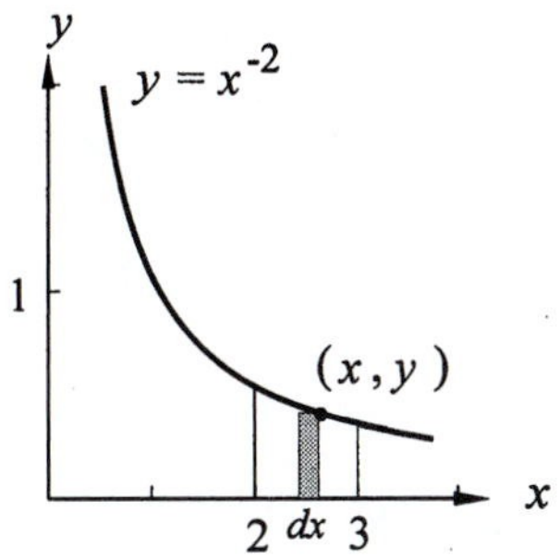

13. $y=\frac{2}{\sqrt{x}};\ x=0,\ y=1,\ y=4$

$\sqrt{x}=\frac{2}{y};\ x=\frac{4}{y^2}$

$A=\int_1^4 x\,dy;\ A=4\int_1^4 y^{-2}dy=-4y^{-1}\Big|_1^4$

$=-\frac{4}{y}\Big|_1^4=-\frac{4}{4}-\left(-\frac{4}{1}\right)$

$=-1+4=3$

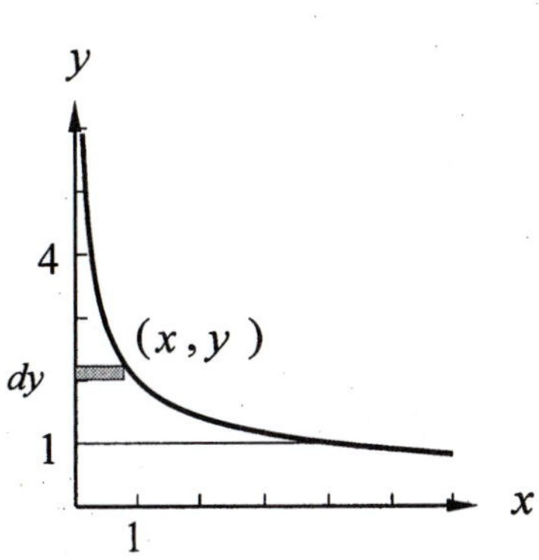

17. $A = \int_0^4 \left(0 - \left(x - 2\sqrt{x}\right)\right)dx$

$A = -\frac{x^2}{2} + \frac{2x^{3/2}}{\frac{3}{2}}\Big|_0^4$

$A = \frac{8}{3}$

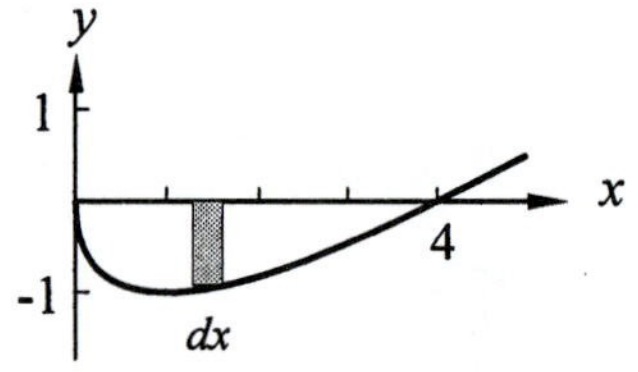

21. $y = x^4 - 8x^2 + 16,\ y = 16 - x^4$

$A = 2\int_0^2 \left(16 - x^4 - \left(x^4 - 8x^2 + 16\right)\right)dx$

$A = 2\int_0^2 \left(8x^2 - 2x^4\right)dx = 2\left(\frac{8}{3}x^3 - \frac{2}{5}x^5\right)\Big|_0^2 = \frac{256}{15}$

$\int_{-2}^{2} = 2\int_0^2$ because of symmetry with respect to y-axis.

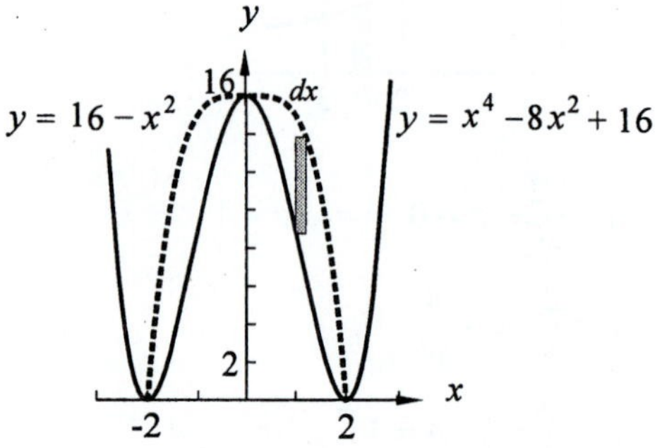

25. $y = x^5;\ x = -1,\ x = 2,\ y = 0$

$A_T = A_{(-1\to 0)} + A_{(0\to 2)}$

$= \text{negative} + \text{positive}$

$= -\left(A\big|_{-1}^{0}\right) + A\big|_0^2$

$A = -\int_{-1}^{0} y\,dx + \int_0^2 y\,dx$

$= -\int_{-1}^{0} x^5\,dx + \int_0^2 x^5\,dx = -\frac{x^6}{6}\Big|_{-1}^{0} + \frac{x^6}{6}\Big|_0^2$

$= 0 - \left(-\frac{1}{6}\right) + \frac{64}{6} - 0 = \frac{65}{6}$

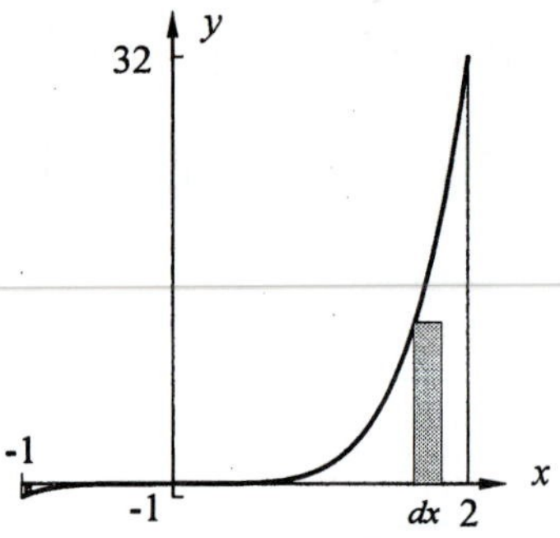

29. $A = \int_1^2 \left(4x - x^2 - \left(4 - x^2\right)\right)dx$

$= \int_1^2 (4x - 4)\,dx = 2x^2 - 4x\big|_1^2$

$= 2$

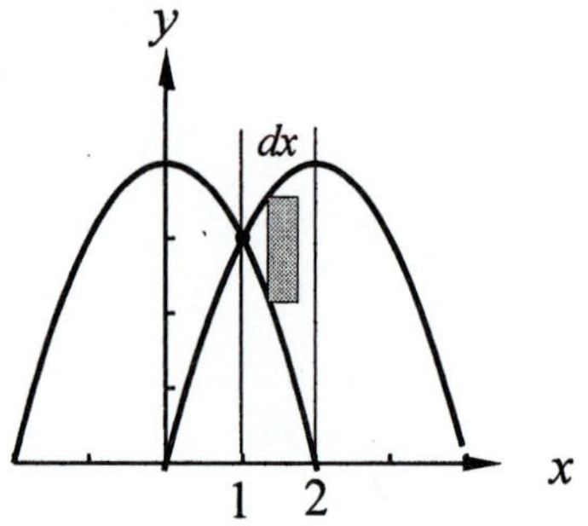

33.

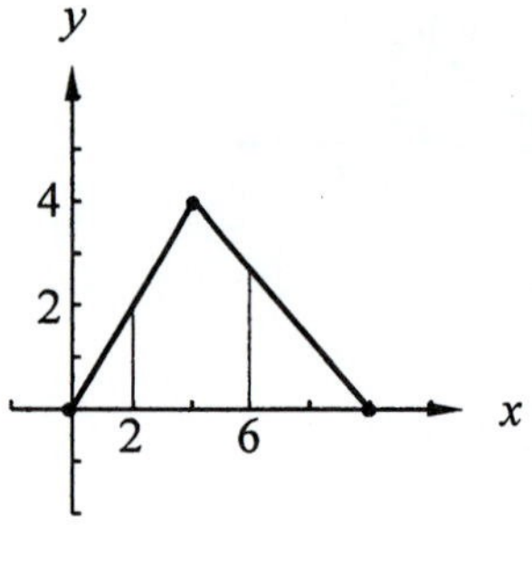

$A = \int_0^4 x\,dx + \int_4^{10}\left(-\frac{2}{3}(x-4) + 4\right)dx = \int_4^{10}\left(-\frac{2}{3}x + \frac{20}{3}\right)dx$

$A = \frac{x^2}{2}\Big|_0^4 + \left(-\frac{x^2}{3} + \frac{20x}{3}\right)\Big|_4^{10}$

$= \frac{4^2}{2} + \left(-\frac{10^2}{3} + \frac{20(10)}{3} + \frac{4^2}{3} - \frac{20(4)}{3}\right)$

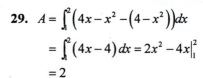

$A = 20$

37. $\int_0^2 \left(4-x^2\right)dx = 2\int_0^{\sqrt{c}} \left(c-x^2\right)dx$

$$4x-\frac{x^3}{3}\Bigg|_0^2 = 2\left(cx-\frac{x^3}{3}\right)\Bigg|_0^2$$

$$8-\frac{8}{3} = 2\left(c\sqrt{c}-\frac{c\sqrt{c}}{3}\right)$$

$$\frac{16}{3} = \frac{4}{3}c^{3/2}$$

$$c = 4^{2/3}$$

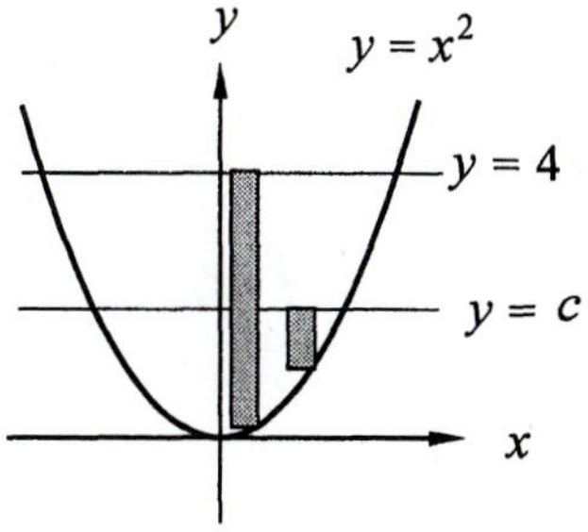

41. (a) $\int_0^2 \left(8x-x^4\right)dx = 9.6 = \frac{48}{5}$

(b) $\int_0^{16}\left(\sqrt[4]{y}-\frac{y}{8}\right)dy = 9.6 = \frac{48}{5}$

(a)

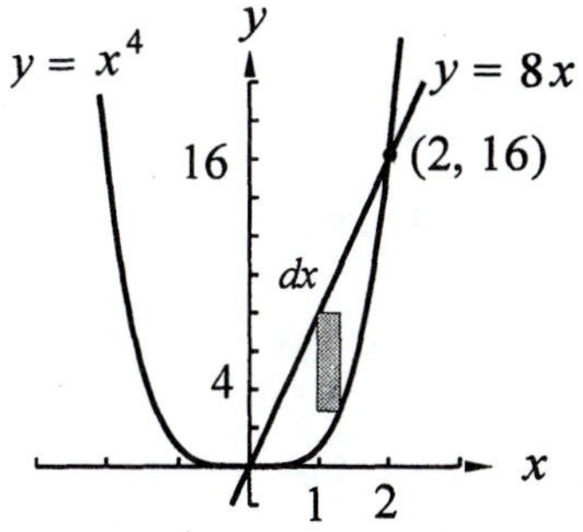

(b)

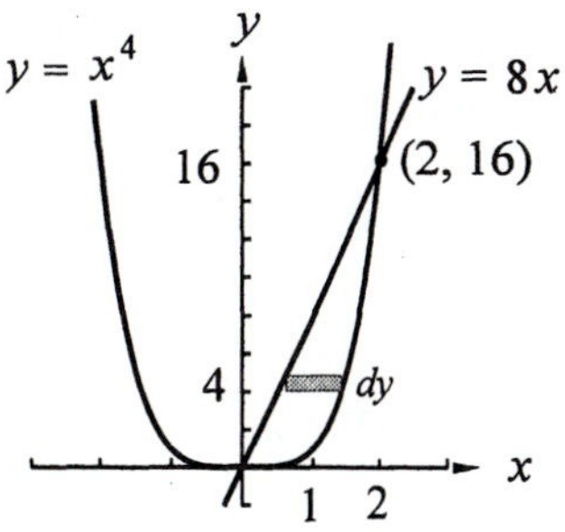

45. $v = 1-0.01\sqrt{2t+1}$; $t = 10$ s to $t = 100$ s

$$\Delta s = \int_{10}^{100}\left(1-0.01\sqrt{2t+1}\right)dt$$

$$= \int_{10}^{100} dt - 0.01\int_{10}^{100}(2t+1)^{1/2}\,dt$$

$$\Delta s = \int_{10}^{100} dt - \frac{0.01}{2}\int(2t+1)^{1/2}\,2\,dt$$

$$= t-\frac{0.01}{2}\cdot\frac{2}{3}(2t+1)^{3/2}\Bigg|_{10}^{100}$$

$$= \left[t-\frac{0.01}{3}(2t+1)^{3/2}\right]_{10}^{100}$$

$= 90.501 - 9.679 = \Delta s = 80.8$ km,
change in position from $t = 10$ s to $t = 100$ s

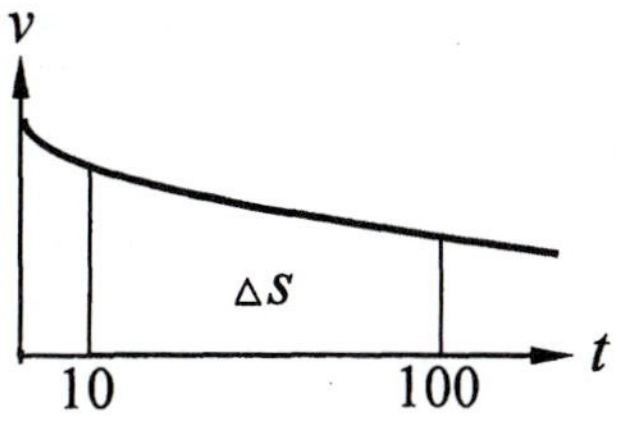

49. $y = 0.25x^4$ and $y = 12 - 0.25x^4$ intersect when

$$0.25x^4 = 12 - 0.25x^4$$
$$0.50x^4 = 12$$
$$x^4 = 24$$
$$x = \sqrt[4]{24}$$
$$y = 0.25\left(\sqrt[4]{24}\right)^4 = 6$$

$$A = 2\int_0^{\sqrt[4]{24}} \left(12 - 0.25x^4 - 0.25x^4\right)dx$$
$$A = 2\int_0^{\sqrt[4]{24}} \left(12 - 0.50x^4\right)dx$$
$$A = 2\left(12x - \frac{0.50x^5}{5}\right)\Bigg|_0^{\sqrt[4]{24}}$$
$$A = 42.5 \text{ dm}^2$$

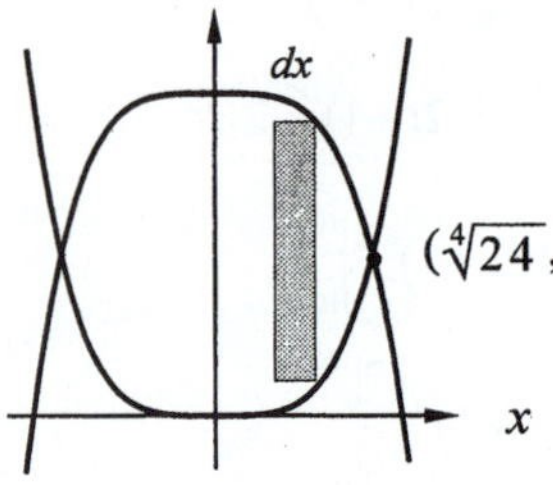

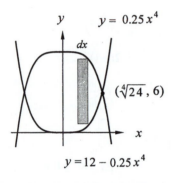

26.3 Volumes by Integration

1. $y = x^3, x = 2, y = 0$ about x-axis.

$$V = \int_0^2 \pi y^2 dx = \pi \int_0^2 (x^3)^2 dx$$
$$V = \pi \left.\frac{x^7}{7}\right|_0^2 = \frac{128\pi}{7}$$

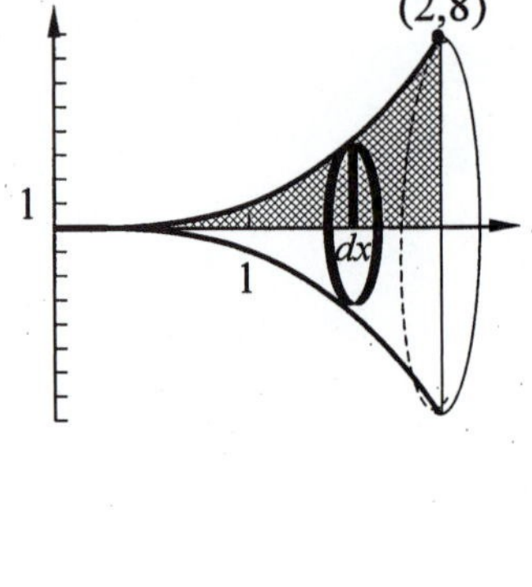

5. $V = \int_0^2 2\pi x(2-x)\,dx = \dfrac{8\pi}{3}$

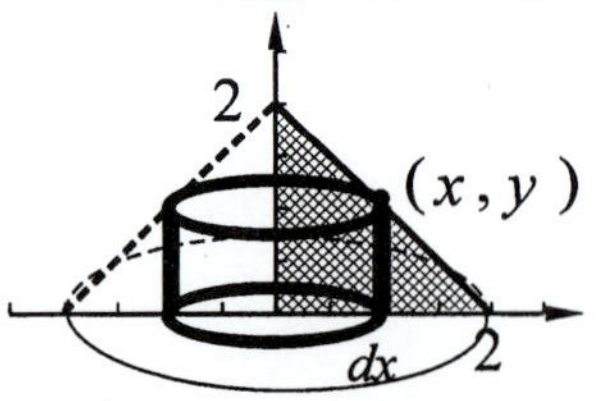

9. $y = 3\sqrt{x}, y = 0, x = 4$
Disk: $dV = \pi y^2 dx$

$$V = \pi \int_0^4 y^2 dx = \int_0^4 9x\, dx$$
$$= \pi \left(\frac{9}{2}x^2\right)\Bigg|_0^4$$
$$= \pi \left[\frac{9}{2}(4)^2 - 0\right] = 72\pi$$

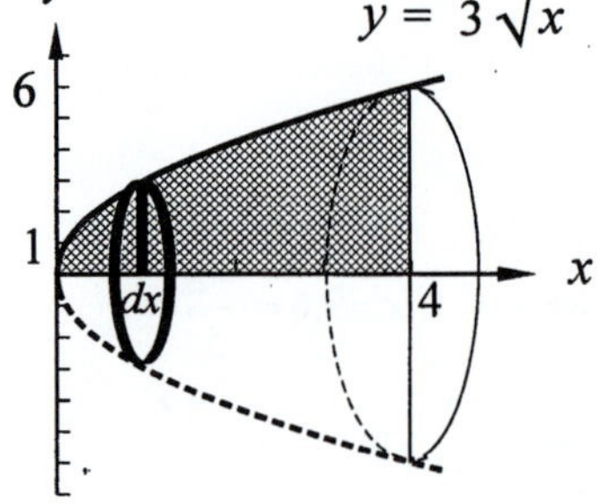

13. $y = x^2 + 1, x = 0, , x = 3, y = 0$
Disk: $dV = \pi y^2 dx$

$$V = \pi \int_0^3 (x^2+1)^2 dx$$
$$= \pi \int_0^3 (x^4 + 2x^2 + 1) dx$$
$$= \pi \left(\frac{1}{5}x^5 + \frac{2}{3}x^3 + x\right)\Bigg|_0^3$$
$$= \pi \left[\frac{1}{5}(3)^5 + \frac{2}{3}(3)^3 + 3 - 0\right]$$
$$= \frac{348}{5}\pi$$

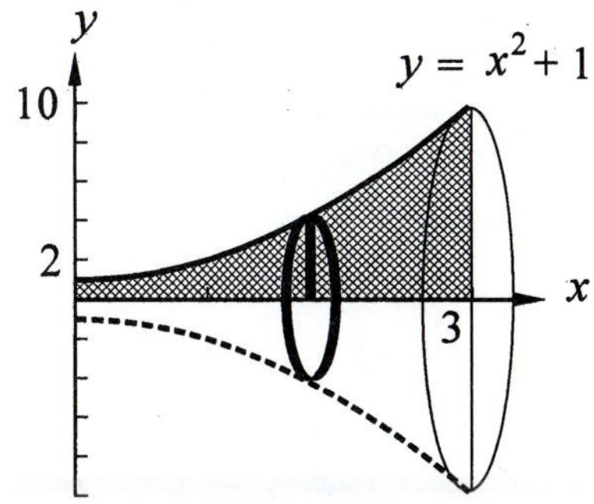

17. $y = x^{1/3}, x = 0, y = 2$
$x = y^3;\ x^2 = y^6$
Disk: $dV = \pi x^2 dy$

$$V = \pi \int_0^2 y^6 dy$$
$$= \frac{\pi}{7} y^7 \Big|_0^2 = \frac{128\pi}{7}$$

21. $x^2 - 4y^2 = 4, x = 3, h = 2y$
Shell: $dV = 2\pi x(2y)dx$

$$V = 4\pi \int_2^3 x\sqrt{\frac{x^2-4}{4}}dx$$
$$= \frac{2\pi}{2}\int_2^3 (x^2-4)^{1/2} 2x\, dx$$

$u = x^2 - 4, du = 2x\, dx$

$$V = \pi \cdot \frac{2}{3}(x^2-4)^{3/2}\Big|_2^3 = \frac{2\pi}{3}(5^{3/2}) - 0$$
$$= \frac{10\sqrt{5}}{3}\pi$$

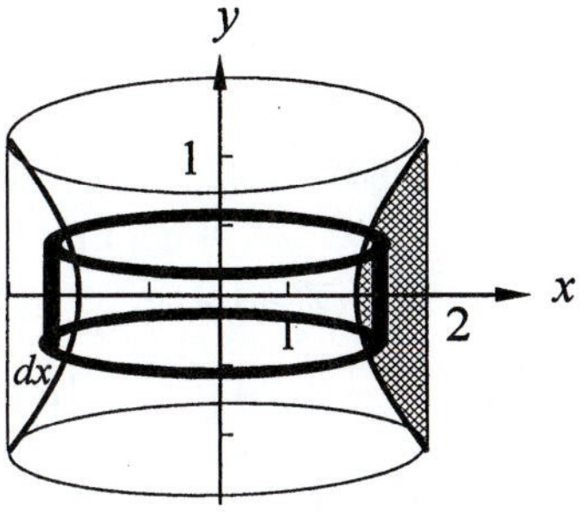

25. $y = \sqrt{4 - x^2}$, Quad I
Shell: $dV = 2\pi xy\, dx$

$$V = 2\pi \int_0^2 x\sqrt{4-x^2}dx$$

$u = 4 - x^2, du = -2x\, dx$

$$V = -\pi \int_0^2 (4-x^2)^{1/2}(-2x\, dx)$$
$$= -\pi \frac{2}{3}(4-x^2)^{3/2}\Big|_0^2$$
$$= -\frac{2\pi}{3}(0-8) = \frac{16\pi}{3}$$

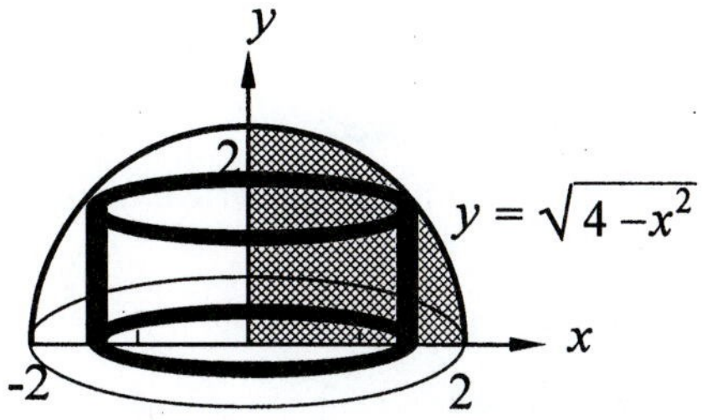

29. $y = 2x - x^2$, $y = 0$, rotated around
$x = 2$, using shells $r = 2 - x$
$h = y, t = dx$
$dV = 2\pi(2-x)y\, dx$

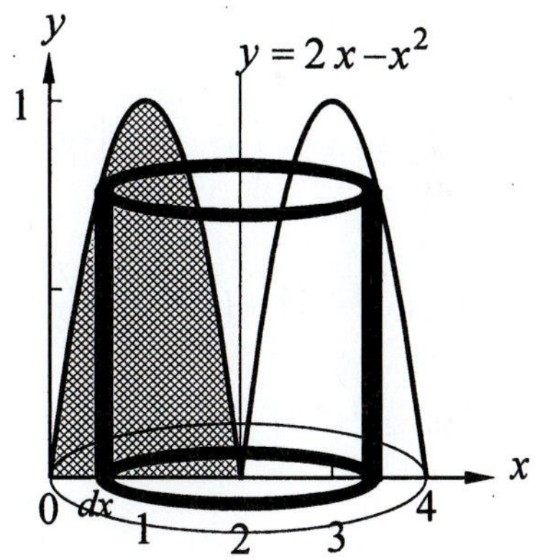

$$V = 2\pi \int_0^2 (2-x)(2-x^2)\, dx$$
$$= 2\pi \int_0^2 (4x - 4x^2 + x^3)\, dx$$
$$= 2\pi \left[2x^2 - \frac{4}{3}x^3 + \frac{1}{4}x^4\right]\Big|_0^2$$
$$= 2\pi \left[8 - \frac{32}{3} + 4\right] = \frac{8}{3}\pi$$

33.

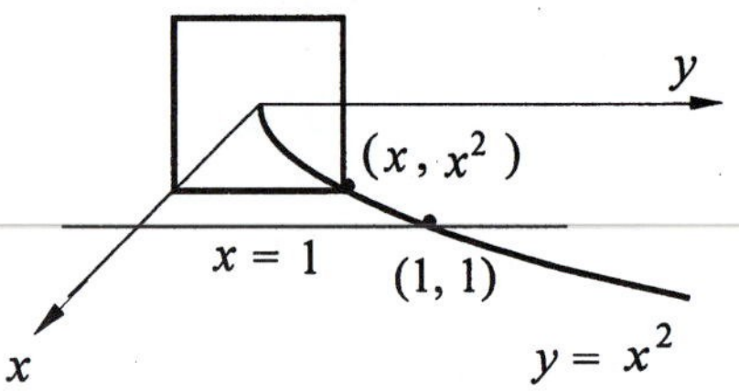

$$V = \int_0^1 (x^2)^2 dx = \frac{x^5}{5}\bigg|_0^1 = \frac{1}{5}$$

37. $x^2 + y_2^2 = 9;\ y_1 = 1$

Volume of lead removed:

V = volume of cylinder + spherical caps

$$dV = \pi y_1^2 dx + \pi y_2^2 dx$$

$$V = 2\left[\pi \int_0^{2\sqrt{2}} (1)dx\right] + 2\pi \int_{2\sqrt{2}}^3 y^2 dx$$

$$= 2\pi x|_0^{2\sqrt{2}} + 2\pi \int_{2\sqrt{2}}^3 (9 - x^2)dx$$

$$= 2\pi(2\sqrt{2}) + 2\pi\left(9x - \frac{x^3}{3}\right)\bigg|_{2\sqrt{2}}^3$$

$$= 2\pi(2\sqrt{2}) + 2\pi\left(27 - 9 - \left[18\sqrt{2} - \frac{16\sqrt{2}}{3}\right]\right)$$

$$= 2\pi\left[2\sqrt{2} + 18 - 18\sqrt{2} + 16\frac{\sqrt{2}}{3}\right]$$

$V = 18.3\ \text{cm}^3$

26.4 Centroids

1. $y = |x|$ is symmetric with respect to y-axis $\Rightarrow \bar{x} = 0$

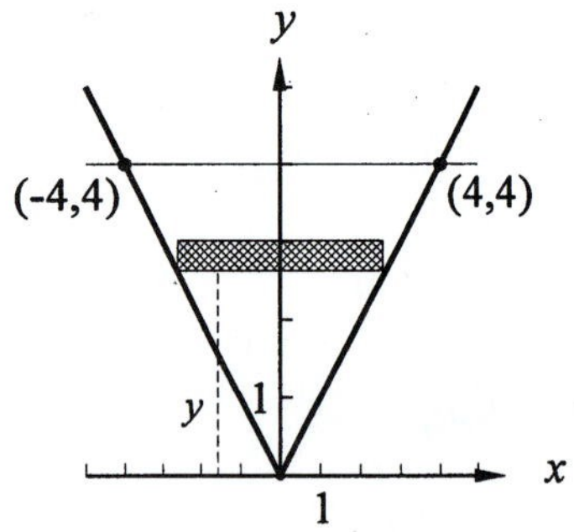

$$\bar{y} = \frac{\int_0^4 y(2x)dy}{\int_0^4 2x\,dy}$$

$$\bar{y} = \frac{\int_0^4 2y^2 dy}{\int_0^4 2y\,dy} = \frac{\frac{y^3}{3}\Big|_0^4}{\frac{y^2}{2}\Big|_0^4}$$

$$= \frac{\frac{4^3}{3}}{\frac{4^2}{2}} = \frac{8}{3}$$

$$(\bar{x}, \bar{y}) = \left(0, \frac{8}{3}\right)$$

5. $M\bar{x} = m_1x_1 + m_2x_2 + m_3x_3 + m_4x_4$

$$(42 + 24 + 15 + 84)\bar{x} = 42(-3.5) + 24(0) + 15(2.6) + 84(3.7)$$

$$\bar{x} = 1.2\ \text{cm}$$

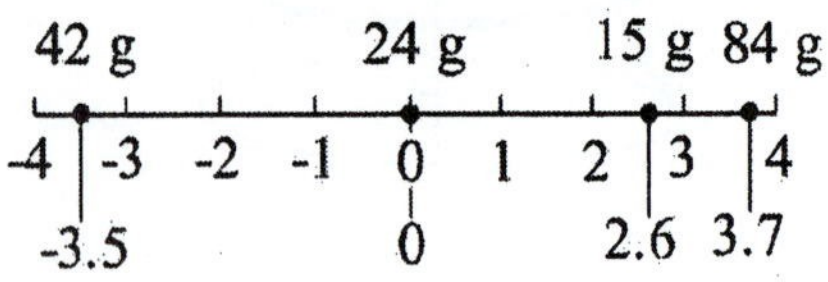

9. Break area into three rectangles.

First: center $(-1,-1)$; $A_1 = 4$

Second: center $(0, \frac{1}{2})$; $A_2 = 4$

Third: center $(2\frac{1}{2}, 1\frac{1}{2})$; $A_3 = 3$

Therefore,

$4(-1) + 4(0) + 3\left(2\frac{1}{2}\right) = (4+4+3)\bar{x}$

$\bar{x} = 0.32$

Therefore, $4(-1) + 4\left(\frac{1}{2}\right) + 3\left(1\frac{1}{2}\right) = 11\bar{y}$

$\bar{y} = 0.23$

Therefore, (0.32 in, 0.23 in) is the center of mass.

13. $y = 4 - x$, and axes

$$\bar{x} = \frac{\int_0^4 xy\,dx}{\int_0^4 y\,dx} = \frac{\int_0^4 x(4-x)dx}{\int_0^4 (4-x)dx} = \frac{\int_0^4 (4x - x^2)dx}{\int_0^4 (4-x)dx}$$

$$= \frac{\left(2x^2 - \frac{1}{3}x^3\right)\Big|_0^4}{\left(4x - \frac{1}{2}x^2\right)\Big|_0^4} = \frac{\frac{32}{3}}{8} = \frac{32}{24} = \frac{4}{3}$$

$$\bar{y} = \frac{\int_0^4 y(x)dy}{\int_0^4 x\,dy} = \frac{\int_0^4 y(4-y)dy}{\int_0^4 (4-y)dy} = \frac{\int_0^4 (4y - y^2)dy}{\int_0^4 y(4-y)dy}$$

$$= \frac{\left(2y^2 - \frac{1}{3}y^3\right)\Big|_0^4}{\left(4y - \frac{1}{2}y^2\right)\Big|_0^4} = \frac{4}{3}$$

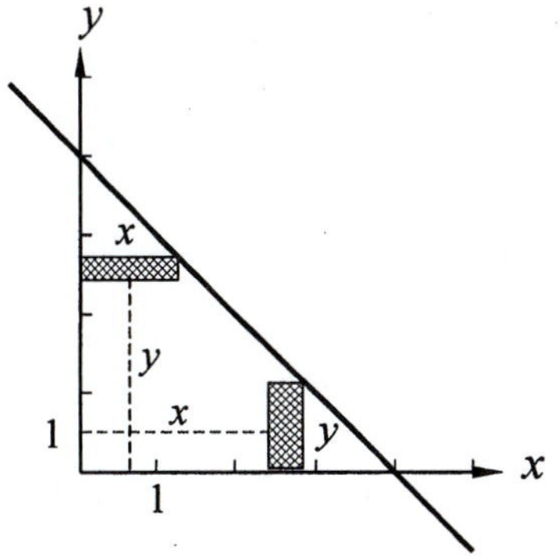

$$(\bar{x}, \bar{y}) = \left(\frac{4}{3}, \frac{4}{3}\right).$$

17. $A = \int_0^2 x\,dx + \int_2^3 (6-2x)dx = 3$

$$\bar{x} = \frac{\int_0^2 x(3x-2x)dx + \int_2^3 x(6-2x)dx}{3} = \frac{5}{3}$$

$$\bar{y} = \frac{\int_0^6 y\left(\frac{y}{2} - \frac{y}{3}\right)dy}{3} = \frac{12}{3} = 4$$

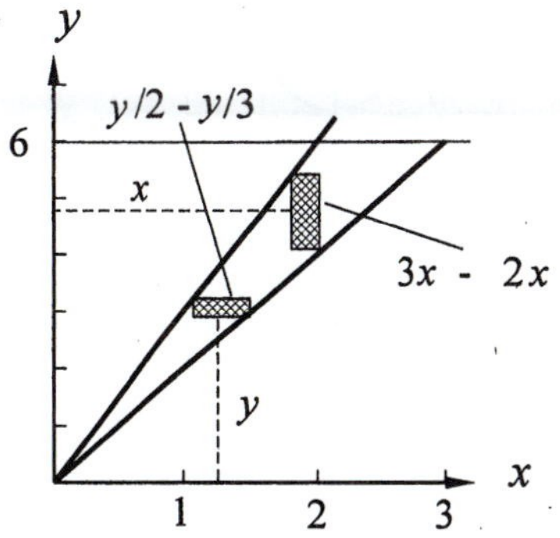

$$(\bar{x}, \bar{y}) = \left(\frac{5}{3}, 4\right)$$

21.

Curves intersect when $\frac{x^2}{4p} = a$

$x = \pm 2\sqrt{pa}$

$y = a$

Region is symmetric with respect to y-axis, $\bar{x} = 0$.

$$\bar{y} = \frac{\int y(2x)dy}{\int 2x\,dy}$$

$$= \frac{\int_0^a 2y(2\sqrt{py})dy}{\int_0^a 2(2\sqrt{py})dy}$$

$$\bar{y} = \frac{4\sqrt{p}\int_0^a y^{3/2}dy}{4\sqrt{p}\int_0^a y^{1/2}dy}$$

$$= \frac{\frac{2}{5}y^{5/2}\Big|_0^a}{\frac{2}{3}y^{3/2}\Big|_0^a}$$

$$= \frac{3}{5}\frac{a^{5/2}}{a^{3/2}} = \frac{3}{5}a$$

$$(\bar{x}, \bar{y}) = \left(0, \frac{3}{5}a\right)$$

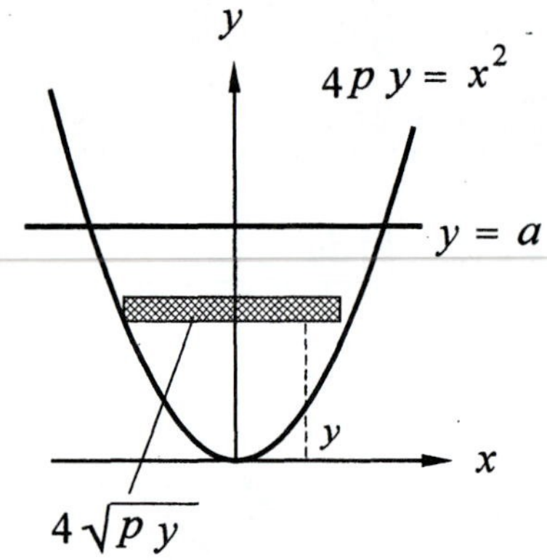

25. $y^2 = 4x,\ y = 0,\ x = 1,$

$y^2 = 4x,\ x = \dfrac{y^2}{4}$

Rotate about y-axis, $\overline{x} = 0$

$$\overline{y} = \frac{\int_0^2 y\left(-\left(\frac{y^2}{4}\right)^2 + 1\right)dy}{\int_0^2 \left(\left(-\frac{y^2}{4}\right)^2 + 1\right)dy}$$

$$\overline{y} = \frac{\int_0^2 \left(-\frac{1}{16}y^5 + y\right)dy}{\int_0^2 \left(-\frac{1}{16}y^4 + 1\right)dy} = \frac{-\frac{1}{96}y^6 + \frac{1}{2}y^2\Big|_0^2}{-\frac{1}{80}y^5 + y\Big|_0^2}$$

$$= \frac{-\frac{64}{96} + 2}{-\frac{32}{80} + 2} = \frac{\frac{128}{96}}{\frac{128}{80}} = \frac{5}{6}$$

$$\left(\overline{x}, \overline{y}\right) = \left(0, \frac{5}{6}\right)$$

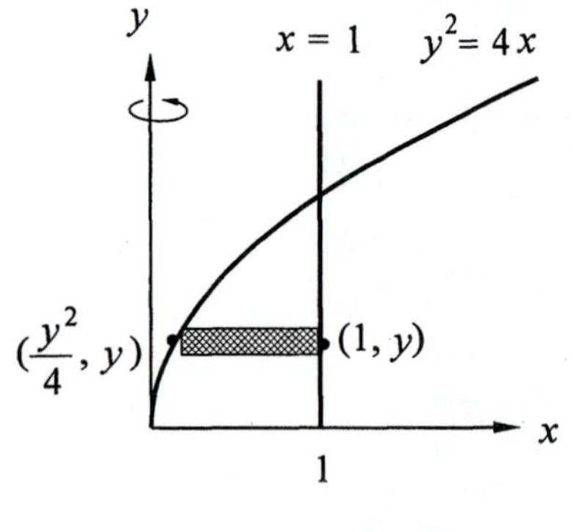

29. Triangle is area bounded by $y = \dfrac{a}{b}x$, x-axis, $x = b$

$$\overline{x} = \frac{\int_0^b xy\,dx}{\int_0^b y\,dx} = \frac{\int_0^b x\frac{a}{b}x\,dx}{\int_0^b \frac{a}{b}x\,dx}$$

$$= \frac{\frac{a}{b}\int_0^b x^2dx}{\frac{a}{b}\int_0^b x\,dx} = \frac{\frac{x^3}{3}\Big|_0^b}{\frac{x^2}{2}\Big|_0^b} = \frac{\frac{b^3}{3}}{\frac{b^2}{2}} = \frac{2}{3}b$$

$$\overline{y} = \frac{\int_0^b y(b-x)dy}{\int_0^b (b-x)dy} = \frac{\int_0^b y\left(b - \frac{b}{a}y\right)dy}{\int_0^b \left(b - \frac{b}{a}y\right)dy}$$

$$= \frac{\int_0^b \left(by - \frac{b}{a}y^2\right)dy}{\int_0^b \left(b - \frac{b}{a}y\right)dy} = \frac{\frac{b}{2}y^2 - \frac{b}{3a}y^3\Big|_0^a}{by - \frac{b}{2a}y^2\Big|_0^a}$$

$$= \frac{\frac{a^2b}{2} - \frac{a^2b}{3}}{ab - \frac{ab}{2}} = \frac{a}{3}$$

$$\left(\overline{x}, \overline{y}\right) = \left(\frac{2}{3}b, \frac{1}{3}a\right)$$

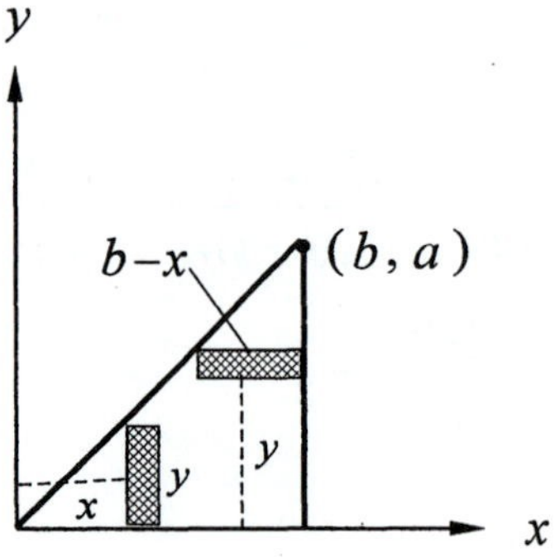

33.

Bounded area: $y = -4x + 80, y = 60, y = 0,$ $x = 0, \overline{x} = 0$

$y = -4x + 80;\ 4x = 80 - y;\ x = 20 - \dfrac{1}{4}y$

$$\overline{y} = \frac{\int_0^{60} yx^2dy}{\int_0^{60} x^2dy} = \frac{\int_0^{60} y\left(20 - \frac{1}{4}y\right)^2 dy}{\int_0^{60}\left(20 - \frac{1}{4}y\right)^2 dy}$$

$$\overline{y} = \frac{\int_0^{60}\left(400y - 10y^2 + \frac{1}{16}y^3\right)dy}{\int_0^{60}\left(400 - 10y + \frac{1}{16}y^2\right)dy}$$

$$= \frac{200y^2 - \frac{10}{3}y^3 + \frac{1}{64}y^4\Big|_0^{60}}{400y - 5y^2 + \frac{1}{48}y^3\Big|_0^{60}}$$

$$= \frac{720\ 000 - 720\ 000 + 202\ 500}{24\ 000 - 18\ 000 + 4\ 500}$$

$= 19.3$ cm from larger base

$$\left(\overline{x}, \overline{y}\right) = (0, 19.3)$$

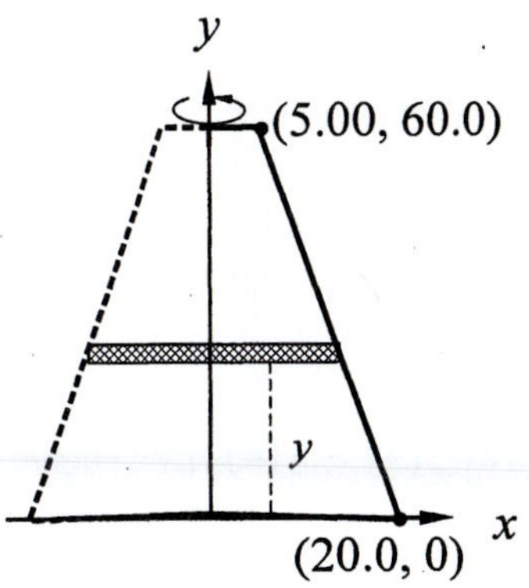

26.5 Moments of Inertia

1. $I_y = k\int_0^1 x^2 y\,dx$

$I_y = k\int_0^1 x^2(4x)dx$

$I_y = 4k\left.\frac{x^4}{4}\right|_0^1$

$I_y = k$

$m = k\int_0^1 y\,dx = k\int_0^1 4x\,dx$

$m = 4k\left.\frac{x^2}{2}\right|_0^1 = 2k$

$R_y^2 = \frac{I_y}{m} = \frac{k}{2k}$

$R_y = \frac{\sqrt{2}}{2}$

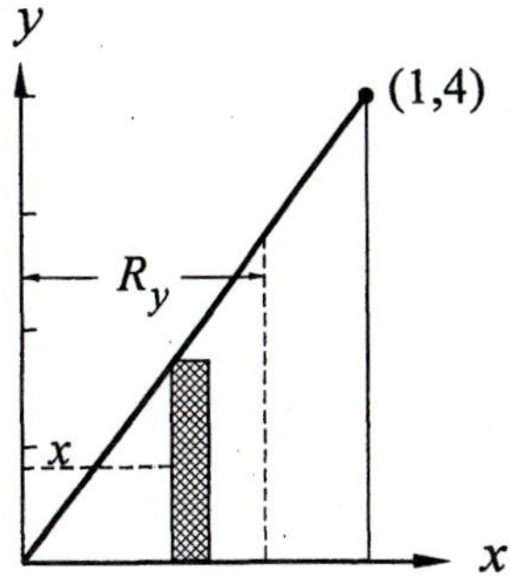

5. $I = m_1x_1^2 + m_2x_2^2 + m_3x_3^2$
$I = 45.0(-3.80)^2 + 90.0(0.00)^2 + 62.0(5.50)^2$
$I = 2530\text{ g}\cdot\text{cm}^2$

$I = MR^2.$
$2530 = (45.0 + 9.0 + 62.0)R^2$
$R = 3.58\text{ cm}$

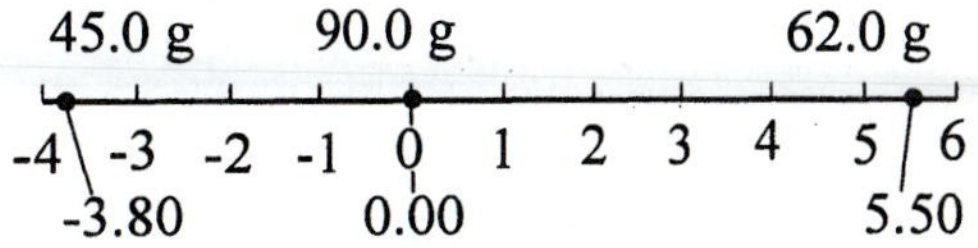

9. $y^2 = x$, $x = 9$, x-axis, with respect to the x-axis

$I_x = k\int_0^3 y^2(9-y^2)dy = k\int_0^3(9y^2-y^4)dy$

$= k\left.\left(3y^3-\frac{1}{5}y^5\right)\right|_0^3 = k\left(81-\frac{243}{5}\right) = \frac{162}{5}k$

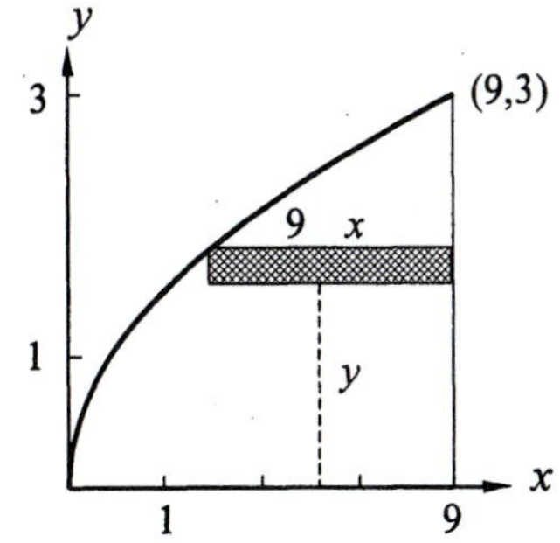

13. $y = \frac{b}{a}x;\ x = a, y = 0$

$I_x = k\int_0^b y^2(a-x)\,dy = k\int_0^b y^2\left(a-\frac{ay}{b}\right)dy$

$= ka\int_0^b\left(y^2-\frac{y^3}{b}\right)dy = ka\left.\left(\frac{1}{3}y^3-\frac{1}{4b}y^4\right)\right|_0^b$

$= \frac{kab^3}{12}$

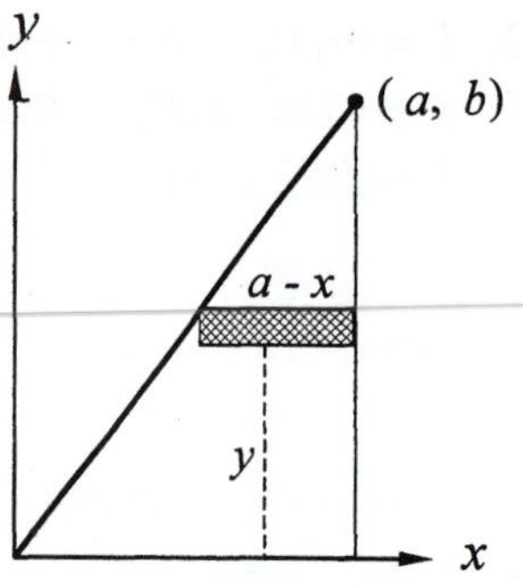

For $k = 1$; $m = \frac{1}{2}ab$ and

$$I_x = ab\left(\frac{b^2}{12}\right) = 2m\left(\frac{b^2}{12}\right) = \frac{1}{6}mb^2$$

17. $y^2 = x^3, y = 8$, y-axis, with respect to the x-axis

$$I_x = k\int_0^8 y^2 x\,dy = k\int_0^8 y^2(y^{2/3})\,dy$$

$$= k\int_0^8 y^{8/3}dy = k\frac{3}{11}y^{11/3}\Big|_0^8$$

$$= \frac{3}{11}(8)^{11/3}k = \frac{3}{11}(2)^{11}k = \frac{6144}{11}k$$

$$m = k\int_0^8 x\,dy = k\int_0^8 y^{2/3}dy = k\left(\frac{3}{5}y^{5/3}\right)\Big|_0^8$$

$$= \frac{3}{5}(8)^{5/3}k = \frac{3}{5}(2)^5 k = \frac{96k}{5}$$

$$R^2 = \frac{I_x}{m} = \frac{6144}{11} \div \frac{96k}{5} = \frac{64(5)}{11}$$

$$R = \sqrt{\frac{64(5)}{11}} = \frac{8}{11}\sqrt{55}$$

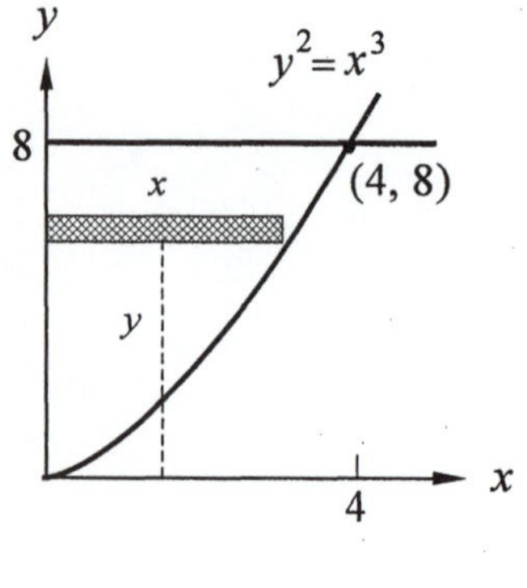

21. $y = 4x - x^2$, $y = 0$, rotated about y-axis

$$I_y = 2\pi k\int_0^4 (4x - x^2)\,dx$$

$$= 2\pi k\int_0^4 (4x^4 - x^5)dx$$

$$= 2\pi k\left[\frac{4}{5}x^5 - \frac{1}{6}x^6\right]_0^4$$

$$= 2\pi k\left[\frac{4}{5}(4)^5 - \frac{1}{6}(4)^6\right] = \frac{4096\pi k}{15}$$

$$m = 2\pi k\int_0^4 (4x - x^2)(x)\,dx$$

$$= 2\pi k\int_0^4 (4x^2 - x^3)dx$$

$$= 2\pi k\left[\frac{4}{3}x^3 - \frac{1}{4}x^4\right]_0^4$$

$$= 2\pi k\left[\frac{4}{3}(4)^3 - \frac{1}{4}(4)^4\right] = \frac{128\pi k}{3}$$

$$R_y^2 = \frac{I_y}{m} = \frac{\frac{4096\pi k}{15}}{\frac{128\pi k}{3}} = \frac{32}{5}$$

$$R_y = \sqrt{\frac{32}{5}\left(\frac{5}{5}\right)} = \frac{4\sqrt{10}}{5}$$

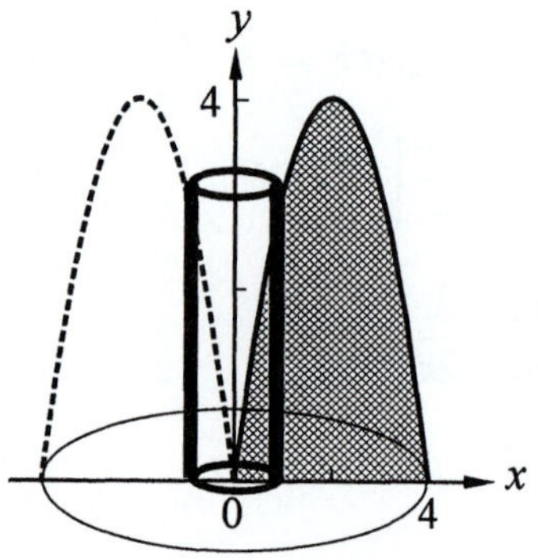

25. $r = 0.600$ cm, $h = 0.800$ cm, $m = 3.00$ g

$$y = \frac{0.600}{0.800}x = 0.750x;\ x = 1.333y$$

$$I_x = 2\pi k \int_0^{0.600} (0.800 - 1.333y)y^3\,dy$$

$$= 2\pi k(0.200y^4 - 0.2667y^5)\Big|_0^{0.600}$$

$$= 2\pi k(0.005\,181)$$

$$m = \frac{k}{3}\pi r^2 h,\ 2\pi k = \frac{6m}{r^2 h} = \frac{6(3.00)}{(0.600^2)(0.800)}$$

$$= 62.5\ \text{g/cm}^3$$

$$I_x = (62.5)(0.005\,181) = 0.324\ \text{g}\cdot\text{cm}^2$$

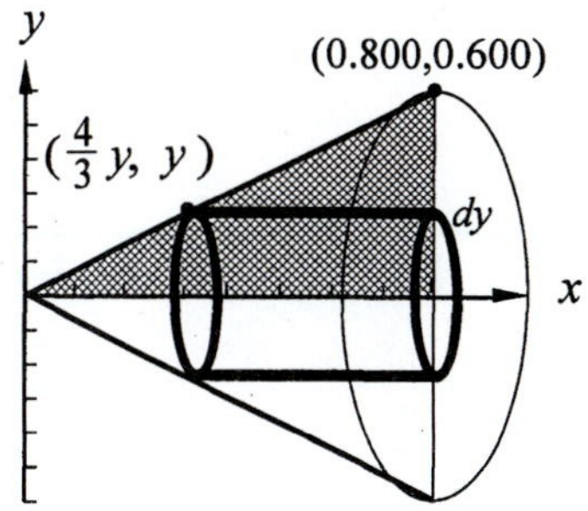

26.6 Other Applications

1. $f(x) = kx$

$6.0 = k(2.0)$

$k = 3.0$ lb/in.

$$W = \int_{3.0}^{6.0} 3.0x\,dx - 1.5x^2\Big|_{3.0}^{6.0}$$

$W = 41$ lb · in.

5. $f(x) = kx$; $6.0 = k(1.5)$; $k = 4.0$ lb/in

$$w = \int_0^{2.0} 4.0x\,dx = 2.0x^2\Big|_0^{2.0} = 2.0(2.0^2) - 0$$

$= 8.0$ lb · in

9. $f(x) = \dfrac{kq_1q_2}{x^2}$, $1.0\text{ pm} = 1.0 \times 10^{-12}$,

$4.0\text{ pm} = 4.0 \times 10^{-12}$

$$W = \int_{1.0\times10^{-12}}^{4.0\times10^{-12}} \frac{9.0\times10^9(1.6\times10^{-19})^2}{x^2}dx$$

$$= 9.0\times10^9(1.6\times10^{-19})^2\left(-\frac{1}{x}\right)\Bigg|_{1.0\times10^{-12}}^{4.0\times10^{-12}}$$

$$= -23.04\times10^{-29}[0.25\times10^{12} - 10^{12}]$$

$$= -23.04\times10^{-29}(-0.75\times10^{12})$$

$$= 1.7\times10^{-16}\ \text{J}$$

13. $W = 1500(24) + \int_0^{24} 12(24 - x)\,dx$

$$W = 36{,}000 + 12\left(24x - \frac{x^2}{2}\right)\Bigg|_0^{24}$$

$W = 36{,}000 + 3456$

$W = 39{,}4556$ lb · ft

17.

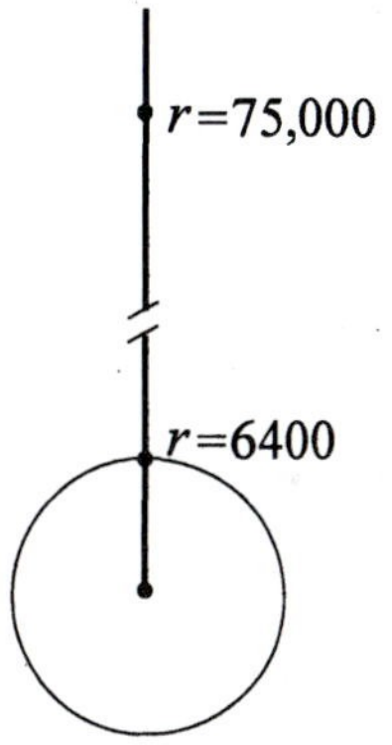

$$F = \frac{Gm_1m_2}{r^2}$$

$$160 = \frac{Gm_1m_2}{6400^2} \Rightarrow Gm_1m_2 = 160(6400^2)$$

$$W = \int_{75,000}^{6400} -\frac{Gm_1m_2}{r^2}dr = \frac{Gm_1m_2}{r}\Bigg|_{75,000}^{6400}$$

$$W = 160(6400^2)\left(\frac{1}{6400} - \frac{1}{75{,}000}\right)$$

$W = 940{,}000$ N · km

21. $F = 62.4\int_0^{2.50}(12.0x)\,dx = 62.4\left(6.00x^2\right)\Big|_0^{2.50}$

$= 62.4(37.5 - 0) = 2340 \text{ lb}$

25.

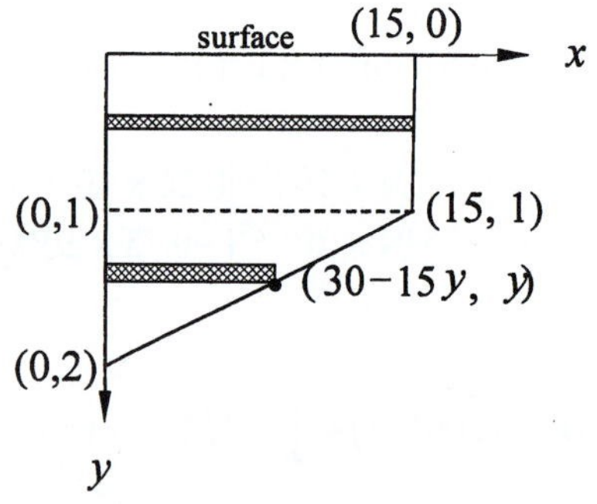

$F = \int_0^1 9800y(15dy) + \int_1^2 9800y(30-15y)\,dy$

$F = 9800(15)\dfrac{y^2}{2}\Big|_0^1 + 9800\left(15y^2 - 5y^3\right)\Big|_1^2$

$F = 172{,}000 \text{ N}$

29. $F_{\text{TOP}} = 9800(2.00)^2(1.00) = 3.92 \times 10^4 \text{ N}$

$F_{\text{BOTTOM}} = 9800(2.00)^2(3.00) = 1.18 \times 10^5 \text{ N}$

The difference in the two forces is the buoyant force.

33. $e = 0.768s - 0.00004s^3$, $s_1 = 30.0$ km/h, $s_2 = 90.0$ km/h

$e_{av} = \displaystyle\int_{30.0}^{90.0}\frac{(0.768s - 0.00004s^3)}{60.0}ds$

$e = \dfrac{1}{60.0}\left(\dfrac{0.768s^2}{2} - 0.00001s^4\right)\Big|_{30.0}^{90.0}$

$= \dfrac{1}{60.0}\left\{\dfrac{0.768}{2}(90.0)^2 - 0.00001(90.0)^4 - \left(\dfrac{0.768}{2}(30.0)^2 - 0.00001(30.0)^4\right)\right\}$

$= \dfrac{1}{60.0}(2.454 \times 10^3 - 3.375 \times 10^2)$

$= 35.3\%$

37. $y = \dfrac{r}{h}x$

$\dfrac{dy}{dx} = \dfrac{r}{h};\ \left(\dfrac{dy}{dx}\right)^2 = \dfrac{r^2}{h^2}$

$S = 2\pi\displaystyle\int_0^h \frac{r}{h}x\sqrt{1 + \frac{r^2}{h^2}}dx$

$= 2\pi\dfrac{r}{h}\dfrac{1}{h}\sqrt{h^2 + r^2}\displaystyle\int x\,dx$

$= \dfrac{2\pi r}{h^2}\sqrt{h^2 + r^2}\dfrac{x^2}{2}\Big|_0^h$

$= \dfrac{\pi r}{h^2}\sqrt{h^2 + r^2}x^2\Big|_0^h$

$= \dfrac{\pi r}{h^2}\sqrt{h^2 + r^2}h^2$

$= \pi r\sqrt{h^2 + r^2}$

Chapter 26 Review Exercises

1. $t = \dfrac{s}{v} = \dfrac{17.1}{45.0}$

$\text{drop} = \dfrac{1}{2}(9.8)\left(\dfrac{17.1}{45.0}\right)^2 = 0.708 \text{ m}$

5. $a = -0.750$

$v = -0.750t + 2.50 = 0 \Rightarrow t = \dfrac{2.50}{0.750}$

$s = -\dfrac{0.750}{2}t^2 + 2.50t\Big|_{t=\frac{2.50}{0.750}} = 4.17 < 4.20$

The ball does not make it to the hole.

9. $V_c = \dfrac{1}{c}\int i\,dt$

$V_c = \dfrac{i}{c}\int dt = \dfrac{it}{c} + V_0 = \dfrac{it}{c} + 0$

$V_c = \dfrac{12\times10^3\left(25\times10^{-6}\right)}{5.5\times10^{-9}} = 55 \text{ V}$

13. $A = \int_0^9 y\,dx = \int_0^9 \sqrt{9-x}\,dx$

$A = -\int_0^9 (9-x)^{1/2}(-dx)$

$A = -\dfrac{2}{3}(9-x)^{3/2}\Big|_0^9 = 18$

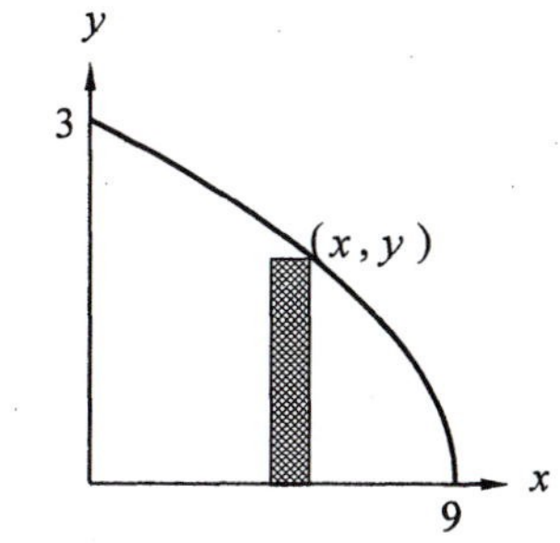

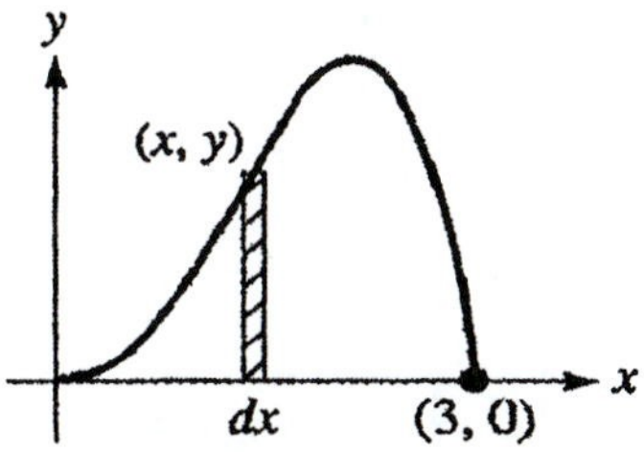

17. $A = \int_0^3 \left(x^2 - \left(x^3 - 2x^2\right)\right)dx$

$A = \int_0^3 \left(3x^2 - x^3\right)dx$

$A = x^3 - \frac{x^4}{4}\Big|_0^3$

$A = \frac{27}{4}$

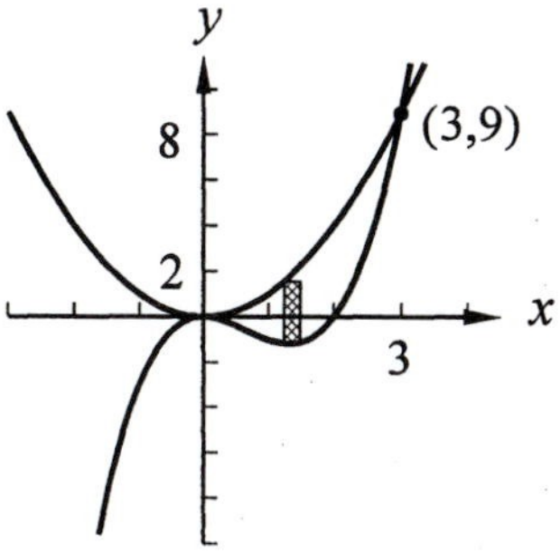

21.

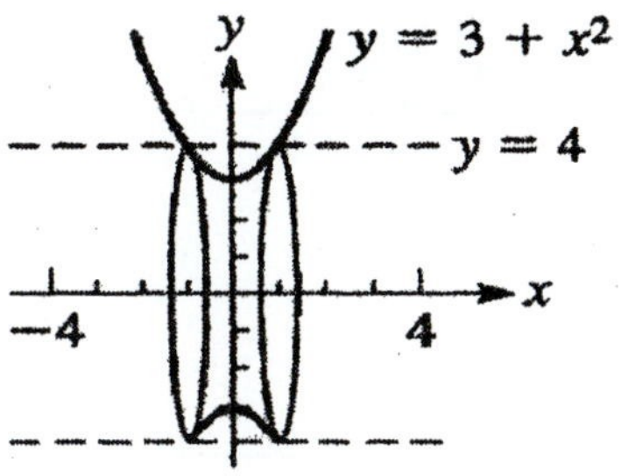

$$V = \pi\int_{-1}^{1} 4^2\,dx - \pi\int_{-1}^{1} y^2\,dx = \pi\int_{-1}^{1}\left(4^2 - y^2\right)dx$$

$$= \pi\int_{-1}^{1}\left[16 - \left(3 + x^2\right)^2\right]dx$$

$$= \pi\int_{-1}^{1}\left(16 - 9 - 6x^2 - x^4\right)dx = \pi\int_{-1}^{1}\left(7 - 6x^2 - x^4\right)dx$$

$$= \pi\left(7x - 2x^3 - \frac{1}{5}x^5\right)\Big|_{-1}^{1}$$

$$= \pi\left(7 - 2 - \frac{1}{5}\right) - \pi\left(-7 + 2 + \frac{1}{5}\right) = \frac{24\pi}{5} + \frac{24\pi}{5}$$

$$= \frac{48\pi}{5}$$

25.

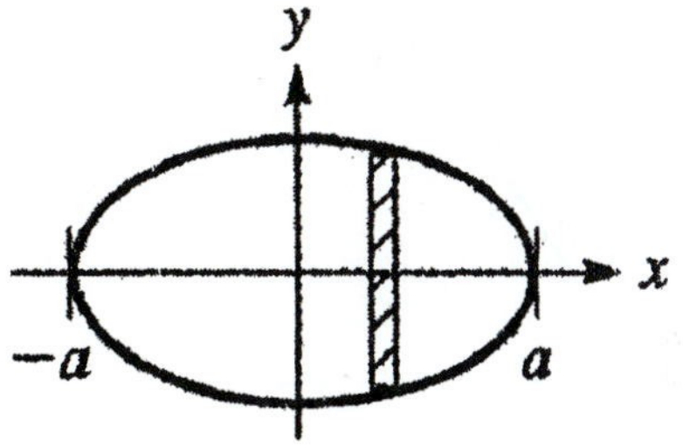

$$V = \pi\int_{-a}^{a} y^2\,dx = \pi\int_{-a}^{a}\left(\frac{a^2b^2 - b^2x^2}{a^2}\right)dx = \pi\int_{-a}^{a}\left(b^2 - \frac{b^2}{a^2}x^2\right)dx$$

$$= \pi\left(b^2x - \frac{b^2}{3a^2}x^3\right)\Big|_{-a}^{a}$$

$$= \pi\left[\left(ab^2 - \frac{ab^2}{3}\right) - \left(-ab^2 + \frac{ab^2}{3}\right)\right] = \pi\left(2ab^2 - \frac{2ab^2}{3}\right)$$

$$= \pi\left(\frac{4ab^2}{3}\right) = \frac{4\pi ab^2}{3} = \frac{4}{3}\pi ab^2$$

29. $y^2 = x^3$ and $y = 3x$ intersect at (0, 0) and (9, 27).

$$\bar{x} = \frac{\int_0^9 x\left(3x - x^{3/2}\right)dx}{\int_0^9 \left(3x - x^{3/2}\right)dx}$$

$$= \frac{\int_0^9 \left(3x - x^{5/2}\right)dx}{\int_0^9 \left(3x - x^{3/2}\right)dx}$$

$$\bar{x} = \frac{x^3 - \frac{2}{7}x^{7/2}\Big|_0^9}{\frac{3}{2}x^2 - \frac{2}{5}x^{5/2}\Big|_0^9} = \frac{30}{7}$$

$$\bar{y} = \frac{\int_0^{27} y\left(y^2 - \frac{1}{3}y\right)dy}{\frac{243}{10}} = \frac{\int_0^{27}\left(y^{5/3} - \frac{y^2}{2}\right)dy}{\frac{243}{10}}$$

$$\bar{y} = \frac{\frac{3}{8}y^{8/3} - \frac{1}{9}y^3\Big|_0^{27}}{\frac{243}{10}} = \frac{45}{4}$$

$$(\bar{x}, \bar{y}) = \left(\frac{30}{7}, \frac{45}{4}\right)$$

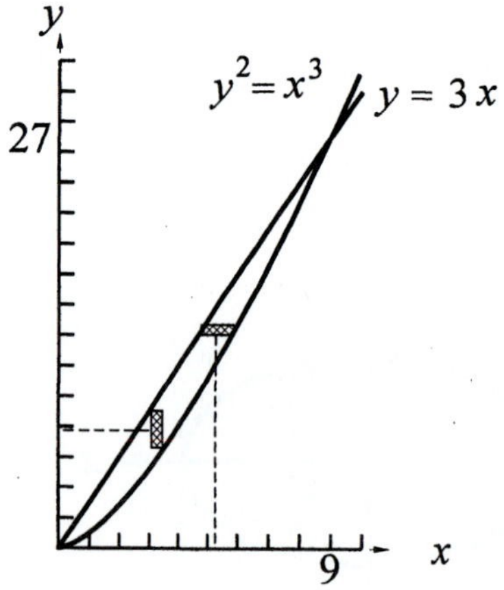

33. $I_y = k\int_0^2 x^2\left(\left(3x - x^2\right) - x\right)dx$

$I_y = k\int_0^2\left(2x^3 - x^4\right)dx$

$I_y = k\left(\frac{x^4}{2} - \frac{x^5}{5}\right)\Big|_0^2$

$I_y = \frac{8k}{5}$

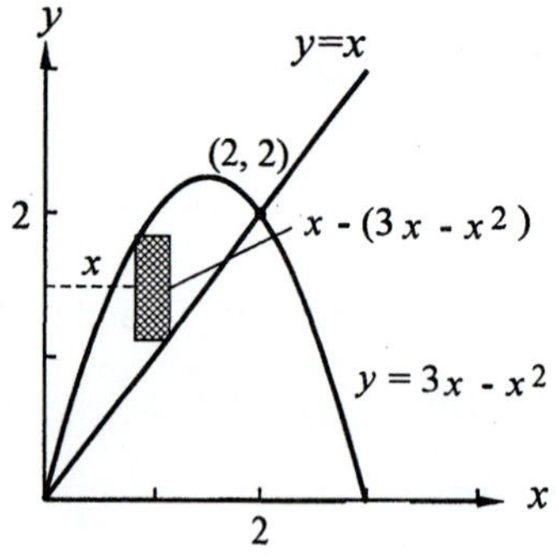

37. rope has a mass per unit length $= \frac{1}{10}$ lb/ft

$W = W_{\text{bucket}} + W_{\text{rope}}$

$W = 80(100) + \int_0^{100} \frac{1}{10}(100 - x)\,dx$

$W = 800 + \left(10x - \frac{x^2}{20}\right)\Big|_0^{100}$

$= 8500 \text{ ft} \cdot \text{lb}$

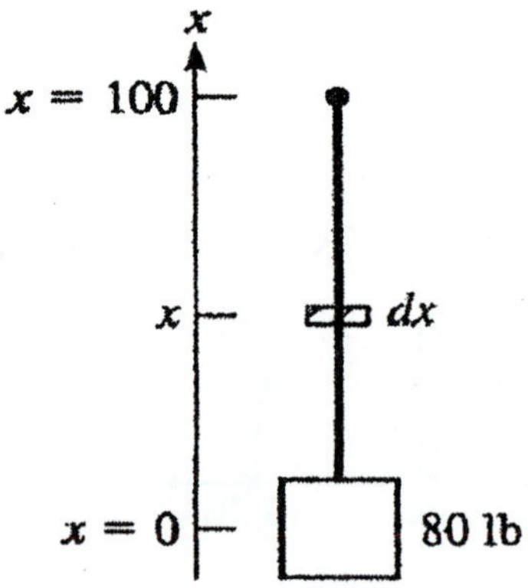

41. $a = -18$

$v = -18t + 28\big|_{t=5} = -62$ m/s

After 5.0 s the velocity of rock is 62 m/s down slope.

45. Using disks and $\frac{x^2}{1.5^2} + \frac{y^2}{10^2} = 1$ as equation of ellipse.

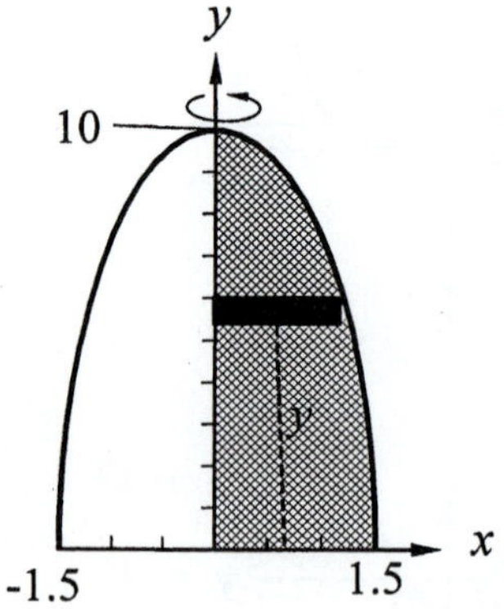

49. The circumference of the bottom, $c = 2\pi r = 9\pi$, equates to l, the length of the vertical surface area.

$$F = 68.0\int_0^{3.25}(9\pi h)\,dh = 68.0\left[4.50\pi h^2\right]_0^{3.25}$$

$$= 68.0\left[4.50\pi(3.25)^2 - 0\right] = 10{,}200 \text{ lb}$$

53. $a = 5\times10^{14}$

$v = 5\times10^{14}t$

$s = 2.5\times10^{14}t^2 = 0.025 \Rightarrow t = 1\times10^{-8}$

$v = 5\times10^{14}\left(1\times10^{-8}\right) = 5\times10^{6} \text{ m/s}$

Chapter 27

DIFFERENTIATION OF TRANSCENDENTAL FUNCTIONS

27.1 Derivatives of the Sine and Cosine Functions

1. $r = \sin^2 2\theta^2$

$\frac{dr}{d\theta} = 2\sin 2\theta^2 \cos 2\theta^2(2(2\theta))$

$\frac{dr}{d\theta} = 8\theta \sin 2\theta^2 \cos 2\theta^2$

$\frac{dr}{d\theta} = 4\theta \sin 4\theta^2$

5. $y = 2\sin(2x^3 - 1)$

$\frac{dy}{dx} = 2\cos(2x^3 - 1)(6x^2) = 12x^2\cos(2x^3 - 1)$

9. $y = 2\cos(3x - \pi)$

$\frac{dy}{dx} = 2[-\sin(3x - \pi)(3)] = -6\sin(3x - \pi)$

13. $y = 3\cos^3(5x + 2)$;

$\frac{dy}{dx} = 3.3\cos^2(5x+2)[-\sin(5x+2)(5)]$

$\frac{dy}{dx} = -45\cos^2(5x+2)\sin(5x+2)$

17. $y = 3x^3\cos 5x$;

$\frac{dy}{dx} = 3[x^3(-5\sin 5x) + \cos 5x(3x^2)]$

$\frac{dy}{dx} = 9x^2\cos 5x - 15x^3\sin 5x$

21. $y = \sqrt{1 + \sin 4x} = (1 + \sin 4x)^{1/2}$

$\frac{dy}{dx} = \frac{1}{2}(1 + \sin 4x)^{-1/2}(4\cos 4x)$

$\frac{dy}{dx} = \frac{2\cos 4x}{\sqrt{1 + \sin 4x}}$

25. $y = \frac{2\cos x^2}{3x - 1}$

$\frac{dy}{dx} = \frac{(3x-1)(2)(-\sin x^2)(2x) - 2\cos x^2(3)}{(3x-1)^2}$

$\frac{dy}{dx} = \frac{-4x(3x-1)\sin x^2 - 6\cos x^2}{(3x-1)^2}$

$\frac{dy}{dx} = \frac{4x(1-3x)\sin x^2 - 6\cos x^2}{(3x-1)^2}$

29. $s = \sin(\sin 2t)$

$\frac{ds}{dt} = \cos(\sin 2t)\cos(2t)(2)$

$\frac{ds}{dt} = 2\cos 2t\cos(\sin 2t)$

33. $p = \frac{1}{\sin s} + \frac{1}{\cos s}$

$\frac{dp}{ds} = \frac{\sin s(0) - \cos s}{\sin^2 s} + \frac{\cos s(0) - (-\sin s)}{\cos^2 s}$

$\frac{dp}{ds} = \frac{-\cos s}{\sin^2 s} + \frac{\sin s}{\cos^2 s}$

37. (a) $\cos 1.0000 = 0.5403023059$, calculator

Represents $\frac{d}{dx}(\sin x)$ at $x = 1$ (derivative, slope of T.L. to sine curve at $x = 1$).

(b) $(\sin 1.001 - \sin 1.0000)/0.0001$
$= 0.5402602315$, calculator = slope of secant line through the two points on sine curve where $x = 1.000$ and $x = 1.0001$.

41. $\sin(xy)\cos 2y = x^2$

$$\cos(xy)\left(x\frac{dy}{dx}+y\right)+(-\sin 2y)\left(2\frac{dy}{dx}\right)=2x$$

$$x\frac{dy}{dx}\cos(xy)+y\cos(xy)-2\frac{dy}{dx}\sin 2y=2x$$

$$x\frac{dy}{dx}\cos(xy)-2\frac{dy}{dx}\sin 2y=2x-y\cos(xy)$$

$$\frac{dy}{dx}(x\cos(xy)-2\sin 2y)=2x-y\cos(xy)$$

$$\frac{dy}{dx}=\frac{2x-y\cos xy}{x\cos xy-2\sin 2y}$$

45. $\cos 2x = 2\cos^2 x - 1;$

$$-\sin 2x(2)=2(2)\cos x(-\sin x)-0$$

$$-2\sin 2x=-4\sin x\cos x$$

$$\sin 2x=2\sin x\cos x$$

49. $y = x\cos 2x;\ x = 1.20$

$$\frac{dy}{dx}=x(-2\sin 2x)+\cos 2x$$

$$m_{\text{TL}}=\cos 2x-2x\sin x$$

$$m_{\text{TL}}|_{x=1.20}=-2.36$$

53. $y = 1.85\sin 36\pi t;$

$$v=\frac{dy}{dx}=(1.85\cos 36\pi t)(36\pi)$$

$$v|_{t=0.0250}=[1.85\cos(36\pi\cdot 0.025)][36\pi]=-199\text{ cm/s}$$

27.2 Derivatives of the Other Trigonometric Functions

1. $y = 3\sec^2 x^2$

$$\frac{dy}{dx}=3(2)(\sec x^2)\frac{d}{dx}(\sec x^2)$$

$$\frac{dy}{dx}=6\sec x^2\sec x^2\tan x^2(2x)$$

$$\frac{dy}{dx}=12x\sec^2 x^2\tan x^2$$

5. $y = 5\cot(0.25\pi-\theta)$

$$\frac{dy}{d\theta}=-5\csc^2(0.25\pi-\theta)\cdot(-1)$$

$$=5\csc^2(0.25\pi-\theta)$$

9. $y=-3\csc\sqrt{2x+3}$

$$\frac{dy}{dx}=-3[-\csc\sqrt{2x+3}\cot\sqrt{2x+3}\cdot\frac{1}{2}(2x+3)^{-1/2}(2)]$$

$$\frac{dy}{dx}=\frac{3\csc\sqrt{2x+3}\cot\sqrt{2x+3}}{\sqrt{2x+3}}$$

13. $y=2\cot^4\frac{1}{2}x$

$$\frac{dy}{dx}=2(4)\cot^3\frac{1}{2}x\left[-\csc^2\frac{1}{2}x\left(\frac{1}{2}\right)\right]$$

$$=-4\cot^3\frac{1}{2}x\csc^2\frac{1}{2}x$$

17. $y=3\csc^4 7x$

$$\frac{dy}{dx}=3\cdot 4\csc^3 7x[-\csc 7x\cot 7x(7)$$

$$=-84\csc^4 7x\cot 7x$$

21. $y=4\cos x\csc x^2$

$$\frac{dy}{dx}=4[\cos x(-\csc x^2\cot x^2\cdot 2x)+\csc x^2(-\sin x)]$$

$$\frac{dy}{dx}=-4\csc x^2(2x\cos x\cot x^2+\sin x)$$

25. $y=\dfrac{2\cos 4x}{1+\cot 3x}$

$$\frac{dy}{dx}=\frac{(1+\cot 3x)[-2\sin 4x(4)]-2\cos 4x(-\csc^2 3x)(3)}{(1+\cot 3x)^2}$$

$$\frac{dy}{dx}=\frac{-8\sin 4x(1+\cot 3x)+6\cos 4x\csc^2 3x}{(1+\cot 3x)^2}$$

$$\frac{dy}{dx}=\frac{2(-4\sin 4x-4\sin 4x\cot 3x+3\cos 4x\csc^2 3x)}{(1+\cot 3x)^2}$$

29. $r=\tan(\sin 2\pi\theta)$

$$\frac{dr}{d\theta}=\sec^2(\sin 2\pi\theta)\cos(2\pi\theta)(2\pi)$$

$$=2\pi\cos 2\pi\theta\sec^2(\sin 2\pi\theta)$$

33. $x\sec y-2y=\sin 2x$

$$x\sec y\tan y\frac{dy}{dx}+\sec y-2\frac{dy}{dx}=2\cos 2x$$

$$x\sec y\tan y\frac{dy}{dx}-2\frac{dy}{dx}=2\cos 2x-\sec y$$

37. $y = \tan 4x \sec 4x$

$$\frac{dy}{dx} = \tan 4x \cdot 4 \sec 4x \tan 4x + \sec 4x \cdot 4 \sec^2 4x$$

$$dy = 4 \sec 4x(\tan^2 4x + \sec^2 4x)dx$$

41. (a)

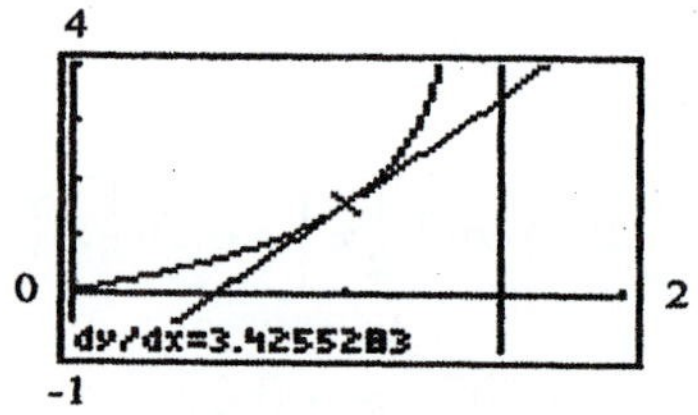

(b)

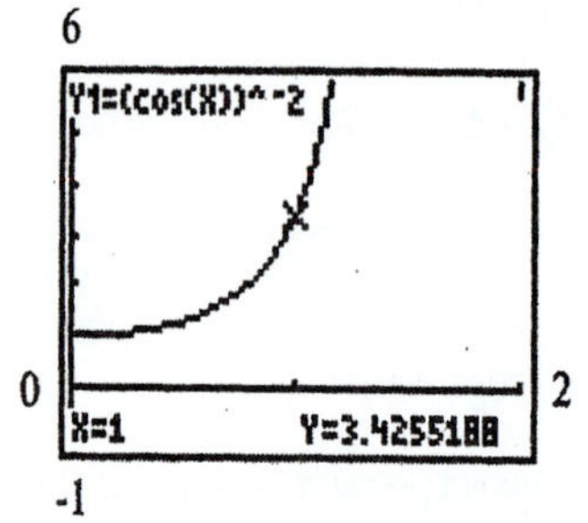

The values are the same in the first four decimal places.

45. $y = 2 \cot 3x;\ x = \frac{\pi}{12};$

$$\frac{dy}{dx} = 2(-\csc^2 3x)(3) = -6 \csc^2 3x;$$

$$\left.\frac{dy}{dx}\right|_{x=\pi/12} = -6 \csc^2 \frac{\pi}{4} = -6(\sqrt{2})^2 = -12$$

49. $y = 2t^{1.5} - \tan 0.1t;\ v = \frac{dy}{dt} = 3t^{0.5} - 0.1 \sec^2 0.1t;$

$$v|_{t=15} = 3(15)^{0.5} - 0.1 \sec^2[0.1(15)] = -8.4 \text{ cm/s}$$

27.3 Derivatives of the Inverse Trigonometric Functions

1. $y = \sin^{-1} x^2$

$$\frac{dy}{dx} = \frac{1}{\sqrt{1-(x^2)^2}}(2x) = \frac{2x}{\sqrt{1-x^4}}$$

5. $y = 2 \sin^{-1} 3x^3$

$$\frac{dy}{dx} = 2\frac{1}{\sqrt{1-9x^6}}(9x^2) = \frac{18x^2}{\sqrt{1-9x^6}}$$

9. $y = 2 \cos^{-1} \sqrt{2-x}$

$$\frac{dy}{dx} = 2 \cdot \frac{-1}{\sqrt{1-(2-x)}} \cdot \frac{1}{2}(2-x)^{-1/2}(-1)$$

$$\frac{dy}{dx} = \frac{1}{\sqrt{x-1}\sqrt{2-x}} = \frac{1}{\sqrt{(x-1)(2-x)}}$$

13. $= 6 \tan^{-1}\left(\frac{1}{x}\right)$

$$\frac{dy}{dx} = 6\frac{1}{1+\frac{1}{x^2}}\left(-\frac{1}{x^2}\right) = \frac{-\frac{6}{x^2}}{(x^2+1)/x^2} = \frac{-6}{x^2+1}$$

17. $y = 0.4u \tan^{-1} 2u$

$$\frac{dy}{du} = 0.4u \cdot \frac{2}{1+4u^2} + 0.4 \tan^{-1} 2u$$

$$\frac{dy}{du} = \frac{0.8u}{1+4u^2} + 0.4 \tan^{-1} 2u$$

21. $y = \frac{\sin^{-1} 2x}{\cos^{-1} 2x}$

$$\frac{dy}{dx} = \frac{\cos^{-1} 2x \frac{1}{\sqrt{1-4x^2}}(2) - \sin^{-1} 2x \frac{-1}{\sqrt{1-4x^2}}(2)}{(\cos^{-1} 2x)^2}$$

$$\frac{dy}{dx} = \frac{2}{\sqrt{1-4x^2}}\left[\frac{\cos^{-1} 2x + \sin^{-1} 2x}{(\cos^{-1} 2x)^2}\right]$$

$$\frac{dy}{dx} = \frac{2(\cos^{-1} 2x + \sin^{-1} 2x)}{\sqrt{1-4x^2}(\cos^{-1} 2x)^2}$$

25. $u=\left[\sin^{-1}(4t+3)\right]^2$

$$\frac{du}{dt}=2\left[\sin^{-1}(4t+3)\right]\frac{4}{\sqrt{1-(4t+3)^2}}$$

$$=\frac{2\sqrt{2}\sin^{-1}(4t+3)}{\sqrt{-2t^2-3t-1}}$$

29. $y=\dfrac{1}{1+4x^2}-\tan^{-1}2x$

$$=(1+4x^2)^{-1}-\tan^{-1}2x$$

$$\frac{dy}{dx}=-(1+4x^2)^{-2}(8x)-\frac{1}{1+4x^2}(2)$$

$$\frac{dy}{dx}=\frac{-8x}{(1+4x^2)^2}-\frac{2}{1+4x^2}=\frac{-8x-2(1+4x^2)}{(1+4x^2)^2}$$

$$\frac{dy}{dx}=\frac{-8x-2-8x^2}{(1+4x^2)^2}=\frac{-2(1+4x+4x^2)}{(1+4x^2)^2}$$

$$=\frac{-2(1+2x)^2}{(1+4x^2)^2}$$

33. $2\tan^{-1}xy+x=3$

$$2\frac{1}{1+x^2y^2}\left(x\frac{dy}{dx}+y\right)+1=0$$

$$\frac{2x}{1+x^2y^2}\frac{dy}{dx}+\frac{2y}{1+x^2y^2}=-1$$

$$\frac{2x}{1+x^2y^2}\frac{dy}{dx}=-1-\frac{2y}{1+x^2y^2}$$

$$\frac{2x}{1+x^2y^2}\frac{dy}{dx}=\frac{-1-x^2y^2-2y}{1+x^2y^2}$$

$$2x\frac{dy}{dx}=-1-x^2y^2-2y;\ \frac{dy}{dx}=\frac{-(x^2y^2+2y+1)}{2x}$$

37. $y=(\sin^{-1}x)^3$

$$\frac{dy}{dx}=3(\sin^{-1}x)^2\frac{1}{\sqrt{1-x^2}}$$

$$dy=\frac{3(\sin^{-1}x)^2dx}{\sqrt{1-x^2}}$$

41. $y=x\tan^{-1}x$

$$\frac{dy}{dx}=\frac{x}{1+x^2}+\tan^{-1}x$$

$$\frac{d^2y}{dx^2}=\frac{-x}{\left(1+x^2\right)^2}(2x)+\frac{1}{1+x^2}+\frac{1}{1+x^2}$$

$$\frac{d^2y}{dx^2}=\frac{2}{\left(1+x^2\right)^2}$$

45. $y=\tan^{-1}2x;\ \dfrac{dy}{dx}=\dfrac{1}{1+(2x)^2}(2)=\dfrac{2}{1+4x^2};$

$$\frac{d^2y}{dx^2}=\frac{(1+4x^2)(0)-2(8x)}{(1+4x^2)^2}=\frac{-16x}{(1+4x^2)^2}$$

49. $t=\dfrac{1}{\omega}\sin^{-1}\dfrac{A-E}{mE}$

$$=\frac{1}{\omega}\sin^{-1}\left(\frac{A-E}{E}\right)\left(\frac{1}{m}\right)$$

$$=\frac{1}{\omega}\sin^{-1}\left(\frac{A-E}{E}\right)m^{-1}$$

$$u=\left(\frac{A-E}{E}\right)m^{-1};\ \frac{du}{dm}=-\left(\frac{A-E}{E}\right)m^{-2}$$

$$\frac{dt}{dm}=\frac{1}{\omega\sqrt{1-\left(\frac{-A+E}{E}\right)^2m^{-2}}}\left(\frac{-A+E}{Em^2}\right)$$

$$=\frac{E-A}{\omega Em^2\sqrt{1-\frac{(A-E)^2}{E^2m^2}}}$$

$$=\frac{E-A}{\omega Em^2\sqrt{\frac{E^2m^2-(A-E)^2}{E^2m^2}}}$$

$$\frac{dt}{dm}=\frac{E-A}{\omega m\sqrt{E^2m^2-(A-E)^2}}$$

53. $a^2=b^2+c^2-2bc\cos A$ (Law of Cosines)
$x^2=5^2+8^2-2(5)(8)\cos A$

Therefore,

$$\cos A=\frac{5^2+8^2-x^2}{80}$$

$$\cos A=\frac{89-x^2}{80}$$

$$A=\cos^{-1}\frac{89-x^2}{80}$$

$$\frac{89-x^2}{80}=\frac{89}{80}-\frac{x^2}{80};\ \frac{d}{dx}\frac{89-x^2}{80}=-\frac{x}{40}$$

27.4 Applications

1. Sketch the curve $y = \sin x - \frac{x}{2}, 0 \le x \le 2\pi$.

$x = 0 \Rightarrow y = 0, (0,0)$ is both x-intercept and y-intercept. Using Newton's method or zero feature on graphing calculator $(1.90, 0)$ is the only x-intercept for $0 \le x \le 2\pi$. For $x = 2\pi$
$y = \sin 2\pi - \frac{2\pi}{2} = -\pi \Rightarrow (2\pi, -\pi)$ is right hand end point.

$\frac{dy}{dx} = \cos x - \frac{1}{2} = 0$ for $x = \frac{\pi}{3}, \frac{5\pi}{3}$

$\frac{d^2y}{dx^2} = -\sin x = \frac{-\sqrt{3}}{2}$ for $x = \frac{\pi}{3} \Rightarrow \max\left(\frac{\pi}{3}, 0.34\right)$

$\frac{d^2y}{dx^2} = -\sin x = \frac{\sqrt{3}}{2}$ for $x = \frac{5\pi}{3} \Rightarrow \min\left(\frac{5\pi}{3}, -3.48\right)$

$\frac{d^2y}{dx^2} = -\sin x = 0$ for $x = 0, \pi$

$\frac{d^2y}{dx^2} < 0$ for $0 < x < \pi \Rightarrow$ infl $(0,0)$

$\frac{d^2y}{dx^2} > 0$ for $\pi < x < \frac{\pi}{2} \Rightarrow$ infl $\left(\pi, -\frac{\pi}{2}\right)$

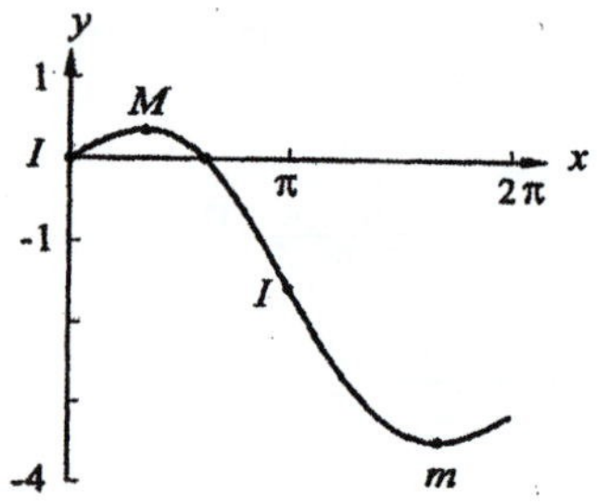

5. $y = \tan^{-1} x$

$\frac{dy}{dx} = \frac{1}{1+x^2} > 0$ for all x.

9. $y = x\sin^{-1} x\big|_{x=0.5} = 0.26179939$

$\frac{dy}{dx} = \frac{x}{\sqrt{1-x^2}} + \sin^{-1} x\big|_{x=0.5} = 1.1009497$

$y - 0.26179939 = 1.1009497(x - 0.5)$

$y = 1.1x - 0.29$

13. $y = 6\cos x - 8\sin x$; minimum value occurs when $f'(x) = 0$.

$f'(x) = -6\sin x - 8\cos x = 0$

$\sin x = -\frac{8}{6}\cos x$

$\tan x = -\frac{8}{6} = -\frac{4}{3}$

$\alpha = 0.927$ (a 3, 4, 5 triangle)

$Q_1 = 2.214,\ Q_2 = 5.356$

$f''(x) = -6\cos x + 8\sin$

$f''(2.214) > 0$, min; $f''(5.356) < 0$, max.

Minimum occurs where $x = 2.214$ rad.

$f(2.214) = 6\left(\frac{-3}{5}\right) - 8\left(\frac{4}{5}\right)$

$= \frac{-18}{5} + \frac{-32}{5} = \frac{-50}{5} = -10$

Minimum value of $y = -10$.

17. $y = 0.50\sin 2t + 0.30\cos t$;

$v = \frac{dy}{dt} = 1.00\cos 2t - 0.30\sin t$

$v|_{t=0.40s} = 1.00\cos 0.80 - 0.30\sin 0.40$
$= 0.58$ ft/s

$a = \frac{d^2y}{dt^2} = -2.00\sin 2t - 0.30\cos t$

$a|_{t=0.40s} = -200\sin 0.80 - 0.30\cos 0.40$
$= -1.7$ ft/s^2

21.

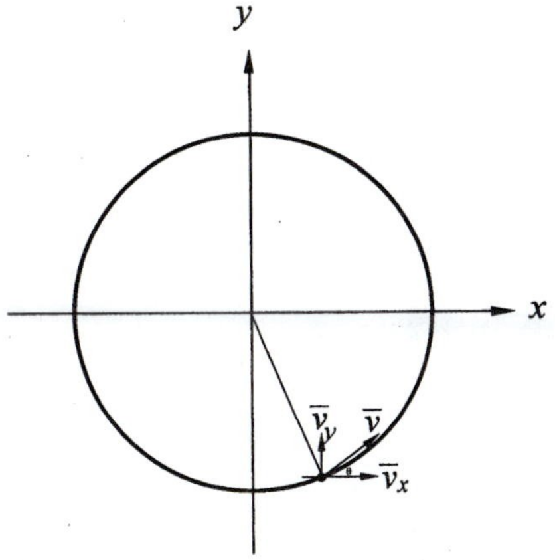

$$v = \sqrt{v_x^2 + v_y^2} = \sqrt{\left(\frac{dx}{dt}\right)^2 + \left(\frac{dy}{dt}\right)^2}$$

$$v = \sqrt{\left(-19(6\pi)\sin 6\pi t\right)^2 + \left(19(6\pi)\cos 6\pi t\right)^2}\,\Big|_{t=0.600}$$

$$v = 358 \text{ cm/s}$$

$$\tan\theta = \frac{v_y}{v_x} = \frac{19(6\pi)\cos 6\pi t}{19(6\pi)\sin 6\pi t}\Big|_{t=0.600}$$

$$\theta = 18.0^\circ$$

25.

$$s = 16t^2$$

$$\tan\theta = \frac{200 - s}{100} = \frac{200 - 16t^2}{100}$$

$$= 2 - 0.16t^2$$

$$\theta = \tan^{-1}(2 - 0.16t^2)$$

$$\frac{d\theta}{dt} = \frac{1}{1 + (2 - 0.16t^2)^2}(-0.32t)$$

$$= \frac{-0.32t}{5 - 0.64(t)^2 + 0.0256(t)^4}$$

$$\frac{d\theta}{dt}\Big|_{t=1.0} = \frac{-0.32(1.0)}{5 - 0.64(1.0)^2 + 0.0256(1.0)^4}$$

$$= -0.073 \text{ rad/s}$$

29. $F = \dfrac{0.25w}{0.25\sin\theta + \cos\theta}$

$$\frac{dF}{d\theta}$$

$$= \frac{(0.25\sin\theta + \cos\theta)(0) - 0.25w(0.25\cos\theta - \sin\theta)}{(0.25\sin\theta + \cos\theta)^2}$$

$$= 0$$

$$\tan\theta = 0.25 \Rightarrow \theta = 14^\circ$$

33. $V = 0.48(1.2 - \cos 1.26t)$

$$\frac{dV}{dt} = 0.48(1.26)\sin 1.26t = 0$$

$$1.26t = 0, \pi$$

$$t = 0, \frac{\pi}{1.26}$$

$$\frac{d^2V}{dt^2} = 0.48(1.26)^2\cos 1.26t\Big|_{t=\frac{\pi}{1.26}} < 0, \text{ max at}$$

$$t = \frac{\pi}{1.26}$$

$$V_{\max} = 0.48\left(1.2 - \cos 1.26\frac{\pi}{1.26}\right) = 1.056$$

$$V_{\max} = 1.1 \text{ L/s}$$

37. $y = 6.0\csc\theta + 4.0\sec\theta$

$$\frac{dy}{d\theta} = 6.0(-\csc\theta\cot\theta) + 4(\sec\theta\tan\theta)$$

$$\frac{dy}{d\theta} = -6.0\csc\theta\cot\theta + 4\sec\theta\tan\theta$$

$$\frac{dy}{d\theta} = -6.0\frac{1}{\sin\theta}\cdot\frac{\cos\theta}{\sin\theta} + 4\frac{1}{\cos\theta}\cdot\frac{\sin\theta}{\cos\theta}$$

$$\frac{dy}{d\theta} = \frac{-6.0\cos^3\theta + 4\sin^3\theta}{\sin^2\theta\cos^2\theta}$$

$$-6.0\cos^3\theta + 4\sin^3\theta = 0;\ 4\sin^3\theta = 6\cos^3\theta$$

$$\sin^3\theta = \frac{6}{4}\cos^3\theta;\ \sin\theta = \sqrt[3]{\frac{3}{2}}\cos\theta$$

$$\tan\theta = \sqrt[3]{\frac{3}{2}};\ \theta = \tan^{-1}\sqrt[3]{\frac{3}{2}} = 0.853 \text{ rad}$$

$$y_{\theta=0.853} = 14 \text{ ft}$$

27.5 Derivative of the Logarithmic Function

1. $y = \ln \cos 4x$

$$\frac{dy}{dx} = \frac{1}{\cos 4x}(-\sin 4x)(4)$$

$$\frac{dy}{dx} = -4\tan x$$

5. $y = 4\log_5(3-x)$

$$\frac{dy}{dx} = 4\frac{1}{3-x}\log_5 e(-1) = \frac{4}{x-3}\log_5 e$$

9. $y = 2\ln \tan 2x$

$$\frac{dy}{dx} = 2\frac{1}{\tan 2x}\sec^2 2x(2) = \frac{4\sec^2 2x}{\tan 2x} = \frac{4\sec^2 2x}{\frac{\sec 2x}{\csc 2x}}$$

$$\frac{dy}{dx} = 4\sec 2x \csc 2x$$

13. $y = \ln(4x-3)^2$

$$\frac{dy}{dx} = \frac{1}{(4x-3)^2}\cdot 2(4x-3)(4)$$

$$= \frac{8}{4x-3}$$

17. $y = 3x\ln(6-x)$

$$\frac{dy}{dx} = 3x\cdot\frac{-1}{6-x} + 3\ln(6-x)$$

$$= \frac{3x}{x-6} + 3\ln(6-x)$$

21. $y = \ln\frac{2x}{1+x} = \ln 2x - \ln(1+x)$

$$\frac{dy}{dx} = \frac{1}{2x}(2) - \frac{1}{1+x} = \frac{1+x-x}{x(1+x)}$$

$$= \frac{1}{x(1+x)} = \frac{1}{x^2+x}$$

25. $u = 3v\ln^2 2v$

$$\frac{du}{dv} = 3v\cdot 2\ln 2v\cdot\frac{1}{2v}\cdot 2 + 3\ln^2 2v$$

$$\frac{du}{dv} = 6\ln 2v + 3\ln^2 2v$$

29. $r = \ln\frac{v^2}{v+2}$

$$= \ln v^2 - \ln(v+2)$$

$$\frac{dr}{dv} = \frac{1}{v^2}(2v) - \frac{1}{v+2}$$

$$= \frac{2}{v} - \frac{1}{v+2}$$

$$= \frac{2v+4-v}{v(v+2)}$$

$$\frac{dr}{dv} = \frac{v+4}{v(v+2)}$$

33. $y = x - \ln^2(x+y)$

$$\frac{dy}{dx} = 1 - 2\ln(x+y)\frac{1}{x+y}\left(1+\frac{dy}{dx}\right)$$

$$\frac{dy}{dx} = 1 - \frac{2\ln(x+y)}{x+y} - \frac{2\ln(x+y)}{x+y}\frac{dy}{dx}$$

$$\frac{dy}{dx} + \frac{2\ln(x+y)}{x+1} = 1 - \frac{2\ln(x+y)}{x+1}$$

$$= \frac{x+y-2\ln(x+y)}{x+y}$$

$$\frac{dy}{dx}\left[\frac{x+y+2\ln(x+y)}{x+y}\right] = \frac{x+y-2\ln(x+y)}{x+y}$$

$$\frac{dy}{dx} = \frac{x+y-2\ln(x+y)}{x+y+2\ln(x+y)}$$

37. $y = (1+x)^{(1/x)}$

(a)

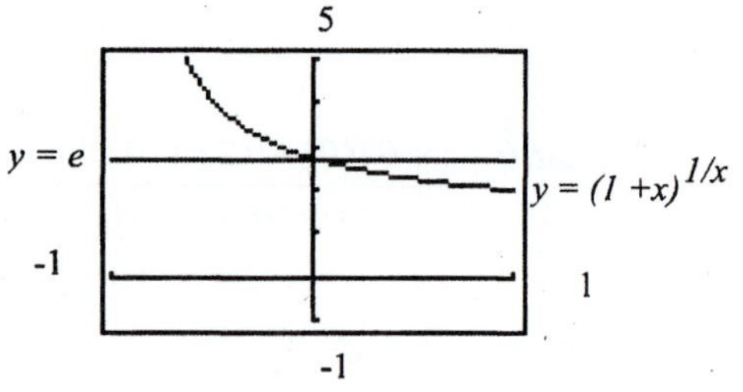

(b) See table.

X	Y_1	Y_2
.1	2.5937	2.7183
.01	2.7048	2.7183
.001	2.7169	2.7183
1E-4	2.7181	2.7183
1E-5	2.7183	2.7183

X=

41. $y = \sin^{-1} 2x + \sqrt{1-4x^2};\ x = 0.250;$

$$\frac{dy}{dx} = \frac{2}{\sqrt{1-(2x)^2}} + \frac{1}{2}(1-4x^2)^{-1/2}(-8x)$$

$$= \frac{2-4x}{\sqrt{1-4x^2}}$$

$$\left.\frac{dy}{dx}\right|_{x=0.250} = \frac{2-4(0.250)}{\sqrt{1-4(0.250)^2}} = 1.15$$

45. $y = \tan^{-1} 2x + \ln(4x^2+1);$

$$m_{\tan} = \frac{dy}{dx} = \frac{1}{4x^2+1}(2) + \frac{1}{4x^2+1}(8x)$$

$$= \frac{2}{1+4x^2} + \frac{8x}{4x^2+1} = \frac{2+8x}{1+4x^2}$$

When $x = 0.625$,

$$m_{\tan} = \frac{2+8(0.625)}{1+4(0.625)^2}$$

$$= \frac{7}{2.5625}$$

$$= 2.73$$

49. $y_1 = \ln(x^2),\ x^2 \neq 0$

$$\frac{dy_1}{dx} = \frac{1}{x^2}(2x) = \left.\frac{2}{x}\right|_{x=-1} = -2$$

$y_2 = 2\ln x,\ x > 0$

$$\frac{dy_2}{dx} = 2\left(\frac{1}{x}\right)$$

$$= \left.\frac{2}{x}\right|_{x=-1} \text{ is not defined since}$$

$-1 < 0$

53. $y = \ln\left(\dfrac{1+\sqrt{1+x^2}}{x}\right) - \sqrt{1-x^2}$

$$\frac{dy}{dx} = \frac{1}{\frac{1+\sqrt{1+x^2}}{x}}\left(\frac{x\left(0+\frac{2x}{2\sqrt{1+x^2}}\right)-(1+\sqrt{1+x^2})(1)}{x^2}\right) - \frac{-2x}{2\sqrt{1-x^2}}$$

$$\frac{dy}{dx} = \frac{x}{1+\sqrt{1+x^2}}\left(\frac{\frac{x^2}{\sqrt{1+x^2}}-1-\sqrt{1+x^2}}{x^2}\right) + \frac{x}{\sqrt{1-x^2}}$$

$$\frac{dy}{dx} = \frac{x^2-\sqrt{1+x^2}-1-x^2}{x(1+\sqrt{1+x^2})\sqrt{1+x^2}} + \frac{x}{\sqrt{1-x^2}}$$

$$\frac{dy}{dx} = \frac{-1-\sqrt{1+x^2}}{x\sqrt{1+x^2}+x+x^3} + \frac{x}{\sqrt{1-x^2}}$$

$$\frac{dy}{dx} = \frac{x^2-x^2-1-\sqrt{1+x^2}}{x(x^2+1+\sqrt{1+x^2})} + \frac{x}{\sqrt{1-x^2}}$$

$$\frac{dy}{dx} = \frac{x^2-(x^2+1+\sqrt{1+x^2})}{x(x^2+1+\sqrt{1+x^2})} + \frac{x}{\sqrt{1-x^2}}$$

$$\frac{dy}{dx} = \frac{x}{x^2+1+\sqrt{1+x^2}} - \frac{1}{x} + \frac{x}{\sqrt{1-x^2}}$$

which may also be simplified to $\dfrac{dy}{dx} = \dfrac{-1}{x\sqrt{1+x^2}} + \dfrac{x}{\sqrt{1-x^2}}$

27.6 Derivatives of the Exponential Function

1. $y = \ln \sin e^{2x}$

$$\frac{dy}{dx} = \frac{1}{\sin e^{2x}}(\cos e^{2x})(2e^{2x})$$

$$\frac{dy}{dx} = 2e^{2x}\cot e^{2x}$$

5. $y = 6e^{\sqrt{x}}$

$$\frac{dy}{dx} = 6e^{\sqrt{x}} \cdot \frac{1}{2\sqrt{x}} = \frac{3e^{\sqrt{x}}}{\sqrt{x}}$$

9. $R = Te^{-T}$

$$\frac{dR}{dT} = T(e^{-T})(-1) + (1)(e^{-T})$$
$$= e^{-T} - Te^{-T} = e^{-T}(1 - T)$$

13. $r = \dfrac{2(e^{2s} - e^{-2s})}{e^{2s}}$

$$r = 2(1 - e^{-4s})$$
$$\frac{dr}{ds} = 2(4e^{-4s})$$
$$\frac{dr}{ds} = 8e^{-4s}$$

17. $y = \dfrac{2e^{3x}}{4x+3}$;

$$\frac{dy}{dx} = \frac{(4x+3)(2e^{3x})(3) - (2e^{3x})(4)}{(4x+3)^2}$$
$$\frac{dy}{dx} = \frac{(12x+9)(2e^{3x}) - 8e^{3x}}{(4x+3)^2}$$
$$= \frac{2e^{3x}(12x+5)}{(4x+3)^2}$$

21. $y = (2e^{2x})^3 \sin x^2$
$= 8e^{6x} \sin x^2$

$$\frac{dy}{dx} = 8e^{6x}(\cos x^2)(2x) + \sin x^2(8e^{6x})(6)$$
$$= 16e^{6x}(x \cos x^2 + 3 \sin x^2)$$

25. $y = xe^{xy} + \sin y$

$$\frac{dy}{dx} = x(e^{xy})\left(x\frac{dy}{dx} + y\right) + (1)e^{xy} + \cos y\frac{dy}{dx}$$
$$= x(e^{xy})\left(x\frac{dy}{dx}\right) + x(e^{xy})(y) + e^{xy} + \cos y\frac{dy}{dx}$$
$$\frac{dy}{dx} - x(e^{xy})\left(x\frac{dy}{dx}\right) - \cos y\frac{dy}{dx} = x(e^{xy})y + e^{xy}$$
$$\frac{dy}{dx}(1 - x(e^{xy})(x) - \cos y) = x(e^{xy})y + e^{xy}$$
$$\frac{dy}{dx} = \frac{xy(e^{xy}) + e^{xy}}{1 - x^2e^{xy} - \cos y} = \frac{e^{xy}(xy+1)}{1 - x^2e^{xy} - \cos y}$$

29. $I = \ln \sin 2e^{6t}$

$$\frac{dI}{dt} = \frac{1}{\sin 2e^{6t}} \cdot \cos 2e^{6t} \cdot 12e^{6t}$$

33. (a) $e = e^x = 2.7182818$ when $x = 1.0000$. This is the slope of a tangent line to the curve $f(x) = e^x$ when $x = 1.0000$. It is the value of $f'(x) = e^x$, since $\frac{de^x}{dx} = e^x$.

(b) $\dfrac{e^{1.0001} - e^{1.0000}}{0.0001} = 2.7184178$ This is the slope of a secant line through the curve

$f(x) = e^x$ at $x = 1.0000$, where $\Delta x = 0.0001$.

$$\lim_{\Delta x \to 0} \frac{e^{(x+\Delta x)} - e^x}{\Delta x} = \frac{de^x}{dx} = e^x$$

For $\Delta x = 0.0001$, the slope of the tangent line is approximately equal to the slope of the secant line.

37. $y = e^{-x/2} \cos 4x$

$$\frac{dy}{dx} = -4 \sin 4xe^{-x/2} - \frac{1}{2}e^{-x/2} \cos 4x\Big|_{x=0.625}$$
$$= -1.458341701$$

numerical derivative feature gives -1.458339582

41.

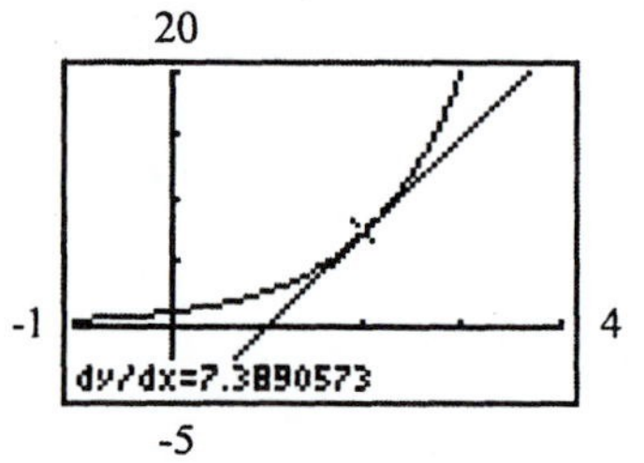

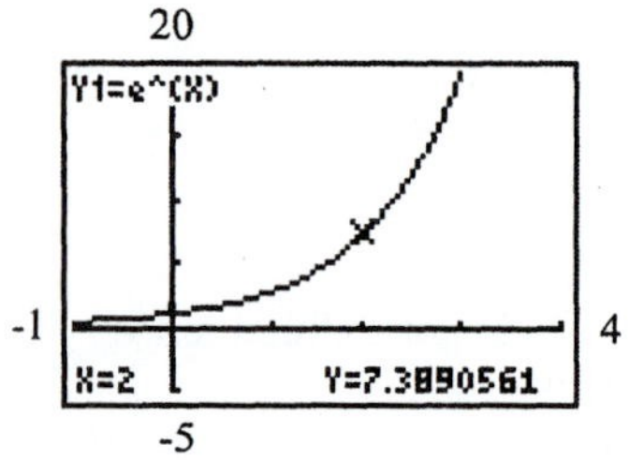

45. $y = \dfrac{e^{2x}-1}{e^{2x}+1}$

$\dfrac{dy}{dx} = 1 - y^2$

$\dfrac{dy}{dx} = \dfrac{(e^{2x}+1)2e^{2x} - (e^{2x}-1)2e^{2x}}{(e^{2x}+1)^2}$

$\dfrac{dy}{dx} = \dfrac{2e^{2x}(e^{2x}+1-e^{2x}+1)}{(e^{2x}+1)^2} = \dfrac{4e^{2x}}{(e^{2x}+1)^2}$

$1 - y^2 = 1 - \dfrac{(e^{2x}-1)^2}{(e^{2x}+1)^2}$

$= \dfrac{(e^{2x}+1)^2 - (e^{2x}-1)^2}{(e^{2x}+1)^2}$

$1 - y^2 = \dfrac{e^{4x}+2e^{2x}+1-e^{4x}+2e^{2x}-1}{(e^{2x}+1)^2}$

$= \dfrac{4e^{2x}}{(e^{2x}+1)^2}$

49. $y = Ae^{kx} + Be^{-kx}$

$y' = Ake^{kx} - Bke^{-kx}$

$y'' = Ak^2e^{kx} + Bk^2e^{-kx}$

$= k^2\left(Ae^{kx} + Be^{-kx}\right)$

$= k^2 y$

53. $i = 4.42e^{-66.7t}\sin(226t)$

$\dfrac{di}{dt} = 4.42e^{-66.7t}\cos(226t)\cdot 226 - 294.8e^{-66.7t}\sin(226t)$

$\dfrac{di}{dt} = e^{-66.7t}(999\cos(226t) - 295\sin(226t))$

57. $\dfrac{d}{dx}\sinh u = \dfrac{d}{dx}\dfrac{1}{2}\left(e^u - e^{-u}\right) = \dfrac{1}{2}\left(e^u\dfrac{du}{dx} - e^{-u}\cdot -\dfrac{du}{dx}\right)$

$\dfrac{d}{dx}\sinh u = \dfrac{1}{2}e^u\dfrac{du}{dx} + \dfrac{1}{2}e^{-u}\dfrac{du}{dx} = \left[\dfrac{1}{2}\left(e^u + e^{-u}\right)\right]\dfrac{du}{dx}$

$\dfrac{d}{dx}\sinh u = \cosh\dfrac{du}{dx}$

$\dfrac{d}{dx}\cosh u = \dfrac{d}{dx}\dfrac{1}{2}\left(e^u + e^{-u}\right) = \dfrac{1}{2}\left(e^u\dfrac{du}{dx} + e^{-u}\cdot -\dfrac{du}{dx}\right)$

$\dfrac{d}{dx}\cosh u = \dfrac{1}{2}e^u\dfrac{du}{dx} - \dfrac{1}{2}e^{-u}\dfrac{du}{dx} = \left[\dfrac{1}{2}\left(e^u - e^{-u}\right)\right]\dfrac{du}{dx}$

$\dfrac{d}{dx} = \cosh u = \sinh u\dfrac{du}{dx}$

27.7 Applications

1. $y = e^{-x}\sin x, 0 \le x \le 2\pi$

$y = 0$ for $x = 0, \pi, 2\pi$. Intercepts: $(0,0)$, $(\pi,0), (2\pi,0)$

$\dfrac{dy}{dx} = e^{-x}\cos x - e^{-x}\sin x = 0$ for $\sin x = \cos x$

$x = \dfrac{\pi}{4}, \dfrac{5\pi}{4}$

$\dfrac{d^2y}{dx^2} = e^{-x}(-\sin x - \cos x) - e^{-x}(\cos x - \sin x)$

$\dfrac{d^2y}{dx^2} = e^{-x}(-2\cos x) = 0$ for $x = \dfrac{\pi}{2}, \dfrac{3\pi}{2}$

$\left.\dfrac{d^2y}{dx^2}\right|_{x=\frac{\pi}{4}} = -2e^{-\pi/4}\cos\dfrac{\pi}{4}$

$= -0.64 < 0 \Rightarrow \left(\dfrac{\pi}{4}, 0.322\right)$ is a max

$\left.\dfrac{d^2y}{dx^2}\right|_{\frac{5\pi}{4}} = -2e^{-5\pi/4}\cos\dfrac{5\pi}{4}$

$= 0.03 > 0 \Rightarrow \left(\dfrac{5\pi}{4}, -0.014\right)$ is a min

$\dfrac{d^2y}{dx^2}$ changes from negative to positive at $x = \dfrac{\pi}{2}$

$\left(\dfrac{\pi}{2}, 0.208\right)$ is an infl. point

$\dfrac{d^2y}{dx^2}$ changes from positive to negative at $x = \dfrac{3\pi}{2}$

$\left(\dfrac{3\pi}{2}, -0.009\right)$ is an infl. point

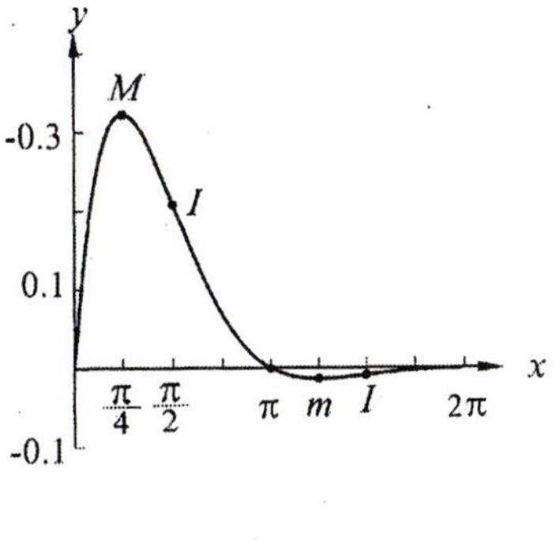

5. $y = xe^{-x} = \dfrac{x}{e^x}$

$$\frac{dy}{dx} = x(-e^{-x}) + e^{-x} = e^{-x}(1-x)$$

$$\frac{d^2y}{dx^2} = e^{-x}(-1) + (1-x)(-e^{-x})$$
$$= e^{-x} - e^{-x} + xe^{-x} = e^{-x}(x-2)$$

(1) Intercepts: $x = 0, y = 0$ (origin)

(2) Symmetry: none

(3) As $x \to +\infty, y \to 0$ positively. Horizontal asymptote $y = 0$. As $x \to -\infty, y \to -\infty$.
(4) Vertical asymptote: none

(5) Domain: all x, range to be determined.

(6) $\dfrac{dy}{dx} = 0$; $e^{-x}(1-x) = 0$; $x = 1$; $f''(1) < 0$,

max. $\left(1, \dfrac{1}{e}\right)$

$\dfrac{d^2y}{dx^2} = 0$; $e^{-x}(x-2) = 0$; $x = 2$; infl. $\left(2, \dfrac{2}{e^2}\right)$

9. $y = 4e^{-x^2}$; $\dfrac{dy}{dx} = 4e^{-x^2}(-2x) = \dfrac{-8x}{e^{x^2}}$

$$\frac{d^2y}{dx^2} = \frac{e^{x^2}(-8) + 8x(2xe^{x^2})}{e^{2x^2}} = 8e^{x^2}(2x^2 - 1)$$

(1) Intercepts: $x = 0, y = 4$, $(0, 4)$ intercept.

(2) Symmetry: yes, with respect to y-axis.

(3) As $x \to \pm\infty, y \to 0$ positively; x-axis is a horizontal asymptote.

(4) No vertical asymptote.

(5) Domain: all x; range: to be determined.

(6) $\dfrac{dy}{dx} = 0$; $\dfrac{-8x}{e^{x^2}} = 0$; $f''(0) < 0$, max. $(0, 4)$

$\dfrac{d^2y}{dx^2} = 0$; $2x^2 - 1 = 0$; $x = \pm\sqrt{\dfrac{1}{2}} = \pm\dfrac{\sqrt{2}}{2}$;

$\left(-\dfrac{\sqrt{2}}{2}, \dfrac{4}{\sqrt{e}}\right), \left(\dfrac{\sqrt{2}}{2}, \dfrac{4}{\sqrt{e}}\right)$ are inflection points.

13. $y = \dfrac{1}{2}(e^x - e^{-x})$; $\dfrac{dy}{dx} = \dfrac{1}{2}(e^x + e^{-x})$

$$\frac{d^2y}{dx^2} = \frac{1}{2}(e^x - e^{-x})$$

(1) Intercepts: $x = 0, y = 0, (0, 0)$

(2) No symmetry.

(3) As $x \to +\infty, y \to +\infty, x \to -\infty, y \to -\infty$

(4) No vertical asymptote.

(5) Domain: all x; range: all y.

(6) $\dfrac{dy}{dx} = 0$; $e^x = -e^{-x} = -\dfrac{1}{e^x}$; $(e^x)^2 = -1$;

$e^x = \sqrt{-1}$. $\dfrac{dy}{dx} > 0$, inc for all x.

(imaginary), no max,. no min.

$\dfrac{d^2y}{dx^2} = 0, e^x = e^{-x} = \dfrac{1}{e^x}$; $(e^x)^2 = 1$; $e^x = \pm 1$

$e^x \neq -1$; $e^x = 1$; $x = 0$, therefore infl. at $(0, 0)$

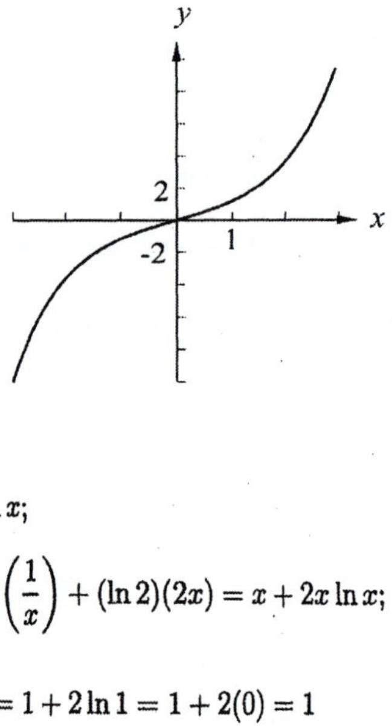

17. $y = x^2 \ln x$;

$$\frac{dy}{dx} = x^2\left(\frac{1}{x}\right) + (\ln 2)(2x) = x + 2x\ln x;$$

$$\left.\frac{dy}{dx}\right|_{x=1} = 1 + 2\ln 1 = 1 + 2(0) = 1$$

Slope is 1, $x = 1, y = 0$; using slope intercept form of the equation and substituting gives $0 = 1(1) + b$ or $b = 1$. The equation is $y = (1)x - 1$ or $y = x - 1$.

21. $f(x) = x^2 - 3 - \ln 4x$

pick $x = 2$ and use NEWTON calculator program

$x_2 = 2.308411873$

$x_3 = 2.283096216$

$x_4 = 2.282926297$

$x_5 = 2.282926289$

$x_6 = 2.282926289$

zero feature gives 2.2829263

25. $P = 100e^{-0.005t}; t = 100$ days;

$$\frac{dP}{dt} = 100e^{-0.005t}(-0.005)$$

$$\frac{dP}{dt} = -0.5e^{-0.005t}\Big|_{t=100} = -0.303 \text{ W/day}$$

29. $\ln p = \frac{a}{T} + b \ln T + c;\ p = e^{(a/T + b\ln T + c)}$

$$\frac{dp}{dT} = e^{(a/T+b\ln T+c)}\left(-aT^{-2} + \frac{b}{T}\right)$$

$$= p\left(\frac{-a+bT}{T^2}\right)$$

$$= e^{(a/T+b\ln T+c)}\left(\frac{-a}{T^2} + \frac{b}{T}\right)$$

$$= \frac{p(-a+bT)}{T^2}$$

33. $y = \ln \sec x;\ -1.5 \le x \le 1.5;\ u = \sec x;$

$\frac{dy}{dx} = \frac{1}{\sec x} \cdot \sec x \tan x = \tan x = 0$ at $x = 0$; $x = 0$ is a critical value; also, multiples of 2π. $\frac{d^2y}{dx^2} = \sec^2 x$; $\sec^2(0) = 1$ so the curve is concave up and there is a minimum point at $x = 0$, $y = \ln \sec 0 = \ln 1 = 0$ recurring at multiples of $x = 2\pi$; $(2\pi, 0), (4\pi, 0), \ldots (0, 0)$ is an intercept.

Asymptotes occur where $\frac{dy}{dx} = \tan x$ is undefined.

These values are odd multiples of $\frac{\pi}{2}$; $-\frac{\pi}{2}, \frac{\pi}{2}, \frac{3\pi}{2} \ldots$

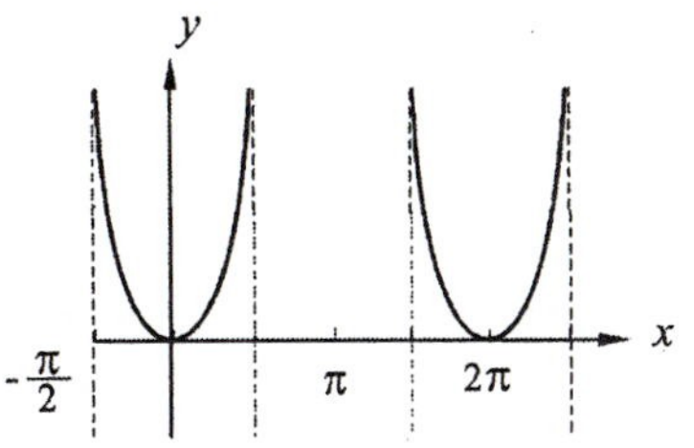

37. $y = 6.0e^{-0.020x}\sin(0.20x), 0 \le x \le 60$

$$\frac{dy}{dx} = e^{-0.020x}\left[\frac{6\cos(0.2x)}{5} - \frac{3\sin(0.2x)}{25}\right] = 0$$

$$\tan 0.2x = 10$$

$$0.2x = \tan^{-1} 10 + k\pi$$

$$x = 5\tan^{-1} 10 + 5k\pi$$

$k = 0$ $x_1 = 7.355638372, y_1 = 5.153476505$

$k = 1$ $x_2 = 23.06360164, y_2 = -3.764113107$

$k = 2$ $x_3 = 38.77156491, y_3 = 2.749318343$

$k = 3$ $x_4 = 54.47952818, y_4 = -2.008109516$

$(x_1, y_1) = 117.6°$ W $50.2°$N, maximum

$(x_2, y_2) = 101.9°$ W $41.2°$ N, minimum

$(x_3, y_3) = 86.2°$ W $47.7°$ N, maximum

$(x_4, y_4) = 70.5°$ W $43.0°$ N, minimum

41. $s = kx^2 \ln \frac{1}{x} = k[x^2(\ln 1 - \ln x)] = -kx^2 \ln x$

$$\frac{ds}{dx} = -k\left(x^2\frac{1}{x} + \ln x\, 2x\right) = -k(x + 2x\ln x)$$

$$\frac{ds}{dx} = -kx(1 + 2\ln x) = 0$$

For max., min.:

$$x = 0;\ \ln x = -\frac{1}{2};\ x = e^{-1/2} = \frac{1}{\sqrt{e}} = 0.607$$

Chapter 27 Review Exercises

1. $y = y\cos(4x-1);\ \frac{dy}{dx} = [-3\sin(4x-1)][4]$
$= -12\sin(4x-1)$

5. $y = \csc^2(3x+2);\ \frac{dy}{dx}$
$= 2\csc(3x+2)[-\csc(3x+2)\cot(3x+2)](3)$
$= -6\csc^2(3x+2)\cot(3x+2)$

9. $y = (e^{x-3})^2;\ \frac{dy}{dx} = (e^{x-3})(e^{x-3})(1) = 2e^{2(x-3)}$

13. $y = 10\tan^{-1}\left(\frac{x}{5}\right);\ \frac{dy}{dx} = 10\left[\frac{1}{1+\left(\frac{x}{5}\right)^2}\right]\frac{1}{5}$

$$= \frac{2}{1+\left(\frac{x}{5}\right)^2} = \frac{2}{1+\frac{x^2}{25}} = \frac{50}{25+x^2}$$

17. $y = \sqrt{\csc 4x + \cot 4x} = (\csc 4x + \cot 4x)^{1/2}$

$$\frac{dy}{dx} = \frac{1}{2}(\csc 4x + \cot 4x)^{-1/2}$$
$$(-4\csc 4x\cot 4x - 4\csc^2 4x)$$
$$= \frac{1}{2}(\csc 4x + \cot 4x)^{-1/2}$$
$$(-4\csc 4x\cot 4x - 4\csc^2 4x)$$
$$= \frac{1}{2}(\csc 4x + \cot 4x)^{-1/2}(-4\csc 4x)$$
$$(\csc 4x + \cot 4x)$$
$$= -2\csc 4x(\csc 4x + \cot 4x)^{1/2}$$
$$= (-2\csc 4x)\sqrt{\csc 4x + \cot 4x}$$

21. $y = \frac{\cos^2 x}{e^{3x}+2}$

$$\frac{dy}{dx} = \frac{(e^{3x}+\pi^2)[2\cos x(-\sin x)] - (\cos^2 x)(e^{3x})(3)}{(e^{3x}+\pi^2)^2}$$
$$= \frac{(e^{3x}+\pi^2)[-2\sin x\cos x] - 3e^{3x}\cos^2 x}{(e^{3x}+\pi^2)^2}$$
$$= \frac{-\cos x\left[(e^{3x}+\pi^2)^2(2\sin x) + 3e^{3x}\cos x\right]}{(e^{3x}+\pi^2)^2}$$
$$= \frac{-\cos x\left[2e^{3x}\sin x + 2\pi^2\sin x + 3e^{3x}\cos x\right]}{(e^{3x}+\pi^2)^2}$$
$$= \frac{-\cos x(2e^{3x}\sin x + 3e^{3x}\cos x + 2\pi^2\sin x)}{(e^{3x}+\pi^2)^2}$$

25. $y = \ln(\csc x^2)^2;\ \frac{dy}{dx} = \frac{1}{\csc x^2}(-\csc x^2\cot x^2)(2x)$
$= -2x\cot x^2$

29. $L = 0.1e^{-2t}\sec(\pi t)$

$$\frac{dL}{dt} = 0.1e^{-2t}\sec(\pi t)\cdot\pi + \sec(\pi t)\cdot 0.1(-2)e^{-2t}$$
$$\frac{dL}{dt} = 0.1\pi e^{-2t}\sec(\pi t)\tan(\pi t) - 0.2e^{-2t}\sec(\pi t)$$
$$\frac{dL}{dt} = e^{-2t}\sec(\pi t)\cdot[0.1\pi\tan(\pi t) - 0.2]$$

33. $\tan^{-1}\frac{y}{x} = x^2e^y;\ u = \frac{y}{x} = yx^{-1};\ \frac{du}{dx} = -yx^{-2} + x^{-1}\frac{dy}{dx}$

$$\frac{1}{1+(yx^{-1})^2}\left(-yx^{-2}+x^{-1}\right)\frac{dy}{dx} = x^2e^y\frac{dy}{dx} + 2xe^y$$
$$\frac{\frac{-y}{x^2}+\frac{1}{x}\frac{dy}{dx}}{1+y^2x^{-2}} = x^2e^y\frac{dy}{dx} + 2xe^y$$
$$\frac{-y}{x^2} + \frac{1}{x}\frac{dy}{dx} = \left(x^2e^y\frac{dy}{dx} + 2xe^y\right)(1+y^2x^{-2})$$
$$\frac{-y}{x^2} + \frac{1}{x}\frac{dy}{dx} = x^2e^y\frac{dy}{dx} + 2xe^y + y^2e^y\frac{dy}{dx}$$
$$+2x^{-1}y^2e^y$$
$$\frac{1}{x}\frac{dy}{dx} - x^2e^y\frac{dy}{dx} - y^2e^y\frac{dy}{dx} = 2xe^y + 2x^{-1}y^2e^y + \frac{y}{x^2}$$
$$\frac{dy}{dx}\left(\frac{1}{x} - x^2e^y - y^2e^y\right) = 2xe^y + \frac{2y^2ey}{x} + \frac{y}{x^2}$$
$$\frac{dy}{dx}\left(\frac{1-x^3e^y - xy^2e^y}{x}\right) = \frac{2xe^y + 2xy^2e^y + y}{x^2}$$

$$\frac{dy}{dx}=\frac{2x^3e^y+2xy^2e^y+y}{x^2}\cdot\frac{x}{1-x^3e^y-xy^2e^y}$$

$$\frac{dy}{dx}=\frac{2x^3e^y+2xy^2e^y+y}{x-x^4e^y-x^2y^2e^y}$$

37. $\ln xy+ye^{-x}=1$

Using implicit differentiation,

$$\frac{d\ln xy}{dx}+\frac{d\ ye^{-x}}{dx}=\frac{(1)}{dx}$$

$$\frac{1}{xy}\left(x\frac{dy}{dx}+y\frac{dx}{dy}\right)+y\frac{de^{-x}}{dx}+e^{-x}\frac{dy}{dx}=0$$

$$\frac{1}{xy}\left(x\frac{dy}{dx}+y\right)+ye^{-x}(-1)+e^{-x}\frac{dy}{dx}=0$$

$$\frac{1}{y}\frac{dy}{dx}+\frac{1}{x}-ye^{-x}+e^{-x}\frac{dy}{dx}=0$$

$$\frac{1}{y}\frac{dy}{dx}+e^{-x}\frac{dy}{dx}=ye^{-x}-\frac{1}{x}$$

$$\frac{dy}{dx}\left(\frac{1}{y}+e^{-x}\right)=ye^{-x}-\frac{1}{x}$$

$$\frac{dy}{dx}\left(\frac{1+ye^{-x}}{y}\right)=\frac{x\,ye^{-x}-1}{x}$$

$$\frac{dy}{dx}=\left(\frac{x\,ye^{-x}-1}{x}\right)\left(\frac{y}{1+y^{e-x}}\right)=\frac{y\left(xye^{-x}-1\right)}{x\left(1+ye^{-x}\right)}$$

41. $y=x-\cos 0.5x\big|_{x=0}=-1\Rightarrow$ y-int: $(0,\ -1)$

$0=x-\cos 0.5x\Rightarrow x=0.9\Rightarrow$ x-int: $(0,\ 0.9)$

$\frac{dy}{dx}=1+0.5\sin 0.5x=0\Rightarrow \sin x=-2$, no solution

$\Rightarrow$ no critical points

$\frac{d^2y}{dx^2}=0.25\cos 0.5x=0\Rightarrow x=\pi+k(2\pi)$

$\frac{d^2y}{dx^2}=$ changes sign at $x=\pi+k(2\pi)$

$\Rightarrow x=\pi+k(2\pi)$ are inflection points

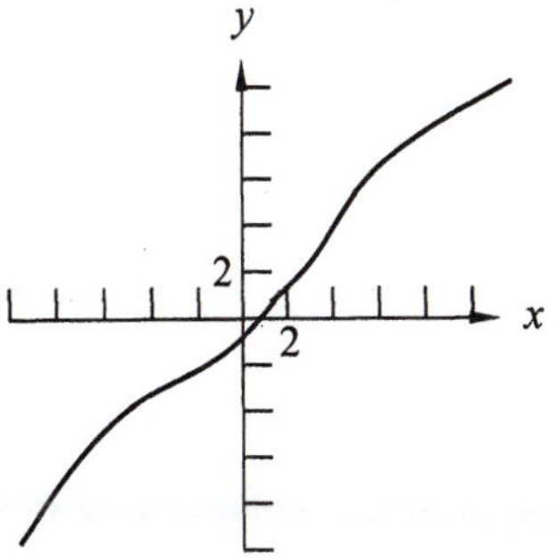

45. $y=4\cos^2\left(x^2\right)$; slope $=\frac{dy}{dx}$

$=2\left[4\cos\left(x^2\right)\right]\left[-\sin\left(x^2\right)\right](2x)$

$=-16x\cos x^2\sin x^2$

$\left.\frac{dy}{dx}\right|_{x=1}=-16\cos\left(1^2\right)\sin\left(1^2\right)$

$=-16(0.5403)(0.8415)$

$=-7.27$

$f(1)=4\cos^2\left(1^2\right)=4(0.5403)^2=1.168$

$y=-7.27x+b;\ 1.168=-7.27(1)+b,\ b=8.44;$

$y=-7.27x+8.44;\ 7.27x+y-8.44=0$

49. $\sin^2 x+\cos^2 x=1$

$$\frac{d\left(\sin^2 x+\cos^2 x\right)}{dx}=\frac{d(1)}{dx}$$

$$\frac{d\sin^2 x}{dx}+\frac{d\cos^2 x}{dx}=0$$

$2\sin x\cos x+2\cos x(-\sin x)=0$

$2\sin x\cos x-2\cos x\sin x=0;\ 0=0$

53. $y=e^x-x^2=0$

Let $x_1=-1$ and use Newton's calculator program

$x_2=-0.7330436122$

$x_3=-0.7038077877$

$x_4=-0.7034674683$

$x_5=-0.7034674225$

$x_6=-0.7034674225$

57. $\mu=\tan\theta,\ 18^\circ\left(\frac{\pi}{180^\circ}\right)=\frac{\pi}{10},\ \Delta\theta=2^\circ\left(\frac{\pi}{180^\circ}\right)=\frac{\pi}{90}$

$\Delta\mu\approx d\mu=\sec^2\theta d\theta=\sec^2\frac{\pi}{10}\left(\frac{\pi}{90}\right)=0.03859$

61. $y = 0.75\left(\sec\sqrt{0.15t} - 1\right)$

$$\frac{dy}{dt} = 0.75\sec\sqrt{0.15t}\tan\sqrt{0.15t}\,\frac{0.15}{2\sqrt{0.15t}}\Bigg|_{t=5.0}$$

$$\frac{dy}{dt}\bigg|_{t=5.0} = 0.12 \text{ m/s}$$

65. $W = 25\sin^2 2t$

$$P = \frac{dW}{dt} = 50\sin 2t\cos 2t(2)$$

$$P = 100\sin 2t\cos 2t$$

69. $I = kE_0^2\cos^2\frac{1}{2}\theta$

$$\frac{dI}{d\theta} = 2kE_0^2\cos^2\frac{1}{2}\theta\left(-\sin\frac{1}{2}\theta\right)\left(\frac{1}{2}\right)$$

$$\frac{dI}{d\theta} = -kE_0^2\cos\frac{1}{2}\theta\sin\frac{1}{2}\theta$$

73. $w = \sqrt{\frac{g}{l\cos\theta}}$, $g = 9.800 \text{ m/s}^2$, $l = 0.6375$ m

$\theta = 32.50^\circ$, $d\theta = 0.25^\circ$

$$\Delta w \approx dw = \frac{1}{2}\sqrt{\frac{g}{l\cos\theta}}\tan\theta d\theta$$

$$\Delta w \approx \frac{1}{2}\sqrt{\frac{9.800}{0.6375\cos 32.5^\circ}}\tan 32.5^\circ\left(\frac{\pi}{180^\circ}\right)$$

$$\Delta w \approx 0.005934 \text{ rad/s}$$

77. $n = 160 - 140e^{-0.30t}$

$$\frac{du}{dt} = 42e^{-0.30t}\Big|_{t=10} = 2.091056871$$

In 2010 the annual rate of increase of Internet users will be 2,100,000/year.

81.

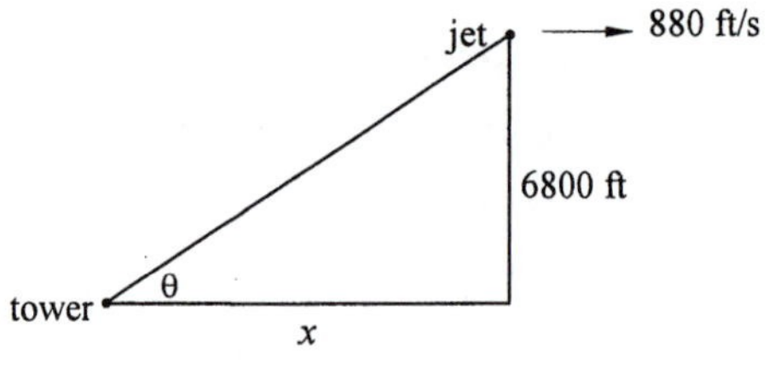

$$\theta = \tan^{-1}\frac{6800}{x}, \quad \frac{d\theta}{dt} = \frac{d\theta}{dx}\frac{dx}{dt}$$

$$\frac{d\theta}{dt} = \frac{1}{1+\left(\frac{6800}{x}\right)^2}\left(\frac{-6800}{x^2}\right)(880)\bigg|_{x=\frac{6800}{\tan 13^\circ}}$$

$$\frac{d\theta}{dt} = -0.0065486205$$

When the angle of elevation of the jet from the contral tower is 13°, the angle of elevation is changing -0.0065 rad/s.

89a.

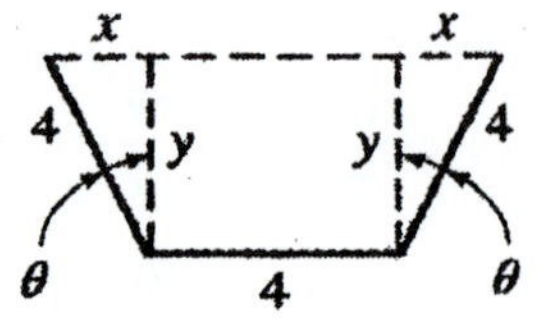

89b.

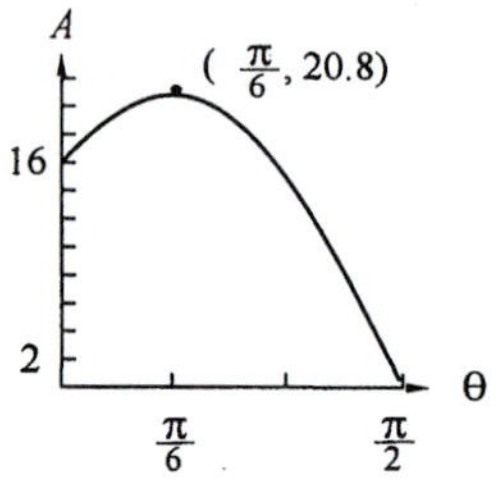

93.

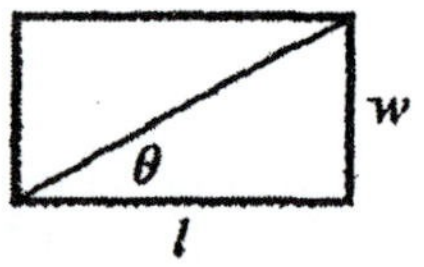

$$P = 2l + 2w = 40$$

$$l + w = 20$$

$$l = 20 - w$$

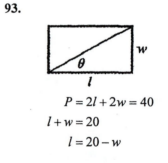

(a) Using algebra, $A = lw = (20 - w)w = 20w - w^2$

$$\frac{dA}{dw} = 20 - 2w = 0 \Rightarrow w = 10 \text{ and } l = 10$$

The area is maximum when the rectangle is a square.

(b) Using trigonometry $\tan\theta = \frac{w}{l} = \frac{w}{20-x}$ from which

$w = \frac{20\tan\theta}{1+\tan\theta}$, then $A = \frac{20^2\tan\theta}{1+\tan\theta} - \frac{20^2\tan^2\theta}{(1+\tan\theta)^2}$ and

taking $\frac{dA}{d\theta} = 0$ gives a maximum for $\theta = 45^\circ$; again, a square.

Chapter 28

METHODS OF INTEGRATION

28.1 The General Power Formula

1. Change $\cos x$ to $-\sin x$, then

$$\int \cos^3 x(-\sin x\,dx) = \frac{1}{4}\cos^4 x + C$$

5. $u = \cos x$; $n = \frac{1}{2}$; $du = -\sin x\,dx$

$$0.4\int \sqrt{\cos x}\sin x\,dx = 0.4\int(\cos x)^{1/2}(-\sin x\,dx)$$
$$= -\frac{0.8}{3}(\cos x)^{3/2} + C$$

9. Let $u = \cos 2x$; $n = 1$; $du = -2\sin 2x\,dx$

$$\int_0^{\pi/8} \cos 2x \sin 2x\,dx$$
$$= -\frac{1}{2}\int_0^{\pi/8}(\cos 2x)^{-1}(-2\sin 2x\,dx)$$
$$= -\frac{1}{2}\,\frac{(\cos 2x)^2}{2}\Big|_0^{\pi/8}$$
$$= -\frac{1}{4}\cos^2 2x\Big|_0^{\pi/8}$$
$$= -\frac{1}{4}\left(\cos^2\frac{\pi}{4} - \cos^2 0\right)$$
$$= -\frac{1}{4}\left(\frac{1}{2} - 1\right) = \frac{1}{8}$$

13. $u = \tan^{-1}5x$; $du = \frac{1}{1+25x^2}(5\,dx)$; $n = 1$

$$\int \frac{5\tan^{-1}5x}{1+25x^2}dx = \int(\tan^{-1}5x)^1\frac{5\,dx}{1+25x^2}$$
$$= \frac{1}{2}(\tan^{-1}5x)^2 + C$$

17. $u = \ln(2x+3)$; $du = \frac{1}{2x+3}(2\,dx)$; $n = 1$

$$\frac{1}{2}\int_0^{1/2}[\ln(2x+3)]\frac{2\,dx}{2x+3} = \frac{1}{2}\,\frac{[\ln(2x+3)]^2}{2}\Big|_0^{1/2}$$
$$= \frac{1}{4}\ln^2(2x+3)\Big|_0^{1/2}$$
$$= \frac{1}{4}(\ln^2 4 - \ln^2 3)$$
$$= 0.179$$

21. $\int \frac{e^{2t}}{(1-e^{2t})^3}dt$. Let $u = 1 - e^{2t}$, $du = -2e^{2t}dt$.

$$\int \frac{e^{2t}}{(1-e^{2t})^3}dt = -\frac{1}{2}\int(1-e^{2t})^{-3}(-2e^{2t}dt)$$
$$= -\frac{1}{2}\int u^{-3}du = -\frac{1}{2}\cdot\frac{u^{-3+1}}{-3+1} + C$$
$$= -\frac{1}{2}\cdot\frac{1}{-2(u^2)} + C$$
$$= \frac{1}{4(1-e^{2t})^2 + C}$$

25. $\int_{\pi/6}^{\pi/4}(1+\cot x)^2\csc^2 x\,dx$

$$= \int_{\pi/6}^{\pi/4}(1+\cot x)^2(-\csc^2 x\,dx)$$
$$= -\frac{(1+\cot x)^3}{3}\Big|_{\pi/6}^{\pi/4}$$
$$= -\frac{(1+\cot\frac{\pi}{4})^3}{3} + \frac{(1+\cot\frac{\pi}{6})^3}{3}$$
$$= 2\sqrt{3} + \frac{2}{3}$$
$$= 4.1308$$

29. $\int\frac{dx}{x\ln^2 x} = \int\ln^{-2}x\left(\frac{1}{x}dx\right) = \int u^{-2}du$ with

$u = \ln x,\ du = \frac{1}{x}dx,\ n = -2$

33. $A = \int_0^2 \frac{1+\tan^{-1}2x}{1+4x^2}dx;\ u = \tan^{-1}2x;$

$du = \frac{1}{1+(2x)^2}\times 2$

$$A = \frac{1}{2}\int_0^2 1+u\,du = \frac{1}{2}\left(u+\frac{1}{2}u^2\right)\Big|_0^2$$

$$= \frac{1}{2}\left[\tan^{-1}2x + \frac{1}{2}(\tan^{-1}2x)^2\right]\Big|_0^2$$

$$= \left[\frac{1}{2}\tan^{-1}4 + \frac{1}{2}(\tan^{-1}2x)^2 - \tan^{-1}0 - \frac{1}{2}(\tan^{-1}0)^2\right]$$

$$= \frac{1}{2}[1.326 + 0.879 - 0 - 0] = 1.102$$

37. $P = mnv^2\int_0^{\pi/2}\sin\theta\cos^2\theta\,d\theta;\ n = 2;$

$\mu = \cos\theta;\ du = -\sin\theta\,d\theta$

$$P = mnv^2\int_0^{\pi/2}\cos^2\theta(-\sin\theta\,d\theta)$$

$$= -mnv^2\left[\frac{\cos^3\theta}{3}\right]\Big|_0^{\pi/2}$$

$$= -mnv^2\left[\frac{1}{3}\left(\cos^3\frac{\pi}{2} - \cos^3 0\right)\right]$$

$$= -mnv^2\left[\frac{1}{3}(0-1)\right]$$

$$= -mnv^2\left(-\frac{1}{3}\right) = \frac{1}{3}mnv^2$$

28.2 The Basic Logarithmic Form

1. Add $2x$ to the integrand, then

$$\int\frac{2x\,dx}{x^2+1} = \ln|x^2+1| + C$$

5. $\int\frac{2x\,dx}{4-3x^2};\ u = 4-3x^2;\ du = -6x\,dx$

$$-\frac{1}{3}\int\frac{-6x\,dx}{4-3x^2} = -\frac{1}{3}\ln|4-3x^2| + C$$

9. $0.4\int\frac{\csc^2 2\theta\,d\theta}{\cot 2\theta};\ u = \cot 2\theta;\ du = -2\csc^2 2\theta\,d\theta$

$$\frac{0.4}{2}\int\frac{2\csc^2 2\theta\,d\theta}{\cot 2\theta} = -0.2\ln|\cot 2\theta| + C$$

13. $u = 1-e^{-x};\ du = e^{-x}dx$

$$\int\frac{e^{-x}dx}{1-e^{-x}} = \ln|1-e^{-x}| + C$$

17. $u = 1+4\sec x;\ du = 4\sec x\tan x\,dx$

$$\int\frac{\sec x\tan x\,dx}{1+4\sec x} = \frac{1}{4}\int\frac{4\sec x\tan x\,dx}{1+4\sec x}$$

$$= \frac{1}{4}\ln|1+4\sec x| + C$$

21. $u = \ln r;\ du = \frac{dr}{r}$

$$0.5\int\frac{dr}{r\ln r} = 0.5\int\frac{\frac{dr}{r}}{\ln r} = 0.5\ln|\ln r| + C$$

25. $n = -\frac{1}{2};\ u = 1-2x;\ du = -2\,dx$

$$\int\frac{2\,dx}{\sqrt{1-2x}} = -\int(1-2x)^{-1/2} = 2\,dx$$

$$= -(1-2x)^{1/2}\cdot 2 + C$$

$$= -2\sqrt{1-2x} + C$$

29. $u = 4+\tan 3x;\ du = 3\sec^2 3x\,dx$

$$\int_0^{\pi/12}\frac{\sec^2 3x}{4+\tan 3x}dx = \frac{1}{3}\int_0^{\pi/12}\frac{3\sec^2 3x\,dx}{(4+\tan 3x)}$$

$$= \frac{1}{3}\ln|4+\tan 3x|\Big|_0^{\pi/12}$$

$$= \frac{1}{3}(\ln 5 - \ln 4) = \frac{1}{3}\ln\frac{5}{4}$$

$$= 0.0744$$

33.

$$\begin{array}{r} 1 \\ x+4\overline{)x-4} \\ \underline{x+4} \\ -8 \end{array}$$

$$\int \frac{x-4}{x+4}\,dx = \int dx - \int \frac{8}{x+4}\,dx$$
$$= x - 8\ln|x+4| + C$$

37. $m = \dfrac{dy}{dx} = \dfrac{\sin x}{3+\cos x};\ y = \displaystyle\int \frac{1}{3+\cos x} \times \sin x\,dx$

$$y = -\int \frac{1}{3+\cos x}(-\sin x)\,dx;$$

let $u = 3+\cos x$; $du = -\sin x\,dx$

$$y = -\int \frac{1}{u}\,du = -\ln|u| + C$$
$$= -\ln(3+\cos x) + C$$

$2 = -\ln\left(3+\cos\dfrac{\pi}{3}\right) + C$; substitute values of x

and y

$2 = -\ln(3+0.5) + C$; $C = 2 + \ln 3.5$

$y = -\ln(3+\cos x) + \ln 3.5 + 2$; substituting for C

$$y = \ln\frac{3.5}{3+\cos x} + 2$$

41. $a(t) = (t+4)^{-1}$

$v(t) = \ln|t+4| + c$

$v(0) = 0 = \ln|0+4| + c \Rightarrow c = -\ln 4$

$v(t) = \ln|t+4| - \ln 4$

$v(4) = \ln|4+4| - \ln 4 = \ln 8 - \ln 4 = \ln\dfrac{8}{4}$

$v(4) = \ln 2 \text{ m/s}^2$

45. $t = L\displaystyle\int \frac{di}{E-iR};\ u = E - iR;\ du = -R\,di$

$$t = -\frac{L}{R}\int \frac{-R\,di}{E-iR} = \frac{-L}{R}\ln|E-iR| + C;$$

$t = 0$ for $i = 0$

$$0 = -\frac{L}{R}\ln E + C;\ C = \frac{L}{R}\ln|E|$$

$$t = \frac{L}{R}(-\ln|E-iR| + \ln|E|)$$

$$t = \frac{L}{R}\ln\frac{E}{E-iR};\ \frac{R}{L}t = \ln\frac{E}{E-iR};\ e^{Rt/L} = \frac{E}{E-iR}$$

$$i = \frac{E}{R} - \frac{E}{R}e^{-Rt/L};\ i = \frac{E}{R}\left(1 - e^{-Rt/L}\right)$$

28.3 The Exponential Form

1. $\displaystyle\int 3x^2 e^{x^3}\,dx = \int e^{x^3}(3x^2\,dx)$

$$= e^{x^3} + C$$

5. $u = 2x+5$; $du = 2\,dx$

$$\int e^{2x+5}dx = \frac{1}{2}\int e^{2x+5}(2\,dx) = \frac{1}{2}e^{2x+5} + C$$

9. $y = x^3$; $du = 3x^2dx$

$$\int 6x^2 e^{x^3}dx = 6\int e^{x^3}(x^2dx) = \frac{6}{3}\int e^{x^3}(3x^2dx)$$
$$= 2e^{x^3} + C$$

13. $u = 2\sec\theta$; $du = 2\sec\theta\tan\theta\,d\theta$

$$\int 4(\sec\theta\tan\theta)e^{2\sec\theta}d\theta = \frac{4}{2}\int e^{2\sec\theta}2\sec\theta\tan\theta\,d\theta$$
$$= 2e^{2\sec\theta} + C$$

17. $u = -2x$; $du = -2\,dx$

$$\int_1^3 3e^{2x}(e^{-2x} - 1)dx = 3\int (e^0 - e^{2x})dx$$
$$= 3\int dx - \frac{3}{2}\int e^{2x}(2\,dx)$$
$$= 3x - \frac{3}{2}e^{2x}\Big|_1^3$$

$$= \left[9 - \frac{3}{2}e^6 - \left(3 - \frac{3}{2}e^2\right)\right]$$

$$= 9 - 3 - \frac{3}{2}e^6 + \frac{3}{2}e^2$$

$$= 6 - \frac{3}{2}(e^6 - e^2) = -588.06$$

21. $u = \tan^{-1} x;\ du = \frac{1}{1+x^2}dx$

$$\int \frac{e^{\tan^{-1} x}}{x^2+1}dx = \int e^{\tan^{-1} x}\frac{dx}{x^2+1} = e^{\tan^{-1} x} + C$$

25. $u = \cos^2 x;$
$du = 2\cos x(-\sin x)dx = -2\sin x\cos x\,dx$
$= -2\sin 2x\,dx$

$$\int_0^{\pi} (\sin 2x)e^{\cos^2 x}dx = -\frac{1}{2}\int_0^{\pi} e^{\cos^2 x}(-2\sin 2x\,dx)$$

$$= -\frac{1}{2}e^{\cos^2 x}\Big|_0^{\pi}$$

$$= -\frac{1}{2}[(\cos\pi)^2 - (\cos 0)^2]$$

$$= -\frac{1}{2}[(-1)^2 - 1^2] = 0$$

29. $A = \int_0^2 3e^x dx = 3e^x\big|_0^2 = 3e^2 - 3e^0 = 19.2$

33. $1+e^x\overline{)1}$ ⇒ $\frac{1}{1+e^2}1 - \frac{e^x}{1+e^x}$

$\underline{1+e^x}$

$-e^x$

$$\int\frac{dx}{1+e^x} = \int dx - \int\frac{e^x}{1+e^x}dx$$

$= x - \ln\left|1+e^x\right| + c$ and since $1+e^x > 0$

$= x - \ln\left(1+e^x\right) + c$

37. $$y_{av} = \frac{\int_0^4 e^{2x}dx}{4-0} = \frac{\frac{1}{2}\int e^{2x}(2\,dx)}{4}$$

$$= \frac{\frac{1}{2}e^{2x}\Big|_0^4}{4} = \frac{1}{8}e^{2x}\Big|_0^4$$

$$= \frac{1}{8}(e^8 - 1) = 372$$

41. $qe^{t/RC} = \frac{E}{R}\int e^{t/RC}dt;\ u = \frac{t}{RC};\ du = \frac{1}{RC}dt$

$$qe^{t/RC} = RC\cdot\frac{E}{R}\int e^{t/RC}\left(\frac{1}{RC}\right)dt$$

$qe^{t/RC}EC(e^{t/RC}) + C_1$, where C_1 is the constant of integration.

$q = 0$ for $t = 0$; $0 = EC + C_1$; $C_1 = -EC$;

$$qe^{t/RC} = EC(r^{t/RC}) - EC$$

$$q = EC - \frac{EC}{e^{t/RC}};\ q = EC(1 - e^{-t/RC})$$

28.4 Basic Trigonometric Forms

1. Change $x\,dx$ to $3x^2dx$ then

$$\int \sec^2 x^3(3x^2dx) = \tan x^3 + C$$

5. $u = 3\theta;\ du = 3\,d\theta$

$$\int 0.3\sec^2 3\theta\,d\theta = 0.3\int \sec^2 3\theta(3\,d\theta)$$

$$= 0.1\tan 3\theta + C$$

9. $u = x^3;\ du = 3x^2dx$

$$\int_{0.5}^{1} x^2\cot x^3dx = \frac{1}{3}\int_{0.5}^{1}\cot x^3(3x^2dx)$$

$$= \frac{1}{3}\ln\left|\sin x^3\right|\Big|_{0.5}^{1}$$

$$= \frac{1}{3}\left(\ln|\sin 1| - \ln\left|\sin\frac{1}{8}\right|\right.$$

$$= 0.6365$$

13. $u = \frac{1}{x} = x^{-1};$

$$du = -1x^{-2}dx = -\frac{dx}{x^2}$$

$$\int\frac{\sin\left(\frac{1}{x}\right)}{x^2}dx = -\int\sin\left(\frac{1}{x}\right)\left(-\frac{dx}{x^2}\right)$$

$$= -\left[-\cos\left(\frac{1}{x}\right)\right] + C$$

$$= \cos\left(\frac{1}{x}\right) + C$$

17. $\int\sqrt{\frac{1-\cos x}{2}}dx = \int\sqrt{\sin^{-2}\frac{x}{2}}dx = \int\left|\sin\frac{x}{2}\right|dx$

$$= 2\int \sin\frac{x}{2}\left(\frac{1}{2}dx\right) \text{ for } 0 \le x < 2\pi$$

$$= -2\cos\frac{x}{2} + c$$

21. $\int\frac{2\tan T}{1-\tan^2 T}dT = \frac{1}{2}\int \tan 2T(2dT)$

$$= -\frac{1}{2}\ln|\cos 2T| + c$$

25. $\sin 3x\left(\frac{1}{\sin 3x} + \frac{1}{\cos 3x}\right) = 1 + \tan 3x;$

$u = 3x;\ du = 3\,dx$

$$\int_0^{\pi/9} \sin 3x(\csc 3x + \sec 3x)dx$$

$$= \int_0^{\pi/9} (1 + \tan 3x)dx$$

$$= \int_0^{\pi/9} dx + \int_0^{\pi/9} \tan 3x\,dx$$

$$= \int_0^{\pi/9} dx + \frac{1}{3}\int_0^{\pi/9} \tan 3x(3\,dx)$$

$$= \left(x - \frac{1}{3}\ln|\cos 3x|\right)\Big|_0^{\pi/9}$$

$$= \frac{\pi}{9} - \frac{1}{3}\ln\left|\cos\frac{\pi}{3}\right| - \left(0 - \frac{1}{3}\ln|\cos 0|\right)$$

$$= \frac{\pi}{9} - \frac{1}{3}\ln\left(\frac{1}{2}\right) = \frac{\pi}{9} + \frac{1}{3}\ln 2$$

$$= 0.580$$

29. $\int\frac{dx}{1+\sin x} = \int\frac{dx}{1+\sin x}\cdot\frac{1-\sin x}{1-\sin x}$

$$= \int\frac{(1-\sin x)}{1-\sin^2 x}dx$$

$$= \int\frac{1-\sin x}{\cos^2 x}dx$$

$$= \int\frac{1}{\cos^2 x}dx + \int\frac{-\sin x\,dx}{\cos^2 x}$$

$$= \int(\sec^2 x\,dx - \sec x\tan x)dx$$

$$= \tan x - \sec x + C$$

33. $y = \sec x;\ x = 0;\ x = \frac{\pi}{3}$

$y = 0$; rotated about x-axis, disks.

$dV = \pi r^2 dx;\ r = y$

$$V = \int_0^{\pi/3} \pi y^2 dx = \pi\int_0^{\pi/3} \sec^2 x\,dx$$

$$V = \pi\tan x\Big|_0^{\pi/3} = \pi\sqrt{3} = 5.44$$

37. $y = \tan x^2;\ y = 0, x = 1$

$$\overline{x} = \frac{\int_0^1 xy\,dx}{\int_0^1 y\,dx} = \frac{\int_0^1 (\tan x^2)x\,dx}{\int_0^1 \tan x^2 dx}$$

$u = x^2;\ du = 2x\,dx$

$$\overline{x} = \frac{\frac{1}{2}\int_0^1 \tan x^2(2x\,dx)}{0.3984} = \frac{-\frac{1}{2}\ln\left|\cos x^2\right|\Big|_0^1}{0.3984}$$

$$\overline{x} = \frac{-\frac{1}{2}(\ln\cos 1 - \ln\cos 0)}{0.3984} = \frac{0.3078}{0.3984}$$

$$= 0.7726\text{ m}$$

28.5 Other Trigonometric Forms

1. Change dx to $\cos 2x\,dx$, then

$$\begin{aligned}\int \sin^2 2x(\cos 2x\,dx) &= \frac{1}{2}\int \sin^2 2x(2\cos 2x\,dx)\\ &= \frac{1}{2}\cdot\frac{\sin^3 2x}{3}+C\\ &= \frac{\sin^3 2x}{6}+C\end{aligned}$$

5. $u = 2x$; $du = 2\,dx$; $u = \cos 2x$; $du = -2\sin 2x\,dx$

$$\begin{aligned}\int \sin^3 2x\,dx &= \int \sin^2 2x\sin 2x\,dx = \int (1-\cos^2 2x)\sin 2x\,dx = \int \sin 2x\,dx - \int \cos^2 2x\sin 2x\,dx\\ &= \frac{1}{2}\int \sin 2x(2\,dx) + \frac{1}{2}\int \cos^2 2x(-2\sin 2x\,dx) = -\frac{1}{2}\cos 2x + \frac{1}{2}\frac{\cos^3 2x}{3}+C\\ &= -\frac{1}{2}\cos 2x + \frac{1}{6}\cos^3 2x + C\end{aligned}$$

9. $u = \cos x$; $du = -\sin x\,dx$

$$\begin{aligned}\int_0^{\pi/4} 5\sin^5 x\,dx &= 5\int_0^{\pi/4} \sin x\sin^4 x\,dx = 5\int_0^{\pi/4} \sin x(1-\cos^2 x)^2 dx = 5\int_0^{\pi/4} \sin x(1-2\cos^2 x+\cos^4 x)dx\\ &= 5\left(\int \sin x\,dx - 2\int \cos^2 x\sin x\,dx + \int \cos^4 x\sin x\,dx\right) = 5(-\cos x) + 10\frac{\cos^3 x}{3} - 5\,\frac{\cos^5 x}{5}\Bigg|_0^{\pi/4}\\ &= 5\left(-\frac{1}{\sqrt{2}}\right) + \frac{10}{3}\left(\frac{1}{\sqrt{2}}\right)^3 - \left(\frac{1}{\sqrt{2}}\right)^5 - \left[5(-1)+\frac{10}{3}(1)-1\right]\\ &= -\frac{5}{\sqrt{2}} + \frac{10}{3}\frac{1}{2\sqrt{2}} - \frac{1}{4\sqrt{2}} - \left(-5+\frac{10}{3}-1\right) = \frac{-60+10(2)-3}{12\sqrt{2}} - \left[\frac{-15+10-3}{3}\right]\\ &= \frac{-43+32\sqrt{2}}{12\sqrt{2}} = \frac{64-43\sqrt{2}}{24} = 0.1329\end{aligned}$$

13. $$\begin{aligned}\int 2(1+\cos 3\phi)^2 d\phi &= \int 2(1+2\cos 3\phi+\cos^2 3\phi)d\phi = \int (2+4\cos 3\phi+2\cos^2 3\phi)d\phi\\ &= 2\phi + \frac{4\sin 3\phi}{3} + \int (1+\cos 6\phi)d\phi = 2\phi + \frac{4\sin 3\phi}{3} + \phi + \frac{1}{6}\int \cos 6\phi(6\,d\phi)\\ &= 3\phi + \frac{4\sin 3\phi}{3} + \frac{\sin 6\phi}{6} + C = \frac{1}{3}(9\phi + 4\sin 3\phi + \sin 3\phi\cos 3\phi) + C\end{aligned}$$

17. $u = \tan x$; $du = \sec^2 x\,dx$

$$\int_0^{\pi/4} \tan x \sec^4 x\,dx = \int_0^{\pi/4} \tan x \sec^2 x(1+\tan^2 x)dx = \int_0^{\pi/4} (\tan x)^1 \sec x\,dx + \int_0^{\pi/4} \tan^3 x \sec^2 x\,dx$$

$$= \frac{1}{2}(\tan x)^2 + \frac{1}{4}(\tan x)^4\Big|_0^{\pi/4} = \frac{1}{2}(1)^2 + \frac{1}{4}(1)^4 = \frac{1}{2} + \frac{1}{4} = \frac{3}{4}$$

21. $$\int 0.5 \sin s \sin 2s\,ds = \int 0.5 \sin s \cdot 2 \sin s \cos s\,ds = \int \sin^2 s \cos s\,ds = \frac{\sin^3 s}{3} + C$$

25. $$\int \frac{1-\cot x}{\sin^4 x}dx$$

$$= \int (1-\cot x)\csc^4 x\,dx = \int (1-\cot x)\csc^2 x(1+\cot^2 x)dx = \int (1-\cot x+\cot^2 x-\cot^3 x \csc^2 x\,dx)$$

$$= \int \csc^2 x\,dx - \int \cot x \csc^2 x\,dx + \int \cot^2 x \csc^2 x\,dx - \int \cot^3 x \csc^2 x\,dx$$

$$= \int \csc^2 x\,dx + \int \cot x(-\csc^2 x\,dx) - \int \cot^2 x(-\csc^2 x\,dx) + \int \cot^3 x(-\csc^2 x\,dx)$$

$$= -\cot x + \frac{\cot^2 x}{2} - \frac{\cot^3 x}{3} + \frac{\cot^4 x}{4} + C = \frac{1}{4}\cot^4 x - \frac{1}{3}\cot^3 x + \frac{1}{2}\cot^2 x - \cot x + C$$

29. $$\int \sec^6 x\,dx = \int \sec^4 x(1+\tan^2 x)dx = \int \sec^4 x\,dx + \int \sec^4 x \tan^2 x\,dx$$

$$= \int \sec^2 x(1+\tan^2 x)dx + \int \sec^2 x(1+\tan^2 x)x\,dx$$

$$= \int \sec^2 x\,dx + \int \tan^2 x \sec^2 x\,dx + \int \tan^2 x \sec^2 x\,dx + \int \tan^4 x \sec^2 x\,dx$$

$$= \tan x + \frac{1}{3}\tan^3 x + \frac{1}{3}\tan^3 x + \frac{1}{5}\tan^5 + C = \frac{1}{5}\tan^5 x + \frac{2}{3}\tan^3 x + \tan x + C$$

33. $u = e^{-x} \Rightarrow du = -e^{-x} \Rightarrow -du = \dfrac{dx}{e^x}$

$$\int \frac{\sec e^{-x}}{e^x}dx = -\int \sec u\,du = -\ln|\sec u + \tan u| + c$$

$$= -\ln\left|\sec e^{-x} + \tan e^{-x}\right| + c$$

37. Rotate about x-axis, disks

$$V = \pi\int_0^{\pi} y^2 dx = \pi\int_0^{\pi} \sin^2 x\,dx = \pi\int_0^{\pi} \frac{1}{2}(1-\cos 2x)\,dx = \frac{\pi}{2}\int_0^{\pi} dx - \frac{\pi}{2}\int_0^{\pi} \cos 2x\,dx$$

$$= \frac{\pi}{2}x - \frac{\pi}{2} \times \frac{1}{2}\sin 2x\Big|_0^{\pi} = \frac{\pi^2}{2} - \frac{\pi}{4}\sin 2\pi - 0 + \frac{\pi}{4}\sin 0 = \frac{1}{2}\pi^2 = 4.935$$

41. $\int \sin x \cos x\, dx;\ u = \sin x;\ du = \cos x\, dx$

$$\int u\, du = \frac{1}{2}u^2 + C = \frac{1}{2}\sin^2 x + C_1$$

Let $u = \cos x;\ du = -\sin x\, dx$

$$-\int \cos x(-\sin x)\, dx = -\frac{1}{2}\cos^2 x + C_2$$

$$\frac{1}{2}\sin^2 x + C_1 = \frac{1}{2}(1 - \cos^2 x + C_1)$$
$$= \frac{1}{2} - \frac{1}{2}\cos^2 x + C_1$$
$$-\frac{1}{2}\cos^2 x + C_2 = \frac{1}{2} - \frac{1}{2}\cos^2 x + C_1$$

$$C_2 = C_1 + \frac{1}{2}$$

45. $a = \sin^2 t \cos t$

$$v = \int \sin^2 t \cos t dt = \frac{\sin^3 t}{3} + v_0$$

$$v(0) = 6 = \frac{\sin^3 0}{3} + v_0 \Rightarrow v_0 = 6$$

$$v = \frac{\sin^3 t}{3} + 6$$

$$s = \int \frac{\sin^3 t}{3} dt + \int 6 dt$$

$$s = \frac{1}{3}\int \sin^2 t \sin t dt + \int 6 dt$$

$$s = \frac{1}{3}\int (1 - \cos^2 t)\sin t dt + 6\int dt$$

$$s = \frac{1}{3}\int \sin t dt - \frac{1}{3}\int \cos^2 t \sin t dt + 6\int dt$$

$$s = \frac{1}{3}\cos t + \frac{1}{9}\cos^3 t + 6t + s_0$$

$$s(0) = 0 = -\frac{1}{3}\cos 0 + \frac{1}{9}\cos^3 0 + 6(0) + s_0 \Rightarrow s_0 = \frac{2}{9}$$

$$s = -\frac{1}{3}\cos t + \frac{1}{9}\cos^3 t + 6t + \frac{2}{9}$$

49. $$V_{rms}=\sqrt{\frac{1}{1/60.0}\int_0^{1/60.0}(340\sin 120\pi t)^2dt}=\sqrt{60}\sqrt{\int_0^{1/60.0}340^2\frac{1-\cos 240\pi t}{2}dt}$$
$$=\sqrt{60}\sqrt{\frac{340^2}{2}\left(t-\frac{1}{240\pi}\sin 240\pi t\right)\Big|_0^{1/60.0}}=240\text{ V}$$

28.6 Inverse Trigonometric Forms

1. Change dx to $-x\,dx$, then

$$\int\frac{-x\,dx}{\sqrt{9-x^2}}=\frac{1}{2}\int(9-x^2)^{-1/2}(-2x\,dx)=\frac{1}{2}\frac{(9-x^2)^{1/2}}{\frac{1}{2}}+C=\sqrt{9-x^2}+C$$

5. $a=8; u=x; du=dx;$

$$\int\frac{dx}{64+x^2}=\frac{1}{8}\tan^{-1}\frac{x}{8}+C$$

9. $$\int_0^2\frac{3e^{-t}dt}{1+9e^{-2t}}$$
$$=-\int_0^2\frac{-3e^{-t}dt}{1+(3e^{-t})^2}\text{ which has form }\int\frac{dx}{1+x^2}$$
$$=-\tan^{-1}(3e^{-t})\Big|_0^2$$
$$=-\tan^{-1}\frac{3}{e^2}+\tan^{-1}3$$
$$=0.8634$$

13. $u=9x^2+16; du=18x\,dx$

$$\int\frac{8x\,dx}{9x^2+16}=\frac{8}{18}\int\frac{18x\,dx}{9x^2+16}$$
$$=\frac{4}{9}\ln\left|9x^2+16\right|+C$$

17. $a=1; u=e^x; du=e^x dx$

$$\int\frac{e^x dx}{\sqrt{1-e^{2x}}}=\sin^{-1}e^x+C$$

21. $a = 2$; $u = x + 2$; $du = dx$

$$\int \frac{4\,dx}{\sqrt{-4x - x^2}} = \int \frac{4\,dx}{\sqrt{4-(x+2)^2}}$$
$$= 4\int \frac{dx}{\sqrt{4-(x+2)^2}}$$
$$= 4\sin^{-1}\left(\frac{x+2}{2}\right) + C$$

25. $a = 2$; $u = x$; $du = dx$; $n = -\frac{1}{2}$; $u = 4 - x^2$;

$du = -2x\,dx$

$$\int \frac{2-x}{\sqrt{4-x^2}}dx = \int \frac{2\,dx}{\sqrt{4-x^2}} - \int \frac{x\,dx}{\sqrt{4-x^2}}$$
$$= 2\int \frac{dx}{\sqrt{4-x^2}} + \frac{1}{2}\int \frac{(-2x\,dx)}{(4-x^2)^{1/2}}$$
$$= 2\sin^{-1}\frac{x}{2} + \frac{1}{2}(4-x^2)^{1/2}\cdot 2 + C$$
$$= 2\sin^{-1}\frac{x}{2} + \sqrt{4-x^2} + C$$

29. $$\int \frac{x^2+3x^5}{1+x^6}dx = \int \frac{x^2}{1+x^6}dx + \int \frac{3x^5}{1+x^6}dx$$
$$\int \frac{x^2}{1+x^6}dx = \frac{1}{3}\int \frac{3x^2}{1+(x^3)^2}dx = \frac{1}{3}\tan^{-1}x^3$$
$$\int \frac{3x^5}{1+x^6}dx = \frac{1}{2}\int \frac{6x^5\,dx}{1+x^6} = \frac{1}{2}\ln\left(1+x^6\right)$$
$$\int \frac{x^2+3x^5}{1+x^6}dx = \frac{1}{3}\tan^{-1}x^3 + \frac{1}{2}\ln\left(1+x^6\right) + c$$

33. (a) General power, $\int u^{-1/2}du$ where $u = 4 - 9x^2$.

$du = -18x\,dx$; numerator can fit du of denominator.
Square root becomes $-1/2$ power.
Does not fit inverse sine form.

(b) Inverse sine; $a = 2$; $u = 3x$; $du = 3\,dx$

(c) Logarithmic; $u = 4 - 9x$; $du = -9\,dx$

37. $y = \frac{1}{1+x^2}$; $A = \int_0^2 \frac{1}{1+x^2}dx$;

$a = 1$; $u = x$; $du = dx$

$$A = \frac{1}{1}\tan^{-1}\frac{x}{1}\bigg|_0^2 = \tan^{-1}2 - \tan^{-1}0 = 1.11$$

41. $\int \frac{dx}{\sqrt{A^2-x^2}} = \int \sqrt{\frac{k}{m}}dt$; $\sin^{-1}\frac{x}{A} = \sqrt{\frac{k}{m}}t + C$

Solve for C by letting $x = x_0$ and $t = 0$.

$\sin^{-1}\frac{x_0}{A} = \sqrt{\frac{k}{m}}(0) + C$; $C = \sin^{-1}\frac{x_0}{A}$;

therefore, $\sin^{-1}\frac{x}{A} = \sqrt{\frac{k}{m}}t + \sin^{-1}\frac{x_0}{A}$

28.7 Integrations by Parts

1. $u = \sqrt{1-x}$, $dv = x\,dx$

$$du = \frac{-1}{2\sqrt{1-x}}dx, v = \frac{x^2}{2}$$

$$\int x\sqrt{1-x}\,dx = \frac{x^2}{2}\sqrt{1-x} - \int \frac{x^2}{2}\frac{-dx}{2\sqrt{1-x}}$$
$$= \frac{x^2}{2}\sqrt{1-x} + \frac{1}{4}\int \frac{x^2dx}{\sqrt{1-x}}$$

No, the substitution $u = \sqrt{1-x}$, $dv = x\,dx$ does not work since $\int \frac{x^2dx}{\sqrt{1-x}}$ is more complex than $\int x\sqrt{1-x}\,dx$.

5. $\int xe^{2x}dx$; $u = x$; $du = dx$; $dv = e^{2x}dx$

$$v = \frac{1}{2}\int e^{2x}(2\,dx) = \frac{1}{2}e^{2x}$$
$$\int xe^{2x}dx = \frac{1}{2}xe^{2x} - \frac{1}{2}\int e^{2x}dx$$
$$= \frac{1}{2}xe^{2x} - \frac{1}{4}\int e^{2x}(2\,dx)$$
$$= \frac{1}{2}xe^{2x} - \frac{1}{4}e^{2x} + C$$

9. $\int 2\tan^{-1} x\, dx$; $u = \tan^{-1} x$; $du = \frac{1}{1+x^2}dx$;

$dv = dx$;

$v = x$

$$2\int \tan^{-1} x\, dx$$

$$= 2\left(x\tan^{-1} x - \int \frac{x\, dx}{1+x^2}\right)$$

$u = x^2$; $du = 2x\, dx$

$$= 2\left(x\tan^{-1} x - \frac{1}{2}\int \frac{2x\, dx}{1+x^2}\right)$$

$$= 2x\tan^{-1} x - \ln\left|1+x^2\right| + C$$

OR

$$= 2x\tan^{-1} x - 2\ln\sqrt{1+x^2} + C$$

13. $\int x\ln x\, dx$; $u = \ln x$; $du = \frac{dx}{x}$; $dv = x\, dx$; $v = \frac{x^2}{2}$

$$\int x\ln x\, dx = \frac{1}{2}x^2\ln x - \frac{1}{2}\int x^2\frac{dx}{x}$$

$$= \frac{1}{2}x^2\ln x - \frac{1}{2}\int x\, dx$$

$$= \frac{1}{2}x^2\ln x - \frac{1}{2}\cdot\frac{x^2}{2} + C$$

$$= \frac{1}{2}x^2\ln x - \frac{1}{4}x^2 + C$$

17. $\int_0^{\pi/2} e^x\cos x\, dx$; $u = e^x$; $du = e^x dx$; $dv = \cos x\, dx$;

$v = \sin x$

$$\int_0^{\pi/2} e^x\cos x\, dx = e^x\sin x - \int \sin x e^x dx$$

$u = e^x$; $du = e^x dx$; $dv = \sin x\, dx$; $v = -\cos x$

$$\int_0^{\pi/2} e^x\cos x\, dx = e^x\sin x - \left(-e^x\cos x + \int e^x\cos x\, dx\right)$$

$$\int_0^{\pi/2} e^x\cos x\, dx = e^x\sin x + e^x\cos x - \int e^x\cos x\, dx$$

$$2\int e^x\cos x\, dx = e^x\sin x + e^x\cos x$$

$$\int_0^{\pi/2} e^x\cos x\, dx = \frac{1}{2}e^x(\sin x + \cos x)\Big|_0^{\pi/2} = \frac{1}{2}e^{\pi/2}(1+0) - \frac{1}{2}e^0(0+1)$$

$$= \frac{1}{2}e^{\pi/2} - \frac{1}{2} = \frac{1}{2}(e^{\pi/2} - 1) = 1.91$$

21. $u = \cos(\ln x)$, $dv = dx$

$v = x$

$$du = -\sin(\ln x)\cdot\frac{1}{x}dx$$

$$\int\cos(\ln x)\, dx = x\cos(\ln x) + \int x\sin(\ln x)\frac{1}{x}dx$$

$u = \sin(\ln x)$, $dv = dx$

$v = x$

$$du = \cos(\ln x)\cdot\frac{1}{x}dx$$

$$\int\cos(\ln x)\, dx = x\cos(\ln x) + x\sin(\ln x) - \int x\cos(\ln x)\cdot\frac{1}{x}dx$$

$$2\int\cos(\ln x)\, dx = x\cos(\ln x) + x\sin(\ln x)$$

$$\int\cos(\ln x)\, dx = \frac{x}{2}(\cos(\ln x) + \sin(\ln x)) + c$$

25. $A = \int_0^2 xe^{-x}dx;\ u = x;\ du = dx;\ dv = e^{-x}dx;$

$v = \int e^{-x}dx = -e^{-x}$

$A = -xe^{-x}\Big|_0^2 = \int_0^2 -e^{-x}dx = -xe^{-x} - e^{-x}\Big|_0^2$

$= 2e^{-2} - e^{-2} - (0-1) = 1 - \dfrac{3}{e^2} = 0.594$

29. $\bar{x} = \dfrac{\int_0^{\pi/2} x(\cos x)\,dx}{\int_0^{\pi/2} \cos x\,dx}$

Let $u = x;\ du = dx;\ dv = \cos x;\ v = \sin x$

$\bar{x} = \dfrac{x\sin x\big|_0^{\pi/2} - \int_0^{\pi/2} \sin x\,dx}{\sin x\big|_0^{\pi/2}}$

$= \dfrac{x\sin x\big|_0^{\pi/2} - (-\cos x)_0^{\pi/2}}{1}$

$= x\sin x + \cos x\big|_0^{\pi/2} = \dfrac{\pi}{2} - 1 = 0.571$

33. $v = \dfrac{ds}{dt} = \dfrac{t^3}{\sqrt{t^2+1}};\ s = \int \dfrac{t^3dt}{\sqrt{t^2+1}}$

Let $u = t^2;\ du = 2t\,dt;\ dv = \dfrac{t\,dt}{(t^2+1)^{1/2}}$

$v = \dfrac{1}{2}\int \dfrac{2t\,dt}{(t^2+1)^{1/2}} = \dfrac{1}{2}(2)(t^2+1)^{1/2} = (t^2+1)^{1/2}$

$s = t^2(t^2+1)^{1/2} - \int (t^2+1)^{1/2}(2t\,dt) = t^2(t^2+1)^{1/2} - \dfrac{2}{3}(t^2+1)^{3/2} + C$

$s = 0$ for $t = 0$; $0 = -\dfrac{2}{3} + C;\ C = \dfrac{2}{3}$

$s = \dfrac{1}{3}[3t^2(t^2+1)^{1/2} - 2(t^2+1)^{3/2} + 2] = \dfrac{1}{3}[(t^2-2)(t^2+1)^{1/2} + 2]$

28.8 Integration by Trigonometric Substitution

1. Delete the x^2 before the radical in the denominator.

$\int \dfrac{dx}{\sqrt{1-x^2}} = \sin^{-1} x + C$

5. $\int \dfrac{dx}{x^2\sqrt{x^2+1}}$. Let $x = \tan\theta,\ dx = \sec\theta\tan\theta\,d\theta$

$\int \dfrac{dx}{x^2\sqrt{x^2+1}} = \int \dfrac{\sec\theta\tan\theta\,d\theta}{\tan^2\theta\sqrt{\tan^2\theta+1}}$

$= \int \dfrac{\sec\theta\tan\theta\,d\theta}{\tan^2\theta\sqrt{\sec^2\theta}} = \int \dfrac{\sec\theta\,d\theta}{\tan\theta\sec\theta}$

$= \int \dfrac{d\theta}{\tan\theta} = \int \cot\theta\,d\theta$

9. Let $x = \sin\theta;\ dx = \cos\theta\,d\theta$

$\int \dfrac{\sqrt{1-x^2}}{x^2}dx = \int \dfrac{\sqrt{1-\sin^2\theta}}{\sin^2\theta}\cos\theta\,d\theta = \int \dfrac{\cos^2\theta}{\sin^2\theta}d\theta$

$= \int \cot^2\theta\,d\theta = \int (\csc^2\theta - 1)\,d\theta$

$= \int \csc^2\theta\,d\theta - \int d\theta = -\cot\theta - \theta + C$

$= \dfrac{-\sqrt{1-x^2}}{x} - \sin^{-1} x + C$

13. Let $z = 3\tan\theta$; $dz = 3\sec^2\theta\, d\theta$

$$\int \frac{6\,dz}{z^2\sqrt{z^2+9}} = 6\int \frac{3\sec^2\theta\, d\theta}{9\tan^2\theta\sqrt{9\tan^2\theta+9}} = 6\int \frac{3\sec^2\theta\, d\theta}{27\tan^2\theta\sqrt{\tan^2\theta+1}} = \frac{6}{9}\int \frac{\sec\theta\, d\theta}{\tan^2\theta} = \frac{6}{9}\int \frac{\cos\theta\, d\theta}{\sin^2\theta}$$

$$= \frac{6}{9}\int \csc\theta\cot\theta\, d\theta = -\frac{6}{9}\csc\theta + C = \frac{-6}{9\sin\theta} + C$$

$\tan\theta = \frac{z}{3}$; $\sin\theta = \frac{z}{\sqrt{9+z^2}}$

$$\frac{-6}{9\sin\theta} + C = \frac{-6}{\frac{9z}{\sqrt{9+z^2}}} + C = -\frac{2\sqrt{z^2+9}}{3z} + C$$

17. $\int_0^{0.5} \frac{x^3\,dx}{\sqrt{1-x^2}}$, $x = \sin\theta$; $dx = \cos\theta\, d\theta$

$$\int \frac{\sin^3\theta\cos\theta\, d\theta}{\sqrt{1-\sin^2\theta}} = \int \sin^3\theta\, d\theta = \int \sin\theta\sin^2\theta\, d\theta = \int \sin\theta(1-\cos^2\theta)\, d\theta = \int \sin\theta\, d\theta - \int \cos^2\theta\sin\theta\, d\theta$$

$$= -\cos\theta + \frac{\cos^3\theta}{3}$$

21. $\int \frac{dy}{y\sqrt{4y^2-9}}$; $2y = 3\sec\theta$; $y = \frac{3}{2}\sec\theta$; $dy = \frac{3}{2}\sec\theta\tan\theta\, d\theta$

$$\int \frac{\frac{3}{2}\sec\theta\tan\theta\, d\theta}{\frac{3}{2}\sec\theta\sqrt{4\left(\frac{3}{2}\sec\theta\right)^2-9}} = \int \frac{\tan\theta\, d\theta}{\sqrt{9\sec^2\theta-9}} = \int \frac{\tan\theta\, d\theta}{3\sqrt{\sec^2\theta-1}} = \int \frac{\tan\theta\, d\theta}{3\tan\theta} = \frac{1}{3}\int d\theta$$

$$= \frac{1}{3}\theta + C = \frac{1}{3}\sec^{-1}\frac{2}{3}y + C$$

$$\int_{2.5}^{3} \frac{dy}{y\sqrt{4y^2-9}} = \frac{1}{3}\sec^{-1}\left(\frac{2y}{3}\right)\Bigg|_{2.5}^{3} = \frac{1}{3}\cos^{-1}\left(\frac{3}{2y}\right)\Bigg|_{2.5}^{3} = 0.03997$$

25. (a) $\int x\sqrt{1-x^2}\,dx = -\frac{1}{2}\int (1-x^2)^{1/2}(-2x\,dx)$

$$= -\frac{1}{2}\frac{(1-x^2)^{3/2}}{3/2} + C$$

$$= -\frac{1}{3}(1-x^2)^{3/2} + C$$

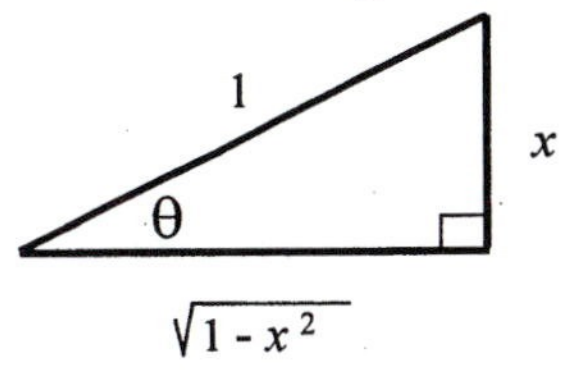

$x = \sin\theta$

$dx = \cos\theta\, d\theta$

$$\int x\sqrt{1-x^2}\,dx = \int \sin\theta\sqrt{1-\sin^2\theta}\cos\theta d\theta$$
$$= -\int \cos^2\theta(-\sin\theta d\theta)$$
$$= -\frac{\cos^3\theta}{3} + C$$
$$= -\frac{\left(1-x^2\right)^{3/2}}{3} + C$$

29. $x^2 + y^2 = a^2$; $y = \sqrt{a^2 - x^2}$

$$I_y = \int_0^a kx^2 y\,dx = k\int_0^a x^2\sqrt{a^2-x^2}dx$$

$x = a\sin\theta$; $dx = a\cos\theta\, d\theta$

$$I_y = k\int a^2\sin^2\theta\sqrt{a^2 - a^2\sin^2\theta}a\cos\theta\,d\theta = ka^4\int \sin^2\theta\cos^2\theta\,d\theta = ka^4\int \sin^2\theta(1-\sin^2\theta)d\theta$$
$$= ka^4\left(\int \sin^2\theta - \sin^4\theta)d\theta = ka^4\left[\frac{1}{2}\int(1-\cos 2\theta)d\theta - \int\frac{1}{2}^2(1-\cos\theta)^2 d\theta\right]$$
$$= ka^4\left[\frac{1}{2}d\int \theta - \frac{1}{2}\int \cos 2\theta\,d\theta - \int\frac{1}{4}(1-2\cos 2\theta + \cos^2 2\theta)d\theta\right]$$
$$= ka^4\left[\frac{1}{2}\int d\theta - \frac{1}{2}\int \cos 2\theta\,d\theta - \frac{1}{4}\int d\theta + \frac{1}{2}\int \cos 2\theta - \frac{1}{4}\cos^2 2\theta\,d\theta\right]$$
$$= ka^4\left[\frac{1}{2}\int d\theta - \frac{1}{4}d\theta - \frac{1}{4}\int\frac{1}{2}(1+\cos 4\theta)d\theta\right] = ka^4\left[\frac{1}{4}\int d\theta - \frac{1}{8}\int d\theta - \frac{1}{4}\int \cos 4\theta\,d\theta\right]$$
$$= ka^4\left[\frac{1}{8}\int d\theta - \frac{1}{32}\int \cos 4\theta\,d\theta\right]$$
$$I_y = ka^4\left[\frac{1}{8}\theta - \frac{1}{32}\sin 4\theta\right]$$

$\sin 4A = 2\sin 2A\cos 2A = 2[2\sin A\cos A(1-2\sin^2 A)]$

$$I_y = ka^4\left[\frac{1}{8}\sin^{-1}\frac{x}{a} - \frac{1}{32}(4\sin\theta\cos\theta)(1-2\sin^2\theta)\right] = ka^4\left[\frac{1}{8}\sin^{-1}\frac{x}{a} - \frac{1}{8}\cdot\frac{x}{a}\cdot\frac{\sqrt{a^2-x^2}}{a}\left(1-2\frac{x^2}{a^2}\right)\right]\Bigg|_0^a$$
$$= ka^4\left[\frac{1}{8}\sin^{-1}1\right] = \frac{1}{8}ka^4\frac{\pi}{2}$$

33. $\bar{x} = 0$ by symmetry.

37. $u = \sqrt{x+1} \Rightarrow u^2 = x+1 \Rightarrow x = u^2 - 1$

$$du = \frac{1}{2\sqrt{x+1}}dx$$

$2udu = dx$

$$\int x\sqrt{x+1}d = \int\left(u^2-1\right)u\left(2udu\right) = \int\left(2u^4-2u^2\right)du$$
$$= \frac{2u^5}{5} - \frac{2u^3}{3} + C$$
$$= \frac{2(x+1)^{5/2}}{5} - \frac{2(x+1)^{3/2}}{3} + C$$

28.9 Integration by Partial Fractions: Nonrepeated Linear Factors

1. $$\frac{10-x}{x^2+x-2} = \frac{10-x}{(x-1)(x+2)} = \frac{A}{x-1} + \frac{B}{x+2}$$
$$10 - x = A(x+2) + B(x-1)$$

for $x = -2$:

$$10 - (-2) = A(-2+2) + B(-2-1)$$
$$12 = -3B$$
$$B = -4$$

for $x = 1$:

$$10 - 1 = A(1+2) + B(1-1)$$
$$9 = 3A$$
$$A = 3$$

$$\frac{10-x}{x^2+x-2} = \frac{3}{x-1} + \frac{-4}{x+2}$$

5. $$\frac{x^2-6x-8}{x^3-4x} = \frac{x^2-6x-8}{x(x^2-4)}$$
$$= \frac{x^2-6x-8}{x(x+2)(x-2)}$$
$$= \frac{A}{x} + \frac{B}{x+2} + \frac{C}{x-2}$$

9. $$\int\frac{dx}{x^2-4} = \int\frac{dx}{(x+2)(x-2)}$$
$$= \int\frac{-\frac{1}{4}}{x+2}dx + \int\frac{\frac{1}{4}}{x-2}dx$$
$$= -\frac{1}{4}\ln|x+2| + \frac{1}{4}\ln|x-2| + C$$
$$= \frac{1}{4}\ln\left|\frac{x-2}{x+2}\right| + C$$

13. $$\int_0^1 \frac{2t+4}{3t^2+5t+2}dt$$
$$= \int_0^1 \frac{8}{3t+2}dt - \int_0^1 \frac{2}{t+1}dt$$
$$= \frac{8\ln(3t+2)}{3}\Bigg|_0^1 - 2\ln(t+1)\Bigg|_0^1$$
$$= \frac{8\ln 5}{3} - \frac{8\ln 2}{3} - 2\ln 2 + 2\ln 1$$
$$= 1.057$$

17. $$\int\frac{6x^2-2x-1}{4x^3-x}dx$$
$$= \int\frac{6x^2-2x-1}{x(4x^2-1)}dx = \int\frac{6x^2-2x-1}{x(2x+1)(2x-1)}dx$$
$$= \int\frac{dx}{x} + \int\frac{\frac{3}{2}}{2x+1}dx - \frac{\frac{1}{2}}{2x-1}dx$$
$$= \ln|x| + \frac{3\ln|2x+1|}{4} - \frac{\ln|2x-1|}{4} + C$$
$$= \frac{4\ln|x|}{4} + \frac{3\ln|2x+1|}{4} - \frac{\ln|2x-1|}{4} + C$$
$$= \frac{1}{4}\ln\left|\frac{x^4(2x+1)^3}{2x-1}\right| + C$$

21. $$\int\frac{dV}{(V^2-4)(V^2-9)}$$
$$= \int\frac{dV}{(V-2)(V+2)(V+3)(V-3)}$$
$$= \int\frac{\frac{1}{30}}{V-3}dV - \int\frac{\frac{1}{30}dV}{V+3} - \int\frac{\frac{1}{20}dV}{V-2} + \int\frac{\frac{1}{20}dV}{V+2}$$
$$= \frac{1}{30}\ln|V-3| - \frac{1}{30}\ln|V+3|$$
$$-\frac{1}{20}\ln|V-2| + \frac{1}{20}\ln|V+2| + C$$

$$= \frac{2}{60}\ln|V-3| - \frac{2}{60}\ln|V+3| - \frac{3}{60}\ln|V-2|$$

$$+ \frac{3}{60}\ln|V+2| + C$$

$$= \frac{1}{60}\ln\left|\frac{(V+2)^3(V-3)^2}{(V-2)^3(V+3)^2}\right| + C$$

25. $\frac{1}{u(a+bu)} = \frac{A}{u} + \frac{B}{a+bu}$

$$1 = A(a+bu) + Bu$$

for $u = 0$: $1 = aA$

$$A = \frac{1}{a}$$

for $u = -\frac{a}{b}$: $1 = B\left(-\frac{a}{b}\right)$

$$B = -\frac{b}{a}$$

$$\int \frac{du}{u(a+bu)}$$

$$= \int \frac{\frac{1}{a}}{u}\,du + \int \frac{-\frac{b}{a}}{a+bu}\,du$$

$$= \frac{1}{a}\ln|u| - \frac{1}{a}\int \frac{b\,du}{a+bu}$$

$$= \frac{1}{a}\ln|u| - \frac{1}{a}\ln|a+bu| + C$$

$$= -\frac{1}{a}(-\ln|u| + \ln|a+bu| + C$$

$$= -\frac{1}{a}\ln\frac{|a+bu|}{|u|} + C$$

$$= -\frac{1}{a}\ln\left|\frac{a+bu}{u}\right| + C$$

29. $V = \int_1^3 \frac{2\pi x\,dx}{(x^3+3x^2+2x)}$

$$= \int_1^3 \frac{2\pi\,dx}{x+1} - \int_1^3 \frac{2\pi\,dx}{x+2}$$

$$V = 2\pi\ln(x+1)|_1^3 - 2\pi\ln(x+2)|_1^3$$

$$= 2\pi[\ln 4 - \ln 2 - \ln 5 + \ln 3]$$

$$V = 2\pi\ln\frac{4\cdot 3}{2\cdot 5} = 2\pi\ln\frac{6}{5}$$

$$= 1.146$$

33. $w = \int_0^{0.5} \frac{4x\,dx}{x^2+3x+2} = 0.1633\ \text{N}\cdot\text{cm}$

28.10 Integration by Partial Fractions: Other Cases

1. $\frac{2}{x(x+3)^2} = \frac{A}{x} + \frac{B}{x+3} + \frac{C}{(x+3)^2}$

5. $\int \frac{x-8}{x^3-4x^2+4x} = \int \frac{-2}{x}\,dx + \int \frac{2}{x-2}\,dx - \int \frac{3}{(x-2)^2}\,dx$

$$= -2\ln|x| + 2\ln|x-2| + \frac{3}{x-2} + C$$

$$= 2\ln\left|\frac{x-2}{x}\right| + \frac{3}{x-2} + C$$

9. $\frac{2s}{(s-3)^3} = \frac{A}{s-3} + \frac{B}{(s-3)^2} + \frac{C}{(s-3)^3}$

$$= \frac{A(s-3)^2 + B(s-3) + C}{(s-3)^3}$$

$$2s = A(s-3)^2 + B(s-3) + C$$

$$s = 3,\ 6 = C$$

$$2s = A(s-3)^2 + B(s-3) + 6$$

$$\left.\begin{array}{l} s = 1,\quad 2 = 4A - 2B + 6 \\ s = 2,\quad 4 = A - B + 6 \end{array}\right\} A = 0, = 2$$

$$\int_1^2 \frac{2s}{(s-3)}\,ds = \int_1^2 \frac{2}{(s-3)^3}\,ds + \int_1^2 \frac{6}{(s-3)^3}\,dx$$

$$= 2\cdot\frac{(s-3)^{-2+1}}{-2+1} + 6\cdot\frac{(s-3)^{-3+1}}{-3+1}\Bigg|_1^2$$

$$= \frac{-2}{(s-3)} - \frac{3}{(s-3)^2}\Bigg|_1^2 = -\frac{5}{4}$$

13. $$\frac{x^2+x+5}{(x+1)(x^2+4)} = \frac{A}{x+1} + \frac{Bx+C}{x^2+4} = \frac{A(x^2+4)+(Bx+C)(x+1)}{(x+1)(x^2+4)}$$

$$x^2+x+5 = A(x^2+4)+(Bx+C)(x+1)$$

$$\left.\begin{array}{lr} x=0, & 5 = 4A + \phantom{2B+{}} C \\ x=1, & 7 = 5A+2B+2C \\ x=2, & 11 = 8A+6B+3C \end{array}\right\} A=1,\ B=0,\ C=1$$

$$\int_0^2 \frac{x^2+x+5}{(x+1)(x^2+4)}\,dx = \int_0^2 \frac{dx}{x+1} + \int_1^2 \frac{1}{x^2+4}\,dx\ \ln|x+1| + \left.\frac{\tan^{-1}\frac{x}{2}}{2}\right|_0^2 = 1.491$$

17. $$\frac{10x^3+40x^2+22x+7}{(4x^2+1)(x^2+6x+10)} = \frac{Ax+B}{4x^2+1} + \frac{Cx+D}{x^2+6x+10}$$

$$10x^3+40x^2+22x+7 = (Ax+B)(x^2+6x+10)+(Cx+D)(4x^2+1)$$
$$10x^3+40x^2+22x+7 = Ax^3+6Ax^2+10Ax+Bx^2+6Bx+10B+4Cx^3+Cx+4Dx^2+D$$
$$10x^3+40x^2+22x+7 = (A+4C)x^3+(6A+B+4D)x^2+(10A+6B+C)x+10B+D$$

$$\left.\begin{array}{ll} (1)\ A+4C & = 10 \\ (2)\ 6A+B+4D & = 40 \\ (3)\ 10A+6B+C & = 22 \\ (4)\ 10B+D & = 7 \end{array}\right\} A=2, B=0, C=2, D=7$$

$$\int \frac{10x^3+40x^2+22x+7}{(4x^2+1)(x^2+6x+10)}\,dx = \int \frac{2x}{4x^2+1}\,dx + \int \frac{2x+7}{x^2+6x+10}\,dx$$

$$\int \frac{2x}{4x^2+1}\,dx = \frac{1}{4}\int \frac{8x}{4x^2+1}\,dx = \frac{\ln(4x^2+1)}{4}$$

$$\int \frac{2x+7}{x^2+6x+10}\,dx = \int \frac{2x+7}{x^2+6x+9+1}\,dx = \int \frac{2x+7}{(x+3)^2+1}\,dx \quad \text{let} \quad u = x+3,\ x = u-3, \quad du = dx$$

$$\int \frac{2x+7}{x^2+6x+10}\,dx = \int \frac{2(u-3)+7}{u^2+1}\,du$$

$$\int \frac{2x+7}{x^2+6x+10}\,dx = \int \frac{2u-6+7}{u^2+1}\,du$$

$$\int \frac{2x+7}{x^2+6x+10}\,dx = \int \frac{2u}{u^2+1}\,du + \int \frac{1}{u^2+1}\,du$$

$$\int \frac{2x+7}{x^2+6x+10}\,dx = \ln(u^2+1) + \tan^{-1}u + C$$

$$\int \frac{2x+7}{x^2+6x+10}\,dx = \ln(x^2+6x+10) + \tan^{-1}(x+3) + C$$

$$\int \frac{10x^3+40x^2+22x+7}{(4x^2+1)(x^2+6x+10)}\,dx = \frac{\ln(4x^2+1)}{4} + \ln(x^2+6x+10) + \tan^{-1}(x+3) + C$$

21. $\frac{x}{(x-2)^3}=\frac{A}{(x-2)}+\frac{B}{(x-2)^2}+\frac{C}{(x-2)^3}=\frac{A(x-2)^2+B(x-2)+C}{(x-2)^3}$

$x=A(x^2-4x+4)+B(x-2)+C$

$x=Ax^2-4Ax+4A+Bx-2B+C$

$x^2:\ A=0$

$x:\ 1=-4A+B\Rightarrow B=1$

Constant: $0=4A-2B+C\Rightarrow 0=-2+C\Rightarrow C=2$

$$\int\frac{xdx}{(x-2)^3}=\int\frac{dx}{(x-2)^2}+\int\frac{2dx}{(x-2)^3}$$
$$=\frac{(x-2)^{-2+1}}{-2+1}+2\frac{(x-2)^{-3+1}}{-3+1}+C$$
$$=\frac{-1}{(x-2)}-\frac{1}{(x-2)^2}+C$$
$$=\frac{-1(x-2)-1}{(x-2)^2}+C$$
$$=\frac{1-x}{(x-2)^2}+C$$

25. $V=\int_0^2 2\pi x\cdot\frac{4}{x^4+6x^2+5}dx=\pi\ln\frac{25}{9}=3.210$

29. $\bar{x}=\frac{\int_1^2 x\cdot\frac{4}{x^3+x}dx}{\int_1^2\frac{4}{x^3+x}dx}=\frac{\pi-4\tan^{-1}\frac{1}{2}}{2\ln\frac{8}{5}}$

$\bar{x}=1.369$

28.11 Integration by Use of Tables

1. $\int\frac{xdx}{(2+3x)^2}$ is formula 3 with $u=x$, $du=dx$, $a=2$, $b=3$.

5. $u=x^2$, $du=2xdx$

$$\int\frac{xdx}{(4-x^4)^{3/2}}=\int\frac{xdx}{\left(2^2-(x^2)^2\right)^{3/2}}=\int\frac{\frac{1}{2}du}{(2^2-u^2)^{3/2}}$$

which is formula 25 with $a=2$.

9. Formula #1; $u=x$; $a=2$; $b=5$; $du=dx$

$$\int\frac{3xdx}{2+5x}=3\int\frac{3xdx}{2+5x}$$
$$=3\left\{\frac{1}{25}(2+5x)-2\ln|2+5x|\right\}+C$$
$$=\frac{3}{25}\left[2+5x-2\ln|2+5x|\right]+C$$

13. Formula #24; $u = y$, $a = 2$

$$\int \frac{dy}{\left(y^2+4\right)^{3/2}} = \frac{y}{4\sqrt{y^2+4}} + C$$

17. Formula #17; $u = 2x$; $du = 2dx$; $a = 3$

$$\int \frac{\sqrt{4x^2-9}}{x}dx = \int \frac{\sqrt{(2x)^2-3^2}}{2x}dx$$

$$= \sqrt{4x^2-9} - 3\sec^{-1}\left(\frac{2x}{3}\right) + C$$

21. Formula #52; $u = r^2$; $du = 2r\,dr$

$$6\int \tan^{-1} r^2 (r\,dr)$$

$$= 3\int \tan^{-1} r^2 (2r\,dr)$$

$$= 3\left[r^2 \tan^{-1} r^2 - \frac{1}{2}\ln(1+r^4)\right] + C$$

$$= 3r^2 \tan^{-1} r^2 - \frac{3}{2}\ln(1+r^4) + C$$

25. Formula #11; $u = 2x$; $du = 2\,dx$; $a = 1$

$$\int \frac{dx}{x\sqrt{4x^2+1}} = \int \frac{2\,dx}{2x\sqrt{(2x)^2+1^2}}$$

$$= -\ln\left(\frac{1+\sqrt{4x^2+1}}{2x}\right) + C$$

29. Formula #40; $a = 1$; $u = x$; $du = dx$; $b = 5$

$$\int_0^{\pi/12} \sin\theta \cos 5\theta \, d\theta = -\frac{\cos(-4\theta)}{2(-4)} - \frac{\cos 6\theta}{12}$$

$$= \frac{1}{8}\cos 4\theta - \frac{1}{12}\cos 6\theta \bigg|_0^{\pi/12}$$

$$= 0.0208$$

33. let $u = x^2$, $du = 2x\,dx$
$u^2 = x^4$

$$\int \frac{2x\,dx}{(1-x^4)^{3/2}} = \int \frac{du}{(1-u^2)^{3/2}}$$

Formula #25: $a = 1$

$$\int \frac{2x\,dx}{(1-x^4)^{3/2}} = \frac{u}{\sqrt{1-u^2}} + C;$$

$$\int \frac{2x\,dx}{(1-x^4)^{3/2}} = \frac{x^2}{\sqrt{1-x^4}} + C$$

37. Formula #46; $u = x^2$; $du = 2x\,dx$; $n = 1$

$$\int x^3 \ln x^2 dx = \frac{1}{2}\int x^2 \ln x^2 (2x\,dx)$$

$$= \frac{1}{2}\left[(x^2)^2\left(\frac{\ln x^2}{2} - \frac{1}{4}\right)\right]$$

$$= \frac{1}{2}\left[\frac{x^4}{2}\left(\ln x^2 - \frac{1}{2}\right)\right]$$

$$= \frac{1}{4}x^4\left(\ln x^2 - \frac{1}{2}\right) + C$$

41. Let $u = t^3$, $du = 3t^2\,dt$

$$\int t^2\left(t^6+1\right)^{3/2} dt = \int \frac{1}{3}\left(u^2+1\right)^{3/2} du, \ \#19, \ a = 1$$

$$= \frac{1}{3}\left[\frac{u}{4}\left(u^2+1\right)^{3/2} + \frac{3u}{8}\sqrt{u^2+1} + \frac{3}{8}\ln\left(u+\sqrt{u^2+1}\right)\right]$$

$$= \frac{1}{12}t^3\left(t^6+1\right)^{3/2} + \frac{t^3}{8}\sqrt{t^6+1} + \frac{1}{8}\ln\left(t^3+\sqrt{t^6+1}\right) + C$$

45. From Exercise 17 of Section 26-6,

$$s = \int_a^b \sqrt{1+\left(\frac{dy}{dx}\right)^2}\,dx; \ y = x^2; \ \frac{dy}{dx} = 2x;$$

$$s = \int_0^1 \sqrt{1+(2x)^2}\,dx)$$

$$= \frac{1}{2}\int_0^1 \sqrt{(2x)^2+1}(2dx$$

Formula #14; $u = 2x$; $du = 2\,dx$

$$s=\frac{1}{2}\left[\frac{2x}{2}\sqrt{4x^2+1}+\frac{1}{2}\ln(2x+\sqrt{4x^2+1})\right]\Bigg|_0^1$$

$$=\frac{1}{2}\left[\left(1\sqrt{5}+\frac{1}{2}\ln(2+\sqrt{5})\right)-\frac{1}{2}\ln 1\right]$$

$$=\frac{1}{4}[2\sqrt{5}+\ln(2+\sqrt{5})]=1.479$$

49. $F=w\int_0^3 lhdh=w\int_0^3 x(3-y)\,dy=w\int_0^3\frac{3-y}{\sqrt{1+y}}dy$

Formula #6

$$\int\frac{3-y}{\sqrt{1+y}}dy=3\int\frac{dy}{\sqrt{1+y}}-\int\frac{ydy}{\sqrt{1+y}}$$

$$=3\frac{(1+y)^{1/2}}{\frac{1}{2}}-\left[\frac{-2(2-y)\sqrt{1+y}}{3(1)^2}\right]+C$$

$$F=w\int_0^3\frac{3-y}{\sqrt{1+y}}dy$$

$$=w\left[6(1+y)^{1/2}+\frac{2}{3}(2-y)(1+y)^{1/2}\right]\Bigg|_0^3$$

$$=w\left[6(2)+\frac{2}{3}(-1)(2)-6(1)-\frac{2}{3}(2)(1)\right]$$

$$F=w\left(12-\frac{4}{3}-6-\frac{4}{3}\right)=\frac{10w}{3}=\frac{10(62.4)}{3}=208\text{ lb}$$

Chapter 28 Review Exercises

1. $u=-8x,\ du=-8dx$

$$\int e^{-8x}dx=-\frac{1}{8}\int e^{-8x}(-8dx)=-\frac{1}{8}e^{-8x}+C$$

5. $\int_0^{\pi/2}\frac{4\cos\theta\ d\theta}{1+\sin\theta}=4\int_0^{\pi/2}\frac{\cos\theta\ d\theta}{1+\sin\theta};\ u$

$=1+\sin\theta,$

$du=\cos\theta$

$$=4\ln(1+\sin\theta)\Big|_0^{\pi/2}$$

$$=2.77$$

9. $\int_0^{\pi/2}\cos^3 2\theta\ d\theta=\int_0^{\pi/2}\cos^2 2\theta\cos 2\theta\ d\theta$

$$=\int_0^{\pi/2}\left(-\sin^2 2\theta\right)\cos 2\theta\ d\theta$$

$$=\int_0^{\pi/2}\cos 2\theta\ d\theta-\int_0^{\pi/2}\sin^2 2\theta\cos 2\ d\theta$$

$$=\frac{1}{2}\int_0^{\pi/2}\cos 2\theta(2d\theta)-\frac{1}{2}\int_0^{\pi/2}\sin^2 2\theta\cos 2\theta(2d\theta)$$

$$=\frac{1}{2}\left[\sin 2\theta-\frac{1}{3}\sin^3 2\theta\right]\Bigg|_0^{\pi/2}$$

$$=\frac{1}{2}\left[\left(\sin\pi-\frac{1}{3}\sin^3\pi\right)=\left(\sin 0-\frac{1}{3}\sin^3 0\right)\right]$$

$$=\frac{1}{2}(0)=0$$

13. $\int(\sin t+\cos t)^2\cdot\sin t\,dt$

$$=\int\left(\sin^2 t+2\sin t\cos t+\cos^2 t\right)\cdot\sin t dt$$

$$=\int\left(1+2\sin t\cos t\right)\cdot\sin t dt$$

$$=\int\left(\sin t+2\sin^2 t\cos t\right)dt$$

$$=\int\sin t\,dt+2\int\sin^2 t(\cos t\,dt)$$

$$=-\cos t+\frac{2\sin^3 t}{3}+C$$

17. $\int\sec^4 3xdx=\int\sec^2 3x\sec^2 3xdx$

$$=\int\left(1+\tan^2 3x\right)\sec^2 3xdx$$

$$=\frac{1}{3}\int\sec^2 3x(3dx)+\frac{1}{3}\int\tan^2 3x\sec^2 3x(3dx)$$

$$=\frac{1}{3}\tan 3x+\frac{1}{3}\frac{\tan^3 3x}{3}+C$$

$$=\frac{1}{9}\tan^3 3x+\frac{1}{3}\tan 3x+C$$

21. $\int\frac{3xdx}{4+x^4}=3\int\frac{xdx}{4+x^4}=3\int\frac{1}{2^2+\left(x^2\right)^2}xdx$

$$=\frac{3}{2}\int\frac{1}{2^2+\left(x^2\right)^2}2xdx$$

$$=\frac{3}{2}\left(\frac{1}{2}\tan^{-1}\frac{x^2}{2}+C_1\right)$$

$$=\frac{3}{4}\tan^{-1}\frac{x^2}{2}+C\text{ where }C=\frac{3}{2}C_1.$$

25. $u = e^{2x},\ du = e^{2x}(2dx)$

$$\int \frac{e^{2x}dx}{\sqrt{e^{2x}+1}} = \frac{1}{2}\int \left(e^{2x}+1\right)^{-1/2} e^{2x}(2dx)$$
$$= \frac{1}{2}\left(e^{2x}+1\right)^{1/2}(2) + C = \sqrt{e^{2x}+1} + C$$

29. $$\int_0^{\pi/6} 3\sin^2 3\phi\, d\phi = \int_0^{\pi/6} 3\cdot\frac{(1-\cos 6\phi)}{2}d\phi$$
$$= \int_0^{\pi/6}\frac{3}{2}d\phi - \frac{1}{4}\int_0^{\pi/6}\cos 6\phi(6d\phi)$$
$$= \frac{3}{2}\phi\Big|_0^{\pi/6} - \frac{1}{4}\sin 6\phi\Big|_0^{\pi/6}$$
$$= \frac{3}{2}\left[\frac{\pi}{6}-0\right] - \frac{1}{4}[\sin\pi - \sin 0] = \frac{\pi}{4}$$

33. $$\frac{3u^2-6u-2}{u^2(3u+1)} = \frac{Au+B}{u^2} + \frac{C}{3u+1}$$
$$= \frac{(Au+B)(3u+1)+Cu^2}{u^2(3u+1)}$$

$3u^2-6u-2 = 3Au^2 + Au + 3Bu + B + Cu^2$

$3u^2-6u-2 = (3A+C)u^2 + (A+3B)u + B$

(1) $3A+C=3,\ 3(0)+C=3,\ C=3$

(2) $A+3B=-6;\ A+3(-2)=-6,\ A=0$

(3) $B=-2$

$$\int\frac{3u^2-6u-2}{u^2(3u+1)}du = \int\frac{-2}{u^2}du + \int\frac{3}{3u+1}du$$
$$= \frac{2}{u} + \ln|3u+1| + C$$

37. $$\int_1^e 3\cos(\ln x)\cdot\frac{dx}{x} = 3\sin(\ln x)\Big|_1^e$$
$$= 3\sin(\ln e) - 3\sin(\ln 1)$$
$$= 3\sin(1) - 3\sin(0)$$
$$= 3\sin 1 - 3\cdot 0 = 3\sin 1 \approx 2.52$$

41. $u = \cos x,\ du = -\sin x dx$

$$\int\frac{\sin x\cos^2 s}{5+\cos^2 x}dx = -\int\frac{u^2du}{5+u^2} = -\int\left(1-\frac{5}{5+u^2}\right)du$$
$$= -u + \frac{5}{\sqrt{5}}\tan^{-1}\frac{u}{\sqrt{5}} + C$$
$$= -\cos x + \frac{5}{\sqrt{5}}\tan^{-1}\frac{\cos x}{\sqrt{5}} + C$$

45. $$\int e^{\ln 4x}dx = \int 4x dx = 2x^2 + C$$
$$\int \ln e^{4x}dx = \int 4x\cdot\ln e\, dx = \int 4x\, dx = 2x^2 + C$$

49. Use the general power formula. Let $u = e^x + 1$, $du = e^x dx,\ n = 2$.

$$\int e^x\left(e^x+1\right)^2 dx = \int e^x\left(e^{2x}+2e^x+1\right)dx$$
$$= \int\left(e^{3x}+2e^{2x}+e^x\right)dx$$
$$= \int e^{3x}dx + \int 2e^{2x}dx + \int e^x dx$$
$$= \frac{1}{3}\int e^{3x}(3dx) + 2\left(\frac{1}{2}\right)\int e^{2x}(2dx) + \int e^x dx$$
$$= \frac{1}{3}e^{3x} + e^{2x} + e^x + C_2;\ C_2 = C_1 + \frac{1}{3}$$

53. (a) $u = x^2+4$

$du = 2xdx$

$$\int\frac{x}{\sqrt{x^2+4}}dx = \int\frac{\frac{1}{2}du}{\sqrt{u}} = \frac{1}{2}\frac{u^{-1/2+1}}{-\frac{1}{2}+1} + C$$
$$= \frac{u^{1/2}+C}{\sqrt{x^2+4}}$$

(b) $x = 2\tan\theta$

$dx = 2\sec^2\theta d\theta$

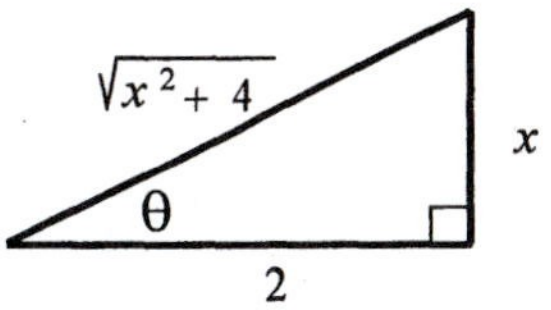

$$\int\frac{xdx}{\sqrt{x^2+4}} = \int\frac{2\tan\theta\left(2\sec^2\theta d\theta\right)}{\sqrt{4\tan^2\theta+4}}$$
$$= \int\frac{2\tan\theta\sec^2\theta d\theta}{\sec\theta}$$
$$= \int 2\tan\theta\sec\theta d\theta$$
$$= 2\sec\theta + C$$
$$= 2\frac{\sqrt{x^2+4}}{2} + C$$
$$= \sqrt{x^2+4} + C$$

(a) is simpler

57. $$A = \int_0^{1.5} ydx = \int_0^{1.5} 4e^{2x}dx$$
$$= 2\int_0^{1.5} e^{2x}(2dx) = 2e^{2x}\Big|_0^{1.5}$$
$$= 2\left(e^3 - e^0\right) = 38.17$$

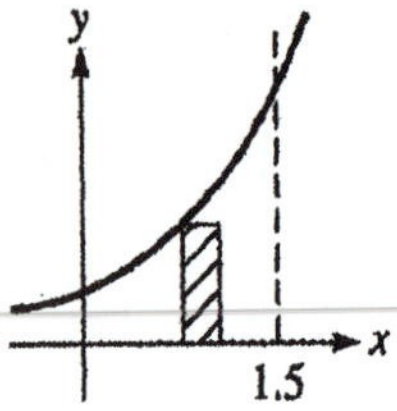

61. $\int \tan^{-1} 2x\,dx$

$u = \tan^{-1} 2x, \qquad dv = dx$

$du = \dfrac{1}{1+4x^2}(2dx) \qquad v = x$

$$\int \tan^{-1}(2x)\,dx = x\tan^{-1} 2x - \int x \frac{2dx}{1+4x^2}$$

$$= x\tan^{-1} 2x - \frac{1}{4}\int \frac{8x\,dx}{1+4x^2}$$

$$= x\tan^{-1} 2x - \frac{1}{4}\ln\left(1+4x^2\right)$$

$$A = \int_0^2 \tan^{-1} 2x\,dx = x\tan^{-1} 2x - \frac{1}{4}\ln\left(1+4x^2\right)\Big|_0^2$$

$$A = 2\tan^{-1} 4 - \frac{1}{4}\ln 17 = 1.94$$

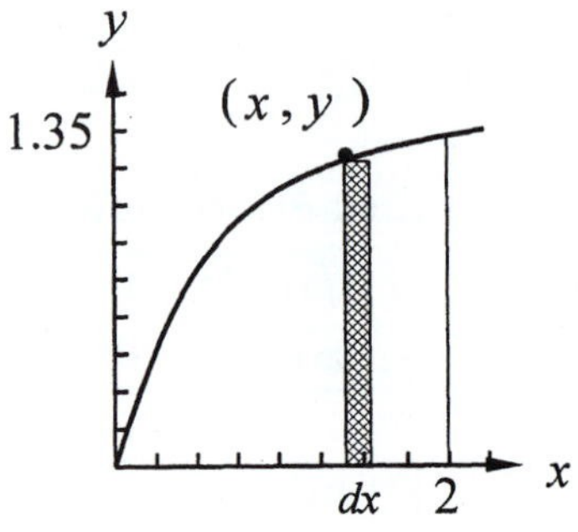

65. $V = \int_0^{\pi} \pi\left(e^x \sin x\right)^2 dx$

$$= \frac{\pi e^{2x}}{8}\left(2\sin^2 x - 2\sin x\cos x + 1\right)\Big|_0^{\pi}$$

$$= \frac{\pi}{8}\left(e^{2\pi} - 1\right) = 209.89$$

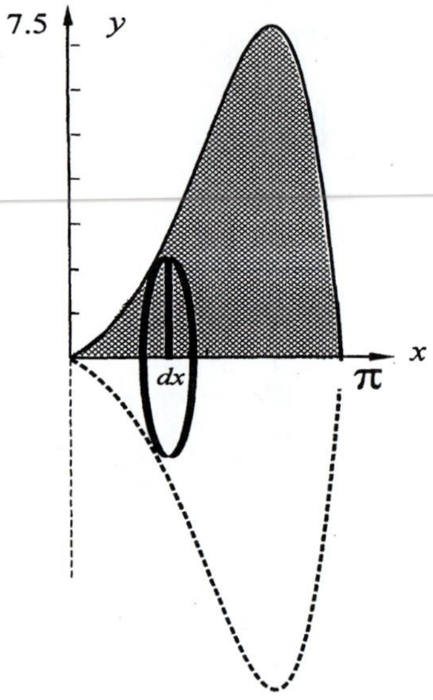

69. $I = \dfrac{1}{2}\int_0^{0.25} \dfrac{5}{1+2t}(2dt)$

$$= \frac{5}{2}\ln(1+2t)\Big|_0^{0.25} = 1.01$$

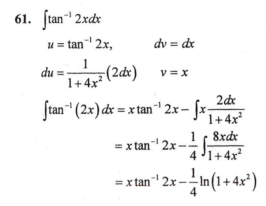

73. $y = 16.0\left(e^{x/32} + e^{-x/32}\right)$, $x = -25.0$ to $x = 25.0$

$$\frac{dy}{dx} = \frac{1}{2}\left(e^{x/32} - e^{-x/32}\right)$$

$$1 + \left(\frac{dy}{dx}\right)^2 = \frac{e^{x/16} - e^{-x/16} + 2}{4}$$

$$\sqrt{1 + \left(\frac{dy}{dx}\right)^2} = \frac{e^{x/32} - e^{-x/32}}{2}$$

$$L = 2\int_0^{25} \frac{e^{x/32} - e^{-x/32}}{2}dx = 32\,e^{x/32} - e^{-x/32}\Big|_0^{25}$$

$$L = 32\left(e^{25/32} - e^{-25/32}\right) = 55.24375836$$

$$L = 55.2 \text{ m}$$

77. $$y_{rms} = \sqrt{\frac{1}{T}\int i^2\,dt} = \sqrt{\frac{1}{T}(2\sin t)^2\,dt}$$

$$\int_0^{2\pi}(2\sin t)^2\,dt = \int_0^{2\pi}4\sin^2 t\,dt = 4\int_0^{2\pi}\sin^2 t\,dt$$

$$= 4\left(\frac{t}{2} - \frac{1}{2}\sin t\cos t\right)\Big|_0^{2\pi}$$

Formula 29 in table of integrals.

$$= 2t - 2\sin t\cot t\Big|_0^{2\pi} = 4\pi;\ T = 2\pi$$

$$y_{rms} = \sqrt{\frac{1}{2\pi}(4\pi)} = \sqrt{2}$$

81. $V = \pi\int_{2.00}^{4.00} y^2\,dx = \pi\int_{2.00}^{4.00} e^{-0.2x}\,dx$; and

$u = 0.2x\ du = -0.2dx$

$$= -\frac{\pi}{0.2}\int_{2.00}^{4.00} e^{-0.2x}(-0.2)\,dx = \frac{-\pi}{0.2}e^{-0.2x}\Big|_{2.00}^{4.00}$$

$$= -\frac{\pi}{2}\left[e^{-0.8} - e^{-0.4}\right] = 3.47\text{ cm}^3$$

85. (a) $A = \int_0^{\pi}(4-y)\,dx$

$$= \int_0^{\pi}\left(4 - 4\cos^2 x\right)dx = \int_0^{\pi}4\left(1-\cos^2 x\right)dx$$

$$= \int_0^{\pi}\left(4 - 4\frac{1+\cos 2x}{2}\right)dx = 2\pi$$

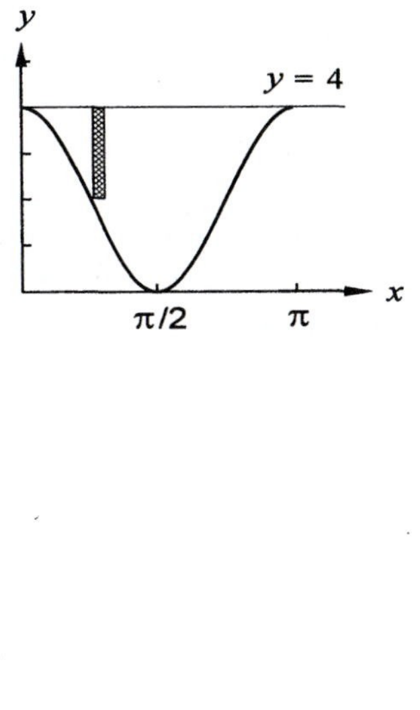

Chapter 29

PARTIAL DERIVATIVES AND DOUBLE INTEGRALS

29.1 Functions of Two Variables

1. $f(x_1, y) = 3x^2 + 2xy - y^3$

$f(-2. 1) = 3(-2)^2 + 2(-2)(1) - (1)^3$

$= 7$

5. From geometry:

$A = 2\pi rh + 2\pi r^2;\ V = \pi r^2 h, h = \dfrac{V}{\pi r^2}$

$A = 2\pi r\left(\dfrac{V}{\pi r^2}\right) + 2\pi r^2 = \dfrac{2V}{r} + 2\pi r^2$

9. $f(x, y) = 2x - 6y$

$f(0, -4) = 2(0) - 6(-4) = 24$

13. $Y(y,t) = \dfrac{2-3y}{t-y} + 2y^2t$

$Y(y,2) = \dfrac{2-3y}{t-y} + 2y^2(2)$

$= 2 - 3y + 4y^2$

17. $H(p,q) = p - \dfrac{p - 2q^2 - 5q}{p+q}$

$H(p, q+k)$

$= p - \dfrac{p - 2(q+k)^2 - 5(q+k)}{p+q+k}$

$= \dfrac{p(p+q+k) - p + 2(q^2+2kq+k^2) + 5(q+k)}{p+q+k}$

$= \dfrac{p^2+pq+pk-p+2q^2+4kq+2k^2+5q+5k}{p+q+k}$

21. $f(x,y) = xy + x^2 - y^2$

$f(x,x) - f(x,0)$

$= x(x) + x^2 - x^2 - [x(0) + x^2 - 0^2]$

$= x^2 + x^2 - x^2 - x^2$

$= 0$

25. $f(x,y) = \frac{\sqrt{y}}{2x}$; considering $\sqrt{y}$, $y \geq 0$ for real values of $f(x,y)$; considering $2x, x \neq 0$ to avoid division by zero. Thus, $y \geq 0$ and $x \neq 0$.

29. $v = iR;\ i = 3\ A,\ R = 6\ \Omega$

$v = 3(6) = 18\ \text{V}$

33. For a, b, T with the same sign: circle if $a = b$, ellipse if $a \neq b$.
For a and b with different signs: hyperbola.

37. $p = 2\ell + 2w;\ \ell = \dfrac{p-2w}{2}$

$A = \ell w = \dfrac{p-2w}{2}w = \dfrac{pw - 2w^2}{2}$

$p = 250\ \text{cm},\ w = 55\ \text{cm}$

$A = \dfrac{250(55) - 2(55)^2}{2} = 3850\ \text{cm}^2$

29.2 Curves and Surfaces in Three Dimensions

1. $3x - y + 2z + 6 = 0$

intercepts: $(0, 0, -3), (0, 6, 0), (-2, 0, 0)$

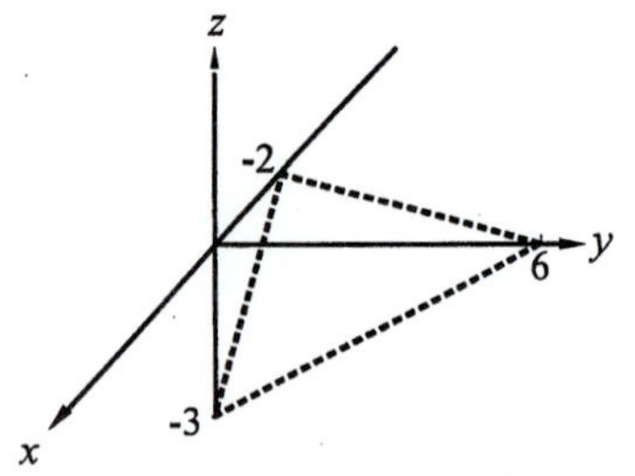

5. $x + y + 2z - 4 = 0$; plane

Intercepts: $(4, 0, 0), (0, 4, 0), (0, 0, 2)$

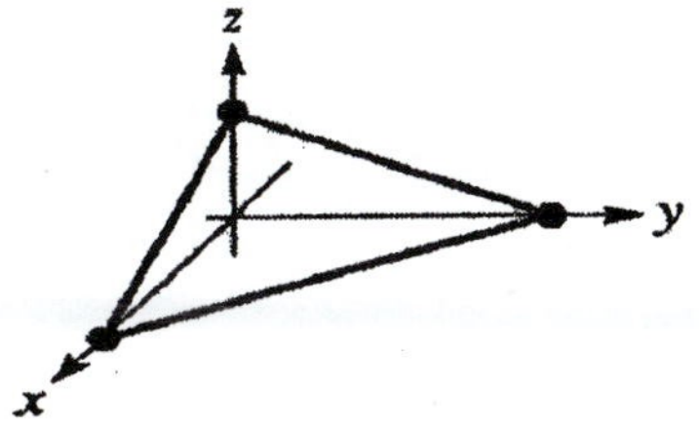

9. $z = y - 2x - 2$; plane

Intercepts: $(-1, 0, 0), (0, 2, 0), (0, 0, -2)$

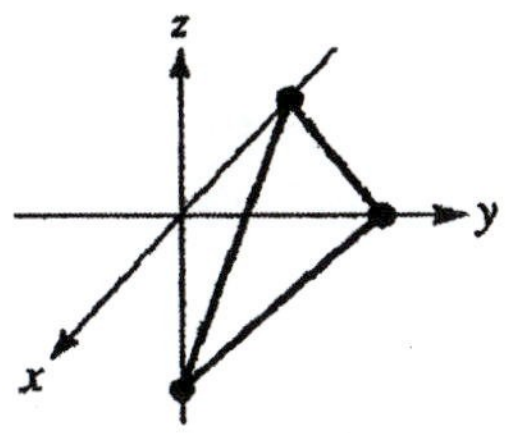

13. $x^2 + y^2 + z^2 = 4$

Intercepts: $(\pm 2, 0, 0), (0, \pm 2, 0), (0, 0, \pm 2)$

Traces:
yz-plane: $y^2 + z^2 = 4$, circle, $r = 2$
xz-plane: $x^2 + z^2 = 4$, circle, $r = 2$
xy-plane: $x^2 + y^2 = 4$, circle, $r = 2$

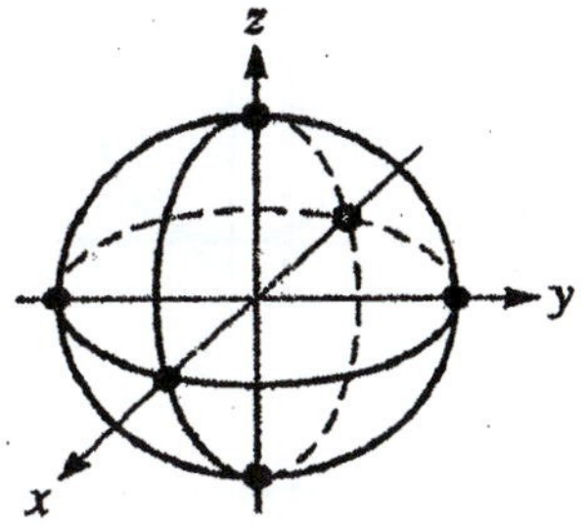

17. $z = 2x^2 + y^2 + 2$

Intercepts: No x-int., no y-int., $(0, 0, 2)$

Traces:

yz-plane:: $z = y^2 + 2$; parabola, $V(0, 0, 2)$
xz-plane: $z = 2x^2 + 2$, parabola, $V(0, 0, 2)$
xy-plane: No trace, $(2x^2 + y^2 + 2 \neq 0)$

Section: For $z = 4$, $2x^2 + y^2 = 2$, ellipse

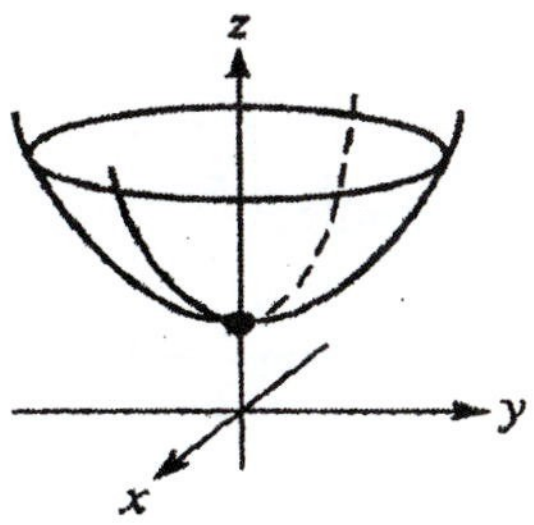

21. $x^2 + y^2 = 16$

Intercepts: $(\pm 4, 0, 0), (0, \pm 4, 0)$, no z-int.

Traces and Sections:

Since z is not present in the equation, the trace and sections are circles $x^2 + y^2 = 16$, with $r = 4$, for all z. This is a cylindrical surface.

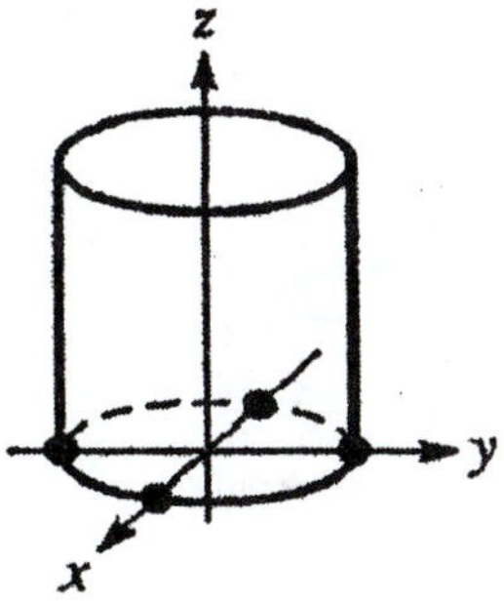

25. (a) $\left(3, \frac{\pi}{4}, 5\right)$

$$x = r\cos\theta = 3\cos\frac{\pi}{4} = \frac{3\sqrt{2}}{2}$$

$$y = r\sin\theta = 3\cos\frac{\pi}{4} = \frac{3\sqrt{2}}{2}$$

$$z = z = 5$$

$$\left(\frac{3\sqrt{2}}{2}, \frac{3\sqrt{2}}{2}, 5\right)$$

(b) $\left(2, \frac{\pi}{2}, 3\right)$

$$x = r\cos\theta = 2\cos\frac{\pi}{2} = 0$$

$$y = r\sin\theta = 2\sin\frac{\pi}{2} = 2$$

$$z = z = 3$$

$(0, 2, 3)$

(c) $\left(4, \frac{\pi}{3}, 2\right)$

$$x = r\cos\theta = 4\cos\frac{\pi}{3} = 4\left(\frac{1}{2}\right) = 2$$

$$y = r\sin\theta = 4\sin\frac{\pi}{3} = 4\left(\frac{\sqrt{3}}{2}\right) = 2\sqrt{3}$$

$$z = z = 2$$

$(2, 2\sqrt{3}, 2)$

29.
$$r^2 = 4z$$
$$x^2 + y^2 = 4z$$

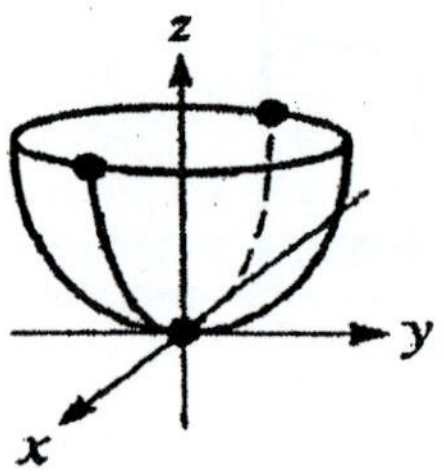

33. $2x^2 + 2y^2 + 3z^2 = 6$

Intercepts: $(\pm\sqrt{3}, 0, 0), (0, \pm\sqrt{3}, 0), (0, 0, \pm\sqrt{2})$

Traces:

yz-plane: $2y^2 + 3z^2 = 6$,
ellipse, $a = \sqrt{3}, b = \sqrt{2}$
xz-plane: $2x^2 + 3z^2 = 6$,
ellipse, $a = \sqrt{3}, b = \sqrt{2}$
xy-plane: $2x^2 + 2y^2 = 6$,
circle, $x^2 + y^2 = 3, r = \sqrt{3}$

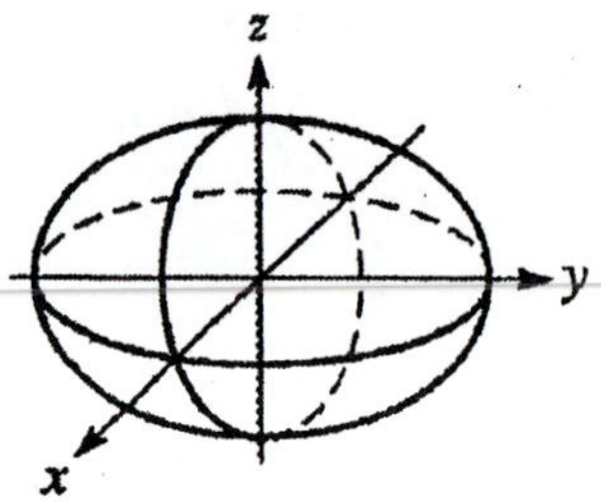

37. $x^2 + y^2 - 2y = 0$

Since z does not appear in the equation, all the traces and sections are circles,

$$x^2 + y^2 - 2y + 1 = 1$$
$$x^2 + (y-1)^2 = 1$$

with center $(0, 1, z)$ for z and $r = 1$. This is a cylindrical surface.

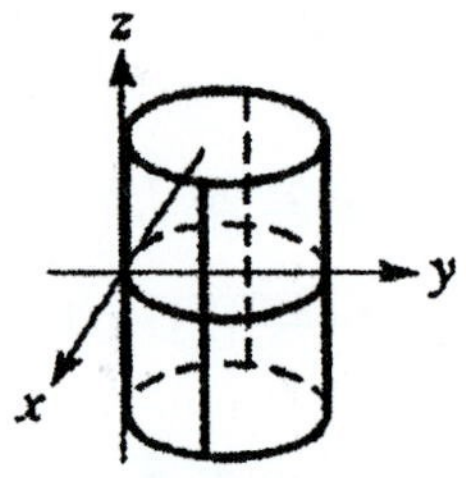

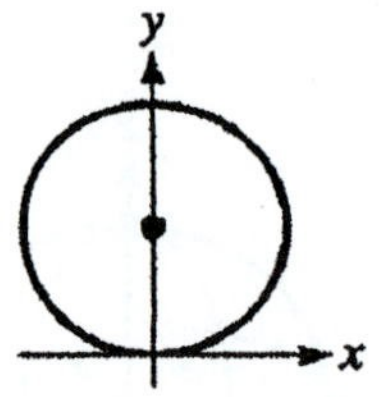

29.3 Partial Derivatives

1. $z = 4y^2 + xy - y^2 = 3y^2 + xy$

$$\frac{\partial z}{\partial x} = y$$

$$\frac{\partial z}{\partial y} = 6y + x$$

5. $f(x,y) = xe^{2y}$

$$\frac{\partial f}{\partial x} = (1)e^{2y} = e^{2y}$$

$$\frac{\partial f}{\partial y} = xe^{2y}(2) = 2xe^{2y}$$

9. $\phi = r\sqrt{1+2rs}$

$$\frac{\partial \phi}{\partial r} = r\left(\frac{1}{2}\right)(1+2rs)^{-1/2}(2s) + (1+2rs)^{1/2}$$

$$= \frac{rs}{(1+2rs)^{1/2}} + (1+2rs)^{1/2}$$

$$= \frac{1+3rs}{\sqrt{1+2rs}}$$

$$\frac{\partial \phi}{\partial s} = r\left(\frac{1}{2}\right)(1+2rs)^{-1/2}(2r)$$

$$= \frac{r^2}{\sqrt{1+2rs}}$$

13. $z = \sin xy$

$$\frac{\partial z}{\partial x} = (\cos xy)(y)$$

$$= y\cos xy$$

$$\frac{\partial z}{\partial y} = (\cos xy)(x)$$

$$= x\cos xy$$

17. $f(x,y) = \dfrac{2\sin^3 2x}{1-3y}$

$$\frac{\partial f}{\partial x} = \frac{2(3)(\sin^2 2x)(\cos 2x)(2)}{1-3y}$$

$$= \frac{12\sin^2 2x\cos 2x}{1-3y}$$

$$\frac{\partial f}{\partial y} = -(2\sin^3 2x)(1-3y)^{-2}(-3)$$

$$= \frac{6\sin^3 2x}{(1-3y)^2}$$

21. $z = \sin x + \cos xy - \cos y$

$$\frac{\partial z}{\partial x} = \cos x - (\sin xy)(y)$$

$$= \cos x - y\sin xy$$

$$\frac{\partial z}{\partial y} = -(\sin xy)(x) + \sin y$$

$$= -x\sin xy + \sin y$$

25. $z = 3xy - x^2$

$$\frac{\partial z}{\partial x} = 3y - 2x$$

$$\left.\frac{\partial z}{\partial y}\right|_{(1,-2,-7)} = 3(-2) - 2(1) = -8$$

29. $z = 2xy^3 - 3x^2y$

$$\frac{\partial z}{\partial x} = 2y^3 - 6xy$$

$$\frac{\partial z}{\partial y} = 6xy^2 - 3x^2$$

$$\frac{\partial^2 z}{\partial x^2} = -6y, \frac{\partial^2 z}{\partial x^2} = 12xy$$

$$\frac{\partial^2 z}{\partial x \partial y} = \frac{\partial^2 z}{\partial x \partial y} = 6y^2 - 6x$$

33. $z = 9 - x^2 - y^2$

$$\frac{\partial z}{\partial y} = -2y$$

$$\left.\frac{\partial z}{\partial y}\right|_{(1,2,4)} = -4$$

$$\left.\frac{\partial z}{\partial y}\right|_{(2,2,1)} = -4$$

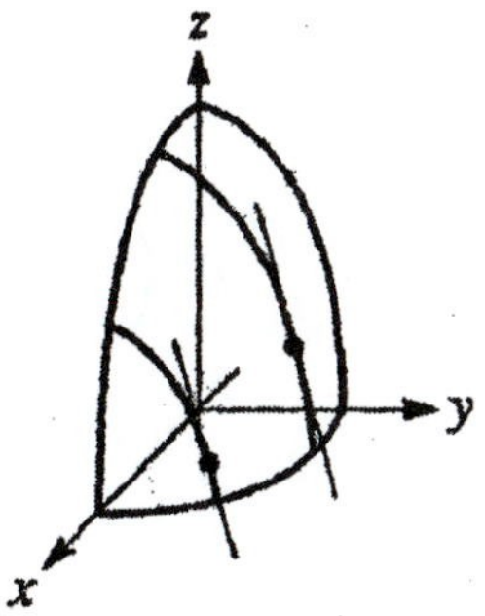

37. $V = \pi r^2 h + \frac{1}{2}\left(\frac{4}{3}\pi r^3\right)$

$$\frac{\partial V}{\partial R} = 2\pi r h + 2\pi r^2 \Big|_{r=2.65,h=4.20}$$

$$\frac{\partial V}{\partial R} = 2\pi(2.65)(4.20) + 2\pi(2.65)^2$$

$$\frac{\partial V}{\partial R} = 114\ \text{cm}^2$$

41. $i_b = 50(e_b + 5e_c)^{1.5}$

$$\frac{\partial i_b}{\partial e_c} = 50(1.5)(e_b + 5e_c)^{0.5}(5)$$
$$= 375(e_b + 5e_c)^{0.5}$$

For $e_b = 200$ V and $e_c = -20$ V

$$\frac{\partial i_b}{\partial e_c} = 375(200 - 100)^{0.5}$$
$$= 3750\ \mu\text{A/V} = 3.75\ 10^{-3}\ 1/\Omega$$

29.4 Double Integrals

1. $\int_0^1 \int_{x^2}^{x} (x+y)\,dy\,dx = \int_0^1 xy + \frac{y^2}{2}\Big|_{x^2}^{x}\ dx$

$$= \int_0^1 \left(x^2 + \frac{x^2}{2} - \left(x^3 + \frac{x^4}{2}\right)\right)dx$$

$$= \int_0^1 \left(-\frac{x^4}{2} - x^3 + \frac{3x^2}{2}\right)dx$$

$$= -\frac{x^5}{10} - \frac{x^4}{4} + \frac{3x^2}{6}\Big|_0^1 = \frac{3}{20}$$

5. $\int_1^2 \int_0^{y^2} xy^2\,dx\,dy = \int_1^2 y^2\left(\frac{1}{2}x^2\right)\Big|_0^{y^2} dy$

$$= \int_1^2 y^2\left(\frac{1}{2}y^4\right) dy$$

$$= \frac{1}{2}\int_1^2 y^6\,dy$$

$$= \frac{1}{14}y^7\Big|_1^2$$

$$= \frac{127}{14}$$

9. $\int_0^{\pi/6} \int_{\pi/3}^{y} \sin x\,dx\,dy$

$$= \int_0^{\pi/6} (-\cos x)\Big|_{\pi/3}^{y} dy$$

$$= -\int_0^{\pi/6} \left(\cos y - \cos\frac{\pi}{3}\right) dy$$

$$= -\int_0^{\pi/6} \left(\cos y - \frac{1}{2}\right) dy$$

$$= -\sin y + \frac{1}{2}y\Big|_0^{\pi/6}$$

$$= -\sin\frac{\pi}{6} + \frac{\pi}{12}$$

$$= \frac{\pi}{12} - \frac{1}{2}$$

$$= \frac{\pi - 6}{12}$$

13.

$$\int_1^2 \int_0^x yx^3 e^{xy^2}\,dy\,dx = \frac{1}{2}\int_1^2 \int_0^x x^2\left(2xye^{xy^2}\right)dx$$

$$= \frac{1}{2}\int_1^2 x^2\left(e^{xy^2}\right)\Big|_0^x dx = \frac{1}{2}\int_1^2 x^2\left(e^{xy^2} - 1\right)dx$$

$$= \frac{1}{6}\int_1^2 3x^2 e^{x^3}\,dx - \frac{1}{2}\int_1^2 x^2\,dx$$

$$= \frac{1}{6}e^{x^3} - \frac{1}{6}x^3\Big|_1^2 = \frac{1}{6}\left[e^8 - 8 - (e-1)\right]$$

$$= 495.2$$

17. $V = \int_0^4 \int_x^{4-x} z\,dy\,dx = \int_0^4 \int_x^{4-x} (4-x-y)\,dy\,dx$

$$= \int_0^4 \left(4y - xy - \frac{1}{2}y^2\right)\Big|_0^{4-x} dx$$

$$= 8x - 2x^2 + \frac{1}{6}x^3\Big|_0^4 = 32 - 32 + \frac{64}{6}$$

$$= \frac{32}{3}$$

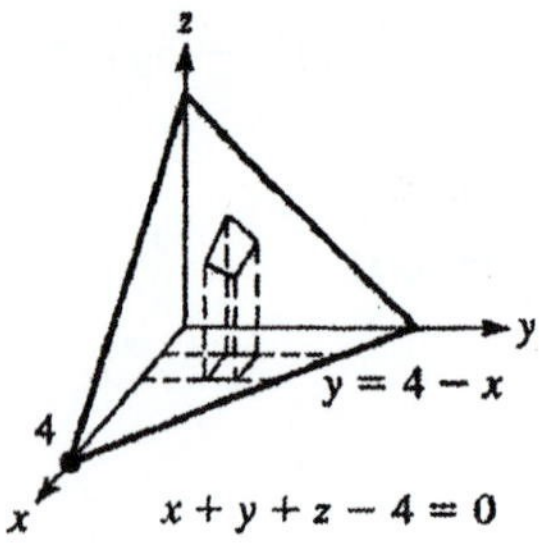

21. $V = \int_0^2 \int_x^2 z\, dy\, dx \int_0^2 \int_x^2 \left(2 + x^2 + y^2\right) dy\, dx$

$$= \int_x^2 \left(2y + x^2 y + \frac{1}{3}y^3\right)\Bigg|_x^2 dx$$

$$= \int_x^2 \left(4 + 2x^2 + \frac{8}{3} - 2x - x^3 - \frac{1}{3}x^3\right) dx$$

$$= \int_x^2 \left(\frac{20}{3} - 2x + 2x^2 - \frac{4}{3}x^3\right) dx$$

$$= \frac{20}{3}x - x^3 + \frac{2}{3}x^3 - \frac{1}{3}x^4 \Bigg|_0^2 = \frac{40}{3} - 4 + \frac{16}{3} - \frac{16}{3}$$

$$= \frac{28}{3}$$

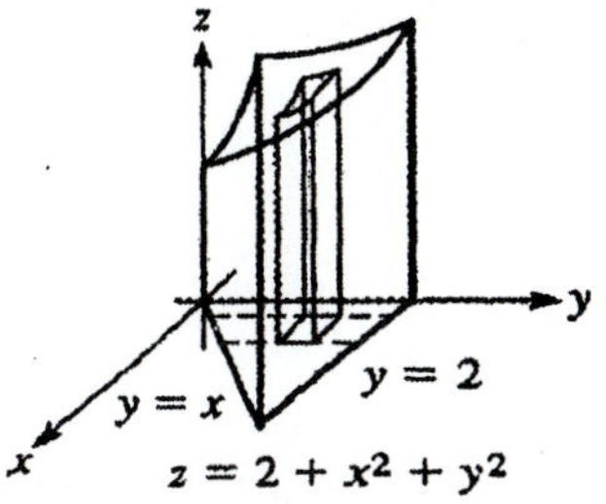

25. $\int_0^5 \int_0^{12} z\, dx\, dy = \int_0^5 \int_0^{12} (10 - 2y)\, dx\, dy$

$$= \int_0^5 (10 - 2y)\Big|_0^{12} dy = 12 \int_0^5 (10 - 2y) dy$$

$$= 12\left(10y - y^2\right)\Big|_0^5 = 12(50 - 25)$$

$$= 300 \text{ cm}^2$$

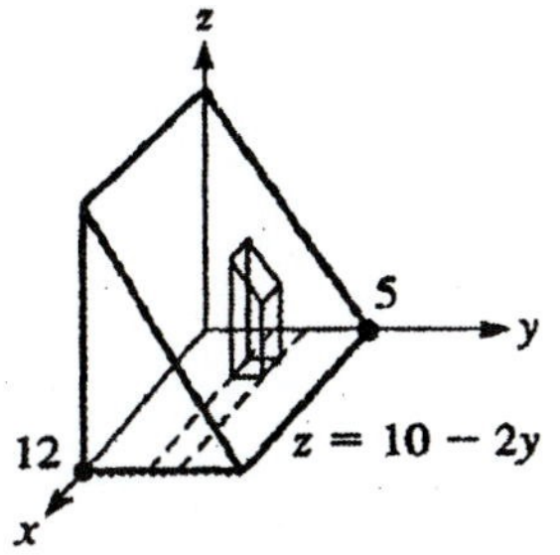

Chapter 29 Review Exercises

1. $f(x, y) = 3x^2 y - y^3$

$f(-1, 4) = 3(-1)^2(4) - 4^3 = -52$

5. $x - y + 2z - 4 = 0$, plane

intercepts: $(0, 0, 2), (0, -4, 0), (4, 0, 0)$

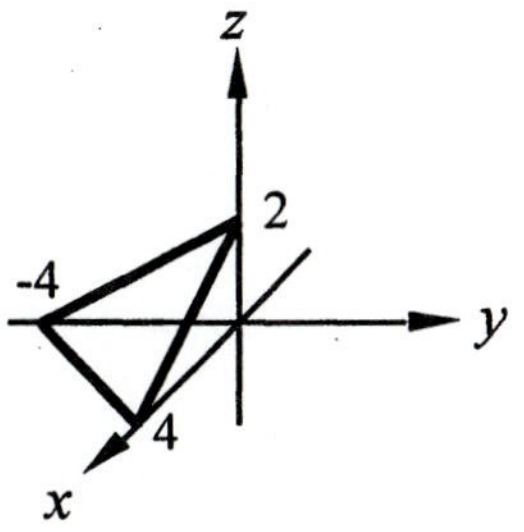

9. $z = 5x^3 y^2 - 2xy^4$

$$\frac{\partial z}{\partial x} = 5y^2\left(3x^2\right) - 2y^4(1)$$

$$= 15x^2 y^2 - 2y^4$$

$$\frac{\partial z}{\partial y} = 5x^3(2y) - 2x\left(4y^3\right)$$

$$= 10x^3 y - 8xy^3$$

13. $z = \dfrac{2x - 3y}{x^2 y + 1}$

$$\frac{\partial z}{\partial x} = \frac{\left(x^2 y + 1\right)(2) - (2x - 3y)(2xy)}{\left(x^2 y + 1\right)^2}$$

$$= \frac{2 - 2x^2 y + 6xy^2}{\left(x^2 y + 1\right)^2}$$

$$\frac{\partial z}{\partial y} = \frac{\left(x^2 y + 1\right)(-3) - (2x - 3y)\left(2x^2\right)}{\left(x^2 y + 1\right)^2}$$

$$= \frac{-\left(3 + 2x^3\right)}{\left(x^2 y + 1\right)^2}$$

17. $z = \sin^{-1}\sqrt{x + y}$

$$\frac{\partial z}{\partial x} = \frac{1}{\sqrt{1 - (x + y)}}\left(\frac{1}{2}\right)(x + y)^{-1/2}(1)$$

$$= \frac{1}{2\sqrt{(x + y)(1 - x - y)}}$$

$$\frac{\partial z}{\partial y} = \frac{1}{\sqrt{1 - (x + y)}}\left(\frac{1}{2}\right)(x + y)^{-1/2}(1)$$

$$= \frac{1}{2\sqrt{(x + y)(1 - x - y)}}$$

21. $\int_0^2\int_1^2 (3y+2xy)\,dx\,dy = \int_0^2 \left(3xy + x^2y\right)\Big|_1^2\,dy$

$= \int_0^2 (6y+4y-3y-y)\,dy = \int_0^2 6y\,dy$

$= 3y^2\Big|_0^2 = 12$

25. $\int_0^1\int_0^{2x} x^2 e^{xy}\,dy\,dx = \int_0^1\int_0^{2x} x\left(e^{xy}x\,dy\right)dx =$

$\int_0^1 x\,e^{xy}\Big|_0^{2x} dx = \int_0^1 x\left(e^{2x^2}-1\right)dx$

$= \frac{1}{4}e^{2x^2} - \frac{1}{2}x^2\Big|_0^1 = \frac{1}{4}e^2 - \frac{1}{2} - \frac{1}{4}$

$= \frac{1}{4}\left(e^2-3\right) = 1.097$

29. $z = \sqrt{x^2+4y^2}$

Intercepts: $(0, 0, 0)$

Traces:

$z \ge 0$ for all x and y

(defined by positive square root)

In yz-plane: $z = \pm 2y$

In xz-plane: $z = \pm x$

In xy-plane: $(0, 0, 0)$

Section: For $z = 2$

$x^2 + 4y^2 = 4$ (ellipse)

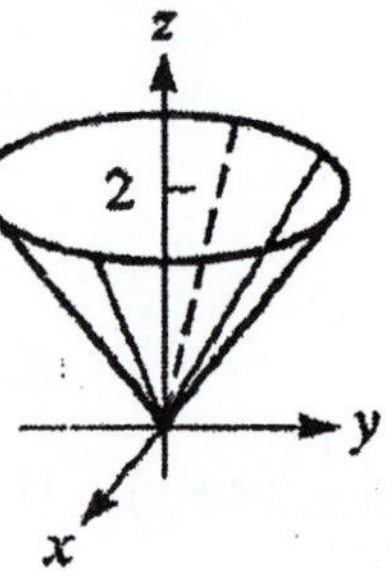

33. (a) $\theta = 3$ is a vertical half-plane

(b) $z = r^2 = x^2 + y^2$

intercepts: $(0, 0, 0)$

traces are ellipses, sections are circles

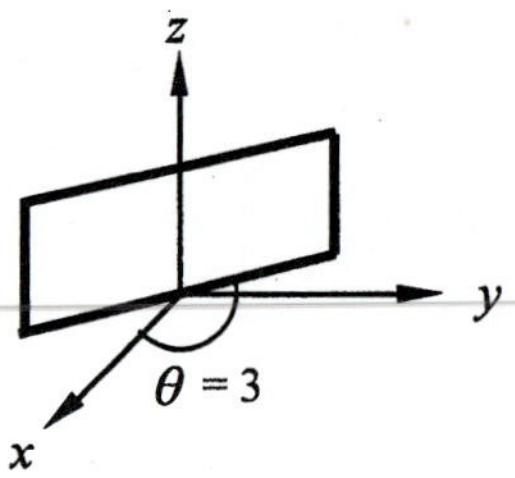

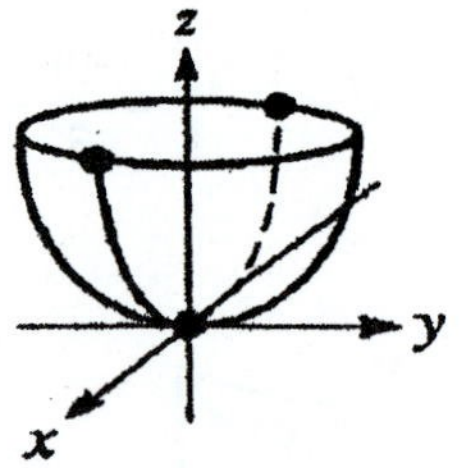

37. $i_c = i_e\left(1-e^{-2v_c}\right)$

$\alpha = \frac{\partial i_c}{\partial i_e} = 1 - e^{-2v_c}$

For $v_c = 2$ V, $\alpha = 1 - e^{-4} = 0.982$

41. $\alpha = \frac{1}{L}\frac{\alpha L}{\alpha T}$

$\alpha = \frac{k_2 + 2k_3FT}{L_0 + k_1F + k_2T + k_3FT^2}$

45. $V = \int_0^4 \int_0^{\sqrt{16-x^2}} z\,dy\,dz = \int_0^4 \int_0^{\sqrt{16-x^2}} (8-x)\,dy\,dz$

$= \int_0^4 (8-x)y\Big|_0^{\sqrt{16-x^2}} dx = \int_0^4 (8-x)\sqrt{16-x^2}\,dx$

$= 8\int_0^4 \sqrt{16-x^2}\,dx - \int_0^4 x\sqrt{16-x^2}\,dx$

$= 8\left[\frac{x}{2}\sqrt{16-x^2} + \frac{16}{2}\sin^{-1}\frac{x}{4}\right] + \frac{1}{3}\left(16-x^2\right)^{3/2}\Big|_0^4$

$= 8\left[2(0) + 8\sin^{-1}1\right] + \frac{1}{3}(0) - 8\left[0 + 8\sin^{-1}0\right]$

$-\frac{1}{3}(16)^{3/2} = 64\sin^{-1}1 - \frac{64}{3}$

$= 64\left(\frac{\pi}{2}\right) - \frac{64}{3} = 32\left(\pi - \frac{2}{3}\right)$

$= 79.2$

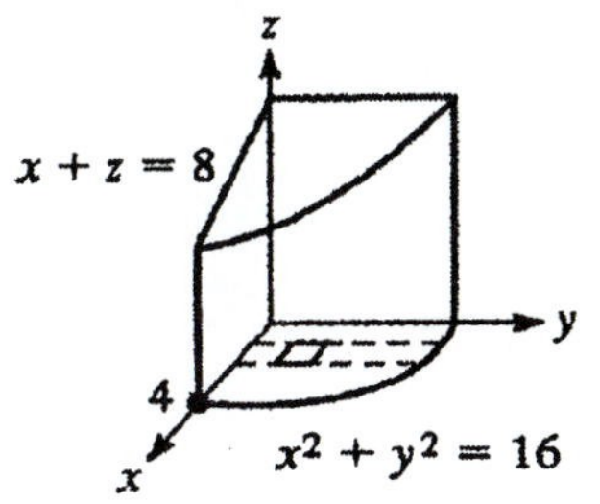

Chapter 30

EXPANSION OF FUNCTIONS IN SERIES

30.1 Infinite Series

1. $\sum_{u=1}^{\infty} 0.5^n = 0.5 + 0.5^2 + 0.5^3 + 0.5^4 + \cdots + 0.5^n + \cdots$

$s_1 = 0.5, s_2 = 0.75, s_3 = 0.875, s_4 = 0.9375.$
Converges

5. $a_n = \dfrac{1}{n+2};\ n = 0, 1, 2, 3, \ldots$

$$a_0 = \frac{1}{0+2} = \frac{1}{2} \qquad a_2 = \frac{1}{2+2} = \frac{1}{4}$$
$$a_1 = \frac{1}{1+2} = \frac{1}{3} \qquad a_3 = \frac{1}{3+2} = \frac{1}{5}$$

9. $a_n = \cos\dfrac{n\pi}{2}, n = 0, 1, 2, 3, \ldots$

$$a_0 = \cos\frac{0\cdot\pi}{2} = 1$$
$$a_1 = \cos\frac{1\cdot\pi}{2} = 0$$
$$a_2 = \cos\frac{2\cdot\pi}{2} = -1$$
$$a_3 = \cos\frac{3\cdot\pi}{2} = 0$$

(a) $1, 0, -1, 0$

(b) $1 + 0 - 1 + 0 - \cdots$

13. $\dfrac{1}{2\times3} - \dfrac{1}{3\times4} + \dfrac{1}{4\times5} - \dfrac{1}{5\times6} + \cdots$

$$n = 1,\ a_1 = \frac{1}{(1+1)(1+2)} = \frac{1}{2\times3}$$
$$n = 2,\ a_2 = \frac{-1}{(2+1)(2+2)} = \frac{-1}{3\times4}$$
$$a_n = \frac{(-1)^{n+1}}{(n+1)(n+2)}$$

17. $1 + \dfrac{1}{2} + \dfrac{2}{3} + \dfrac{3}{4} + \dfrac{4}{5} + \cdots$

$$S_0 = 1;\ S_1 = 1 + \frac{1}{2} = \frac{3}{2} = 1.5$$
$$S_2 = 1 + \frac{1}{2} + \frac{2}{3} = \frac{13}{6} = 2.1666667$$
$$S_3 = 1 + \frac{1}{2} + \frac{2}{3} + \frac{3}{4} = \frac{35}{12} = 2.9166667$$
$$S_4 = 1 + \frac{1}{2} + \frac{2}{3} + \frac{3}{4} + \frac{4}{5} = \frac{223}{60} = 3.7166667$$

Divergent

21. $\sum_{n=1}^{\infty} \dfrac{2n+1}{n^2(n+1)^2}$

First five terms:

$$a_1 = \frac{3}{4};\ a_2 = \frac{5}{36};\ a_3 = \frac{7}{144};\ a_4 = \frac{9}{400};\ a_5 = \frac{11}{900}$$

First five partial sums:

$$S_1 = 0.75;$$
$$S_2 = \frac{3}{4} + \frac{5}{36} = 0.8888889$$
$$S_3 = \frac{3}{4} + \frac{5}{36} + \frac{7}{144} = 0.9375000$$
$$S_4 = \frac{3}{4} + \frac{5}{36} + \frac{7}{144} + \frac{9}{400} = 0.9600000$$
$$S_5 = \frac{3}{4} + \frac{5}{36} + \frac{7}{144} + \frac{9}{400} + \frac{11}{900} = 0.9722222$$

Convergent, converging to 1 (approx. sum)

25. $1 + 2 + 4 + \cdots + 2^n + \cdots;\ n = 0, 1, 2, 3, \cdots$

$$S_0 = 1$$
$$S_1 = 3$$
$$S_2 = 7$$
$$S_3 = 15$$
$$\vdots$$
$$S_n = 2^{n+1} - 1$$

$\lim_{n\to\infty} S_n = \lim_{n\to\infty} (2^{n+1} - 1) = \infty$, divergent

29. $10 + 9 + 8.1 + 7.29 + 6.561 + \cdots + 10(0.9)^n + \cdots;$
$n = 0, 1, 2, 3, \ldots S_n = 100 - 100(0.9)^{n+1};$
$n = 0, 1, 2, 3, \ldots$

$\lim_{n\to\infty} S_n = \lim_{n\to\infty} 100 - 100(0.9)^{n+1} = 100$, convergent

33. $\sum_{n=0}^{\infty}(x-4)^n$ is a GS with $a_1 = 1$, $r = x - 4$ which converges for $|x-4| < 1 \Leftrightarrow -1 < x - 4 < 1$ or $-3 < x < 5$.

37. $S_n = \dfrac{a_1(1-r^n)}{(1-r)}$; $r \neq 1$; geometric series

Series: $\dfrac{1}{2} + \dfrac{1}{4} + \dfrac{1}{8} + \cdots;\ a_n = \dfrac{1}{2^n}, a = \dfrac{1}{2}, r = \dfrac{1}{2}$

$f(x) = \dfrac{a_1(1-r^x)}{(1-r)};\ f(x) = \dfrac{\frac{1}{2}(1-r^x)}{(1-\frac{1}{2})} = (1-r^x)$

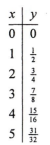

x	y
0	0
1	$\frac{1}{2}$
2	$\frac{3}{4}$
3	$\frac{7}{8}$
4	$\frac{15}{16}$
5	$\frac{31}{32}$

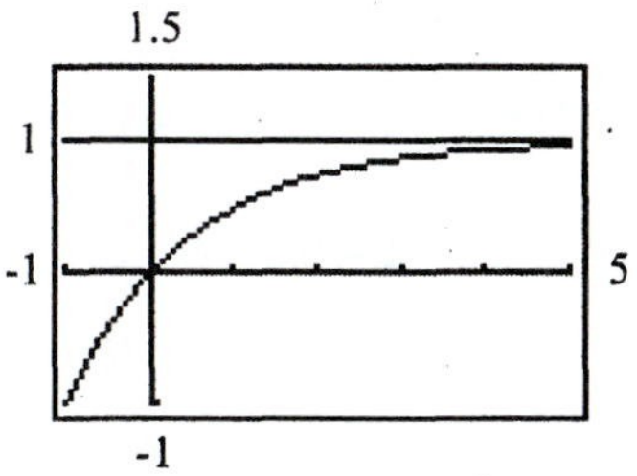

The infinite series approaches 1.

41. $\sum_{n=0}^{\infty} x^n = 1 + x + x^2 + \cdots + x^n + \cdots$

For $|x| < 1, a_1 = 1, r = x$,

$$S = \frac{1}{1-x}$$

$$\sum_{n=0}^{\infty} x^n = \frac{1}{1-x}$$

30.2 Maclaurin's Series

1. $f(x) = \dfrac{2}{2+x}, f(0) = 1$

$f'(x) = \dfrac{-2}{(2+x)^2}, f'(0) = -\dfrac{1}{2}$

$f''(x) = \dfrac{4}{(2+x)^3}, f''(0) = \dfrac{1}{2}$

$f'''(x) = \dfrac{-12}{(2+x)^4}, f'''(0) = -\dfrac{3}{4}$

$f(x) = \dfrac{2}{2+x} = 1 - \dfrac{1}{2}x + \dfrac{1}{2}x^2 - \dfrac{3}{4}x^3 + \cdots$

5.
$f(x) = \cos x \qquad f(0) = 1$
$f'(x) = -\sin x \qquad f'(0) = 0$
$f''(x) = -\cos x \qquad f''(0) = -1$
$f'''(x) = \sin x \qquad f'''(0) = 0$
$f^{iv}(x) = \cos x \qquad f^{iv}(0) = 1$

$$f(x) = \cos x = f(0) + f''(0)\frac{x^2}{2!} + f^{iv}(0)\frac{x^4}{4!} - \cdots$$

$$\cos x = 1 - 1\frac{x^2}{2} + 1\frac{x^4}{24} - \cdots$$

$$\cos x = 1 - \frac{1}{2}x^2 + \frac{1}{24}x^4 - \cdots$$

9.
$f(x) = e^{-2x} \qquad f(0) = 1$
$f'(x) = -2e^{-2x} \qquad f'(0) = -2$
$f''(x) = 4e^{-2x} \qquad f''(0) = 4$

$$e^{-2x} = 1 - 2x + 4\frac{x^2}{2} - \cdots = 1 - 2x + 2x^2 - \cdots$$

13.
$f(x) = \dfrac{1}{(1-x)} \qquad f(0) = 1$

$f'(x) = \dfrac{1}{(1-x)^2} \qquad f'(0) = 1$

$f''(x) = \dfrac{2}{(1-x)^3} \qquad f''(0) = 2$

$$\frac{1}{(1-x)} = 1 + x + \frac{2x^2}{2} + \cdots = 1 + x + x^2 + \cdots$$

17.
$$\begin{aligned} f(x) &= \cos^2 x & f(0) &= 1\\ f'(x) &= -2\sin x\cos x & f'(0) &= 0\\ f''(x) &= 2-4\cos^2 x & f''(0) &= -2\\ f'''(x) &= 8\sin x\cos x & f'''(0) &= 0\\ f^{iv}(x) &= 16\cos^2 x - 8 & f^{iv}(0) &= 8. \end{aligned}$$

$$\cos^2 x = 1 - 2\frac{x^2}{2!} + 8\frac{x^4}{4!} - \cdots$$
$$= 1 - x^2 + \frac{1}{3}x^4 - \cdots$$

21. $f(x) = \tan^{-1} x$

$$f'(x) = \frac{1}{1+x^2} = (1+x^2)^{-1}$$
$$f''(x) = -(1+x^2)^{-2}2x = -2x(1+x^2)^{-2}$$
$$f'''(x) = -2x[-2(1+x^2)^{-3}(2x)] + (1+x^2)^{-2}(-2)$$
$$f(0) = 0$$
$$f'(0) = 1$$
$$f''(0) = 0$$
$$f'''(0) = -2$$

$$f(x) = 0 + 1x + \frac{0x^2}{2!} - \frac{2x^3}{3!} + \cdots = x - \frac{1}{3}x^3 + \cdots$$

25. $f(x) = \ln\cos x$

$$f'(x) = -\frac{1}{\cos x}\sin x = -\tan x$$
$$f''(x) = -\sec^2 x$$
$$f'''(x) = -2\sec x\sec x\tan x = -2\sec^2 x\tan x$$
$$f^{iv}(x) = -2\sec^2 x\sec^2 x - 2\tan x(2\sec x\sec x\tan x)$$
$$f(0) = \ln 1 = 0$$
$$f'(0) = 0$$
$$f''(0) = -1$$
$$f'''(0) = 0$$
$$f^{iv}(0) = -2 - 0 = -2$$

$$f(x) = 0 + 0x - \frac{1x^2}{2!} + \frac{0x^3}{3!} - \frac{2x^4}{4!} + \cdots$$
$$= -\frac{1}{2}x^2 - \frac{1}{12}x^4 - \cdots$$

29. (a) It is not possible to find a Maclaurin's expansion for $f(x) = \csc x$ since the formula is not defined when $x = 0$.

(b) $f(x) = \ln x$ is not defined when $x = 0$.

33. The Maclaurin's expansion of $f(x) = e^{3x}$ is

$f(x) = 1 + 3x + \frac{9}{2}x^2 + \frac{9}{2}x^3 + \cdots$. The linearization is $L(x) = 1 + 3x$, the first two terms of the expansion.

37. $f(x) = x^2, f'(x) = 2x, f''(x) = 2, f^{(n)}(x) = 0,$ $n \geq 3$

$f(0) = 0, f'(0) = 0, f''(0) = 2, f^{(n)}(x) = 0, n \geq 3$

$$x^2 = 0 + 0(x) + 2\frac{x^2}{2!} + 0\frac{x^3}{3!} + \cdots$$
$$x^2 = x^2$$

41. $0 \leq R \leq 1$; $R = e^{-0.001t}$

$$\frac{dR}{dt} = -0.001e^{-0.001t};\ \frac{d^2R}{dt^2} = 1\times 10^{-6}e^{-0.001t}$$

$f(0) = 1$; $f'(0) = -0.001$; $f''(0) = 1\times 10^{6}$

$$e^{-0.001t} = 1 - 0.001t + 1\times 10^{-6}\frac{t^2}{2!} - \cdots$$
$$= 1 - 0.001t + 10\times 10^{-7}\frac{t^2}{2} - \cdots$$
$$= 1 - 0.001t + (5\times 10^{-7})t^2 - \cdots$$

30.3 Operations with Series

1.
$$e^x = 1 + x + \frac{x^2}{2!} + \frac{x^3}{3!} + \cdots$$
$$e^{2x^2} = 1 + 2x^2 + \frac{(2x^2)^2}{2!} + \frac{(2x^2)^3}{3!} + \cdots$$
$$e^{2x^2} = 1 + 2x^2 + 2x^4 + \frac{4}{3}x^6 + \cdots$$

5. $f(x) = \sin\left(\frac{1}{2}x\right)$; $\sin x = x - \frac{x^3}{3!} + \frac{x^5}{5!} - \frac{x^7}{7!} + \cdots$

$$f(x) = \sin\left(\frac{1}{2}x\right)$$
$$= \frac{1}{2}x - \frac{\left(\frac{1}{2}x\right)^3}{6} + \frac{\left(\frac{1}{2}x\right)^5}{120} - \frac{\left(\frac{1}{2}x\right)^7}{7!} + \cdots$$
$$= \frac{1}{2}x - \frac{x^3}{2^3 3!} + \frac{x^5}{2^5 5!} - \frac{x^7}{2^7 7!} + \cdots$$

9. $f(x)=\ln(1+x^2)$; $\ln(1+x)=x-\frac{x^2}{2}+\frac{x^3}{3}-\frac{x^4}{4}+\cdots$

$$\ln(1+x^2)=x^2-\frac{(x^2)^2}{2}+\frac{(x^2)^3}{3}-\frac{(x^2)^2}{4}+\cdots$$
$$=x^2-\frac{1}{2}x^4+\frac{1}{3}x^6-\frac{1}{4}x^8+\cdots$$

13. $$\int_0^{0.2}\cos\sqrt{x}dx=\int_0^{0.2}\left(1-\frac{(\sqrt{x})^2}{2}+\frac{(\sqrt{x})^4}{24}\right)dx$$
$$=\int_0^{0.2}\left(1-\frac{1}{2}x-\frac{1}{24}x^2\right)dx$$
$$=\left(x-\frac{1}{4}x^2+\frac{1}{72}x^3\right)\Big|_0^{0.2}$$
$$=0.2-\frac{1}{4}(0.2)^2+\frac{1}{72}(0.2)^3$$
$$=0.1901$$

17. $e^x\sin x=f(x)$

$$e^x=1+x+\frac{x^2}{2!}+\frac{x^3}{3!}+\cdots \quad (1)$$
$$\sin x=x-\frac{x^3}{3!}+\frac{x^5}{5!}-\cdots \quad (2)$$

Multiply (1) by x from (2): $x+x^2+\frac{x^3}{2!}+\frac{x^4}{3!}+\cdots$

Multiply (1) by $-\frac{x^3}{3!}$ from (2): $-\frac{x^3}{3!}-\frac{x^4}{3!}-\cdots$

Combine:

$$e^x\sin x=x+x^2+\frac{x^3}{2!}+\frac{x^4}{3!}-\frac{x^3}{3!}-\frac{x^4}{3!}$$
$$=x+x^2+\frac{1}{2}x^3-\frac{1}{6}x^3$$
$$=x+x^2+\frac{2}{6}x^3+\cdots$$
$$=x+x^2+\frac{1}{3}x^3+\cdots$$

21. Replacing x in $\ln(1+x)=x-\frac{1}{2}x^2+\frac{1}{3}x^3-\frac{1}{4}x^4+\ldots$ with $\sin x$ gives

$$\ln(1+\sin x)=\sin x-\frac{1}{2}\sin^2 x+\frac{1}{3}\sin^{3x}-\frac{1}{4}\sin^4 x+\ldots$$

from which, replacing $\sin x$ with

$$\sin x=x-\frac{1}{6}x^3+\frac{1}{120}x^5-\ldots,$$
$$\ln(1+\sin x)=x-\frac{1}{6}x^3+\frac{1}{120}x^5-\ldots$$
$$-\frac{1}{2}\left(x-\frac{1}{6}x^3+\frac{1}{120}x^5-\ldots\right)^2$$
$$+\frac{1}{3}\left(x-\frac{1}{6}x^3+\frac{1}{120}x^5-\ldots\right)^3-\ldots$$
$$\ln(1+\sin x)=x-\frac{1}{6}x^3+\frac{1}{120}x^5-\ldots$$
$$-\frac{1}{2}x^2+\frac{1}{6}x^4-\frac{1}{45}x^6+\frac{1}{720}x^8-\frac{1}{28,800}x^{10}+\ldots$$
$$+\frac{1}{3}x^3-\frac{1}{6}x^5+\frac{13}{360}x^7-\frac{7}{1620}x^9+\frac{13}{43,200}x^{11}$$
$$-\frac{1}{86,400}x^{13}+\frac{1}{5,184,000}x^{15}-\ldots$$
$$\ln(1+\sin x)=x-\frac{1}{2}x^2+\frac{1}{6}x^3-\ldots$$

25. $\cos x=1-\frac{x^2}{2!}+\frac{x^4}{4!}-\frac{x^6}{6!}+\cdots$

$$\int\cos x\,dx$$
$$=\int dx-\frac{1}{2}\int x^2dx+\frac{1}{4!}\int x^4dx$$
$$-\frac{1}{6!}\int x^6dx+\cdots$$
$$=x-\frac{1}{2}\frac{x^3}{3}+\frac{1}{4!}\frac{x^5}{5}-\frac{1}{6!}\frac{x^7}{7}+\cdots$$
$$=x-\frac{x^3}{3!}+\frac{x^5}{5!}-\frac{x^7}{7!}+\cdots=\sin x$$

29. $$\int_0^1 e^x dx=e^x\Big|_0^1=e-e^0=e-1$$
$$=2.7182818-1=1.7182818$$

$f(x)=e^x$ $\quad f(0)=e^0=1$

$f'(x)=e^x$ $\quad f'(0)=1$

$f''(x)=e^x$ $\quad f''(0)=1$

$f'''(x)=e^x$ $\quad f'''(0)=1$

$$e^x=1+x+\frac{x^2}{2!}+\frac{x^3}{3!}+\cdots$$

$$\int_0^1 \left(1 + x + \frac{x^2}{2} + \frac{x^3}{6}\right) dx$$
$$= x + \frac{x^2}{2} + \frac{x^3}{6} + \frac{x^4}{24}\Big|_0^1$$
$$= 1 + \frac{1}{2} + \frac{1}{6} + \frac{1}{24}$$
$$= 1.7083333$$

33. $y = x^2e^x$; $x = 0.2$, x-axis

$$A_{0,0.2} = \int_0^{0.2} x^2e^x dx$$
$$= \int_0^{0.2} x^2\left(1 + x + \frac{x^2}{2}\right) dx$$
$$= \int_0^{0.2} \left(x^2 + x^3 + \frac{1}{2}x^4\right) dx$$
$$= \frac{x^3}{3} + \frac{x^4}{4} + \frac{x^5}{10}\Big|_0^{0.2}$$
$$= \frac{1}{3}(0.2)^3 + \frac{1}{4}(0.2)^4 + \frac{1}{10}(0.2)^5$$
$$= 0.003099$$

37. $K = \left[\left(1 - \frac{v^2}{c^2}\right)^{-1/2} - 1\right] mc^2$

$$= \left[1 - \frac{1}{2}\left(\frac{-v^2}{c^2}\right) + \frac{-\frac{1}{2}\left(-\frac{1}{2} - 1\right)}{2!}\left(\frac{-v^2}{c^2}\right)^2 + \cdots - 1\right] mc^2$$
$$= \left[1 + \frac{1}{2}\frac{v^2}{c^2} + \frac{3}{8}\frac{v^4}{c^4} + \cdots\right] mc^2$$
$$= \frac{1}{2}mv^2 + \frac{3}{8}\frac{mv^4}{c^4}$$
$$= \frac{1}{2}mv^2 \text{ for } v \text{ much smaller than c.}$$

41. $y_1 = \ln(1 + x)$,

$y_2 = x$,

$y_3 = x - \frac{1}{2}x^2$,

$y_4 = x - \frac{1}{2}x^2 + \frac{1}{3}x^3$

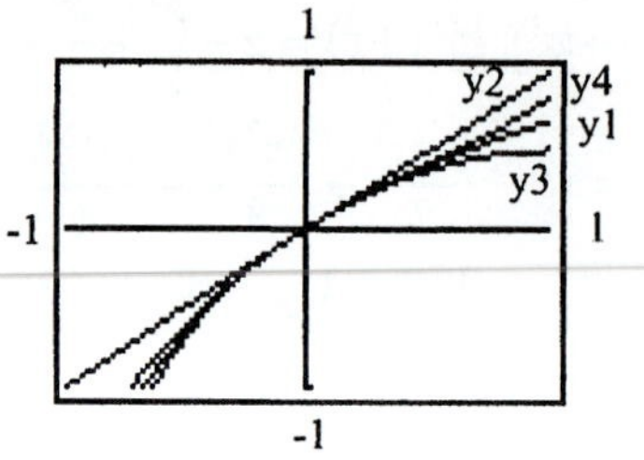

30.4 Computations by Use of Series Expansions

1. $e^x = 1 + x + \frac{x^2}{2!} + \cdots$

$$e^{-0.1} = 1 + (-0.1) + \frac{(-0.1)^2}{2!} + \cdots$$
$$e^{-0.1} = 0.905$$

5. $\sin 0.1$, (2 terms); $\sin x = x - \frac{x^3}{3!}$

$$\sin 0.1 = 0.1 - \frac{(0.1)^3}{6} = 0.09983333$$

(0.0998334 calculator)

9. $\cos \pi^\circ$, (2 terms); $\pi^\circ = \frac{\pi^2}{180}$ radians

$$\cos x = 1 - \frac{x^2}{2!}$$
$$\cos 3^\circ = 1 - \frac{\left(\frac{\pi^2}{180}\right)^2}{2} = 0.9984967733$$

(0.9984791499 calculator)

13. $\sin 0.3625$, (3 terms): $\sin x = x - \frac{x^3}{3!} + \frac{x^5}{5!}$

$$\sin 0.3625 = 0.3625 - \frac{(0.3625)^3}{6} + \frac{(0.3625)^5}{5!}$$
$$= 0.3546130$$

(0.3546129 calculator)

17. $(1+x)^6 = 1+6x+15x^2+20x^3+15x^4 + 6x^5+x^6$

$(1.032)^6 = 1+6(0.032)+15(0.032)^2 = 1.20736$

$(1.032)^6 = 1.20803$, calculator

21. $\sqrt{1.1076} = 1.1076^{1/2} = (1+0.1076)^{1/2}$

$(1+x)^n = 1+nx+\frac{n(n-1)x^2}{2!}+\cdots$

$x = 0.1076$ and $n = \frac{1}{2}$

$$\sqrt{1.1076} = 1+\frac{1}{2}(0.1076)+\frac{\frac{1}{2}\left(-\frac{1}{2}\right)(0.1076)^2}{2}+\cdots$$
$$= 1+0.0538000-0.0014472+\cdots$$
$$= 1.0523528$$

25. From Exercise 5, $\sin(0.1) = 0.1+\frac{0.1^3}{6} = 0.1001667$

The maximum possible error is the value of the first term omitted,

$$\frac{x^5}{5!} = \left|\frac{0.1^5}{120}\right| = 8.3\times 10^{-8}$$

29. $(1+x)^n = 1+nx+\frac{n(n-1)}{2!}x^2$

$$\sqrt{3.92} = 2(1+(-0.02))^{1/2}$$
$$= 2\left[1+\frac{1}{2}(-0.02)+\frac{\frac{1}{2}\left(\frac{1}{2}-1\right)}{2!}(-0.02)^2\right]$$
$$= 1.9799$$

33. $e^x = 1+x+\frac{x^2}{2}+\frac{x^3}{3!}+\frac{x^4}{4!}+\cdots > 1+x+\frac{x^2}{2}$

for $x>0$ since the terms of the expansion for e^x after those on right hand side of the inequality have a positive value.

37. $f(t) = \frac{E}{R}(1-e^{-Rt/L})$; $e^x = 1+x+\frac{x^2}{2}+\cdots$

$$e^{-Rt/L} = 1-\frac{Rt}{L}+\frac{R^2t^2}{2L^2}+\cdots$$

$$i = \frac{E}{R}\left[1-\left(1-\frac{Rt}{L}+\frac{R^2t^2}{2L^2}\right)\right] = \frac{E}{L}\left(t-\frac{Rt^2}{2L}\right)$$

The approximation will be valid for small values of t.

30.5 Taylor Series

1. $f(x) = x^{1/2}, f(1) = 1$

$f'(x) = \frac{1}{2x^{1/2}}, f'(1) = \frac{1}{2}$

$f''(x) = -\frac{1}{4x^{3/2}}, f''(1) = -\frac{1}{4}$

$f'''(x) = \frac{3}{8x^{5/2}}, f'''(1) = \frac{3}{8}$

$$\sqrt{x} = 1+\frac{1}{2}(x-1)+\frac{-\frac{1}{4}(x-1)^2}{2!}+\frac{\frac{3}{8}(x-1)^3}{3!}+\cdots$$
$$\sqrt{x} = 1+\frac{1}{2}(x-1)-\frac{1}{8}(x-1)^2+\frac{1}{16}(x-1)^3-\cdots$$

5. $\sqrt{4.2}$; $\sqrt{x} = 2+\frac{(x-4)}{4}-\frac{(x-4)^2}{64}+\frac{(x-3)^3}{512}$

$$\sqrt{4.2} = 2+\frac{(4.2-4)}{4}-\frac{(4.2-4)^2}{64}+\frac{(4.2-4)^3}{512}$$
$$= 2.049;\ (2.04939 \text{ calculator})$$

9. $\sin x = \frac{1}{2}+\frac{\sqrt{3}}{2}\left(x-\frac{\pi}{6}\right)-\frac{1}{4}\left(x-\frac{\pi}{6}\right)^2$

$\sin 29.53°$

$$= \frac{1}{2}+\frac{\sqrt{3}}{2}\left(\frac{29.53\pi}{180}-\frac{\pi}{6}\right) - \frac{1}{4}\left(\frac{29.53\pi}{180}-\frac{\pi}{6}\right)^2$$
$$= 0.49288;\ (0.4928792 \text{ calculator})$$

13. $\sin x$; $a = \frac{\pi}{3}$

$f(x) = \sin x \qquad f\left(\frac{\pi}{3}\right) = \frac{\sqrt{3}}{2}$

$f'(x) = \cos x \qquad f'\left(\frac{\pi}{3}\right) = \frac{1}{2}$

$f''(x) = -\sin x \qquad f''\left(\frac{\pi}{3}\right) = -\frac{\sqrt{3}}{2}$

$$\sin x = \frac{\sqrt{3}}{2}+\frac{1}{2}\left(x-\frac{\pi}{3}\right)-\frac{\sqrt{3}}{2!}\left(x-\frac{\pi}{3}\right)^2-\cdots$$
$$= \frac{1}{2}\left[\sqrt{3}+\left(x-\frac{\pi}{3}\right)-\frac{\sqrt{3}}{2!}\left(x-\frac{\pi}{3}\right)^2-\cdots\right]$$

17. $\tan x;\ a = \frac{\pi}{4}$

$$f(x) = \tan x \qquad f\left(\frac{\pi}{4}\right) = 1$$

$$f'(x) = \sec^2 x \qquad f'\left(\frac{\pi}{4}\right) = (\sqrt{2})^2 = 2$$

$$f''(x) = 2\sec x \sec x \tan x = 2\sec^2 x \tan x$$
$$f''(x) = 2(\sqrt{2})^2(1) = 4$$

$$\tan x = 1 + 2\left(x - \frac{\pi}{4}\right) + \frac{4\left(x - \frac{\pi}{4}\right)^2}{2!} + \cdots$$
$$= 1 + 2\left(x - \frac{\pi}{4}\right) + 2\left(x - \frac{\pi}{4}\right)^2 + \cdots$$

21. $f(x) = \frac{1}{x+2},\ f(3) = \frac{1}{5}$

$$f'(x) = -\frac{1}{(x+2)^2},\ f'(3) = -\frac{1}{25}$$

$$f''(x) = \frac{2}{(x+2)^3},\ f''(3) = \frac{2}{125}$$

$$\frac{1}{x+2} = \frac{1}{5} - \frac{1}{25}(x-3) + \frac{1}{125}(x-3)^2$$

25. $\sqrt{9.3};\ a = 9$

$$f(x) = \sqrt{x} \qquad f(9) = 3$$

$$f'(x) = \frac{1}{2\sqrt{x}} \qquad f'(9) = \frac{1}{6}$$

$$f''(x) = -\frac{1}{4x^{3/2}} \qquad f''(9) = -\frac{1}{108}$$

$$\sqrt{x} = 3 + \frac{1}{6}(x-9) - \frac{1}{108}\frac{(x-9)^2}{2!}$$

$$\sqrt{9.3} = 3 + \frac{1}{6}(0.3) - \frac{1}{108}\frac{(0.3)^2}{2} = 3.0496$$

29. $\sin x = \frac{1}{2}\left[\sqrt{3} + \left(x - \frac{\pi}{3}\right) - \frac{\sqrt{3}}{2}\left(x - \frac{\pi}{3}\right)^2\right];\ a = \frac{\pi}{3}$

$$61° = 60° + 1° = \frac{\pi}{3} + \frac{\pi}{180}$$

$$\sin 61° = \frac{1}{2}\left[\sqrt{3} + \frac{\pi}{180} - \frac{\sqrt{3}}{2}\left(\frac{\pi}{180}\right)^2\right] = 0.87462$$

33. Expand $f(x) = 2x^3 + x^2 - 3x + 5$ about $x = 1$

$$f'(x) = 6x^2 + 2x - 3$$
$$f''(x) = 12x + 1$$
$$f'''(x) = 12$$

$$f(x) = f(1) + f'(1)(x-1) + \frac{f''(1)(x-1)^2}{2!} + \frac{f'''(1)(x-1)^2}{3!}$$
$$= 2(1)^3 + 1^2 - 3(1) + 5 + \left(6(1)^2 + 2(1) - 3\right)(x-1)$$
$$+ \frac{(12(1)+2)(x-1)^2}{2!} + \frac{12(x-1)^3}{3!}$$
$$= 5 + 5(x-1) + 7(x-1)^2 + 2(x-1)^3$$

37. $i = 6\sin \pi t, \qquad i(\pi/2) = 6\sin \pi^2/2$

$$i' = 6\pi\cos \pi t, \qquad i'(\pi/2) = 6\pi\cos \pi^2/2$$
$$i'' = -6\pi^2 \sin \pi t, \quad i''(\pi/2) = -6\pi^2 \sin \pi^2/2$$

$$i = 6\sin\frac{\pi^2}{2} + 6\pi\cos\frac{\pi^2}{2}\left(t - \frac{\pi}{2}\right)$$
$$-3\pi^2 \sin\frac{\pi^2}{2}\left(t - \frac{\pi}{2}\right)^2 + \cdots$$

41. $f(x) = \frac{1}{x};\ x = 0$ to $x = 4$

(a) $y_1 = \frac{1}{x}$

(b) $y_2 = \frac{1}{2} - \frac{1}{4}(x-2)$

Graph of part (b) will fit the graph of part (a) well for values of x close to $x = 2$.

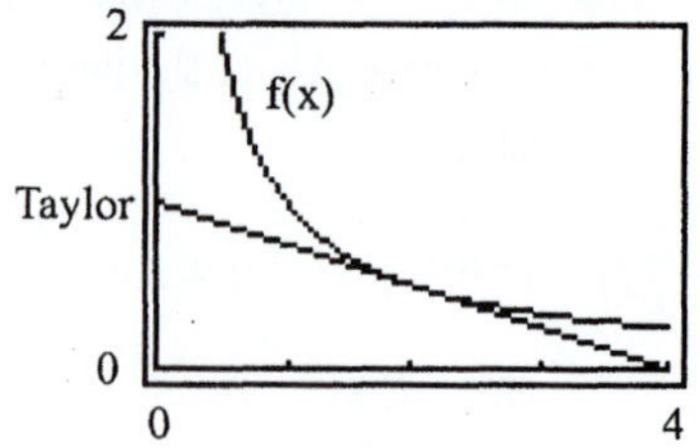

30.6 Introduction to Fourier Series

1. $f(x) = \begin{cases} -2, -\pi \le x < 0 \\ 2, \quad 0 \le x < \pi \end{cases}$

$$a_0 = \frac{1}{2\pi}\int_{-\pi}^{0} (-2)dx + \frac{1}{2\pi}\int_{0}^{\pi} 2\,dx = 0$$

$$a_n = \frac{1}{\pi}\int_{-\pi}^{0} -2\cos nx\,dx + \frac{1}{\pi}\int_{0}^{\pi} 2\cos nx\,dx = 0$$

$$b_n = \frac{1}{\pi}\int_{-\pi}^{0} -2\sin nx\,dx + \frac{1}{\pi}\int_{0}^{\pi} 2\sin nx\,dx$$

$$= \frac{4}{\pi}\left(\frac{1}{n} - \frac{\cos(\pi n)}{n}\right)$$

$$b_n = \frac{4}{\pi}(1 - \cos\pi n) = \begin{cases} \dfrac{8}{n\pi}, n \text{ odd} \\ 0, n \text{ neven} \end{cases}$$

$$b_1 = \frac{8}{\pi}, b_3 = \frac{8}{3\pi}, b_5 = \frac{8}{5\pi}$$

$$f(x) = \frac{8}{\pi}\sin x + \frac{8}{3\pi}\sin 3x + \frac{8}{5\pi}\sin 5x + \cdots$$

$$f(x) = \frac{8}{\pi}\left(\sin x + \frac{1}{3}\sin 3x + \frac{1}{5}\sin 5x + \cdots\right)$$

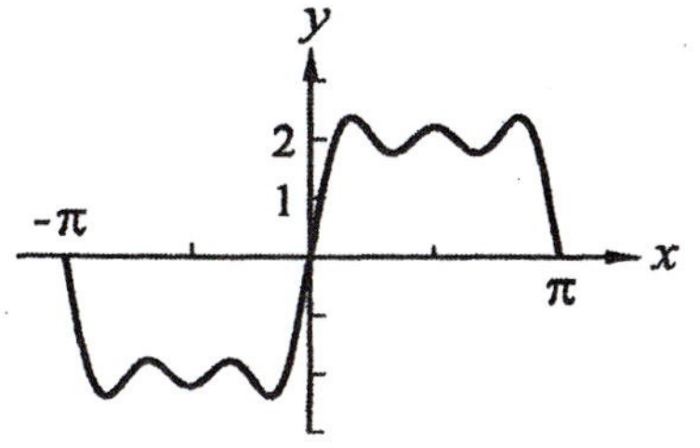

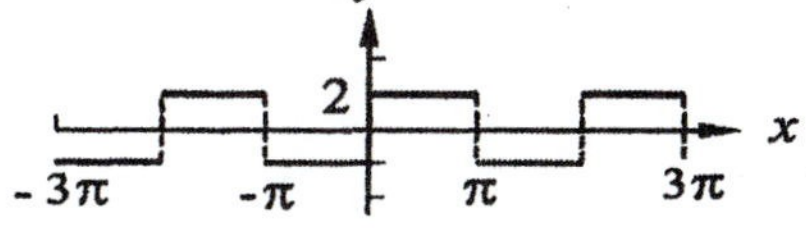

5. $f(x) = \begin{cases} 1 & -\pi \le x < 0 \\ 2 & 0 \le x < \pi \end{cases}$

$$a_0 = \frac{1}{2\pi}\int_{-\pi}^{0} 1\,dx + \frac{1}{2\pi}\int_{0}^{\pi} 2\,dx$$

$$= \frac{x}{2\pi}\Big|_{-\pi}^{0} + \frac{2x}{2\pi}\Big|_{0}^{\pi}$$

$$= 0 + \frac{\pi}{2\pi} + \frac{2\pi}{2\pi} - 0 = \frac{1}{2} + 1 = \frac{3}{2}$$

$$a_1 = \frac{1}{\pi}\int_{-\pi}^{0} 1\cos x\,dx + \frac{1}{\pi}\int_{0}^{\pi} 2\cos x\,dx$$

$$= \frac{1}{\pi}\sin x\Big|_{-\pi}^{0} + \frac{2}{\pi}\sin x\Big|_{0}^{\pi}$$

$$= \frac{1}{\pi}(0-0) + \frac{2}{\pi}(0-0) = 0$$

$a_n = 0$ since $\sin n\pi = 0$

$$b_1 = \frac{1}{\pi}\int_{-\pi}^{0} 1\sin x\,dx + \frac{1}{\pi}\int_{0}^{\pi} 2\sin x\,dx$$

$$= -\frac{1}{\pi}\cos x\Big|_{-\pi}^{0} - \frac{2}{\pi}\cos x\Big|_{0}^{\pi}$$

$$= -\frac{1}{\pi}(1+1) - \frac{2}{\pi}(-1-1)$$

$$= -\frac{2}{\pi} + \frac{4}{\pi} = \frac{2}{\pi}$$

$$b_2 = \frac{1}{\pi}\int_{-\pi}^{0} 1\sin 2x\,dx + \frac{1}{\pi}\int_{0}^{\pi} 2\sin x\,dx$$

$$= -\frac{1}{2\pi}\cos 2x\Big|_{-\pi}^{0} - \frac{1}{\pi}\cos 2x\Big|_{0}^{\pi}$$

$$= -\frac{1}{2\pi}(1-1) - \frac{1}{\pi}(1-1) = 0$$

$$b_3 = \frac{1}{\pi}\int_{-\pi}^{0} \sin 3x\,dx + \frac{1}{\pi}\int_{0}^{\pi} 2\sin 3x\,dx$$

$$= -\frac{1}{3\pi}\cos 3x\Big|_{-\pi}^{0} - \frac{2}{3\pi}\cos 3x\Big|_{0}^{\pi}$$

$$= -\frac{1}{3\pi}(1+1) - \frac{2}{3\pi}(-1-1) = \frac{2}{3\pi}$$

Therefore, $b_n = 0$ for n even; $b_n = \dfrac{2}{n\pi}$ for n odd.

Therefore, $f(x) = \dfrac{3}{2} + \dfrac{2}{\pi}\sin x + \dfrac{2}{3\pi}\sin 3x + \cdots$

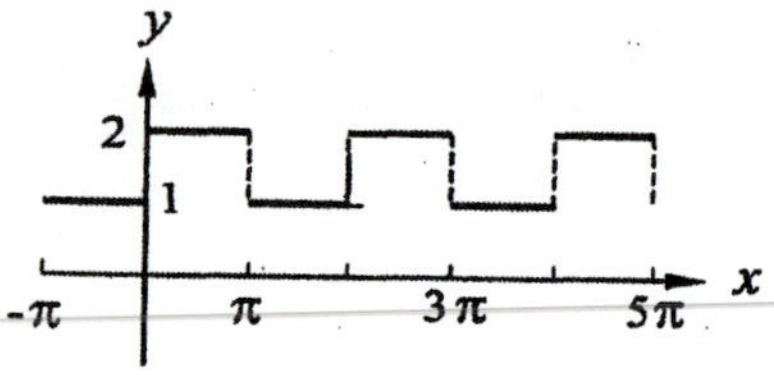

9. $f(x) = \begin{cases} -1 & -\pi \le x < 0 \\ 0 & 0 \le x < \frac{\pi}{2} \\ 1 & \frac{\pi}{2} \le x < \pi \end{cases}$

$$a_0 = \frac{1}{2\pi}\int_{-\pi}^{0} -dx + \frac{1}{2\pi}\int_{\pi/2}^{\pi} dx$$
$$= \frac{1}{2\pi}x\Big|_{-\pi}^{0} + \frac{1}{2\pi}x\Big|_{\pi/2}^{\pi}$$
$$= -\frac{1}{2\pi}\left(x\Big|_{-\pi}^{0} - x\Big|_{\pi/2}^{\pi}\right)$$
$$= -\frac{1}{2\pi}\left[\pi - \left(\pi - \frac{\pi}{2}\right)\right] = -\frac{1}{4}$$
$$a_1 = \frac{1}{\pi}\int_{-\pi}^{0} -\cos x\, dx + \frac{1}{\pi}\int_{\pi/2}^{\pi} c\cos x\, dx$$
$$= -\frac{1}{\pi}\sin x\Big|_{-\pi}^{0} + \frac{1}{\pi}\sin x\Big|_{\pi/2}^{\pi}$$
$$= -\frac{1}{\pi}\left(\sin x\Big|_{-\pi}^{0} - \sin x\Big|_{\pi/2}^{\pi}\right) = -\frac{1}{\pi}$$

$$a_2 = \frac{1}{\pi}\int_{-\pi}^{0} -\cos 2x\, dx + \frac{1}{\pi}\int_{\pi/2}^{\pi} \cos 2x\, dx$$
$$= -\frac{1}{2\pi}\sin 2x\Big|_{-\pi}^{0} + \frac{1}{2\pi}\sin 2x\Big|_{\pi/2}^{\pi}$$
$$= -\frac{1}{2\pi}\left(\sin 2x\Big|_{-\pi}^{0} - \sin 2x\Big|_{\pi/2}^{\pi}\right) = 0$$

$$a_3 = \frac{1}{\pi}\int_{-\pi}^{0} -\cos 3x\, dx + \frac{1}{\pi}\int_{\pi/2}^{\pi} \cos 3x\, dx$$
$$= -\frac{1}{3\pi}\sin 3x\Big|_{-\pi}^{0} + \frac{1}{3\pi}\sin 3x\Big|_{\pi/2}^{\pi}$$
$$= -\frac{1}{3\pi}\left(\sin 3x\Big|_{-\pi}^{0} - \sin 3x\Big|_{\pi/2}^{\pi}\right) = \frac{1}{3\pi}$$

Therefore, $a_n = \pm\frac{1}{n\pi}$ for n odd; $a_n = 0$ for n even.

$$b_1 = \frac{1}{\pi}\int_{-\pi}^{0} -\sin x\, dx + \frac{1}{\pi}\int_{\pi/2}^{\pi} \sin x\, dx$$
$$= \frac{1}{\pi}\cos x\Big|_{-\pi}^{0} - \frac{1}{\pi}\cos x\Big|_{\pi/2}^{\pi}$$
$$= \frac{1}{\pi}\left(\cos x\Big|_{-\pi}^{0} - \cos x\Big|_{\pi/2}^{\pi}\right) = \frac{3}{\pi}$$
$$b_2 = \frac{1}{\pi}\int_{-\pi}^{0} -\sin 2x\, dx + \frac{1}{\pi}\int_{\pi/2}^{\pi} \sin 2x\, dx$$
$$= \frac{1}{2\pi}\cos 2x\Big|_{-\pi}^{0} - \frac{1}{2\pi}\cos 2x\Big|_{\pi/2}^{\pi}$$
$$= \frac{1}{2\pi}\left(\cos x\Big|_{-\pi}^{0} - \cos 2x\Big|_{\pi/2}^{\pi}\right) = -\frac{1}{\pi}$$
$$b_3 = \frac{1}{\pi}\int_{-\pi}^{0} -\sin 3x\, dx + \frac{1}{\pi}\int_{\pi/2}^{\pi} \sin 3x\, dx$$
$$= \frac{1}{3\pi}\cos 3x\Big|_{-\pi}^{0} - \frac{1}{3\pi}\cos 3x\Big|_{\pi/2}^{\pi}$$
$$= \frac{1}{3\pi}\left(\cos 3x\Big|_{-\pi}^{0} - \cos 3x\Big|_{\pi/2}^{\pi}\right) = -\frac{1}{\pi}$$
$$b_n = \pm\frac{1}{\pi} \text{ for } n > 1$$
$$f(x) = -\frac{1}{4} - \frac{1}{\pi}\cos x + \frac{1}{3\pi}\cos 3x - \cdots$$
$$+ \frac{3}{\pi}\sin x - \frac{1}{\pi}\sin 2x + \frac{1}{\pi}\sin 3x - \cdots$$

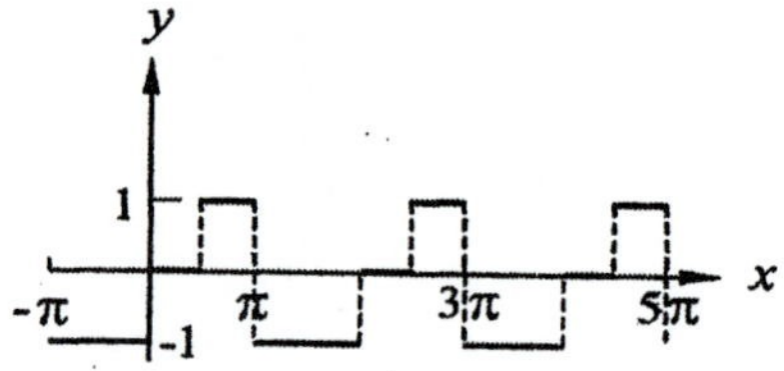

13. $a_0 = \frac{1}{2\pi}\int_{-\pi}^{\pi} e^x\,dx = \frac{e^{\pi}-e^{-\pi}}{2\pi}$

$a_1 = \frac{1}{\pi}\int_{-\pi}^{\pi} e^x \cos x\,dx = -\frac{e^{\pi}-e^{-\pi}}{2\pi}$

$a_2 = \frac{1}{\pi}\int_{-\pi}^{\pi} e^x \cos 2x\,dx = \frac{e^{\pi}-e^{-\pi}}{5\pi}$

$b_1 = \frac{1}{\pi}\int_{-\pi}^{\pi} e^x \sin x\,dx = -\frac{e^{\pi}-e^{-\pi}}{2\pi}$

$b_2 = \frac{1}{\pi}\int_{-\pi}^{\pi} e^x \sin 2x\,dx = -\frac{2(e^{\pi}-e^{-\pi})}{5\pi}$

$$e^x = \frac{e^{\pi}-e^{-\pi}}{2\pi} - \frac{e^{\pi}-e^{-\pi}}{2\pi}\cos x + \frac{e^{\pi}-e^{-\pi}}{5\pi}\cos 2x + \cdots + \frac{e^{\pi}-e^{-\pi}}{2\pi}\sin x - \frac{2(e^{\pi}-e^{-\pi})}{5\pi}\sin 2x + \cdots$$

$$e^x = \frac{e^{\pi}-e^{-\pi}}{\pi}\left(\frac{1}{2} - \frac{1}{2}\cos x + \frac{1}{5}\cos 2x + \cdots + \frac{1}{2}\sin x - \frac{2}{5}\sin 2x + \cdots\right)$$

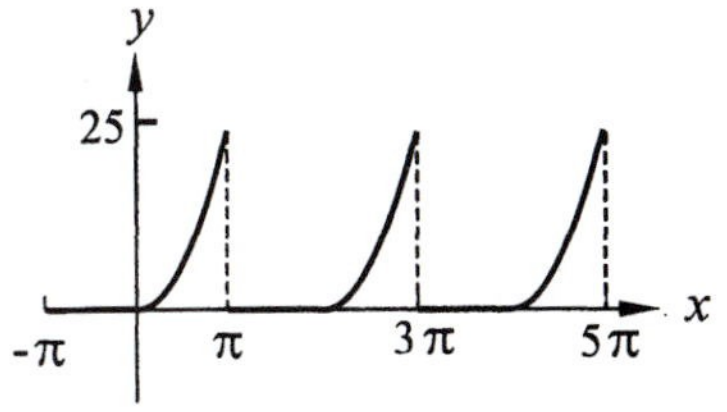

17. $f(x) = \begin{cases} 1, -\pi \le x < 0 \\ 2, 0 \le x < \pi \end{cases}$

Graph $y_1 = \frac{3}{2} + \frac{2}{\pi}\sin x + \frac{2}{3\pi}\sin 3x$

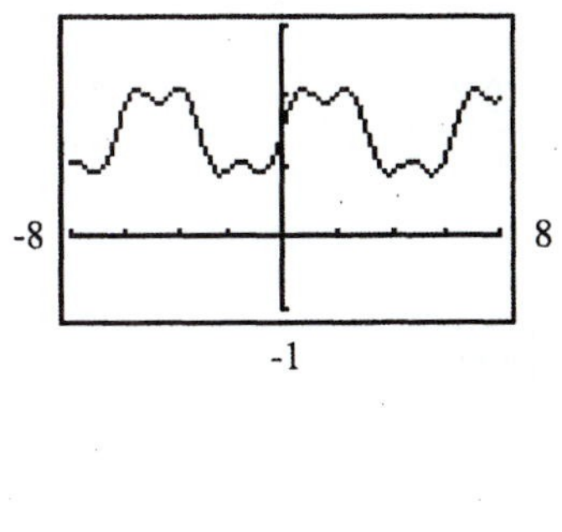

21. $F(t) = \begin{cases} 0, -\pi \le t < 0 \\ t^2 + t, \ 0 < t < \pi \end{cases}$

$a_0 = \frac{1}{2\pi}\int_0^{\pi}\left(t^2+t\right)dt = \frac{\pi^2}{6} + \frac{\pi}{4}$

$a_1 = \frac{1}{\pi}\int_0^{\pi}\left(t^2+t\right)\cos t\,dt = -\frac{2}{\pi} - 2$

$a_2 = \frac{1}{\pi}\int_0^{\pi}\left(t^2+t\right)\cos 2t\,dt = \frac{1}{2}$

$a_3 = \frac{1}{\pi}\int_0^{\pi}\left(t^2+t\right)\cos 3t\,dt = \frac{-2-2\pi}{9\pi}$

$b_1 = \frac{1}{\pi}\int_0^{\pi}\left(t^2+t\right)\sin t\,dt = \pi - \frac{4}{\pi} + 1$

$b_2 = \frac{1}{\pi}\int_0^{\pi}\left(t^2+t\right)\sin 2t\,dt = \frac{-\pi-1}{2}$

$b_3 = \frac{1}{\pi}\int_0^{\pi}\left(t^2+t\right)\sin 3t\,dt = \frac{\pi}{3} - \frac{4}{27\pi} + \frac{1}{3}$

$$F(t) = \frac{\pi^2}{6} + \frac{\pi}{4} - \left(\frac{2}{\pi}+2\right)\cos t + \frac{1}{2}\cos 2t - \left(\frac{2+2\pi}{9\pi}\right)\cos 3t + \cdots + \left(\pi - \frac{4}{\pi} + 1\right)\sin t - \left(\frac{\pi+1}{2}\right)\sin 2t + \left(\frac{\pi}{3} - \frac{4}{27\pi} + \frac{1}{3}\right)\sin 3t + \cdots$$

30.7 More About Fourier Series

1. $f(x) = \begin{cases} 2 & -\pi \le x < -\frac{\pi}{2}, \frac{\pi}{2} \le x < \pi \\ 3 & -\frac{\pi}{2} \le x < \frac{\pi}{2} \end{cases}$

From Example 2,

$$f(x) = \frac{1}{2} + \frac{2}{\pi}\left(\cos x - \frac{\cos 3x}{3} + \frac{\cos 5x}{5} - \cdots\right) + 2.$$

$$= \frac{5}{2} + \frac{2}{\pi}\left(\cos x - \frac{\cos 3x}{3} + \frac{\cos 5x}{5} - \cdots\right)$$

5. $f(x)=\begin{cases}5 & -3\le x<0\\ 0 & 0\le x<3\end{cases}$

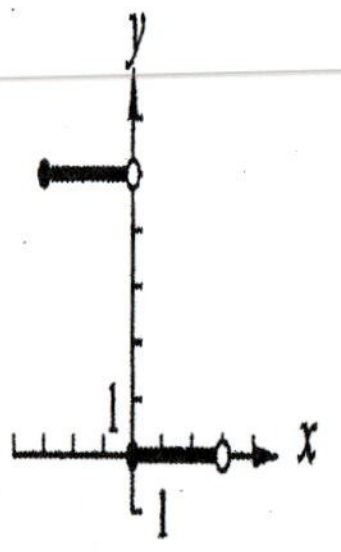

from the graph $f(x)$ is neither odd nor even.

9. $f(x)=|x| \quad -4\le x<4$

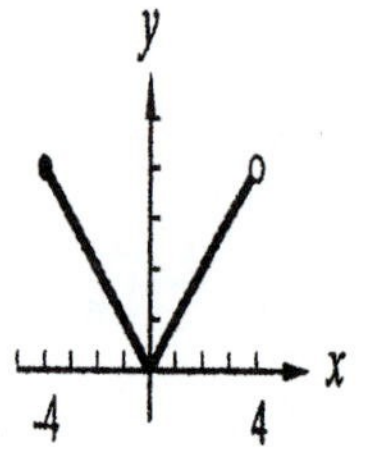

is even from the graph.

13. From the graph, $f(x)=2-x,\ -4\le x<4$ is not odd or even. Fourier series may contain both sine and cosine terms. In fact, the expansion contains only sine terms.

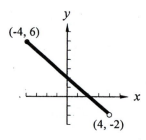

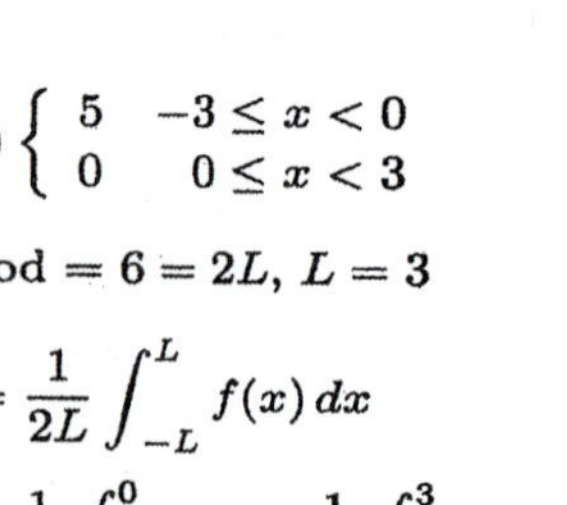

17. $f(x)\begin{cases}5 & -3\le x<0\\ 0 & 0\le x<3\end{cases}$

period $= 6 = 2L,\ L = 3$

$$a_0=\frac{1}{2L}\int_{-L}^{L} f(x)\,dx$$

$$=\frac{1}{6}\int_{-3}^{0}5\,dx+\frac{1}{6}\int_{0}^{3}0\cdot dx=\frac{5}{2}$$

$$a_n=\frac{1}{L}\int_{-L}^{L} f(x)\cos\frac{n\pi x}{L}\,dx$$

$$=\frac{1}{3}\int_{-3}^{0}5\cos\frac{n\pi x}{3}dx+\frac{1}{3}\int_{0}^{3}0\cdot\cos\frac{n\pi x}{3}\,dx$$

$$a_n=\frac{5\sin(n\pi)}{n\pi}=0, n=1,2,3\cdots$$

$$b_n=\frac{1}{L}\int_{-L}^{L} f(x)\sin\frac{n\pi x}{L}\,dx$$

$$=\frac{1}{3}\int_{-3}^{0}5\sin\frac{n\pi x}{3}dx+\frac{1}{3}\int_{0}^{3}0\cdot\sin\frac{2\pi x}{3}\,dx$$

$$b_n=\frac{5\cos(n\pi)-5}{n\pi}=\frac{5}{\pi}\left(\frac{\cos(n\pi)-1}{n}\right)$$

n	b_n
1	$\frac{5}{\pi}\cdot(-2)=\frac{-10}{\pi}$
2	0
3	$\frac{5}{\pi}\left(-\frac{2}{3}\right)=\frac{-10}{3\pi}$
4	0
5	$\frac{5}{\pi}\left(-\frac{2}{5}\right)=\frac{-10}{5\pi}$

$$f(x)=a_0+a_1\cos\frac{\pi x}{L}+a_2\cos\frac{2\pi x}{L}+a_3\cos\frac{3\pi x}{L}+\cdots$$
$$+b_1,\sin\frac{\pi x}{L}+b_2\frac{2\pi x}{L}+b_3\frac{3\pi x}{L}+\cdots$$

$$f(x)=\frac{5}{2}$$
$$-\frac{10}{\pi}\left(\sin\frac{\pi x}{3}+\frac{1}{3}\sin\frac{3\pi x}{3}+\frac{1}{5}\sin\frac{5\pi x}{3}+\cdots\right)$$

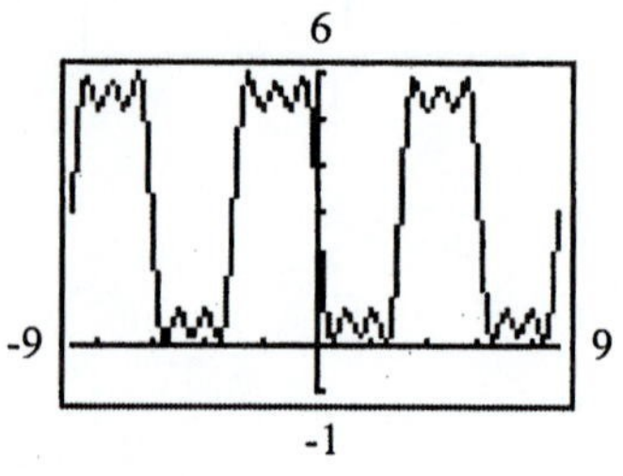

21. $f(x) = \begin{cases} -x & -4 \le x < 0 \\ x & 0 \; \le x < 4 \end{cases}$

$$a_0 = \frac{1}{8}\int_{-4}^{0} -x\,dx + \frac{1}{8}\int_0^4 x\,dx = -\frac{1}{16}x^2\Big|_{-4}^{0} + \frac{1}{16}x^2\Big|_0^4 = 2$$

$$a_n = \frac{1}{4}\int_{-4}^{0} -x\cos\frac{n\pi x}{4}\,dx + \frac{1}{4}\int x\cos\frac{n\pi x}{4}\,dx$$

$$= -\frac{1}{4}\frac{16}{(n\pi)^2}\int_{-4}^{0}\frac{n\pi}{4}x\cos\frac{n\pi}{4}x\frac{n\pi}{4}\,dx + \frac{1}{4}\frac{16}{(n\pi)^2}\int_0^4 \frac{n\pi}{4}x\cos\frac{n\pi}{4}x\frac{n\pi}{4}\,dx$$

$$= -\frac{4}{(n\pi)^2}\left(\cos\frac{n\pi x}{4} + \frac{n\pi x}{4}\sin\frac{n\pi x}{4}\right)\Bigg|_{-4}^{0} + \frac{4}{(n\pi)^2}\left(\cos\frac{n\pi x}{4} + \frac{n\pi x}{4}\sin\frac{n\pi x}{4}\right)\Bigg|_0^4$$

$$-\frac{4}{(n\pi)^2}(\cos 0 - [\cos(-n\pi) + n\pi\sin n\pi]) + \frac{4}{(n\pi)^2}(\cos n\pi + n\pi\sin n\pi - [\cos 0])$$

$$= -\frac{4}{(n\pi)^2}(1 - \cos n\pi - n\pi\sin n\pi) + \frac{4}{(n\pi)^2}(\cos n\pi + n\pi\sin n\pi - 1)$$

$$= -\frac{4}{(n\pi)^2}(1 - \cos n\pi - n\pi\sin n\pi - \cos n\pi - n\pi\sin n\pi + 1)$$

$$= -\frac{4}{(n\pi)^2}(2 - 2\cos n\pi - 2n\pi\sin n\pi)$$

$$a_1 = -\frac{16}{\pi^2};\; a_2 = 0;\; a_3 = -\frac{16}{9\pi^2}$$

$$b_n = \frac{1}{4}\int_{-4}^{0} -x\sin\frac{n\pi x}{4}\,dx + \frac{1}{4}\int_0^4 x\sin\frac{n\pi x}{4}\,dx$$

$$= -\frac{1}{4}\frac{16}{(n\pi)^2}\int_{-4}^{0}\frac{n\pi x}{4}\sin\frac{n\pi x}{4}\left(\frac{n\pi}{4}\,dx\right) + \frac{1}{4}\frac{16}{(n\pi)^2}\int_0^4 \frac{n\pi x}{4}\sin\frac{n\pi x}{4}\cdot\frac{n\pi\,dx}{4}$$

$$= -\frac{4}{(n\pi)^2}\left(\sin\frac{n\pi x}{4} - \frac{n\pi x}{4}\cos\frac{n\pi x}{4}\right)\Bigg|_{-4}^{0} + \frac{4}{(n\pi)^2}\left(\sin\frac{n\pi x}{4} - \frac{n\pi x}{4}\cos\frac{n\pi x}{4}\right)\Bigg|_0^4$$

$$= -\frac{4}{(n\pi)^2}\{-[\sin(-n\pi) + n\pi\cos(-n\pi)]\} + \frac{4}{(n\pi)^2}(\sin n\pi - n\pi\cos n\pi)$$

$$= -\frac{4}{(n\pi)^2}(\sin n\pi - n\pi\cos n\pi) + \frac{4}{(n\pi)^2}(\sin n\pi - n\pi\cos n\pi) = 0, \text{ for all } n.$$

Therefore,

$$f(x) = 2 - \frac{16}{\pi^2}\cos\frac{\pi x}{4} - \frac{16}{9\pi^2}\cos\frac{3\pi x}{4} = 2 - \frac{16}{\pi^2}\left(\cos\frac{\pi x}{4} + \frac{1}{9}\cos\frac{3\pi x}{4} + \cdots\right)$$

and comparing with calculator,

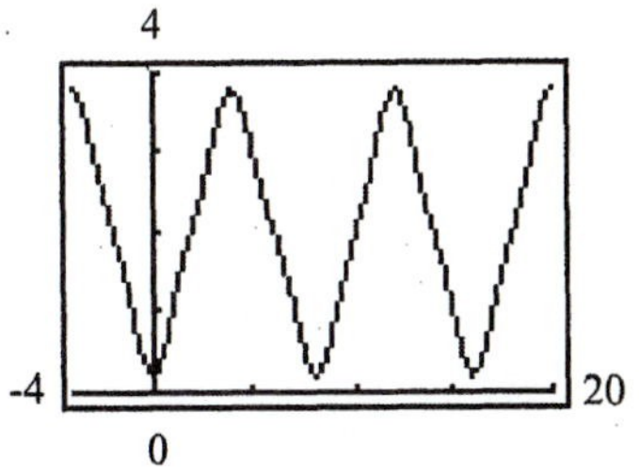

25. Expand $f(x) = x^2$ in a half-range cosine series for $0 \le x < 2$.

$$a_0 = \frac{1}{L}\int_0^L f(x)dx = \frac{1}{2}\int_0^2 x^2dx$$

$$= \frac{1}{2}\frac{x^3}{3}\Big|_0^2 = \frac{1}{6}(2^3 - 0) = \frac{4}{3}$$

$$a_n = \frac{2}{L}\int_0^L f(x)\cos\frac{n\pi x}{L}\,dx, (n = 1, 2, 3, \ldots)$$

$$a_n = \frac{2}{2}\int_0^2 x^2\frac{n\pi x}{2}\,dx$$

$$= \frac{2x}{\frac{n^2\pi^2}{4}}\cos\frac{n\pi x}{2} + \left(\frac{x^2}{\frac{n\pi}{2}} - \frac{2}{\frac{n^3\pi^3}{8}}\right)\sin\frac{n\pi x}{2}\Bigg|_0^2$$

$$a_n = \frac{8x}{n^2\pi^2}\cos\frac{n\pi x}{2} + \left(\frac{2x^2}{n\pi} - \frac{16}{n^3\pi^3}\right)\sin\frac{n\pi x}{2}\Bigg|_0^2$$

$$a_n = \frac{16}{n^2\pi^2}\cos n\pi + \left(\frac{8}{n\pi} - \frac{16}{n^3\pi 3}\right)\sin n\pi$$

$$a_n = \frac{16}{n^2\pi^2}\cos n\pi$$

$$a_1 = \frac{16}{1^2\pi^2}\cos\pi = \frac{-16}{\pi^2}$$

$$a_2 = \frac{16}{2^2\pi^2}\cos 2\pi = \frac{4}{\pi^2}$$

$$a_3 = \frac{16}{3^2\pi^2}\cos 3\pi = \frac{-16}{9\pi^2}$$

$$f(x) = \frac{4}{3} - \frac{16}{\pi^2}\cos\frac{\pi x}{2} + \frac{4}{\pi^2}\cos\frac{2\pi x}{2} - \frac{16}{9\pi^2}\cos\frac{3\pi x}{2} + \cdots$$

$$f(x) = \frac{4}{3} - \frac{16}{\pi^2}\left(\cos\frac{\pi x}{2} - \frac{1}{4}\cos\pi x + \frac{1}{9}\cos\frac{3\pi x}{2} - \cdots\right)$$

Chapter 30 Review Exercises

1. $f(x) = \frac{1}{1+e^x} = (1+e^x)^{-1}$

$f'(x) = -(1+e^x)^{-2}(e^x) = e^x(1+e^x)^{-2}$

$f''(x) = -(1+e^x)^{-2}e^x + e^x(2)(1+e^x)^{-3}(e^x)$

$f'''(x) = -(1+e^x)^{-2}e^x + e^x(2)(1+e^x)^{-3}(e^x)$
$+ 2e^{2x}(-3)(1+e^x)^{-4}e^x$
$+ (1+e^x)^{-3}(2e^{2x})(2)$

$f(0) = \frac{1}{1+1} = \frac{1}{2}$

$f'(0) = -1(2^{-2}) = -\frac{1}{4}$

$f''(0) = -\frac{1}{4} + \frac{1}{4} = 0$

$f'''(0) = -\frac{1}{4} + \frac{1}{4} - \frac{3}{8} + \frac{1}{2} = \frac{1}{8}$

$f(x) = \frac{1}{2} - \frac{1}{4}x + \frac{0x^2}{2!} + \left(\frac{1}{8}\right)\frac{x^3}{3!} + \cdots$

$= \frac{1}{2} - \frac{1}{4}x + \frac{1}{48}x^3 - \cdots$

5. $f(x) = (x+1)^{1/3}$ $\quad f(0) = 1$

$f'(x) = \frac{1}{3}(x+1)^{-2/3}$ $\quad f'(0) = \frac{1}{3}$

$f''(x) = -\frac{2}{9}(x+1)^{-5/3}$ $\quad f''(0) = -\frac{2}{9}$

$f(x) = 1 + \frac{1}{3}x - \frac{2x^2}{9(2)} + \cdots$

$= 1 + \frac{1}{3}x - \frac{1}{9}x^2 + \cdots$

9. $f(x) = \cos(a+x),$ $\quad f(0) = \cos a$

$f'(x) = -\sin(a+x),$ $\quad f'(0) = -\sin a$

$f''(x) = -\cos(a+x),$ $\quad f''(0) = -\cos a$

$f(x) = \cos(a+x) = \cos a - \sin a \cdot x - \frac{\cos a}{2!}x^2 + \cdots$

13. See Exercise 5.

$\sqrt[3]{1+x} = 1 + \frac{1}{3}x - \frac{1}{9}x^2 + \cdots$

$\sqrt[3]{1+0.3} = 1 + \frac{1}{3}(0.3) - \frac{1}{9}(0.3)^2$

$\sqrt[3]{1.3} = 1.09$

17. $\ln(1+x) = x - \frac{x^2}{2} + \frac{x^3}{3} - \cdots$

$\ln[1+(-0.1828)] = -0.1828 - \frac{(-0.1828)^2}{2} + \frac{(-0.1828)^3}{3} - \cdots$

$\ln 0.8172 = -0.2015$

21. $f(x) = \sqrt{x}, a = 144$

$f(x) = \sqrt{x}, f(144) = 12$

$f'(x) = \frac{1}{2}x^{-1/2}, f'(144) = \frac{1}{24}$

$f''(x) = -\frac{1}{4}x^{-3/2}, f''(144) = -\frac{1}{6912}$

$\sqrt{x} = 12 + \frac{1}{24}(x-144) - \frac{1}{6912}\frac{(x-144)^2}{2} + \cdots$

$\sqrt{148} = 12 + \frac{4}{24} - \frac{4^2}{13,824} = 12.1655$

25. $f(x) = \cos x, a = \frac{\pi}{3}$

$f(x) = \cos x, f\left(\frac{\pi}{3}\right) = \frac{1}{2}$

$f'(x) = -\sin x, f'\left(\frac{\pi}{3}\right) = \frac{1}{2}\sqrt{3}$

$f''(x) = -\cos x, f''\left(\frac{\pi}{3}\right) = -\frac{1}{2}$

$f(x) = \frac{1}{2} + \frac{1}{2}\sqrt{3}\left(x - \frac{\pi}{3}\right) - \frac{1}{4}\left(x - \frac{\pi}{3}\right)^2 + \cdots$

29. $f(x) = \begin{cases} \pi - 1, & -4 \le x < 0 \\ \pi + 1, & 0 \le x < 4 \end{cases}$ is Example 7

$f(x) = \pi - 1 + 1 + \frac{4}{\pi}\sin\frac{\pi x}{4} + \frac{4}{3\pi}\sin\frac{3\pi x}{4} + \cdots$

$f(x) = \pi + \frac{4}{\pi}\left(\sin\frac{\pi x}{4} + \frac{1}{3}\sin\frac{3\pi x}{4} + \cdots\right)$

33. $f(x)=x \quad -2\le x<2$, period $=4$, $L=2$; $a_0=\frac{1}{4}\int_{-2}^{2}x\,dx=\frac{1}{8}x^2\Big|_{-2}^{2}$

$$a_n=\frac{1}{2}\int_{-2}^{2}x\cos\frac{n\pi x}{2}\,dx=\frac{1}{2}\left(\frac{2}{n\pi}\right)^2\left(\cos\frac{n\pi x}{2}+\frac{n\pi x}{2}\sin\frac{n\pi x}{2}\right)\Big|_{-2}^{2}$$

$$=\frac{2}{n^2\pi^2}[\cos n\pi+n\pi\sin n\pi-\cos(-n\pi)+n\pi\sin(-n\pi)]=0 \text{ for all } n$$

$$b_n=\frac{1}{2}\int_{-2}^{2}x\sin\frac{n\pi x}{2}\,dx=\frac{1}{2}\left(\frac{2}{n\pi}\right)^2\left(\sin\frac{n\pi x}{2}-\frac{n\pi x}{2}\cos\frac{n\pi x}{2}\right)\Big|_{-2}^{2}$$

$$=\frac{2}{n^2\pi^2}[\sin n\pi-n\pi\cos n\pi-\sin(-n\pi)-n\pi\cos(-n\pi)]$$

$$=\frac{2}{n^2\pi^2}(-n\pi\cos n\pi-n\pi\cos n\pi)=\frac{-4}{n\pi}\cos n\pi;\ b_1=-\frac{4}{\pi}\cos\pi=\frac{4}{\pi},$$

$$b_2=-\frac{4}{2\pi}\cos 2\pi=-\frac{2}{\pi},\ b_3=\frac{-4}{3\pi}\cos 3\pi=\frac{4}{3\pi}$$

37. It is a geometric series for which $|r|<1=0.75$.
Therefore the series converges.

$$S=\frac{64}{1-0.75}=256$$

41. $f(x)=\tan x,\qquad f\left(\frac{\pi}{4}\right)=1$

$f'(x)=1+\tan^2 x,\qquad f'\left(\frac{\pi}{4}\right)=2$

$f''(x)=2\tan x\left(+\tan^2 x\right),\qquad f''(x)=4$

$$f(x)=\tan x=1+2\left(x-\frac{\pi}{4}\right)+\frac{4\left(x-\frac{\pi}{4}\right)^2}{2!}+\cdots$$

$$=1+2\left(x-\frac{\pi}{4}\right)+2\left(x-\frac{\pi}{4}\right)^2+\cdots$$

45. $\sin x=x-\frac{x^3}{3!}+\cdots$

$\sin(x+h)-\sin(x-h)$

$$=(x+h)-\frac{(x+h)^3}{3!}+\cdots-(x-h)+\frac{(x+h)^3}{3!}-\cdots$$

$$=x+h-\frac{x^3+3x^2h+3xh^2+h^3}{3!}+\cdots-x+h+\frac{x^3-3x^2h+3xh^2-h^3}{3!}-\cdots$$

$$=2h-\frac{6x^2h}{3!}-\frac{2h^3}{3!}+\cdots=2h\left(1-\frac{x^2}{2}+\cdots\right)-\frac{2h^3}{3!}+\cdots$$

$=2h\cos x$ for small h

49. $\sin^2 x=\frac{1}{2}(1-\cos 2x)$

$$=\frac{1}{2}\left(1-\left(1-\frac{(2x)^2}{2!}\right)+\frac{(2x)^4}{4!}-\frac{(2x)^6}{6!}+\cdots\right)$$

$$=\frac{1}{2}\left(1-1+2x^2-\frac{2}{3}x^4+\frac{4}{45}x^6-\cdots\right)$$

$$=x^2-\frac{1}{3}x^4+\frac{2}{45}x^6-\cdots$$

53. $f(x)=\frac{1}{1-x}=1+x+x^2+\cdots$

$$\frac{1}{1+x}=\frac{1}{1-(-x)}=1+(-x)+(-x)^2+\cdots$$

$$=1-x+x^2-\cdots$$

57. $e^x=1+x+\frac{x^2}{2!}+\frac{x^3}{3!}$

$$e^{0.9}=1+(0.9)+\frac{0.9^2}{2}+\frac{0.9^3}{6}=2.4265$$

$$e^x=e\left[1+(x-1)+\frac{(x-1)^2}{2}\right]$$

$$e^{0.9}=e\left[1+(0.9-1)+\frac{(0.9-1)^2}{2}\right]=2.4600$$

$e^{0.9}=2.459603$ directly from the calculator.

61. $\tan^{-1}x=\int\frac{1}{1+x^2}dx$

$$=\int\left(1-x^2+x^4-x^6+\cdots\right)dx$$

$$\tan^{-1}x=x-\frac{x^3}{3}+\frac{x^5}{5}-\frac{x^7}{7}+\cdots$$

65. $N = N_0 e^{-\lambda t} \cdot N$

$$= N_0\left[1+(-\lambda t)+\frac{(-\lambda t)^2}{2!}+\frac{(-\lambda t)^3}{3!}+\cdots\right]$$

$$= N_0\left[1-\lambda t+\frac{\lambda^2 t^2}{2}-\frac{\lambda^3 t^3}{6}+\cdots\right]$$

69. $f(x) = \ln\frac{1+x}{1-x},\ f(0) = \ln\frac{1+0}{1-0} = \ln 1 = 0$

$$f'(x) = -\frac{2}{x^2-1},\ f'(0) = 2$$

$$f''(x) = \frac{4x}{(x+1)^2(x-1)^2},\ f''(0) = 0$$

$$f'''(x) = \frac{-4(3x^2+1)}{(x+1)^3(x-1)^3},\ f'''(0) = 4$$

$$f^{iv}(x) = \frac{48x(x^2+1)}{(x+1)^4(x-1)^4},\ f^{iv}(0) = 0$$

$$f^{(5)}(x) = \frac{-48(5x^4+10x^2+1)}{(x+1)^5(x-1)5},\ f^{(5)}(0) = 48$$

$$f^{(6)}(x) = \frac{480x(3x^4+10x^2+3)}{(x+1)^6(x-1)^6},\ f^{(6)}(0) = 0$$

$$f^{(7)}(x) = \frac{-1440(7x^6+35x^4+21x^2+1)}{(x+1)^7(x-1)^7},\ f^{(7)}(0) = 1440$$

$$V = \ln\frac{1+x}{1-x} = 2x + \frac{4x^3}{3!} + \frac{48x^5}{5!} + \frac{1440x^7}{7!}$$

$$V = \ln\frac{1+x}{1-x} = 2x + \frac{2}{3}x^3 + \frac{2}{5}x^5 + \frac{2}{7}x^7$$

Chapter 31

DIFFERENTIAL EQUATIONS

31.1 Solutions of Differential Equations

1. $y = c_1e^{-x} + c_2e^{2x}$

$\frac{dy}{dx} = -c_1e^{-x} + 2c_2e^{2x}$

$\frac{d^2y}{dx^2} = c_1e^{-x} + 4c_2e^{2x}$

$$\begin{aligned}\frac{d^2y}{dx^2} - \frac{dy}{dx} &= c_1e^{-x} + 4c_2e^{2x} - (-c_1e^{-x} + 2c_2e^{2x})\\ &= c_1e^{-x} + 4c_2e^{2x} + c_1e^{-x} - 2c_2e^{2x}\\ &= 2c_1e^{-x} + 2c_2e^{2x} = 2(c_1e^{-x} + c_2e^{2x})\\ &= 2y\end{aligned}$$

$y = 4e^{-x}, \frac{dy}{dx} = -4e^{-x}, \frac{d^2y}{dx^2} = 4e^{-x}$

$$\begin{aligned}\frac{d^2y}{dx^2} - \frac{dy}{dx} &= 4e^{-x} - (-4e^{-x}) = 8e^{-x}\\ &= 2(4e^{-x}) = 2y\end{aligned}$$

5. $y'' + 3y' - 4y = 3e^x;\ y = c_1e^x + c_2e^{-4x} + \frac{3}{5}xe^x$

$$\begin{aligned}y' &= c_1e^x - 4c_2e^{-4x} + \frac{3}{5}(xe^x + e^x)\\ &= c_1e^x - 4c_2e^{-4x} + \frac{3}{5}xe^x + \frac{3}{5}e^x\end{aligned}$$

$$\begin{aligned}y'' &= c_1e^x + 16c_2e^{-4x} + \frac{3}{5}e^x + \frac{3}{5}(xe^x + e^x)\\ &= c_1e^x + 16c_2e^{-4x} + \frac{3}{5}e^x + \frac{3}{5}xe^x + \frac{3}{5}e^x\\ &= c_1e^x + 16c_2e^{-4x} + \frac{3}{5}xe^x + \frac{6}{5}e^x\end{aligned}$$

Substitute y, y', y'' into differential equation.

$$c_1e^x + 16c_2e^{-4x} + \frac{3}{5}xe^x + \frac{6}{5}e^x$$
$$+3\left(c_1e^x - 4c_2e^{-4x} + \frac{3}{5}xe^x + \frac{3}{5}e^x\right)$$
$$-4\left(c_1e^x + c_2e^{-4x} + \frac{3}{5}xe^x\right) = 3e^x$$

$$c_1e^x + 16c_2e^{-4x} + \frac{3}{5}xe^x + \frac{6}{5}e^x$$
$$+3c_1e^x - 12c_2e^{-4x} + \frac{9}{5}xe^x + \frac{9}{5}e^x$$
$$-4c_1e^x - 4c_2e^{-4x} - \frac{12}{5}xe^x = 3e^x$$

$\frac{15}{5}e^x = 3e^x;\ 3e^x = 3e^x$ identity

General solution; order two, c_1 and c_2

9. $y = 3\cos 2x;\ y'' = -12\cos 2x$

$y'' + 4y = -12\cos 2x + 4(3\cos 2x) = 0$

$y = c_1\sin 2x + c_2\cos 2x;$

$y'' = -4c_1\sin 2x - 4c_2\cos 2x$

$$\begin{aligned}y'' + 4y &= -4c_1\sin 2x - 4c_2\cos 2x\\ &\quad + 4(c_1\sin 2x + c_2\cos 2x)\\ &= 0\end{aligned}$$

13. $y = 2 + x - x^3$

$\frac{dy}{dx} = 1 - 3x^2$

17. $y'' + 9y = 4\cos x;\ 2y = \cos x;\ y = \frac{1}{2}\cos x;$

$y' = -\frac{1}{2}\sin x;\ y'' = -\frac{1}{2}\cos x$

Substitute y and y''.

$-\frac{1}{2}\cos x + 9\left(\frac{1}{2}\cos x\right) = 4\cos x$

$-\frac{1}{2}\cos x + \frac{9}{2}\cos x = 4\cos x$

$\frac{8}{2}\cos x = 4\cos x;\ 4\cos x = 4\cos x$ identity

21. $x\frac{d^2y}{dx^2} + \frac{dy}{dx} = 0;\ y = c_1\ln x + c_2$

$\frac{dy}{dx} = \frac{c_1}{x} = c_1x^{-1};\ \frac{d^2y}{dx^2} = -c_1x^{-2} = -\frac{c_1}{x^2}$

Substitute $\frac{dy}{dx}$ and $\frac{d^2y}{dx^2}$.

$x\left(\frac{-c_1}{x^2}\right) + c_1x^{-1} = 0;\ -\frac{c_1}{x} + \frac{c_1}{x} = 0$

$0 = 0$ identity

25. $y = c_1e^x + c_2e^{2x} + \frac{3}{2}$

$y' = c_1e^x + 2c_2e^{2x}$

$y'' = c_1e^x + 4c_2e^{2x}$

$$y'' - 3y' + 2y = c_1e^x + 4c_2e^{2x} - 3\left(c_1e^x + 2c_2e^{2x}\right) + 2\left(c_1e^x + c_2e^{2x} + \frac{3}{2}\right)$$

$$= c_1e^x + 4c_2e^{2x} - 3c_1e^x - 6c_2e^{2x} + 2c_1e^x + 2c_2e^{2x} + 3$$

$$= 3$$

29. $(y')^2 + xy' = y$

$y = cx + c^2;\ y' = c$

Substitute.

$(c)^2 + x(c) = y;\ c^2 + cx = y;\ y = y$ identity

33. $y = x^3 + c_1x^2 + c_2$

$-4 = 0^3 + c_1(0)^2 + c_2 \Rightarrow c_2 = -4$

$y = x^3 + c_1x^2 - 4$

31.2 Separation of Variables

1. $2xy\,dx + (x^2+1)dy = 0$

$$\frac{2x\,dx}{x^2+1} + \frac{dy}{y} = 0$$

$$\ln(x^2+1) + \ln y = \ln c$$

$$\ln(y(x^2+1)) = \ln c$$

$$y(x^2+1) = c$$

5. $y^2dx + dy = 0$; divide by y^2;

$dx + \frac{dy}{y^2} = 0$; integrate

$x + \frac{y^{-1}}{-1} = c;\ x - \frac{1}{y} = c$

9. $x^2 + (x^3+5)y' = 0;\ (x^3+5)\frac{dy}{dx} = -x^2$

$(x^3+5)dy = -x^2dx;\ dy = \frac{-x^2dx}{x^3+5}$; integrate

$y = -\frac{1}{3}\ln(x^3+5) + c$

$3y + \ln(x^3+5) = 3c_1$; $3c_1$ is constant

$3y + \ln(x^3+5) = c$

13. $e^{x^2}dy = x\sqrt{1-y}\,dx;\ \frac{dy}{\sqrt{1-y}} = \frac{x\,dx}{e^{x^2}}$

$\frac{dy}{(1-y)^{1/2}} = e^{-x^2}x\,dx$; integrate

$-\frac{(1-y)^{1/2}}{\frac{1}{2}} = -\frac{1}{2}e^{-x^2} + c$

$-2\sqrt{1-y} = -\frac{1}{2}e^{-x^2} + c$; multiply by -2

$4\sqrt{1-y} = e^{-x^2} - 2c_1;\ -2c_1 = c$

$4\sqrt{1-y} = e^{-x^2} + c$

17. $y' - y = 4;\ \frac{dy}{dx} = 4 + y;\ \frac{dy}{4+y} = dx$; integrate

$\ln(4+y) = x + c$

21. $y\tan x\,dx + \cos^2 x\,dy = 0;\ \frac{\tan x\,dx}{\cos^2 x} + \frac{dy}{y} = 0$

$(\tan x)^1\sec^2 x\,dx + \frac{dy}{y} = 0$; integrate

$\frac{1}{2}\tan^2 x + \ln y = c_1;\ 2c_1 = c;\ \tan^2 x + 2\ln y = c$

25. $e^{\cos\theta}\tan\theta\,d\theta + \sec\theta\,dy = 0$

$e^{\cos\theta}\sin\theta\,d\theta + dy = 0$

$-e^{\cos\theta} + y = c$

29. $2\ln t\,dt + t\,di = 0;$

$2\ln t\frac{dt}{t} + di = 0$; integrate

$\frac{2(\ln t)^2}{2} + i = 0;\ (\ln t)^2 + i = c;$

$i = c - (\ln t)^2$

33. $\frac{dy}{dx}+yx^2=0;\ \frac{dy}{y}+x^2dx=0$

Integrate: $\ln y+\frac{x^3}{3}+c$

Substitute $x=0, y=1;\ \ln 1=c;\ c=0$

$3\ln y+x^3=0$

37. $y^2e^x dx+e^{-x}dy=y^2dx;\ x=0, y=2$

$y^2dx-y^2e^xdx=e^{-x}dy;\ y^2(1-e^x)dx=e^{-x}dy$

$\frac{(1-e^x)}{e^{-x}}dx=\frac{dy}{y^2};\ e^xdx-e^{2x}=y^{-2}dy;$ integrate

$e^x-\frac{1}{2}e^{2x}=-\frac{1}{y}+c$

Substitute $x=0, y=2$

$e^0-\frac{1}{2}e^0=-\frac{1}{2}+c;$ therefore, $c=1$

$e^x-\frac{1}{2}e^{2x}=-\frac{1}{y}+1;\ 2e^x-e^{2x}=-\frac{2}{y}+2$

$e^{2x}-\frac{2}{y}=2(e^x-1)$

41. $dT+0.15(T-10)dt=0,\qquad T(0)=10$

$\int\frac{dT}{T-10}=\int -0.15dt$

$\ln(T-10)=-0.15t+\ln c$

$T-10=ce^{-0.15t}$

$T=ce^{-0.15t}+10$

$T(0)=100=c+10\Rightarrow c=90$

$T(t)=90e^{-0.15t}+10$

31.3 Integrating Combinations

1. $x\,dy+y\,dx+2xy^2dy=0$

$\frac{x\,dy+y\,dx}{xy}+2y\,dy=0$

$\frac{d(xy)}{xy}+2y\,dy=0$

$\ln xy+y^2=c$

5. $y\,dx-x\,dy+x^3dx=2\,dx;$

$x\,dy-y\,dx-x^3dx=-2\,dx$

$\frac{(x\,dy-y\,dx)}{x^2}-x\,dx=-\frac{2\,dx}{x^2}$

$\frac{y}{x}-\frac{1}{2}x^2=2x^{-1}=\frac{2}{x}+c_1;\ y-\frac{1}{2}x^3=2+c_1x$

$2y-x^3=4+2c_1x;\ x^3-2y=-2c_1x-4;\ -2c_1=c$

$x^3-2y=cx-4$

9. $\sin x\,dy=(1-y\cos x)\,dx$

$\sin x\,dy+y\cos x\,dx=dx$

$d(y\sin x)=dx$

$y\sin x=x+c$

13. $\tan(x^2+y^2)dy+x\,dx+y\,dy=0;$

$dy+\frac{x\,dx+y\,dy}{\tan(x^2+y^2)}=0$

$d(x^2+y^2)=2x\,dx+2y\,dy=2(x\,dx+y\,dy)$

$dy+\cot(x^2+y^2)(x\,dx+y\,dy)=0$

$dy+\frac{1}{2}\cot(x^2+y^2)^2d(x^2+y^2)$

$y+\frac{1}{2}\ln\sin(x^2+y^2)=c;\ y=c-\frac{1}{2}\ln\sin(x^2+y^2)$

17. $10x\,dy+5y\,dx+3y\,dy=0$

$5(2x\,dy+y\,dx)+3y\,dy=0;$ multiply by y

$5(2xy\,dy+y^2dx)+3y^2dy=0;\ 5d(xy^2)+3y^2dy=0$

$5xy^2+y^3=c$

21. $y\,dx-x\,dy=y^3dx+y^2x\,dy;\ x=2, y=4$

$\frac{y\,dx-x\,dy}{y^2}=y\,dx+x\,dy;\ d\left(\frac{x}{y}\right)=d(xy)$

$\frac{x}{y}=xy+c;\ x=2, y=4;\ \frac{2}{4}=2(4)+c;\ c=-\frac{15}{2}$

$\frac{x}{y}=xy-\frac{15}{2};$ multiply by $2y$

$2x=2xy^2-15y$

25. $e^{-x}\,dy - 2y\,dy = ye^{-x}\,dx$

$e^{-x}\,dy - ye^{-x}\,dx = 2y\,dy$

$d\left(ye^{-x}\right) = d\left(y^2\right)$

$ye^{-x} = y^2 + c$

31.4 The Linear Differential Equation of the First Order

1. $dy + \left(\frac{2}{x}\right) y\,dx = 3\,dx$

$ye^{\int \frac{2}{x}dx} = \int 3e^{\int \frac{2}{x}dx}dx + c$

$ye^{2\ln x} = \int 3e^{2\ln x}dx + c$

$ye^{\ln x^2} = \int 3e^{\ln x^2}\,dx + c$

$yx^2 = \int 3x^2 dx + c$

$yx^2 = x^3 + c$

$y = x + cx^{-2}$

5. $dy + 2y\,dx = e^{-4x}dx;\ P = 2, Q = e^{-4x};$

$e^{\int 2dx} = e^{2x}$

$ye^{2x} = \int e^{-4x}e^{2x}dx = -\frac{1}{2}\int e^{-2x}(-2\,dx)$

$= -\frac{1}{2}e^{-2x} + c$

$y = -\frac{1}{2}e^{-2x}e^{-2x} + ce^{-2x} = -\frac{1}{2}e^{-4x} + ce^{-2x}$

9. $\sin x\,dy = \left(1 - y\cos x\right)dx$

$dy + \cot x\left(y\,dy\right) = \csc x\,dx$

$ye^{\int \cot x\,dx} = \int \csc x e^{\int \cot x\,dx}dx$

$ye^{\ln \sin x} = \int \csc x e^{\ln \sin x}dx$

$y\sin x = \int \csc x \sin x\,dx$

$y\sin x = x + c$

13. $dr + r\cot\theta\,d\theta = d\theta;\ dr + \cot\theta r\,d\theta = d\theta$

$P = \cot\theta, Q = 1, e^{\int \cot\theta\,d\theta} = e^{\ln \sin\theta} = \sin\theta$

$r\sin\theta = \int \sin\theta\,d\theta + c;\ r\sin\theta = -\cos\theta + c$

$r = -\frac{\cos\theta}{\sin\theta} + \frac{c}{\sin\theta} = -\cot\theta + c\csc\theta$

17. $y' + y = x + e^x$

$dy + y\,dx = \left(x + e^x\right)dx, \qquad P(x) = 1$

$ye^{\int dx} = \int \left(x + e^x\right)e^{\int dx}dx$

$ye^x = \int \left(xe^x + e^{2x}\right)dx$

$ye^x = xe^x - e^{-x} + \frac{1}{2}e^{2x} + c$

21. $y' = x^3(1 - 4y);\ \frac{dy}{dx} = x^3 - 4x^3y;$

$dy = x^3dx - 4x^3y\,dx$

$dy + 4x^3y\,dx = x^3dx;\ P = 4x^3, Q = x^3;$

$e^{4\int x^3dx} = e^{x^4}$

$ye^{x^4} = \int x^3e^{x^4}dx + c = \frac{1}{4}e^{x^4}4x^3dx + c$

$= \frac{1}{4}e^{x^4} + c$

$y = \frac{1}{4} + ce^{-x^4}$

25. $\sqrt{1+x^2}\,dy + x(1+y)\,dx = 0$

$\sqrt{1+x^2}\,dx + xdx + xy\,dx = 0$

$dy + \frac{x}{\sqrt{1+x^2}}y\,dx = -\frac{x}{\sqrt{1+x^2}}dx, \quad P(x) = \frac{x}{\sqrt{1+x^2}}$

$ye^{\int \frac{x}{\sqrt{1+x^2}}dx} = \int \frac{-x}{\sqrt{1+x^2}}e^{\int \frac{x}{\sqrt{1+x^2}}dx}dx$

$ye^{\sqrt{1+x^2}} = \int \frac{-x}{\sqrt{1+x^2}}e^{\sqrt{1+x^2}}dx$

$ye^{\sqrt{1+x^2}} = -e^{\sqrt{1+x^2}} + c$

29. $y' = 2(1-y)$; solve by separation of variables.

$\frac{dy}{1-y} = 2\,dx; -\ln(1-y) = 2x - \ln c;$

$\ln\frac{c}{1-y} = 2x;$

$c = (1-y)e^{2x}; 1-y = ce^{-2x}; y = 1-ce^{-2x}$

Solve as a first order equation.

$dy = 2\,dx - 2y\,dx; dy + 2y\,dx = 2\,dx$

$ye^{\int 2dx} = \int e^{\int 2dx}dx; ye^{2x} = \int 2e^{2x}dx$

$ye^{2x} = e^{2x} + c; y = 1 + ce^{-2x}$

33. $\frac{dy}{dx} + 2y\cot x = 4\cos x; x = \frac{\pi}{2}, y = \frac{1}{3};$

$dy + 2y\cot x\,dx = 4\cos x\,dx; P = 2\cot x;$

$Q = 4\cos x$

$e^{\int P dx} = e^{\int 2\cot x dx} = 2^{2\ln|\sin x|} = e^{\ln|\sin x|^2}$

$= |\sin x|^2$

$y(\sin x)^2 = \int 4\cos x(\sin x)^2 dx + c = \frac{4(\sin x)^3}{3} + c$

$y = \frac{4}{3}\sin x + c(\csc^2 x)$

$x = \frac{\pi}{2}$ when $y = \frac{1}{3}; \frac{1}{3} = \frac{4}{3}\sin\frac{\pi}{2} + c; c = -1$

$y = \frac{4}{3}\sin x - \csc^2 x$

37. $y' + P(x)y = Q(x)y^2$

$dy + P(x)y\,dx = Q(x)y^2\,dx$ which is not linear because $Q(x)y^2$ is not a function of x only.

Let $u = \frac{1}{y} \Rightarrow dy = -y^2\,du$, then $dy + P(x)y\,dx$
$= Q(x)y^2 dx$ is $-y^2 du + P(x)y\,dx = Q(x)y^2\,dx$
$du - P(x)u\,dx = -Q(x)\,dx$ which is linear.

31.5 Numerical Solutions of First-Order Equations

1. $\frac{dy}{dx} = x + 1$

x	y	$x+1$	dy	y(correct)
0.0	1.00	1.0	0.20	1.00
0.2	1.20	1.2	0.24	1.22
0.4	1.44	1.4	0.28	1.48
0.6	1.72	1.6	0.32	1.78
0.8	2.04	1.8	0.36	2.12
1.0	2.40	2.0	0.40	2.50

$y = \frac{1}{2}x^2 + x + c$

$y = 1$ when $x = 0$

$c = 1$

$y = \frac{1}{2}x^2 + x + 1$

5.

x	y approximate	y exact
0.0	1	1
0.1	$1.0+(0.0+1)(0.1)=1.10$	0.105
0.2	$1.10+(0.1+1)(0.1)=1.21$	1.220
0.3	$1.21+(0.2+1)(0.1)=1.33$	1.345
0.4	$1.33+(0.3+1)(0.1)=1.46$	1.480
0.5	$1.46+(0.4+1)(0.1)=1.60$	1.625
0.6	$1.60+(0.5+1)(0.1)=1.75$	1.780
0.7	$1.75+(0.6+1)(0.1)=1.91$	1.945
0.8	$1.91+(0.7+1)(0.1)=2.08$	2.120
0.9	$2.08+(0.8+1)(0.1)=2.26$	2.305
1.0	$2.26+(0.9+1)(0.1)=2.45$	2.500

9. $\frac{dy}{dx} = xy + 1$, $x = 0$ to $x = 0.4$, $\Delta x = 0.1$, $(0, 0)$

x	y	y to 4 places
0	0	0
0.1	0.1003339594	0.1003
0.2	0.20268804	0.2027
0.3	0.3091639819	0.3092
0.4	0.42203172548	0.4220

13. $\frac{dy}{dx} = \cos(x+y)$, $x = 0$ to $x = 0.5$, $\Delta x = 0.1$, $\left(0, \frac{\pi}{2}\right)$

x	y
0	$\frac{\pi}{2} = 1.5708$
0.1	1.5660
0.2	1.5521
0.3	1.5302
0.4	1.5011
0.5	1.4656

17. $\frac{di}{dt} = 2i = \sin t$

t	i	$\sin t - 2i$	di
0.0	0.0000	0.0000	0.0000
0.1	0.0000	0.0998	0.0100
0.2	0.0100	0.1787	0.0179
0.3	0.0279	0.2398	0.0240
0.4	0.0518	0.2857	0.0286
0.5	0.0804	0.3186	0.0319

$i = 0.0804A$ for $t = 0.5$ s

$di + 2i\,dt = \sin t\,dt;\ e^{\int 2dt} = e^{2t}$

$ie^{2t} = \int e^{2t}\sin t\,dt = \frac{e^{2t}(2\sin t - \cos t)}{4+1} + c$

$i = \frac{1}{5}(2\sin t - \cos t) + ce^{-2t}$ (Formula 49)

$i = 0$ for $t = 0$, $0 = \frac{1}{5}(0-1) + c$, $c = \frac{1}{5}$

$i = \frac{1}{5}\left(2\sin t - \cos t + e^{-2t}\right)$

$i = 0.0898A$ for $t = 0.5t$

31.6 Elementary Applications

1. $y^2 = cx$

$2y\frac{dy}{dx} = c = \frac{y^2}{x}$

$\frac{dy}{dx} = \frac{y}{2x}$ for slope of any member of family.

$\frac{dy}{dx} = -\frac{2x}{y}$ for slope of orthogonal trajectories.

$y\,dy = -2x\,dx$

$\frac{y^2}{2} = -x^2 + \frac{c}{2}$

$y^2 + 2x^2 = c$

5. $\frac{dy}{dx} = \frac{2x}{y};\ y\,dy = 2x\,dx;\ \frac{1}{2}y^2 = x^2 + c$

Substitute $x = 2$, $y = 3$; $\frac{1}{2}(9) = 4 + c$; $c = 0.5$

$\frac{1}{2}y^2 = x^2 + 0.5;\ y^2 = 2x^2 + 1;\ y = \pm\sqrt{2x^2+1}$

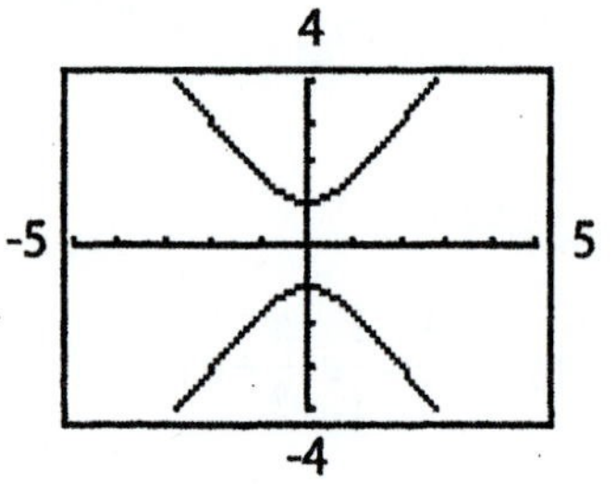

9. See Example 2; $\frac{dy}{dx} = ce^x;\ y = ce^x;\ c = \frac{y}{e^x}$

Substitute for c in the equation for the derivative.

$\frac{dy}{dx} = \frac{y}{e^x}e^x = y;\ \left.\frac{dy}{dx}\right|_{OT} = -\frac{1}{y};\ y\,dy = -dx$

Integrating, $\frac{y^2}{2} = -x + \frac{c}{2};\ y^2 = c - 2x$

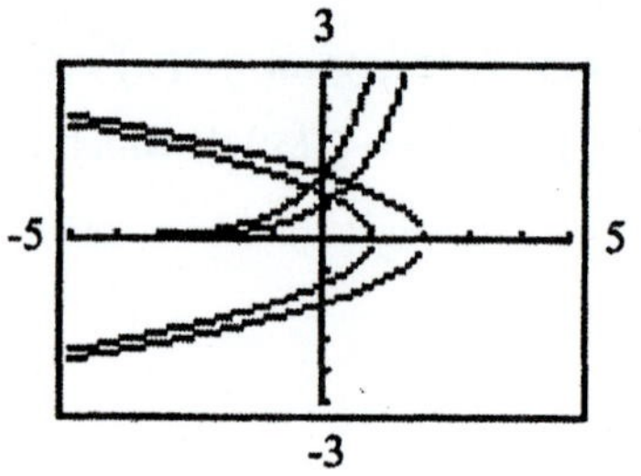

13. From Example 3, $N = N_0(0.5)^{t/5.27}$

$N = N_0(0.5)^{2.00/5.27} = N_0(0.769)$

76.9% of the initial amount remains

17. $\frac{dN}{dt} = r - kN$

$\frac{dN}{r - kN} = dt$

$-\frac{1}{k}\ln(r - kN) = t + c$

$N = 0$ for $t = 0$

$c = -\frac{1}{k}\ln r$

$-\frac{1}{k}\ln(r - kN) = t - \frac{1}{k}\ln r$

$\ln\frac{r - kN}{r} = -kt$

$r - kN = re^{-kt}$

$N = \frac{r}{k}(1 - e^{-kt})$

21. $r\frac{dS}{dr} = 2(a - S)$; $r\,dS = 2a\,dr - 2S\,dr$;

$dS + \frac{2}{r}S\,dx = 2a\frac{dr}{r}$; $y = S$, $x = r$, $P = \frac{2}{r}$,

$Q = \frac{2a}{r}$

$e^{\int P dx} = e^{\int (2/r)dr} = e^{2\ln x} = e^{\ln r^2} = r^2$

$Sr^2 = \int\left(2a\frac{dr}{r}\right) - 2a\ln r + c = \int 2ar\,dr$

$Sr^2 = ar^2 + c$; $S = a + \frac{c}{r^2}$

25. $\frac{dT}{dt} = k(T - 80.0)$; $\frac{dT}{T - 80.0} = k\,dt$;

$\ln(T - 80.0) = kt + \ln c$; $T = 80.0 + ce^{kt}$;

$200 = 80.0 + c$; $T = 80.0 + 120e^{kt}$;

$140 = 80.0 + 120e^{5k}$; $e^{5k} = 0.5$; $e^{k} = (0.5)^{1/5}$;

$T = 80.0 + 120(0.5)^{t/5}$; $100 = 80.0 + 120(0.5)^{t/5}$;

$0.5^{t/5} = \frac{1}{6.0}$; $t = 5\frac{\ln\frac{1}{6.0}}{\ln\frac{1}{2}} = 13$ min

29. $\frac{dc}{dt} = k(c_o - c) \Rightarrow \frac{dc}{c - c_0} = -kdt$

$\ln(c - c_0) = -kt + \ln c_1$

$\ln\frac{c - c_0}{c_1} = -kt \Rightarrow \ln\frac{0 - c_0}{c_1} = 0 \Rightarrow c_1 = -c_0$

$c - c_0 = -c_0 e^{-kt}$

$c = c_0\left(1 - e^{-kt}\right)$

33. $V = E\sin\omega t$; $t = 0, i = 0$

$L\frac{di}{dt} + Ri = E\sin\omega t$; $L\,di + Ri\,dt = E\sin\omega t\,dt$

$di + \frac{R}{L}i\,dt = \frac{E}{L}\sin\omega t\,dt$

$P = \frac{R}{L}$; $Q = \frac{E}{L}\sin\omega t$; $e^{\int (R/L)dt} = e^{Rt/L}$

$ie^{Rt/L} = \int\frac{E}{L}\sin\omega t\,e^{Rt/L}dt = \frac{E}{L}\int e^{Rt/L}\sin\omega t\,dt$

Integration by parts:

$$ie^{Rt/L} = \frac{E}{L}\frac{L^2\omega^2}{L^2\omega^2 + R^2} \times \left(\frac{R}{L\omega^2}e^{Rt/L}\sin\omega t - \frac{1}{\omega}e^{Rt/L}\cos\omega t + c\right)$$

$$0 = \frac{EL\omega^2}{L^2\omega^2 + R^2}\left(0 - \frac{1}{\omega} + c\right) = -\frac{1}{\omega} + c$$

Therefore, $c = \frac{1}{\omega}$

$$ie^{Rt/L} = \frac{EL\omega^2}{R^2 + L^2\omega^2} \times \left(\frac{R}{L\omega^2}e^{Rt/L}\sin\omega t - \frac{1}{\omega}e^{Rt/L}\cos\omega t + \frac{1}{\omega}\right)$$

$$ie^{Rt/L} = \frac{E}{R^2+L^2\omega^2}(\mathrm{R}e^{Rt/L}\sin\omega t - L\omega e^{Rt/L}\cos\omega t + L\omega)$$

$$i = \frac{E}{R^2 + L^2\omega^2}(R\sin\omega t - L\omega\cos\omega t + L\omega e^{-Rt/L})$$

37. $\frac{dv}{dt} = 32 - v;\ \frac{dv}{32 - v} = dt;\ -\frac{-dv}{32 - v} = dt$

Integrating, $(-1)\ln(32 - v) = t - \ln c$;

$\ln\frac{32 - v}{c} = -t$;

$32 - v = ce^{-t};\ -v = -32 + ce^{-t};\ v = 32 - ce^{-t}$

Starting from rest means $v = 0$ when $t = 0$;
$0 = 32 - c;\ c = 32;\ v = 32 - 32e^{-t} = 32(1 - e^{-t})$

$\lim_{t\to\infty} 32(1 - e^{-t}) = 32$

41. $\frac{dx}{dt} = 6t - 3t^2;\ dx = (6t - 3t^2)$;
$x = \frac{6t^2}{2} - \frac{3t^3}{3} + c;\ x = 3t^2 - t^3 + c$;
$x = 0$ for $t = 0;\ 0 = c;\ x = 3t^2 - t^3$;
$y = 2(3t^2 - t^3) - (3t^2 - t^3)^2$;
$y = -t^6 + 6t^5 - 9t^4 - 2t^3 + 6t^2$

45. $\frac{dv}{dt} = kv;\ \frac{dv}{v} = k\ dt$; integrating, $\ln v = kt + \ln c$
Let $v = 33,000$ when $t = 0$; $33,000 = c$;
$\ln v = kt + 33,000;\ \ln v - \ln 33,000 - kt$;
$\ln\frac{v}{33,000} = kt;\ \frac{v}{33,000} = e^{kt};\ v = 33,000e^{kt}$
Let $v = 19,700$ when $t = 3$; $19,700 = 33,000e^{3k}$;
$e^k = (0.597)^{1/3};\ v = 33,000(0.597)^{1/3}$
After 11 years, $v = 33,000(0.597)^{11/3} = \$4977$

49. $y = e^{x/2} + c$
$\frac{dy}{dx} = \frac{1}{2}e^{x/2}$
$\left.\frac{dy}{dx}\right|_{OT} = -2e^{-x/2}$
$y = 4e^{-x/2} + c$
isobars: $y = e^{x/2}$, $y = e^{x/2} + 3$
wind direction: $y = 4e^{-x/2} + 1$, $y = 4e^{-x/2} - 2$

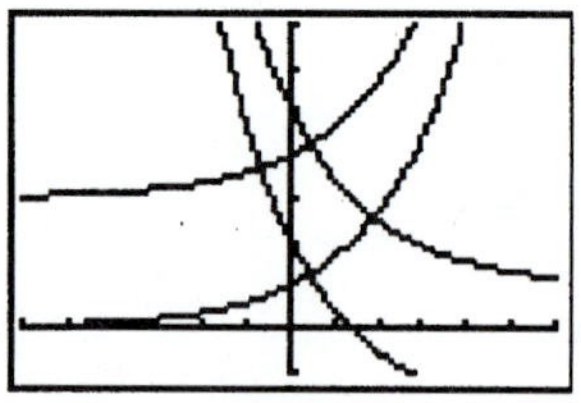

31.7 Higher-Order Homogeneous Equations

1. $D^2y - 5Dy = 0$
$m^2 - 5m = 0$
$m(m - 5) = 0$
$m = 0,\ m = 5$
$y = c_1 + c_2e^{5x}$

5. $3\frac{d^2y}{dx^2} + 4\frac{dy}{dx} + y = 0$
$3D^2y + 4Dy + y = 0;;\ 3m^2 + 4m + 1 = 0$
$(3m + 1)(m + 1) = 0;\ m_1 = -\frac{1}{3}, m_2 = -1$
$y = c_1e^{-(1/3)x} + c_2e^{-x}$

9. $2D^2y - 3y = Dy$
$2D^2y - Dy - 3y = 0$
$2m^2 - m - 3 = 0$
$(2m - 3)(m + 1) = 0$
$m = \frac{3}{2},\ m = -1$
$y = c_1e^{\frac{3x}{2}} + c_2e^{-x}$

13. $3D^2y + 8Dy - 3y = 0$; $3m^2 + 8m - 3 = 0$;
$(3m - 1)(m + 3) = 0$; $m_1 = \frac{1}{3}$ and $m_2 = -3$;
$y = c_1e^{x/3} + c_2e^{-3x}$

17. $2\frac{d^2y}{dx^2} - 4\frac{dy}{dx} + y = 0$; $2D^2y - 4Dy + y = 0$;
$2m^2 - 4m + 1 = 0$

Quadratic formula: $m = \frac{4 \pm \sqrt{16 - 8}}{4}$;

$m_1 = 1 + \frac{\sqrt{2}}{2}$,

$m_2 = 1 - \frac{\sqrt{2}}{2}$

$y = c_1e^{(1+(\sqrt{2}/2))x} + c_2e^{(1-(\sqrt{2}/2))x}$;
$y = c_1e^xe^{(\sqrt{2}/2)x} + c_2e^xe^{-(\sqrt{2}/2)x}$
$= e^x(c_1e^{x(\sqrt{2}/2)} + c_2e^{-x(\sqrt{2}/2)})$

21. $y'' = 3y' + y$; $D^2y - 3Dy - y = 0$; $m^2 - 3m - 1 = 0$

Quadratic formula:

$m = \frac{3 \pm \sqrt{9+4}}{2}$; $m_1 = \frac{3}{2} + \frac{\sqrt{13}}{2}$; $m_2 = \frac{3}{2} - \frac{\sqrt{13}}{2}$

$y = c_1e^{((3/2)+(\sqrt{13}/2))x} + c_2e^{((3/2)-(\sqrt{13}/2))x}$;
$y = e^{3x/2}(c_1e^{x(\sqrt{13}/2)} + c_2e^{-x(\sqrt{13}/2)})$

25. $2D^2y + 5aDy - 12a^2 = 0,\ a > 0$

$2m^2 + 5am - 12a^2 = 0$

$(2m - 3a)(m + 4a) = 0$

$m = \frac{3a}{2}, \qquad m = -4a$

$y = c_1e^{\frac{3a}{2}x} + c_2e^{-4ax}$

29. $D^2y - Dy = 12y$
$D^2y - Dy - 12y = 0$; $y = 0$ when $x = 0$; $y = 1$ when $x = 1$
$m^2 - m - 12 = 0$; $(m - 4)(m + 3) = 0$; $m_1 = 4$;
$m_2 = -3$
$y = c_1e^{4x} + c_2e^{-3x}$

Substituting given values: $0 = c_1 + c_2$;
therefore, $c_1 = -c_2$;
$1 = c_1e^4 + c_2e^{-3}$

$1 = -c_2e^4 + c_2e^{-3} = -c_2e^4 + \frac{c_2}{e^3} = \frac{-c_2e^7 + c_2}{e^3}$

$e^3 = c_1(1 - e^7)$; therefore,

$c_2 = \frac{e^3}{(1 - e^7)}, c_1 - \frac{e^3}{(1 - e^7)}$

$y = -\frac{e^3}{(1 - e^7)}e^{4x} + \frac{e^3}{(1 - e^7)}e^{-3x}$

$= \frac{e^3}{e^7 - 1}e^{4x} - \frac{e^3}{e^7 - 1}e^{-3x}$

$y = \frac{e^3}{e^7 - 1}(e^{4x} - e^{-3x})$

33. $D^4y - 5D^2y + 4y = 0, m^4 - 5m^2 + 4 = 0$
$m = -2, -1, 1, 2$
$y = c_1e^x + c_2e^{-x} + c_3e^{2x} + c_4e^{-2x}$

31.8 Auxiliary Equations with Repeated or Complex Roots

1. $\frac{d^2y}{dx^2} + 10\frac{dy}{dx} + 25y = 0$

$$D^2y + 10Dy + 25y = 0$$
$$m^2 + 10m + 25 = 0$$
$$(m + 5)^2 = 0$$
$$m = -5,\ -5$$
$$y = e^{-5x}(c_1 + c_2x)$$

5. $D^2y - 2Dy + y = 0$; $m^2 - 2m + 1 = 0$;
$(m - 1)^2 = 0$; $m = 1, 1$
$y = e^x(c_1 + c_2x)$; $y = (c_1 + c_2x)e^x$

9. $D^2y + 9y = 0$; $m^2 + 9 = 0$; $m_1 = 3j$ and $m_2 = -3j$, $\alpha = 0, \beta = 3$

$y = e^{0x}(c_1 \sin 3x + c_2 \cos 3x)$;
$y = c_1 \sin 3x + c_2 \cos 3x$

13. $D^4y - y = 0$
$m^4 - 1 = 0$
$(m^2 - 1)(m^2 + 1) = 0$
$(m - 1)(m + 1)(m^2 + 1) = 0$
$m = \pm 1,\ m = \pm j$
$y = c_1e^x + c_2e^{-x} + c_3 \sin x + c_4 \cos(-x)$
$y = c_1e^x + c_2e^{-x} + c_3 \sin x + c_4 \cos x$

17. $16D^2y - 24Dy + 9y = 0;\ 16m^2 - 24m + 9 = 0;$

$(4m - 3)^2 = 0;\ m = \frac{3}{4}, \frac{3}{4}$

21. $2D^2y + 5y = 4Dy;\ 2D^2y + 5y - 4Dy = 0;$
$2m^2 - 4m + 5 = 0$

Quadratic formula:

$$m = \frac{4 \pm \sqrt{16 - 40}}{4} = \frac{4 \pm 2\sqrt{-6}}{4}$$

$$m_1 = 1 + \frac{\sqrt{6}}{2}j;\ m_2 = 1 - \frac{\sqrt{6}}{2}j;\ \alpha = 1, \beta = \frac{1}{2}\sqrt{6}$$

$$y = e^x\left(c_1 \cos \frac{1}{2}\sqrt{6}x + c_2 \sin \frac{1}{2}\sqrt{6}x\right)$$

25. $2D^2y - 3Dy - y = 0;\ 2m^2 - 3m - 1 = 0$

By the quadratic formula, $m = \dfrac{3 \pm \sqrt{9 + 8}}{4}$

$$m_1 = \frac{3}{4} + \frac{\sqrt{17}}{4},\ m_2 = \frac{3}{4} - \frac{\sqrt{17}}{4}$$

$y = c_1e^{((3/4)+(\sqrt{17}/4))x} + c_2e^{((3/4)+(\sqrt{17}/4))x};$
$y = e^{(3/4)x}(c_1e^{x(\sqrt{17}/4)} + c_2e^{-x(\sqrt{17}/4)})$

29. $D^3y - 6D^2y + 12Dy - 8y = 0$
$m^3 - 6m^2 + 12m - 8 = 0$
$(m - 2)(m^2 - 4m + 4) = 0$
$(m - 2)(m - 2)(m - 2) = 0$
$m = 2, 2, 2$ repeated root
$y = e^{2x}(c_1 + c_2x + c_3x^2)$

33. $D^2y + 2Dy + 10y = 0;\ m^2 + 2m + 10 = 0$

By the quadratic formula, $m = \dfrac{-2 \pm \sqrt{4 - 40}}{2}$

$m_1 = -1 + 3j;\ m_2 = -1 - 3j,\ \alpha = -1,\ \beta = 3;$
$y = e^{-x}(c_1 \sin 3x + c_2 \cos 3x)$

Substituting $y = 0$ when $x = 0$;

$0 = e^0(c_1 \sin 0 + c_2 \cos 0);\ c_2 = 0$

Substituting $y = e^{-\pi/6}$, $x = \frac{\pi}{6}$;

$e^{-\pi/6} = e^{-\pi/6}\left(c_1 \sin \frac{\pi}{2}\right);\ e^{-\pi/6} = e^{-\pi/6}c_1;$
$c_1 = 1$
$y = e^{-x} \sin 3x$

37. $y = c_1e^{3x} + c_2e^{-3x}$
$(m - 3)(m + 3) = 0$
$m^2 - 9 = 0;\ (D^2 - 9)y = 0$

41. $y(x) = c_1 \cos 2x + c_2 \sin 2x$
$y(0) = 0 = c_1 \cos 0 + c_2 \sin 0 \Rightarrow c_1 = 0$
$y(1) = 0 = c_2 \sin 2 \Rightarrow c_2 = 0$
$y = 0$

31.9 Solutions of Nonhomogeneous Equations

1. b: $x^2 + 2x + e^{-x}$
$y_p = A + Bx + Cx^2 + Ee^{-x}$

5. $D^2y - Dy - 2y = 4;\ m^2 - m - 2 = 0$
$(m - 2)(m + 1) = 0;\ m_1 = 2,\ m_2 = -1$
$y_c = c_1e^{2x} + c_2e^{-x}$
$y_p = A;\ Dy_p = 0;\ D^2y_p = 0$

Substituting in diff. equation, $0 - 0 - 2A = 4;$
$A = -2$
Therefore, $y_p = -2;\ y = c_1e^{2x} + c_2e^{-x} - 2$

9. $y'' - 3y' = 2e^x + xe^x;\ D^2y - 3Dy = 2e^x + xe^x$
$m^2 - 3m = 0;\ m(m - 3) = 0;\ m_1 = 0,\ m_2 = 3$
$y_c = c_1e^0 + c_2e^{3x} = c_1 + c_2e^{3x};\ y_p = Ae^x + Bxe^x$
$Dy_p = Ae^x + B(xe^x + e^x) = Ae^x + Bxe^x + Be^x$

$$D^2y_p = Ae^x + Be^x + B(xe^x + e^x)$$
$$= Ae^x + Be^x + Bxe^x + Be^x$$
$$D^2y_p = Ae^x + 2Be^x + Bxe^x$$

Substituting in diff. equation:

$$Ae^x + 2Be^x + Bxe^x - 3(Ae^x + Bxe^x + Be^x)$$
$$= 2e^x + xe^x$$
$$Ae^x + 2Be^x + Bxe^x - 3Ae^x - 3Bxe^x - 3Be^x$$
$$= 2e^x + xe^x$$
$$-2Ae^x - Be^x - 2Bxe^x$$
$$= 2e^x + xe^x$$
$$e^x(-2A - B) + xe^x(-2B)$$
$$2e^x + xe^x$$
$$-2A - B = 2;\ -2A + \frac{1}{2} = 2;\ -2A = \frac{3}{2};$$
$$A = -\frac{3}{4};\ -2B = 1;\ B = -\frac{1}{2}$$
$$y = c_1 + c_2e^{3x} - \frac{3}{4}e^x - \frac{1}{2}xe^x$$

13. $\dfrac{d^2y}{dx^2} - 2\dfrac{dy}{dx} + y = 2x + x^2 + \sin 3x$

$$D^2y - 2Dy + y = 2x + x^2 + \sin 3x$$

$$m^2 - 2m + 1 = 0;\ (m-1)^2 = 0;\ m = 1,\ 1$$
$$y_c = e^x(c_1 + c_2x);$$
$$y_p = A + Bx + Cx^2 + E\sin 3x + F\cos 3x$$
$$Dy_p = B + 3Cx + 3E\cos 3x - 3F\sin 3x$$
$$D^2y_p = 2c - 9E\sin 3x - 9F\cos 3x$$

Substituting in diff. equation:

$$2C - 9E\sin 3x - 9F\cos 3x$$
$$-2(B + 2Cx + 3E\cos 3x - 3F\sin 3x)$$
$$+A + Bx + Cx^2 + E\sin 3x + F\cos 3x$$
$$= 2x + x^2 + \sin 3x$$
$$2C - 9E\sin 3x - 9F\cos 3x - 2B - 4Cx - 6E\cos 3x$$
$$+6F\sin 3x + A + Bx + Cx^2 + E\sin 3x + F\cos 3x$$
$$= 2x + x^2 + \sin 3x$$
$$2C - 2B + A - 8E\sin 3x + 6F\sin 3x - 8F\cos 3x$$
$$-6E\cos 3x - 4Cx + Bx + Cx^2$$
$$= 2x + x^2 + \sin 3x$$
$$(2C - 2B + A) + \sin 3x(6F - 8E) + \cos 3x(-8F - 6E)$$
$$+x(B-4) + Cx^2$$
$$= 2x + x^2 + \sin 3x$$
$$2C - 2B + A = 0;\ A = 10$$

$$6F - 8E = 1;\ -8F - 6E = 0;\ E = -\frac{2}{25};$$
$$F = \frac{3}{50};\ B - 4 = 2;\ B = 6;\ C = 1$$
$$y = e^x(c_1 + c_2x) + 10 + 6x + x^2$$
$$-\frac{2}{25}\sin 3x + \frac{3}{50}\cos 3x$$

17. $D^2y - Dy - 30y = 10;\ m^2 - m - 30 = 0;$

$$(m-6)(m+5) = 0;\ m_1 = -5;\ m_2 = 6$$
$$y_c = c_1e^{-5x} + c_2e^{6x};\ y_p = A;\ Dy_p = 0;$$
$$D^2y_p = 0;\ 0 = 0 - 30A = 10;\ A = -\frac{1}{3};$$
$$y_p = -\frac{1}{3};\ y = c_1e^{-5x} + c_2e^{6x} - \frac{1}{3}$$

21. $D^2y - 4y = \sin x + 2\cos x;\ m^2 - 4 = 0;$

$$m_1 = 2;\ m_2 = -2$$
$$y_c = c_1e^{2x} + c_2e^{-2x}$$
$$y_p = A\sin x + B\cos x;\ Dy_p = A\cos x - B\sin x$$
$$D^2y_p = -A\sin x - B\cos x$$
$$-A\sin x - B\cos x - 4A\sin x - 4B\cos x$$
$$= \sin x + 2\cos x$$
$$-5A\sin x - 5B\cos x$$
$$= \sin x + 2\cos x$$
$$-5A = 1;\ A = -\frac{1}{5};\ -5B = 2;\ B = -\frac{2}{5}$$
$$y_p = -\frac{1}{5}\sin x - \frac{2}{5}\cos x$$
$$y = c_1e^{2x} + c_2e^{-2x} - \frac{1}{5}\sin x - \frac{2}{5}\cos x$$

25. $D^2y + 5Dy + 4y = xe^x + 4;\ m^2 + 5m + 4 = 0;$

$$(m+1)(m+4) = 0;\ m_1 = -1,\ m_2 = -4$$
$$y_c = c_1e^{-x} + c_2e^{-4x};\ y_p = Ae^x + Bxe^x + C$$
$$Dy_p = Ae^x + B(xe^x + e^x) = Ae^x + Bxe^x + Be^x$$
$$D^2yp = Ae^x - B(xe^x + e^x) + Be^x$$
$$= Ae^x + 2Be^x + Bxe^x$$
$$Ae^x + 2Be^x + Bxe^x + 5(Ae^x + Bxe^x + Be^x)$$
$$+4(Ae^x + Bxe^x + C) = xe^x + 4$$

$(10A+7B)e^x + 10Bxe^x + 4C = xe^x + 4$

$10A+7B=0;\ 10B=1;\ B=\frac{1}{10}$

$4C=4;\ C=1;\ 10A+7\left(\frac{1}{10}\right)=0$

$A=-\frac{7}{100};\ y_p=-\frac{7}{100}e^x+\frac{1}{10}e^x+1$

$y=c_1e^{-x}+c_2e^{-4x}-\frac{7}{100}e^x+\frac{1}{10}xe^x+1$

29. $D^2y+y=\cos x$

$m^2+1=0$

$m=\pm i$

$y_c=c_1\sin x+c_2\cos x$

let $y_p=x(A\sin x+B\cos x)$

$Dy_p=x(A\cos x-B\sin x)+A\sin x+B\cos x$

$D^2y_p=x(-A\sin x-B\cos x)+A\cos x$
$-B\sin x+A\cos x-B\sin x$

$D^2y_p+y_p=\cos x$

$-x(A\sin x+B\cos x)+2A\cos x-2B\sin x$
$+x(A\sin x+B\cos x)=\cos x$
$2A\cos x-2B\sin x=\cos x$

$2A=1, B=0$

$A=\frac{1}{2}$

$y_p=\frac{1}{2}x\sin x$

$y=y_c+y_p$

$y=c_1\sin x+c_2\cos x+\frac{1}{2}x\sin x$

33. $D^2y-Dy-6y=5-e^x;\ m^2-m-6=0;$
$(m-3)(m+2)=0;\ m_1=3,\ m_2=-2$
$y_c=c_1e^{3x}+c_2e^{-2x}$
$y_p=A+Be^x;\ Dy_p=Be^x;\ D^2y_p=Be^x$
$Be^x-Be^x-6(A+Be^x)=5-e^x$

$-6A-6Be^x=5-e^x;\ -6A=5;\ A=-\frac{5}{6};$

$-6B=-1;\ B=\frac{1}{6}$

$y_p=-\frac{5}{6}+\frac{1}{6}e^x;\ y=c_1e^{3x}+c_2e^{-2x}+\frac{1}{6}e^x-\frac{5}{6}$

Substituting $x=0$ when $y=2$; $3c_1+3c_2=8$

$Dy=3c_1e^{3x}-2c_2e^{-2x}+\frac{1}{6}e^x$

Substituting, $Dy=4$ when $x=0$, $18c_1-12c_2=23$

Solving the two linear equations simultaneously,

$c_1=\frac{11}{6},\ c_2=\frac{5}{6}$

$y=\frac{11}{6}e^{3x}+\frac{5}{6}e^{-2x}+\frac{1}{6}e^x-\frac{5}{6}$

$=\frac{1}{6}(11e^{3x}+5e^{-2x}+e^x-5)$

37. $D_y-y=x^2$

$m-1=0\Rightarrow m=1$

$y_c=c_1e^x$

$y_p=A+Bx+Cx^2$

$Dy_p=B+2Cx$

$B+2Cx-A-Bx-Cx^2=x^2$

$-C=1\Rightarrow C=-1$

$2C-B=0\Rightarrow 2(-1)-B=0\Rightarrow B=-2$

$A-B=0\Rightarrow A=B=-2$

$y=c_1e^x-2-2x-x^2$

31.10 Applications of Higher-Order Equations

1. $x=c_1\sin 4t+c_2\cos 4t;\ x=0, Dx=2$ for $t=0$
$0=c_1\sin(4(0))+c_2\cos(4(0))\Rightarrow c_2=0$
$x=c_1\sin 4t$
$Dx=4c_1\cos 4t$

$4c_1\cos(4(0))=2\Rightarrow c_1=\frac{1}{2}$

$x=\frac{1}{2}\sin 4t$

5. $D^2\theta+\frac{g}{l}\theta=0;\ g=9.8\text{ m/s}^2,\ l=0.1\text{ m};$
$D^2\theta+9.8\theta=0;\ m^2+9.8=0$

$m_1=\sqrt{9.8}j,\ m_2=-\sqrt{9.8}j;\ \alpha=0, \beta=\sqrt{9.8}$
$x=c_1\sin\sqrt{9.8}t+c_2\cos\sqrt{9.8}t$

Substituting $\theta = 0.1$ when $t = 0$;
$0.1 = c_1 \sin 0 + c_2 \cos 0$; $0.1 = c_2$;
$D_x = c_1 \cos \sqrt{9.8}t - c_2 \sin \sqrt{9.8}t$
Substituting $D\theta = 0$ when $t = 0$;
$0 = c_1 \cos 0 - c_2 \sin 0$; $c_1 = 0$;
$\theta = 0.1 \cos \sqrt{9.8}t = 0.1 \cos 3.1t$

$\sqrt{9.8}t$	t	$\cos \sqrt{9.8}t$	$0.1 \cos \sqrt{9.8}t$
0	0	1	0.1
$\frac{\pi}{2}$	0.50	0	0
π	1.00	−1	−0.1
$\frac{3\pi}{2}$	1.50	0	0
2π	2.00	1	0.1

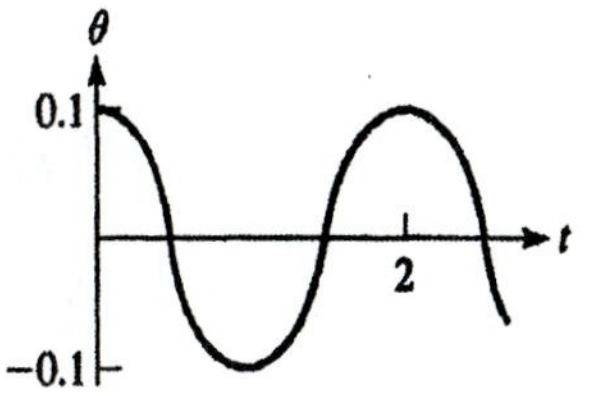

9. $m\dfrac{d^2s}{dt^2} = -\dfrac{mg}{e}(S - L)$

$$\frac{d^2s}{dt^2} + \frac{g}{e}s = \frac{g}{e}L$$

$$m^2 + \frac{g}{e} = 0 \Rightarrow m = \pm j\sqrt{\frac{g}{e}}$$

$$S_c = c_1 \sin\sqrt{\frac{g}{e}}t + c_2 \cos\sqrt{\frac{g}{e}}t$$

$$S_p = A + Bt$$

$$DS_p = B$$

$$D^2S_p = 0$$

$$0 + \frac{g}{e}(A + Bt) = \frac{g}{e}L$$

$$\frac{g}{e}A = \frac{g}{e}L \Rightarrow A = L$$

$$\frac{g}{e}B = 0 \Rightarrow B = 0$$

$$S = c_1 \sin\sqrt{\frac{g}{e}}t + c_2 \cos\sqrt{\frac{g}{e}}t + L$$

$$S_0 = c_1 \sin 0 + c_2 \cos 0 + L \Rightarrow c_2 = S_0 - L$$

$$\frac{ds}{dt} = c_1\sqrt{\frac{g}{e}}\cos\sqrt{\frac{g}{e}}t - (S_0 - L)\sqrt{\frac{g}{e}}\sin\sqrt{\frac{g}{e}}t$$

$$0 = c_1\sqrt{\frac{g}{e}}\cos(0) - (S_0 - L)\sqrt{\frac{g}{e}}\sin(0) \Rightarrow c_1 = 0$$

$$S = (S_0 - L)\cos\sqrt{\frac{g}{e}}t + L$$

13. $m = \dfrac{4.00 \text{ lb}}{32.0 \text{ ft/s}^2} = 0.125 \text{ slug}$

$F = kx, 4.00 = k(0.125)$
$k = 32.0 \text{ lb/ft}$
$0.125D^2x = 4 \sin 2t - 32.0x$
$D^2x + 256x = 32.0 \sin 2t$
$m^2 + 256 = 0$; $m = \pm 16.0j$
$x_c = c_1 \sin 16.0t + c_2 \cos 16.0t$
$x_p = A \sin 2t + B \cos 2t$
$Dx_p = 2A \cos 2t - 2B \sin 2t$
$D^2x_p = -4A \sin 2t - 4B \cos 2t$
$-4A \sin 2t - 4B \cos 2t + 256(A \sin 2t + B \cos 2t)$
$= 32.0 \sin 2t$

$252A \sin 2t + 252B \cos 2t = 32.0 \sin 2t$
$A = \dfrac{32.0}{252} = 0.127, B = 0$
$x_p = 0.127 \sin 2.00t$
$x = c_1 \sin 16.0t + c_2 \cos 16.0t + 0.127 \sin 2.00t$
$Dx = 16.0c_1 \cos 16.0t - 16.0c_2 \sin 16.0 + 0.254 \cos 2.00t$
$x = 0.250$ ft, $Dx = 0$, when $t = 0$
$0.250 = c_1(0) + c_2 + 0.127(0)$
$c_2 = 0.250$
$0 = 16.0c_1 - 16.0c_2(0) + 0.254$
$c_1 = -0.016$
$x = 0.250 \cos 16.0t + 0.127 \sin 2.00t - 0.016 \sin 16.0t$

17. $L = 0.100H, R = 0, C = 100\ \mu F = 10^{-4}, E = 100$ V

$$0.100\frac{d^2q}{dt^2} + 0\frac{dq}{dt} + \frac{q}{10^{-4}} = 100$$

$$\frac{d^2q}{dt^2} + 10^5q = 1000;\ m^2 + 10^5 = 0;\ m = \pm 316j$$

$q_c = c_1 \sin 316t + c_2 \cos 316t$; $q_p = A$; $q_p' = 0$; $q_p'' = 0$

$0 + 0 + 10^5 A = 1000$; $A = 0.01$

$q = c_1 \sin 316t + c_2 \cos 316t + 0.01$; $q = 0$ when $t = 0$; $0 = 0 + c_2(1) + 0.01$

$c_2 = -0.01$; $\frac{dq}{dt} = 316c_1 \cos 316t + 3.16 \sin 316t$;

$i = \frac{dq}{dt} = 0$ when $t = 0$

$0 = 316c_1 + 0$; $c_1 = 0$;

$y = 0 \sin 316t - 0.01 \cos 316t + 0.01$

$q = 0.01(1 - \cos 316t)$

21. $L = 8.00$ mH $= 8.00 \times 10^{-3}$ H; $R = 0$; $C = 0.50$ μF

$C = 5.00 \times 10^{-7}$ F

$E = 20.0e^{-200t}$ mV $= 2.00 \times 10^{-2}e^{-200t}$ V

$$8.00 \times 10^{-3}\frac{d^2q}{dt^2} + 0\frac{dq}{dt} + \frac{q}{5.00 \times 10^{-7}}$$
$$= 2.00 \times 10^{-2}e^{-200t}$$

$$\frac{d^2q}{dt^2} + 2.50 \times 10^8 q = 2.50e^{-200t}$$

$m^2 + 2.50 \times 10^8 = 0$; $m_1 = 1.58 \times 10^4 j$ and $m_2 = -1.58 \times 10^4 j$

$q_c = c_1 \sin 1.58 \times 10^4 t + c_2 \cos 1.58 \times 10^4 t$

$q_p = Ae^{-200t}$; $q_p' = -200Ae^{-200t}$;

$q_p'' = 4.00 \times 10^4 Ae^{-200t}$

Substituting into the differential equation,

$4.00 \times 10^4 Ae^{-200t} + 2.50 \times 10^8 Ae^{-200t} = 2.50e^{-200t}$

$4.00 \times 10^4 A + 2.50 \times 10^8 A = 2.50$; $A = 10^{-8}$

$q = c_1 \sin 1.58 \times 10^4 t + c_2 \cos 1.58 \times 10^4 + 10^{-8}e^{-200t}$

$q = 0$ when $t = 0$; $0 = c_1(0) + c_2(1) + 10^{-8}$; $c_2 = -10^{-8}$

$$i = \frac{dq}{dt} = 1.58 \times 10^4 c_1 \cos 1.58 \times 10^4 t$$
$$- 1.58 \times 10^4 c_2 \sin 1.58 \times 10^4 t - 200 \times 10^{-8}e^{-200t}$$

$i = 0$ when $t = 0$;

$0 = 1.58 \times 10^4 c_1(1) - 2.00 \times 10^{-6}(1)$

$$c_1 = \frac{2.00 \times 10^{-6}}{1.58 \times 10^4}$$

25. $$EI\frac{d^4y}{dx^4} = w$$

$$y = c_1 + c_2 + c_3x^2 + c_4x^3 + \frac{w}{24EI}x^4$$

@ $x = 0, y = 0 \Rightarrow c_1 = 0$

$$y = c_2x + c_3x^2 + c_4x^3 + \frac{w}{24EI}x^4$$

$$y' = c_2 + 2c_3x + 3c_4x^2 + \frac{4w}{24EI}x^3$$

@ $x = 0, y' = 0 \Rightarrow c_2 = 0$

$$y = c_3x^2 + c_4x^3 + \frac{w}{24EI}x^4$$

$$y'' = 2c_3 + 6c_4 + \frac{12w}{24EI}x^2$$

@ $x = L, y'' = 0$

$$0 = 2c_3 + 6c_4L + \frac{12w}{24EI}L^2$$

$$y''' = 6c_4 + \frac{24w}{24EI}x$$

@ $x = L, y''' = 0$

$$0 = 6c_4 + \frac{24w}{24EI} \cdot L \Rightarrow c_4 = \frac{-wL}{6EI}$$

$$0 = 2c_3 + 6 \cdot \frac{-wL}{6EI} \cdot L + \frac{12w}{24EI}L^2 \Rightarrow c_3 = \frac{wL^2}{4EI}$$

$$y = \frac{wL^2}{4EI}x^2 + \frac{-wL}{6EI}x^3 + \frac{w}{24EI}x^4$$

$$y = \frac{w}{24EI}(6L^2x^2 - 4Lx^3 + x^4)$$

31.11 Laplace Transforms

1. $f(t) = 1, t > 0$

$$L(f) = \int_0^\infty e^{-st} \cdot 1\,dt$$
$$= \lim_{c\to\infty} \frac{-1}{s}\int_0^c e^{-st}(-s\,dt)$$
$$L(f) = -\frac{1}{s} \lim_{c\to\infty} e^{-st}\Big|_0^c$$
$$= -\frac{1}{s}\left[\lim_{c\to\infty} (e^{-sc} - e^{-s(0)})\right]$$
$$= -\frac{1}{s} \lim_{c\to\infty} (0 - 1)$$
$$L(f) = \frac{1}{s}$$

5. $f(t) = e^{3t}$; from transform (3) of the table,

$a = -3$; $L(3t) = \dfrac{1}{s-3}$

9. $f(t) = \cos 2t - \sin 2t$; $L(f) = L(\cos 2t) - L(\sin 2t)$

By transforms (5) and (6),

$L(f) = \dfrac{s}{s^2+4} - \dfrac{2}{s^2+4}$; $L(f) = \dfrac{s-2}{s^2+4}$

13. $y'' + y'$; $f(0) = 0$; $f'(0) = 0$

$$\begin{aligned} &L[f''(y) + f'(y)] \\ &= L(f'') + L(f') \\ &= s^2L(f) - sf(0) - f'(0) + sL(f) - f(0) \\ &= s^2L(f) - s(0) - 0 + sL(f) - 0 \\ &= s^2L(f) + sL(f) \end{aligned}$$

17. $L^{-1}(F) = L^{-1}\left(\dfrac{2}{s^3}\right) = 2L^{-1}\left(\dfrac{1}{s^3}\right)$;

$L^{-1}(F) = \dfrac{2t^2}{2} = t^2$; transform (2)

21. $L^{-1}(F) = L^{-1}\dfrac{1}{(s+1)^3} = L^{-1}\dfrac{1}{2}\left[\dfrac{2}{(s+1)^3}\right]$

$= \dfrac{1}{2}t^2e^{-t}$;

transform (12)

25. $F(s) = \dfrac{4s^2-8}{(s+1)(s-2)(s-3)} = \dfrac{-\frac{1}{3}}{s+1} + \dfrac{-\frac{8}{3}}{s-2} + \dfrac{7}{s-3}$

$$L^{-1}(F) = -\frac{1}{3}L^{-1}\left(\frac{1}{s+1}\right) - \frac{8}{3}L^{-1}\left(\frac{1}{s-2}\right) + 7L^{-1}\frac{1}{s-3}$$

$f(t) = -\dfrac{1}{3}e^{-t} - \dfrac{8}{3}e^{2t} + 7e^{3t}$

29.
$$\begin{aligned} L\{tf(t)\} = L\left(te^{-at}\right) &= -\frac{d}{ds}\left(\frac{1}{s+a}\right) \\ &= -\left(\frac{-1}{(s+a)^2}\right) \\ &= \frac{1}{(s+a)^2} \end{aligned}$$

31.12 Solving Differential Equations by Laplace Transforms

1. $2y' - y = 0$; $y(0) = 2$

$$\begin{aligned} L(2y') - L(y) &= L(0) \\ 2L(y') - L(y) &= 0 \\ 2sL(y) - 2(2) - L(y) &= 0 \end{aligned}$$

$L(y) = \dfrac{4}{2s-1} = 2\dfrac{1}{s-\frac{1}{2}}$, from transform 3,

$y = 2e^{\frac{1}{2}t}$

5. $y' + y = 0$; $y(0) = 1$; $L(y') + L(y) = L(0)$;
$L(y') + L(y) = 0$; $sL(y) - y(0) + L(y) = 0$
$sL(y) - 1 + L(y) = 0$; $(s+1)L(y) = 1$;

$L(y) = \dfrac{1}{s+1}$; $a = -1$, transforms (3); $y = e^{-t}$

9. $y' + 3y = e^{-3t}$; $y(0) = 1$; $L(y') + L(3y) = L(e^{-3t})$;

$L(y') + 3L(y) = L(3^{-3t})$;

$[sL(y-1] + 3L(y) = \dfrac{1}{s+3}$

$(s+3)L(y) = \dfrac{1}{s+3} + 1$; $L(y) = \dfrac{1}{(s+3)^2} + \dfrac{1}{s+3}$

The inverse is found from transforms (11) and (3).

$y = te^{-3t} + e^{-3t} = (1+t)e^{-3t}$

13. $4y''+4y'+5y=0,\ y(0)=1,\ y'(0)=-\frac{1}{2}$

$4L(y'')+4L(y')+5L(y)=0$

$4(s^2L(y)-sy(0)-y'(0))+4(sL(y)-y(0))$

$+5L(y)=0$

$4s^2L(y)-4s+2+4sL(y)-4+5L(y)=0$

$(4s^2+4s+5)L(y)=4s+2$

$$L(y)=\frac{4s+2}{4s^2+4s+5}=\frac{4(s+\frac{1}{2})}{4(s^2+s+\frac{1}{4})+4}$$

$$L(y)=\frac{s+\frac{1}{2}}{(s+\frac{1}{2})^2+1},\quad \#20,\ a=\frac{1}{2},\ b=1$$

$y=e^{-1/2t}\cos t$

17. $y''+y=1;\ y(0)=1;\ y'(0)=1;$

$L(y'')+L(y)=L(1);$

$$s^2L(y)-s-1+L(y)=\frac{1}{s}$$

$$(s^2+1)L(y)=\frac{1}{s}+s+1;$$

$$L(y)=\frac{1}{s(s^2+1)}+\frac{s}{s^2+1}+\frac{1}{s^2+1}$$

By transforms (7), (5), and (6),

$y=1-\cos t+\cos t+\sin t;\ y=1+\sin t$

21. $y''-4y=10e^{3t},\ y(0)=5,\ y'(0)=0;$

$L(y'')-4L(y)=10\cdot L(e^{3t})$

$$s^2L(y)-s\cdot y(0)-y'(0)-4L(y)=\frac{10}{s-3};$$

$$s^2L(y)-5s-0-4L(y)=\frac{10}{s-3}$$

$$(s^2-4)L(y)=5s+\frac{10}{s-3}$$

$$L(y)=\frac{5s}{(s+2)(s-2)}+\frac{10}{(s+2)(s-2)(s-3)}$$

$$L(y)=\frac{\frac{5}{2}}{s+2}+\frac{\frac{5}{2}}{s-2}+\frac{\frac{1}{2}}{s+2}+\frac{-\frac{5}{2}}{s-2}+\frac{2}{s-3}$$

$$L(y)=\frac{3}{s+2}+\frac{2}{s-3}$$

$y=3e^{-2t}+2e^{3t}$

25. $2v'=6-v$; since the object starts from rest,

$f(0)=0,\ f'(0)=0$

$$2L(v')+L(v)=6L(1);\ 2sL(v)-0+L(v)=\frac{6}{s};$$

$$(2s+1)L(v)=\frac{6}{s}$$

$$L(v)=\frac{6}{s(2s+1)}=6\left[\frac{\frac{1}{2}}{s(s+\frac{1}{2})}\right]$$

By transforms (4), $v=6(1-e^{-t/2})$

29. $L\frac{d^2q}{dt^2}+R\frac{dq}{dt}+\frac{q}{i}=E$

$$0(q'')+50q'+\frac{q}{4\times10^{-6}}=40$$

$$50q'+\frac{10^6}{4}q=40$$

$$50L(q')+\frac{10^6}{4}L(q)=L(40)$$

$$50[sL(q)-q(0)]+\frac{10^6}{4}L(q)=\frac{40}{5}$$

$$L(q)\left(50s+\frac{10^6}{4}\right)=\frac{40}{s}$$

$$L(q)=\frac{40}{s\left(50s+\frac{10^6}{4}\right)}$$

$$L(q)=F(s)=\frac{40}{50s(s+5000)}$$

$$=\frac{40}{50(5000)}\frac{5000}{s(s+5000)}$$

$$=0.00016\cdot\frac{5000}{s(s+5000)}$$

$L^{-1}[F(s)=f(t)=\frac{1.60}{10^4}(1-e^{-5000t})=9$; from transform (4), $a=5000$

$q=1.60\times10^{-4}(1-e^{-5000t})$

33. $D^2y+9y=18\sin 3t;\ y=0, Dy=0, t=0$

$y''+9y=18\sin 3t;\ L(y'')+9L(y)=18L(\sin 3t)$

$$s^2L(y)-sy(0)-y'(0)+9L(y)=18\frac{3}{s^2+9}=\frac{54}{s^2+9}$$

$$L(y)(s^2+9)=\frac{54}{s^2+9}$$

$$L(y) = \frac{54}{(s^2+9)^2} = F(s); \text{ from transform (15),}$$
$a = 3;$

$2a^3 = 54$

$$y = L^{-1}\left[\frac{54}{(s^2+9)^2}\right] = L^{-1}\left[\frac{2(27)}{(s^2+9)^2}\right]$$

$y = \sin 3t - 3t\cos 3t$

37. $L\frac{d^2q}{dt^2} + R\frac{dq}{dt} + \frac{q}{C} = E;\ q(0) = 0, q'(0) = 0$

$$Lm^2 + Rm + \frac{1}{C} = 0$$

$$m = \frac{-R}{2L} \pm \sqrt{\frac{1}{LC} - \frac{R^2}{4L^2}}j = -66.7 \pm 226j$$

$q = e^{-66.7t}(c_1\sin(226t) + c_2\cos(226t) + CE$
@ $t = 0, q = 0 \Rightarrow c_2 = -CE$
$q = e^{-66.7t}(c_1\sin(226t) - CE\cos(226t) + CE$

$$i = \frac{dq}{dt}$$
$= e^{-66.7t}(226c_1\cos(226t) + 226CE\sin(226t)$
$- 66.7(c_1\sin(226t) - CE\cos(226t))e^{-66.7t}$
@ $t = 0, i = 0$
$$0 = 226c_1 + 66.7CE \Rightarrow c_1 = \frac{-66.7CE}{226}$$

$$i = e^{-66.7t}[(-66.7CE\cos(226t) + 226E\sin(226t)$$
$$+ \frac{66.7^2CE}{226}\sin(226t) + 66.7CE\cos(226t)]$$

$$i = \left(\frac{66.7^2CE}{226} + 226CE\right)e^{-66.7t}\sin(226t)$$

$$i = \left(\frac{66.7^2(300\times10^{-6})(60)}{226}\right.$$
$$\left. +226(300\times10^{-6})(60)\right)e^{-66.7t}\sin(226t)$$
$i = 4.42e^{-66.7t}\sin(226t)$

Chapter 31 Review Exercises

1. $4xy^3\,dx + (x^2+1)\,dy = 0$; divide by y^3 and x^2+1

$$\frac{4x}{x^2+1}dx + \frac{dy}{y^3} = 0;\ \text{integrating,}$$

$$2\ln(x^2+1) - \frac{1}{2y^2} = c$$

5. $2D^2y + Dy = 0$. The auxiliary equation is

$2m^2 + m = 0$

$m(2m+1) = 0;\ m_1 = 0$ and $m_2 = -\frac{1}{2}$

$y = c_1e^0 + c_2e^{-1/2};\ y = c_1 + c_2e^{-x/2}$

9. $(x+y)\,dx + (x+y^3)\,dy = 0;\ x\,dx + y\,dx + x\,dy +$ $y^3\,dy = 0;\ x\,dx + d(xy) + y^3\,dy = 0$

Integrating, $\frac{1}{2}x^2 + xy + \frac{1}{4}y^4 = c_1;$

$2x^2 + 4xy + y^4 = c$

13. $dy = (2y+y^2)\,dx;\ \frac{dy}{(2y+y^2)} = dx$

$$-\frac{1}{2}\ln\left(\frac{2+y}{y}\right) = x + \ln c_1;\ \ln\frac{2+y}{y}$$

$$= -2x - \ln c_1^2\left(\frac{2+y}{y}\right) = -2x;\ \frac{2+y}{y} = \frac{e^{-2x}}{c_1^2};$$

$y = c_1^2(y+2)e^{2x};\ y = c(y+2)e^{2x}$

17. $y' + 4y = 2e^{-2x};\ \frac{dy}{dx} + 4y = 2e^{-2x};\ dy + 4\cdot y\,dx$

$= 2e^{-2x}\,dx$

$e^{\int 4dx} = e^{4x};\ e^{4x}dy + 4ye^{4x}dx = 2e^{-2x}\cdot e^{4x}dx;$

$d(ye^{4x}) = 2e^{2x}dx;\ \int d(ye^{4x}) = \int e^{2x}\cdot(2dx);$

$ye^{4x} = e^{2x} + c;$

$y = e^{-2x} + ce^{-4x}$

21. $2D^2s + Ds - 3s = 6;\ 2m^2 + m - 3 = 0;$

$(m-1)(2m+3) = 0;\ m_1 = 1,\ m_2 = -\frac{3}{2}$

$s_c = c_1e^t + c_2e^{-3t/2};\ s_p = A;\ s_p' = 0;\ s_p'' = 0$

Substituting into the differential equation,

$2(0) + 0 - 3A = 6;\ A = -2;\ s_p = -2.$

$s = c_1e^t + c_2e^{-3t/2} - 2$

25. $9D^2y - 18Dy + 8y = 16 + 4x$

$D^2y - 2Dy + \frac{8}{9}y = \frac{16}{9} + \frac{4}{9}x$

$m^2 - 2m + \frac{8}{9} = 0;\ m = \frac{2 \pm \sqrt{4 - 4\left(\frac{8}{9}\right)}}{2};\ m_1 = \frac{2}{3};\ m_2 = \frac{4}{3}$

$y_c = c_1e^{2x/3} + c_2e^{4x/3};\ y_p = A + Bx;\ y_p' = B;\ y_p'' = 0$

Substituting into the differential equation.

$0 - 2B + \frac{8}{9}(A + Bx) = \frac{16}{9} + \frac{4}{9}x$

$\left(-2B + \frac{8}{9}A\right) + \frac{8}{9}Bx = \frac{16}{9} + \frac{4}{9}x;\ -2B + \frac{8}{9}A = \frac{16}{9};$

$\frac{8}{9}B = \frac{4}{9};\ B = \frac{1}{2};\ -2\left(\frac{1}{2}\right) + \frac{8}{9}A = \frac{16}{9};$

$A = \frac{25}{8};\ y_p = \frac{1}{2}x + \frac{25}{8}$

$y = c_1e^{2x/3} + c_2e^{4x/3} + \frac{1}{2}x + \frac{25}{8}$

29. $y'' - 7y' - 8y = 2e^{-x}$

$D^2y_c - 7Dy_c - 8y_c = 0$

$m^2 - 7m - 8 = 0$

$(m+1)(m-8) = 0$

$m = -1,\ m = 8$

$y_c = c_1e^{-x} + c_2e^{8x}$

Let $y_p = Axe^{-x},\ y_p' = Ae^{-x}(1-x),\ y_p'' = Ae^{-x}(x-2)$

$Ae^{-x}(x-2) - 7Ae^{-x}(1-x) - 8Axe^{-x} = 2e^{-x}$

$A(x-2) - 7A(1-x) - 8Ax = 2$

$-9A = 2$

$A = \frac{-2}{9}$

$y_p = \frac{-2}{9}xe^{-x}$

$y = y_c + y_p$

$y = c_1e^{-x} + c_2e^{8x} + \frac{-2}{9}xe^{-x}$

33. $3y' = 2y\cot x;\ \frac{dy}{dx} = \frac{2}{3}y\cot x;\ \frac{dy}{y} = \frac{2}{3}\cot x\,dx;$

$\ln y = \frac{2}{3}\ln\sin x + \ln c;\ \ln y - \ln\sin^{2/3}x = \ln c;$

$\ln\frac{y}{\sin^{2/3}x} = \ln c;\ \frac{y}{\sin^{2/3}x} = c;\ y = c\sin^{2/3}x$

Substituting $y = 2$ when $x = \frac{\pi}{2}$, $2 = c\sin^{2/3}\frac{\pi}{2};\ c = 2$

Therefore $y = 2\sin^{2/3}x = 2\sqrt[3]{\sin^2 x};\ y^3 = 8\sin^2 x$

37. $D^2v + Dv + 4v = 0.\ m^2 + m + 4 = 0;\ m = \frac{-1 \pm \sqrt{-15}}{2}$

$m_1 = \frac{1}{2} + \frac{\sqrt{15}}{2}j;\ m_2 = -\frac{1}{2} - \frac{\sqrt{15}}{2}j;\ \alpha = -\frac{1}{2},\ \beta = \frac{\sqrt{15}}{2};$

by Eq. (30-17)

$v = e^{-t/2}\left(c_1\sin\frac{\sqrt{15}}{2}t + c_2\cos\frac{\sqrt{15}}{2}t\right)$

$Dv = e^{-t/2}\left(\frac{\sqrt{15}}{2}c_1\cos\frac{\sqrt{15}}{2}t - c_2\sin\frac{\sqrt{15}}{2}t\right)$

$+\left(c_1\sin\frac{\sqrt{15}t}{2}c_2\cos\frac{\sqrt{15}}{2}t\right)\left(-\frac{1}{2}e^{-t}\right)$

Substituting $Dv = \sqrt{15},\ v = 0$ when $t = 0;\ c_2 = 0;\ c_1 = 2.$

$y = 2e^{-t/2}\sin\left(\frac{1}{2}\sqrt{15}t\right)$

41. $L(4y')-L(y)=0;\ 4L(y')-L(y)$
$=0;\ 4sL(y)-1-L(y)$
by Eq. (30-24)

$(4s-1)L(y)=1;\ L(y)=\frac{1}{4s-1}=\frac{\frac{1}{4}}{s-\frac{1}{4}}$

By transform (3), $y=e^{t/4}$.

45. $L(y'')+L(y)=0;\ y(0)=0,$
$y'(0)=-4$

$s^2L(y)+4+L(y)=0;\ (s^2+1)L(y)=-4;$

$L(y)=-4\left(\frac{1}{s^2+1}\right)$

By transform (6), $y=-4\sin t$.

49. Using EULRMETH in Appendix C for

$\frac{dy}{dx}=1+y^2,\ x=0$ to $x=0.4,\ \Delta x=0.1,\ (0,\ 0)$

x	y
0	0
0.1	0.1
0.2	0.201
0.3	0.3050401
0.4	0.4143450463

53. $\frac{dx}{dt}=2t$

$x=t^2+c$

$1=0^2+c\Rightarrow c=1$

$x=t^2+1$

$xy=1$

$x\frac{dy}{dt}+y\frac{dx}{dt}=0$

$x\frac{dy}{dt}+y(2t)=0$

$\frac{x}{y}dy+2t\,dt=0$

$\frac{1}{y^2}dy+2t\,dt=0$

$-\frac{1}{y}+t^2=c$

$-\frac{1}{1}+0^2=c\Rightarrow=-1$

$-\frac{1}{y}+t^2=-1$

$-1+yt^2=-y$

$y(t^2+1)=1$

$y=\frac{1}{t^2+1}$

In terms of t, $x=t^2+1$ and $y=\frac{1}{t^2+1}$.

57. $\frac{dm}{dt}=km$

$\frac{dm}{m}=kdt$

$\ln m=kt+c$

$\ln m_0=k(0)+c\Rightarrow c=\ln m_0$

$\ln m=kt+\ln m_0$

$\ln\frac{m}{m_0}=kt$

$\frac{m}{m_0}=e^{kt}$

$m=m_0e^{kt}$

61. $\frac{dy}{dx}=\frac{y}{y-x},\ (-1,\ 2),\ y>0$

$ydy=ydx+xdy=d(xy)$

$\frac{y^2}{2}=xy+C$

$\frac{2^2}{2}=(-1)(2)+C\Rightarrow C=4$

$\frac{y^2}{2}=xy+4$

$y^2=2xy+8\Rightarrow y=x\pm\sqrt{x^2-8}$ from which the path consists of two functions

$f_1(x)=x+\sqrt{x^2-8},\ x\ge\sqrt{8}$ and

$f_2(x)=x-\sqrt{x^2-8},\ x\ge\sqrt{8}$

65. $N = N_0 e^{-kt}$

$$\frac{N_0}{2} = N_0 e^{-kt(1.28\times 10^9)}$$

$$e^{k(1.28\times 10^9)} = 2$$

$$k(1.28\times 10^9) = \ln 2$$

$$k = \frac{\ln 2}{1.28\times 10^9}$$

$$N = N_0 e^{-\frac{\ln 2}{1.28\times 10^9}\cdot t}$$

$$0.75N_0 = N_0 e^{\frac{-\ln 2}{1.28\times 10^9}\cdot t}$$

$$e^{\frac{\ln 2}{1.28\times 10^9}\cdot t} = \frac{4}{3}$$

$$\frac{\ln 2}{1.28\times 10^9}\cdot t = \ln\frac{4}{3}$$

$$t = \frac{\ln\frac{4}{3}}{\frac{\ln 2}{1.28\times 10^9}}$$

$t = 5.31\times 10^8$ years

69. See Example 2, Section 31.6

$$y = cx^5,\ c = \frac{y}{x^5}$$

$$y' = 5cx^4 = 5\left(\frac{y}{x^5}\right)x^4 = \frac{5y}{x}$$

$$y'|OT = -\frac{x}{5y};\ 5y\,dy = -x\,dx$$

$$\frac{5}{2}y^2 = -\frac{1}{2}x^2 + \frac{1}{2}c$$

$$5y^2 + x^2 = c$$

73. $L\frac{di}{dt} + Ri = E$

$$2\frac{di}{dt} + 40i = 20$$

$$\frac{di}{dt} + 20i = 10$$

$$di + 20i\,dt = 10\,dt$$

$$ie^{\int 20\,dt} = \int 10e^{\int 20\,dt} + c$$

$$ie^{20t} = 0.5e^{20t} + c$$

$$(0)e^{20(0)} = 0.5e^{20(0)} + c$$

$$c = -0.5$$

$$ie^{20t} = 0.5e^{20t} - 0.5 = 0.5\left(e^{20t} - 1\right)$$

$$i = 0.5\left(1 - e^{-20t}\right)$$

77. $LD^2q + RDq + \frac{q}{C} = E \quad \left(D = \frac{d}{dt}\right)$

$L = 0.5$ H, $R = 6\,\Omega$, $C = 20$ mF, $E = 24\sin 10t$

$$0.5D^2q + 6Dq + 50q = 24\sin 10t,\ D^2q + 12Dq + 100q = 48\sin 10t$$

$$m^2 + 12m + 100 = 0,\ m = \frac{-12\pm\sqrt{144-400}}{2} = -6\pm 8j$$

$$q_e = e^{-6t}\left(c_1\sin 8t + c_2\cos 8t\right),\ q_p = A\sin 10t + B\cos 10t$$

$$Dq_p = 10A\cos 10t - 10B\sin 10t,\ D^2q_p = -100A\sin 10t - 100B\cos 10t$$

$$-100A\sin 10t - 100B\cos 10t + 12(10A\cos 10t - 10B\sin 10t) + 100(A\sin 10t + B\cos 10t) = 48\sin 10t$$

$$-120B = 48,\ B = -0.4;\ 120A = 0,\ A = 0$$

$$q = e^{-6t}\left(c_1\sin 8t + c_2\cos 8t\right) - 0.4\cos 10t$$

$$q = 0 \text{ when } t = 0,\ 0 = c_2 - 0.4,\ c_2 = 0.4$$

$$q = e^{-6t}\left(c_1\sin 8t + 0.4\cos 8t\right) - 0.4\cos 10t$$

$$Dq = e^{-6t}\left(-6c_1\sin 8t - 2.4\cos 8t + 8c_1\cos 8t - 3.2\sin 8t\right) + 4\sin 10t$$

$$Dq = 0 \text{ when } t = 0;\ 0 = -2.4 + 8c_1,\ c_1 = 0.3$$

$$q = e^{-6t}\left(0.3\sin 8t + 0.4\cos 8t\right) - 0.4\cos 10t$$

81. $2\frac{di}{dt} + i = 12,\ i(0)0$

$$2L(i') + L(i) = L(12)$$

$$2\left[sL(i) - 0\right] + L(i) = \frac{12}{s}$$

$$L(i) = \frac{12}{s(2s+1)} = \frac{12(1/2)}{s(s+1/2)}$$

$$i = 12\left(1 - e^{-t/2}\right)$$

$$i(0.3) = 12\left(1 - e^{-0.3/2}\right)$$

$i = 1.67$ A

85. $m = 0.25$ slug, $F = 8$ lb, $x = 6$ in. $= 0.5$ ft; $8 = 0.5k$,

$k = 16$ lb/ft

$$0.25D^2y + 16y = \cos 8t \quad (D = d/dt)$$

$$D^2y + 64y = 4\cos 8t,\ y(0) = 0,\ Dy(0) = 0$$

$$L(D^2y) + 64L(y) = 4L(\cos 8t);$$

$$s^2L(y) - s(0) - 0 + 64L(y) = 4\left(\frac{s}{s^2+64}\right)$$

$$L(y) = \frac{4s}{(s^2+64)^2} = \frac{1}{4}\left[\frac{2(8s)}{(s^2+64)^2}\right];\ y = 0.25t\sin 8t$$

89. $EI\frac{d^2y}{dx^2} = M,\ M = 2000x - 40x^2$

$$EID^2y = 2000x - 40x^2$$

$$D^2y = \frac{1}{EI}(2000x - 40x^2)$$

$$Dy = \frac{1}{EI}\left(1000x^2 - \frac{40}{3}x^3\right) + c_1$$

$$y = \frac{1}{EI}\left(\frac{1000}{3}x^3 - \frac{10}{3}x^4\right) + c_1x + c_2$$

$y = 0$ for $x = 0$ and $x = L$

$$0 = \frac{1}{EI}(0-0) + 0 + c_2,\ c_2 = 0$$

$$0 = \frac{1}{EI}\left(\frac{1000}{3}L^3 - \frac{10}{3}L^4\right) + c_1L$$

$$c_1 = \frac{1}{EI}\left(\frac{10}{3}L^3 - \frac{1000}{3}L^2\right)$$

$$y = \frac{1}{EI}\left(\frac{1000}{3}x^3 - \frac{10}{3}x^4\right) + \frac{1}{EI}\left(\frac{1000}{3}L^3 - \frac{1000}{3}L^2\right)x$$

$$= \frac{10}{3EI}(100x^3 - x^4 + L^3x - 100L^2x)$$